# Wetlands: Ecology, Conservation and Management

Volume 15

The recognition that wetlands provide many values for people and are important foci for conservation worldwide has led to an increasing amount of research and management activity. This has resulted in an increased demand for high quality publications that outline both the value of wetlands and the many management steps necessary to ensure that they are maintained and even restored. Recent research and management activities in support of conservation and sustainable development provide a strong basis for the book series. The series presents current analyses of the many problems afflicting wetlands as well as assessments of their conservation status. Current research is described by leading academics and scientists from the biological and social sciences. Leading practitioners and managers provide analyses based on their vast experience.

The series provides an avenue for describing and explaining the functioning and processes that support the many wonderful and valuable wetland habitats, such as swamps, lagoons and marshes, and their species, such as waterbirds, plants and fish, as well as the most recent research directions. Proposals cover current research, conservation and management issues from around the world and provide the reader with new and relevant perspectives on wetland issues.

Rahul Kumar • R.K. Gupta • Kiran Yodha
Editors

# Ecology, Conservation of Aquatic Birds

## Aquatic Birds, Wetland, Sustainability

Springer

*Editors*
Rahul Kumar
Department of Zoology
Chaudhary Charan Singh Haryana
Agricultural University, College
of Agriculture
Bawal, Haryana, India

R.K. Gupta
Department of Zoology, College of Basic
Sciences and Humanities
Chaudhary Charan Singh Haryana
Agricultural University
Hisar, Haryana, India

Kiran Yodha
Department of Zoology, College of Basic
Sciences and Humanities
Chaudhary Charan Singh Haryana
Agricultural University
Hisar, Haryana, India

ISSN 1875-1261 ISSN 1875-127X (electronic)
Wetlands: Ecology, Conservation and Management
ISBN 978-3-032-17124-5 ISBN 978-3-032-17125-2 (eBook)
https://doi.org/10.1007/978-3-032-17125-2



This Springer imprint is published by the registered company Springer Nature Switzerland AG
The registered company address is: Gewerbestrasse 11, 6330 Cham, Switzerland

# Acknowledgments

It is with immense gratitude that we acknowledge the support and encouragement we received during the preparation and completion of this book, *Ecology, Conservation of Aquatic Birds: Aquatic Birds, Wetland, Sustainability*.

First and foremost, we would like to express our sincere thanks to the Hon'ble Vice Chancellor Prof. B.R. Kamboj, CCSHAU, Hisar, Haryana, for providing a visionary academic environment and constant encouragement throughout this journey. Our heartfelt appreciation also goes to Dr. S.K. Pahuja, Dean COA, and Dr. Rajesh Gera, Dean COBS&H, for their invaluable guidance and institutional support that facilitated this work.

We are equally grateful to Dr. Naresh Kaushik, Principal, COA Bawal, and Dr. R.K. Gupta, HOD Zoology, for their consistent motivation and infrastructural support and for fostering a spirit of academic excellence.

A special word of thanks to my colleagues and faculty members, whose insights and discussions have enriched the content and direction of this book. Their cooperation and camaraderie made this endeavor both enjoyable and fulfilling.

We would also like to extend our deepest gratitude to our family for their unconditional love, patience, and support, without which this work would not have been possible. To our friends, thank you for your encouragement, inspiration, and unwavering belief in our capabilities.

Thanks to the authors and co-authors for sharing their expertise and time with the current and future of *Ecology, Conservation of Aquatic Birds: Aquatic Birds, Wetland, Sustainability*. We also thank technical reviewers for providing suggestions to improve the manuscript chapters; unfortunately, space precludes us from listing everyone.

Lastly, we thank Springer Nature for the opportunity to contribute to the growing body of knowledge in the field of Ecology, Conservation of Aquatic Birds and for their professional support throughout the publication process.

This book is a collective achievement, and we remain ever thankful to everyone who contributed to its realization.

Warm regards,
Editors

Bawal, Haryana, India Rahul Kumar
Hisar, Haryana, India R.K. Gupta
Hisar, Haryana, India Kiran Yodha

# Contents

# Chapter 1
# Wetlands: A Vital Ecosystem for Aquatic Birds

**Nibha Singh, Shweta Sharma, Dilbag Singh, and Mayur Mukut Murlidhar Sharma**

**Abstract** Wetlands are among the most critical ecosystems since they not only support biologically rich life but also offer a wide range of services. One of the most important uses is as a habitat to both migratory and residential waters' birds species. The purpose of this chapter is to assess the role wetlands play in sustaining the population of aquatic birds, focusing on the impact of water, vegetation, and prey availability on bird species richness and abundance. Recent works in ecology and ornithology highlight the complex interactions between wetland water dynamics, nutrient balance, and the population cycles of aquatic birds. We examine the consequences of human impacts like land use practices, pollution, and climate change on the wetland ecosystems and the birds. Findings show that habitat loss and wetland degradation pose the greatest risk to aquatic birds for maintaining populations in a global context. Subsequently, provides policies for the restoration of habitats and wetlands, improved community-based monitoring, and conservational measures as effective frameworks to sustain damaged ecosystems. This chapter stresses that without developed management plans for wetlands and targeted for bird species, the ecological functions of wetlands will not be fulfilled, making it essential to change the direction of the strategies used to preserve the ecosystem.

**Keywords** Diversity · Ecosystem services · Biodiversity · Ornithological · Hydrology · Migratory birds

---

N. Singh
Department of Endocrinology and Metabolism, AIIMS, New Delhi, India

S. Sharma (✉)
Department of Pharmaceutical Sciences, University of California, Irvine, CA, USA
e-mail: Shwetas4@uci.edu

D. Singh
School of Agriculture Sciences, Galgotias University, New Delhi, India

M. M. M. Sharma
Department of Agriculture and Life Industry, Kangwon National University, Chuncheon, South Korea



R. Kumar et al. (eds.), *Ecology, Conservation of Aquatic Birds*, Wetlands: Ecology, Conservation and Management 15, https://doi.org/10.1007/978-3-032-17125-2_1

## 1.1 Introduction

Wetlands are mid-point environments between water and land, with periodic or permanent flooding, hydric soils, and hydrophytic plants [30]. Wetland ecosystems are a wide range of ecosystems such as marshes, swamps, bogs, fens, and mangroves, each with unique biotic assemblages. Due to their high productivity and biodiversity, they are ideal reservoirs for biodiversity, especially for birds. Wetlands cover nearly 12.1 million km$^2$ around the world, covering around 6.2% of the Earth's surface, but 35% have been estimated to have disappeared since 1970 because of human pressures and climate change [8, 9, 43]. India has over 757,000 wetlands covering a total area of 15.3 million hectares [1], out of which 75 have been recognized as Ramsar wetlands of global significance covering 1.33 million hectares. Foremost among these are Chilika Lake, Keoladeo National Park, and Loktak Lake, which are important habitats for resident and migratory water bird species.

Wetlands are critical to migratory birds and waterbirds in that they provide nutritious food, secure breeding and nesting habitats, refuge for predators, and major stopover points during long-distance flight. The Greater Flamingo (*Phoenicopterus roseus*), Bar-headed Goose (*Anser indicus*), and Siberian Crane (*Leucogeranus leucogeranus*) heavily depend on wetland habitats for feeding on significant flyways. The complex bird diversity and wetland dynamics tell it all regarding their keystone habitat role they play, putting emphasis on the cascading impact of habitat loss on ecosystem stability, bird migration, and bird communities. Wetland conservation is thus critical to maintaining bird diversity worldwide and ecological stability. They are marked by hydrophytic vegetation, an untypical hydrologic regime, and hydric soils that support dense biotic aggregations.

### *1.1.1 Classification of Wetlands*

Ramsar Convention (1971) and Cowardin classification system [7] categorize wetlands into the following broad categories:

- **Marshes**—Herb wetlands with reeds, grasses, and sedges predominant, seasonally or permanently flooded.
- **Swamps**—Woody plant-dominated wetlands, for instance, trees and shrubs.
- **Bogs**—Malnourished, acid wetlands dominated by sphagnum mosses.
- **Fens**—Peat-forming wetlands with nutrient supply through groundwater, less acidic than bogs.
- **Mangroves**—Coastal wetlands with woody vegetation that is salt-tolerant, occurring in the tropics and subtropics.

### *1.1.2 Significance of Wetlands to Migratory and Aquatic Birds*

Wetlands are vital homing areas for waterbirds, such as rich food sources, breeding and nesting grounds, predator refuge, and crucial stopover places for migratory birds [30]. Seasonal migratory species like Siberian Cranes (*Leucogeranus leucogeranus*), Greater Flamingos (*Phoenicopterus roseus),* and Bar-headed Geese (*Anser indicus*) rely on wetlands for seasonal migration on global flyways. These landscapes also function very importantly to sustain avian population processes and accommodate species diversity along ecological gradients. Wetlands are among the most biologically productive landscapes because they support significant resources that subsidize rich bird communities, especially aquatic and migratory birds. These environments provide consistent and prolific food sources like fish, mollusks, crustaceans, water plants, and invertebrates supporting various avian feeding guilds like piscivores, insectivores, herbivores, and omnivores [30]. Some migratory birds like Flamingos, Cranes, and Ducks. The high primary productivity of wetlands provides birds with ample energy-rich diets to aid in reproduction, molting, and migration.

For migratory birds, wetlands are key stopover and fueling points along global flyways. Transcontinental long-distance migrants like the Bar-tailed Godwit (*Limosa lapponica*), Northern Pintail (*Anas acuta*), and Siberian Crane (*Leucogeranus leucogeranus*) rely on such habitats to replenish their energy reserves during transcontinental flights. Range coverage of wetlands along migratory corridors often determines the existence and breeding success of the species [8, 9].

Wetlands are also safe breeding and nesting sites, especially in shallow vegetation with nest structural defences and land predator refuge. Herons, egrets, and grebes breed in emergent plants, while ducks and geese breed in shoreline formations. Wetlands also provide cover and roosting structures that protect environmental stressors like storms and temperature. The ecological worth of wetlands transcends the immediate provision of habitat. By supporting bird populations, they play a role in nutrient cycling, seed dispersal, and trophic regulation and hence guarantee overall ecosystem stability. Wetland loss or deterioration deprives these functions, triggering decreases in bird abundance, modifications in bird migration patterns, and even extinction. Wetlands are therefore irreplaceable ecological commodities in international bird conservation.

## 1.2 Wetland Ecosystems and Their Ecological Functions

Wetlands are likely the most ecologically diverse and nature-rich ecosystems on Earth, serving a vast range of functions critical for sustaining ecosystem biodiversity, controlling water budgets, and hosting a truly vast array of fauna, especially aquatic birds. Wetlands are transitional ecotones between aquatic and terrestrial ecosystems with critical roles in nutrient cycling, sediment trapping, and biotic interactions, as argued by Mitsch & Gosselink [30]. Their structural differences and functional diversity render them essential to the ecology of birds, specifically for migratory and resident waterbird populations. Wetlands hold a variety of habitat structures like open water, mudflats, marsh vegetation, and reed beds as a source of overall breeding, nesting, foraging, and roosting requirements of waterbirds [10]. It supports diverse trophic resources, such as invertebrates, aquatic, fish, amphibians, and seeds, and a dynamic avian food web that supports specialist and generalist bird species. The microhabitat mosaics of wetlands augment ecological niches, thus species richness and abundance.

From an ecosystem services viewpoint, wetlands act as natural filters for water by capturing sediments and purifying pollutants like heavy metals, excess nutrients, and pathogens through plant and microbial process led biogeochemical processes. This filtering process indirectly ensures bird health by maintaining aquatic environment quality. In addition, wetlands control hydrological regimes through floodwater storage, aquifer recharging, and river base flows maintenance, and therefore they ensure hydrological stability for the water bird breeding cycles [57]. Wetlands also regulate carbon sequestration and climate. Mangrove wetlands and peat wetlands have high organic carbon content and therefore are very important in the regulation of greenhouse gas emissions. This climatic control influences seasonally migratory and seasonality of the breeding period in bird species, which are equally very sensitive to temperature and hydrological fluctuation [18]. Furthermore, ecological health of wetlands preserves intra-species interactions like predator-prey and symbiotic interactions that underpin ecosystem balance. For instance, having bird predators like egrets and herons will determine the structure of populations of fish and amphibians and thereby indirectly influence aquatic vegetation through trophic cascades [3].

Nonetheless, pressures caused by humans such as land use change, pollution, hydrological alteration, and climate change are degrading wetland habitats globally. Direct effects of wetland loss or degradation to aquatic birds are reduced habitat quantity and quality. Thus, conservation strategies that are multi-faceted including restoration of habitats, pollution management, and planning for sustainable land use are important to ensure the sustenance of wetland ecological processes and bird populations relying on them. Wetlands are intricate ecological networks that deliver fundamental services for aquatic birds. Their structural complexity, nutrient cycling function, and diversity of habitats are all essential to the very existence and well-being of bird communities on temporal and spatial scales. Conservation and management of these systems is as much an imperative of biodiversity conservation as it is the foundation of sustainable ecological management.

**Aquatic Birds: Diversity and Dependence on Wetlands**

Water birds are a vast group of birds living in and around bodies of water. They can be broadly categorized into:

- **Migratory species**: A few examples include Northern Pintail (*Anas acuta*), Common Teal (*Anas crecca*), and Bar-headed Goose (*Anser indicus*). They migrate long distances from breeding to wintering sites and rely significantly on stopover wetlands as resting and refueling habitats during migration [62, 63].

- **Resident species:** i.e., Indian Spot-billed Duck (*Anas poecilorhyncha*), Purple Moorhen (*Porphyrio poliocephalus*), Little Grebe (*Tachybaptus ruficollis*). These stay continuously in wetlands and use them for the whole life cycle [20].

## 1.3 The Value of Wetlands to Migratory Birds

### *1.3.1 Food Sources*

Wetlands are one of the most biologically productive habitats with high primary and secondary productivity based on nutrient-rich sediments and hydrological shallowness [30]. Such habitats provide supportive rich trophic levels like phytoplankton, aquatic macrophytes, benthic algae, invertebrates (mollusks, crustaceans, insect larvae), and small fish that offer housing to avian populations as a whole. These have migratory birds to rely on for breeding and migratory high energetic requirements. An example is the greater flamingo (*Phoenicopterus roseus*) in Chilika Lake, Odisha, whose diet consists of intense concentrations of cyanobacteria, diatoms, and planktonic crustaceans based on its lamellae-filtering beak specializations [12].

### *1.3.2 Breeding and Nesting Sites*

Semi-aquatic edge wetlands dominated by emergent plants like reeds (*Phragmites australis*) and sedges (*Cyperaceae*) provide the best nesting environment with the addition of structural support, predator safety, and microclimatic stability [35]. The water surrounding the wetlands also provides physical protection from ground predators, improving breeding success in herons (*Ardeidae*), storks (*Ciconiidae*), and jacanas (*Jacanidae*). Colonial nesters especially maximize these protective advantages, which minimize predation threat and enable cooperative defence.

### *1.3.3 Stopover Sites Along Migration*

Wetlands are essential stopover points over major flyways such as the Central Asian Flyway, enabling birds to restore lipid and glycogen reserves necessary for maintaining long-distance flight [33]. Wetland loss of essential stopover habitats may break migratory connectivity, decrease survival, and threaten reproductive success. Avian species such as the bar-headed goose (*Anser indicus*) and northern pintail (*Anas acuta*) utilize sequential wetland habitats for resting and refueling on transcontinental migrations.

### 1.3.4 Shelter and Rest

In addition to breeding and feeding sites, wetlands offer important shelter for rest during migration, molting, and storms. Close reed thickets, floating vegetation mats, and open water spaces provide refuge from predators and human impacts [53]. Migration life cycle and migratory birds' dependence on wetlands, illustrating key phases like breeding, nesting, feeding, migration, and resting, wherein wetlands offer critical resources for survival and reproduction. Heterogeneity in structure of wetland vegetation also shields the birds from wind and temperature fluctuation, lowering metabolic stress, which is of greater significance in juveniles and molts.

## 1.4 Aquatic Bird Ecosystem Services

Aquatic birds play an essential function in preserving the ecological value of wetlands. All of their functions help preserve biodiversity, nutrient cycling, pest regulation, and socio-economic value in terms of ecotourism and education [21]. Some aquatic birds, especially water birds (e.g., duck, goose, and swan), assist in the dispersal of wetland plants' seeds by endozoochory (transport in the gut by ingestion) and ectozoochory (transport on the feathers or feet). Seeds that have travelled through the digestive tract show enhanced germination activity through partial scarification. Figure 1.1 depicts life cycle and dependence of migratory birds on wetlands, highlighting vital phases such as breeding, nesting, feeding, migration, and resting, where wetlands play an important role in their reproduction and survival. This facilitates the colonization of new wetland environments and the conservation of plant genetic diversity [13, 48]. Dispersal by migratory ducks, for instance, has been found to disperse *Potamogeton pectinatus* seeds hundreds of kilometers from one another and thus connect otherwise fragmented wetland systems.

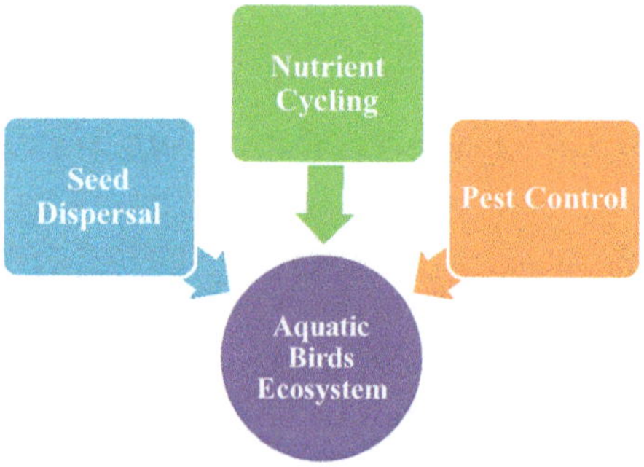


**Fig. 1.1** Diagrammatic representation of Aquatic Bird Ecosystem Services. (Source: Author)

- **Nutrient Cycling**: Waterbirds are significant contributors to nutrient cycling in wetlands through guano deposition. Waterbird feces contain high concentrations of nitrogen (N) and phosphorus (P), which fuel primary productivity by facilitating phytoplankton and macrophyte growth [14, 28]. Seabird colonies in coastal wetlands transfer marine-derived nutrients onto land, fueling terrestrial plant and invertebrate assemblages. Over-enrichment of small wetlands with nutrients can induce eutrophication, supporting arguments for balanced population dynamics.
- **Pest Control**: The insect population, larvae, mollusks, and other invertebrate pests are kept at manageable levels by aquatic birds through top-down control actions. Herons (*Ardea* spp.), terns (*Sterna* spp.), and shorebirds consume large quantities of insect larvae (like mosquito larvae), thus keeping surrounding human populations at low risk for vector-borne diseases like malaria and Japanese encephalitis [6, 47]. This environmental service is most valuable in agro-ecosystems around wetlands, where birds also serve to control pests that have an impact on crops.
- **Cultural Services:** Wetlands and avifauna have deep cultural, recreational, and educational value. Ecotourism and bird tourism contribute significantly to the economic benefits of local communities and promote environmental protection [49]. Signature birds like the Siberian crane (*Leucogeranus leucogeranus*) of Keoladeo National Park receive thousands of visitors every year. Wetlands also provide living laboratories for ornithological research, citizen science projects, and environmental education programs, and thus increase public awareness of conservation requirements [29].

## 1.5 Impacts on Wetland Aquatic Birds

Wetlands worldwide are increasingly threatened by human activity of major consequence for waterbirds. Habitat loss due to urbanization, intensive agriculture, and construction such as dam and canal schemes leads to degradation and fragmentation of high-quality breeding and feeding areas [17, 55, 56, 62]. Additionally, industrial effluent, agricultural runoff with pesticides and fertilizers, and plastic litter contaminate wetland ecosystems and affect the health and breeding quality of birds [30, 52]. Climate change is a pressing driver as well since altered conditions of temperature and precipitation increase result in wetland desiccation, shifting migratory timing, and exacerbated food deficiency [11, 18]. Apart from that, alien invasion of plants such as water hyacinth (*Eichhornia crassipes*) displaces indigenous biodiversity and breeding grounds, whereas human invasions such as uncontrolled tourism, disturbance, and invasion stress and displace vulnerable bird populations [16, 26]. In total, all these multifaceted threats establish the sheer need for sustainable conservatory practice for the conservation of both wetlands and their avian dependent communities. The major effects on wetland aquatic birds, where it describes how habitat destruction, food deficiency, disruption of breeding, migration obstacles, pollution, and global warming affect their survival and reproduction in

**Table 1.1** Impacts of wetland loss on aquatic birds

| Effect area | Description | Examples |
|---|---|---|
| Loss of habitat | Draining, encroachment, and land conversion decrease nesting and feeding habitat. | Decline in India's Sarus cranes' population. |
| Food shortage | Pollution and eutrophication shift prey numbers, decreasing feeding efficiency. | Less fish for cormorants and herons. |
| Breeding disturbance | Disturbance and water fluctuation disrupt nesting and chick survival. | Decreased storks' and pelicans' nesting success. |
| Migration barriers | Wetland contraction limitations restrict stopover/ resting places for long-distance migration. | Decline of migratory birds such as Bar-headed goose. |
| Pollution-induced death | Ingestion of plastic, toxin, and pesticide residue causes disease and death. | Duck and goose mass mortality events. |
| Climate change stress | Changes in rainfall and temperature regimes induce wetland hydrology modification. | Flamingo and pintail route modification. |

general show in (Table 1.1). Wetlands are some of the most productive ecosystems on earth, but they are faced with numerous anthropogenic and natural stresses that have led to serious effects on aquatic birds worldwide. The main threats are as follows:

I. **Habitat loss due to Urbanization and Agriculture:** Increased urbanization and agricultural expansion have led to the large-scale reclamation of wetlands, drainage, and conversion into agricultural land, industrial estates, and settlements. It reduces the quantity of breeding, foraging, and roosting habitats for waterbirds [8, 9, 23, 24]. For example, paddy field expansion and aquaculture activities displaced significant wetland habitats in Asia, directly impacting species such as the Painted Stork (*Mycteria leucocephala*) and Black-tailed Godwit (*Limosa limosa*). Fragmentation of wetland habitats also isolates bird populations, making them less genetically diverse and resilient.

II. **Water Pollution:** Industrial effluents, raw sewage, and pesticide-laden agricultural runoff pollute wetland water to cause eutrophication, toxic buildup, and changed food web composition [27]. Heavy metals such as lead and mercury bioconcentrate in fish and invertebrates and are ingested by piscivorous and omnivorous birds, resulting in reproductive collapse, neurotoxicity, and increased mortality [46]. Eutrophication also causes low dissolved oxygen levels, which impact the populations of fish, thereby affecting the aquatic bird prey base.

III. **Egg collection and overhunting:** Unsustainable waterbird hunting for food, recreation, and trade is a general threat for most developing regions [63]. Egg collection for sale or consumption prevents breeding success and causes local population reduction, especially in colonial breeding species like terns and gulls. Overhunting directly reduces numbers and also causes disturbance leading to the abandonment of nests or feeding sites.

IV. **Climate Change and Regime Shift of Water:** Global climate change is modifying precipitation regimes, temperature conditions, and hydrology, leading to seasonal wetland desiccation, sea-level rise, and modification of flood pulse regimes [17]. The alterations may shift the timing of food, thereby causing mismatches between the arrival times of migratory birds and food availability. Additionally, elimination of breeding habitats in the Arctic through melting permafrost and shifting tundra vegetation impacts long-distance migrants like plovers and sandpipers.

V. **Invasive Species Altering Food Webs:** Non-native like water hyacinth (Eichhornia crassipes), zebra mussel (Dreissena polymorpha), and predatory fish like Nile perch (Lates niloticus) may significantly alter wetland communities. Exotic aquatic vegetation overrides native vegetation, decreasing habitat complexity for nesting and foraging [50]. Invasive predators and competitors may decrease native fish and invertebrate populations, depriving aquatic birds of key food sources [50].

## 1.6 Wetland Loss Consequences on Aquatic Bird Communities

Wetland degradation and loss lead to direct reductions in habitat carrying capacity critical to the survival of migratory birds. These include migratory species like the Bar-headed Goose (*Anser indicus*) and Northern Pintail (*Anas acuta*), which rely on a series of intact wetland stopovers upon which to rest and fuel during long-distance flights. Loss of a single large wetland on these corridors could create a bottleneck effect, which leads to abrupt population crashes and decreased species richness [26, 31].

I. **Disturbance of migration cycles**: Other bird aquatic species such as terns, herons, and storks nest in extremely large colonies with particular habitat requirements such as shallow water bodies, reed stands, or mud flats. Physical destruction of breeding sites from the loss of wetlands due to drainage, land reclamation, or hydrology changes can lead to reduced opportunities for breeding. This means decreased reproductive success that in turn destabilizes population trends [25]. Seasonal cues like water levels, temperature, and food availability control migratory water birds' migratory timing. Climate change and wetland loss alter hydrological regimes, thereby disorganizing these cues. This might lead to temporal mismatches between peak food and migratory times to the detriment of survival and reproductive success [32, 39].

II. **Species-specific extinction risk:** Specialized species like the Siberian Crane (*Leucogeranus leucogeranus*) and Baer's Pochard (*Aythya baeri*) possess highly restricted ecological niches, relying solely on particular wetland habitats for searching and breeding. Their loss or alteration raises the threat to extinction since they are unable to rapidly adapt to substitute habitats (BirdLife International 2021; [8, 9]).

### 1.6.1 Research Methods in Wetland and Bird Studies

Wetland ecosystems and their avifaunal communities necessitates a multidisciplinary framework of integrated field and laboratory protocols. Application of field survey methods such as point counts, line transects, and mist netting are critical in establishing bird abundance, diversity, and behavioral ecology across different wetland habitats [5, 41]. In the meantime, water and habitat quality surveys with physicochemical analysis (dissolved oxygen, pH, nutrient levels), vegetation mapping, and hydrological profiling are important in determining ecological well-being and its impact on bird populations [30]. The use of remote sensing coupled with Geographic Information Systems (GIS) has greatly enhanced wetland monitoring as it has provided the spatial examination of land use change, wetland loss, as well as habitat fragmentation over time [34, 45]. In addition, population modeling with programs like MARK and statistical modeling with programs like generalized linear models (GLMs) enables the quantification of population trends, survival, and habitat use, enabling evidence-based conservation policy [51, 54]. These methods collectively provide the backbone to modern wetland ornithology research, enabling us to strengthen our knowledge of ecological processes and guide effective management and policy interventions. The human action and policy management system driving wetlands, with unlimited population expansion, water use, and poor action leading to wetland degradation, but managed finance, coordination, and community engagement leading to wetland rehabilitation are required to maintain water bird populations (Fig. 1.2).

## 1.7 Conservation Strategies

Wetland conservation is essential for the protection of waterbird populations and their ecological roles. Successful management is a four-dimensional process involving the balancing of legislative frameworks, habitat protection, community involvement, scientific monitoring, and coordination of wetland conservation within overall policy agendas.

I. **Legislative Protection:** National laws and international treaties have an essential role to play in the protection of wetlands and their bird-of-wetland communities. The Ramsar Convention on Wetlands (1971) is an international agreement offering a global framework for the conservation and sustainable. Utilization of wetlands are Legal protection for many bird species under the Wildlife Protection Act (1972) in India and forbids acts that would result in wetland destruction. Bilateral agreements like the India & Japan and India & Australia Migratory Bird Agreements also promote joint research collaboration, habitat maintenance, and the conservation of mutual migratory birds species. Legal protection not only forestalls instant exploitation but also guarantees adherence to sustainable development objectives [38, 61].

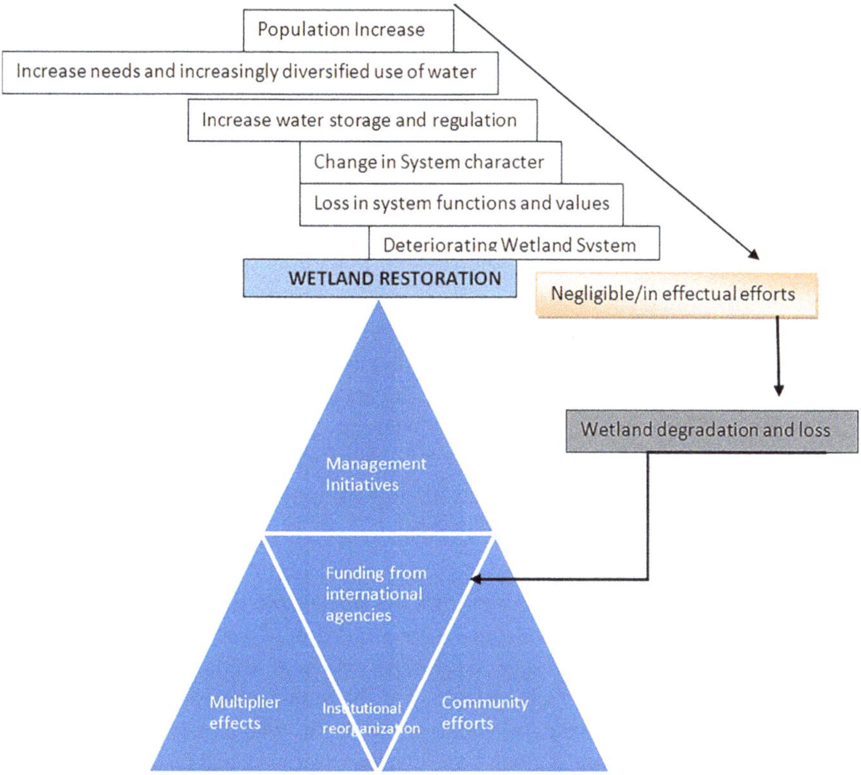


**Fig. 1.2** Framework of human activities and policies for wetland degradation versus restoration. (Source: Author)

II. **Wetland restoration:** Restoration of wetlands is central to reviving lost ecosystem functions and bird populations. Restoration involves re-wetting drained wetlands, natural hydrological regimes restoration, river-floodplain reconnection, and reintroduction of indigenous aquatic vegetation. These measures promote waterbird feeding and nesting habitats and enhance climate change resilience of wetlands [57]. For instance, rehabilitation efforts at Keoladeo National Park (Rajasthan) and Chilika Lake (Odisha) have shown encouraging results in the enrichment of waterbirds [36]. Wetland conservation should not be conducted in isolation but incorporated into national climate change adaptation and sustainable development policy. Wetlands are carbon biological sinks, flood buffers, and reservoirs of biodiversity and thus significantly contribute to adaptation resilience to global warming. Institutionalization of wetland management in policy plans such as India's National Action Plan on Climate Change (NAPCC) fosters inter-sectoral convergence of environment, agriculture, and water resources ministries [8, 9]. Integration thus makes the aims of conservation align with socio-economic development to the benefit of long-term sustainability.

III. **Community Engagement:** Community participation is of utmost significance in wetland preservation. Ecotourism projects, bird-watching festivals organized by the community, and public awareness campaigns socialize the populace to appreciate the wetlands as a livelihood source and measure of biodiversity. Participatory methods empower the populace to turn into custodians of the wetlands and restrict non-sustainable utilization like poaching or over-fishing [40]. In India, conservation in and around Loktak Lake (Manipur) has proven that a convergence of socio-economic dividends with conservation objectives can make a significant contribution to mitigating human-wildlife conflict [10]. Technological advancements of modern times, i.e., satellite telemetry, GIS-based habitat mapping, and drone monitoring, are transforming wetland conservation. Satellite tracking of bird flyways during bird migration identifies priority stopover sites and potential threats along flyways [22]. Long-term tracking of bird numbers offers early warning signs of habitat degradation for proactive management. The Asian Waterbird Census (AWC) conducted by Wetlands International has yielded sound long-term data on trend in migratory and resident waterbirds throughout Asia.

International stimulus to wetland conservation has arisen through agreements such as the Ramsar Convention which promotes the sustainable use and conservation of internationally important wetlands and their protection as waterfowl habitats [62]. There are a good number of successful experiences of wetland restoration, like Chilika Lake in India, in which hydrological rehabilitation restored bird life and fishery livelihood [37], Doñana wetlands in Spain, in which restoration was augmented by water reallocation and land-use management [44], and Okavango Delta in Botswana, in which watershed management that incorporated has maintained biodiversity even under developmental pressure [42]. Local society and citizen science efforts are important contributors to wetland stewardship by means of bird tracking, monitoring of ecological processes, and eco-restoration, as in the case of India's Loktak Lake model of wetland conservation and European Wetland Bird Survey [63]. Strong policy and legal measures, e.g., India's Wetlands (Conservation and Management) Rules, EU's Water Framework Directive, and country-level biodiversity strategies, all working to avoid loss of habitat, enable wise use, and include wetlands in planning for climate resilience [29, 59]. Collectively, these multilevel initiatives bring into focus the need for coordination among international agreements, science-driven restoration, and local action in protecting wetland ecosystems and their avifaunal diversity.

## 1.8 Future Prospects

New technologies like artificial intelligence (AI) and unmanned aerial vehicles (drones) are changing wetland monitoring through the offering of high-resolution mapping, machine-based species identification, and real-time estimation of habitat change. They offer efficient and reliable data for migratory bird population analysis and wetland dynamics.

As migratory birds fly across more than one geopolitical border, international collective action must be taken in order for them to be conserved. Cross-border science and multilateral conventions consolidate flyway networks and provide protection to key stopover points and coordinate conservation efforts at an international level.

Because climate change has substantive impacts on wetland hydrology and bird habitats, adaptive management strategies must be used. Actions like restoration of natural hydrological processes, enhancement of the carbon sequestration capacity, and inclusion of wetlands in climate resilience planning will yield long-term biodiversity conservation as well as ecosystem services.

## 1.9 Conclusion

Wetlands are specialized ecosystems that offer significant wetland habitat for waterbirds in facilitating breeding, foraging, and migratory activities. The chapter highlights the significance of biodiversity in wetlands, avian community life history dependency, and extensive threats of habitat loss, pollution, and climate change [8, 9, 60]. Conservation practices embedded through international frameworks such as the Ramsar Convention, participatory approaches, restoration of habitats, and policy implementation are necessary to provide resilience to wetland ecosystems as well as avifaunal species. Adaptation has to be ecosystem-based, community-based, and adaptive monitoring-based to provide effective wetland management [18, 58]. However, critical research gaps remain in understanding long-term ecological processes, climate-induced shifts in avian migration, and the effectiveness of current restoration strategies. Upcoming studies ought to focus on interdisciplinary studies, utilizing technologies like remote sensing, predictive modeling, and citizen science for tracking and controlling wetland-bird interaction under changing environmental conditions [15, 19]. A science-based, integrated strategy is needed to secure future generations of the ecological and socio-economic services of wetlands.

Wetlands are amongst the most productive systems on the planet and are essential as habitats of bird migration and aquatic food chains. Wetlands offer essential ecosystem services such as feeding, breeding, roosting, and stopover sites in long-distance migration. Structural wetland complexity from shallow vegetation to open-water sites supports a variety of trophic interactions and maintains a diversified richness of bird guilds. As a result, migratory birds play an important role in wetland processes including seed dispersal, nutrient cycling, and the regulation of invertebrate populations, thus strengthening the two-way ecological link between bird species and wetland ecosystems.

The high rate of wetland loss due to urbanization, agricultural encroachment, industrial pollution, climate change, and biological invasions poses a severe threat to avian populations. The stressors not only reduce bird numbers and diversity but also change migratory patterns, ruin breeding colonies, and pose a threat to species that live in wetlands with extinction. Downstream effects of wetland loss extend beyond avian diversity to affect ecosystem health, human well-being, and global ecological balance. Conservation of wetlands thus calls for an integrated strategy involving legal protection, habitat restoration, scientific monitoring, policy direction, and public involvement. Legal frameworks such as international agreements and national policy form the basis, but successful implementation hinges on stakeholders' active engagement and adaptive management measures. Integrating wetland conservation into climate change adaptation planning and sustainable development policy is especially important to protect avian diversity and human health in the Anthropocene age. Wetlands are not just important freshwater and migratory bird ecosystems but also keystone landscapes for global biodiversity and ecological resilience maintenance. Conservation of these habitats will be enhanced by interdisciplinarity of research, participatory governance, and transboundary cooperation to ensure continued survival and maintenance of the migratory bird communities reliant on them.

## References


1. Bhardwaj, E. A. (2011). SPECTRAL CALIBRATION OF 64 BAND HYPERSPECTRAL IMAGER OF CHANDRAYAAN-1 KR MURALI, GIRISH M. GOUDA, SWATI BISHT, HIMAVANTKUMAR JHAKA and JITENDRA SHARMA Sensors Development Area, Space Applications Centre (ISRO). Planetary Science (PS), 25, 133.
2. Newton, A., Icely, J., Cristina, S., Perillo, G. M., Turner, R. E., Ashan, D.,... & Kuenzer, C. (2020). Anthropogenic, direct pressures on coastal wetlands. Frontiers in Ecology and Evolution, 8, 144.
3. Batzer DP, Sharitz RR (eds) (2014) Ecology of freshwater and estuarine wetlands. University of California Press
4. Sharma, S., Phartiyal, M., Madhav, S., & Singh, P. (2021). Global wetlands: categorization, distribution and global scenario. Wetlands conservation: current challenges and future strategies, 1-16.
5. Bibby CJ, Burgess ND, Hill DA, Mustoe SH (2000) Bird census techniques. Academic Press
6. Buchanan GM, Donald PF, Butchart SH (2011) Identifying priority areas for conservation: a global assessment for forest-dependent birds. PLoS One 6(12):e29080
7. Cowardin LM, Carter V, Golet FC, LaRoe ET (1979) Classification of wetlands and deepwater habitats of the United States. U.S. Fish and Wildlife Service
8. Davidson NC (2014) How much wetland has the world lost? Long-term and recent trends in global wetland area. Mar Freshw Res 65(10):934–941
9. Davidson NC (2014) How much wetland has the world lost? Mar Freshw Res 65(10):934–941. https://doi.org/10.1071/MF14173
10. Jain, M., Garg, V. K., & Kadirvelu, K. (2011). Investigation of Cr (VI) adsorption onto chemically treated Helianthus annuus: optimization using response surface methodology. Bioresource technology, 102(2), 600-605
11. Davidson NC, Dinesen L, Fennessy S, Finlayson CM, Grillas P, Grobicki A, McInnes RJ, Stroud DA (2019) A review of the adequacy of reporting to the Ramsar Convention on change in the ecological character of wetlands. Mar Freshw Res 71(1):117–126

12. Kundu, D., Vajargah, S. H., Wan, L., Adams, B., Prendergast, D., & Nazar, L. F. (2018). Aqueous vs. nonaqueous Zn-ion batteries: consequences of the desolvation penalty at the interface. Energy & Environmental Science, 11(4), 881-892.
13. Figuerola J, Green AJ (2002) Dispersal of aquatic organisms by waterbirds: a review of past research and priorities for future studies. Freshw Biol 47(3):483–494
14. Hahn, O., Carollo, C. M., Porciani, C., & Dekel, A. (2007). The evolution of dark matter halo properties in clusters, filaments, sheets and voids. Monthly Notices of the Royal Astronomical Society, 381(1), 41-51.
15. Gawlik DE (2006) The role of wildlife science in wetland ecosystem restoration: lessons from the Everglades. Ecol Eng 26(1):70–83
16. Gopal B (2013) Future of wetlands in tropical and subtropical Asia, especially in the face of climate change. Aquat Sci 75:39–61
17. Finlayson, C. M., Davies, G. T., Moomaw, W. R., Chmura, G. L., Natali, S. M., Perry, J. E.,... & Sutton-Grier, A. E. (2019). The second warning to humanity–providing a context for wetland management and policy. Wetlands, 39(1), 1-5.
18. Junk WJ, An S, Finlayson CM, Gopal B, Květ J, Mitchell SA, Mitsch WJ, Robarts RD (2013) Current state of knowledge regarding the world's wetlands and their future under global climate change: a synthesis. Aquat Sci 75:151–167
19. Kingsford RT, Walker KF, Lester RE, Young WJ, Fairweather PG, Sammut J, Geddes MC (2011) A Ramsar wetland in crisis–the Coorong, lower lakes and Murray mouth, Australia. Mar Freshw Res 62(3):255–265
20. Kiran, Singh D, Kour A, Priya, Delu V, Kumar R (2022) Community composition and status of avian diversity at Campus and Agricultural landscapes of Chaudhary Charan Singh Haryana Agricultural University, Hisar (Haryana). J Appl Nat Sci 14(4):1130–1140. https://doi.org/10.31018/jans.v14i4.3784
21. Kiran, Singh D, Kour A, Priya, Delu V, Kumar R, Punia N, Sunita (2022) Different strategies adopted by birds to sustain ecosystem: a review. Pharm Innov SP-11(9):412–422
22. Kölzsch A, Müskens GJ, Kruckenberg H, Glazov P, Weinzierl R, Nolet BA, Wikelski M (2016) Towards a new understanding of migration timing: slower spring than autumn migration in geese reflects different decision rules for stopover use and departure. Oikos 125(10):1496–1507
23. Kumar R, Singh D, Yadav R (2025) Toxic effects of chlorantraniliprole and fipronil on *Eisenia fetida*: from growth and reproductive parameters to avoidance behavior. Environ Eng Manag J 24(7):1499–1508. Accepted manuscript
24. Kumar R, Singh D (2022) Insecticide- tolerant bacterial population in *Eisenia fetida*'s gut and Vermicast exposed to Chlorantraniliprole and Fipronil. Appl Biol Res 24(3):273–279
25. Kushlan JA (1997) The conservation of wading birds. Colon Waterbirds 20:129–137
26. Ma, X., Geiser-Lee, J., Deng, Y., & Kolmakov, A. (2010). Interactions between engineered nanoparticles (ENPs) and plants: phytotoxicity, uptake and accumulation. Science of the total environment, 408(16), 3053-3061.
27. Li, D., Hsu, S., Purushotham, D., Sears, R. L., & Wang, T. (2019). WashU epigenome browser update 2019. Nucleic acids research, 47(W1), W158-W165.
28. Manny BA, Johnson WC, Wetzel RG (1994) Nutrient additions by waterfowl to lakes and reservoirs: predicting their effects on productivity and water quality. Hydrobiologia 279(1): 121–132
29. Assessment, M. E. (2005). Ecosystems and human well-being: wetlands and water. World resources institute.
30. Mitsch WJ, Gosselink JG (2015) Wetlands. John wiley & sons
31. Murray, N. J., Clemens, R. S., Phinn, S. R., Possingham, H. P., & Fuller, R. A. (2014). Tracking the rapid loss of tidal wetlands in the Yellow Sea. Frontiers in Ecology and the Environment, 12(5), 267-272.
32. Both, C., Bouwhuis, S., Lessells, C. M., & Visser, M. E. (2006). Climate change and population declines in a long-distance migratory bird. Nature, 441(7089), 81-83.
33. Newton I (2023) The migration ecology of birds. Elsevier

34. Ozesmi SL, Bauer ME (2002) Satellite remote sensing of wetlands. Wetl Ecol Manag 10:381–402
35. Paracuellos M (2006) How can habitat selection affect the use of a wetland complex by waterbirds? Biodivers Conserv 15(14):4569–4582
36. Parikh P, Diep L, Hofmann P, Tomei J, Campos LC, Teh TH et al (2021) Synergies and trade-offs between sanitation and the sustainable development goals. UCL Open Environ 3:e016
37. Pattnaik AK (2024) Reversing the degeneration of an ecological hotspot through an ecosystem approach: a case study of Chilka Lake. In: Regenerative ecosystems in the anthropocene: a transdisciplinary ecosystemic framework for regenerativeness. Springer Nature Switzerland, Cham, pp 129–150
38. Prasad SN, Ramachandra TV, Ahalya N, Sengupta T, Kumar A, Tiwari AK et al (2002) Conservation of wetlands of India-a review. Trop Ecol 43(1):173–186
39. Lehikoinen, P., Santangeli, A., Jaatinen, K., Rajasärkkä, A., & Lehikoinen, A. (2019). Protected areas act as a buffer against detrimental effects of climate change—Evidence from large-scale, long-term abundance data. Global change biology, 25(1), 304-313.
40. Pretty J, Smith D (2004) Social capital in biodiversity conservation and management. Conserv Biol 18(3):631–638
41. Ralph CJ, Sauer JR, Droege S (1995) Monitoring bird populations by point counts. Pacific Southwest Research Station
42. Ramberg MR (2014) Teacher change in an era of neo-liberal policies: a neo-institutional analysis of teachers' perceptions of their professional change. Eur Educ Res J 13(3):360–379
43. Ramsar Convention Secretariat (2022) The Ramsar list of wetlands of international importance. Ramsar Convention Secretariat, Gland
44. Montes, M. F. (1998). The currency crisis in Southeast Asia (Vol. 218, No. 2114). Institute of Southeast Asian Studies
45. Adam, B. (2010). History of the future: Paradoxes and challenges. Rethinking History, 14(3), 361-378
46. Scheuhammer AM, Meyer MW, Sandheinrich MB, Murray MW (2007) Effects of environmental methylmercury on the health of wild birds, mammals, and fish. AMBIO 36(1):12–19
47. Shaalan EAS, Canyon DV (2009) Aquatic insect predators and mosquito control. Trop Biomed 26:223–261
48. Soons MB, Brochet AL, Kleyheeg E, Green AJ (2016) Seed dispersal by dabbling ducks: an overlooked dispersal pathway for a broad spectrum of plant species. J Ecol 104(2):443–455
49. Strayer, D. L. (2010). Alien species in fresh waters: ecological effects, interactions with other stressors, and prospects for the future. Freshwater Biology, 55, 152–174. Publisher: Wiley-Blackwell (John Wiley & Sons).
50. Strayer DL (2010) Alien species in fresh waters: ecological effects, interactions with other stressors, and prospects for the future. Freshw Biol 55:152–174
51. Sutherland WJ, Newton I, Green R (2004) Bird ecology and conservation: a handbook of techniques, vol 1. Oxford University Press
52. Urmila, Gupta RK, Sharma S, Kumar R (2025) Earthworms: the environmentally preferred bioengineers for managing heavy metals. In: Dey S, Bhattacharya S (eds) Biotechnological interventions in the removal of emerging pollutants. Springer Nature Singapore, Singapore, pp 365–384
53. Weller MW (1999) Wetland birds: habitat resources and conservation implications. Cambridge University Press
54. White GC, Burnham KP (1999) Program MARK: survival estimation from populations of marked animals. Bird Study 46(sup1):S120–S139
55. Yadav R, Gupta RK, Kumar R, Kaur T, Moun R (2022) Biochemical response of earthworm, *Eisenia fetida* to heavy metals toxicity. J Appl Nat Sci 14(3):990–998
56. Yadav R, Gupta RK, Kumar R, Kaur T (2022) Assessment of toxicity of Lead and nickel on the biochemical and immunological parameters of earthworm, Eudrilus eugeniae. Agric Sci Dig 44(5):922–929

57. Zedler JB, Kercher S (2005) Wetland resources: status, trends, ecosystem services, and restorability. Annu Rev Environ Resour 30(1):39–74
58. Prigent, C., Warren, J. M., Kohli, A. H., & Teyssier, C. (2020). Fracture-mediated deep seawater flow and mantle hydration on oceanic transform faults. Earth and Planetary Science Letters, 532, 115988.
59. Sinha, D., & Mehta, D. (2017). Environmental Clearances and Monitoring in India: Report Card for the Ministry for Environment, Forest and Climate Change. Vidhi Centre for Policy Change, 1-34.
60. Moomaw, W. R., Chmura, G. L., Davies, G. T., Finlayson, C. M., Middleton, B. A., Natali, S. M.,... & Sutton-Grier, A. E. (2018). Wetlands in a changing climate: science, policy and management. Wetlands, 38(2), 183-205.
61. Ramsar, P. (2013). The Ramsar Convention Manual: a guide to the Convention on Wetlands (Ramsar, Iran, 1971) Ramsar Convention Secretariat.
62. Davidson, N. C. (2016). Ramsar convention on wetlands: scope and implementation. In The Wetland Book (pp. 1-9). Springer, Dordrecht.
63. Squires, T. M., Yuda, P., Akbar, P. G., Collar, N. J., Devenish, C., Taufiqurrahman, I.,... & Marsden, S. J. (2021). Citizen science rapidly delivers extensive distribution data for birds in a key tropical biodiversity area. Global Ecology and Conservation, 28, e01680.

# Chapter 2
# Waterbird Population: Trends, Challenges, Conservation, and Recommendations

**S. K. Gharde, Mukul Machhindra Barwant, Baikunth Kumar Yadav, Geetanjli Sahni, Mohit Deswal, Pooja Kadyan, and Sudhir Kumar Kataria**

**Abstract** Waterbirds are an integral part of the health and functioning of aquatic ecosystems and serve as ecological indicators, and contribute to biodiversity through their roles in nutrient cycling, pest control, and seed dispersal. This chapter examines the global and regional trends in the waterbird populations, revealing alarming declines of many species due to anthropogenic pressures. The key challenges include wetland degradation, habitat fragmentation, climate change, pollution, invasive species, and unregulated hunting. The migratory waterbirds face an additional threat along with flyways due to the inconsistent conservation measures across the countries. This chapter evaluates the current conservation efforts, such as the Ramsar Convention, AEWA, and community-driven wetland management strategies, highlighting the successes and the persistent gaps. It also outlines strategic recommendations like enhanced habitat protection, adaptive management practices, policy integration, long-term monitoring, and public engagement. The present chapter explains that more dedicated work should be put into ensuring that the threats are reduced and enhancing the recovery of waterbirds around the world.

**Keywords** Conservation · Habitat loss · Migratory species · Waterbirds · Wetland

---

S. K. Gharde
Department of Entomology, School of Agriculture, Lovely Professional University, Jalandhar, Punjab, India

M. M. Barwant
Department of Botany, Sanjivani Rural Education Society's Sanjivani Arts, Commerce and Science College, Kopargaon, Maharashtra, India

B. K. Yadav
Department of Zoology, Mahatma Gandhi Central University, Motihari, Bihar, India

G. Sahni
Department of Zoology, Arya P.G. College, Panipat, Haryana, India

M. Deswal · P. Kadyan (✉) · S. K. Kataria
Department of Zoology, Maharshi Dayanand University, Rohtak, Haryana, India

 
R. Kumar et al. (eds.), *Ecology, Conservation of Aquatic Birds*, Wetlands: Ecology, Conservation and Management 15, https://doi.org/10.1007/978-3-032-17125-2_2

## 2.1 Introduction

Waterbirds are an ecologically diverse group of birds, which is dependent on the freshwater, coastal, and wetland environments. They occupy a unique niche in the operation of ecosystems and biodiversity in the whole world. All these birds occur on all continents and they belong to the families such as the Anatidae (ducks, geese, swans), Ardeidae (herons, egrets), Scolopacidae (sandpipers, curlews) and so on. They make an extensive range of water bodies their migrating, foraging, and nesting as well as roosting places [75]. They have many more ecological values than serving as charismatic wildlife; waterbirds are important indicators of the health of the ecosystem, seed dispersers and distributors of nutrients, essential members of aquatic food webs.

The global population of waterbirds has however reduced drastically in the last few decades. The Living Planet Index showed that freshwater vertebrates which represented most of waterbird species have declined by an average of 83% since the year 1970 [78]. An indicator of more environmental degradation is pollution, climate-induced changes in the habitats, and outlandish conversion of wetlands. Migratory waterbirds are of particular concern to local and global issues, because of their reliance on intact networks of breeding, staging, and wintering habitats across continents [40, 82].

Waterbirds form one of the groups of most closely monitored birds around the globe in large part owing to their ecological significance and prominence. Examples of long term large scale monitoring initiatives include the International Waterbird Census (IWC) and Asian Waterbird census as well as regional atlas bird projects, which have produced a rich source of data on population trends, distributional changes and conservation needs [71]. Through such endeavours, it has been established that a significant number of species continue to decline, particularly in Africa, Asia and the Pacific, whilst in some cases within certain species in Europe and North America, they have been stable or even increased due to specialised intensive conservation measures [74]. Migratory shorebirds and island endemics are among the most threatened as they are easily losing a habitat in a significant site or can be threatened by invasion of the foreign species or hunting [61, 64].

Whereas the Ramsar Convention on Wetlands and other international agreements have provided policy bases that can be used to protect these areas, in most cases, there is a hiatus between scientific knowledge and practice. One of the reasons why conservation is hampered in most parts of the world is due to socioeconomic limitations, ineffective governance, an inadequate amount of data, as well as the failure to integrate all the sectors of a given region, such as agriculture, fishery and urban development [22].

A thorough analysis of the dynamics of waterbird populations worldwide is given in this chapter. It is designed to initially examine temporal and spatial changes in waterbird populations, emphasizing species-specific patterns and regional variations. The methods and instruments used to monitor waterbirds are covered in detail in a separate section, with a focus on the contributions of telemetry, citizen research, and cutting-edge technology like drones and machine learning. The following parts concentrate on conservation strategies, which include international collaboration through flyway projects, habitat restoration, and protected area designation. The presentation of a number of case studies offers useful insights into the complexities of waterbird conservation by illuminating achievements and difficulties in actual settings. To give researchers, conservationists, politicians, and students a basic understanding, this chapter synthesizes contemporary scientific literature, global conservation initiatives, and practical methods. For waterbird populations and the ecosystems, they represent to survive, cross-sectoral, multidisciplinary, and transboundary methods are critically needed. The destiny of wetlands and water-dependent communities around the world is similar to that of waterbirds, which serve as mobile environmental health sentinels. In addition to maintaining avian diversity, their conservation is essential to maintaining the ecological underpinnings that support human well-being.

## 2.2 Importance of Waterbirds in Ecosystems

Waterbirds are an integral part of aquatic and semi-aquatic ecosystems globally. Their ecological importance transcends their biological value, contributing significantly to nutrient cycling, seed dispersal, pest control, and the maintenance of wetland health. In addition, waterbirds serve as reliable bioindicators, reflecting the overall state of the environments they inhabit [31].

### *2.2.1 Nutrient Cycling and Energy Flow*

Many waterbirds influence the nutrient dynamics of wetland ecosystems through their feeding and excretion. For example, geese and swans graze on aquatic vegetation, influencing primary production and nutrient availability [54]. Their droppings contribute significant amounts of nitrogen and phosphorus, which can fertilize plant communities or, in some cases, contribute to eutrophication depending on population densities [38, 41]. Colonial waterbirds such as herons, cormorants, and pelicans, which roost and nest in large numbers, can drastically alter soil chemistry beneath nesting trees through guano deposition [3].

### *2.2.2 Seed Dispersal and Vegetation Dynamics*

Waterbirds contribute to seed dispersal both internally (endozoochory) and externally (epizoochory), affecting plant community composition and colonization of remote wetlands. Ducks and other dabbling birds ingest seeds of aquatic and riparian plants, which may remain viable after gut passage and be deposited in new habitats [59]. This behavior enhances plant genetic exchange across fragmented landscapes and supports the recovery of degraded wetlands [70].

### 2.2.3 *Trophic Regulation and Pest Control*

Waterbirds regulate food webs by preying on fish, invertebrates, and insects. Piscivorous birds such as grebes, herons, and terns can control fish populations, influencing fish community structure and reducing the dominance of certain species [77]. In agricultural landscapes, insectivorous waterbirds like storks and ibises contribute to pest control by feeding on grasshoppers, caterpillars, and beetles, indirectly supporting crop production [48]. Shorebirds foraging in intertidal zones can also influence benthic invertebrate abundance and distribution, thereby affecting sediment dynamics and nutrient cycling [69].

### 2.2.4 *Ecosystem Engineering and Habitat Maintenance*

Some waterbirds act as ecosystem engineers by modifying habitats. Flamingos stir up sediments while feeding, redistributing nutrients. They maintain open mudflats that support diverse invertebrate communities [10]. The nesting colonies of large birds can change the structure of vegetation through trampling, guano deposition, or by tree mortality, which can create a new microhabitat for the other organisms [26]. These modifications sometimes seen as locally destructive and are the natural processes that promotes the dynamic wetland ecosystems.

### 2.2.5 *Bioindicators of Environmental Health*

Waterbirds are most widely used as bioindicators in conservation and the ecological research. Because of their visibility, migratory behavior, habitat specificity, and trophic diversity, they are sensitive to changes in habitat quality, pollution levels, and hydrological regimes [20, 39]. The changes in their abundance can be the signal which underly problems such as heavy metals contamination, eutrophication, or water scarcity. As such, long-term monitoring of the waterbird populations is a central tool for wetland assessment and management [37, 52, 79].

### *2.2.6 Cultural, Economic, and Aesthetic Value*

Waterbirds also contribute substantially to human society. Many cultures regard specific waterbird species as sacred, symbolic and totemic [47]. The birdwatching and eco-tourism centred around the wetland birds generate a significant revenue and promote the local conservation efforts [60]. Major Sustainable hunting practices involve some waterfowl species and also contribute to livelihoods and food security in rural areas, particularly where traditional knowledge governs the harvest levels [68].

## 2.3 Global Distribution Overview

Waterbirds occupy a wide range of habitats across all continents and biomes, from the tundra of the Arctic to tropical mangroves, highland lakes, and coastal wetlands. Their global distribution is shaped by ecological needs such as water availability, breeding and feeding habitats, and migratory behavior. As a group, waterbirds are highly mobile and often migratory, exploiting seasonal resources and spanning vast distances between breeding, stopover, and wintering sites [12].

### *2.3.1 Geographic Patterns of Richness and Diversity*

Waterbird diversity is unevenly distributed across the globe. Tropical and subtropical regions—particularly Southeast Asia, sub-Saharan Africa, and parts of South America—tend to host high species richness due to a variety of wetland types, climatic stability, and habitat productivity [24]. For instance, the Amazon basin, Congo River system, and Mekong Delta are hotspots of waterbird diversity and endemism [82].

Inland wetlands, estuaries, deltas, lakes, floodplains, and mangrove forests all serve as vital habitats. Some species are highly specialized, depending on narrow habitat types such as saline lakes (e.g., Lesser Flamingo *Phoeniconaias minor* in East Africa), while others are generalists capable of adapting to human-altered landscapes, such as the Cattle Egret (*Bubulcus ibis*), which has expanded its range globally [48].

### *2.3.2 Flyways and Migration Corridors*

The migratory behavior of many waterbirds creates distinctive flyway systems—broad corridors linking breeding, staging, and wintering areas. There are nine major

**Table 2.1** Major global flyways and their corresponding migratory routes for waterbirds

| S. no. | Global flyways | Migratory routes |
|---|---|---|
| 1 | East Atlantic flyway | Arctic Europe and Greenland to West Africa |
| 2 | Black Sea–Mediterranean flyway | Northern and Eastern Europe, Western and Central Asia to Sub-Saharan Africa |
| 3 | Central Asian flyway | Siberia and Central and South Asia, to East Africa |
| 4 | East Asian–Australasian flyway (EAAF | Siberia and Alaska to Southeast Asia and Australia |
| 5 | West Asian–East African flyway | Europe, Central and Western Asia, to the Middle East and East Africa |
| 6 | Americas flyway | Atlantic and Pacific routes |
| 7 | Central Pacific flyway | Pacific Ocean, Alaska, and Northeast Asia to Pacific islands and parts of Oceania |
| 8 | East Pacific flyway | North America, from Alaska to Central and South America. |
| 9 | African–Eurasian flyway | Europe and parts of Asia to Africa and vice versa |

global flyways, recognized by international conservation frameworks, which encompass the migratory routes of most waterbird species [18] (Table 2.1 and Fig. 2.1).

Among these, the East Asian Australasian Flyway is one of the most species-rich and most threatened, especially for shorebirds dependent on tidal mudflats in the Yellow Sea [49, 61] (Table 2.1 and Fig. 2.1).

### 2.3.3 Key Wetland Systems and Important Bird and Biodiversity Areas (IBAs)

Several large wetland systems are globally significant for waterbird populations (Table 2.2) due to their ecological productivity, vast size, and strategic location along major flyways:

### 2.3.4 Species with Broad vs. Restricted Distributions

Some waterbird species exhibit cosmopolitan distributions, found across continents and a range of habitats. Examples include the Mallard (*Anas platyrhynchos*), Great Egret (*Ardea alba*), and Black-winged Stilt (*Himantopus himantopus*). These generalists tend to be resilient to moderate habitat change and often benefit from artificial wetlands, reservoirs, and rice paddies [37].

Many species have localised or endemic ranges, often limited to island wetlands, highland lakes, or saline lagoons. Species such as the Madagascar Pochard (*Aythya innotata*), Juan Fernández Petrel (*Pterodroma externa*), and Philippine Duck (*Anas

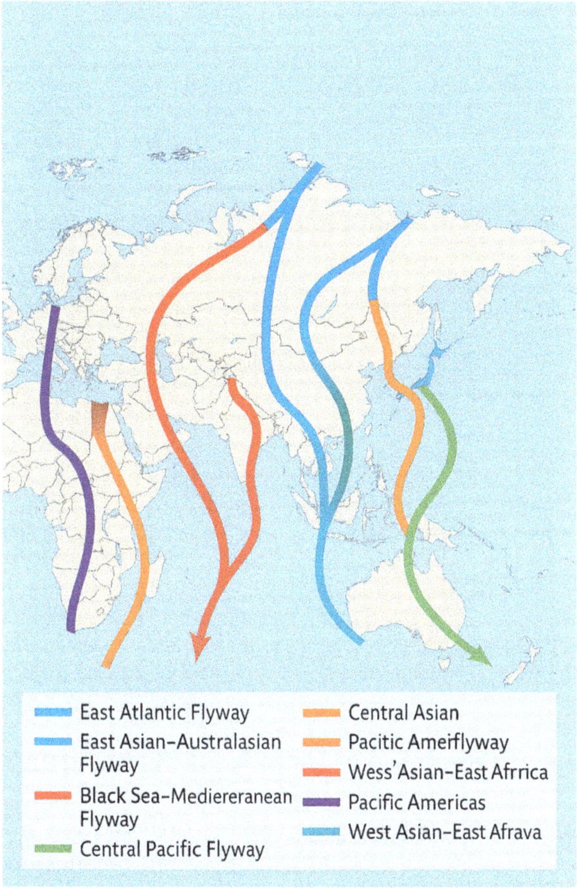


**Fig. 2.1** Overview of major flyway systems for migratory waterbirds

*luzonica*) are highly vulnerable to habitat degradation, invasive species, and hunting due to their narrow ecological tolerances and small population sizes [11].

### *2.3.5 Seasonal Variability and Dynamic Ranges*

Waterbird distribution is not static. Seasonal migrations, drought cycles, and climatic fluctuations cause annual shifts in waterbird abundance and presence at specific sites. In floodplain ecosystems such as the Inner Niger Delta or Australia's Murray–Darling Basin, bird numbers fluctuate dramatically depending on rainfall patterns and river flows [27]. Climate change leads to poleward shifts in breeding ranges and alters the phenology for many migratory species [43, 80].

**Table 2.2** Wetland hubs in major waterbird migration corridors

| Wetland | Location | Key features | Bird importance | Designation |
|---|---|---|---|---|
| Pantanal | Brazil–Bolivia–Paraguay | The world's largest tropical wetland, a highly productive ecosystem | Supports over 650 bird species, including jabirus and ibises | Ramsar Site, IBA |
| Sundarbans | Bangladesh–India | Largest mangrove forest; tidal estuaries and mudflats | Crucial for storks, egrets, kingfishers, and other waterbirds | Ramsar Site, UNESCO World Heritage Site |
| Okavango delta | Botswana | Inland delta with seasonal flooding; high biodiversity | Breeding ground for herons, pelicans, and cranes | Ramsar Site, UNESCO World Heritage Site |
| Yellow Sea | China–Korea | Intertidal mudflats, part of the East Asian–Australasian flyway | Stopover site for 50+ million migratory shorebirds annually | IBA, Ramsar Sites, UNESCO Natural Heritage (China) |
| Danube delta | Europe (mainly Romania & Ukraine) | Largest European wetland; diverse aquatic habitats | Key for breeding/migrating waterbirds in the Black Sea–Mediterranean flyway | Ramsar Site, UNESCO Biosphere Reserve & World Heritage Site |

All these wetlands are recognised as Ramsar Sites, Important Bird and Biodiversity Areas (IBAs), or UNESCO World Heritage Sites, reflecting their global ecological value

## 2.4 Regional Assessments of Waterbird Populations

Waterbird population trends vary significantly and are influenced by the regional ecological contexts, land-use changes, conservation measures, and socio-economic pressures. Asia and Africa face the steepest waterbird declines due to the rapid habitat transformation and the limited enforcement. Europe and North America achieved notable recoveries due to integrated policy frameworks. America and Oceania remain understudied in these parts but show regional variable trends depending on the wetland protection and the climate resilience (Table 2.3).

## 2.5 Migratory and Resident Waterbird Species

Waterbirds can be categorized as migratory and resident species with different ecological roles. Migratory waterbirds undertake extensive seasonal journeys across international boundaries and rely on a chain of wetlands for breeding, staging, and overwintering. Migrants birds include species such as the Bar-tailed Godwit (*Limosa lapponica*) which travels up to 12,000 km non-stop, and the Northern Pintail (*Anas acuta*) which breeds in the Arctic and winters in the tropics [5, 51]. These birds are important for nutrient cycling, seed dispersal, and linking distant ecosystems across

**Table 2.3** Comparative analysis of waterbird population dynamics and management efforts by regions

| Region | Trends | Key threats/ challenges | Conservation efforts | References |
|---|---|---|---|---|
| Asia | Severe decline (>70%) in migratory shorebirds like Spoon-billed Sandpiper and Great Knot. | Coastal reclamation, pollution, overfishing, and hunting. Degradation of Yellow Sea tidal flats. | Protected area designations in the Yellow Sea (China). Migratory Birds Framework Act (South Korea). Ramsar Sites in Japan. | [1, 44, 49, 82] |
| Africa | Variable regional trends. Decline of Palearctic migrants like Black-tailed Godwits due to Sahel wetland drying. | Wetland conversion for agriculture. Damming, hunting. 90% reduction in Lake Chad. | Community-based wetland management (East Africa). Ramsar Sites. Transboundary conservation under AEWA. | [8, 19, 23] |
| Europe | Stabilisation or increase in species like Greylag Goose and Common Crane. | Habitat loss in Mediterranean wetlands. Intensive agriculture, pollution, and climate change. | Birds Directive & Natura 2000 network. LIFE program for habitat restoration. EU agri-environment schemes. | [6, 42] |
| North America | Increases in duck species like Mallards and Northern Pintails. Effective breeding ground conservation. | Urbanization, pesticide use. Wetland conversion in the U.S. and Canada. | Ducks Unlimited. North American Waterfowl Management Plan (NAWMP). | [2, 66] |
| Latin America | Many species are under-studied. Habitat loss impacting herons, ibises, storks. | Agriculture in floodplains. Hydroelectric dams, deforestation. | Protected areas like Pantanal Matogrossense National Park. Local NGO networks. | [34] |
| Oceania | Major declines in Australian inland waterbirds due to drought and river regulation. | Invasive predators on islands. Habitat loss (esp. for island-endemics). | Ramsar sites in Australia. Indigenous-led "Caring for Country" programs. Predator control in New Zealand (e.g., NZ Dotterel). | [25, 37] |

**Table 2.4** Ecological and conservation aspects of migratory vs. resident waterbirds

| Sl. no | Aspects | Migratory waterbirds | Resident waterbird |
|---|---|---|---|
| 1. | Movement range | International, intercontinental | Local or regional |
| 2. | Conservation scale | Requires international cooperation | Effective within local/national frameworks |
| 3. | Vulnerability | Cumulative threats across flyways | Habitat-specific and range-restricted threats |
| 4. | Data gaps | Complex to monitor; high mobility | Easier to study but often understudied in remote areas |
| 5. | Example species | Red Knot, Eurasian Curlew | Black Swan, Purple Swamphen |

continents. However, their reliance on multiple habitats exposes them to numerous threats like habitat loss at stopover sites climate-driven phenological mismatches and pressures like hunting and pollution in various regions [1, 13, 61].

Resident waterbirds remain within the same general area throughout the year, making them more reliant on the quality and consistency of local wetland environments. Species like African Jacana (*Actophilornis africanus*) and Australasian Grebe (*Tachybaptus novaehollandiae*) play a continuous role in local food webs predating on aquatic invertebrates and contributing to the control of mosquito populations [45]. Some residents birds act as keystone wetland indicators with their presence reflecting the ecological health and water quality of their habitats [36]. Although they are less affected by the transboundary threats, resident waterbirds are extremely vulnerable to localized disturbances such as wetland drainage, water pollution, invasive species, and introduced predators particularly in isolated island ecosystems.

The effective conservation strategies must be tailored to these distinct life histories. Migratory species benefited from international cooperation, flyway-scale initiatives (like AEWA and EAAFP), satellite tracking, and citizen science platforms that monitor and model their vast movements [12, 62]. In contrast, the resident species conservation often hinges on the protected area management, sustainable water use, and community-based stewardship programs that foster local engagement and long-term habitat restoration. A dual approach addressing both global and local pressures is essential for the protection of all waterbird species (Table 2.4) [4].

## 2.6 Drivers of Population Change in Waterbirds

Waterbird populations are influenced by both natural and human-induced factors. Natural drivers include climatic variability, disease outbreaks, and ecological interactions [16, 32, 76]. Seasonal fluctuations in rainfall and temperature affect the availability of wetland habitats, which influence the breeding success and migratory behaviour [46]. Events like El Niño can severely impact food resources, especially in arid and semi-arid regions such as sub-Saharan Africa and Australia [35, 46]. Predation by the natural enemies and outbreaks of the diseases such as avian influenza can cause significant mortality, particularly when compounded by extreme weather. Additionally, competition for resources and fluctuations in prey abundance shape the waterbird distribution and the reproductive success.

Anthropogenic drivers, however, are the dominant forces behind the widespread decline in waterbird populations globally. Habitat loss due to wetland drainage, agricultural expansion, dam construction, and urban development is a critical concern, especially in biodiversity-rich regions like Southeast Asia and East Asia's coastal zones [22]. Coastal reclamation in the Yellow Sea has eliminated over 60% of tidal flats, critical for migratory shorebirds [49]. Mangrove destruction and peatland fires in Southeast Asia have decimated waterbird breeding sites [29]. Pollution from agricultural runoff, heavy metals, and oil spills degrades water quality and compromises the health of birds. Heavy metals, pesticides, and oil spills accumulate in the food chain, impairing reproduction and immunity in species like herons and cormorants [65]. Plastic ingestion and entanglement have also been documented across many species [56]. Overhunting, egg collection, and illegal trade further exacerbate pressures on already vulnerable species [15]. Plastic pollution and toxic contamination continue to affect reproductive success and increase mortality across species and habitats. Fluctuations in invertebrate prey or aquatic vegetation can impact species like dabbling ducks and shorebirds [53].

Climate change acts as both a direct and compounding threat, with sea-level rise, altered hydrological regimes, and temperature shifts affecting the availability and suitability of habitats [33, 72]. It disrupts migratory cues, leading to mistimed arrivals and departures and increased exposure to predation and unsuitable conditions. Invasive species, introduced either intentionally or accidentally, compete with native waterbirds or prey on them directly, especially on islands. The habitat destruction and fragmentation caused by dams, roadways, and power lines result in habitat destruction and a direct source of death due to collision or electrocution [9]. These drivers collectively highlight the urgent need for integrated, proactive conservation strategies. The most severe impacts often result from cumulative interactions, like habitat loss and climate change can reduce habitat area, weakening resilience to flooding or drought. Pollution and disease may reduce immune response, making birds more susceptible to outbreaks.

## 2.7 Challenges in Monitoring and Conservation Efforts

Despite the awareness about the importance of waterbirds as ecological indicators and their vulnerability to human-induced threats there are significant challenges which hinder effective monitoring, data integration and conservation implementation. These challenges are complex, multidimensional, and often vary depending on geography, political will, and funding availability. This chapter details the core issues which undermine the conservation success for waterbirds worldwide.

Waterbird conservation is hampered by a range of inter-related challenges that undermine the efforts to monitor populations, understand ecological trends, and implement effective protection measures. A major barrier is the lack of consistent and standardized monitoring where financial and logistical constraints lead to the irregular data collection [73]. Existing programs like the International Waterbird Census often have patchy geographic coverage and rely on the non-standardized methods, limiting data comparability and hindering the use of global conservation indices [24]. Although remote sensing technologies and satellite tagging are promising for monitoring, their high cost, limited accessibility, and technical challenges have prevented widespread application, especially in lower-income regions [7].

The weak policy frameworks and the governance issues significantly affect the conservation success [14]. Many countries either lack a robust authorized protection for wetlands and waterbirds or struggle with enforcement due to the limited capacity and competing land-use demands [21, 22]. International collaboration remains fragmented, and migratory species that cross multiple borders are particularly vulnerable when conservation efforts are un-coordinated [12]. Although agreements like AEWA, CMS, and Ramsar Convention exist, their implementation is inconsistent especially in the areas with high biodiversity but low institutional capacity [12]. The socio-economic pressures such as dependence on wetlands for livelihoods and traditional hunting practices further strain the waterbird populations. Compounding these issues is the ambiguity posed by climate change which disrupts habitat availability and the migration patterns in impulsive ways. Current conservation models remain poorly equipped to adapt these dynamic and complex challenges emphasizing the need for integrative, data-driven and community-inclusive approaches [17, 28, 63].

## 2.8 Conservation Strategies and Success Stories

Waterbird conservation has evolved significantly from isolated, species-specific efforts to the broader integrated strategies that include habitat protection landscape-scale planning and community engagement. Central to these efforts is the protection and restoration of wetlands which serve as essential habitats for both migratory and resident waterbirds. Global designations such as Ramsar Sites and Important Bird and Biodiversity Areas (IBAs) have played a critical role in safeguarding key wetland areas. Over 2400 Ramsar sites globally safeguard over 250 million hectares of wetlands critical to migratory and resident waterbirds [55]. Integration of Important Bird and Biodiversity Areas (IBAs) within national park networks has helped protect species like the Siberian Crane in Iran and India.

Restoration initiatives in regions like the Danube Delta revived the populations of herons and pelicans [58], and North America has restored over six million hectares of wetlands which benefit millions of migratory ducks and geese. Both have shown that rewetting and habitat rehabilitation can effectively revive the declining waterbird populations.

Given that many waterbird species migrate across the continents and conservation approaches have shifted to emphasize international coordination through flyway-based frameworks. Agreements such as the African-Eurasian Waterbird Agreement (AEWA) and the East Asian-Australasian Flyway Partnership (EAAFP) foster collaborative monitoring, habitat protection, and policy harmonization like the Yalu Jiang wetlands in China [50]. The site network approaches like the Western Hemisphere Shorebird Reserve Network (WHSRN) ensure that birds have access to the suitable habitats throughout their migration cycle, protecting millions of hectares across key flyways.

Incredible recoveries have also been achieved through species-oriented conservation initiatives. Examples of successful critically endangered species are Whooping Crane in North America [67], that have been brought back after being on the brink of extinction by using combination of captive breeding and re-introduction and Spoon-billed Sandpiper species [81] where population is stabilized through breeding programs and habitat protection in Asia. These achievements indicate the power of targeted interventions which can be facilitated by the use of scientific research and international cooperation.

Community-based conservation and the technological innovations are gaining significance in the ability of maintaining the gains in conservation. The active involvement of local citizens in management decisions as at Lake Natron and the Hakaluki Haor enables the locals to be involved in its conservation as well as enjoying the sustainable exploitation. Novel technologies such as satellite tracking [30] gt4cgescode. These tools and frameworks are vital in dealing with new threats and in the long-term conservation success together with other legal tools such as the Convention on Biological Diversity (CBD) and national wetland policies.

## 2.9 Future Prospectus

There is a growing change in the environment and it is these changes that need to have an adaptative science based and inclusive approach in conservation of waterbird populations. There should also be intensified long-term tracking and cross-disciplinary research to provide intelligent conservation planning. A broader coverage of standardized surveys, particularly in poorly studied areas, such as the tropics, and involvement of citizen science, such as of eBird and iNaturalist, has the potential to increase data collection and international cooperation. Encouraging studies that connect the ecology of birds to hydrology and land use will enhance knowledge on dynamics of the ecosystems particularly in the wetlands that are endowed in biodiversity.

Protection and linkage of important habitats takes precedence. These include incorporation of the extension of the networks of protection areas, which include the neglected micro-habitats like ephemeral wetlands and saltpans. The inclusion of ecological corridors and the landscape planning will guarantee the connection of the defense across the migration routes. Agriculture, aquaculture, and urban areas should be sometimes utilized with conservation-friendly processes that prevent loss and fragmentation of habitats.

Technological advances present effective decision-making and monitoring tools. Population estimation, mapping of reacting habitats and threats can be simplified with the help of AI and remote observation, whereas the satellite tracking will uncover previously unknown migration patterns with knowledge that can be used on the flyway conservation. It is also important to improve governance. The threats such as illegal hunting and encroachment can be dealt with by harmonising international conservation policies across flyways and improving national laws, and enhancing the capacity to enforce them.

The two keys to successfulness include empowering local communities and solving climate issues. The community welfare can be integrated with the conservation objectives through participatory conservation, option of sustainable livelihoods and the acknowledgment of indigenous stewardship. The incorporation of the climate resilience into the process of wetlands management and its harmonization to the overall climate policies will support the ecosystem and species adaptability. Advocacy, education, and awareness, especially by means of international events and localized activities, will serve as central to the development of long-term conservation ethic by creating support and promoting the aspect of future generations.

Communicate success stories to motivate action to show the effects of conservation.

Measures to conserve waterbirds in the twenty-first century have to move beyond single-dimensional approaches, and into collaborative, adaptive and inclusive systems. Through innovation, incorporation of traditional cultures together with scaling the models that have been successful, we could create a sustainable future of the water birds and the wetlands they live on.

## 2.10 Conclusion

Waterbirds are not only the charismatic species, they are also essential indicators of ecosystem health and predictors of climate change and also a crucial aspect of wetland ecology all over the world. Their high levels of diversity, their interesting migration pattern and their close association with the lives of man make them very effective materials in the quest of conservation. However, the continuous decrease in most populations is indicative of an overall inability to protect both freshwater and coastal ecosystems all over the globe.

These and other waterbird conservation success stories have been achieved over the past decades through local wetland rehabilitation and species recovery efforts, transboundary flyway projects, and even systems of international policy. All these initiatives are in support of the significant fact that conservation works when science, communities, policy and technology are joined. Other examples of case studies, such as the recovery of the Whooping Crane, the conservation of the Yellow Sea intertidal zones and empowerment of the Ramsar sites, show that there is a way forward by reversing negative trends with well-placed and adequate measures.

But there are numerous questions. Through the mounting pressure of habitat deterioration, climate shifts, pollution due to water, and human interference along with the insufficiency of funds and the lack of governance, many species are in danger of becoming extinct. The lack of data in some areas and some of the less-studied species also prevents us to make informed decisions.

Looking to the future, several key priorities emerge:

- Incorporation of waterbird conservation into mainstream wetland and other water resource management so as to have integrated and sustainable solutions.
- Increase and coordinate long-term monitoring schemes, utilising novel technologies, citizen science, and data-sharing web sites.
- Expand locally based, people-centered conservation that avoids the top-down approaches; allows community to make their own choices and follow local wisdom; and uses livelihood practices that benefit local people.
- Enforce worldwide collaboration in migratory flyways where waterbirds do not recognize borders and need collective efforts on a larger scale.
- Integrate the planning of climate adaptation and resilience in conservation measures so that wetlands and populations of waterbirds will withstand the changing world more effectively.

The transcontinental migration patterns, as well as the dependence of waterbirds on interrelated ecosystems, allows providing a unique perspective according to which the environment change can be understood and addressed. We safeguard them and with them we save the precious wetlands that feed millions of other animals not to mention human beings. The future way needs farsightedness, collaboration, and firm determination. In the event that these factors come together, there is hope that generations to come would still continue to enjoy the magnificent visions and the noise of successful breeding waterbirds in different parts of the world.

## References

1. Amano T, Székely T, Koyama K, Amano HE, Sutherland WJ (2018) Successful conservation of global waterbird populations depends on effective governance. Nature 553(7687):199–202
2. Anderson MG, Padding PI (2015) The North American waterfowl management plan: 1986–2015. Waterbirds 38(2):325–332
3. Anderson WB, Polis GA (1999) Nutrient fluxes from water to land: seabirds affect plant nutrient status on Gulf of California islands. Oecologia 118(3):324–332
4. Bamford M, Watkins D, Bancroft W, Tischler G, Wahl J (2008) Migratory shorebirds of the East Asian-Australasian Flyway: Population estimates and internationally important sites. Wetlands International Oceania, Canberra, Australia
5. Battley PF, Conklin JR, Parody-Merino ÁM, Langlands PA, Southey I, Burns T, ... & Potter MA (2020) Interacting roles of breeding geography and early-life settlement in godwit migration timing. Front Ecol Evol 8:52

6. Bauer HG, Bezzel E, Fiedler W (2018) Birds of Europe: their migration and ecology. Princeton University Press
7. Bayraktarov E, Ehmke G, O'Mara J, Possingham HP, Maron M (2019) Measuring and achieving cost-effective biodiversity conservation: a review. Biol Conserv 237:264–273
8. Beale CM, Baker NE, Brewer MJ, Lennon JJ (2013) Protected area networks and savannah bird biodiversity in the face of climate change and land-use pressures. Ecol Lett 16(8):1061–1068
9. Bernitz Z, Rushworth I, Taylor MR (2021) Avian electrocution and collision with power infrastructure in southern Africa: an emerging conservation concern. Ostrich 92(3):211–223
10. Bildstein KL, Bancroft GT, Dugan PJ, Gordon DH, Erwin RM, Nol E et al (1991) Approaches to the conservation of coastal wetlands in the Western hemisphere. Wilson Bull 103(2):218–254
11. BirdLife International (2023) Data zone: species factsheets. https://datazone.birdlife.org
12. Boere GC, Stroud DA (2006) The flyway concept: what it is and what it isn't. In: Waterbirds around the world. The Stationery Office, Edinburgh, pp 40–47
13. Both C, Bouwhuis S, Lessells CM, Visser ME (2006) Climate change and population declines in a long-distance migratory bird. Nature 441(7089):81–83. https://doi.org/10.1038/nature04539
14. Bridge ES, Thorup K, Bowlin MS, Chilson PB, Diehl RH, Fléron RW et al (2011) Technology on the move: recent and forthcoming innovations for tracking migratory birds. Bioscience 61(9):689–698
15. Brochet AL, Van Den Bossche W, Jbour S, Ndang'Ang'A PK, Jones VR, Abdou WALID et al (2016) Preliminary assessment of the scope and scale of illegal killing and taking of birds in the Mediterranean. Bird Conserv Int 26(1):1–28
16. Burger J, Gochfeld M (1994) Predation and effects of humans on Island-nesting seabirds. Birds N Am 19(2):39–47
17. Chen IC, Hill JK, Ohlemüller R, Roy DB, Thomas CD (2011) Rapid range shifts of species associated with high levels of climate warming. Science 333(6045):1024–1026
18. CMS (Convention on Migratory Species) (2014) A strategic framework for the conservation of migratory species. UNEP/CMS Secretariat
19. Coe MT, Foley JA (2001) Human and natural impacts on the water resources of the Lake Chad basin. J Geophys Res Atmos 106(D4):3349–3356
20. Cuthbert RJ, Green RE, Ranwala SMW, Pain DJ (2021) Waterbirds as bioindicators of wetland quality: global perspectives and challenges. Ecol Indic 132:108263
21. Danielsen F, Burgess ND, Jensen PM, Pirhofer-Walzl K (2014) Environmental monitoring: the scale and speed of implementation varies according to the degree of people's involvement. J Appl Ecol 47(6):1166–1178
22. Davidson NC (2014) How much wetland has the world lost? Long-term and recent trends in global wetland area. Mar Freshw Res 65(10):934–941
23. Davidson NC, Stroud DA (2021) African-Eurasian migratory waterbirds: current challenges and future opportunities. Conserv Biol 35(1):20–30
24. Delany S, Scott D (2006) Waterbird population estimates, 4th edn. Wetlands International
25. Dowding JE, Davis AM (2007) New Zealand dotterel recovery plan, 2004–14. Department of Conservation, Wellington
26. Ellis JC, Fariña JM, Witman JD (2006) Nutrient transfer from sea to land: the role of seabirds in coastal ecosystem functioning. Glob Chang Biol 12(3):510–516. https://doi.org/10.1111/j.1365-2486.2006.01102.x
27. Finlayson CM, D'Cruz R, Davidson NC (2018) The Ramsar convention and ecosystem-based approaches to the wise use and sustainable development of wetlands. Routledge
28. Finlayson CM, Everard M, Irvine K, McInnes RJ, Middleton BA, van Dam AA, Davidson NC (2017) The wetland and waterbird nexus. In: The wetland book. Springer, pp 1–17
29. Giam X, Bradshaw CJA, Tan HTW, Sodhi NS (2011) Future habitat loss and the conservation of plant biodiversity. Biol Conserv 144(1):159–166

30. Gill RE, Tibbitts TL, Douglas DC, Handel CM, Mulcahy DM, Gottschalck JC et al (2009) Extreme endurance flights by landbirds crossing the Pacific Ocean: ecological corridor rather than barrier? Proc R Soc B Biol Sci 276(1656):447–457
31. Green AJ, Elmberg J (2014) Ecosystem services provided by waterbirds. Biol Rev 89(1): 105–122. https://doi.org/10.1111/brv.12045
32. Hill SC, Manvell RJ, Schulenburg B, Shell W, Wikramaratna PS, Brown IH (2015) Outbreaks of H5N8 HPAI in waterfowl in Europe and Asia in 2014–2015. Avian Dis 59(3):436–439
33. Jetz W, Wilcove DS, Dobson AP (2007) Projected impacts of climate and land-use change on the global diversity of birds. PLoS Biol 5(6):e157
34. Junk WJ, Piedade MTF, Lourival R, Wittmann F, Kandus P, Lacerda LD et al (2020) Classification, ecology, and human impacts of floodplain wetlands in South America. Ecosystems 23(4):700–722
35. Kingsford RT, Thomas RF (2004) Destruction of wetlands and waterbird populations by dams and irrigation on the Murrumbidgee River in arid Australia. Environ Manag 34(3):383–396
36. Kingsford RT, Porter JL, Halse SA (2010) Ecological character of Australian inland wetlands and waterbird responses to water availability. In: Scott DA, Clemens CM, Minton J (eds) Status of migratory shorebirds in the east Asian-Australasian flyway (pp. 17–33). Wetlands International
37. Kingsford RT, Bino G, Porter JL (2017) Continental impacts of water development on waterbirds, contrasting two Australian river basins: global implications for sustainable water use. Glob Chang Biol 23(11):4958–4969
38. Kiran Singh D, Kour A, Priya Delu V, Kumar R (2022) Community composition and status of avian diversity at Campus and Agricultural landscapes of Chaudhary Charan Singh Haryana Agricultural University, Hisar (Haryana). J Appl Nat Sci 14(4):1130–1140. https://doi.org/10.31018/jans.v14i4.3784
39. Kiran Singh D, Kour A, Priya Delu V, Kumar R, Punia N, Sunita (2022) Different strategies adopted by birds to sustain ecosystem: a review. The Pharma Innovation SP-11(9):412–422
40. Kirby JS, Stattersfield AJ, Butchart SH, Evans MI, Grimmett RF, Jones VR et al (2008) Key conservation issues for migratory land- and waterbird species on the world's major flyways. Bird Conserv Int 18(S1):S49–S73
41. Kitchell JF, Schindler DE, Herwig BR, Post DM, Olson MH (1999) Nutrient cycling and fish in aquatic ecosystems. In: Ecosystems and human well-being: wetlands and water. Island Press, pp 133–145
42. Koffijberg K, Hornman M, Laursen K (2020) Trends and numbers of wintering waterbirds in the European Union. Wetlands International Report
43. Lehikoinen A, Lindén A, Byholm P, Ranta E, Saurola P, Valkama J, Högmander H (2021) Poleward shifts in ranges of Finnish birds: causes and consequences. Ecography 44(2): 285–297. https://doi.org/10.1111/ecog.05327
44. Ma Z, Melville DS, Liu J, Chen Y, Yang H, Ren W et al (2019) Rethinking China's new great wall. Science 366(6465):534–536
45. Marchant S, Higgins PJ (eds) (1990) Handbook of Australian, New Zealand and Antarctic Birds (Vols. 1–7). Oxford University Press, Melbourne
46. McKechnie AE, Hockey PAR, Wolf BO (2012) Feeling the heat: Australian arid-zone birds and climate change. Emu Austral Ornithol 112(1):i–vii
47. Moser M (2015) The cultural significance of wetland birds. Wetl Ecol Manag 23(5):807–820. https://doi.org/10.1007/s11273-015-9446-6
48. Munyenyembe FEC, Harris J, Safalaoh ACL, Chaguluka JL (1989) The feeding ecology of the Cattle Egret Bubulcus ibis in Malawi. Malawi J Sci Technol 1(2):27–32
49. Murray NJ, Ma Z, Fuller RA (2014) Tidal flats of the Yellow Sea: a review of ecosystem status and anthropogenic threats. Austral Ecol 40(4):472–481
50. Murray NJ, Clemens RS, Phinn SR, Possingham HP, Fuller RA (2017) The global distribution and trajectory of tidal flats. Nat Commun 8:1–9
51. Newton I (2010) Bird migration (New Naturalist, No. 113). HarperCollins, London

52. Nisha A, Rani S, Kumar S, Kumar R, Kumar A (2025) Recent technologies and strategies for bioremediation of toxic and hazardous waste. In: Dubey KK, Shah MP, Yadav A (eds) Challenges and sustainable solutions in bioremediation. CRC Press, pp 1–16. https://doi.org/10.1201/9781003407317-1
53. Piersma T, Lok T, Chen Y, Hassell CJ, Yang H, Boyle A et al (2016) Simultaneous declines in summer survival of three shorebird species signals a flyway at risk. J Appl Ecol 53(2):479–490
54. Post DM, Taylor JP, Kitchell JF, Olson MH, Schindler DE, Herwig BR (1998) The role of migratory waterfowl as nutrient vectors in a managed wetland. Conserv Biol 12(4):910–920
55. Ramsar Secretariat. (2022). Global Wetland Outlook: Special Edition 2021. Gland, Switzerland: Secretariat of the Convention on Wetlands. https://www.global-wetland-outlook.ramsar.org/report-1
56. Roman L, Schuyler QA, Wilcox C, Hardesty BD (2019) Plastic ingestion by marine birds. Mar Ecol Prog Ser 513:199–208
57. Runge CA, Martin TG, Possingham HP, Willis SG, Fuller RA (2014) Conserving mobile species. Front Ecol Environ 12(7):395–402. https://doi.org/10.1890/130237
58. Schneider-Jacoby M, Schwarz U, Sackl P, Stumberger B, Saveljic D (2009) The importance of the Neretva Delta for birds. Ornis Svec 19(4):217–223
59. Soons MB, Brochet AL, Kleyheeg E, Green AJ (2008) Seed dispersal by dabbling ducks: a quantitative review. Funct Ecol 22(3):582–589. https://doi.org/10.1111/j.1365-2435.2008.01418.x
60. Steven R, Morrison C, Castley JG (2015) Birdwatching and avitourism: a global review of research into its participant markets, distribution and impacts, highlighting future research directions. J Sustain Tour 23(8–9):1257–1276. https://doi.org/10.1080/09669582.2014.924955
61. Studds CE, Kendall BE, Murray NJ, Wilson HB, Rogers DI, Clemens RS et al (2017) Rapid population decline in migratory shorebirds relying on Yellow Sea tidal mudflats as stopover sites. Nat Commun 8:14895. https://doi.org/10.1038/ncomms14895
62. Sullivan BL, Wood CL, Iliff MJ, Bonney RE, Fink D, Kelling S (2014) eBird: A citizen-based bird observation network in the biological sciences. Biol Conserv 169:31–40
63. Sutherland WJ, Pullin AS, Dolman PM, Knight TM (2004) The need for evidence-based conservation. Trends Ecol Evol 19(6):305–308
64. Symes, A., Butchart, S. H. M., Donald, P. F., & Hoffmann, M. (2018). A global assessment of the status of migratory species (UNEP/CMS Technical Series No. 45). Bonn, Germany: Secretariat of the Convention on the Conservation of Migratory Species of Wild Animals (CMS) & United Nations Environment Programme
65. Tsipoura N, Burger J, Niles LJ, Dey AD, Gochfeld M, Peck M, Mizrahi D (2011) Metal levels in shorebird feathers and blood during migration through Delaware Bay, New Jersey: a preliminary assessment. Sci Total Environ 409(4):956–964
66. U.S. Fish & Wildlife Service (2023) Waterfowl population status, 2023. Department of the Interior
67. Urbanek RP, Fondow LEA, Zimorski SE, Wellington MA, Nipper MA (2010) Winter release and management of reintroduced migratory Whooping Cranes Grus americana. Bird Conserv Int 20(1):43–54
68. Väisänen RA, Lammi E, Koskimies P (2020) Monitoring game bird populations for sustainable use. Ann Zool Fenn 57(1–6):49–66. https://doi.org/10.5735/086.057.0110
69. Van de Koppel J, Rietkerk M, Dankers N, Herman PM (2005) Scale-dependent feedback and regular spatial patterns in young mussel beds. Am Nat 165(3):E66–E77. https://doi.org/10.1086/427734
70. Van Leeuwen CH, Lovas-Kiss Á, Ovegård M, Green AJ (2012) Great cormorants reveal overlooked dispersal of aquatic organisms via endozoochory by piscivorous birds. Proc R Soc B Biol Sci 279(1749):2221–2229. https://doi.org/10.1098/rspb.2011.2405
71. Van Roomen M, Van Winden E, Van Turnhout C, Kleefstra R (2015) Trends of waterbird populations in the East Atlantic flyway. Wetlands International, EAFR Report Series

72. Villamagna AM, Murphy BR (2010) Ecological and socio-economic impacts of invasive water hyacinth (*Eichhornia crassipes*): a review. Freshw Biol 55(2):282–298
73. Wetlands International (2020) State of the world's waterbirds 2020. Ede
74. Wetlands International (2021) Waterbird population estimates – Fifth edition. https://wpe.wetlands.org
75. Wetlands International (2023) Global waterbird monitoring and conservation. https://www.wetlands.org
76. Wobeser G (2006) Essentials of disease in wild animals. Wiley-Blackwell
77. Wold A, Hershey AE (1999) Effects of salmon carcass decomposition on benthic macroinvertebrates in an Alaskan stream. Freshw Biol 42(3):561–575
78. World Wide Fund for Nature (WWF) (2022) Living planet report 2022: building a nature-positive society. WWF International
79. Yadav R, Gupta RK, Kumar R, Kaur T (2022) Assessment of toxicity of lead and nickel on the biochemical and immunological parameters of earthworm, Eudrilus eugeniae. Agric Sci Dig 44(5):922–929
80. Yodha K, Singh D, Kour A, Yadav R, Kumar R (2024) Species diversity and abundance of birds in ber (Zizyphus mauritiana) crop. Asian J Conserv Biol 13(1). https://doi.org/10.53562/ajcb.83683
81. Zöckler C, Syroechkovskiy EE Jr, Bunting G (2016) Spoon-billed sandpiper: a review of progress toward conservation of the most threatened migratory shorebird in the East Asian–Australasian flyway. BirdingASIA 25:45–52
82. Zöckler C, Syroechkovskiy EE, Atkinson PW (2020) Rapid declines in the spoon-billed sandpiper Calidris pygmaea indicate the need for urgent conservation action. Bird Conserv Int 30(4):433–450

# Chapter 3
# Waterbirds: Specialized Traits, Community Structure, and Foraging Preferences

**Farhat Jahan and Haroon Ali**

**Abstract** A diverse group of avian species that are uniquely adapted to aquatic habitats ranging from freshwater wetlands to saltwater marine ecosystems. They have specialised morphological traits such as webbed feet, well-formed waterproof plumage, salt glands and a variety of beaks that enable these birds to survive and utilisation of resources in aquatic environments. Ecological factors like habitat quality, food availability, migration and predation as well as anthropogenic pressures including wetland drainage, pollution, agriculture and urbanisation greatly influence the composition of waterbird communities. Feeding ecology reflects both the morphological adaptations of waterbirds and their habitat-specific strategies. Waterbirds occupy a wide range of dietary guilds. These guilds include mostly piscivorous, insectivorous, molluscivorous, crustacean-eating, herbivorous, omnivorous and scavenging aquatic birds. The foraging techniques of these birds include snipping, diving, probing, sifting, filtering and surface feeding. Their foraging habits are also determined by the type of habitat like mudflats, vegetated wetlands, coastal areas and open water. These ecological specialisations and their adaptations are indicative of biodiversity, highlighting the important role of waterbirds as an integral component of aquatic ecosystems.

**Keywords** Aquatic adaptations · Feeding ecology · Foraging strategies · Ecosystem indicators · Habitat disturbance · Piscivores · Waterbirds

---

F. Jahan
Department of Zoology, Hansraj College, University of Delhi, Delhi, India

H. Ali (✉)
Department of Public Health, College of Applied Medical Sciences, Khamis Mushait Campus, King Khalid University (KKU), Abha, Kingdom of Saudi Arabia
e-mail: haalahmad@kku.edu.sa

 
R. Kumar et al. (eds.), *Ecology, Conservation of Aquatic Birds*, Wetlands: Ecology, Conservation and Management 15, https://doi.org/10.1007/978-3-032-17125-2_3

## 3.1 Introduction

Waterbirds are an assorted group of avian species, that adapted to aquatic ecosystems and exhibiting specialized morphological, physiological, and behavioral traits. These characteristics facilitate them to take advantage of freshwater lakes, coastal marshes, estuaries, and wetlands habitats efficiently. These Water birds species play a significant role in maintaining ecological balance and serve as bioindicators of aquatic environmental health ecosystem output. Understanding their community structure, inter-specific interactions and foraging preferences of waterbirds is essential for conservation of wetland and their management.

This chapter explores the adaptive characteristics of waterbirds and their niche differentiation, trophic dynamics and strategies use in foraging, extensive research representation and the case studies. It also highlights the threats faced by populations of water birds and the need for conservatory measures to protect these vital species. Waterbirds inhabit a range of wetland ecosystems like rivers, lakes, marshes, estuaries and coastal regions. These water birds have evolved a variety of specialized characteristics features to survive in the aquatic environments. This enabling them to utilize diverse food resources and co-exist with other species in structured communities. Waterbird ecology, foraging strategies, and community interactions provide valuable insights into wetland ecosystem functioning and conservation needs [125].

This is consistent with building upon the foundational principles of bird community ecology. This chapter delves into the unique structure and intricate interactions within waterbird communities. In our observations, these assemblages of water bird species are intrinsically linked with aquatic environments, relying on ponds, lakes, rivers, wetlands, estuaries and coastal zones for essential resources such as food, nesting sites, and refuge. The adaptive features of water birds possess for life in and around water profoundly shape their community organization and the ecological roles they fulfils. Understanding the dynamics of waterbird communities has become increasingly critical, and now in the face of escalating anthropogenic pressures on aquatic ecosystems [56].

Waterbirds exhibit a remarkable diversity of foraging preferences and strategies, influenced the availability of food and characteristics of prey within lakes, rivers, wetlands, estuaries and coastal zones. This chapter explores the diverse foraging preferences of waterbirds and their morphological, behavioral and ecological adaptations. That allows them to make the most of a wide spectrum of aquatic bird resources.

### *3.1.1 The Importance of Water Bird Species in Ecosystem*

Water birds are essential for understanding the interactions between species and their surroundings. These birds provide valuable insights into the ecosystem dynamics. They also play a crucial role in regulating fish, invertebrate, and plant populations, fulfilling essential ecological functions. Waterbirds' presence, abundance, and behavior indicate the health of wetland habitats, making them reliable indicators for environmental monitoring and conservation (Green and Elmberg in 2014). They aid in seed dispersal and promoting plant growth in wetland regions [107]. Waterbirds also serve as bioindicators, reflecting wetland health and the levels of pollution present in the aquatic ecosystem [21]. Waterbirds serve as vital components of aquatic and semi-aquatic ecosystems, performing numerous ecological roles that support both biodiversity and ecosystem functioning. Aquatic birds are essential for nutrient cycling and energy flow in these ecosystems. Water birds forage in various habitats, moving nutrients between land and water [125]. They help to spread organic matter around. Their guano is rich in nitrogen and phosphorus, work as a fertilizer to wetland soils and stimulates the primary productivity of the ecosystem [4]. This ecological service is significant in nutrient-poor areas like coastal lagoons, salt marshes, and high-altitude wetlands. Water birds also act as seed dispersers to help spread seeds and influence plant communities. Many aquatic bird species eat seeds and then poop them out at various locations, sometimes spanning hundreds of kilometers. This process facilitates long-distance dispersal of seeds [107]. This promotes plant gene flow and colonization of fragmented or newly formed habitats especially in the floodplain ecosystems and seasonal wetlands. Moreover, waterbirds contribute to the regulation of aquatic invertebrate and fish populations. Herons, cormorants, and kingfishers, among other species, exert top-down control over prey communities. This influence shapes food web dynamics and contributes to maintaining ecological balance [120]. By doing so, they support trophic regulation and assist in preventing overpopulation of specific aquatic organisms, which could otherwise result in habitat degradation.

In terms of ecosystem connectivity, migratory waterbirds create natural linkages between distant wetlands across continents. Their seasonal movements link different ecological areas, increasing gene flow for the associated biota and contributing to global biodiversity conservation [57]. Wetlands that serve as migratory flyways also generate economic benefits through ecotourism, particularly in regions that have well-established bird watching industries. The abundance and variety of waterbird species serve as a testament to the long-term ecological stability of the region. Waterbird population declines often signify wetland degradation caused by pollution, habitat loss, invasive species and climate change [21]. Consequently, preserving the water bird habitats translates in to broader initiatives for ecosystem conservation, health, sustainability and climate resilience.

### *3.1.2 Specialised Traits, Community Structure and Foraging Strategies of Water Birds*

Ecological research, conservation of biodiversity, and habitat management, all are rely on understanding the unique traits of water birds, their social dynamics, and how they hunt. In our observations, we found that the waterbirds thrive in a diverse range of aquatic environments, from fresh water lakes to saline coastal marshes, due to their specialized characteristics. These characteristics are morphological and physiological traits such as shape of beak, foot and leg structure, water proofing nature of feathers, and salt gland function [4]. The adaptations found in water birds define their ecological niches and facilitate resource partitioning. Consequently, inter-specific competition is reduced enabling the coexistence of diverse species within the same habitat [19].

Studying the community structure of water birds in wetland ecosystems offers valuable insights in to species richness, functional diversity, and trophic interactions. Community patterns reveal how water birds species assemble based on habitat, feeding groups, and migration [56]. According to Green and Elmberg in 2014, providing data is vital for identifying keystone species, assessing ecosystem health and detecting changes due to human activities like pollution and habitat loss. The study of species composition and evenness also helps to understand how the seasonal changes, wetland hydrology and climate change affect populations of birds [104]. The observations regarding foraging strategies reveals how birds exploit food resources. These strategies from exploring the surface to venturing deep in to the ocean, show how our physical and behavioural traits are closely connected with their ecological roles. The knowledge also helps to predict water bird population in responses to habitat changes and food web disruptions [127]. This data of foraging helps in wetland restoration by showing the resources of food and micro-habitats are needed to support specific bird communities.

The foraging data also inform restoration of wetland practices by indicating what types of resources of food and micro-habitats need to be preserved or re-established to support specific communities of birds. A deeper understanding of bird traits, structure, and foraging skills can really improve. Ecological theory is the backbone of evidence-based conservation planning. This helps to design the bird-friendly wetlands. It assess impacts, and meet international commitments like the Ramsar Convention and the Convention on Migratory Species.

### *3.1.3 Scope and Objectives*

Water birds are adapted to both freshwater and marine habitats, and they play pivotal ecological roles in wetlands globally. The unique physical traits, behaviors and social interactions of aquatic birds reveal how ecosystems function, food chains work, and evolution unfolds. In this chapter we delves in to the unique anatomical and physiological adaptations of waterbirds, that enabling them to thrive in diverse environments, from shallow marshes to vast oceans. Here, this chapter classifies waterbird species in detail, showcasing their taxonomic diversity up to orders and families. It also studies their physical traits, like beak shape, foot, leg structure, plumage, and salt glands. According to Raikow [98] and Denny [25], all these traits are closely linked to their aquatic lifestyles. This chapter also explores the behavioral and physiological adaptations that help individual migration, foraging, regulate their body temperature, and reproduce successfully in the changing aquatic environments. This chapter explores the community structure of aquatic bird populations, encompassing species richness, habitat utilization, zonation patterns and inter-specific interactions beyond the individual characteristics. This chapter summarizes the responses of water bird communities to environmental gradients, seasonal changes and the activities done by humans [42, 116].

The foraging guilds of water birds are comprehensively explained and demonstrating how various types of species mitigate competition by effectively partitioning time, space, and dietary resources. The main aims of this chapter is to synthesise existing knowledge on foraging guilds and foraging preferences and strategies among aquatic birds. Specifically, it examines how diverse beak morphologies in the water birds species and feeding behaviors correspond to types of habitat and prey availability. This chapter also explores how different organisms interact and feed on each other which helps us understand how ecosystems work. It looks at how nutrients are cycled, energy is transferred, and how certain organisms control pests and diseases.

In this chapter, we will explore the significance of aquatic birds beyond their role in our ecosystem. They play a pivotal role in conservation initiatives as well. This content provides a multi-disciplinary approach to safeguarding water birds from threats such as habitat degradation, pollution and climate change. It is a valuable content to the reference for ecologists, conservation biologists, wetland managers, ornithologists and researchers who are studying wetlands.

## 3.2 Taxonomic Classification and Diversity of Water Birds

Water birds constitute a taxonomically diverse and specific group of avian species that are ecologically dependent on aquatic habitats for foraging, nesting and breeding. They belong to various orders of class Aves, mainly Anseriformes (ducks, geese, swans), Charadriiformes (gulls, terns, shorebirds, waders), Pelecaniformes (pelicans, herons, ibises), Gruiformes (cranes, rails, coots), and Podicipediformes (grebes). This classification is based on morphological characteristics, including beak shape, foot structure, plumage adaptations and behavioral specialisations that associated with aquatic life.

For instance, members of the Anseriformes order, such as the mallard (*Anas platyrhynchos*) have broad and lamellated bills which are adapted for dabbling and filtering aquatic invertebrates and vegetations. Charadriiformes is an order particularly diverse group both freshwater and marine species. Sandpipers (*Calidris* spp.) have long legs and slender bill which is adapted for probing soft substrates. Pelecaniformes order e.g., herons (*Ardea* spp.) possess spear-like beak for capturing fish. In the Gruiformes order including *Rallus aquaticus* exhibit strong legs and toes which enable them to walk effortlessly on floating vegetation.

Podicipediformes order such as the great crested grebe (e.g. *Podiceps cristatus*) show the exceptional swimming abilities due to their lobed toes. Phylogenetic studies conducted using mitochondrial and nuclear DNA have revolutionized our understanding of the inter-relationships between these groups. Molecular tools have demonstrated that water bird traits have evolved independently due to ecological convergence, rather than shared ancestry [45, 68]. Therefore, the classification of water birds is not strictly linear but rather reflects a complex history of adaptive radiation in response to diverse wetland ecosystems.

These taxonomic frameworks are super important for conservatory purposes. They help us to find out which species are different from each other and what each species does in its environment. This information guides us to habitat management and species recovery efforts [24, 105].

## *3.2.1 Types of Aquatic Birds*

Waders, swimmers and divers are the water sports enthusiasts as follows:

### 3.2.1.1 Waders

Waders are commonly referred as shorebirds. They constitute a diverse avian order characterised by their elongated legs, which enable them to excel in foraging activities within shallow water bodies, mudflats and wetlands. They have slender and elongated bills that are typically designed for probing in to soft substrates to extract invertebrates for feeding. Waders are easily distinguished by their exceptionally long legs and light weight bodies. Through which they facilitate their swift movement through marshes, flooded fields and estuarine environments. Herons (*Ardea* spp.), egrets (*Egretta* spp.), stilts (*Himantopus* spp.), sandpipers (*Calidris* spp.), and ibises (*Plegadis* spp.) are the common examples. Herons (*Ardea* spp.), egrets (*Egretta* spp.), stilts (*Himantopus* spp.), sandpipers (*Calidris* spp.) and ibises (*Plegadis* spp.) are the common examples. Waders mostly hunt by looking or feeling around and they have different ways of doing it. Some just wait patiently while others actively chase their prey.

### 3.2.1.2 Swimmers

Swimmers are water birds that have adapted to move efficiently on water. Usually, they have fully webbed feet and buoyant bodies that enabling them to glide smoothly across the lakes, rivers and coastal lagoons. These birds are typically forage by dabbling (feeding at or just below the water surface) or filter-feeding. Their beaks are typically broad, flat and well equipped with lamellae (comb-like structures) that enable them to strain small organisms from water. Ducks (*Anas* spp.), geese (*Branta* spp.) and swans (*Cygnus* spp.) are well-known swimmers. Some swimmers also dive into shallow depths. Their plumage is waterproof and coated with oils, which are essential for thermoregulation and buoyancy.

#### 3.2.1.3 Divers

Divers are highly specialized group of waterbird species that can forage underwater by actively swimming. These birds have sleek bodies and strong legs at the back, which makes them excellent swimmers. But they are not the best at moving around on land. Usually they eat fish, aquatic creatures and crustaceans. They employ diverse strategies such as pursuit diving like cormorants and loons and plunge diving like terns and gannets. Notable divers species include grebes, loons and cormorants. The waterbirds belong to this group might have smaller air sacs and stronger bones that help them stay submerged underwater. Making them great divers.

### 3.2.2 *Taxonomic Classification Water Birds of Class Aves*

A thorough understanding of the taxonomic classification of water birds is essential for safeguarding biodiversity and preserving these captivating aquatic bird species with in their habitats. By identifying and categorizing these endangered or threatened species, we can make the most of our conservation efforts. A clear taxonomic structure is like a treasure map for scientists, birdwatchers and conservationists. It helps them to spot and keep an eye on waterbird populations. This enables them to monitor migratory species and assess their health. Water birds belong to various taxonomic groups within the aves class, which comprises several distinct orders. The water birds are a specific group of class aves with special traits and have important roles in the ecosystem. They are organized into orders and families. The aquatic avian species demonstrate the taxonomic orders and families of the class Aves, underscoring their specialized traits (Table 3.1).

### 3.2.3 *Global and Regional Distribution Patterns of Water Birds*

Aquatic birds are distributed in all the continents of the world and inhabit a diverse range of ecosystems spanning from polar ice caps to tropical wetlands. This remarkable adaptability showcases their exceptional ecological plasticity and evolutionary success. The global distribution of waterbird species are predominantly determined by the accessibility of freshwater and marine habitats, migratory flyways, seasonal resource availability and climate patterns [9].

**Table 3.1** Water birds classification showing taxonomic orders and families of class aves with their specialized traits and examples

| S. no. | Orders | Families | Specialized traits of water birds | Examples |
|---|---|---|---|---|
| 1 | Anseriformes | Anatidae | Webbed feet and flattened bills and dabbling/diving feeders | Ducks, Geese and Swans |
| 2 | Pelecaniformes | (i) Pelecanidae | Long bill with throat pouch and cooperative fish hunting | Pelicans |
| | | (ii) Ardeidae | Long legs and neck and wading behavior- ambush predation | Herons and Egrets |
| | | (iii)Threskiornithidae | Down-curved/spatulate bills and tactile foraging | Ibises and Spoonbills |
| 3 | Charadriiformes | (i) Scolopacidae | Long, slender bills and probing in mudflats | Sandpipers and Godwits |
| | | (ii) Charadriidae | Short bills, visual hunters on shores and mudflats | Plovers and Lapwings |
| | | (iii) Laridae | Opportunistic feeders, strong fliers and aerial foraging | Gulls and Terns |
| 4 | Gruiformes | Rallidae | Lobate feet, vegetation grazers and good swimmers | Coots and Moorhens |
| 5 | Podicipediformes | Podicipedidae | Lobed toes—excellent divers and nesting on floating plants | Grebes |
| 6 | Gaviiformes | Gaviidae | Heavy body, streamlined and deep diving for fish | Loons |
| 7 | Suliformes | Phalacrocoracidae | Hooked bill, plunge diving and minimal waterproofing | Cormorants |

**References:** [1, 41, 66, 119]

Globally, waterbirds species like ducks, geese, pelicans, gulls, herons and waders are found in various water bodies including lakes, rivers, estuaries, mangroves, salt marshes and coastal lagoons. The Afrotropical, Neotropical, Palearctic and Indo-Malayan realms boast high species richness, particularly in vast wetland complexes such as the Okavango Delta in Botswana, the Pantanal in Brazil and the Sundarbans spanning India and Bangladesh. The Mississippi Flyway and Prairie Pothole Region in the Nearctic realm are crucial migratory stopover sites, supporting millions of waterbirds during their journey.

In Asia, India stands as a prominent sanctuary for both migratory and resident aquatic birds, harboring an astonishing diversity of over 310 wetland-associated avian species. Important habitats include Keoladeo National Park (Bharatpur), Chilika Lake (Asia's largest brackish Water Lagoon), Loktak Lake (The largest natural freshwater lake in the north-eastern region of India), Yamuna Floodplains and the Okhla Bird Sanctuary. These wetlands harbour diverse avian species, including the bar-headed goose (*Anser indicus*), ruddy shelduck (*Tadorna ferruginea*), painted stork (*Mycteria leucocephala*), and black-headed ibis (*Threskiornis melanocephalus*).

**Migratory Water Birds Travel Vast Distances Annually Along Major Flyways, Including**

- East Asian–Australasian Flyway (EAAF)
- Central Asian Flyway (CAF)
- African–Eurasian Flyway (AEWA)
- Pacific Americas Flyway

These flyways are crucial for the survival of species, as they connect breeding, stopover, and wintering sites. The loss of any link in these flyways can significantly impact the viability of the population [12]. Regional factors, including monsoon patterns, water level fluctuations, salinity gradients, and anthropogenic pressures, influence the local abundance and diversity of species. Studies conducted in India have revealed significant variations in the composition of waterbird communities across Himalayan lakes, Gangetic floodplains, and coastal estuaries [112].

Take an example of the Asian openbill is a freshwater-loving bird makes its home in inland wetlands. On the other hand, if we look the greater flamingo (*Phoenicopterus roseus*) is a salt-tolerant species that thrives in saline ecosystems like the Rann of Kutch. Conservation planning for the water birds necessitates an understanding of their distribution patterns, particularly in the context of climate change, wetland degradation and urban expansion which are altering aquatic habitats globally.

## 3.3 Specialised Morphological and Physiological Traits in Water Birds

The water birds are amazing creatures that have adapted to live in all sorts of water bodies from freshwater wetlands to coastal estuaries. They have got special features and ways of working that help them to survive and find food in the different environments. Water birds exhibit distinctive morphological and physiological adaptations that facilitate their success in the aquatic habitats and by analyzing genetic and physical traits across diverse water bird species, researchers can monitor their evolutionary transformations and adaptations over time.

The morphological traits of water birds play a significant role in determining their ecological significance. The beak shapes exhibit substantial variations that influenced by the dietary preferences and foraging techniques employed by the species. The species like herons and egrets have long and spear-like beaks adapted for stabbing fish while ducks and flamingos have broad and lamellate beaks that enable them to filter-feed on planktons and small invertebrates from water [39].

Foot morphology varies across species. For instance, ducks and geese have webbed feet for efficient swimming, while grebes and coots have lobed toes that aid in propulsion through submerged vegetation [97]. Wing morphology also reflects the ecological specialisation of different species. Dabbling ducks and shorebirds generally possess pointed wings, which facilitate swift and sustained flight. In contrast, birds like cormorants have shorter wings, adapted for diving. However, they necessitate wing-drying behaviour due to their plumage's reduced waterproofing capabilities [19]. The plumage of the bird feathers is usually dense and waterproof, with specialised uropygial gland secretions that contribute to buoyancy and insulation.

Physiological Traits Waterbirds have evolved several physiological mechanisms to thrive in the ever-changing aquatic environment. Their efficient osmoregulatory systems enable many species, particularly seabirds such as gulls and pelicans, to excrete excess salt through supraorbital salt glands situated near their eyes [39]. This allows them to consume saltwater without dehydrating.

Another key physiological trait is thermoregulation. Birds such as grebes and swans have a countercurrent heat exchange system in their legs, which minimises heat loss when wading in cold waters. High metabolic rates also support the energy-intensive demands of flight and thermoregulation in variable climates, as highlighted by Klaassen [57]. Migratory species, such as shorebirds, also demonstrate remarkable fat storage capacities and muscle endurance, enabling them to sustain non-stop transcontinental flights [79].

In arid environments, water loss through eggs and nests is minimised through behavioural and physiological incubation strategies.

Gutierrez et al. [44] demonstrated that shorebirds such as stilts and plovers modify their incubation postures to minimise water loss during periods of high temperatures. Different water bird species exhibit distinct morphological traits in their beaks, legs, feet and feathers as follows:

### *3.3.1 Beak Morphology in Water Birds*

The beak shape of aquatic birds is a clear example of evolutionary adaptation that lets species thrive in various ecological niches and food sources. Over the time, the morphology, length and structure of beaks have diversified in response to dietary preferences and foraging techniques. Herons and egrets are the good examples that have long and sharp beaks adapted for grasping slippery fish in shallow water. Kushlan [63] reported the mode of feeding necessitates meticulous precision and is frequently linked with the still-hunting behaviour. Filter-feeding species like flamingos and some dabbling ducks have broad and flat beaks with comb-like structures called lamellae. This allows them to sieve small invertebrates, plankton and organic material from muddy water [48, 128].

Dabbling ducks such as Anas platyrhynchos feed at the water surface or just below by tilting forward without submerging themselves. Their beaks are super sensitive and can spot and sort out yummy food from all the rubbish around them. These beaks often have a spatulate shape which allows them to perform both tactile foraging and filter-feeding. In the interim, species such as ibises and sandpipers possess elongated, decurved beaks that facilitate their penetration of soft sediments, mud, or shallow waters. This enables them to identify concealed prey, including worms, mollusks, and crustaceans. Notably, the beak tips of many of these birds possess sensory pits called Herbst corpuscles, which are capable of detecting vibrations and chemical cues emanating from buried prey [94, 95].

The correlation between beak shape and diet exemplifies the functional design and ecological specialisation of aquatic birds, thereby mitigating interspecific competition and facilitating niche partitioning within shared habitats. The comprehension of these morphological adaptations is crucial for the interpretation of feeding behaviour, habitat selection, and species interactions within aquatic ecosystems.

### 3.3.2 *Beak Adaptation in Water Birds*

- Herons and egrets have elongated, pointed beaks that facilitate their capture of fish.
- Flamingos and dabbling ducks possess filtering beaks that enable them to effectively.
- collect diminutive organisms from the aqueous environment. Cormorants and pelicans are characterised by hooked beaks that are adept at grasping slippery prey (Table 3.2).

### 3.3.3 *Leg and Foot Morphology of Water Birds*

The diverse foot and leg morphologies of aquatic birds are a direct reflection of their ecological roles and habitat preferences. Adaptations within these structures profoundly impact on birds locomotion, foraging strategies and ability to evade predators. Webbed feet a shared trait among ducks, geese, swans and pelicans are fully connected toes that act as paddles. These feet facilitate the powerful propulsion and maintenance of stability during swimming in these birds. The adaptations found in

**Table 3.2** Types of beak modifications in waterbirds

| Type of beak | Examples | Function | Description | Exemplification |
|---|---|---|---|---|
| Spear-like | Herons and Egrets | Spearing fish | Long, and Sharp beak |  |
| Filter-feeding | Flamingos and Ducks | Filtering small organisms | Broad and Lamellate Beak |  |
| Hooked beak | Pelicans and Gulls | Catching slippery prey | Curved Beak |  |

**References:** [1, 39, 98]

these birds facilitate efficient locomotion in open water and contribute to reduced energy expenditure during foraging and migration [71]. On the other hand, grebes and coots have lobed feet where each toe has little extensions or flaps-like structure. During strokes, these lobes expand and contract, enhancing manoeuvrability and precise control which are crucial for diving and pursuing prey underwater [98].

These morphologies are not only super important for how these animals move around but they also help them stay separate from each other and find different places to eat. By occupying diverse water depths and employing distinct leg and foot adaptations, these water birds effectively reduce interspecific competition and optimise resource utilisation within shared habitats.

### *3.3.4 Leg and Foot Adaptations in Water Birds*

Long-legged wading birds, including herons, egrets, storks and ibises possess distinct morphological features. Their elongated legs enable them to elevate their bodies above the water surface and facilitating their wade through deeper wetland areas while maintaining the dryness of their feathers. This adaptation also enables them to maintain stealth and balance while stalking prey in shallow or vegetated aquatic environments [116]. The elongated tarsometatarsus and specialized muscle configurations enable the animals to execute quiet, purposeful movements which are crucial for their sit and wait hunting strategy. Some adaptations in legs and feet of the water birds are as follows:

- Wading birds, such as storks and ibises, have elongated legs that enable them to navigate through shallow aquatic habitats effectively.
- Ducks and pelicans are equipped with webbed feet, which significantly enhance their swimming capabilities (Table 3.3)
- Grebes and coots possess lobed toes, which further enhance their diving proficiency (Table 3.3).

**Table 3.3** Leg and foot modifications in waterbird

| Type of modification | Examples | Functions | Description | Exemplification | References |
|---|---|---|---|---|---|
| Webbed feet | Ducks, Swans and Geese | Efficient swimming and propulsion | Fully webbed feet |  | [1, 39, 98] |
| Lobed toes | Grebes and Coots | Swimming in dense aquatic vegetation | Lobed toes |  | |
| Long legs | Herons and Storks | Wading in shallow waters for prey | Long legs |  | |
| Strong clawed toes | Rails and Moorhens | Walking on floating vegetation and climbing | Long toes with claws | 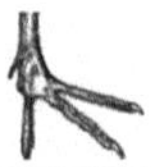 | |

### 3.3.5 Feathers and Plumage Morphology in Water Birds

Water birds rely on feathers and plumage for more than just flight, they also play a crucial role in thermoregulation, waterproofing, and camouflage. One of the most crucial adaptations in water birds is waterproofing. Water birds have a special gland called the uropygial gland that secretes oily substances. During preening, these oils are spread over the bird's feathers, making the plumage water-repellent and preventing water from reaching the skin. The physiological adaptations are not only regulates the body temperature but also facilitates buoyancy and energy efficiency during swimming and diving [25, 29].

Feathers also serve as camouflage, especially for species living in marshes, reed beds and muddy flats. These water birds often have plumage muted brown and grey in colour. These species also have intricate patterns features that blend seamlessly with their surroundings. Hill and McGraw [47] explained the cryptic colouration of birds like bitterns, sandpipers and snipes allows them to remain inconspicuous to both predators and prey. Several water birds species like Ruff, Green Heron, Roseate Spoonbill and Surakav bird undergo colour changes and develop new markings on their plumage during breeding season and seasonally environmental changes like snow or grass. These adaptations confer long-term advantages throughout the year.

The feather structure and colouration patterns of water birds are crucial for their survival. They provide physical protection in harsh environments and enable behavioural adaptations such as stealth for predation and avoidance.

### 3.3.6 Feather Adaptations

- Diving birds, such as cormorants and grebes, possess water-repellent feathers.
- Preen or uropygial gland use for waterproofing.
- Countershading coloration aids in their camouflage within their environment [67]. Various feather adaptations are shown in Table 3.4.

### 3.3.7 Physiological Adaptations of Water Birds

Water birds possess a remarkable suite of physiological adaptations that allow them to thrive in water-rich environments, from freshwater wetlands to saline marine coastal waters. Salt regulation is one of the key adaptations that enables them to survive and forage in these water birds habitats. Avian species such as gulls, pelicans, and cormorants, which possess specialised supraorbital salt glands situated above their eyes. These glands facilitate the excretion of excess salt through their nostrils, enabling them to ingest seawater and marine prey without succumbing to

**Table 3.4** Feather adaptations in waterbirds

| Physical and visual adaptations | Examples | Adaptive functions of feathers in waterbirds |
|---|---|---|
| Preen/Uropygial gland | Most water bird species | Oil secrets—forming a protective layer that enhances buoyancy, maintains body heat and ensures efficient locomotion in water, feather condition, waterproofing and parasite control. |
| Water-repellent feathers | Most water bird species | Prevent water from saturating the plumage and reaching the skin |
| Countershading | Ducks and geese | The dark back blends with the water/wetland surface shows camouflage against predators |

**References:** [25, 48, 67, 104, 105]

dehydration [89]. This adaptation is particularly crucial for marine species and estuarine birds that frequently consume saline water.

Another notable physiological characteristic is the efficient storage and utilisation of oxygen. Diving birds, such as grebes, loons, and auks, have increased myoglobin concentrations in their muscles. This adaptation allows them to slow their heart rates (bradycardia) while submerged, conserving oxygen. This enables them to extend their dive durations and efficiently hunt underwater [108]. Furthermore, their dense plumage offers insulation against frigid water, enabling them to regulate their body temperature through a synergistic combination of waterproofing and trapped air layers.

Waterbirds also have counter-current heat exchange systems in their legs and feet, which help them reduce heat loss in cold water. This physiological system entails the precise alignment of arteries and veins, facilitating the circulation of warm blood from the body's core to the extremities, thereby warming the returning cold blood. Ducks, geese, and swans have particularly well-developed plumage, which enables them to forage in icy wetlands without significant heat loss, as observed by Midtgard [82].

To sustain swimming for long-distance migration and maintain energy output across different climatic conditions. In conclusion, energy-efficient metabolism facilitates high-demand activities, including long-distance migration, sustained swimming, and flying. Numerous aquatic birds accumulate substantial fat reserves prior to migration and demonstrate adaptive thermogenesis to sustain energy output across diverse climatic conditions [49]. These physiological traits illustrate the evolutionary adaptations that have facilitated aquatic birds' colonisation and dominance of diverse aquatic environments.

### 3.3.8 Sensory Adaptations in Water Birds

Water birds possess specialised sensory systems that facilitate hunting, navigation, and avoidance of predators in both aquatic and aerial environments. Among these, vision is highly adapted in numerous species. For instance, ospreys and herons possess binocular depth perception, while kingfishers do Some birds also show ultraviolet vision, which aids in mate selection and foraging efficiency by highlighting prey or vegetation patterns. The cormorants exhibit underwater refraction correction [76]. Some aquatic bird avian species possess ultraviolet vision, which facilitates mate selection and foraging efficiency by illuminating prey or vegetation patterns.

Foraging relies heavily on both vision and tactile sensitivity. For instance, shorebirds such as sandpipers and snipes possess bill-tip organs densely populated with "mechanoreceptors that enable them to discern invertebrates concealed beneath moist sediment- a capability known as "remote touch" [94, 95]. Certain aquatic species possess the ability to perceive ultraviolet light, which facilitates mate selection and enhances foraging efficiency by illuminating prey or vegetation patterns.

Certain aquatic birds, such as penguins and shearwaters, demonstrate remarkable sensitivity to fluctuations in atmospheric pressure and magnetic fields. These adaptations enable them to orient themselves and navigate across vast oceanic regions [121]. These meticulously refined sensory adaptations confer upon aquatic birds the perceptual capabilities essential for effectively navigating and exploiting underwater, wetland, and coastal environments. Notably, these adaptations are frequently associated with their respective feeding guilds and migratory patterns.

Penguins, particularly deep-diving species such as the Emperor Penguin (*Aptenodytes forsteri*), exhibit remarkable tolerance to extreme underwater pressures. Although they lack a singular, unequivocal "pressure receptor organ," they possess mechano-sensitive structures within their inner ear and central nervous system that facilitate their ability:

- To monitor barometric pressure fluctuations during ascent and descent.
- To prevent decompression sickness (the bends) by regulating heart rate and blood flow (bradycardia and vasoconstriction).
- To perceive alterations in hydrostatic pressure, potentially utilising proprioceptors and baroreceptors within their circulatory system and skin (Table 3.5).
- Penguins, and many other migratory and pelagic birds, are believed to use the Earth's geomagnetic field for navigation purposes (Table 3.5).

## 3.4 Behavioral Adaptations in Water Birds

Waterbirds exhibit behavioural flexibility, which is crucial for their survival and reproductive success. Many species migrate seasonally between breeding and wintering grounds to exploit optimal climatic and food conditions [104]. Territorial and flocking behaviour also varies among species. For instance, herons are territorial during breeding, while it is reported that ducks often form large mixed-species flocks for foraging and predator avoidance.

**Table 3.5** Physiological adaptations and their functions in waterbirds

| Physiological adaptations | Examples | Functions | References |
|---|---|---|---|
| Salt glands | Gulls and Pelicans | Excretes excess salt from seawater intake. | [50, 54, 77, 83–86, 91, 109, 122] |
| Dense bones | Cormorants and Grebes | Reduces buoyancy for diving efficiency. | |
| Efficient oxygen utilisation | Cormorants, Grebes and Penguins | Help them stay submerged for long periods. | |
| Strong flight muscles | Arctic Terns, Cranes Greater Flamingo, Northern Pintail and Ruddy Shelducks | Help in long-distance migration. | |
| Sharp vision | Osprey, Great Cormorant, Grey Heron, Kingfisher and Pelican | For seeing underwater/in low-light conditions.<br>To effectively locate, track and capture a prey. | |
| Bill-tip organs | Sandpipers and Snipes | Tactile sensitivity plays a significant role in foraging. | |
| Magnetite-based receptors | Penguins and Northern Shovelers | To detect the earth's geomagnetic field for navigation purposes. | |
| Cryptochromes | Penguins | To develop visual compass for magnetic fields under specific light (blue/UV light) conditions. | |
| Counter-current heat exchange systems | Ducks, Geese and Swans | To forage in icy wetlands without significant heat loss. | |
| Energy-efficient metabolism | Bar—tailed Godwit and Arctic Tern | To sustain swimming for long-distance migration.<br>To maintain energy output across different climatic conditions. | |

Aquatic birds demonstrate a wide range of behavioural adaptations that facilitate their survival, foraging, breeding, and migration within their water-rich environments. These behaviours are frequently influenced by ecological factors such as predation, resource scarcity, and climatic fluctuations, and can exhibit substantial variations across species and geographical regions.

### 3.4.1 Migratory Behaviour in Water Birds

Migration is a vital behavioural adaptation seen in many waterbirds. Bar-tailed Godwit, Greater Flamingo, Northern Pintail, Common cranes, Ruddy Shelducks and Northern Shovelers are common migratory water bird species. The migratory movements are often cued by changes in photoperiod, temperature, and availability of food [7]. Birds like the Bar-tailed Godwit (*Limosa lapponica*) and the Northern Pintail (*Anas acuta*) undertake inter-continental migrations between their breeding and wintering grounds. They often orient themselves using celestial cues, magnetic fields, and landmarks [28, 40]. Migratory birds adapt their timing, duration, and staging to optimise energy expenditure and ensure survival.

#### 3.4.1.1 Long-Distance Migration Strategy and Flyways

Many species of water birds undergo incredible long-distance migrations each year and travel thousands of kilometres across continents to reach their breeding and non-breeding habitats. These migrations adhere to well-defined flyways, including the East Asian–Australasian Flyway, the African–Eurasian Flyway, and the Americas Flyway system [12]. The Arctic Tern (*Sterna paradisaea*) holds the record for the longest bird migration, annually flying over 70,000 km from the Arctic to Antarctica and back [28]. Migration enables water birds to access seasonal resources, minimise competition, and enhance their breeding success.

#### 3.4.1.2 Physiological Endurance in Water Birds

Migratory Behaviour is closely linked to physiological endurance in water birds, which includes fat accumulation, efficient flight mechanics, and the ability to regulate sleep during flight. The Bar-tailed Godwit (*Limosa lapponica*) possess the remarkable ability to undertake non-stop flights across the Pacific Ocean, covering distances of approximately 8000 to 11,000 kilometres. These flights are sustained through the utilisation of stored body fat and muscle proteins as their primary energy sources [40]. Birds demonstrate remarkable navigation skills, relying on celestial cues, magnetic fields, and olfactory signals to orient themselves.

### *3.4.2 Breeding and Nesting Behaviour in Water Birds*

Waterbirds have developed intricate breeding strategies to thrive in dynamic and frequently flooded habitats. Colonial nesting is a prevalent adaptation observed in species such as pelicans, gulls, and terns. This adaptation in waterbirds offers various advantages, including enhanced predator detection and thermoregulation [15]. Several waterbird species including Jacanas and Phalaropes demonstrate a unique parental care pattern characterised by role reversal. In another case study the males assume the responsibility of incubating the eggs and raising the chicks while liberating females to enhance their reproductive output [31].

Several avian species like grebes and ducks return to the same breeding grounds annually showing nest site fidelity. On the other hand, some other water birds species like plovers and stilts use clever distraction displays to trick predators away from their nests.

#### 3.4.2.1 Selection of Sites in Water Birds

The water bird species meticulously select nesting sites that offer safety from predators and proximity to food sources. Other various waterbird species including ducks, coots, and grebes construct floating nests on aquatic vegetation as a protective measure against terrestrial predators. While plovers choose open shorelines and exhibit camouflage to evade detection.

#### 3.4.2.2 Parental Care in Water Birds

Waterbirds have different ways of showing how much they care for their chicks. Notably, in jacanas and phalaropes, a unique phenomenon known as role reversal occurs. In this phenomenon, the males assume the incubation and chick-rearing responsibilities deviating from the traditional parental roles. This observation was first documented by Emlen and Wrege [31]. Mostly waterbirds show pre-cocial development, characterised by the relative maturity and mobility of their chicks shortly after hatching. However, they still need protection and guidance.

#### 3.4.2.3 Colony Nesting in Water Birds

Colonial nesting is a common breeding strategy observed in various aquatic bird species, including herons, gulls, pelicans, and terns. This behaviour presents several advantages, including collective defence against predators, enhanced foraging efficiency due to social facilitation, and thermal benefits [15]. However, it could also increase competition and the risk of disease.

### *3.4.3 Anti-predatory and Social Behaviours of Water Birds*

Waterbirds have evolved a diverse range of anti-predator strategies. Vigilance is a prevalent trait among flocking species, facilitating enhanced threat detection. Mobbing behaviour, characterised by a collective harassment of predators by a group of avian species, is observed in terns, gulls, and lapwings. Numerous bird species, including rails and bitterns, use cryptic plumage and stealthy movements to evade detection within dense vegetation. Furthermore, diving serves as a predator avoidance strategy observed in grebes and coots.

#### 3.4.3.1 Alarm Calls and Mimicry

Aquatic birds have evolved alarm calls to warn their fellow birds of impending danger. The vocalisations of these predators exhibit diverse characteristics, such as pitch, intensity, and urgency, which are influenced by the species of predator [11]. Some of the water birds, like the killdeer, exhibit mimicry or distraction displays to divert predators' attention away from their nests.

#### 3.4.3.2 Flocking and Vigilance

Flocking behaviour is an important social adaptation of water birds, enhancing their ability to detect predators and reducing the risk of individual predation by diluting the chances of being targeted. Large flocks of water birds adopt a cooperative strategy, taking turns being vigilant so that others can forage for food or rest. This phenomenon is observed in ducks, starlings, and gulls during both migration and roosting periods.

#### 3.4.3.3 Aggressive and Territorial Displays

Several waterbirds adopt aggressive postures and vocalizations to safeguard their feeding or nesting habitats. Swans, geese, and coots, for instance, engage in territorial disputes and displays of aggression to prevent intruders and safeguard their resources.

#### 3.4.3.4 Thermoregulation and Roosting

Aquatic birds rely on behavioral thermoregulation to adapt to varying thermal conditions. Many of the waterbird species engage in sun-basking, tucking their bills under their wings, or standing on a single leg to conserve heat. The roosting behavioral adaptation in large groups also facilitates heat conservation and predator avoidance. Cormorants, a group of water birds, are known for their distinctive wing-spreading postures after diving. This behavior helps them dry their feathers and regulate their body temperature [103].

### *3.4.4 Kleptoparasitism Behaviour in Waterbirds*

Kleptoparasitism, commonly known behavior as food theft. It is a captivating foraging strategy employed by various waterbird species. In this phenomenon, individuals engage in the act of stealing prey or food items that have been captured or collected by other birds. This behaviour minimizes the energy expenditure associated with active foraging. Kleptoparasites, including gulls, frigatebirds, skuas, and jaegers, are renowned for their opportunistic exploitation of areas characterised by high bird densities or limited food resources. The Great Frigatebird often harasses seabirds like boobies and terns mid-air until they regurgitate their food, which the frigatebird then catches before it falls into the water [88]. Arctic Skuas also aggressively pursue other birds like puffins and the terns to force them to drop or regurgitate fish [37]. Kleptoparasitism might seem like a clever way to get energy, but it can be quite a competitive game, even among the same species. This behaviour could also affect how coastal and marine bird communities interact especially in breeding and roosting areas where food is scarce.

### 3.4.5 Foraging Specialization in the Water Birds

Water birds exhibit diverse foraging specialisations and facilitating their exploitation of diverse ecological niches and food resources within their aquatic environments. These behavioural adaptations encompass variations in feeding strategies, habitat utilisation and morphological characteristics such as beak shape, length of the neck and limb structure. For example, the Cormorants are specialised divers that engage in fish hunting underwater using powerful legs and webbed feet for propulsion [43].

Like Flamingos, they are filter feeders and well equipped with lamellate beaks that enable them to sieve small invertebrates and algae from shallow waters [77]. Herons and egrets exhibit a patiently stand-and-wait ambush technique that maintaining a state of stillness until prey approaches within striking distance [63]. Other avian aquatic species like Black Skimmer forage by flying low over water surfaces using their elongated lower mandible to skim through the water and capture fish [32]. The highly evolved foraging strategies help aquatic birds avoid competition and find food efficiently. It is very fascinating how their behaviour, environment, and physical characteristics all work together to make them so successful.

Aquatic birds have also adapted their foraging techniques to suit their ecological niches, making them one of the most prominent behavioral adaptations in this group. Ducks and coots dabble and dive, filtering food from water or the seabed. Flamingos, however, have an upside-down head that allows them to filter microorganisms through specialized lamellae in their bills [110]. Cormorants and penguins are skilled pursuers of fish underwater. They have streamlined bodies and powerful feet or flippers for this purpose [72].

Waterbirds have also evolved intricate foraging strategies. Like, Spoonbills employ tactile feeding, sweeping their beaks side-to-side in the shallow waters, while oystercatchers utilise precise pecking to breach molluskan shells. Numerous species synchronise their feeding schedules with tidal or diurnal rhythms to optimise prey accessibility [127]. The parental care behaviours, like precociality in ducklings

**Table 3.6** Behavioural adaptations in waterbirds

| Behavioural adaptations | Description | Examples | References |
|---|---|---|---|
| Migratory behaviour | Seasonal movement to exploit breeding and feeding grounds over long distances. | Arctic Tern (*Sterna paradisaea*) and Bar-tailed Godwit (*Limosa lapponica*) | [28, 40] |
| Breeding and nesting behaviour | Breeding in dense colonies for protection, cooperative vigilance and social interaction. | Gulls (*Larus* spp.) and Pelicans (*Pelecanus* spp.) | [15, 31] |
| Anti -predatory behaviour | Flocking, alarm calls and distraction displays to deter predators. | Lapwings (*Vanellus* spp.) and Terns (*Sterna* spp.) | [11, 104] |
| Kleptoparasitism (food theft) | Stealing food from other the birds to reduce foraging effort. | Frigatebirds (*Fregata* spp.) and Arctic Skua (*Stercorarius parasiticus*) | [37, 89] |
| Foraging specialization | Use of unique hunting and feeding techniques to exploit specific niches. | Cormorants (diving) and flamingos (filter feeding) | [4, 44] |

and cooperative nesting in certain colonial species, which contribute to enhanced chick survival in dynamic wetland environments [19] (Table 3.6).

## 3.5 Community Structure of Water Birds

The water bird community structure encompasses the composition, diversity, and functional organization of coexisting species within a specific aquatic habitat. These communities are shaped by intricate interactions between environmental factors, including habitat type, food availability, water depth, vegetation structure, and the seasonal dynamics [116]. Water bird communities are frequently categorized based on their feeding guilds, like dabblers, divers, filter feeders, and waders. This classification minimizes interspecific competition through resource partitioning [63].

In wetland ecosystems, for instance, a single habitat can support a diverse range of species, including ducks, herons, ibises, stilts, and cormorants. Each of these species occupies a unique trophic niche. Wading birds such as herons and egrets primarily forage in shallow waters, utilizing visual cues to capture fish and amphibians. Conversely, diving ducks and grebes have adapted to deeper aquatic zones, where they specialize in capturing submerged prey. These functional distinctions contribute to the coexistence and complementation of species in aquatic environments [83].

Additionally, the composition of communities of water birds frequently undergoes the seasonal changes, particularly in wetlands that serve as migratory stopovers or breeding grounds. These ecosystems exhibit fluctuations in species richness and abundance due to the periodic influx of migratory species [20]. Anthropogenic pressures, including habitat degradation, water pollution, and urban encroachment, can disrupt the community structures of water birds by diminishing species diversity and favoring more adaptable generalist species over specialized ones [42, 55]. In highly productive and well-managed habitats such as Ramsar wetlands, the aquatic bird communities tend to greater stability, diversity, and functional richness. Analyzing structural and dynamic characteristics of these water bird communities offers valuable insights into the ecological health and biodiversity of wetland ecosystems. The related information enables the development of tailored conservation strategies of water birds that cater to the unique needs of each habitat [73].

### 3.5.1 Species Richness and Evenness in Different Aquatic Habitats

Biodiversity is defined by the structural complexity of water bird communities, which is influenced by the species richness and evenness. Species richness refers to the number of different species of water bird in an area, while evenness describes how evenly individuals are distributed among species. Aquatic habitats like lakes, marshes, estuaries, mangroves, and seasonal wetlands support distinct bird species assemblages based on factors like water depth, vegetation, salinity, and prey availability [20, 116]. For example, freshwater wetlands typically harbour a high diversity of wading and dabbling birds, including herons, storks, ibises, and ducks. The species richness in these wetlands peaks during the migratory season. On the other hand, saltwater lagoons and estuaries support specialized shorebirds and piscivores, such as plovers, sandpipers, terns, and gulls. These water birds have adapted to saline conditions and tidal dynamics. Species evenness is often highest in undisturbed habitats with balanced resource distribution. However, degraded or heavily managed wetlands tend to show low evenness, dominated by a few tolerant or invasive species [42, 73]. The balance between richness and evenness in an aquatic ecosystem reflects its ecological health and influences nutrient cycling and food web dynamics. Monitoring these indices helps in conservation priorities and evaluating wetland restoration programs' effectiveness [38].

### 3.5.2 Guild Structure of Water Birds: Functional Groups Based on Foraging Mode and Trophic Niche

Water bird communities have a guild structure that organizes species into functional groups that exploit similar resources using similar foraging strategies. These trophic guilds contribute to the reduction of interspecific competition and the enhancement of ecosystem stability by facilitating resource partitioning. Aquatic birds are grouped into foraging guilds like dabblers, divers, surface feeders, pursuit feeders, probers, gleaners, and filter feeders, each with unique adaptations for their feeding niche [63, 116]. For instance, dabbling ducks (e.g. *Anas platyrhynchos*) primarily feed at or near the water's surface, often tipping forward without submerging. In contrast, diving ducks (e.g., *Aythya fuligula*) and grebes forage by diving underwater to capture fish or aquatic invertebrates. Shorebirds, such as sandpipers and curlews, possess elongated and highly sensitive bills, uniquely suited for detecting prey within mudflats and shallow substrates. Filter feeders, including flamingos and shovelers, employ specialized beaks to strain plankton and diminutive organisms from the water column [77].

The classification of species into foraging guilds offers a robust framework for comprehending the coexistence of diverse species within a shared habitat. This classification also facilitates the prediction of community responses to environmental perturbations, such as eutrophication or hydrological alterations, which may favour certain guilds over others [20, 73]. From a conservation point of view, it is significant to keep these guilds balanced. That way, we can ensure that the wetlands continue to function properly and support a rich variety of biodiversity.

### *3.5.3 Habitat Utilization and Zoning in Water Birds*

Water birds exhibit remarkable adaptability in exploiting the spatial configuration of wetlands, which generally comprise distinct zones such as the shoreline, open water, and vegetated areas. These zones facilitate species-specific behaviours, particularly pertaining to foraging and nesting strategies [73, 116].

#### 3.5.3.1 Shoreline Zones

Shoreline zones habitats, particularly mudflats and shallow water margins, serve as the primary habitats for wading birds such as sandpipers (*Calidris* spp.), stilts (*Himantopus* spp.) and plovers (*Charadrius* spp.). These avian species exploit the benthic invertebrates inhabiting soft sediments utilizing specialised bills adapted for probing.

#### 3.5.3.2 Open Water Zones

Open water habitats are predominantly inhabited by diving birds, including grebes, cormorants, and diving ducks (e.g., *Aythya* spp.). These avian species engage in fishing and foraging for aquatic invertebrates by submerging their bodies entirely. These adaptations, such as lobed or webbed feet and waterproof plumage, enable them to move efficiently underwater.

#### 3.5.3.3 Vegetation Zones

Emergent and floating vegetation provides habitat for various bird species, including herons (*Ardea* spp.), moorhens (*Gallinula* spp.), and jacanas (*Hydrophasianus* spp.). These aquatic habitats serve as crucial nesting sites, offering camouflage and concealment from potential predators. Nesting platforms are frequently constructed within dense vegetation to mitigate the risk of predation. The wetland zones contribute to resource partitioning, facilitate the coexistence of the species with overlapping ecological niches. The whole process enhances the overall diversity of the ecosystem [20].

### *3.5.4 Factors Affected on Water Bird Communities*

The water bird community structure is influenced by a multiplicity of environmental factors like physical and chemical characteristics. The factors like size, depth, salinity and flow patterns play an important role in determining which species make a home in a particular habitat.

Habitat diversity includes both vegetation and open water gives water birds a wider range of places to live and thrive. The water birds communities are significantly influenced by these factors include habitat heterogeneity, seasonal variations and inter-specific interactions. All of the factors influence species distribution, abundance and coexistence of the community of aquatic birds.

#### 3.5.4.1 Habitat Heterogeneity and Characteristics

According to Green and Elmberg in 2014, The greater aquatic habitat complexity, such as mosaics of open water, mudflats and reed beds supports higher species richness by providing a broader range of foraging and nesting opportunities. Such type of heterogeneity reduces interspecific competition and maximises ecological niche availability.

- Water quality, salinity, and depth impact species distribution.
- Vegetation (e.g., reeds, mangroves) provides nesting and roosting sites.

#### 3.5.4.2 Climate and Seasonality

Waterbird assemblages undergo substantial seasonal turnover owing to their migratory cycles. In regions such as the Indian subcontinent, wetland bird populations experience a significant increase during the winter months. This surge is primarily attributed to the arrival of migratory ducks, geese, and waders from their northern breeding grounds. Seasonal fluctuations in water levels significantly impact the availability of habitats and prey density [59]. Seasonal changes affect migration patterns, breeding, and feeding behaviours. Climate events (e.g., droughts, floods) can alter habitat conditions.

#### 3.5.4.3 Inter-Species Interactions

Species interactions, encompassing competition, facilitation, and mixed-species flocking, profoundly influence the composition of communities. Dominant or aggressive species often take over prime nesting or foraging sites, forcing subordinate species to less suitable areas. Kushlan [63], reported about mixed species flocks. They often have an advantage because they can spot predators more easily and find out food more efficiently. The comprehension of these factors is essential for effective wetland management and water bird conservation particularly in the context of habitat alteration and climate variability.

#### 3.5.4.4 Food Availability

- High food abundance tends to support greater densities in waterbird communities.
- The availability of food must align with key life stages like breeding to ensure successful reproductivity and survival.
- The abundance and accessibility of prey species including fish, invertebrates and plants significantly impact on the composition of a community.

#### 3.5.4.5 Human Impact

- Habitat loss due to agriculture, urbanisation, and wetland drainage disrupts community structure.
- Pollution and climate change (e.g., rising sea levels) affect aquatic ecosystems.

#### 3.5.4.6 Conservation Measures

- Protected wetlands and migratory bird sanctuaries contribute to maintaining aquatic bird populations.
- International agreements, such as the Ramsar Convention, safeguard critical habitats (Table 3.7).

**Table 3.7** Zonations of water bird communities in a wetland ecosystem

| Wetland zones | Dominant group of waterbirds | Species names | Zone characteristics | Ecological functions | References |
|---|---|---|---|---|---|
| Shoreline/ mudflat | Waders and shorebirds | *Charadrius dubius* (Little Ringed Plover) and *Tringa glareola* (Wood Sandpiper) | Exposed mud or sand and shallow water edges | Foraging on benthic invertebrates and nesting on bare ground | [20, 43, 64, 75, 117] |
| Open water | Swimmers and divers | *Aythya fuligula* (Tufted Duck) and *Podiceps cristatus* (Great Crested Grebe) | Deeper water bodies—limited vegetation | Diving and surface foraging territorial displays | |
| Emergent vegetation | Marshs and Reeddwelling birds | *Ardea purpurea* (Purple Heron) and *Gallinula chloropus* (Common Moorhen) | Dominated by reeds, sedges and floating plants | Nesting, roosting and ambush foraging concealment from predators | |
| Floating vegetation | Surface-feeding Ducks and Jacanas | *Anas platyrhynchos* (Mallard) and *Hydrophasianus chirurgus* (Pheasant -tailed Jacana) | Water surface covered with plants like duckweed and water lilies | Walking on floating vegetation foraging on insects and seeds | |
| Riparian/ edge zone | Mixed Forest or Shrubland birds (partial aquatic users) | *Ceryle rudis* (Pied Kingfisher) and *Halcyon smyrnensis* (White-throated Kingfisher) | Transitional zone with trees and shrubs near water | Perching, nesting and diving from edges | |

## 3.6 Foraging Preferences and Feeding Ecology in Water Birds

The water bird species exhibit a wide range of foraging strategies and dietary preferences. These are directly proportional to their morphological adaptations and habitat choices. The feeding ecology of these birds is influenced by various factors including their bill structure, length of the legs, locomotion mode and the availability of food with in specific wetland habitats. Wading birds like herons and ibises usually hunt in shallow water. They use their eyes to spot fish, frogs and other small creatures and swoop down to catch them [63]. In our observation, dabbling ducks (*Anas* spp.) have a unique way of eating. They tilt their bodies forward in shallow waters and munch on aquatic plants and tiny creatures. Conversely, diving ducks

(*Aythya* spp.) and grebes descend to the seabed to capture benthic prey in deeper depths [20].

Flamingos (*Phoenicopterus* spp.) are special birds that use filter-feeding to feed. Their specially designed beaks allow them to scoop up tiny crustaceans and algae from the mud and water [77]. Pelicans also cooperate in foraging, forming semi-circles to gather fish in shallow waters and capture them with their large pouches [2]. Shorebirds are distinguished by their elongated and slender bills such as sandpipers employ their specialised beaks to navigate and explore soft substrates. These substrates serve as habitats for annelids, mollusks and other benthic organisms [93].

Water birds have such unique ways of feeding and it's not just because of where they live. Their feeding habits actually play a significant role in how wetlands work. Their foraging activities contribute to nutrient cycling, trophic dynamics and habitat formation [42]. Lovvorn [69] reported the seasonal changes and how animals forage for food can also affect what they eat. Sometimes these changes lead to changes in their diet or how they share resources to avoid competition with other.

## 3.6.1 Types of Foraging Modes in Water Birds

### 3.6.1.1 Surface Feeding Type

Surface-feeding water birds are like little hunters, hopping and flitting around the water's edge. These creatures do not require extensive exploration to capture their prey. Instead, they swiftly grasp it as it floats just above the water surface. Waterbird species like gulls (*Larus* spp.), terns (*Sterna* spp.), and certain ducks (*Anas* spp.) employ the strategy to catch insects, fish, and plant matter that float on or just below the water's surface [70]. This mode of feeding is energetically efficient, especially in open water bodies with abundant floating or suspended prey. The surface feeders water birds often possess remarkable visual acuity and agile flight capabilities, allowing them to capture prey while in motion.

### 3.6.1.2 Probing Type

Examples of shorebirds such as sandpipers (*Calidris* spp.) and godwits (*Limosa* spp.) frequently engage in probing behaviour. They insert their long, slender bills into soft sediments to locate buried prey such as worms, mollusks, insects and crustaceans [94, 95]. These birds often forage along mudflats and estuarine zones, where prey remains hidden but is abundant. Some species possess mechano-receptors at the tips of their bills, allowing them to detect prey remotely through vibrations or movements in the substrate [85].

#### 3.6.1.3 Visual Foraging in Wading Birds

Wading birds, including herons (*Ardea* spp.), storks (*Ciconia* spp.), and egrets, are visual hunters that heavily rely on sight to detect and capture prey in the shallow waters. These birds show a distinctive patient hunting strategy, standing motionless/ slowly stalking through water before swiftly striking at fish, amphibians, worms or other aquatic invertebrates. Their long legs and necks give them super precise movements and quick strikes no matter how deep the water is present [63]. This approach is particularly effective in clear waters with high prey visibility.

#### 3.6.1.4 Dabbling and Grazing Behaviour

Mallards (*Anas platyrhynchos*) and Northern Pintails (*Anas acuta*) are dabbling ducks, which forage in shallow water by tipping forward to reach submerged vegetation and small aquatic organisms. They feed on plant material, seeds, and invertebrates without diving. Their broad, lamellate bills filter food from water and mud, enabling them to exploit nutrient-rich wetland shallows [20]. These ducks species without diving in water, using their lamellated bills to strain small aquatic invertebrates and plants [4]. Dabblers primarily exploit shallow wetlands, floodplains, and rice fields. Their bills are highly sensitive and allow for tactile feeding, detecting and filtering food particles from mud and water. This mode also supports resource partitioning among sympatric species, reducing interspecific competition.

#### 3.6.1.5 Diving Foraging Strategies

These aquatic birds employ a unique hunting strategy by submerging their entire bodies beneath the water's surface. Their dense plumage, webbed feet and streamlined bodies facilitate exceptional under water propulsion. Cormorants (*Phalacrocorax* spp.) and diving ducks (*Aythya* spp.) are the amazing examples of birds that hunt underwater. They use their feet and wings to propel themselves through the water, making it look like they are flying underwater. They primarily target benthic prey including worms, mollusks, crustaceans and fish. These birds have dense plumage for insulation and compact and hydrodynamic bodies which allow them to move efficiently underwater. Diving is their main way of finding food and they need to be incredibly fit to do it.

Their foraging range frequently extends to deeper water bodies that remain inaccessible to waders or dabblers, as per [69] study. These birds use breath-hold dives and have evolved adaptations like high myoglobin levels to store oxygen efficiently [108].

#### 3.6.1.6 Filter Feeding Specialization

Duck species and flamingos (*Phoenicopterus* spp.) have evolved specialised beaks for filter feeding. These birds possess comb-like structures with in their beaks called lamellae, that facilitate the sifting of algae, plankton and minuscule invertebrates from water and mud. Flamingos stir up sediment using their feet and employ their downward-curved bills upside down in the water to capture food [77]. This technique enables them to feed efficiently in saline or alkaline lakes (Table 3.8).

### *3.6.2 Diet Composition of Water Birds*

#### 3.6.2.1 Invertebrates

Invertebrates form a significant part of the diet for many water birds, particularly shorebirds, dabbling ducks, and waders. Species like sandpipers (*Calidris* spp.), avocets (*Recurvirostra* spp.) and stilts forage for polychaete worms, mollusks, crustaceans, and insect larvae in intertidal zones and soft mudflats. Their slender, probing bills are perfectly adapted to detect and extract these organisms from the sediments. The high protein content of invertebrates is crucial during the breeding seasons and migratory periods, providing the birds with the energy they need and supporting their egg production [92]. Even herbivorous species, such as certain

**Table 3.8** Foraging strategies in water birds

| Foraging strategies | Descriptions | Feeding zones | Examples | References |
|---|---|---|---|---|
| Surface feeding type | Feeding on floating food items from the water surface without submerging. | Water surface | Gulls, Terns and Phalaropes | [16] |
| Probing type | Inserting long bills into mud or sand to extract hidden prey | Mudflats/ shoreline | Sandpipers, Godwits and Curlews | [91] |
| Wading type | Walking through shallow water to capture prey using long legs and beaks | Shallow water | Herons, Egrets and Stilts | [100] |
| Dabbling type | Tipping forward to feed below the water surface without diving | Surface/ subsurface | Mallards and Northern Shovelers | [6] |
| Diving type | Submerging completely to capture prey underwater using feet or wings for propulsion | Underwater | Grebes, Cormorants and Diving Ducks | [50] |
| Filter feeding type | Filtering small organisms from the water using specialized beak structures | Surface/ shallow water | Flamingos and Northern Shovelers | [110] |

ducks, may transition to an invertebrate-rich diet during the early spring to fulfill their nutritional needs.

#### 3.6.2.2 Fish

Fish constitute a primary dietary component for piscivores waterfowl, including cormorants (*Phalacrocorax* spp.), herons (*Ardea* spp.), grebes, kingfishers (*Alcedo* spp.), and pelicans. These avian species possess adaptations such as elongated necks, specialized beaks with sharp or hooked features, and exceptional visual acuity, which facilitate their identification and capture of elusive aquatic prey. While some species engage in submersive hunting techniques, such as cormorants and grebes, others employ aerial hunting or wade slowly through aquatic environments, as exemplified by herons. Fish species that rely on seasonal migrations for spawning grounds are vulnerable to fluctuations in fish populations due to overfishing and raising pollution [60, 61, 75].

#### 3.6.2.3 Aquatic Plants

Aquatic plants and algae are a crucial food source for herbivorous and omnivorous water birds like swans, coots, and certain other duck species like the northern shoveler. These water birds species mostly consume submerged macrophytes (e.g., *Vallisneria*, *Potamogeton*), floating vegetation (e.g., *Lemna*) and emergent plant parts. Aquatic vegetation serves as a vital source of essential nutrients and dietary fibre frequently constituting the primary component of their sustenance during non-breeding or over- wintering periods [58, 62]. Additionally, the grazing pressure exerted by these birds influences the diversity of plants and structure of wetland habitats.

### *3.6.3 Seasonal and Diurnal Variation*

Food availability has changed over time and in different places. The water birds have different ways of finding food and what they feed because of the changing seasons and times of day. These fluctuations arise from changes in water levels, temperature, and prey abundance, which alter the availability of food resources. In temperate regions, numerous waterbird species undertake migratory journeys in pursuit of prey-rich habitats. During the spring and summer months, breeding grounds provide a protein-rich invertebrates diet as a primary source of nutrition. Conversely, wintering sites primarily support diets consisting of seeds or aquatic vegetations [30]. Seasonal shifts in prey distribution influence foraging habitat selection and the composition of community assemblages.

Diurnal variations, on the other hand, exhibit fluctuations in avian activity patterns influenced by light availability, predator presence, and prey behaviour. Many water bird species, including herons and ducks, demonstrate peak foraging activity during dawn and dusk, synchronizing with crepuscular prey movements. Nocturnal foraging is prevalent in the species such as the Black-crowned Night Heron (*Nycticorax nycticorax*), which exploits diminished competition and predation pressure during the nighttime [78]. These temporal adaptations optimize energetic efficiency and survival across a wide range of aquatic ecosystems.

### 3.6.4 Resource Partitioning

**Reducing Competition Through Niche Segregation** Resource partitioning is a fundamental ecological process that facilitates the co-existence of multiple waterbird species within a shared habitat by mitigating inter-specific competition. This is accomplished through niche segregation, where in distinct species exploit distinct foraging areas, dietary resources and temporal feeding patterns. For instance, within a single wetland ecosystem, species such as stilts, egrets, and ducks may forage at varying water depths, specialise in consuming distinct prey items, thereby minimizing resource overlap [100].

Morphological characteristics, such as beak shape and leg length, frequently determine the preferred feeding microenvironment, while behavioural adaptations influence foraging times and techniques. Research conducted in estuarine and marsh ecosystems has demonstrated that even closely related species can spatially or temporally partition their niches to avoid direct competition [66]. These mechanisms contribute to the enhancement of species richness and community stability within the aquatic ecosystems.

## 3.7 Ecological Significance of Aquatic Avian Species

Water birds play a pivotal role in preserving the ecological equilibrium of freshwater and coastal ecosystems. Their diverse strategies for feeding, movement patterns and nesting habits profoundly impact various ecological processes. Water birds play multifaceted and indispensable roles in maintaining the health and functionality of various aquatic environments. Their ecological significance encompasses multiple key areas, such as nutrient cycling, dispersal of seeds, bioindicators, pest control and trophic regulation. Understanding these roles is significant for conservation initiatives and effective ecosystem management.

### 3.7.1 Nutrient Cycling and Seed Dispersal

Water birds contribute to nutrient cycling through their foraging and droppings. The excreta of birds is very rich in nitrogen and phosphorus. These nutrients enhance the productivity of aquatic ecosystems and support the growth of phytoplankton and aquatic vegetation [96]. For example, colonial nesting birds such as cormorants and herons produce concentrated guano thereby enriching the soil and water bodies within their breeding grounds and consequently augmenting primary productivity [74].

Water birds also contribute to seed dispersal, particularly in wetland ecosystems by consuming and excreting seeds over extended distances. Like dabbling ducks and geese have been observed transporting seeds across watersheds, thereby fostering genetic diversity and facilitating the colonization of aquatic plants [34].

Contrary to previous studies, localized nutrient enrichment can impact algal growth, invertebrate communities and fish populations. Aquatic birds especially dabbling ducks and shorebirds play a crucial role in seed dispersal. Interestingly, the aquatic birds ingest seeds from aquatic plants and transport them to new locations either externally or internally. This process is very importance for the colonization of wetlands, facilitating genetic exchange and contributing to the overall aquatic plant biodiversity. The aquatic bird species are important seed dispersers in temporary wetlands which contributing to their resilience and biodiversity.

### 3.7.2 Bioindicators of an Ecosystem

Water birds, due to their position at higher trophic levels and reliance on specific habitats are considered excellent bioindicators. As an example, declines in wader and duck populations have been linked to pollution and hydrological changes indicating environmental stress [21]. Monitoring of these species helps conservationists assess ecosystem health and prioritise protection efforts. The decline of piscivorous birds like bald eagles suggests concerns about fish populations and persistent organic pollutants that accumulate in the food chain [87, 118]. Alterations in waterbird diversity and abundance can serve as indicators of changes in wetland hydrology, water quality and availability of food. Abnormalities and unusual behaviours in aquatic birds can serve as a warning signal. Scientists are much concerned about the potential impact of contaminants and environmental stressors on ecosystems. Long-term monitoring of aquatic bird populations provides valuable data on pollution, habitat degradation and climate change impacts on wetlands [65]. The decline population of Common Loons in specific regions is attributed to mercury contamination rendering them indicators of environmental pollution [33].

### 3.7.3 Pest Control and Trophic Regulation

Aquatic birds play a crucial role in pest control and the overall trophic regulation of aquatic ecosystems. Many species, particularly insectivorous birds such as swallows, terns, and some shorebirds, consume substantial quantities of aquatic insects, including mosquito larvae and other nuisance species. This helps regulate the insect populations and potentially reduces the transmission of diseases to humans and livestock [102]. Purple martins are known to eat flying insects, including mosquitoes. Predatory aquatic birds like pelicans, kingfishers, and cormorants control fish populations by predating on them. While sometimes seen as competitors with human fisheries, their predation can actually benefit fish stocks by removing weaker or diseased individuals, improving their overall fitness [27]. By feeding on insects, mollusks, crustaceans, and small vertebrates, aquatic birds help control pest populations and maintain the trophic level balance in aquatic ecosystem. For example, Herons and egrets regulate fish and amphibian populations, while gulls and terns reduce insect and rodent densities in coastal zones. This natural form of pest control is particularly valuable in agricultural and aquaculture regions adjacent to wetlands and estuaries, as highlighted by Weller [116]. Some birds, like flamingos, filter-feed on small invertebrates and algae, affecting lower trophic levels and water clarity and nutrient dynamics. This complex interplay of predation and consumption by aquatic birds maintains aquatic food web stability and balance.

### 3.7.4 Case Studies and Research Findings

Many case studies and research initiatives have underscored the significant ecological role of aquatic birds. A notable example is the function of great cormorants in regulating fish populations in certain European lakes. Although frequently perceived as a conflict with fisheries, research has demonstrated that cormorant predation can decrease the abundance of small, non-commercial fish species, potentially facilitating the growth of larger, commercially valuable fish by diminishing competition [114].

Wading bird colonies significantly impact wetland ecosystems. For instance, studies in the Everglades show that large colonies of egrets and herons create nutrient hotspots through guano deposition, boosting productivity and biodiversity in the surrounding marshes [36]. The reintroduction of California Condors, a critically endangered species has shown their ability as scavengers to eliminate carcasses prevent disease spread and contribute to nutrient cycling in their vast foraging areas [81].

The Keoladeo National Park (Bharatpur, India) the presence of diverse water bird species has been linked to enhanced fish recruitment and aquatic plant regulation underscoring their ecosystem services [113]. Similar research conducted in Europe and North America has demonstrated that the absence of specific duck species results in heightened algal blooms, primarily due to diminished grazing on submerged vegetation and invertebrates [42]. These case studies underscore ecological functions of aquatic bird species that benefit both biodiversity and wellbeing of human. These diverse examples demonstrate that aquatic birds are not merely components of aquatic ecosystems but they actively contribute to shaping their structure and function. We need to do more research and keep an eye on these amazing species of avian class to fully understand and protect their important roles in our ecosystem.

## 3.8 Threats and Conservation Perspectives of Waterbirds

Habitat loss, pollution and climate change pose significant threats to global biodiversity which causing declines in species populations and ecosystem degradation. These factors operate through distinct mechanisms yet they frequently interact with each other and thereby augmenting their cumulative impact.

**Habitat Loss:** It's a real challenge for water birds all over the world. Wetlands, estuaries, mangroves, and other freshwater and coastal ecosystems, which serve as crucial habitats for feeding, breeding, and migration, are increasingly subjected to drainage, reclamation, or fragmentation due to urban expansion, agricultural activities, and infrastructural development. As reported by Davidson [22], over 64% of the world's wetlands have vanished since 1900, leading to the displacement of numerous waterbird species. This loss not only diminishes nesting and foraging habitats but also isolates populations, thereby enhancing their susceptibility to random events and genetic bottlenecks. Consequently, conservation initiatives such as Ramsar site declarations and community-based wetland management emerge as pivotal strategies for mitigating habitat degradation [99].

**Pollution:** This poses a multifaceted threat to aquatic birds, causing acute toxicity and long-term reproductive and behavioural disturbances. Heavy metals, pesticides, and industrial effluents contaminate water bodies, leading to bioaccumulation in bird tissues and disrupting food chains [17]. Microplastics and macroplastics are increasingly ingested by aquatic birds, causing intestinal blockages and reduced fitness [119, 122, 123]. Oil spills, such as the Deepwater Horizon disaster, have catastrophic effects on seabird populations, causing feather contamination, hypothermia, and impaired buoyancy [46, 124]. Eutrophication is a consequence of agricultural runoff that can result in algal blooms. These blooms diminish prey availability and oxygen levels in the water consequently affecting bird survival indirectly. To safeguard of these aquatic avian species- it is imperative to implement stringent measures such as controlling point-source pollution, enforcing plastic bans and implementing comprehensive waste treatment reforms. Additionally, addressing other environmental challenges is of paramount importance for their conservation.

**Climate change:** It is characterized by shifting climate patterns, poses a multifaceted and escalating threat to aquatic bird species. Changes in temperature, precipitation and hydrological cycles disrupt the distribution and productivity of wetlands.

This disruption impacts migration schedules, breeding outcomes, and food accessibility [101]. Rising sea levels threaten coastal breeding colonies, while droughts dry inland breeding sites used by species like the Black-necked Stilt and the Greater Flamingo. Alterations in wind patterns and an augmentation in storm frequency adversely impact migratory efficiency and survival rates. Several Arctic-breeding waterfowl, such as the Red Knot and Snow Goose, are encountering temporal discrepancies between chick hatching and peak food availability as a result of rapid warming [14]. To safeguard aquatic birds from climate-related stressors, adaptive conservation strategies such as climate-resilient wetland restoration, assisted migration and dynamic habitat protection policies are urgently required.

**Invasive Species**

Invasive species are significant factors contributing to biodiversity loss. These species pose a significant threat to water bird populations, particularly in isolated ecosystems such as islands and protected wetland habitats. Introduced predators, including rats (*Rattus* spp.), feral cats (*Felis catus*) and mongooses (*Herpestes* spp.) have caused substantial damage to ground-nesting bird colonies by preying on eggs, chicks and even adult birds [52]. Vegetative invaders like water hyacinth (*Eichhornia crassipes*) and *Phragmites australis* degrade habitats by reducing open water space and oxygen availability. This alters the structure and food web of aquatic ecosystems [126].

These modifications can substantially diminish the foraging efficiency and reproductive success of native waterbird species. To manage invasive species effectively comprehensive strategies are needed including eradication campaigns biosecurity protocols and habitat restoration to protect vulnerable bird populations [106].

**Human Disturbances**

Human activities, both direct and indirect can significantly disrupt the ecology and behavior of aquatic birds. Recreational activities such as boating, fishing and tourism in proximity to breeding and roosting sites frequently lead to heightened stress levels, displacement and nest abandonment [35]. Urban expansion and infrastructure development such as roads, dams and noise pollution disrupt habitats and reduce the availability of safe nesting and feeding sites. Additionally, unregulated tourism and photographic activities in proximity to sensitive habitats can cause chronic disturbance particularly during the breeding season [109]. These disturbances can diminish reproductive success, augment energy expenditure and modify social behaviors such as flocking and vigilance. To protect our environment from human activities we can create safe zones around sensitive areas, limit access during certain times and encourage eco-tourism that takes care of birds and other wildlife.

## *3.8.1 Conservational Strategies and Management Practices*

Comprehensive and adaptive conservation strategies are significant for protecting and preserving avian water species. These strategies are designed to effectively address to current threats and establish a robust defence mechanism. Resilience against emerging pressures such as climate change and urbanization. Effective management practices must integrate habitat conservation, legislation, involvement of community and scientific monitoring to ensure long-term success.

#### 3.8.1.1 Protected Area Networks and Habitat Restoration

The establishment and effective management of protected areas like Ramsar wetlands and Important Bird and Biodiversity Areas (IBAs) are significant in the conservation of water birds. These areas provide secure breeding, feeding and roosting habitats thereby safeguarding birds from human interference [23]. Despite many protected wetlands facing threats due to encroachment, pollution and inadequate enforcement measures.

Restoring degraded wetlands including reforestation of riparian zones, reintroducing native vegetation and regulating hydrological flows has been shown to enhance biodiversity and increase bird abundance [54].

#### 3.8.1.2 Species-Specific and Nest Management Techniques

In addition to habitat-level strategies, species-specific interventions play a crucial role in conservation efforts. Artificial nest boxes, platforms, and floating islands have been successfully employed to enhance nesting opportunities for various species, including terns, herons, and grebes, particularly in disturbed or modified environments [86]. For endangered or vulnerable species, head-starting programs, such as collecting eggs for incubation and releasing fledglings back into the wild can be effective. However, these programs must be accompanied by long-term habitat protection measures to ensure their long-term sustainability.

#### 3.8.1.3 Controlling of Invasive Species and Pollution

The management of invasive flora and fauna holds paramount importance, particularly in islands and isolated wetlands where native bird species have developed in the absence of substantial predators. The eradication of invasive mammals, such as rats and cats, has resulted in the successful restoration of seabird populations on numerous islands [51]. Through policy enforcement, waste management, and educational campaigns, pollution control can effectively reduce the prevalence of heavy metals, plastics, and pesticides in aquatic environments. This approach not only benefits birds but also contributes to the overall health and resilience of broader ecosystems [17].

#### 3.8.1.4 Monitoring, Research, and Community Involvement

Long-term ecological monitoring, employing bird counts, satellite telemetry, and remote sensing, is paramount for tracking population trends, migration patterns, and habitat utilization. Citizen science initiatives and community-based monitoring programs empower local stakeholders and foster stewardship of wetland resources. Engaging indigenous and local communities in conservation endeavors not only augments success rates but also upholds traditional knowledge and land-use rights [111].

#### 3.8.1.5 International Cooperation and Legal Frameworks

Given the migratory nature of many water birds, international cooperation is crucial. International agreements like the Convention on Migratory Species (CMS), the African-Eurasian Waterbird Agreement (AEWA) and the East Asian-Australasian Flyway Partnership (EAAFP) promote cross-border collaboration on habitat conservation, policy harmonization and joint research [12]. Enacting stronger national laws that align with global frameworks enhances the local enforcement and financial support for conservation efforts.

### *3.8.2 Ramsar Sites, Protected Areas, and Artificial Wetlands*

The effective management of Ramsar sites, protected areas and artificial wetlands are crucial for conserving aquatic bird habitats globally. Ramsar Sites established under the 1971 Ramsar Convention are internationally significant wetlands. Renowned for their ecological value especially as crucial habitats for water birds. These sites are very important for vital breeding, feeding and resting places for various migratory bird species.

Globally, there are over 2400 Ramsar sites spanning an area exceeding 250 million hectares. Notably, many of these sites are harbor of substantial populations of threatened and endemic water birds [22]. Designated wetlands play a crucial role in helping countries prioritize conservation funding, enforce wetland management plans and foster international cooperation in protecting migratory birds. Protected areas, such as national parks, wildlife sanctuaries and bird reserves, complement Ramsar sites by providing legal protection to critical habitats. When effectively managed, these areas substantially minimize human interference, habitat degradation and exploitation [115]. While the mere declaration of protected status is insufficient, sustained monitoring, enforcement and community involvement are crucial.

The success of these areas depends on how we manage the habitats and create buffer zones. These zones help to reduce the impact of changes at the edges and keep the wetlands connected. Artificial wetlands, including constructed lagoons, sewage treatment ponds, and rainwater harvesting systems, have emerged as crucial supplementary habitats for waterfowl, particularly in regions where natural wetlands are diminishing due to climate change or land reclamation. Man-made systems can replicate many ecological functions of natural wetlands, supporting diverse bird communities, including species of conservation concern [84]. For instance, wetlands in Australia and India, specifically wastewater treatment wetlands, have been documented to have high bird richness. These wetlands serve as refuges during periods of drought and migratory periods. To make the most of their ecological importance, careful planning and regular maintenance are crucial. This helps keep the water clean and the plants healthy and human activities in check.

Managing Ramsar sites, protected areas and artificial wetlands is vital for preserving viable aquatic bird populations. This adaptable approach combines with ecological research, climate resilience and stakeholder engagement to ensure its effectiveness in changing environmental conditions (Table 3.9).

## 3.9 Future Prospective

Future research on waterbirds should focus on integrating ecological, behavioural and conservation approaches to better understand their adaptation strategies in rapidly changing environments and to effectively respond to them. Climate change, rising sea level and increasing anthropogenic pressures are expected to cause significant changes in aquatic ecosystems. Which will affect aquatic bird populations, community dynamics and their migration patterns. We can carry out long-term monitoring programmes using modern tools such as satellite telemetry, drone surveys, GIS mapping and eDNA analysis. These tools can provide in-depth information on the habitat use by the waterbirds and their population trends.

It will also be important to examine the effects of pollution, eutrophication, microplastics and emerging contaminants on feeding efficiency and reproductive success. Conservation strategies that prioritise habitat restoration, wetland connectivity and community-based management are key approaches that reconcile human needs with biodiversity conservation. Integrating ornithology, hydrology, climatology and social-ecological studies will enhance our understanding of waterbird ecology. This multidisciplinary approach will also strengthen global conservation policies for aquatic biodiversity. Waterbirds are the reliable indicators of wetland health. Protecting them and their habitats, ensuring sustainable management and integrating them with modern research tools is demand of time. Protecting these species with community participation is essential to secure the future of the aquatic ecosystems.

**Table 3.9** Major threats and conservational status of key water bird species

| Name of the waterbird species | Major threats | Regions affected | Conservation status (IUCN) | References |
|---|---|---|---|---|
| Baer's Pochard (*Aythya baeri*) | Wetland drainage<br>Hunting<br>Pollution | South and Southeast Asia | Critically endangered | [10] |
| Siberian Crane (*Leucogeranus leucogeranus*) | Habitat loss<br>Illegal hunting<br>Disturbance along migratory routes | East Asia<br>Russia<br>India | Critically endangered | [53] |
| Greater Adjutant (*Leptoptilos dubius*) | Habitat destruction<br>Human disturbance<br>Poisoning | India and Southeast Asia | Endangered | [18] |
| Cranes (e.g., Siberian Crane and Sandhill Crane) | Habitat destruction<br>Illegal hunting<br>Collision with powerlines | Asia<br>North America | Endangered (Siberian Crane)<br>Least concern (Sandhill Crane) | [80] |
| Penguins (e.g., Emperor, African and Galápagos) | Climate change<br>Overfishing<br>Oil pollution | Southern hemisphere | Endangered to least concern (varies by species) | [13] |
| Dalmatian Pelican (*Pelecanus crispus*) | Wetland degradation<br>Water pollution<br>Fishing net entanglement | Europe<br>Central Asia | Near threatened | [5] |
| Albatrosses (Family: Diomedeidae) | Bycatch in fisheries<br>Marine pollution<br>Habitat degradation | Southern Oceans<br>Pacific<br>Atlantic | Vulnerable to critically endangered (varies by species) | [90] |
| Flamingos (*Phoenicopterus* spp.) | Habitat degradation<br>Water diversion<br>Pollution | Africa<br>South America<br>Asia | Near threatened to least concern (varies by species) | [8] |
| Common Eider (*Somateria mollissima*) | Oil spills<br>Hunting<br>Climate change impacts | Northern Hemisphere<br>Arctic coasts | Near threatened | [26] |
| Pelicans (e.g. *Pelecanus occidentalis* and *Pelecanus crispus*) | Pollution<br>Fishing net entanglement<br>Habitat loss | Global (coastal and inland) | Least concern to near threatened | [3] |

## 3.10 Conclusion

This chapter explores intricate ecological roles, physiological attributes and behavioural patterns of avian aquatic species that emphasising their pivotal significance in wetland ecosystems globally.

Aquatic birds exhibit a remarkable diversity of specialized adaptations like water-repellent feathers, counter-shading coloration and highly developed beak and foot. These adaptations enable them to occupy a wide range of aquatic habitats. These adaptations facilitate for efficient foraging, navigation in aquatic environments and predation avoidance.

The foraging preferences demonstrate a broad trophic spectrum including fish, aquatic invertebrates, plant materials and detritus. This highlights their adaptability to the environment and their diverse ecological roles. Water birds are categorized in to various functional groups including dabblers, divers, waders, filter feeders and probers. Each group exploits distinct feeding zones and strategies to capture the food. That minimizing interspecies competition and supporting ecosystem balance.

The community structure of avian bird species exhibits diverse patterns depending on the type of habitat, seasonal resource availability and geographical location. Wetlands are characterised by distinct habitat heterogeneity that exhibit a higher species richness and evenness. Water birds establish structured communities known as guilds that based on their trophic niche and foraging strategies. These communities provide valuable insights in to ecosystem functioning and resilience. Water birds serve as indicators of environmental health, agents of seed dispersal, regulators of pest populations and play pivotal role in nutrient cycling. The abundance and behavior of these species often reflect the overall health of the ecosystem. Despite significant progress, major threats continue to endanger populations of water bird globally. Habitat loss, primarily attributed to expansion of agricultural, urbanization and wetland drainage continues to be the primary cause of decline of the waterbird population. Pollution caused by particularly from plastics, heavy metals and pesticides disrupts their survival and affecting their foraging and reproductive process. Climate changes effects like altered migratory patterns, rising of sea levels and shifting of prey availability are intensifying these pressures. Invasive species and disturbances created by human exacerbate these issues particularly for ground-nesting and island-breeding species.

**Pathways and Research Priorities**

To address these complex challenges, we urgently require more scientific research in this field and integrated conservation strategies. Using advanced technologies like GPS telemetry, satellite imaging and eDNA analysis to monitor aquatic bird population trends and their migratory routes.

It is also necessary to examine how pollution and climate change interact to affect the physiology and behavior of water birds. Another way to creating evidence-based management strategies that are tailored to local ecosystems and meet global conservation goals. The chapter also emphasizes the crucial role of protected areas like Ramsar sites and artificial wetlands in preserving aquatic bird habitats. When managed well, these areas act as strong sanctuaries for biodiversity especially in rapidly urbanizing and climate sensitive regions. Conservation of aquatic birds is not just about preserving bird species, it's about safeguard of the entire ecosystems. These birds maintain our environment in balance and healthy. Their decline number serves as a warning signal of broader environmental degradation. By collaborating internationally, engaging with communities, advocating for policies and consistently funding habitat restoration we can ensure the survival of these precious wetland creatures.

**Urgent Need for Continued Research and Conservation Initiatives**

Water birds are known for their ecological significance and ability to indicate ecosystem health. The aquatic avian species are intricately linked to wetland dynamics and playing a crucial role in nutrient cycling, food web regulation, dispersal of seeds and environmental monitoring as shown in this chapter. Despite these challenges they face numerous threats including habitat degradation, pollution, climate change and invasive species which are causing declines population and extinctions.

To effectively address these mounting challenges, ongoing research and targeted conservation efforts are crucial. Despite numerous studies on water birds' biology and ecology several knowledge gaps remain. It is very important to understand how stressors like microplastic pollution changes in water bodies and global warming affect the physiology, behavior and reproductive success of aquatic birds.

**To Conduct Further Research**

- Conduct comprehensive long-term population trend analyses across diverse flyways employing advanced technologies such as satellite telemetry, drone surveys, and environmental DNA (eDNA).
- Simulate the effects of climate change on wetland availability, breeding sites, and migration patterns.
- Develop ecological indicators based on waterbird communities to assess the health of wetlands and evaluate the success of restoration efforts

**To Address Emerging Threats, Conservation Strategies Must Evolve. Key Actions**

- Enhancing Ramsar sites and wetland networks through improved policy implementation, habitat connectivity, and climate resilience.
- Restoring degraded wetlands using nature-based solutions and creating artificial wetlands to compensate for the loss of natural wetlands.
- Controlling invasive species that predate on eggs, compete for food resources, or disrupt native vegetation essential for nesting.
- Mitigating anthropogenic stressors, including pollution, unregulated tourism, and unsustainable fishing methods, in proximity to avian habitats.
- Cultivating local communities as stewards of wetland ecosystems through educational initiatives, incentives, and participatory conservation practices.

Conservation of water birds is vital especially those species migrate internationally requiring international cooperation. Initiatives such as the Convention on Migratory

Species (CMS), the East Asian- Australasian Flyway Partnership (EAAFP) and the African-Eurasian Waterbird Agreement (AEWA) establish frameworks that facilitate coordinated efforts on a large scale. In summary, water birds are not only fascinating subjects for scientific study but also act as guardians of planetary well-being. Preserving these wetlands is crucial for our collective responsibility to protect biodiversity and the vital life support systems they offer. Through sustained research with the integrated policy action and grassroots engagement that the future of aquatic birds can still be secured.

## References


1. Ali S, Ripley SD (1987) Handbook of the birds of India and Pakistan. Oxford University Press
2. Anderson DW (1991) Foraging behavior of pelicans and cormorants. In: Burger J (ed) The behaviour of seabirds. Springer, Boston, pp 231–248
3. Anderson DW, Gress F, Mais KF (1982) Brown pelicans: Influence of food supply on reproduction. Condor 84:322–331. https://doi.org/10.2307/1367377
4. Baldassarre GA, Bolen EG (2006) Waterfowl ecology and management, 2nd edn. Krieger Publishing, Malabar
5. Barov B, Derhé MA (2011) Review of the official status of species and priority levels. In: BirdLife international. Action plans for conservation of vulnerable species in Europe. European Commission, Luxembourg
6. Bellrose FC (1980) Ducks, geese, and swans of North America. Stackpole Books
7. Berthold P (2001) Bird migration: a general survey, 2nd edn. Oxford University Press, Oxford
8. Bildstein KL (1992) Causes and consequences of human-induced alterations to fish-eating bird communities in Western Europe. In: McCullough DR, Barrett RH (eds) Wildlife 2001: Populations. Springer, Dordrecht, pp 825–836. https://doi.org/10.1007/978-94-011-2868-1_64
9. BirdLife International (2021) State of the world's birds 2021. BirdLife International, Cambridge. https://www.birdlife.org
10. BirdLife International (2022) IUCN Red List for birds. Downloaded from http://www.birdlife.org
11. Blumstein DT, Armitage KB (1997) Alarm calling in yellow-bellied marmots: I. The meaning of situationally variable alarm calls. Anim Behav 53(1):143–171
12. Boere GC, Stroud DA (2006) The flyway concept: what it is and what it isn't. In: Boere GC, Galbraith CA, Stroud DA (eds) Waterbirds around the world. The Stationery Office, Edinburgh, pp 40–47
13. Boersma PD, Rebstock GA (2014) Climate change increases reproductive failure in Magellanic penguins. PLoS One 9(1):e85602. https://doi.org/10.1371/journal.pone.0085602
14. Both C, Van Asch M, Bijlsma RG, Van Den Burg AB, Visser ME (2009) Climate change and unequal phenological changes across four trophic levels: constraints or adaptations? J Anim Ecol 78(1):73–83
15. Buckley PA, Buckley FG (1972) The significance of colonial nesting in gulls. Ecology 53(4):682–694
16. Burger J, Gochfeld M (1981) Temporal patterns in foraging behavior of gulls and terns. Condor 83(2):140–146
17. Burger J, Gochfeld M (2004) Marine birds as sentinels of environmental pollution. EcoHealth 1(3):263–274
18. Choudhury A (2000) The Birds of Assam. Gibbon Books and World Wide Fund for Nature, Guwahati

19. Colwell MA (2010) Shorebird ecology, conservation, and management. University of California Press, Berkeley
20. Colwell MA, Taft OW (2000) Waterbird communities in managed wetlands of varying water depth. Waterbirds 23(1):45–55
21. Custer CM, Custer TW, Dummer PM (2012) Exposure and effects of mercury on birds nesting in the St. Louis river area of concern, Minnesota, USA. Arch Environ Contam Toxicol 62: 474–483. https://doi.org/10.1007/s00244-011-9717-1
22. Davidson NC (2014) How much wetland has the world lost? Long-term and recent trends in global wetland area. Mar Freshw Res 65(10):934–941
23. Davidson NC, Fluet-Chouinard E, Finlayson CM (2012) Ramsar wetlands: values, management, and conservation challenges. Aquat Conserv Mar Freshwat Ecosyst 22(3):219–229
24. del Hoyo J, Elliott A, Sargatal J (eds) (1992–2013) Handbook of the birds of the world, vol 1–16. Lynx Edicions
25. Denny M (1993) Air and water: the biology and physics of life's media. Princeton University Press, Princeton
26. Descamps S, Jenouvrier S, Gilchrist HG, Forbes MR (2012) Avian cholera, climate and long-term population dynamics of Common Eiders. Solar Physics 28:299–306. https://doi.org/10.1007/s00300-011-1074-5
27. Dobson AP (1990) Survival rates and their relationship to life-history traits in some common British birds. In: PowerDM (ed) Current Ornithology. Plenum Press, New York, pp 115–146. https://doi.org/10.1007/978-1-4615-6784-4_3
28. Egevang C, Stenhouse IJ, Phillips RA, Petersen A, Fox JW, Silk JRD (2010) Tracking of Arctic terns reveals longest animal migration. Proc Natl Acad Sci U S A 107(5):2078–2081
29. Ehrlich PR, Dobkin DS, Wheye D (1988) The birder's handbook: a field guide to the natural history of north American birds. Simon & Schuster, New York
30. Elphick CS (2008) Landscape effects on waterbird densities in California rice fields: taxonomic differences, scale-dependence, and conservation implications. Ecol Appl 18(2): 483–496
31. Emlen ST, Wrege PH (1986) Forced copulations and intra-specific parasitism: two costs of social living in the white-fronted bee-eater. Ethology 71(1):2–29
32. Erwin RM (1977) Foraging activities of black skimmers on the New Jersey coast. Auk 94(2): 352–358
33. Evers DC, Kaplan JD, Meyer MW, Reaman PS, Braselton WE, Major A, Burgess N, Scheuhammer AM (1998) Geographic trend in mercury measured in Common Loon feathers and blood. Environ Toxicol Chem 17:173–183. https://doi.org/10.1002/etc.5620170205
34. Figuerola J, Green AJ (2002) Dispersal of aquatic organisms by waterbirds: a review of past research and priorities for future studies. Freshw Biol 47:483–494. https://doi.org/10.1046/j.1365-2427.2002.00829.x
35. Finney SK, Pearce-Higgins JW, Yalden DW (2005) The effect of recreational disturbance on an upland breeding bird, the golden plover (Pluvialisapricaria). Biol Conserv 121(1):53–63
36. Frederick PC, Ogden JC (2001) Pulsed resources and the ecology of conservation in the Everglades, Florida. In: Science and Conservation in the Everglades. Springer, pp 435–455. https://doi.org/10.1007/978-1-4615-1353-7_18
37. Furness RW (1987) The Skuas. T & AD Poyser, Calton
38. Gibbs JP (2000) Wetland loss and biodiversity conservation. Conserv Biol 14(1):314–317
39. Gill FB (2007) Ornithology, 3rd edn. W. H. Freeman and Company, New York
40. Gill RE Jr, Tibbitts TL, Douglas DC, Handel CM, Mulcahy DM, Gottschalck JC, Warnock N, McCaffery BJ, Battley PF, Piersma T (2009) Extreme endurance flights by landbirds across the Pacific Ocean: ecological corridor between Alaska and New Zealand. Proc R Soc Lond B Biol Sci 276(1656):447–457
41. Gill F, Donsker D (2023) IOC World Bird List (v13.1). International Ornithologists' Union. https://www.worldbirdnames.org

42. Green AJ, Elmberg J (2014) Ecosystem services provided by waterbirds. Biol Rev 89:105–122. https://doi.org/10.1111/brv.12045
43. Gremillet D, Wright G, Lauder A, Carss DN, Wanless S (2005) Modelling the daily food requirements of wintering great cormorants: a bioenergetics tool for wildlife management. J Appl Ecol 40(2):266–277
44. Gutiérrez JS, Abad-Gómez JM, Sánchez-Guzmán JM et al (2011) Water loss through evaporation in shorebird nests: implications for incubation behaviour under hot desert conditions. J Avian Biol 42:99–106. https://doi.org/10.1111/j.1600-048X.2010.05113.x
45. Hackett SJ, Kimball RT, Reddy S et al (2008) A phylogenomic study of birds reveals their evolutionary history. Science 320:1763–1768. https://doi.org/10.1126/science.1157704
46. Haney JC, Geiger HJ, Short JW (2014) Bird mortality from the Deepwater Horizon oil spill. I. Exposure probability in the offshore Gulf of Mexico. Mar Ecol Prog Ser 513:225–237
47. Hill GE, McGraw KJ (2006) Bird coloration: function and evolution, vol 2. Harvard University Press, Cambridge
48. Jenkin PM (1957) The filter-feeding and food of flamingoes (Phoenicopteri). Philos Trans R Soc Lond Ser B Biol Sci 240(673):401–493
49. Jenni L, Jenni-Eiermann S (1998) Fuel supply and metabolic constraints in migrating birds. J Avian Biol 29(4):521–528. https://doi.org/10.2307/3677171
50. Johnsgard PA (2010) Ducks, geese, and swans of the world. University of Nebraska Press
51. Jon HS, Kim IH, Paek UN (2016) Population and breeding status of the Black-faced Spoonbill Platalea minor in the Democratic People's Republic of Korea. Forktail 32:64–68
52. Jones HP, Tershy BR, Zavaleta ES, Croll DA, Keitt BS, Finkelstein ME, Howald GR (2008) Severity of the effects of invasive rats on seabirds: a global review. Conserv Biol 22(1):16–26
53. Kanai Y, Ueta M, Germogenov N, Nagendran M, Mita N, Higuchi H (2002) Migration routes and important resting areas of Siberian cranes (Grus leucogeranus) between northeastern Siberia and China as revealed by satellite tracking. Biological Conservation 106:339–346. https://doi.org/10.1016/S0006-3207(01)00259-2
54. Keddy PA, Fraser LH, Solomeshch AI, Junk WJ, Campbell DR, Arroyo MT, Alho CJ (2009) Wet and wonderful: the world's largest wetlands are conservation priorities. Bioscience 59(1): 39–51
55. Kiran, Singh D, Kour A, Priya Delu V, Kumar R (2022) Community composition and status of avian diversity at Campus and Agricultural landscapes of Chaudhary Charan Singh Haryana Agricultural University, Hisar (Haryana). J Appl Nat Sci 14(4):1130–1140. https://doi.org/10.31018/jans.v14i4.3784
56. Kiran, Singh D, Kour A, Priya Delu V, Kumar R, Punia N, Sunita (2022) Different strategies adopted by birds to sustain ecosystem: a review. The Pharma Innovation SP-11(9):412–422
57. Klaassen M (1996) Metabolic constraints on long-distance migration in birds. J Exp Biol 199: 57–64. https://doi.org/10.1242/jeb.199.1.57
58. Kloskowski J (2011) Impact of waterfowl on macrophyte vegetation in a shallow lake ecosystem. Freshw Biol 56(10):1983–1995
59. Kumar R, Gupta SK, Sivakumar K, Mathur VB (2005) Waterbird diversity and abundance in relation to habitat and anthropogenic disturbance in selected wetlands of India. Wildlife Institute of India, Dehradun
60. Kumar R, Gupta R, Dahiya T, Yadav R, Saifi R, Yodha K, Kumari B, Singh P, Shefali (2022) Surveillance of *Heteropneustes fossilis* (stinging catfish) diseases and health management practices in Haryana. Appl Ecol Environ Sci 10(4):261–267
61. Kumar R, Gupta RK, Yadav R, Saifi R, Yodha K, Pooja (2022) *Eisenia fetida* as protein source for growth enhancement of *Heteropneustes fossilis*. Egypt J Aquat Biol Fish 26(2):577–588
62. Kumar R, Yadav R, Kumar Gupta R, Pooja (2022) A sustainable way for fish health management by replacement of chemical and drugs by earthworm. In: Environmental degradation in Asia: land degradation, environmental contamination, and human activities. Springer International Publishing, Cham, pp 329–352

63. Kushlan JA (1978) Feeding ecology of wading birds. In: Sprunt AIV, Ogden JC, Winckler S (eds) Wading birds. National Audubon Society Research report no. 7, New York, pp 249–297
64. Kushlan JA, Hancock JA (2005) The herons. Oxford University Press
65. Kushlan JA, Steinkamp MJ, Parsons KC, et al (2002) Waterbird Conservation for the Americas: The North American Waterbird Conservation Plan, Version 1. Waterbird Conservation for the Americas, Washington, DC
66. Lawlor LR (1980) Structure and stability in natural and randomly constructed competitive communities. Am Nat 116(3):394–408
67. Livezey BC (1991) A phylogenetic analysis and classification of recent dabbling ducks (Tribe Anatini) based on comparative morphology. Auk 108:471–507. https://doi.org/10.1093/auk/108.3.471
68. Livezey BC, Zusi RL (2007) Higher-order phylogeny of modern birds (Theropoda, Aves: Neornithes) based on comparative anatomy. II. Analysis and discussion. Zool J Linn Soc 149: 1–95. https://doi.org/10.1111/j.1096-3642.2006.00293.x
69. Lovvorn JR (1994) Biomechanics and foraging profitability: an approach to assessing trophic needs and impacts of diving ducks. Ecol Appl 4(2):253–270
70. Lovvorn JR, Gillingham MP (1996) Food availability and intake capacity as determinants of cellular plant consumption by ducks. Ecology 77:2435–2448. https://doi.org/10.2307/2265742
71. Lovvorn JR, Jones DR (1994) Body mass, volume, and buoyancy of some aquatic birds, and their relation to locomotor strategies. Can J Zool 72(1):32–39. https://doi.org/10.1139/z94-005
72. Lovvorn JR, Grebmeier JM, Cooper LW, Bump JK, Richman SE (1999) Modeling ingestion rate of diving predators from the benthos: a bioenergetics approach for sea ducks. Ecol Appl 9(4):1181–1195
73. Ma Z, Cai Y, Li B, Chen JK (2010) Managing wetland habitats for waterbirds: an international perspective. Wetlands 30(1):15–27
74. Manny BA, Johnson WC, Wetzel RG (1994) Nutrient additions by waterfowl to lakes and reservoirs: predicting their effects on water quality. Hydrobiologia 279:121–132. https://doi.org/10.1007/BF00027125
75. Marques AT, Batalha H, da Silva JP (2014) Understanding the impact of wind farms on shorebird populations: a review of fish availability and piscivorous bird distribution. Mar Ecol Prog Ser 515:239–255
76. Martin GR, Katzir G (1999) Visual fields in short-toed eagles and grey herons. Brain Behav Evol 53(1):33–42. https://doi.org/10.1159/000006586
77. Mascitti V, Kravetz FO (2002) Bill morphology of South American flamingos. Condor 104(1): 73–83
78. McNeil R, Drapeau P, Lachapelle P (1993) Nocturnality in shorebirds and other waterbirds in a tropical lagoon. Auk 110(4):735–743
79. McWilliams SR, Guglielmo C, Pierce B, Klaassen M (2004) Flying, fasting, and feeding in birds during migration: a nutritional and physiological ecology perspective. J Avian Biol 35: 377–393. https://doi.org/10.1111/j.0908-8857.2004.03378.x
80. Meine CD, Archibald GW (1996) The Cranes: Status Survey and Conservation Action Plan. IUCN, Gland, Switzerland
81. Meretsky VJ, Snyder NFR, Beissinger SR, Clendenen DA, Wiley JW (2000) Demography of the California Condor: Implications for Reintroduction. Conservation Biology 14:957–967. https://doi.org/10.1046/j.1523-1739.2000.99114.x
82. Midtgard U (1981) The rete tibiotarsale and arteriovenous association in the hind limb of birds: a comparative morphological study on counter-current heat exchange systems. Acta Zoologica 62(2):67–87. https://doi.org/10.1111/j.1463-6395.1981.tb00715.x
83. Mitsch WJ, Gosselink JG (2015) Wetlands, 5th edn. Wiley, Hoboken
84. Murray BR, Hose GC, Eamus D (2010) Constructed wetlands in arid and semi-arid regions: a review of their potential for habitat provision and wastewater treatment. Ecol Eng 36(12): 1681–1690

85. Nebel S, Thompson GJ, Humphrey BW, et al (2005) Do shorebirds use dietary triglycerides as a cue for site selection during migration? Condor 107:662–670. https://doi.org/10.1093/condor/107.3.662
86. Newton I (2010) The migration ecology of birds. Academic Press, London
87. Nisha A, Rani S, Kumar S, Kumar R, Kumar A (2025) Recent technologies and strategies for bioremediation of toxic and hazardous waste. In: Dubey KK, Shah MP, Yadav A (eds) Challenges and sustainable solutions in bioremediation. CRC Press, pp 1–16. https://doi.org/10.1201/9781003407317-1
88. Osorno JL, Garcia CM, Drummond H (1992) Why do frigatebirds attack masked booby chicks? Anim Behav 44(4):703–707
89. Peaker M, Linzell JL (1975) Salt glands in birds and reptiles. Cambridge University Press, Cambridge
90. Phillips RA, Silk JRD, Croxall JP, Afanasyev V (2006) Year-round distribution of white-chinned petrels Procellaria aequinoctialis from South Georgia. Mar Ecol Prog Ser 319:287–300. https://doi.org/10.3354/meps319287
91. Pienkowski MW (1983) Differences in habitat requirements and distribution patterns of plovers and sandpipers as a function of bill length. J Anim Ecol 52(1):147–158
92. Piersma T (2007) Using the power of comparison to explain habitat use and migration strategies of shorebirds worldwide. J Ornithol 148(Suppl 1):S45–S59
93. Piersma T, De Goeij P, Tulp I (1993) An evaluation of intertidal feeding habitats from a shorebird perspective: towards relevant comparisons between temperate and tropical mudflats. Neth J Sea Res 31(4):503–512
94. Piersma T, van Aelst R, Kurk K, Berkhoudt H, Maas LR (1998) A new pressure sensory mechanism for prey detection in birds: the tactile foraging of shorebirds. Proc R Soc Lond B 265(1402):1377–1383. https://doi.org/10.1098/rspb.1998.0435
95. Piersma T, van Aelst R, Kurk K, Berkhoudt H, Maas LRM (1998) A new pressure sensory mechanism for prey detection in birds: the use of the bill-tip organ by Calidris sandpipers. J Exp Biol 201(23):3255–3263
96. Post ES, Forchhammer MC, Stenseth NC, Langvatn R (1998) Extrinsic modification of vertebrate phylogeny and the evolution of complex life cycles. Nature 394:568–571. https://doi.org/10.1038/29053
97. Proctor NS, Lynch PJ (1993) Manual of ornithology: avian structure and function. Yale University Press, New Haven
98. Raikow RJ (1994) Structure and function of the hindlimb in birds. In: King AS, McLelland J (eds) Form and function in birds, vol 4. Academic, London, pp 57–147
99. Ramsar Convention Secretariat (2013) The Ramsar convention manual: a guide to the convention on wetlands
100. Recher HF, Recher JA (1969) Some aspects of the ecology of migrant shorebirds. Ecology 50(2):198–206
101. Rehfisch MM, Crick HQP (2003) Predicting the impact of climatic change on Arctic-breeding waders. Wader Study Group Bull 100:86–95
102. Reiter J, Sprenger B (1987) Ecological and ethological studies on the Black Stork Ciconia nigra in the northern part of Switzerland. Ornithologische Beobachter 84:115–130
103. Rijke AM (1968) The water repellency and feather structure of cormorants, Phalacrocoracidae. J Exp Biol 48(1):185–189
104. Roshier DA, Robertson AI (2003) Seasonal movements and habitat use by waterbirds in the lower Murrumbidgee River floodplain, southeastern Australia. Wildl Res 30:281–290. https://doi.org/10.1071/WR01090
105. Sibley CG, Monroe BL Jr (1990) Distribution and taxonomy of birds of the world. Yale University Press, New Haven
106. Simberloff D, Martin JL, Genovesi P, Maris V, Wardle DA, Aronson J et al (2013) Impacts of biological invasions: what's what and the way forward. Trends Ecol Evol 28(1):58–66

107. Soons MB, Brochet AL, Kleyheeg E, Green AJ (2008) Seed dispersal by waterbirds: an overlooked dispersal pathway for terrestrial plants. J Ecol 96:570–582. https://doi.org/10.1111/j.1365-2745.2008.01354.x
108. Stephenson R, Turner DL, Butler PJ (1989) The relationship between diving activity and oxygen storage capacity in tufted ducks (Aythya fuligula). J Exp Biol 141:265–275
109. Steven R, Pickering C, Castley JG (2011) A review of the impacts of nature-based recreation on birds. J Environ Manag 92(10):2287–2294
110. Studer-Thiersch A (1975) Basal feeding rhythm in greater flamingos Phoenicopterus ruber roseus maintained under artificial conditions. Ornis Scand 6(2):137–146
111. Sullivan BL et al (2014) The eBird enterprise: an integrated approach to development and application of citizen science. Biol Conserv 169:31–40
112. Sundar KG, Kittur S (2012) On the importance of agricultural landscapes for the conservation of Sarus cranes (Grus antigone) in India. Bird Conserv Int 23(4):467–478. https://doi.org/10.1017/S0959270912000120
113. Sundar KSG, Subramanya S (2010) Sarus Crane Grus antigone in India: A Review of Biology, Threats, and Conservation. In: Waterbirds around the world. Edinburgh: The Stationery Office, pp 367-371
114. Van Eerden MR, Gregersen J (1994) Long-term changes in the northwest European population of Cormorants Phalacrocorax carbo sinensis. Ardea 83:61–79
115. Watson JEM, Dudley N, Segan DB, Hockings M (2014) The performance and potential of protected areas. Nature 515(7525):67–73
116. Weller MW (1999) Wetland birds: habitat resources and conservation implications. Cambridge University Press, Cambridge
117. Wetlands International (2020) Asian waterbird census report. Wetlands International South Asia, New Delhi
118. Wiemeyer SN, Lamont TG, Bunck CM, et al (1984) Organochlorine pesticide, polychlorobiphenyl, and mercury residues in bald eagle eggs—1969–79—and their relationships to shell thinning and reproduction. Archives of Environmental Contamination and Toxicology 13:529–549. https://doi.org/10.1007/BF01056331
119. Wilcox C, Van Sebille E, Hardesty BD (2015) Threat of plastic pollution to seabirds is global, pervasive, and increasing. Proc Natl Acad Sci 112(38):11899–11904
120. Williams JB, Siegfried WR (1980) Foraging ecology and time budgets of the crowned, black and grey-headed herons. Ardea 68:135–147
121. Wiltschko R, Wiltschko W (2005) Magnetic orientation and magnetoreception in birds and other animals. J Comp Physiol A 191:675–693. https://doi.org/10.1007/s00359-005-0627-7
122. Yadav R, Gupta RK, Kumar R, Kaur T, Moun R (2022) Biochemical response of earthworm, *Eisenia fetida* to heavy metals toxicity. J Appl Nat Sci 14(3):990–998
123. Yadav R, Gupta RK, Kumar R, Kaur T (2022) Assessment of toxicity of lead and nickel on the biochemical and immunological parameters of earthworm, Eudrilus eugeniae. Agric Sci Dig 44(5):922–929
124. Yadav R, Kumar R, Gupta RK, Kaur T, Yodha K, Kour A, Kaur S, Rajput A (2023) Heavy metal toxicity in earthworms and its environmental implications: a review. Environ Adv 12: 100374
125. Yodha K, Singh D, Kour A, Yadav R, Kumar R (2024) Species diversity and abundance of birds in ber (*Zizyphus mauritiana*) crop. Asian J Conserv Biol 13(1). https://doi.org/10.53562/ajcb.83683
126. Zedler JB, Kercher S (2004) Causes and consequences of invasive plants in wetlands: opportunities, opportunists, and outcomes. Crit Rev Plant Sci 23(5):431–452
127. Zwarts L, Ens BJ, Goss-Custard JD, Hulscher JB (1996) Why do waders reach high feeding densities? In: Goss-Custard JD (ed) The ecology of birds. Springer, Dordrecht, pp 237–258
128. Zweers GA, Gerritsen AFC, Vanden Berg A (1995) Filter feeding in flamingos (Phoenicopterus ruber). Condor 97(2):297–324

# Chapter 4
# Water Bird Migration: A Conservation Perspective

**Anshu, Jyoti, Deepanshu Kumar, Rahul Kumar, and Sindhu Sheoran**

**Abstract** Migration is a periodic phenomenon exhibited by birds. Waterbirds travel distances of up to thousands of kilometres to reach their wintering or breeding grounds, following mostly fixed flyways. This chapter discusses the major flyways of the world, including the Central American, Central Asian, Pacific, Atlantic, Black Sea, Mediterranean flyways, East Asian–Australasian Flyway, and Mississippi flyway, highlighting their ecological importance for aquatic migrant birds. Flyways can be intracontinental or transboundary, later requiring birds to cross vast oceans to reach their destinations. Stopover sites are refuelling points along the route between the starting point and destination, with wetlands mainly serving as stopover sites. Wetlands provide temporary habitats, food, and resting places. Therefore, wetland conservation is vital to protect migratory birds. Data from various studies confirm that bird populations on major flyways are declining rapidly due to human activities such as habitat loss, poaching, illegal killing, overfishing at stopover sites, and pollution. The routes of the Siberian crane, Northern Pintail, Bar-headed Goose, Blue-throat, and flamingo are shown in this chapter, including several localities in India where these birds arrive. To cope with migration-related challenges, birds undergo many morphological and physiological changes. Morphological features include the shape of the body, wings, feet, and beak, while physiological adaptations include changes in the skeletal system, such as fused pneumatic bones, keel, well-developed flight muscles, and changes in the digestive, sensory, and respiratory systems.

This chapter particularly highlights the importance of wetlands, conservation regulations, and their impact on flyways. It emphasizes the need to identify

Anshu · Jyoti · D. Kumar
Department of Zoology and Environmental Sciences, Punjabi University, Patiala, India

R. Kumar
Department of Zoology, Chaudhary Charan Singh Haryana Agricultural University, College of Agriculture, Bawal, Haryana, India

S. Sheoran (✉)
Western Sydney University, Sydney, NSW, Australia
e-mail: S.sheoran@westernsydney.edu.au

 
R. Kumar et al. (eds.), *Ecology, Conservation of Aquatic Birds*, Wetlands: Ecology, Conservation and Management 15, https://doi.org/10.1007/978-3-032-17125-2_4

"vulnerable" and "threatened" flyways and to develop appropriate protective measures. The chapter covers the main flyways, the role of wetlands in waterbird conservation, conservation regulations, the need for new cooperative legislation, migratory bird routes, and bird adaptations.

**Keywords** Conservation flyway · Migration · Stopover · Wetland

## 4.1 Introduction

Bird migration is one of the most incredible events in nature and causes major changes in the distribution of birds across Earth's surface every spring and autumn [55]. This phenomenon involves animals traveling back and forth from wintering sites to their reproductive grounds. They typically move to breeding grounds, often in multiple springtime flights, pausing at various locations to rest and replenish because they cannot cover the entire distance in a single journey. Birds' choice of stopover sites is influenced by their ecological features, as well as by their location and visibility along major migratory routes [79]. The term 'flyway' was initially introduced to describe the pathways taken by birds during their seasonal migration [47]. Recently, a widely cited definition of a flyway is "the totality of the ecological systems that are necessary to enable a migratory waterbird to survive and fulfill its annual cycle" [11].

Most birds in the northern hemisphere migrate through a process called "latitudinal migration," which involves traveling long distances and explains how birds move from north to south and vice versa [35]. These birds originate from the northern temperate and subarctic regions, where they breed in the summer, and migrate southward during winter. Species such as the Bar-tailed Godwit *(Limosa lapponica)*, Northern Pintail (*Anas acuta*), Eurasian Teal (*Anas crecca*), Eurasian Wigeon (*Mareca penelope*), Ruddy Turnstone (*Arenaria interpres*), Pallid Harrier (*Circus macrourus*), Common Greenshank (*Tringa nebularia*), Sandpiper (*Tringa stagnatilis*), Eurasian Whimbrel (*Numenius phaeopus*), and Pallas's Gull *(Ichthyaetus ichthyaetus)* are known for covering long distances [62]. Although some species travel shorter latitudinal distances, such as from the Himalayas to peninsular India.

Wetlands play a crucial role in supporting migratory birds by providing essential stopover sites to rest, refuel, and recover during their long journeys. These habitats offer abundant food resources, shelter from predators, and safe resting areas, especially for waterbirds and shorebirds migrating across continents. Wetlands serve as vital nodes in the flyway network, and the loss of even a single site can disrupt entire migration routes [29, 92]. Figure 4.1 depicts the migratory route. Currently, India has 1243 wetlands, which are categorized into 1053 other wetlands, 115 notable wetlands, and 75 Ramsar wetlands. India joined the Ramsar Convention—also known as the "Convention on Wetlands"—on February 1st, 1982. The total area of India's Ramsar sites is 13266.77 $km^2$.

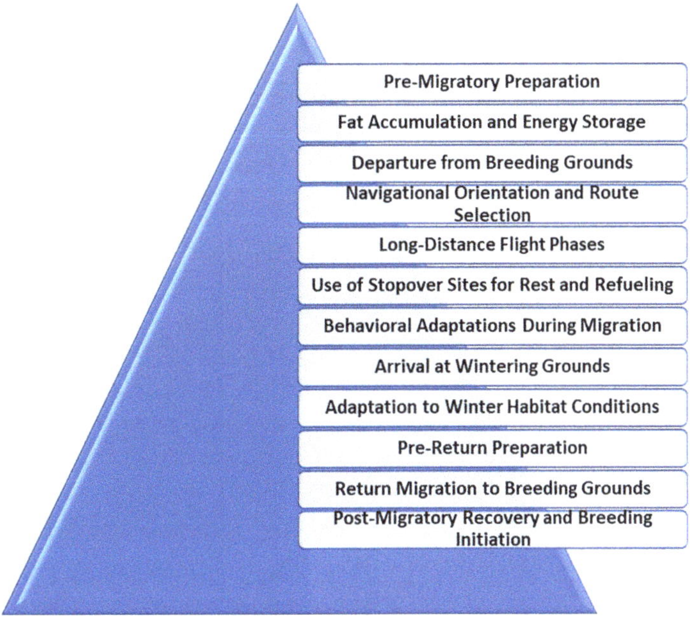


**Fig. 4.1** Illustration of the general migratory route adopted by birds

**Route of Migration**

## 4.2 Migration of Birds and Their Flyways in India

Migratory bird species travel south for wintering and return to their habitats when conditions are suitable [2, 8, 56, 73]. About 2000 bird species participate in this bidirectional seasonal movement worldwide [10], some of which are depicted in Fig. 4.2. These birds follow specific routes due to natural obstacles like the Caribbean Sea, the Mediterranean Sea, and the towering Himalayan peaks. These routes are known as migratory flyways [75]. Many migratory birds face numerous threats, which lead to declines in their populations and species diversity across the globe, mainly due to human-related factors such as habitat loss and degradation, illegal and unsustainable hunting, pollution, climate change, infrastructure development, and disease [4, 9, 82]. Coordinated international efforts are crucial for their conservation. Various approaches, including numerous multilateral agreements, have been implemented at the intergovernmental level to promote conservation and ensure the sustainability of these migratory birds. These include the Neotropical Migratory Bird Conservation Act, the Convention on Migratory Species, the State of the World's Birds, the Convention for the Protection of Migratory Birds, and the Ramsar Convention.

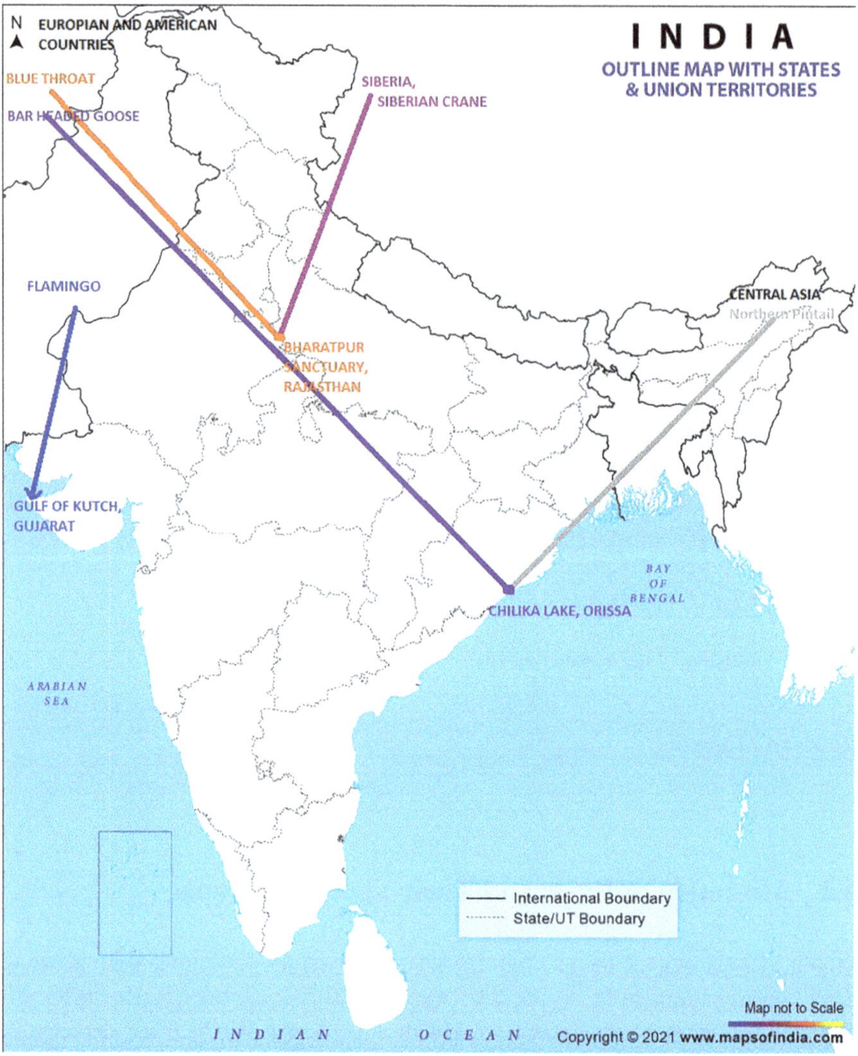


**Fig. 4.2** Showing the arrival of various migratory birds at distinct destinations in India

## 4.3 Problems Faced by Birds During Migration

The population of migratory birds is decreasing due to various natural calamities, including sandstorms, thunderstorms, diseases, and predation, as well as human activities such as wind farms, oil spills, pollution, habitat degradation, radiation, electric lights, and hunting [93]. Migratory birds are important indicators of ecosystem health because their population trends often reflect changes in habitat quality and environmental conditions [25]. The decline is usually caused by multiple interconnected factors. Some of the main challenges faced by migratory birds in India are as follows:

### 4.3.1 Fish Harvesting

Residents of nearby wetlands often fish in the marshes. This human activity can cause serious problems for migratory bird species. Birds are easily frightened by human presence and noise, which increases their stress, disrupts their foraging, and can even lead them to abandon nests or resting sites. These disruptions can harm their energy balance, which is crucial for completing long migratory journeys. Studies have shown that repeated human disturbance in wetlands lowers the suitability of these habitats for migratory birds [24, 43–45].

### 4.3.2 Human Disturbance

It is observed that herders and tourists often wander around nearby wetlands. In addition to grazing on natural resources, their cattle disturb wildlife. Such grazing worsens lake siltation by causing soil erosion around the lakes. Tourist activity adds another layer of stress. Human presence, noise, and vehicle movements can cause birds to flush frequently, leading to increased energy use and decreased feeding efficiency. This is especially harmful during migration, when conserving energy is vital for long-distance flight [7, 89, 91].

### 4.3.3 Agricultural Encroachment

Wetland areas often encroach on surrounding regions, clearing native vegetation. It is common to irrigate crops with water from lakes. These irrigation systems divert water from lakes and marshes, lowering water levels and harming aquatic ecosystems. Fertilizer runoff from farms adds excess nutrients to wetlands, causing eutrophication, algal blooms, and oxygen depletion, which harm aquatic food webs [40, 94]. Migratory birds that depend on aquatic invertebrates and submerged plants for food are also affected. Habitat loss due to encroachment is one of the main reasons for the global decline in migratory bird populations [34, 87].

### 4.3.4 *Bird Poaching*

Migratory birds are poached for various reasons. They are hunted for meat, used in traditional medicine, or captured for the illegal pet trade and exotic bird markets. Bird poaching also has ecological impacts. Many migratory birds play essential roles in ecosystems, such as controlling insects, dispersing seeds, and cycling nutrients in wetlands. Removing even a small number of these birds can disrupt food web dynamics and cause ecological imbalance [77]. Some local poachers often take migratory birds from marshes. However, forest officials respond quickly to stop these activities.

### 4.3.5 *Pollution*

Wetlands cannot clean themselves and support fragile, complex ecosystems, which makes them susceptible to pollution. It is common to see people discarding bags, plastic wrappers, and water bottles near marshes' banks. Residents living near wetlands are often caught dumping their trash in these areas, polluting both the landscape and water. Recent studies have shown that light pollution greatly affects nocturnally migrating passerines [15].

Light pollution is another rising concern, particularly for nocturnal migratory birds. Artificial lights disorient birds during their nighttime flights, causing them to deviate from migratory paths, collide with structures, or become trapped in urban areas. Recent studies show that light pollution significantly disrupts the migratory behavior of nocturnally migrating passerines, leading to increased mortality and altered stopover behavior.

Similarly, pesticides, plastics, and chemical contaminants are among the pollutants [42, 48, 90] that directly threaten migratory birds. The health and reproductive abilities of birds are affected by the accumulation of pesticides and other toxins in their bodies. As a passage migrant, the yellow-billed cuckoo (*Coccyzus americanus*) travels through Costa Rica on its migratory route but does not stay there for the winter. However, migrating birds passing through croplands may die mysteriously after being poisoned by pesticides, while some migratory birds simply perish from exhaustion during migration.

Perhaps the biggest threat to migratory species is human development; our infrastructure has turned migratory routes into dangerous obstacle courses; migratory bird populations are heavily impacted by collisions with transparent windows, buildings, communication towers, and power lines; light and noise pollution can disorient birds and cause them to collide with these structures, increasing the stressors they already face.

The population of migratory birds is declining at an alarming rate worldwide compared to resident birds. Migratory birds that breed in Europe and spend their winters in Sub-Saharan Africa and central and South America are declining at a considerable rate, and the probable reasons behind this are suggested to be habitat destruction, illegal hunting or killing, and climate change [3].

### 4.3.6 Illegal Killing

According to Brochet et al. [14], approximately 11 million to 36 million birds are illegally captured each year in the Mediterranean region. The Eurasian Chaffinch, Song Thrush, and Blackcap are among the most affected by hunting and killing. Passenger pigeons, once the most common migratory bird in North America during the nineteenth century, were driven to extinction by the end of the twentieth century due to illegal killing for cheap meat [32].

### 4.3.7 Habitat Destruction and Loss

Many European migratory birds spend their winters in the dry savannas of Sub-Saharan Africa. In this region, changes in rainfall, land use, and land cover have significantly impacted the populations of migratory birds. Over a 25-year span (1975–2000), agriculture grew by 57%, resulting in the loss of about 5 million hectares of vegetation that served as habitat for some specialist species. Consequently, European migratory bird populations declined in this area [84].

Residential and commercial development, along with activities such as drilling, mining, quarrying, and building transportation and utility routes, are identified as major threats causing widespread and serious impacts on Numeniini populations. Numeniini, a group of 13 wader species including seven that are Near Threatened or internationally threatened, two of which are Critically Endangered, are especially affected in coastal non-breeding habitats where significant land use changes are occurring. Besides habitat loss, roads can decrease local breeding wader densities in nearby fields [23], leading to population-level impacts, especially when roads cut through large parts of breeding grounds. Among various flyways, the East Asian-Australian Flyway presents the greatest number of breeding threats to Numeniini, primarily due to coastal reclamation; similar threats are seen across the Central and Atlantic Americas and East Atlantic flyways. The most significant dangers to breeding grounds are found in the Central and Atlantic Americas, East Atlantic, and West Asian flyways [57].

Studies in the Midwestern region have shown that many Neotropical Migratory Birds of forests and grasslands are area-sensitive and prone to reproductive failure in small, fragmented habitats. Cowbird parasitism, habitat heterogeneity, and nest predation are also regarded as the main causes of the decline of these birds [65].

### 4.3.8 Climatic Variation

Climate change significantly influences biodiversity changes by impacting migration timing and duration. For example, higher temperatures driven by climate change cause insects (prey) to arrive earlier at the breeding grounds of the Dutch Pied Flycatcher (predator) in sub-Saharan Africa. Still, Dutch Pied Flycatchers arrive at different times. This mismatch in prey and predator arrival times at the same site has caused ecological imbalance and decreased the Dutch Pied Flycatcher population by up to 90% [12, 92].

### 4.3.9 Degradation of Stopover Sites

Baker et al. [4] reported a decline in the population of red knot birds. These birds winter in Tierra del Fuego and migrate to the Arctic polar deserts for breeding. In between, they make a major stopover at Delaware Bay in the USA, where they feed on horseshoe crab eggs to store energy for long migratory journeys to the Arctic. However, from 1997 to 2002, their population decreased rapidly due to overharvesting of crabs, which reduced their nutrient storage and fitness. This led to the death of many during migration. Between 2000 and 2002, the population in Tierra del Fuego dropped sharply from 51,000 to 27,000, severely threatening the survival of this species.

The migratory paths of the oriental white stork and evaluated the potential risk of the population collapsing by modeling the impact of removing key stopover sites. Their results indicated that if the storks could not use stopover locations like Bohai Bay during their southward migration, they would be unable to complete their wintering habitats along the Yangtze River.

## 4.4 Adaptations

The primary method birds in temperate regions use to escape harsh winter conditions is migration. As a result, migratory birds' bones and wings have developed unique morphological, structural, and microstructural features that assist with various environmental adaptations. These traits vary greatly among species depending on their habitat and, most importantly, their ability to fly [80]. Different types of adaptations are listed in Table 4.1.

**Table 4.1** Diverse adaptations of migratory water birds

| Adaptation type | Adaptation | Example(s) | References |
|---|---|---|---|
| **Morphological** | Streamlined body | Bar-headed goose (*Anser indicus*) | [71] |
| | Long, pointed wings | Godwit (*Limosa limosa*), Pintail | [26] |
| | Webbed feet | Ducks, Geese | [33] |
| | Lightweight, pneumatic bones | All migratory birds | [69] |
| | Specialized bills (filter feeding, probing) | Flamingo, Avocet, Sandpiper | [67] |
| **Physiological** | High-efficiency lungs & unidirectional airflow | Bar-headed goose | [70] |
| | High fat reserves for fuel | Red Knot, Arctic Tern | [27] |
| | Salt glands to excrete saline | Flamingos, Gulls | [31] |
| | Thermoregulation in extreme environments | Snow goose | [36] |
| **Behavioral** | Zugunruhe (migratory restlessness) | Warblers, captive waterfowl | [8] |
| | Use of flyways and stopover sites | Central Asian Flyway species | |
| | Navigation using magnetic, solar, and star cues | Cranes, Ducks | [88] |
| | V-formation flying to save energy. | Geese, Cranes | [58] |
| **Reproductive** | Seasonal breeding is synchronized with the habitat | Arctic-breeding shorebirds | [51] |
| | Site fidelity | Bar-headed goose, cranes | [5] |

## 4.5 Factors Responsible for Bird Migration

Bird migration involves regular, seasonal movement of birds from one habitat to another, typically driven by changes in temperature, food availability, or breeding conditions. The word migration has been derived from the Latin word "migrare," which means movement from one spatial unit to another.

### *4.5.1 Breeding/Feeding*

Birds travel long distances, often across continents, to find more suitable environments for feeding, breeding, or surviving during harsh seasons [68]. Seabirds, particularly those migrating over vast open oceans, use the sun, moon, and stars to help them orient themselves. This ability to navigate by celestial bodies is especially useful for long-distance migratory birds that need to maintain a consistent direction over long periods. Some species, like the Arctic Tern, European Herring Gull and Wandering Albatross, are thought to use the position of the sun during the day and the stars at night to maintain their migratory route, especially when flying in unfamiliar or featureless oceanic environments [64].

### 4.5.2 Photoperiod/Circadian Rhythm

Changes in day length acts as a zeitgeber for migratory preparations, involving increases in body mass, fat deposition, muscle hypertrophy and increased locomotor activity [28]. Increasing day length also triggers an increase in androgen production, which induces the physiological and behavioural changes linked with transitioning to a migratory state. Photoperiods serve as environmental cue for tuning of circadian rhythm [86].

### 4.5.3 Hormonal Control

Corticosterone, leptin and prolactin are hormones that indirectly regulates departure to migratory journey. These hormones help in energy accumulation and metabolic adjustments required before migration; it is possible that they have a permissive role or otherwise influence the probability of departure [66].

## 4.6 Conservation

### 4.6.1 Role of Wetlands During Migration

Conservation of migratory birds is a complex process that requires understanding how species move across large distances—often hundreds to thousands of kilometers—and the different habitats they depend on for breeding, wintering, and resting. Climate and land-use changes are major threats to these migratory habitats, significantly increasing the global risk of species decline [74]. The yearly life cycle needs of migratory waterbirds, such as shorebirds, wading birds, and ducks, are supported by a broad continental network of wetland habitats. Currently, 40% of the world's food production comes from water-related areas like urban zones, agricultural regions, and irrigated lands that support metropolitan centers [19, 21, 78]. This growth has caused water birds to move across the landscape, using agricultural food sources to compensate for the loss of historic wetlands. Climate change stresses the wetland network, and water scarcity affects land use, disrupting ecological processes and complicating waterbird habitat needs [20, 30].

India currently ranks first among South Asian countries and third globally in terms of Ramsar sites. Most of these locations are situated in the Gangetic plains and the northwest Himalayas. Several key wetlands across India serve as essential wintering grounds for migratory birds. Notable among them are the Asan Conservation Reserve in Uttarakhand, Pong Dam in Himachal Pradesh [38, 39], Wular Lake in Jammu and Kashmir, and the Tso Kar Wetland Complex in Ladakh. Punjab's Harike Lake, Sultanpur National Park, and Bhindawas Wildlife Sanctuary in Haryana are also significant. In Uttar Pradesh, Sur Sarovar Bird Sanctuary played a vital role, alongside Keoladeo National Park and Sambhar Lake in Rajasthan. Other important sites include Kabartal Wetland in Bihar, Bhoj Wetland in Madhya Pradesh, Deepor Beel in Assam, the Bhitarkanika Mangroves, and Hirakud Reservoir in Odisha, as well as Loktak Lake in Manipur. In southern India, the East Kolkata Wetlands and Sunderbans Wetland in West Bengal are major wetlands, along with Kolleru Lake in Andhra Pradesh, Point Calimere Bird Sanctuary, and Pichavaram Mangrove in Tamil Nadu, as well as Ashtamudi Wetland and Sasthamkotta Lake in Kerala. Additionally, Nanda Lake in Goa, Nalsarovar Bird Sanctuary, Thane Creek in Maharashtra, Khijadia Wildlife Sanctuary, and Wadhvana Wetland in Gujarat are important sites for the conservation of migratory birds.

Wetlands are vital areas for migrating birds, offering resting and feeding spots. These areas must be protected from drainage, and woody vegetation should be shielded from grazing or even replanted. Such efforts would boost local biodiversity, support migratory birds, and benefit the livelihoods of nearby farmers and herders. Additionally, illegal hunting and capture can be stopped. The necessary political frameworks, including the African-Eurasian Waterbird Agreement and the African–Eurasian Migratory Landbird Action Plan [3], are already in place.

To develop more effective conservation strategies, it is essential to understand ecological connectivity and key stopover sites that act as critical midpoints between breeding and non-breeding areas, where birds rely heavily on refueling. Not all stopover sites are equally important for migrating birds; some are vital, while others have less impact on the journey. For example, birds relying on stopover sites in the Yellow Sea region have experienced significant population declines, as this area was a major refueling point for them. Therefore, prioritizing the conservation of this site is crucial, as it plays a key role in supporting migration [76].

The four bird species (swan goose, white-fronted goose, bar-headed goose, and whooper goose) along the Asian flyway to analyze their migration networks and key stopover sites. Migration networks with fewer stopover sites are more vulnerable to habitat loss because these sites are crucial for migratory birds to rest and refuel during their journeys. When only a few stopover locations exist, the destruction or degradation of even one site can severely disrupt the entire migration route. Species relying on these limited stopovers may have difficulty completing their journeys, leading to declines in their survival and reproduction due to a lack of resources. Conversely, migration networks with more stopover sites tend to be more resilient, as birds have alternative options, reducing the risk of severe outcomes if a single site is lost. Therefore, maintaining or restoring a well-connected network of stopover

**Table 4.2** Conservation strategies for migratory waterbirds in India

| Strategy | Description | References |
|---|---|---|
| Wetland Protection & Monitoring | Identifying Ramsar sites or key wetlands using remote sensing tools (e.g., NDWI) to monitor hydrology and habitat health. | [53] |
| Water Availability & Habitat Maintenance | Ensuring water supply in seasonal tanks and coastal wetlands; preventing tree/nest-site loss and invasive vegetation. | [16] |
| Sustainable Use & Community Engagement | Integrating aquaculture, agro-wetland practices, and ecotourism with conservation efforts to reduce habitat pressures and promote coexistence. | [49] |
| Hunting Regulation & Site Protection | Research on agricultural landscapes indicates that hunting is a major factor affecting bird abundance; restrictions are essential in vulnerable wetlands. | [61] |
| Standardized Monitoring Protocols | Nationwide adoption of AEWA waterbird monitoring guidelines for consistent counts across key sites. | [18] |
| Remote Sensing & GIS | Monitoring and tracking wetland changes over time to inform restoration and conservation efforts. | [53] |
| **International Agreements & Frameworks** | | |
| **Agreement/Instrument** | **Scope & Mechanism** | **References** |
| Ramsar Convention | Wetlands of International Importance; India has 91 sites (e.g., Chilika, Sambhar). Supports site management and national wetland planning. | [72] |
| AEWA (CMS technical agreement) | Connects 84 countries and protects 255 waterbird species. Focuses on flyway connectivity, monitoring, and national action plans. | [18] |
| AEWA Waterbird Monitoring Guidelines | Provides standardized protocols for waterbird surveys across flyways, including India. | [18] |
| Siberian Crane MoU (CMS) | Non-binding, multi-country conservation plan focused on Siberian crane habitats, including Indian sites like Keoladeo and Bharatpur. | [85] |

locations is essential for the survival and health of migratory species. Conservation strategies are outlined in Table 4.2.

### *4.6.2 Conservation Policies*

In conservation policymaking, the importance of a site for migratory species is typically evaluated by considering factors like the presence of appropriate habitat, the vulnerability of that habitat, the degree of habitat loss, the site's protection status, and the abundance or diversity of species [50, 52].

## 4.7 Migration Corridors

Birds follow specific fixed routes during intracontinental and transboundary migration. These routes begin at their homing ground and end at a destination, which could be a wintering or breeding ground. Figures 4.3 and 4.4 illustrate the main migratory pathways of birds worldwide.

### *4.7.1 East Atlantic Flyway*

Extends from northern Europe and Arctic Canada down to West Africa. The majority of wading birds, or waders, migrate from Nearctic, Palaearctic, and temperate breeding regions to the Wadden Sea, a vital stopover site on the East Atlantic

**Fig. 4.3** Major waterbird migration flyways

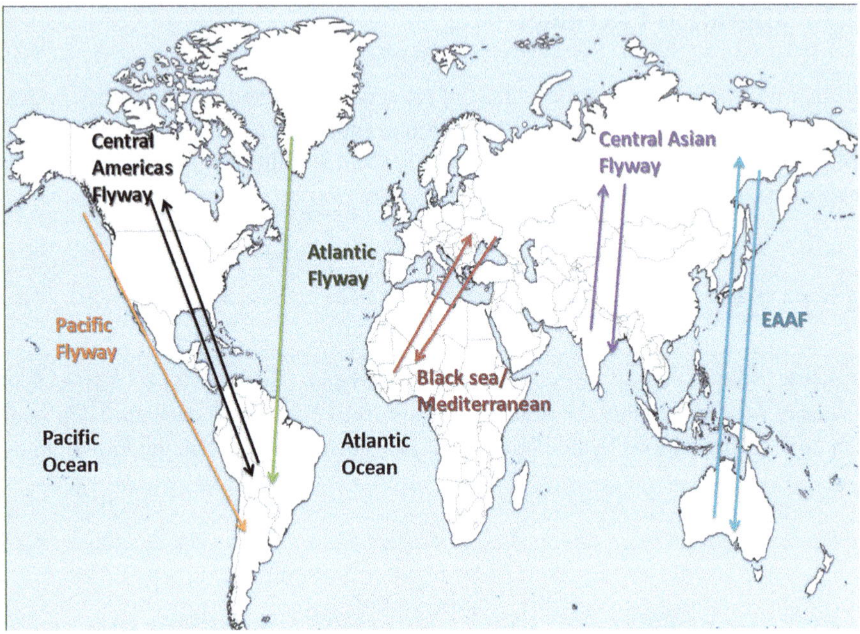


**Fig. 4.4** Demonstration of major global flyways

Flyway. During migration, birds stop here once or twice a year. The main difference with the Mediterranean Flyway is the feeding habitat. Food availability in the Wadden Sea is mainly limited by harsh winter temperatures and depends on tidal fluctuations. In contrast, wind-driven changes in water levels affect food availability in the Black Sea lagoons. More waders choose this route because the Wadden Sea's tidal mudflats support larger feeding populations. This explains why the Mediterranean Flyway has fewer wader populations [37]. Within the Iberian Peninsula, the east Atlantic flyway is important for shorebirds. The Santona and Noja Marshes are wetlands in the Iberian Peninsula that serve as wintering grounds for ducks, especially for the Eurasian spoonbill. According to a study by Navedo et al. [54], Santona hosts 19,848 overwintering waterbirds, and population analysis shows an increase in eight aquatic bird species and a decrease in Ardea cinerea and Pluvialis squatarola.

## *4.7.2 Black Sea–Mediterranean Flyway*

Connects the Palearctic region of northern Europe and Asia with southern Europe and North Africa. The Dalmatian pelican and great white pelican are members of Pelicaniformes, and these birds rely heavily on the Black Sea/Mediterranean flyway for migration and breeding. Human disturbances, nest flooding, and power lines are major threats to their populations. Loss of wetlands and prey fish at stopover sites along this flyway critically affects the pelican population. This highlights the importance of the Black Sea flyway as a migration corridor and underscores the urgent need to conserve wetlands and nesting sites along this route [17].

The European turtle dove is a bird native to the Palearctic region, and its population is declining. It is classified as 'Vulnerable' on the IUCN Red List. Bea et al. [6] conducted a study on this bird and discovered that it uses a new migratory flyway that passes through the Black Sea, Caspian Sea, Turkey, the Middle East, Egypt, and ultimately Africa. This flyway is critically important for the conservation of this species.

### 4.7.3 East Asian–Australasian Flyway (EAAF)

Spanning from Alaska and Siberia to Southeast Asia and Australasia. By identifying biodiversity hotspots throughout breeding, non-breeding, and migration times, the East Asian–Australasian Flyway (EAAF) can be used to conserve migratory wader species. Species richness patterns obtained from species distribution models are compared with conventional BirdLife range maps to determine if the current network of protected areas (PAs) is adequate to safeguard these important habitats. The model results, which are strongly supported by eBird field data, especially in the Yellow Sea region, demonstrate a larger and spatially distinct species richness than previously anticipated. It is concerning to note that just 5% of the most species-rich zones are located within protected areas, and less than 10% of the terrestrial habitat that waders use is protected. The results emphasize how critical it is to increase study and conservation activities in regions such as the Russian Far East and the Yellow Sea.

### 4.7.4 Atlantic Americas Flyway

It links breeding grounds in Canada/Alaska with wintering sites in South America. The goal of this study is to forecast how human activities, such as fishing, gas extraction, recreation, pollution, land reclamation, and sea level rise, may affect the number of waders and waterfowl in the East Atlantic Flyway and their migratory patterns. Study of Ens [22] highlights the necessity of robust, practical mathematical models backed by empirical evidence in order to direct conservation. The study emphasizes how variations in prey availability and competition for nesting and wintering locations impact bird populations.

### 4.7.5 Central Asian Flyway (CAF)

The Central Asian Flyway stretches across Central and South Asia, from northern Siberia to the Indian Ocean. Over 300 migrating bird species, many of which depend on fragile wetland habitats, rely on this route. Human disturbance, pollution, and habitat loss threaten it. Under CAF, international cooperation among 30 countries is the primary focus of conservation efforts.

### 4.7.6 Mississippi Flyway

One of the main bird migration routes in North America is the Mississippi Flyway, which follows the course of the Mississippi River. It is home to millions of ducks and songbirds and extends from Canada to the Gulf of Mexico. There are many wetlands and wooded areas along the route that are perfect for stopping points. It is actively managed for bird conservation and has been well studied.

### 4.7.7 Pacific Flyway

Stretching from Alaska to South America, the Pacific Flyway traces the western coast of the Americas. This route exposes birds to diverse geography, winds, and resource availability throughout the year. Mountain valleys, estuaries, and coastal wetlands are key habitats. For many shorebirds and ducks, the flyway is vital, even though it is underrepresented in studies.

## 4.8 Future Prospectus

Further study of flyways and surveys of wetland areas can provide valuable information to support conservation policy-making. Additional research is recommended for ecologists to develop more cautious policies and identify key stopover sites. Global collaboration will be essential to protect flyways, as they extend across multiple provinces. Merely focusing on habitat conservation will not suffice for the protection of vulnerable and threatened species; a multidisciplinary approach is necessary.

## 4.9 Conclusion

Wetlands function as refueling sites for birds during their migrations worldwide. Data from previous decades have demonstrated that the reduction and loss of wetlands globally have significantly impacted the population dynamics of migratory birds. Many action plans have already been implemented by different national governments, and further amendments to existing conservation laws are needed to enhance their effectiveness. Firstly, the pathways of 'vulnerable' and 'threatened' categories must be identified, and protection measures should be developed accordingly. In simple terms, birds are graceful travelers of the skies and are living jewels of nature; therefore, prioritizing their conservation is essential to maintaining ecological balance and preserving the biodiversity they embody.

## References


1. Aarif KM, Nefla A, Muzaffar SB, Musammilu KK, Prasadan PK (2017) Traditional fishing activities enhance the abundance of selected waterbird species in a wetland in India. Avian Res 8:1–10
2. Alerstam T (1990) Bird migration. Cambridge University Press, Cambridge
3. Bairlein F (2016) Migratory birds are under threat. Science 354(6312):547–548
4. Baker AJ, Gonzalez PM, Piersma T, Niles LJ, de Lima Serrano do Nascimento I, Atkinson PW, Clark NA, Minton CD, Peck MK, Aarts G (2004) Rapid population decline in red knots: fitness consequences of decreased refuelling rates and late arrival in Delaware Bay. Proc R Soc Lond Ser B Biol Sci 271(1541):875–882
5. Batbayar N, Takekawa JY, Natsagdorj T, Spragens KA, Xiao X (2014) Site selection and nest survival of the Bar-headed Goose (Anser indicus) on the Mongolian Plateau. Waterbirds 37(4): 381–393
6. Bea A, Olano I, Svazas S, Henry J, Yanenko V, Grishanov G (2023) The Black Sea-Eastern Mediterranean flyway of the globally threatened European turtle dove (*Streptopelia turtur*). Turk J Zool 47(1):48–52
7. Beale CM, Monaghan P (2004) Human disturbance: people as predation-free predators? J Appl Ecol 41(2):335–343
8. Berthold P (2001) Bird migration: a general survey. Oxford University Press
9. Biber JP, Salathé T (1991) Threats to migratory birds. In: Salathé T (ed) Conserving migratory birds. International Council for Bird Preservation (Technical Publication No. 12), Cambridge, U.K, pp 17–35
10. BNHS (2020). http://www.bnhsenvis.nic.in/Database/Migratory-Birds-and Flyways_17325.aspx
11. Boere GC (2006) The flyway concept: what it is and what it isn't. In: Boere GC, Galbraith CA, Stroud DA (eds) Waterbirds around the world. The Stationery Office, Edinburgh, pp 40–47
12. Both C, Bouwhuis S, Lessells CM, Visser ME (2006) Climate change and population declines in a long-distance migratory bird. Nature 441(7089):81–83
13. Brink AB, Eva HD (2009) Monitoring 25 years of land cover change dynamics in Africa: a sample-based remote sensing approach. Appl Geogr 29(4):501–512
14. Brochet AL, Van den Bossche W, Jbour S, Ndang'ang'a PK, Jones VR, Abdou WALI, Al-Hmoud AR, Asswad NG, Atienza JC, Atrash I, Barbara N (2016) Preliminary assessment of the scope and scale of illegal killing and taking of birds in the Mediterranean. Bird Conserv Int 26(1):1–28

15. Burt CS, Kelly JF, Trankina GE, Silva CL, Khalighifar A, Jenkins-Smith HC, Fox AS, Fristrup KM, Horton KG (2023) The effects of light pollution on migratory animal behavior. Trends Ecol Evol 38:355–368
16. Byju H, Maitreyi H, Natarajan R, Vijayan R, Maran BAV (2025) The avifauna of Ramanathapuram, Tamil Nadu along the Southeast coast of India: waterbird assessments and conservation implications across key sanctuaries and Ramsar sites. PeerJ 13:e18899
17. Catsadorakis G, Onmuş O, Bugariu S, Gül O, Hatzilacou D, Hatzofe O et al (2015) Current status of the Dalmatian pelican and the great white pelican populations of the Black Sea/Mediterranean flyway. Endanger Species Res 27(2):119–130
18. Dodman T (2022) Agreement on the conservation of African-Eurasian Migratory Waterbirds (AEWA)
19. Donnelly JP, King SL, Knetter J, Gammonley JH, Dreitz VJ, Grisham BA, Nowak MC, Collins DP (2021) Migration efficiency sustains connectivity across agroecological networks supporting sandhill crane migration. Ecosphere 12(6):e03543
20. Donnelly JP, King SL, Silverman NL, Collins DP, Carrera-Gonzalez EM, Lafón-Terrazas A, Moore JN (2020) Climate and human water use diminish wetland networks supporting continental waterbird migration. Glob Chang Biol 26(4):2042–2059
21. Elphick CS, Oring LW (2003) Conservation implications of flooding rice fields on winter waterbird communities. Agric Ecosyst Environ 94(1):17–29
22. Ens BJ, Piersma T, Drent RH (1994) The dependence of waders and waterfowl migrating along the East Atlantic flyway on their coastal food supplies: what is the most profitable research programme? Ophelia:127–151
23. Fikenscher A, Hooijmeijer J, Kentie R, Piersma T (2015) Black-tailed Godwits avoid traffic-intense roads less in high-quality breeding habitat. De Levende Natuur 116:51–56
24. Fox AD, Madsen J (2017) Threatened species to super-abundance: the unexpected international implications of successful goose conservation. Ambio 46:179–187
25. Frick WF, Kingston T, Flanders J (2020) A review of the major threats and challenges to global bat conservation. Ann NewYork Acad Sci 1469(1):5–25
26. Gill FB (2007) Ornithology, 3rd edn. W. H. Freeman and Company, New York
27. Guglielmo CG (2010) Move that fatty acid: fuel selection and transport in migratory birds and bats
28. Gwinner E (1996) Circadian and circannual programmes in avian migration. J Exp Biol 199(1): 39–48
29. Haig SM, Mehlman DW, Oring LW (1998) Avian movements and wetland connectivity in landscape conservation. Conserv Biol 12(4):749–758
30. Haig SM, Murphy SP, Matthews JH, Arismendi I, Safeeq M (2019) Climate-altered wetlands challenge waterbird use and migratory connectivity in arid landscapes. Sci Rep 9(1):4666
31. Hughes MR (2003) Regulation of salt gland, gut, and kidney interactions. Comp Biochem Physiol A Mol Integr Physiol 136(3):507–524
32. Hung CM, Shaner PJL, Zink RM, Liu WC, Chu TC, Huang WS, Li SH (2014) Drastic population fluctuations explain the rapid extinction of the passenger pigeon. Proc Natl Acad Sci 111(29):10636–10641
33. Kent GC (1987) Comparative anatomy of the vertebrates. Times Mirror
34. Kiran SD, Kour A, Priya DV, Kumar R (2022) Community composition and status of avian diversity at Campus and Agricultural landscapes of Chaudhary Charan Singh Haryana Agricultural University, Hisar (Haryana). J Appl Nat Sci 14(4):1130–1140. https://doi.org/10.31018/jans.v14i4.3784
35. Kirby JS, Stattersfield AJ, Butchart SHM, Evans MI, Grimmett RFA, Jones VR, O'Sullivan J, Tucker GM, Newton I (2008) Key conservation issues for migratory land- and waterbird species on the world's major flyways. Bird Conserv Int 18:S49–S73
36. Klaassen M (1996) Metabolic constraints on long-distance migration in birds. J Exp Biol 199(1):57–64

37. Kube J, Korzyukov AI, Nankinov DN, M'tinster OAG, Weber P (1998) The northern and western Black Sea region - the Wadden Sea of the Mediterranean flyway for wader populations. Int Wader Stud 10:379–393
38. Kumar A, Paliwal R (2015) Galloanserae and aquatic Neoaves of Pong Dam Wetland, Himachal Pradesh: *status and conservation issues*. In: Aquatic ecosystem: biodiversity, ecology and conservation. Springer, New Delhi, pp 307–321
39. Kumar A (2011) Pong dam wetlands and zoological survey of India. Curr Sci 85(10):1398
40. Kumar R, Singh D, Yadav R (2025) Toxic effects of chlorantraniliprole and fipronil on *Eisenia fetida*: from growth and reproductive parameters to avoidance behavior. Environ Eng Manag J 24(7):1499–1508. Accepted manuscript
41. Kumar R, Pattnaik AK (2012) Chilika: an integrated management planning framework for conservation and wise use
42. Kumar R, Singh D (2022) Insecticide-tolerant bacterial population in *Eisenia fetida*'s Gut and Vermicast exposed to Chlorantraniliprole and Fipronil. Appl Biol Res 24(3):273–279
43. Kumar R, Gupta R, Dahiya T, Yadav R, Saifi R, Yodha K, Kumari B, Singh P, Shefali. (2022a) Surveillance of *Heteropneustes fossilis* (Stinging Catfish) diseases and health management practices in Haryana. Appl Ecol Environ Sci 10(4):261–267
44. Kumar R, Gupta RK, Yadav R, Saifi R, Yodha K, Pooja. (2022b) *Eisenia fetida* as protein source for growth enhancement of *Heteropneustes fossilis*. Egypt J Aquat Biol Fish 26(2): 577–588
45. Kumar R, Yadav R, Kumar Gupta R, Pooja (2022c) A sustainable way for fish health management by replacement of chemical and drugs by earthworm. In: Environmental degradation in Asia: land degradation, environmental contamination, and human activities. Springer International Publishing, Cham, pp 329–352
46. Lewis M (2016) AEWA at twenty: an appraisal of the African-Eurasian Waterbird Agreement and its unique place in international environmental law. J Int Wildl Law Policy 19(1):22–61
47. Lincoln FC (1935) The waterfowl flyways of North America. US Department of Agriculture, No. 342: 16
48. Longcore T, Rich C (2004) Ecological light pollution. Front Ecol Environ 2(4):191–198
49. Mathevet R, Tourenq C, Mesléard F (2002) Agricultural policies, land-use and waterbird conservation: the case study of a major Mediterranean wetland, the Camargue. Cybergeo: Eur J Geograph
50. Mehlman DW, Mabey SE, Ewert DN, Duncan C, Abel B, Cimprich D, Sutter RD, Woodrey M (2005) Conserving stopover sites for forest-dwelling migratory landbirds. Auk 122(4): 1281–1290
51. Meltofte H (ed) (2007) Effects of climate variation on the breeding ecology of Arctic shorebirds, vol 59. Museum Tusculanum Press
52. Merken R, Deboelpaep E, Teunen J, Saura S, Koedam N (2015) Wetland suitability and connectivity for trans-Saharan migratory waterbirds. PLoS One 10(8):e0135445
53. Naik R, Sharma LK (2022) Monitoring migratory birds of India's largest shallow saline Ramsar site (Sambhar Lake) using geospatial data for wetland restoration. Wetl Ecol Manag 30(3): 477–496
54. Navedo JG, Masero JA, Juanes JA (2007) Updating waterbird population estimates within the East Atlantic Flyway: status and trends of migratory waterbirds in Santoña marshes. Ardeola 54(2):237–249
55. Newton I (2007) Weather-related mass-mortality events in migrants. Ibis 149:453–467
56. Newton I (2003) Geographical patterns in bird migration. In: Berthold P, Gwinner E, Sonnenschein E (eds) Avian migration. Springer, Berlin/Heidelberg
57. Pearce-Higgins JW, Brown DJ, Douglas DJ, Alves JA, Bellio M, Bocher P, Buchanan GM, Clay RP, Conklin J, Crockford N, Dann P (2017) A global threats overview for Numeniini populations: synthesising expert knowledge for a group of declining migratory birds. Bird Conserv Int 27(1):6–34

58. Portugal SJ, Hubel TY, Fritz J, Heese S, Trobe D, Voelkl B, Usherwood JR (2014) Upwash exploitation and downwash avoidance by flap phasing in ibis formation flight. Nature 505(7483):399–402
59. Prasad SN, Ramachandra TV, Ahalya N, Sengupta T, Kumar A, Tiwari AK, Vijayan L (2002) Conservation of wetlands of India review. Trop Ecol 43(1):173–186
60. Quader S, Raza RH MigrantWatch: a citizen science programme
61. Ramachandran R, Kumar A, Gopi Sundar KS, Bhalla RS (2017) Hunting or habitat? Drivers of waterbird abundance and community structure in agricultural wetlands of southern India. Ambio 46:613–620
62. Rasmussen PC, Anderton JC (2005) Birds of South Asia. The Ripley guide. Smithsonian Institution and Lynx Edicions, Washington DC/Barcelona, 378(1)
63. Richardson PL (2015) Upwind dynamic soaring of albatrosses and UAVs. Prog Oceanogr 130: 146–156
64. Ritchison G (2023) Navigation and Orientation. In: In a Class of Their Own. Fascinating Life Sciences. Springer, Cham. https://doi.org/10.1007/978-3-031-14852-1_14
65. Robinson SK (1996) Threats to breeding Neotropical migratory birds in the Midwest. United States Department Of Agriculture Forest Service General Technical Report Nc,:1–21
66. Romero LM, Holt DW, Maples M, Wingfield JC (2006) Corticosterone is not correlated with nest departure in snowy owl chicks (*Nyctea scandiaca*). Gen Comp Endocrinol 149(2):119–123
67. Rubega MA (1996) Sexual size dimorphism in red-necked phalaropes and functional significance of nonsexual bill structure variation for feeding performance. J Morphol 228(1):45–60
68. Salewski V, Bruderer B (2007) The evolution of bird migration—a synthesis. Naturwissenschaften 94:268–279
69. Schmidt-Nielsen K (1997) Animal physiology: adaptation and environment. Cambridge University Press, Cambridge
70. Scott GR, Milsom WK (2006) Flying high: a theoretical analysis of the factors limiting exercise performance in birds at altitude. Respir Physiol Neurobiol 154(1–2):284–301
71. Scott GR, Egginton S, Richards JG, Milsom WK (2009) Evolution of muscle phenotype for extreme high altitude flight in the bar-headed goose. Proc R Soc B Biol Sci 276(1673): 3645–3653
72. Solanki V, Sharma A (2021) Ramsar wetlands of India-importance and conservation. Int J Human Rights 8:113
73. Somveille M, Manica A, Butchart SHM, Rodrigues ASL (2013) Mapping global diversity patterns for migratory birds. PLoS One 8(8):e70907. https://doi.org/10.1371/journal.pone.0070907
74. Spooner FE, Pearson RG, Freeman R (2018) Rapid warming is associated with population decline among terrestrial birds and mammals globally. Glob Chang Biol 24(10):4521–4531
75. Stroud DA, Boere GC, Galbraith CA, Thompson D (2006) Waterbird conservation in a new millennium—where from and where to? In: Boere GC, Galbraith CA, Stroud DA (eds) Waterbirds around the world. The Stationery Office, Edinburgh, UK
76. Studds CE, Kendall BE, Murray NJ, Wilson HB, Rogers DI, Clemens RS, Gosbell K, Hassell CJ, Jessop R, Melville DS, Milton DA (2017) Rapid population decline in migratory shorebirds relying on Yellow Sea tidal mudflats as stopover sites. Nat Commun 8(1):14895
77. Sutherland WJ, Newton I (2004) Bird ecology and conservation: a handbook of techniques, vol No. 1. Oxford University Press, Oxford
78. Taft OW, Haig SM (2005) The value of agricultural wetlands as invertebrate resources for wintering shorebirds. Agric Ecosyst Environ 110(3–4):249–256
79. Tattoni C, Ciolli M (2019) Analysis of bird flyways in 3D. ISPRS Int J Geo Inf 8(12):535
80. Teryak G, Bakry A, Abuelsaad A, Mohamed AE, Abo-Eleneen R, Mohamed T (2025) Using laser-induced breakdown spectroscopy to explore the functional anatomy of the skeletal appendages in migratory and resident birds. Available at SSRN 5370599
81. Toews DP, Heavyside J, Irwin DE (2017) Linking the wintering and breeding grounds of warblers along the Pacific Flyway. Ecol Evol 7(17):6649–6658

82. Trouwborst A (2012) Transboundary wildlife conservation in a changing climate: adaptation of the Bonn convention on migratory species and its daughter instruments to climate change. Diversity 4:258–300
83. UNESCO-UN-Water (2020) United Nations world water development report 2020:*water and climate change*. United Nations, Paris
84. Walther BA (2016) A review of recent ecological changes in the Sahel, with particular reference to land-use change, plants, birds and mammals. Afr J Ecol 54(3):268–280
85. WARBLERS-JOINT, A.Q.U.A.T.I.C., CMS (2003)
86. Weise CM (1967) Castration and spring migration in the white-throated sparrow. Condor 69(1): 49–68
87. Wilcove DS, Wikelski M (2008) Going, going, gone: is animal migration disappearing. PLoS Biol 6(7):188
88. Wiltschko W, Wiltschko R (2005) Magnetic orientation and magnetoreception in birds and other animals. J Comp Physiol A 191:675–693
89. Yadav R, Gupta RK, Kumar R, Kaur T, Moun R (2022a) Biochemical response of earthworm, *Eisenia fetida* to heavy metals toxicity. J Appl Nat Sci 14(3):990–998
90. Yadav R, Gupta RK, Kumar R, Kaur T (2022b) Assessment of toxicity of lead and nickel on the biochemical and immunological parameters of earthworm. Eudrilus eugeniae Agric Sci Dig 44(5):922–929
91. Yadav R, Kumar R, Gupta RK, Kaur T, Yodha K, Kour A, Kaur S, Rajput A (2023) Heavy metal toxicity in earthworms and its environmental implications: a review. Environ Adv 12: 100374
92. Yodha K, Singh D, Kour A, Yadav R, Kumar R (2024) Species diversity and abundance of birds in ber (*Zizyphus mauritiana*) crop. Asian J Conserv Biol 13(1). https://doi.org/10.53562/ajcb.83683
93. Yodha K, Singh D, Kour A, Yaday R, Dahiya P, Kumar R (2023) Estimation of predatory bird damage to winter guava crop. Indian J Ecol 50(5(SI)):1787–1793. https://doi.org/10.55362/IJE/2023/4138
94. Zedler JB, Kercher S (2005) Wetland resources: status, trends, ecosystem services, and restorability. Annu Rev Environ Resource 30(1):39–74

# Chapter 5
# Migration-Related Morphological and Physiological Adaptations in Seabirds

**Jyoti, Anshu, and Manmeet Pal Kaur**

**Abstract** Migration is a seasonal movement of birds between their breeding and wintering grounds. This movement is driven by both internal and external cues, such as the need to escape harsh conditions, locate food sources, and find nesting locations. The migration period is physiologically challenging for migrants as they encounter fluctuating temperatures, humidity, oxygen levels, air pressure, and pathogenic stress while crossing different types of landscapes. This chapter includes information about various types of migration, the process of migration, migratory costs, Preparation for migration, and morphological and physiological adaptations of birds to cope with challenges in detail. Morphological adaptations include changes in body size/shape, changes in feathers, and changes in wing morphology that influence the flapping, gliding, and soaring during migration. Efficient metabolism and energy conservation are achieved by increasing the storage of fat before migration and dynamic soaring. Muscle mass shows changes before, during, and after migration. Physiological adaptations, such as shrinkage of the digestive system, an improved innate immune system, osmotic balance by salt glands, changes in the respiratory system for efficient oxygen supply in hypoxic terrain, increased circulation to the heart and muscles, increased antioxidant production, and navigation-related structures and mechanisms, are discussed in detail concerning seabirds. This chapter also includes navigation and orientation-related adaptations that include homing through the use of celestial bodies, magnetic field detection by magnetoreceptors, navigation by landmarks, and infrasound in detail.

**Keywords** Migration · Stopover site · Dynamic soaring · Energy conservation · Muscle mass · Navigation, and homing.

Jyoti (✉) · Anshu · M. P. Kaur
Department of Zoology and Environmental Sciences, Punjabi University Patiala, Patiala, India



R. Kumar et al. (eds.), *Ecology, Conservation of Aquatic Birds*, Wetlands: Ecology, Conservation and Management 15, https://doi.org/10.1007/978-3-032-17125-2_5

## 5.1 Introduction

The ability to fly is restricted to specific groups within the animal kingdom, and birds (Aves) are fortunate to possess this capability, which they use to cover vast distances during migration. Flight utilizes more energy than running. Any flight undertaken by a bird will likely involve the following elements: take-off, ascent, gliding, soaring, or sustained flapping flight, and maneuvering, which may include sharp turns and occasional bursts of flight [7]. Bird migration involves regular, seasonal movement of birds from one habitat to another, typically driven by changes in temperature, food availability, or breeding conditions. The word migration has been derived from the Latin word "migrare," which means movement from one spatial unit to another [28]. Birds travel long distances, often across continents, to find more suitable environments for feeding, breeding, or surviving during harsh seasons [52]. The migration distance can be small (a few meters) or large (up to thousands of kilometers). The migration period is physiologically challenging for migrants as they encounter fluctuating temperatures, humidity, oxygen levels, air pressure, and pathogenic stress while crossing different types of landscapes [23].

Birds have lightweight bodies due to flight adaptations like pneumatic or hollow bones, fused bones, only one functional paired reproductive organ, no storage organ like a urinary bladder, and pulmonary sacs associated with the respiratory system [59]. The evolution of birds involves many weight-reducing adaptations, which are reflected in their bone structure, many of which strengthen and stiffen the skeleton. In addition to these, birds undergo physiological and behavioural changes before the migration, including reorganizing body composition to reduce flight cost [41].

## 5.2 Migratory Costs Paid by Birds

### *5.2.1 High Energy Consumption Process*

Migration is a highly expensive task, because it not only relies on a large amount of stored energy but a high mortality rate is also associated with it. Long-distance, nonstop flights can lead to physical exhaustion and muscle depletion. To cover vast migratory distances, birds need to have enough energy molecules to utilize for the production of lift, thrust and drag forces, which should be more than gravitational pull to generate a forward momentum during flight. Wing morphology affects flight endurance and expenses during flight. Wing loading, along with aspect ratio, is an important wing variable that can tell a lot about the flight taken by the bird [2]. Long-distance migrants have a greater aspect ratio, a longer distal wing, pointed wingtips, and lower wing loading, while short-distance migrants or residents have the opposite characteristics [9, 27].

### 5.2.2 Harsh Environmental Variables

Migratory seabirds face violent storms, high winds, and rough waters that can disorient them and increase energy expenditure. Birds migrating from polar regions may encounter freezing temperatures, making survival more difficult during long migrations [56].

### 5.2.3 Lack of Food and Long Periods of Fasting

Seabirds often encounter a shortage of food, especially crossing open oceans and large deserts, leading to periods of fasting and reliance on fat reserves. An extended fasting period can lead to malnutrition or starvation, particularly if prey is sparse or hard to find. Depleted fish stocks due to overfishing can also reduce food availability for seabirds during migration.

### 5.2.4 Navigational Challenges

Migration and homing over vast oceans and large deserts create the most difficult orientation and navigation challenges for seabirds, as no landmark cues are visible. Cloudy weather, storms, and other natural disturbances can further intensify this challenge.

### 5.2.5 Risk of Predators

While migrating, seabirds are vulnerable to predation from other marine animals, such as large predatory fish, squid, and even other seabirds. In addition, some seabirds might be targeted by land-based predators upon arriving at breeding or stopover places. Migrating birds may also face the risk of piracy from other seabirds. For example, skuas are known to harass other birds, stealing food or even attacking their nests, which stresses the migrating species more.

### 5.2.6 Re-Organization of Circadian Rhythms

The reorganization of circadian rhythms in migratory seabirds is crucial to their ability to navigate and perform long-distance migrations. Seabirds, like other migratory birds, rely on highly synchronized biological clocks to ensure that they remain active and oriented at the right times during their journey. This reorganization is fundamental as seabirds travel across different time zones and environmental conditions. When seabirds migrate, they often cross several time zones and may travel through areas with varying lengths of day or constantly changing light conditions (e.g., the Midnight Sun near the poles). These environmental shifts require seabirds to adjust their circadian rhythms. For species like the Albatross or Arctic Tern, which may engage in nonstop flight for days or weeks, circadian rhythms must be reorganized to support long periods of activity. Migratory birds adjust their circadian rhythms to cope with the demand of sleep during their migratory journey. There are two types of sleep pattern based on time duration of sleep- monophasic and polyphasic sleep. In monophasic sleep, animal sleeps for a long period at night but in the case of a polyphasic pattern, sleep is divided into multiple slots throughout the day, which will aid in the reduction of overall sleep time. Migratory birds adopt a polyphasic sleep pattern during migration to maintain alertness and performance. Internal clock is influenced by photoperiodic cues; thus, many seabirds use shortening of day length in autumn as a cue to begin their migration and lengthening of photoperiod in summer as a signal for returning.

## 5.3 Different Types of Migration

Marine birds exhibit different types of migration:

### *5.3.1 Latitudinal Migration*

*Sterna paradisaea* (**Arctic Tern**): A marine bird known for the longest migratory journey in the animal world is the Arctic Tern. This bird breeds in the Arctic region and during winter in the Arctic region, it migrates toward Antarctica, covering up to 80,000 km in one round-trip. It is an example of long-distance migration [12].

*Diomedea exulans* (**The Wandering Albatross**) is popular for having the largest wingspan among birds. It exhibits circumpolar migration [46].

### *5.3.2 Altitudinal Migration*

*Puffinus creatopus* (**Pink-footed Shearwater**): This seabird is commonly found along the coast of Peru and Chile. It moves to high cliffs of islands for breeding and return to lower altitudes during winter [15].

### 5.3.3 *Partial Migration*

*Larus argentatus* (**European Herring Gull**): Not all members of the population of this bird migrate to southward temperate areas in winter but a few of these stays in their breeding ground throughout the year.

### 5.3.4 *Nomadic Migration*

*Oceanites oceanicus* (**Wilson's Storm-petrel**), this bird migrates in an irregular pattern depending on the availability of food. When food becomes scarce at one place, they migrate to another site.

### 5.3.5 *Circumpolar Migration*

*Pagodroma nivea* (**Snow Petrel**): This bird breeds in the Antarctic region and migrates within the polar regions. Migration is restricted to polar and subpolar regions.

### 5.3.6 *Flyway Migration*

*Rissa tridactyla* (**Black-legged Kittiwake**): This bird is commonly found along the Arctic coastline of the North Atlantic Ocean and migrates along established flyways.

Marine birds show fascinating migratory patterns, ranging from long-distance migrations to more confined movements within a region. Each bird has adapted a unique migration strategy to suit its biological, ecological, and environmental needs.

## 5.4 Migration Process

The complex migration process can be divided into various basic steps or stages, including pre-migratory preparation, take-off, flight dynamics, stop-over ecology, and landing at the destination.

### 5.4.1 Take Off

Seabirds start their migratory journey from coastal areas where abundant food is available after completing their pre-migratory preparation. During take-off, many conditions like the health of the bird, fuel storage, wind speed, and direction decide whether the flight will be successful or not.

### 5.4.2 Different Types of Flight Patterns

Based on distance, environmental conditions, and bird species, different flight patterns are adopted as shown in Fig. 5.1.

A. **Dynamic Soaring**: It is common in seabirds like albatrosses. In this technique, wind speeds between the ocean surface and higher altitudes are different, and this difference is used by birds to soar long distances with minimal energy expenditure. Wings appear to show no movement.
B. **Flapping Flight**: Some seabirds rely on sustained flapping flight, especially during the initial stages of migration. Constant wing beating can be seen.
C. **Gliding and Soaring**: Seabirds, particularly those migrating over the ocean, will often glide and soar to conserve energy during long flights. This is typical of species like shearwaters and petrels.

### 5.4.3 Stopovers/Refueling

During long-distance migration, seabirds will often stop at strategic locations to rest and refuel. These stopovers can be islands, coastal regions, or specific points on migratory flyways where food is abundant.

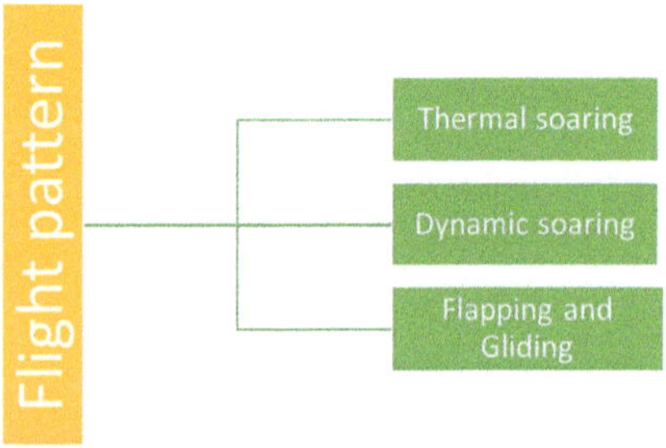


**Fig. 5.1** Types of flight patterns adopted by birds

### 5.4.4 Landing at Destination

Landing after a long migration is crucial for survival, and seabirds typically look for safe, open spaces for landing. Upon reaching their destination, seabirds typically spend time recovering from the long journey. This includes finding food, rehydrating, and potentially re-adjusting their circadian rhythms to match the local environment. If it is breeding migration, seabirds begin engaging in reproductive behaviours and if it is seasonal migration to feeding grounds, then birds continue their feeding patterns.

## 5.5 Preparation of Migration

### 5.5.1 Fuel/Food Storage

Birds exhibit a variety of behaviors in their ecology, but migration is the most energetically costly behavior, as they travel non-stop for thousands of kilometers, pushing their physiological limits. Due to the high energy demand and the absence of frequent refueling on the route, birds need to replenish their energy reserves before starting the migration [20]. Birds store a large amount of lipids and polypeptides to fuel long-distance flights. And amount of fuel/energy stored depends on the following factors-

- Migration distance.
- Wing aspect ratio
- Flight pattern (soaring needs less energy as compared to flapping)
- Weight of body and wing loading [57].

The type of fuel used will greatly influence the endurance performance, along with the quantity of stored fuel. The three main types of biomolecules (lipids, proteins, carbohydrates) can serve as fuel, and the major fuel to be stored will be determined by the following criteria-

(a) transportation and maintenance costs
(b) ease of supply to the muscles
(c) re-fuelling
(d) adverse effects during and after flight.

Lipids fulfil the majority of these criteria and are regarded as the best fuel, except for their supply from adipose tissue to muscles which creates some problems. Long flights of several hours are common and non-stop flights over a thousand kilometers are performed by some species when crossing vast ecological barriers, e.g., Bar-tailed Godwits *Limosa lapponica baueri* crossing the Pacific between China and Australia [39]. For every gram of wet mass, fat in adipose tissue produces eight times as much energy as protein and approximately ten times as much as glycogen.

Protein as well as glycogen usually require more water for storage, although fat can be stored with negligible amounts. Adipose tissue's maintenance cost is around ten times lower than compared of skeletal muscle and up to 45 times lower than that of liver tissue.

### 5.5.2 *Repair of Damaged Feathers*

Feathers have several functions, like flight, courtship display, body insulation, protection from parasites, and signalling. Feathers are shed every year via molting. The barn swallow *Hirundo rustica* is an aerial insectivorous, long-distance migrant bird that replaces body (contour) feathers before the migration period. Birds repair their damaged wing feathers by replacing them with new ones during molting to ensure a smooth flight [51].

### 5.5.3 *Navigation and Homing Organs*

Seabirds are incredible navigators, capable of migrating thousands of kilometers across vast oceans to return to breeding grounds or find suitable feeding areas. Their navigation and orientation abilities are well-developed and allow them to travel with remarkable precision. Without the proper functioning of navigational organs, birds cannot locate their breeding or feeding grounds.

## 5.6 Morphological Adaptations in the Seabirds

### 5.6.1 *Change in Body Size/Shape*

Most bird species experience notable fluctuations in body mass throughout the year. These fluctuations affect both fatty and lean (non-fat) tissue. Lean body mass refers to the bird's body mass after removing all the fat. Structural size refers to a bird's physical dimensions, measured by specific linear traits, without considering the status of nutrition [38]. The reserved nutrients consist of tissues (protein and lipids) on which a bird can rely in case of emergency, such as when facing starvation. These reserves are essential for daily functions and are only used in dire situations. On the other hand, nutrient stores represent the accumulation of nutrients in preparation for specific events, like egg production or long-distance flight [35]. Therefore, when an animal is in a baseline state (i.e., not breeding, migrating, molting, or enduring winter stress), it would have full nutrient reserves but no nutrient stores. A well-known example of geese relying partly or wholly on nutrients stored before departure for the breeding areas [36], there are other explanations as to why birds would increase their fat-free mass before migratory flights:

1. The increase in fat-free mass may not be related to migration but may simply reflect completion of body growth by juveniles [26].
2. Birds may need to increase the size (and power) of their flight muscles (hypertrophy) to be able to fly at optimal speeds. Larger birds have a smaller power margin than smaller birds [21], and we may therefore expect large birds in particular to increase the size of their flight muscles, that is, to add fat-free components along with the fat, before take-off.

Visceral organs play an important role in migratory flight because they contribute to body mass and metabolic rate. Given that the performance of organs increases with their size [40], migrants are expected to have well-developed 'exercise organs' such as flight muscles (pectoral and supracoracoid), heart, and lungs. This should be favoured owing to

1. higher mechanical power production by a larger cross-sectional surface of muscles,
2. better respiration and oxygen supply to sustain mechanical power generation and to compensate for the hypoxic conditions of high-altitude flights,
3. steady energy supply by organs providing protein substrate for catabolism, along with fat, to sustain flight [44].

Besides exercise organs, nutritional organs (gizzard, liver) also contribute to the energy turnover, and hence the physiology of flight, by regulating food processing. While exercise organs are functional during flight, nutritional organs are atrophied during migration to reduce loading and hypertrophied only at stopover refuelling stations. It was shown that migrants have larger bony surfaces for flight muscle attachment [8], but similar pectoral muscle and heart weights as sedentary species, except that heart size was larger in migrant. Besides the size of specific organs, total body mass also matters in optimal migration. Body mass is the most important life-history trait, which can adaptively evolve to meet certain demands or might constrain optimal strategies' realization. Migration is not an exception. Body mass or drag due to mass are integral parts of functions about flight economy. Therefore, body mass is expected to evolve with migration distance [19].

### 5.6.2 Changes in Feathers

Feathers are a defining feature of birds with multiple functions such as flight, insulation, protection against predation and signalling. Table 5.1 shows different types of feathers present on the body of bird and their functions.

Feathers are made of keratin, the same protein that makes up hair and nails. Flight feathers are of prime importance during flight. The size and complex structure of feather components can be changed under the impact of different factors. For

**Table 5.1** Types of feathers with their functional roles

| Types of Feathers | Functions |
|---|---|
| Contour feathers | Feathers cover the entire body of a bird and give it colour and shape. They overlap like shingles, with a fluffy base and a waterproof tip |
| Down feathers | These soft, fluffy feathers are found under the contour feathers. They have little or no shaft and are the simplest type of feather. |
| Flight feathers | These long feathers are found on a bird's wings and tail, and help it fly. They also help the bird steer in flight. |
| Semiplumes | These feathers are a cross between contour and down feathers, and provide insulation and aerodynamics. They also play a role in courtship displays. |
| Filoplumes | These soft, hair-like feathers are scattered throughout the body, especially around the contour feathers. They may have a sensory function, such as helping the bird sense pressure and vibration changes in the contour feathers. |
| Bristle feathers | These feathers are found around the head, mouth, eyelids, and neck. They have a strong rachis and a few barbs at the base, and have both protective and sensory functions. |

example, migratory birds increase their wing span and obtain sharper wingtips by a differential elongation of the outer wing flight feathers (i.e., primaries; [33]). According to a study of de la Hera et al. [10], migration distance favours wing feathers with thicker dorso-ventral rachis and stout barb lengths, along with longer and more concave wings, which confer more stiffness during bending to these feathers.

### 5.6.3 Changes in Wing Morphology

Large Wing Span and Aspect Ratio: Long-distance migratory seabirds like the wandering albatross have exceptionally large wingspans (up to 7.5 feet), which are well-suited for gliding and dynamic soaring. The long wings reduce the energy cost of flight and help maintain lift over long distances.

The wings of migratory seabirds are adapted to provide minimal drag and maximum lift. Grilli et al. [18] studied wing morphology of 3 populations (fully migratory, partially migratory, and resident) of Turkey Vulture (*Cathartes aura*). Wing shape and size influence the maneuverability and efficiency of flapping, gliding, and soaring during flight. According to Watanabe [58], narrow and longer wings (high aspect ratio), with low wing loading and pointed shape, are energetically more efficient for flight. This type of wing is prevalent in long-distance migratory birds. On the other hand, short and wider wings (low aspect ratio) and rounded shapes are found in partially migrant and resident birds [55]. The turkey vulture is a strictly soaring migrant, migrating from Canada in the north to Tierra del Fuego in the south. Studies on this bird revealed that wing loading has more impact than aspect ratio and it is linked with migration distance, speed, length and height of the flight.

## 5.7 Physiological Adaptations in the Seabirds

### *5.7.1 Efficient Metabolism and Energy Generation*

To provide enough energy for flight endurance, birds have an extremely efficient metabolism that generates a large amount of energy. Fat serves as a major energy source and is broken down to meet energy demands. Fatty acids (FA) are stored in adipocytes and the liver. In adipocytes, hormone-sensitive lipase breaks down triacylglycerol into non-esterified fatty acids and glycerol; catecholamines regulate this process [48]. Non-esterified fatty acids bind to albumin in plasma, which has several binding sites, and are transported to the muscles. Besides increasing fat storage, migrant birds may facilitate fatty acid mobilization by altering the physical properties of fat [47]. An intriguing alternative method for delivering exogenous fatty acids to muscles via circulation was suggested by Jenni-Eiermann and Jenni [25], based on their research on migratory passerines. Results indicate that plasma triacylglycerides and very low-density lipoprotein (VLDL) levels are higher in migrant birds caught mid-flight compared to those fasted for several hours or overnight. They proposed that the liver's capacity for fat uptake and processing allows it to act as an alternative reservoir for exogenous fatty acids from adipose tissue. This process enables plasma albumin to transport more fatty acids per unit time. Lipids absorbed by the liver are reconstructed and released as VLDL into the plasma. This mechanism could supply significant fatty acids to muscles without causing osmotic changes from elevated plasma albumin levels, providing a substantial energy source. For short-distance migratory birds, energy may be more vital than time during migration [31].

### *5.7.2 Increased Muscle Mass*

The migratory period causes changes in the mass of skeletal muscles in the majority of species. Flight muscles show phenotypic flexibility, which is an adaptation of flight muscle to change power demand as per changes in body mass. Flights across ecological barriers without any stopover can cause the breakdown of muscle protein beyond the limit necessary for flight, while migration across a "common landscape," where a bird can land to feed and/or drink, may occur without catabolizing flight muscle protein. Phenotypic flexibility is usually associated with three phases of migration:

1. In pre-migratory periods, muscle mass is increased,
2. During flight, muscle mass is decreased
3. At stopover periods, muscle mass is again increased.

The term "flight muscle" refers to the combined pectoralis and supracoracoideus muscles on both the left and right sides, while "leg muscle" denotes the group of muscles attached to the Tibotarsus. Muscle mass is reported as lean dry mass. In passerine birds, changes in the mass of leg and flight muscles can contribute up to twenty-five percent of variations in body mass, with most of these changes taking place after flights over ecological barriers like the Sahara Desert or the Mediterranean Sea. In Grebes, the mass of flight and leg muscles is affected by "use hypertrophy" or "disuse atrophy." During the flightless molting period, flight muscles shrink, while leg muscles increase in size [4].

### 5.7.3 Shrinkage of the Digestive System and Other Organs

Before departure on long-distance flights, birds' exercise organs (pectoral muscle and heart) tend to show hypertrophy, and their nutritional organs (stomach, intestine, and liver) tend to show atrophy [16]. Reductions in nutritional organs appear most prominent in species that are supposed to overfly barren oceans with no opportunities for emergency landings. Birds may shrink their intestines to reduce weight and conserve energy during migration [29]. Gut microbiota may shift to optimize fat processing during long flights. Changes occurring in gut can be seen in into Table 5.2.

### 5.7.4 Immune Organs

The immune system of animals includes innate defences that are present from birth as immediate first lines of defence, and induced immune responses, which are acquired by encountering different pathogens throughout life [6]. The main benefit

**Table 5.2** Showing changes in the digestive system of seabirds

| Species | Changes in organ size | References |
|---|---|---|
| Chaffinch (*Fringilla coelebs*) | Atrophy in the liver because of shift of glycogenesis to lipogenesis before migration. Fat is stored in adipose tissue. | [11] |
| Greater snow goose (*Anser caerulescens atlanticus*) | Atrophy in the gizzard | [17] |
| Eared Grebe (*Podiceps nigricollis*) | Atrophy in the liver, intestine, and stomach was observed before departure. | [16] |
| Red Knot (*Calidris canutus*) | A decrease in mass of both stomach and intestine was observed | [3] |
| Red Knot (*Calidris canutus*); Bar-tailed Godwit (*Limosa lapponica baueri*); Great knot (*Calidris tenuirostris*) | Atrophy of stomach, liver and intestine | [39, 42, 43] |

Changes in the size of the digestive system before migration are reversible and have a great significance in the reduction of overall body size to maintain efficient flight of an improved innate system is probably an increased likelihood of eliminating pathogens before they start multiplying and establishing themselves inside a host, while an innate immune response is mounted when a pathogen starts replicating and establishing itself in the body [22]. Immune responses ensure survival, but these are energetically expensive because they include some sort of self-damage. Hence, a stronger innate system is favoured to avoid the generation of the immune response. Stopovers last only a few days, migrants at stopover are particularly keen on avoiding immune responses and hence rely mainly on the first line of defense mechanism. A strong innate immune system may be relevant for birds, which travel long migratory journeys and visit multiple stopover sites [53] that could hold pathogens that they have never encountered before. The increase in immune function highlights the immune system's sensitivity to changing conditions, even in very short periods. Interestingly, for many individuals, the increase in microbial killing capacity appears stronger than the increase in lysis. This is probably because the bacteria-killing capacity is a broad measure of innate immune function. Understanding stopover ecology is thus is prime significance to understanding bird migration. Stopovers sites are for accumulating fuel, but are also typically thought to allow migrants to 'rest' and recover from migratory endurance flight [13].

### 5.7.5 Hydration Adaptations

Dehydration can occur when the water loss is more than metabolic water production, which may create problems limiting flight range. The lungs serve as the main passage for water loss during flight; the rate of water loss relies on environmental factors such as temperature, relative humidity, and rate of ventilation, and increases with altitude. Synthesis of metabolites leads to increased plasma osmolality, which can threaten the functioning of the organs during flight. The constraints on energy and water balance may interact, influencing various aspects of flight behavior, including flight altitude, flight mode, distance between stops, and stopover duration. To minimize energy loss while avoiding water balance issues, a bird might choose to migrate in short hops rather than long stretches when crossing large ecological barriers [30].

**Salt Glands** Seabirds have salt glands located above their eyes, which allow them to excrete excess salt from seawater. This adaptation is crucial for survival in marine environments, where they may drink seawater during migration. During long flights over the ocean, migratory seabirds can conserve water effectively. Their kidneys are highly efficient at excreting concentrated waste, allowing them to retain water for longer periods. The microscopic structure of a salt gland tubule resembles a bundle of pipes with a small hole in the middle. Each "pipe" consists of five to seven individual cells arranged like wedges. The hole, or lumen, funnels the secretion into the central duct. While the structure and function of the salt gland are essentially the same in all sea birds, its location varies. In gulls and many other birds, the glands are situated on top of the head, above the eye sockets. In cormorants and gannets, however, they are located between the eye and the nasal cavity. Petrel ejects droplets of solution produced by its salt gland through a pair of tubes on top of its beak. The salt-gland secretions of most birds drip from the tip of the beak. The petrel, however, remains in the air almost continuously and has evolved a "water pistol" mechanism as a means of eliminating the fluid while in flight [54].

### *5.7.6 Energy Conservation by Dynamic Soaring*

Many long-distance migratory seabirds use a technique called dynamic soaring, where they take advantage of wind gradients near the ocean's surface to gain lift without constantly flapping their wings. This allows them to travel vast distances while conserving energy. Dynamic soaring is a flying pattern used by birds, like albatrosses, to cover long distances with minimal energy. The key to this technique is the difference in wind speed at different heights in the atmosphere, known as wind shear. The bird begins by flying against the wind, climbing through the wind shear layer where the changes in wind speed occur. Once it reaches a certain point, the bird turns to fly with the wind, taking advantage of the faster winds in that direction. As it flies downwind, the bird drops across the wind shear layer to gather more energy from the wind, then turns back into the wind, climbing again through the wind shear layer. This cycle repeats as the bird soars in an S-shaped path across the wind. This efficient technique allows albatrosses to fly long distances without flapping their wings often, conserving energy while gaining speed [49].

### *5.7.7 Adaptations to Low Oxygen (Hypoxia)*

Birds can fly at very high altitudes during migration. For example, Bar-headed geese cross the Himalayan mountains and fly over Mount Everest. Birds start their migration near sea level and can reach up to the height of 9000 m within a single day which leaves no time for acclimatization. Faraci [14] explained the adaptations in birds shown in Table 5.3.

At higher altitudes, there is low pressure of oxygen that creates hypoxic conditions for organisms. Birds have many physiological mechanisms that contribute to their exceptional tolerance of hypoxia as shown in above table.

### *5.7.8 Oxidative Stress During Migration*

Some migratory birds travel for 200 h long non-stop flight and such intense physical activity is associated with the generation of reactive oxygen and nitrogen species (ROS/RNS). These free radicals are generated in skeletal and cardiac muscles,

**Table 5.3** Adaptations in circulatory and respiratory systems of Seabirds

| Adaptation | Key points |
|---|---|
| Respiratory system (High Gas exchange efficiency of lungs) | Hyperventilation, $P_{50}$ value declines due to Respiratory alkalosis, which increases the affinity for oxygen during hypoxia. |
| Blood oxygen transport | The low P50 observed in high-altitude adapted birds offers a clear advantage for oxygen transport during severe hypoxia. A small drop in blood temperature (due to the cold conditions at high altitudes) notably increases haemoglobin's affinity for oxygen. |
| Regulation of cardiac output | Compared to mammals of similar body mass, birds have larger hearts and a greater stroke volume. The larger size of the avian heart indicates its ability to generate significant increases in cardiac output during hypoxia or exercise. Redistributing the blood toward essential organs such as the heart and brain. Secretion of catecholamines due to increased adrenal blood flow. |
| Cerebral circulation | Bird cerebral circulation is highly sensitive to decreases in arterial $Po_2$, with blood flow beginning to increase when arterial $Po_2$ falls below about 75 mmHg, and it can rise to six times during severe hypoxia. |
| Cellular adaptations | High capillary density in skeleton, and cardiac muscles in birds. High density of mitochondria and oxidative enzymes in muscles. High myoglobin concentration in muscles. |

increasing the oxidative stress in these muscles. If enough anti-oxidative agents will not be present, then free radicals cause tissue damage. Elevated levels of protein carbonyls and glutathione peroxidase in the migrating European robin on its journey toward its non-breeding grounds were found. These two components are markers of excessive oxidation. Carbonyl groups are incorporated into proteins through reactions with free radicals, and protein carbonylation is typically irreversible. The activity of glutathione peroxidase, an antioxidant enzyme, increases when hydrogen peroxide (early byproducts of oxidative damage) levels rise [24].

## 5.8 Adaptations in Navigation and Orientation

### *5.8.1 Celestial Navigation*

Seabirds, particularly those migrating over vast open oceans, use the sun, moon, and stars to help them orient themselves. This ability to navigate by celestial bodies is especially useful for long-distance migratory birds that need to maintain a consistent direction over long periods. Some species, like the Arctic Tern, European Herring Gull and Wandering Albatross, are thought to use the position of the sun during the

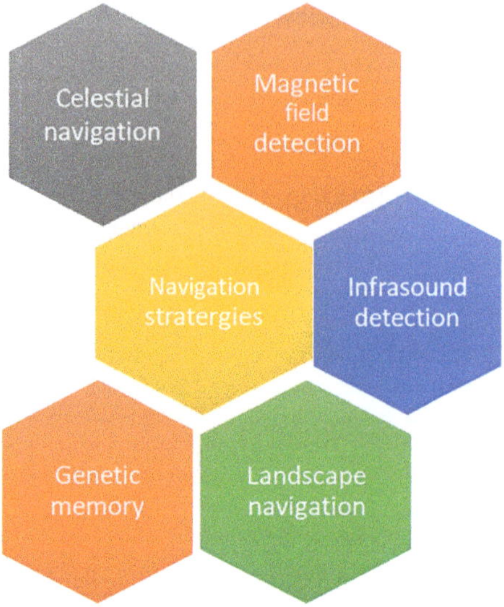


**Fig. 5.2** Illustration of various navigational strategies used by birds

day and the stars at night to maintain their migratory route, especially when flying in unfamiliar or featureless oceanic environments [1, 32, 50]. Navigation strategies of birds is illustrated in Fig. 5.2.

### *5.8.2 Magnetic Field Detection*

Many seabirds can detect Earth's magnetic field, which acts as a compass for navigation. This magnetic sense is thought to be located in the birds' eyes, in specialized cells that respond to magnetic fields (called cryptochromes), or in their beaks, where magnetite particles may play a role. The Earth's magnetic field assists in the long-distance navigation of albatrosses [50]. For example, research shows that albatrosses, despite the lack of obvious landmarks, use a magnetoreceptor on their head to guide them while flying across the vast ocean [34].

### *5.8.3 Odor, or the Sense of Smell*

Using their sense of smell, seabirds may identify chemical clues in their surroundings, including scents linked to specific areas or food supplies. These birds utilize their partners' scent to locate their nests and for homing, even though they can detect it from a significant distance [5]. Seabirds can find food spots over long distances or return to particular nesting sites due to their olfactory abilities.

### 5.8.4 *Navigation Using Landmarks*

Seabirds frequently use visual markers (such as coasts, mountains, or particular island features) to help them find their bearings when they are nearer shore. When birds need to locate their precise nesting sites during breeding seasons, this is very crucial [45].

### 5.8.5 *Infrasound Acquisition*

Certain seabirds can pick up low-frequency sounds, or infrasound, that can travel great distances across the ocean or atmosphere. These noises may originate from faraway beaches or large-scale environmental events like storms. To orient themselves when flying over the wide ocean, seabirds can employ these noises [37].

## 5.9 Future Prospectus

The impact of environmental stressors and climate change on delicate systems of birds can be better understood by studying seabird migration. As ocean currents shift, water temperatures rise, and food becomes scarcer, seabirds may need to modify their flight patterns, migratory routes, or even physiological functions. Their ability to adapt to these environmental changes will be vital for their survival, and understanding these adaptations can help guide conservation efforts. Additionally, advances in tracking and monitoring technologies will enable a deeper understanding of seabird migration patterns, supporting more effective management and protection of seabird populations.

## 5.10 Conclusion

Seabirds have developed a unique set of physiological and morphological adaptations to withstand migration pressures. Their bodies are designed for long-distance flight, with long, thin wings that reduce drag and enable energy-efficient gliding and soaring. This wing shape allows seabirds to cover vast marine areas, especially when combined with specially engineered, sleek, waterproof feathers. Additionally, they can significantly increase their fat reserves before migration to provide energy for the journey and reduce body weight to conserve energy during flight. Their digestive systems undergo a process called "gut atrophy" during migration, but they can quickly re-establish these systems once they reach stopping points. Seabirds' specialized salt glands and efficient kidneys help regulate their water and salt intake—

adaptations that support survival in harsh environments. Furthermore, they possess highly effective respiratory systems that enable optimal oxygen exchange, even over long distances or at high altitudes. Their ability to navigate precisely across the globe relies on their sophisticated knowledge of the Earth's magnetic field, celestial bodies, and solar positions. To save energy and focus on survival, reproductive behaviors are often postponed during migration, with breeding occurring after they reach suitable locations.

## References


1. Åkesson S, Hedenström A (2007) How migrants get there: migratory performance and orientation. Bioscience 57(2):123–133
2. Alerstam T, Rosen M, Backman J, Ericson PGP, Hellgren O (2007) Flight speeds among bird species: Allometric and phylogenetic effects. PLoS Biol 5:e197
3. Battley PF, Piersma T (1997) Body composition of Lesser Knots (*Calidris canutus rogersi*) preparing to take off on migration from northern New Zealand. Notornis 44(3):137–150
4. Bauchinger ULF, Biebach H (2005) Phenotypic flexibility of skeletal muscles during long-distance migration of garden warblers: muscle changes are differentially related to body mass. Ann N Y Acad Sci 1046(1):271–281
5. Bonadonna F (2009) Olfaction in petrels: from homing to self-odor avoidance. Ann N Y Acad Sci 1170(1):428–433
6. Buehler DM, Tieleman BI, Piersma T (2010) How do migratory species stay healthy over the annual cycle? A conceptual model for immune function and for resistance to disease. Integr Comp Biol 50:346–357
7. Butler PJ (2016) The physiological basis of bird flight. Philos Trans R Soc B Biol Sci 371(1704):20150384
8. Calmaestra RG, Moreno E (2000) Ecomorphological patterns related to migration: a comparative osteological study with passerines. J Zool 252:495–501
9. Cheng Z, Ma D, Lu M, Du X, Kong L, Bao X (2025) Variation of migration routes in the Central Asian-breeding Cuculus canorus population influenced by the Qinghai-Tibet Plateau. Avian Res 16:100265
10. de la Hera I, Hernández-Téllez I, Pérez-Rigueiro J, Pérez-Tris J, Rojo FJ, Tellería JL (2020) Mechanical and structural adaptations to migration in the flight feathers of a Palaearctic passerine. J Evol Biol 33(7):979–989
11. Dolnik VR, Blyumental TI (1967) Autumnal premigratory and migratory periods in the Chaffinch (Fringilla coelebs coelebs) and some other temperate-zone passerine birds. Condor 69(5):435–468
12. Egevang C, Stenhouse IJ, Phillips RA, Petersen A, Fox JW, Silk JR (2010) Tracking of Arctic terns Sterna paradisaea reveals longest animal migration. Proc Natl Acad Sci 107(5):2078–2081
13. Eikenaar C, Hessler S, Hegemann A (2020) Migrating birds rapidly increase constitutive immune function during stopover. R Soc Open Sci 7:192031
14. Faraci FM (1991) Adaptations to hypoxia in birds: how to fly high. Annu Rev Physiol 53(1): 59–70
15. Felis JJ, Adams J, Hodum PJ, Carle RD, Colodro V (2019) Eastern Pacific migration strategies of Pink-footed Shearwaters Ardenna creatopus: implications for fisheries interactions and international conservation. Endanger Species Res 39:269–282
16. Gaunt AS, Hikida RS, Jehl JR, Fenbert L (1990) Rapid atrophy and hypertrophy of an avian flight muscle. Auk 107(4):649–659

17. Gauthier G, Bédard J, Huot J, Bedard Y (1984) Protein reserves during staging in greater snow geese. Condor 86(2):210–212
18. Grilli MG, Lambertucci SA, Therrien JF, Bildstein KL (2017) Wing size but not wing shape is related to migratory behavior in a soaring bird. J Avian Biol 48(5):669–678
19. Hedenstro¨m A (2008) Adaptations to migration in birds: behavioural strategies, morphology and scaling effects. Philos Trans R Soc London B 363:287–299
20. Hedenstrom A (2010) Extreme endurance migration: what is the limit to non-stop flight? PLoS Biol 8:e1000362
21. Hedenström A, Alerstam T (1992) Climbing performance of migrating birds as a basis for estimating limits for fuel-carrying capacity and muscle work. J Exp Biol 164(1):19–38
22. Hegemann A, Matson KD, Versteegh MA, Villegas A, Tieleman BI (2013) Immune response to an endotoxin challenge involves multiple immune parameters and is consistent among the annual-cycle stages of a free-living temperate zone bird. J Exp Biol 216:2573–2580. https://doi.org/10.1242/jeb.083147
23. Hegemann A, Fudickar AM, Nilsson JÅ (2019) A physiological perspective on the ecology and evolution of partial migration. J Ornithol 160:893–905. https://doi.org/10.1007/s10336-019-01648-9
24. Jenni-Eiermann S, Jenni L, Smith S, Costantini D (2014) Oxidative stress in endurance flight: an unconsidered factor in bird migration. PLoS One 9(5):e97650
25. Jenni-Eiermann S, Jenni L (1992) High plasma triglyceride levels in small birds during migratory flight: a new pathway for fuel supply during endurance locomotion at very high mass-specific metabolic rates? Physiol Zool 65(1):112–123
26. Johnson OW, Morton ML, Bruner PL, Johnson PM (1989) Fat cyclicity, predicted migratory flight ranges, and features of wintering behavior in Pacific Golden-Plovers. Condor 91(1): 156–177
27. Kaboli M, Aliabadian M, Guillaumet A, Roselaar CS, Prodon R (2007) Ecomorphology of the wheatears (genus Oenanthe). Ibis 149:792–805
28. Kiran SD, Kour A, Priya DV, Kumar R (2022) Community composition and status of avian diversity at Campus and Agricultural landscapes of Chaudhary Charan Singh Haryana Agricultural University, Hisar (Haryana). J Appl Nat Sci 14(4):1130–1140. https://doi.org/10.31018/jans.v14i4.3784
29. Kiran SD, Kour A, Yadav R, Dahiya P, Kumar R, Sharma K (2023) The avian gut microbial community: driving factors and role in wild birds. Latest trend in zoology and entomology science (Volume-14). Akinik Publications, ISBN: 978-93-5570-676-8, pp 15–31
30. Klaassen M (1996) Metabolic constraints on long-distance migration in birds. J Exp Biol 199(1):57–64
31. Klaassen RH, Ens BJ, Shamoun-Baranes J, Exo KM, Bairlein F (2012) Migration strategy of a flight generalist, the Lesser Black-backed Gull *Larus fuscus*. Behav Ecol 23(1):58–68
32. Lincoln FC, Peterson SR (1979) Migration of birds (No. 16). Fish & Wildlife Service, US Department of the Interior
33. Lockwood R, Swaddle JP, Rayner JM (1998) Avian wingtip shape reconsidered: wingtip shape indices and morphological adaptations to migration. J Avian Biol 29:273–292
34. Mouritsen H, Huyvaert KP, Frost BJ, Anderson DJ (2003) Waved albatrosses can navigate with strong magnets attached to their head. J Exp Biol 206(22):4155–4166
35. Murphy MT (1986) Temporal components of reproductive variability in Eastern Kingbirds (Tyrannus tyrannus). Ecology 67(6):1483–1492
36. Newton I (1977) Timing and success of breeding in tundra nesting geese. In: Stonehouse B, Perrins CM (eds) Evolutionary ecology. Macmillan, London
37. Patrick SC, Assink JD, Basille M, Clusella-Trullas S, Clay TA, den Ouden OF, Joo R, Zeyl JN, Benhamou S, Christensen-Dalsgaard J, Evers LG (2021) Infrasound as a cue for seabird navigation. Front Ecol Evol 9:740027
38. Piersma T, Davidson NC (1991) Confusions of mass and size. Auk 108(2):441–443

39. Piersma T, Gill RE Jr (1998) Guts don't fly: small digestive organs in obese bar-tailed godwits. Auk 115(1):196–203
40. Piersma T, Lindstro¨m A (1997) Rapid reversible changes in organ size as a component of adaptive behaviour. Trends Ecol Evol 12:134–138
41. Piersma T, Van Gils JA (2011) The flexible phenotype: a body-centred integration of ecology, physiology, and behaviour. Oxford University Press, Oxford, UK
42. Piersma T, Dietz MW, Dekinga A, Nebel S, Van Gils J, Battley PF, Spaans B (1999) Reversible size-changes in stomachs of shorebirds: when, to what extent, and why. Acta Ornithol 34(2): 175–181
43. Piersma T, Koolhaas A, Dekinga A (1993) Interactions between stomach structure and diet choice in shorebirds. Auk 110(3):552–564
44. Piersma T, Perez-Tris J, Mouritsen H, Bauchinger U, Bairlein F (2005) Is there a "migratory syndrome" common to all migrant birds? Ann N Y Acad Sci 1046:282–293
45. Pollonara E (2015) Landmark-based navigation in birds: learning and using topographical cues
46. Porcasi JF (1999) A statistical method for identification of albatross (Phoebastria) species. Pacific Coast Archaeol Soc Quart 35(6):1
47. Raclot T (2003) Selective mobilization of fatty acids from adipose tissue triacylglycerols. Prog Lipid Res 42(4):257–288
48. Ramenofsky M (1990) Fat storage and fat metabolism in relation to migration. In: Bird migration: physiology and ecophysiology. Springer Berlin Heidelberg, Berlin, Heidelberg, pp 214–231
49. Richardson PL (2015) Upwind dynamic soaring of albatrosses and UAVs. Prog Oceanogr 130: 146–156
50. Ritchison G (2023) Navigation and orientation. In: In a class of their own. Fascinating life sciences. Springer, Cham
51. Rubolini D, Massi A, Spina F (2002) Replacement of body feathers is associated with low pre-migratory energy stores in a long-distance migratory bird, the barn swallow (Hirundo rustica). J Zool 258(4):441–447
52. Salewski V, Bruderer B (2007) The evolution of bird migration—a synthesis. Naturwissenschaften 94:268–279
53. Schmaljohann H, Fox JW, Bairlein F (2012) Phenotypic response to environmental cues, orientation and migration costs in songbirds flying halfway around the world. Anim Behav 84:623–640
54. Schmidt-Nielsen K (1959) Salt glands. Sci Am 200(1):109–119
55. Swaddle JP, Witter MS (1998) Cluttered habitats reduce wing asymmetry and increase flight performance in European starlings. Behav Ecol Sociobiol 42:281–287
56. Thorne LH, Clay TA, Phillips RA, Silvers LG, Wakefield ED (2023) Effects of wind on the movement, behavior, energetics, and life history of seabirds. Mar Ecol Prog Ser 723:73–117
57. Vincze O, Vágási CI, Pap PL, Palmer C, Møller AP (2019) Wing morphology, flight type and migration distance predict accumulated fuel load in birds. J Exp Biol 222(1):jeb183517
58. Watanabe J (2016) Aspects of diversity, paleobiology, and morphology of wing-propelled diving birds. Geobios 90:143
59. Yodha K, Singh D, Kour A, Yadav R, Kumar R (2024) Species diversity and abundance of birds in ber (*Zizyphus mauritiana*) crop. Asian J Conserv Biol 13(1). https://doi.org/10.53562/ajcb.83683

# Chapter 6
# Quantifying the Impact of Wetland Rehabilitation on Bird Abundance in Agricultural Areas

**Shweta Sharma, Dilbag Singh, Varsha Malik, Kiran Kiran, Pushp Lata, and Nibha Singh**

**Abstract** Wetland degradation in agricultural landscapes has significantly impacted biodiversity, particularly bird populations that depend on these ecosystems for habitat and sustenance. Wetland rehabilitation plays a crucial role in enhancing bird abundance, offering both ecological and conservation benefits by restoring degraded wetlands in agricultural areas. This chapter explores the necessity of wetland rehabilitation and its potential to mitigate habitat loss. Multiple methods of bird abundance estimation such as field surveys, remote sensing, and bioacoustic monitoring are presented. Wetland restoration enhances the quality of habitats, species diversity, and population density. Different studies in different regions of the globe present the benefits of wetland rehabilitation on bird assemblages. With these benefits, resource constraints persist, policy lacunae, and land-use conflicts still exist. For the overcoming of these challenges, conservation programs involving integrated ecological monitoring and policy interventions are necessary. This

---

S. Sharma (✉)
Department of Zoology, Chaudhary Charan Singh Haryana Agricultural University, Hisar, India

Department of Pharmaceutical Sciences, University of California, Irvine, Irvine, USA
e-mail: Shwetas4@uci.edu

D. Singh
School of Agriculture Sciences, Galgotias University, Greater Noida, Uttar Pradesh, India
e-mail: dilbag@galgotiasuniversity.edu.in

V. Malik
Department of Zoology, Chaudhary Charan Singh Haryana Agricultural University, Hisar, India

K. Kiran
Department of Botany and Plant Physiology, CCS Haryana Agricultural University, Hisar, Haryana, India

P. Lata
Department of Zoology, Chaudhary Bansilal University, Bhiwani, India

N. Singh
Jamia Hamdard University, New Delhi, India

 
R. Kumar et al. (eds.), *Ecology, Conservation of Aquatic Birds*, Wetlands: Ecology, Conservation and Management 15, https://doi.org/10.1007/978-3-032-17125-2_6

chapter stresses the effective role of wetland restoration in maintaining bird populations as well as ensuring environmentally friendly agricultural patterns.

**Keywords** Bioacoustic · Biodiversity · Degradation · Habitat loss · Wetland rehabilitation

## 6.1 Introduction

Wetlands are the richest semi-aquatic ecosystems on earth, which are seasonally or permanently flooded [17]. Flooding creates anoxic condition, primarily in soil. The wetlands were found to be a transitional area between land and water bodies due to developing oxygen deficit water saturated soil [77]. Also, these are reported to be some of most biologically diverse ecosystem of all, which serves as a habitat to a range of aquatic and semi aquatic flora and fauna, with an improved water quality being a result of filtering efficiency of plants for nitrates and phosphorus [3]. Wetlands provide essential habitat for a variety of species, especially birds [57]. Moreover, in agricultural landscapes, wetland degradation due to anthropogenic activities were observed (Fig. 6.1). This degradation effects both the ecological health and wetland dependent bird population. As agriculture expands and intensifies, the cause of wetland degradation and focussing the requirement for rehabilitation become important in order to conserve biological diversity and regulating ecological balance.

One of the major factors of degrading wetland in agricultural landscapes is drainage for land conversion [6]. Wetlands were majorly identified as areas without productivity and drained for generating space used under crop cultivation and grazing of livestock. This process led to altered wetland hydrology, which declines their capacity to provide habitat for waterbirds and other wildlife [58]. For example, the waterbirds as ducks, herons and cranes depends on wetlands for their food and reproduction. When these ecosystems are drained, water level become insufficient for sustaining aquatic vegetation and the other food sources of these birds. With the

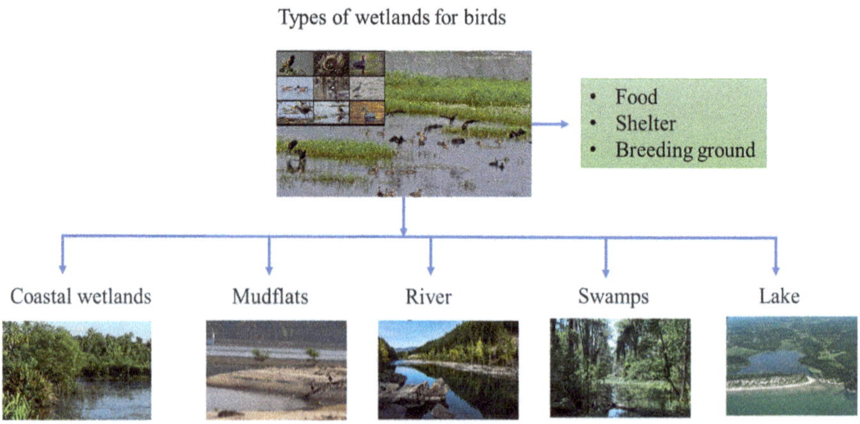


**Fig. 6.1** Representing five major types of wetlands in which birds reside with the reasons for stay

passage of time, the habitat loss contributes to declined bird population, which are already endangered because of other human activities. In combination with the drainage process, the water resource overuse for irrigation effectively alters hydrology of these wetlands [82]. Several agricultural operation cause diversions of water from natural sources causing seasonal drying or even leads a permanent loss of wetlands. This altered water availability effects both residing as well as migratory species of birds. Migratory birds, however, depends on wetlands as a stopping site during long distance flights, where they regain their energy by taking rest [45]. During drying of these wetlands or their shrinkage, competition increases for available resources, which cause a decreased population. Moreover, the disturbance in water regimes can cause an intrusion of land plants to wetland areas, which degrade the quality of habitat for species of birds [70].

Pollution is also an effective pitfall to wetlands in agricultural landscapes [76]. The wastes from fertilizers, pesticides and other agricultural chemicals are source for introduction of harmful wastes of wetlands [42]. This is a major cause of eutrophication, where the excessive nutrients interfere with algae and hence decline water's oxygen levels. This oxygen deficient condition of water is majorly responsible for death of water residing organisms as fishes, invertebrates and other aquatic organisms, which are baseline for wetland food web [17]. Due to this, the water insects preying birds like water fowl faces a lack of food. In addition to it, the biological assimilation of toxic compounds in food chain is a threat to health of birds, which affects survivability and reproduction. The effect of nutrient overload and contaminate highly degraded ecosystems, making them inconvenient for several bird's species [1]. Introduction of invasive communities, mainly by agricultural activities, lead to elevated degradation of wetlands [12]. Non-residing flora and fauna can overcount native species, which alters the balance of these habitats. For instance, invasive flora is dominant wetland vegetation, declined diversity and excess of native plants. For instance, external plants can overcome wetland vegetation, which can decline diversity of native flora, on which birds depend [72]. Also, in relation to this, the residing fishes or invertebrates can disturb the food webs, leading to reduced resource availability for birds. These changes determine the wetland integrity, which makes them support less for different populations of birds [61].

Another major cause in agricultural landscape is habitat fragmentation. As wetlands are changed to smaller, stranded patches bordered by farmland, their relativity and ecological functionality are declined. Several birds need large, adjoining place for completing their life cycles, as making nests and escaping hunting birds. Fragmentation not only defines finding suitable habitats, but it also elevates risk of being hunted and other wildlife conflicts [70]. It also causes a greater reduction in wetlands ability to support bird population. Considering these challenges, the rehabilitation of destructed wetlands is an important process for regaining their ecological functionality and for supporting population of birds. The process of rehabilitation involves restoration of natural hydrology, which includes reintroducing water flow, reducing water diversion, and further addressing other issues of drainage [78]. This helps wetlands regain their ability to support invertebrates, aquatic vegetation and other factors effecting birds' survival. Also, control of agricultural pollution is important. For implementing sustainable farming practices as buffer zones, reduced agrochemicals and waste management improvement, can prevent wetlands from runoff water

[80]. These measures are helpful in improving quality of water causing reduction in eutrophication and contamination risk. Restoration of vegetation is also another important factor for wetland rehabilitation [64]. Growing existing species along with making the food and shelter available for birds, also helps in stabilizing soil, reducing erosion, and filter water [70]. Habitat quality can also be improved by avoiding entry of native species and restoration of already residing plant communities. In addition to it, connection of fragmented wetlands by creating ecological corridors will lead to an increased connectivity of habitat. This allows free movement of birds between and across areas for required resources. These steps can also contribute to other wildlife species by being a greater contribution to ecosystem resilience.

Along with providing ecological benefits, wetland rehabilitation can also can be significantly advantageous for agriculture and local communities. The improved wetland health can provide an improved quality of water, reducing flood risk, and enhances groundwater recharge, these all contribute to sustainable farming [48]. These enhance biodiversity, which help in pollination, pest control, and other ecological services agriculture needs [85]. Rehabilitation wetlands can also prove useful as ecotourism areas, preserving natural functions as well as earning local residents income. Lastly to this, wetland degradation of agricultural landscape is a major threat to bird population and ecosystems' well-being. Through the restoration of wetlands to their natural condition, we are able to create habitats for different bird species and, in turn, indirectly support agriculture [15]. With increasing environmental problems in the current period, the restoration and conservation of wetlands are crucial to long-term biodiversity, ecosystem resilience, and, in turn, ensuring the future of the bird population.

## 6.2 Methods for Quantifying Bird Abundance

For an improved appreciation of the contribution offered by wild birds in wildlife ecology, baseline studies are crucial. This enables one to determine species that can host, transmit, or disperse pathogens. Population studies among wild birds would typically comprise three forms: inventory and monitoring, movement patterns, and behaviour studies. Early activities are in monitoring and stock-taking, with overall goals of: (1) all bird species within a particular area, measurement of the density or abundance of species, and recording dynamics of species and population size that occur seasonally. These activities play an important role in the identification of emergent infectious diseases such as H5N1 avian influenza. Through the monitoring of wild birds, scientists can conceptualize an early warning system to detect sudden rises in mortality rates. These studies provide valuable information to disease control and wildlife conservation to prevent potential epidemics from spreading to domestic animals and man. Species enumeration and population studies are also commonly carried out by biologists through several avian survey methods. The methodology would depend on the aims of the study, the extent of the area, species characteristics, habitat conditions, and logistics. This manual provides a general description of survey methods applied in the field for water birds, shore birds, and those linked with the H5N1 virus. The methods range from complete censuses, where all birds are enumerated, to sampling regimes, where population estimates are obtained over wider areas. Regardless of the method applied, surveys must be conducted by experienced personnel through standardized methods to ensure comparability with repeated surveys. Correct identification of the species is essential because spurious counts and variable methods sully the data. Observers will be seeing various species and conditions and, therefore, should be able to recognize most, if not all, the anticipated species, close relatives, and sex or age differences. Well-run surveys yield useful data to aid conservation and disease-surveillance effort.

### *6.2.1 Techniques and Tools for Monitoring and Measuring Avian Population Changes, Including Surveys, Remote Sensing, and Bioacoustic Monitoring*

(a) **Complete Censuses**

A complete census aims to count all animals within a specified area to obtain an unbiased abundance estimate without relying on statistical inferences. A reliable census depends on the assumption that every individual present is recorded, making it most effective for conspicuous species in well-defined habitats [79]. Examples include counting herons and cormorants nesting along wetland margins, waterbirds in small open wetlands, or shorebirds at high tide roosts. However, in cases where birds are numerous, tightly grouped, or time is limited, estimating rather than counting each individual may be necessary. Experienced counters can quickly estimate groups of 10, 20, 50, or more birds and tally them accordingly. Smaller units (typically 10) are commonly used, while larger units (100 or more) are reserved for birds in flight, colonial nesters, or time-constrained counts. A complete census is particularly effective for large, conspicuous species like swans or geese and is preferred when organized networks of observers are available. This approach is promoted for periodic swan censuses by organizations such as Wetlands International/IUCN/SSC Swan Specialist Group. Large-scale synoptic censuses such as Wetlands International's International Water bird Census annually do focus-species-specific "look-see surveys" to enumerate focus species in appropriate sites, ensuring complete population estimation. Full census is best suited for conspicuous, large animals such as geese or swans where participant networks are in good order. Organizations such as Wetlands International demand regular censuses of swans, while large programs, such as the International Water bird Census, employ "look-see surveys" at important locations [33]. Good censuses need extensive planning and subdivision of large areas into manageable units for systematic surveying. Skilled teams need to adopt standard techniques, methods of species identification, and tools such as spotting scopes and GPS [9]. Precise mapping and clear borders prevent errors and double-counting, and guarantee extensive habitat coverage. Photographic and video surveys, commonly conducted by aerial means, provide abundant material for analysis after the survey. Ground or boat surveys must complement aerial methods to provide species identification and reduce biases.

(b) **Remote Sensing in Avian Ecology**

Remote sensing has revolutionized the field of avian ecology by providing powerful tools for monitoring bird populations, understanding habitat changes, and predicting ecological trends. By using satellite imagery, aerial surveys, and drone technology, researchers can gather large-scale and high-resolution environmental data without the need for extensive fieldwork. Remote sensing is widely used in ecological studies done through Infrared-triggered cameras and video recordings complement fieldwork by monitoring behaviors such as brooding and defense, while infrared imagery provides vast amounts of data [43]. Multi-spectral imaging is frequently applied to habitat mapping and can directly track species when captured at high spatial resolution. Hyperspectral data offers significant potential for classifying objects with similar spectral characteristics. LiDAR data primarily contributes to the study of habitat structure. Radar has been utilized to monitor flying birds over extended periods, with microwave imagery and multi-polarization enhancing the accuracy of complex habitat mapping [4]. However, data scale can influence study outcomes, and certain remote sensing (RS) inversion model parameters may lack ecological relevance. Integrating multi-source data can improve mapping accuracy and provide insights into the relationship between spatial and temporal resolutions in imaging.

- **Applications of Remote Sensing in Avian Ecology**
  - **Habitat Mapping and Monitoring**
    Remote sensing allows ecologists to assess habitat characteristics such as vegetation cover, water availability, and land-use changes. This is critical for understanding the distribution and abundance of bird species, particularly those sensitive to environmental changes.
  - **Tracking Migration Patterns**
    Satellite telemetry and GPS tracking, combined with remote sensing data, help researchers study long-distance migration routes, stopover sites, and breeding grounds. This information is essential for conservation efforts and policy-making.
  - **Climate Change Impact Assessment**
    Remote sensing assists in tracking climate-driven changes in habitats, including deforestation, wetland loss, and desertification. Through the analysis of temperature, rainfall, and land cover patterns, scientists can infer the manner in which bird populations will react to climate change.
  - **Biodiversity and Conservation Planning**
    Conservationists make use of remote sensing to map critical biodiversity hotspots, measure protected area effectiveness, and monitor illicit activities such as deforestation or poaching posing threats to birds.

- **Automated Bird Detection and Population Estimation**
  Artificial intelligence and machine learning have made automatic recognition of bird species possible from all remote sensing data such as thermal imaging and acoustic monitoring. All of these operations involve precise estimates of population and enable direct human observation to a minimum.

(c) **Bioacoustic Monitoring in Measuring Avian Populations**

Bioacoustic monitoring is increasingly an important tool in avian ecology and biodiversity monitoring that provides a non-invasive yet useful tool for investigating bird populations [33]. Using automated recording equipment and sophisticated signal processing techniques, researchers can automatically detect, identify, and count bird species from their sounds [30].

To be effective in natural preservation, bioacoustic techniques ought to complement long-term observations with greater accuracy and efficiency. Monitoring biodiversity is aimed at finding temporal changes in ecological assemblages and ought to be conducted via carefully designed surveys that provide accurate data to analyze trends [47]. Environmental health indicators rely on quantitative surveys providing population trend estimates for target species [34]. Bioacoustic monitoring should involve proper recording procedures, coordinated data collection practices, effective species detection equipment, and population estimates methods. Recent improvements in digital recording technology for audio now enable unsupervised sampling of singing activity over long periods of time, even up to several months. Tailored solutions involving the use of specialized recorders or consumer device-based bespoke techniques are increasingly applied to monitor acoustic activity in animals [54, 81]. Such recordings may be used to estimate bird community structure. When high-sensitivity or highly directional microphones are used, studies have shown minimal differences between the number of species detected in recordings and those observed directly in the field [36, 37]. However, using low-sensitivity microphones tends to yield better results for on-site observers than those analyzing recordings in a laboratory setting [39]. Parallel studies comparing sound recordings with traditional bird census methods, such as point count surveys [81] or line mapping [28], have yielded comparable results. Notably, Wimmer [84] demonstrated that targeted acoustic sensor sampling could even detect significantly more bird species than traditional surveys.

A major challenge in bioacoustic monitoring is estimating the number of individuals from recordings. Stereo or quadraphonic recordings can recreate a spatial impression of the soundscape, helping skilled listeners estimate population sizes [16, 62]. Another approach involves using individual vocal signatures, provided that their acoustic characteristics remain stable throughout the observation period. Gilbert et al. [31] applied this technique to distinguish individuals in Eurasian bittern (*Botaurus stellaris*) and black-throated loon (*Gavia arctica*) populations. Since 1990, sound recordings have been used to monitor and identify individual British bitterns [32].

Call frequency has also been employed to estimate animal numbers, particularly in migration studies. However, variations in call rates between species and environmental influences make absolute population estimates difficult [25, 26, 38]. Still, Buxton and Jones [13] found that nocturnal call activity could provide insights into relative population densities of cave-breeding birds. Comparing various bird survey methods, Buckland [11] found that estimates based on song activity had relatively low precision.

Microphone arrays offer new possibilities for bird monitoring by estimating the locations of vocalizing animals. By analysing the time delay in signal arrival at different microphones, vocalizing birds can be localized with meter-level accuracy, depending on sensor spacing [53, 54] (Blumstein et al. 2011). However, Wahlberg et al. [83] reported a localization accuracy of over 100 meters for booming bitterns, using a four-microphone array with sensor distances ranging from 65 to 294 meters.

- **Bioacoustic Monitoring includes the following steps**
  - **Acoustic Data Collection**—Autonomous recording units (ARUs) are placed in the field to continuously capture bird calls and songs over extended periods.
  - **Signal Processing & Analysis**—Collected audio data is analyzed using machine learning, spectrograms, and pattern recognition software to distinguish species and count vocalizations.
  - **Species Identification**—AI-driven algorithms help classify species based on their unique vocal signatures. Manual verification is often used for accuracy.
  - **Population Estimation**—The frequency and distribution of bird calls are used as proxies for estimating species abundance and monitoring population trends.

  Bioacoustic monitoring offers a non-invasive approach to avian surveys, minimizing human disturbance and stress on birds. ARUs enable continuous data collection across multiple, including remote, locations, facilitating long-term ecological studies. These recordings can be archived for future reanalysis as analytical methods advance. Additionally, bioacoustic monitoring is cost-effective and scalable, reducing the need for extensive fieldwork while allowing simultaneous deployment at multiple sites [60].

## 6.3 Role of Wetland Rehabilitation in Enhancing Bird Habitats

Wetlands serve as critical habitats for a diverse array of avian species, providing essential resources such as food, nesting sites, and protection from predators. However, anthropogenic activities, including habitat destruction, pollution, and climate change, have significantly degraded wetland ecosystems, leading to declines in bird populations worldwide. Wetland rehabilitation aims to restore ecological functions, thereby enhancing habitat availability, diversity, and quality for both aquatic and terrestrial bird species.

### 6.3.1 Habitat Availability and Connectivity

Restored wetlands increase habitat availability through the re-establishment of hydrological connectivity, indigenous vegetation, and water regimes. Many bird species rely on wetlands for feeding, breeding, and stopover grounds during migration. Restoration efforts like controlled flooding, afforestation, and removal of alien invasive species reinforce wetland functioning, providing birds with favorable environments from their early life stages through adulthood [87]. For instance, restoration of India's Chilika Lake wetland included hydrological restoration via dredging and removal of invasive water hyacinth (*Eichhornia crassipes*) which, in turn, effectively enhanced habitat value for migratory birds like the bar-headed goose (*Anser indicus*) and black-tailed godwit (*Limosa limosa*) [24].

### 6.3.2 Enhancement of Biodiversity

Wetland restoration enhances biodiversity through the reintroduction of native vegetation species, enhanced water quality, and food webs due to natural processes. Macroinvertebrate density and aquatic plant productivity enhance top trophic levels that are favorable for herbivorous and insectivorous populations of birds. Research indicates that wetlands that are restored have higher avian species diversity than disturbed areas [29]. Florida Everglades proved that restoration of wetlands through rehydration and the decrease in phosphorus load resulted in enhanced population levels of birds that wade, like the great egret (*Ardea alba*) and the wood stork (*Mycteria americana*), which showed enhanced ecological integrity [27].

### 6.3.3 Improved Habitat Quality for Breeding and Nesting

Restored wetlands offer better nesting habitat and less habitat fragmentation, favoring ground-nesting and cavity-nesting birds. Better structured vegetation and undisturbed nesting space yield greater reproductive success rates [46]. For example, restoration of Chinese wetlands along the Yangtze River floodplain has helped the critically endangered Siberian crane (*Leucogeranus leucogeranus*) through the restoration of shallow water habitats with high cover of dense aquatic plants, essential for foraging and nesting [40].

### 6.3.4 Climate Change Mitigation and Resilience

Restored wetlands serve as natural shields against weather extremes like floods and droughts by controlling water levels and encouraging the uptake by soil. These functions are crucial in the preservation of healthy bird populations, particularly migratory birds that depend on stable wetland conditions along their migration corridors [21]. Wetland restoration in the Mississippi River Delta has enhanced the adaptive potential of waterbird species like mallards (*Anas platyrhynchos*) and northern pintails (*Anas acuta*) via the creation of stable breeding and foraging environments regardless of high sea levels and a heightened frequency of storms [5].

### 6.3.5 Long-Term Monitoring and Conservation Implications

Long-term monitoring and adaptive management approaches are necessary for effective ecological restoration of wetlands. Bioacoustic monitoring, far detection, and people's science projects provide significant contributions to the evaluation of bird population dynamics and wetland quality over time [60]. The Ramsar Convention focuses on the inclusion of wetland rehabilitation into global biodiversity conservation policy agendas due to people's participation and policy-regulated wetland conservation [18]. Wetland rehabilitation is also at the forefront of recovering bird habitats through increasing habitat availability, biodiversity, and ecological fitness. Effective restoration work from around the globe illustrates that specific conservation action can reduce the negative impacts of habitat destruction on resident and migratory bird populations. Conservation planning in the future needs to incorporate ecological restoration with integrated land-use planning for the long-term conservation of wetlands.

## 6.4 Case Studies of Successful Wetland Rehabilitation

Wetlands are one of the most biologically rich ecosystems on the planet. They support a wide variety of species, regulate hydrological cycles, and provide ecosystem services like flood control, water purification, and carbon sequestration [59]. Despite their extremely high value, wetlands have been severely degraded over the past century, primarily as a result of human activities such as agricultural growth, urbanization, and industrialization [55]. Wetland loss has several adverse consequences, such as the loss of birds that use wetlands as breeding grounds, foraging areas, and migration routes [44].

Birds function as crucial bioindicators of the health of ecosystems, conveying information on habitat quality and suitability [22]. Wetland restoration thus has a twofold objective: enhancing ecosystem function as well as raising biodiversity

[71]. In many regions of the world, different schemes of wetland restoration have recorded quantifiable increases in bird population and variety. This chapter offers case studies from around the world that exemplify exemplary wetland re-establishment projects and their positive impacts on bird populations. These examples show the value of integrating scientific research, local people's participation, and policy to guarantee conservation success [41] (Table 6.1).

These case studies reveal the instantaneous impact that wetland restoration has on avifauna population and diversity. Each of these scenarios highlights the importance of restoring hydrology functions, local people engagement, and using region-based methodologies [14].

**Hydrology Restoration:** Restoration of natural water flow is dependent upon the recovery of wetland environments and the related species.

**Participation of the Community:** Local participation, incentives, and sensitization are crucial for sustaining restoration programs.

**Continuity of Monitoring:** Ongoing monitoring provides efficient restoration programs and responsiveness to changing conditions.

**Policy Linkages:** Wetland restoration can maximize environmental and socioeconomic benefits through coordination of policies related to agriculture and climate.

Future restoration activities for wetlands can be designed to produce even more beneficial ecological outcomes based on learning from these examples. Wetlands are not birdhouses but functional ecosystems that are necessary for survival and promoting the well-being of our planet.

## 6.5 Key Challenges and Limitations

### *6.5.1 Allocating Resources and Financial Restraints*

Financial and human resource availability are one of the biggest wetland rehabilitation constraints. Restoration efforts in wetlands are usually costly, with high costs of hydrological engineering, native vegetation planting, land purchase, and ongoing monitoring [55]. Rehabilitation works are normally limited in extent and scope by a lack of finance. Due to the absence of pressing socioeconomic issues like poverty alleviation and food security, wetlands fall behind in developing nations.

Even in developed nations, funding for wetland restoration has to vie with other environmental goals such as avoiding pollution and reforestation. Failing funds can result in poorly thought-out or half-baked restoration plans, with less-than-optimal outcomes. For instance, under-funding can cause projects to disregard fundamental hydrological issues, rendering wetlands unable to be a good support system for biodiversity. Traditional funding mechanisms such as carbon offsetting schemes, public-private partnerships, and ecotourism schemes can be used in addressing budget shortages. Conservation finance and global grants can also be of great significance in large-scale wetland restoration programs.

**Table 6.1** Various wetland restoration case studies, highlighting their ecological significance and rehabilitation strategies

| Sr. No. | Case studies | Rehabilitation | Effects on birds | Important takeaways | References |
|---|---|---|---|---|---|
| 1. | **Restoring North American Prairie Potholes:** Known as the "duck factory" of North America, the Prairie Pothole Region (PPR) is dispersed across both the United States and Canada. Migratory waterfowl use the thousands of small, water-filled depressions in this area as breeding grounds. However, almost half of these wetlands disappeared due to extensive agricultural drainage during the twentieth century, which negatively impacted bird populations. | Large-scale wetland restoration initiatives were initiated in the 1980s to try to reverse this decline. Important strategies included re-flooding former wetlands, replacing drainage tiles, and re-establishing native flora. Municipalities, conservation groups, and local landowners had to work together to implement these programs. Financial incentives and conservation easements were used to entice farmers to participate in restoration projects | The outcomes were astounding. Waterfowl nesting success increased by 70% over 20 years in the restored wetlands. There were population recoveries for species including the Blue-winged Teal, Mallard, and Northern Pintail. Rare species that had all but disappeared from the area, such as Wilson's Phalarope and the Yellow Rail, also started to resurface | The Prairie Potholes story emphasizes how vital it is to engage local people and restore hydrological systems. Offering farmers financial incentives can be a very effective way to strike a balance between agricultural productivity and conservation | [2, 14, 49, 68] |
| 2. | **Restoration of Wetlands in India's Western Ghats:** Many seasonal wetlands of the Western Ghats, a UNESCO World Heritage Site, are home to migratory and unique bird species. But more and more of these wetlands are being used for farming, especially the production of rice. Bird populations have significantly declined as a result of habitat deterioration made worse by the overuse of chemical pesticides and fertilizers. | In the early 2000s, a community-led project was launched to restore wetlands in the Western Ghats. Reintroducing native plant species, establishing buffer zones around wetlands, and switching to organic farming methods were important initiatives. Farmers were largely educated by local NGOs about the ecological significance of wetlands and the advantages of sustainable agriculture. Participation was also encouraged by financial assistance from government initiatives. | The diversity of bird species rose by 40% in a span of 5 years. Improved habitat quality is demonstrated by the return of species like the Painted Stork, Indian Spot-billed Duck, and Lesser Whistling Duck to these wetlands. Along with helping birds recover, the restored wetlands also boosted local economies by attracting eco-tourists and increasing agricultural output. | This case study emphasizes how crucial community participation is to the preservation of wetlands. Financial incentives and education can promote sustainable behaviors that enhance local livelihoods and biodiversity. | [10] |

| | | | | | |
|---|---|---|---|---|---|
| 3. | **Restoration Project in the Everglades, USA:** Many people consider Florida's Everglades, a sizable subtropical wetland, to be among the country's most biologically significant ecosystems. A wide variety of species, including sizable colonies of wading birds, have historically been found in the Everglades. But in the twentieth century, widespread canalization, agricultural expansion, and urbanization seriously harmed the environment, which resulted in sharp drops in bird numbers. | One of the biggest wetland restoration projects worldwide is the Comprehensive Everglades Restoration Plan (CERP), which was introduced in 2000. The project's objectives are to eliminate invasive plant species, restore natural water flow, and reestablish vital wildlife habitats. A lot of work was also done to lessen the nitrogen runoff from agricultural areas, which had led to habitat deterioration and algae blooms. | Populations of wading birds, including the Great Egret, Wood Stork, and Roseate Spoonbill, recovered significantly. Over the course of two decades, nesting success increased by 50%, and species that had been falling started to stabilize. Other fauna, such as fish, amphibians, and reptiles, also benefited from the restoration efforts, which increased the resilience of the environment. | In extensive restoration projects, the Everglades project highlights the value of sustained investment and flexible management. For the initiative to be successful, cooperation between federal and state agencies, conservation organizations, and local stakeholders was crucial. | [41, 73] |
| 4. | **European Peatlands Rewetting:**In Germany and the UK, peatlands have long been drained for farming, which has resulted in the loss of vital habitat for birds that depend on wetlands. Furthermore, peatland drainage exacerbates climate change by making a substantial contribution to carbon emissions. | Rewetting programs aimed to restore water levels by obstructing drainage systems and establishing financial incentives for farmers to switch to sustainable farming methods. In certain places, peatlands were turned into nature reserves, giving animals secure places to live. In order to finance restoration initiatives and encourage climate-friendly land management, carbon-offset systems were implemented. | Bird species richness increased by 30% in just 7 years after peatlands were rewetted. Black-tailed Godwit, Eurasian Curlew, and Common Crane were among the notable species recovered. In addition to preserving species, the restored peatlands started acting as carbon sinks, which helped regulate the climate. | This case study demonstrates how wetland restoration can be used to address several environmental issues at once. Peatland restoration can provide long-term ecological and socioeconomic advantages by combining conservation and climate mitigation techniques. | [2, 35, 44, 68] |
| 5. | **Chilika Lake,** the largest coastal lagoon in India and a designated | Conservation efforts in Chilika Lake have focused on restoring | Anthropogenic activities such as fishing, tourism, and habitat | Rehabilitation efforts, including habitat | [51, 69] |

(continued)

**Table 6.1** (continued)

| Sr. No. | Case studies | Rehabilitation | Effects on birds | Important takeaways | References |
|---|---|---|---|---|---|
| | Ramsar site, is a critical habitat for migratory and resident bird species. Several studies have employed bioacoustic monitoring to assess avian diversity, seasonal variations, and the impact of environmental changes. Research has highlighted the presence of over 230 bird species, with bioacoustic methods aiding in the identification of elusive and nocturnal species that are difficult to survey through traditional means. | wetland ecosystems, improving water quality, and controlling invasive species. The removal of the invasive weed *Eichhornia crassipes* and dredging activities have led to improved habitat conditions, benefiting bird populations. Bioacoustic monitoring has been instrumental in evaluating the success of these rehabilitation measures by tracking changes in bird vocalizations and species diversity. | degradation have influenced bird populations in Chilika Lake. Studies have shown that excessive human interference leads to behavioral changes, including altered vocalization patterns and shifts in habitat usage. However, conservation interventions, such as the establishment of no-fishing zones and regulated ecotourism, have contributed to stabilizing bird populations, as reflected in bioacoustic data. | restoration and invasive species control, positively impact bird populations. Anthropogenic activities influence bird behavior, but conservation strategies can mitigate negative effects. | |

### 6.5.2 Governance Concerns and Policy Gaps

There must be a strong policy and legal framework for effective wetland restoration to promote environmental integrity, manage land use, and regulate conservation activities [49]. The majority of regions lack mechanisms for the enforcement of policies or do not have inclusive wetland conservation policies. Meanwhile, out-of-date or inconsistent policies may not be able to address the new issues facing modern day like climate change and urbanization. Decentralized regulation may result in inefficiencies and conflicts with different institutions handling different aspects of wetlands. Conservation may sometimes be held up due to political resistance or campaigning by vested interests. While a historic global convention, the Ramsar Convention has limited effect on national policies because it lacks legally binding means for enforcement. Wetland degradation also continues to run out of control in some areas of Southeast Asia due to corruption and weak governance.

Weak governance and policy gaps produce uncertainty and hinder the application of effective rehabilitation efforts. They also discourage long-term investment towards the protection of wetlands.

There is also a necessity to revise and reorganize policies on wetlands according to contemporary-day ecological and socioeconomic conditions. Rehabilitation activities can also function better if supported by stakeholder collaboration, capacity development programs, and transparent decision-making procedures.

### 6.5.3 Divergent Land-Use Requirements

As wetlands typically occur in the most desirable places, they are exposed to competing land use demands [73]. Wetland conservation is typically exposed to habitat loss and degradation from industrialization, urbanization, and agricultural development. These involve wetland draining for agricultural purposes, for example, for rice, sugarcane, and palm oil farming, in various parts of the world. Urban growth and development like roads and airports often invade wetland habitats and irreversibly damage them. Policymakers and stakeholders are faced with a dilemma because of the inconsistency of land use needs. It also seems particularly challenging to reconcile economic development and environmental protection in regions where wetlands are seen more as “wastelands” rather than as highly valuable ecosystems. Wetlands need to be prioritized at the top as important ecological assets in land-use planning in order to remedy these issues. Agroforestry and the development of buffer zones are two instances of sustainable land use that can be promoted to mitigate the impact of human activities. Education and awareness campaigns, through highlighting the conservation of wetlands as having economic and environmental benefits, can shift the public’s attitude.

### 6.5.4 Environmental Uncertainty and Climate Change

Efforts to restore wetlands are significantly hampered by climate change. Wetland hydrology and ecology are directly impacted by rising temperatures, changed precipitation patterns, and sea level rise [22]. The biological equilibrium of wetlands that depend on seasonal water flows may be upset by protracted droughts or unexpected flooding. Mangroves and salt marshes are examples of coastal wetlands that are especially susceptible to storm surges and sea level rise. The effectiveness of restoration measures is affected by species distributional changes due to changing climatic conditions. Climate change caused due to global warming is subjecting the Sundarbans, the world's largest mangrove ecosystem, to escalating saline streams as well as heightened cyclones. African sessional wetlands like the Okavango Delta are growing more unpredictable, which is affecting local fauna and human populations as well. Climate change, through inducing uncertain and unfavorable environmental conditions, aggravates the problem of wetland rehabilitation. This makes project planning harder and increases the likelihood of project failure. It is possible that wetland restoration projects need to be made climate resilient. Restoration project success can be enhanced through using adaptive management methods, such as selecting climate-resilient vegetation species and using climate models to inform design.

### 6.5.5 Insufficient Scientific Information and Tracking

Concepts of the hydrology, biodiversity, and biota of an ecosystem need to be thorough in order to effectively restore wetlands. Insufficient monitoring and absence of baseline data afflict most schemes [59]. The absence of investigation on specific wetland habitats prevents effective restoration.

Once the first phase of project execution, monitoring programs either lack adequate finances or are abandoned. The majority of tropical wetlands are poorly understood and therefore poorly informed restoration programs exist. In the absence of continuous observation, projects struggle to measure how successful their work is and adjust accordingly. The project is more likely to fail and effective measures are less likely to be applied elsewhere when there is a lack of scientific data and monitoring. Funding must be provided for ecological research and long-term monitoring projects. Collectively, government agencies, conservation groups, and educational institutions can develop the data that will inform and evaluate restoration work. Finally, wetland restoration is a laborious and complex procedure involving careful consideration of economic, social, and ecological factors. Matters raised in this chapter from loopholes in the law and a lack of funds to alternative uses of land and international climate change highlight the need for adaptive and collaborative approaches to wetland conservation. Parties need to work together to: Get adequate funds through innovative means such as carbon credits and ecotourism. Realign policies and governance frameworks towards wetland preservation. Effective land-use planning can balance ecological sustainability and economic advancement. To address environmental uncertainties, i.e., incorporating climate resilience within restoration processes. To guide and assist the rehabilitation efforts, invest in scientific research and integrated long-term monitoring.

By addressing these challenges, we can ensure the long-term health and well-being of these vital ecosystems and improve the effectiveness of wetland restoration programs. Wetlands are essential to human health and global sustainability, and they are not simply wildlife habitat.

## 6.6 Recommendations for Future Research and Conservation Efforts

For biodiversity preservation and to make bird populations sustainable wetland rehabilitation strategies in agricultural environments have a very important role. In an effort to make such programs more effective, a holistic system involving ecological monitoring, policy guidelines and community participation is needed.

### *6.6.1 Integrating Ecological Monitoring*

A Strong ecological monitoring program is essential to evaluate long-term restoration success of wetland projects. There should be future studies aimed at standardizing monitoring techniques within various wetland types for inter-comparison purposes [23], employing the use of remote sensing and geographic information systems (GIS) to monitor the change in habitat and its effects on the abundance of birds [56], applying bioacoustic monitoring and citizen science platforms to support data gathering on bird abundance [65], performing species-specific reactions to restoration of the habitat in an attempt to apply conservation methods accordingly [74].

### *6.6.2 Strengthening Policy Frameworks*

Policy measures are needed to protect restored wetlands and integrate them into agricultural landscapes. Future action would spur greater environmental policy that promotes wetland restoration and environmentally friendly agriculture [86], promote sector coordination among agricultural stakeholders, conservators, and policymakers to reconcile land use needs [52], construct adaptive management practices that contain ecological feedback mechanisms to improve wetland restoration activities throughout the years [63], increase protection for restored wetlands through the law to avoid future degradation and land conversion [18].

### 6.6.3 Enhancing Community Engagement

Local people have a great influence on the success of wetland restoration schemes. There needs to be further research into the contribution of traditional ecological knowledge (TEK) towards wetland conservation and how it could be incorporated into contemporary rehabilitation strategies [8], economic disincentives to farmers in adopting wetland-friendly agricultural practices such as payment for ecosystem services (PES) [20], educational campaigns emphasizing the environmental and economic advantages of wetland conservation [19], community-based participatory conservation programs that incorporate local communities in decision-making and habitat management [75].

### 6.6.4 Expanding Research on Wetland Functionality and Bird Ecology

More research needs to be performed in order to better understand the ecological processes of restored wetlands. There must be research in studying the implications of climate change on wetland hydrology and its concomitant influence on bird numbers [50], evaluating the contribution of invertebrate abundance and diversity to bird population maintenance in restored wetlands [7], studying the dependence of migratory birds on restored wetlands and their contribution to flyway conservation [66], exploring the influence of agricultural pollutants on wetland integrity and reproductive success of birds [67].

Restoration activities in wetlands can be greatly enhanced through the incorporation of ecological monitoring, policy reinforcement, and community participation. Future studies must take on interdisciplinary strategies that interlink ecological science, policy, and social outreach to promote long-term sustainability of these precious ecosystems.

The restoration of wetlands in agricultural landscapes is a key driver towards improving avian diversity and ecosystem process. In the course of this chapter, it has been proven that restored wetlands highly enhance habitat quality, species number, and are one of the ecologically important services essential for sustainable agriculture. By restoring wetlands, valuable ecosystem services, including enhanced water holding capacity, improved soil quality, and increased supply of food, are regained and provide healthier bird populations. Quantification of abundance and richness of birds in restored wetlands involves both conventional and sophisticated monitoring techniques, such as field surveys, remote sensing, and bioacoustic monitoring.

Integration of classical and high-technology methods presents a possibility to conduct exhaustive and precise assessment of bird community responses to wetland restoration. Effective case studies, including the Prairie Pothole Region of North America, the Danube Delta of Europe, and Chilika Lake in India, demonstrate the potential for reversing habitat degradation and enhancing avian conservation by properly planned rehabilitation schemes. Against this success, however, there are ongoing issues, such as misallocation of resources, policy inconsistency, and competition between agricultural development and wetland protection. Overcoming these challenges calls for a multi-dimensional strategy through ecological surveillance, enhancing policy frameworks, and provoking proactive community engagement. Ongoing ecological monitoring to measure restoration sustainability needs to be given priority in ensuing research, while the potential of wetlands to improve climate resilience needs to be investigated. Wetland rehabilitation is highly crucial for preserving ecological stability in agricultural landscapes and avian biodiversity conservation. By embracing strategic conservation strategies, harmonizing agricultural productivity and ecosystem health, and ensuring stakeholder cooperation, wetland restoration can become a gold standard of sustainable land use. A science-based integrated strategy will prove instrumental in delivering the long-term ecological and socio-economic gains of wetlands and, ultimately, advancing biodiversity conservation and agricultural sustainability.

## 6.7 Conclusion

Restoration of agricultural wetlands is necessary to increase bird abundance, diversity, and ecological resilience in agricultural landscapes. With the improvement in habitat quality, enhanced water holding capacity, and improved food supply, restored wetlands sustain bio-diverse rich avian assemblages of migratory and aquatic birds. Case studies of Chilika Lake, the Everglades, and European floodplains confirm the efficacy of wetland restoration in enhancing bird diversity and ecosystem health. All the same, challenges like land-use conflict, policy loopholes, and global climate change need to be addressed through integrated conservation approaches, support at the policy level, and stakeholder participation. Restoration of wetlands, aside from bird conservation, enhances sustainable agriculture through improved water management and soil quality. Long-term ecological observation, adaptive management, and inter-sectoral cooperation need to be the priority of future interventions for the sustainability and durability of wetland restoration efforts.

## References

1. Alikhani S, Nummi P, Ojala A (2021) Urban wetlands: a review on ecological and cultural values. Water 13(22):3301. https://doi.org/10.3390/w13223301

2. Angelini C, van der Heide T, Griffin JN, Morton JP, Derksen-Hooijberg M, Lamers LP et al (2015) Foundation species' overlap enhances biodiversity and multifunctionality from the patch to landscape scale in southeastern United States salt marshes. Proc R Soc B Biol Sci 282(1811): 20150421. https://doi.org/10.1098/rspb.2015.0421
3. Ansari AA, Khan FA, Gill SS, Varshney J (2011) Aquatic plant diversity in eutrophic ecosystems. In: Eutrophication: causes, consequences and control, pp 247–263. https://doi.org/10.1007/978-90-481-9625-8_12
4. Aplin P (2005) Remote sensing: ecology. Prog Phys Geogr 29(1):104–113. https://doi.org/10.1191/030913305pp437p
5. Baldwin AH, Batzer DP (2012) Wetland habitats of North America. In: Wetland habitats of North America: ecology and conservation concerns, pp 1–9. https://doi.org/10.1525/9780520951419-003
6. Ballut-Dajud GA, Sandoval Herazo LC, Fernández-Lambert G, Marín-Muñiz JL, López Méndez MC, Betanzo-Torres EA (2022) Factors affecting wetland loss: a review. Land 11(3):434. https://doi.org/10.3390/land11030434
7. Batzer D, Boix D (2016) An introduction to freshwater wetlands and their invertebrates. In: Invertebrates in freshwater wetlands: an international perspective on their ecology, pp 1–23. https://doi.org/10.1007/978-3-319-24978-0_1
8. Berkes F, Colding J, Folke C (2000) Rediscovery of traditional ecological knowledge as adaptive management. Ecol Appl 10(5):1251–1262. https://doi.org/10.1890/1051-0761(2000)010[1251:ROTEKA]2.0.CO;2
9. Bibby CJ (2000) Bird census techniques. Elsevier
10. Bo LUAN, Yue LIU, Di CHE, Wenjun ZHOU, Liuliu HU, Yuan LIN (2024) Exploring nature-based solutions on refined Waterbird habitats restoration in high-density urban area: a case study of the Futian mangrove National Important Wetland in Shenzhen, China. Landscap Archit Front 12(3):36
11. Buckland ST (2006) Point-transect surveys for songbirds: robust methodologies. Auk 123(2): 345–357. https://doi.org/10.1093/auk/123.2.345
12. Butt MA, Zafar M, Ahmed M, Shaheen S, Sultana S, Butt MA et al (2021) Threats to biodiversity of wetland. In: Wetland plants: a source of nutrition and ethno-medicines, pp 117–133. https://doi.org/10.1007/978-3-030-69258-2_7
13. Buxton RT, Jones IL (2012) Measuring nocturnal seabird activity and status using acoustic recording devices: applications for Island restoration. J Field Ornithol 83(1):47–60. https://doi.org/10.1111/j.1557-9263.2011.00355.x
14. Byrnes JE, Gamfeldt L, Isbell F, Lefcheck JS, Griffin JN, Hector A et al (2014) Investigating the relationship between biodiversity and ecosystem multifunctionality: challenges and solutions. Methods Ecol Evol 5(2):111–124. https://doi.org/10.1111/2041-210X.12143
15. Casazza ML, McDuie F, Jones S, Lorenz AA, Overton CT, Yee J, Feldheim CL, Ackerman JT, Thorne KM (2021) Waterfowl use of wetland habitats informs wetland restoration designs for multi-species benefits. J Appl Ecol 58(9):1910–1920. https://doi.org/10.1111/1365-2664.13845
16. Celis-Murillo A, Stodola KW, Pappadopoli B, Burton JM, Ward MP (2016) Seasonal and daily patterns of nocturnal singing in the Field Sparrow (Spizella pusilla). J Ornithol 157:853–860. https://doi.org/10.1007/s10336-015-1318-y
17. Chakraborty SK, Sanyal P, Ray R (2023) Biodiversity and its functional significance: case studies from East Kolkata Wetlands. In: Wetlands ecology: eco-biological uniqueness of a Ramsar site (East Kolkata Wetlands, India). Springer International Publishing, Cham, pp 379–520. https://doi.org/10.1007/978-3-031-09253-4_7
18. Davidson NC (2014) How much wetland has the world lost? Long-term and recent trends in global wetland area. Mar Freshw Res 65(10):934–941. https://doi.org/10.1071/MF14173
19. Dudgeon D, Arthington AH, Gessner MO, Kawabata ZI, Knowler DJ, Lévêque C et al (2006) Freshwater biodiversity: importance, threats, status and conservation challenges. Biol Rev 81(2):163–182. https://doi.org/10.1017/S1464793105006950

20. Engel S, Pagiola S, Wunder S (2008) Designing payments for environmental services in theory and practice: an overview of the issues. Ecol Econ 65(4):663–674. https://doi.org/10.1016/j.ecolecon.2008.03.011
21. Erwin KL (2009) Wetlands and global climate change: the role of wetland restoration in a changing world. Wetl Ecol Manag 17(1):71–84. https://doi.org/10.1007/s11273-008-9119-1
22. Estes JA, Terborgh J, Brashares JS, Power ME, Berger J, Bond WJ, Carpenter SR, Essington TE, Holt RD, Jackson JB, Wardle DA (2011) Trophic downgrading of planet earth. Science 333(6040):301–306. https://doi.org/10.1126/science.1205106
23. Fairbairn SE, Dinsmore JJ (2001) Local and landscape-level influences on wetland bird communities of the prairie pothole region of Iowa, USA. Wetlands 21(1):41–47. https://doi.org/10.1672/0277-5212(2001)021[0041:LALLIO]2.0.CO;2
24. Farheen KS, Reyes NJDG, Jeon MS, Kim LH (2022) The Status of Ramsar wetlands in India: a review of ecosystem benefits, threats, and management strategies. J Wetlands Res 24(2): 123–141. https://doi.org/10.17663/JWR.2022.24.2.123
25. Farnsworth A (2005) Flight calls and their value for future ornithological studies and conservation research. Auk 122(3):733–746. https://doi.org/10.1093/auk/122.3.733
26. Farnsworth A, Lunt DJ, O'Brien CL, Foster GL, Inglis GN, Markwick P, Pancost RD, Robinson SA (2019) Climate sensitivity on geological timescales controlled by nonlinear feedbacks and ocean circulation. Geophys Res Lett 46(16):9880–9889. https://doi.org/10.1029/2019GL083574
27. Frederick PC, Ogden JC (2003) Monitoring wetland ecosystems using avian populations: seventy years of surveys in the Everglades. In: Monitoring ecosystems: interdisciplinary approaches for evaluating ecoregional initiatives. Island Press, Washington, DC, pp 321–350
28. Frommolt KH, Bardeli R, Clausen M (2008) Computational bioacoustics for assessing biodiversity. In: Proceedings of the International Expert meeting on IT-based detection of bioacoustical patterns, BfN-Skripten, vol 234
29. Galatowitsch SM, van der Valk AG (1996) Characteristics of recently restored wetlands in the prairie pothole region. Wetlands 16:75–83. https://doi.org/10.1007/BF03160647
30. Gaunt SL, Mccallum DA (2004) Birdsong and conservation. In: Nature's music. Academic, pp 343–362
31. Gilbert G, McGregor PK, Tyler G (1994) Vocal individuality as a census tool: practical considerations illustrated by a study of two rare species. J Field Ornithol:335–348. https://www.jstor.org/stable/4513949
32. Gilbert JL, Lansing CR, Garnsey SM (2012) Seeing facial motion affects auditory processing in noise. Atten Percept Psychophys 74:1761–1781. https://doi.org/10.3758/s13414-012-0375-z
33. Gregory JM, Ingram WJ, Palmer MA, Jones GS, Stott PA, Thorpe RB, Lowe JA, Johns TC, Williams KD (2004) A new method for diagnosing radiative forcing and climate sensitivity. Geophys Res Lett 31(3). https://doi.org/10.1029/2003GL018747
34. Gregory RD, van Strien A (2010) Wild bird indicators: using composite population trends of birds as measures of environmental health. Ornithol Sci 9(1):3–22. https://doi.org/10.2326/osj.9.3
35. Gyurácz J, Bánhidi P, Csuka A (2011) Successful restoration of water level and surface area restored migrant bird populations in a Hungarian wetland. Biologia 66(6):1177–1182. https://doi.org/10.2478/s11756-011-0132-0
36. Haselmayer J, Quinn JS (2000) A comparison of point counts and sound recording as bird survey methods in Amazonian Southeast Peru. Condor 102(4):887–893. https://doi.org/10.1093/condor/102.4.887
37. Hobson KA, Rempel RS, Greenwood H, Turnbull B, Van Wilgenburg SL (2002) Acoustic surveys of birds using electronic recordings: new potential from an omnidirectional microphone system. Wildl Soc Bull:709–720. http://www.jstor.org/stable/3784223
38. Hüppop O, Hilgerloh G (2012) Flight call rates of migrating thrushes: effects of wind conditions, humidity and time of day at an illuminated offshore platform. J Avian Biol 43(1):85–90. https://doi.org/10.1111/j.1600-048X.2011.05443.x

39. Hutto RL, Stutzman RJ (2009) Humans versus autonomous recording units: a comparison of point-count results. J Field Ornithol 80(4):387–398. https://doi.org/10.1111/j.1557-9263.2009.00245.x
40. Jiang H, He C, Luo W, Yang H, Sheng L, Bian H, Zou C (2018) Hydrological restoration and water resource management of Siberian Crane (Grus leucogeranus) stopover wetlands. Water 10(12):1714. https://doi.org/10.3390/w10121714
41. Jones HP, Jones PC, Barbier EB, Blackburn RC, Rey Benayas JM, Holl KD, McCrackin M, Meli P, Montoya D, Mateos DM (2018) Restoration and repair of Earth's damaged ecosystems. Proc R Soc B Biol Sci 285(1873):20172577. https://doi.org/10.1098/rspb.2017.2577
42. Kiran SD, Kour A, Priya DV, Kumar R (2022) Community composition and status of avian diversity at Campus and Agricultural landscapes of Chaudhary Charan Singh Haryana Agricultural University, Hisar (Haryana). J Appl Nat Sci 14(4):1130–1140. https://doi.org/10.31018/jans.v14i4.3784
43. Lei Q, Li J, Ma K (2018) Applications of remote sensing technology in avian ecology. Biodivers Sci 26(8):862–877. https://doi.org/10.17520/biods.2018143
44. Li C, Chen J, Liao X, Ramus AP, Angelini C, Liu L, Silliman BR, Bertness MD, He Q (2023) Shorebirds-driven trophic cascade helps restore coastal wetland multifunctionality. Nat Commun 14(1):8076. https://doi.org/10.1038/s41467-023-43951-3
45. Linscott JA, Senner NR (2021) Beyond refueling: investigating the diversity of functions of migratory stopover events. Condor 123(1):duaa074. https://doi.org/10.1093/ornithapp/duaa074
46. Ma Z, Cai Y, Li B, Chen J (2010) Managing wetland habitats for waterbirds: an international perspective. Wetlands 30:15–27. https://doi.org/10.1007/s13157-009-0001-6
47. Magurran AE, Baillie SR, Buckland ST, Dick JM, Elston DA, Scott EM, Smith RI, Somerfield PJ, Watt AD (2010) Long-term datasets in biodiversity research and monitoring: assessing change in ecological communities through time. Trends Ecol Evol 25(10):574–582. https://doi.org/10.1016/j.tree.2010.06.016
48. Mandal KG, Banerjee KOUSHIK, Purbey SK, Kumar RAVI (2024) Potential measures to enhance ecosystem services of flood-prone and wetland agricultural systems. J Agric Phys 24: S108–S122
49. Manning P, Van Der Plas F, Soliveres S, Allan E, Maestre FT, Mace G, Whittingham MJ, Fischer M (2018) Redefining ecosystem multifunctionality. Nat Ecol Evol 2(3):427–436. https://doi.org/10.1038/s41559-017-0461-7
50. McIntyre NE, Liu G, Gorzo J, Wright CK, Guntenspergen GR, Schwartz F (2019) Simulating the effects of climate variability on waterbodies and wetland-dependent birds in the Prairie Pothole Region. Ecosphere 10(4):e02711. https://doi.org/10.1002/ecs2.2711
51. Mehta P, Chandravanshi S, Yadav SR, Singh A, Jatav SK (2023) Conservation and management of Ramsar Sites in India: challenges and solutions
52. Meli P, Rey Benayas JM, Balvanera P, Martínez Ramos M (2014) Restoration enhances wetland biodiversity and ecosystem service supply, but results are context-dependent: a meta-analysis. PLoS One 9(4):e93507. https://doi.org/10.1371/journal.pone.0093507
53. Mennill DJ, Burt JM, Fristrup KM, Vehrencamp SL (2006) Accuracy of an acoustic location system for monitoring the position of duetting songbirds in tropical forest. J Acoust Soc Am 119(5):2832–2839. https://doi.org/10.1121/1.2184988
54. Mennill DJ, Doucet SM, Ward KAA, Maynard DF, Otis B, Burt JM (2012) A novel digital telemetry system for tracking wild animals: a field test for studying mate choice in a lekking tropical bird. Methods Ecol Evol 3(4):663–672. https://doi.org/10.1111/j.2041-210X.2012.00206.x
55. Moreno-Mateos D, Barbier EB, Jones PC, Jones HP, Aronson J, López-López JA, McCracking ML, Meli P, Montoya D, Rey Benayas JM (2017) Anthropogenic ecosystem disturbance and the recovery debt. Nat Commun 8(1):14163. https://doi.org/10.1038/ncomms14163
56. Murray NJ, Phinn SR, DeWitt M, Ferrari R, Johnston R, Lyons MB et al (2019) The global distribution and trajectory of tidal flats. Nature 565(7738):222–225. https://doi.org/10.1038/s41586-018-0805-8

57. Nayak A, Bhushan B (2022) Wetland ecosystems and their relevance to the environment: importance of wetlands. In: Handbook of research on monitoring and evaluating the ecological health of wetlands. IGI Global, pp 1–16. https://doi.org/10.4018/978-1-7998-9498-8.ch001
58. Orsholm J, Elenius M (2022) Effects of hydrology on wetlandbiodiversity: a literature study and development of hydrological indicators
59. Palmer MA, Ambrose RF, Poff NL (1997) Ecological theory and community restoration ecology. Restor Ecol 5(4):291–300. https://doi.org/10.1046/j.1526-100X.1997.00543.x
60. Pérez-Granados C, Traba J (2021) Estimating bird density using passive acoustic monitoring: a review of methods and suggestions for further research. Ibis 163(3):765–783. https://doi.org/10.1111/ibi.12944
61. Qiu J, Zhang Y, Ma J (2024) Wetland habitats supporting waterbird diversity: conservation perspective on biodiversity-ecosystem functioning relationship. J Environ Manag 357:120663. https://doi.org/10.1016/j.jenvman.2024.120663
62. Rempel RS, Francis CM, Robinson JN, Campbell M (2013) Comparison of audio recording system performance for detecting and monitoring songbirds. J Field Ornithol 84(1):86–97. https://doi.org/10.1111/jofo.12008
63. Reynolds JF, Smith DMS, Lambin EF, Turner BL, Mortimore M, Batterbury SP et al (2007) Global desertification: building a science for dryland development. Science 316(5826): 847–851. https://doi.org/10.1126/science.1131634
64. Rodrigo MA (2021) Wetland restoration with hydrophytes: a review. Plants 10(6):1035. https://doi.org/10.3390/plants10061035
65. Rosenberg KV, Dokter AM, Blancher PJ, Sauer JR, Smith AC, Smith PA et al (2019) Decline of the North American avifauna. Science 366(6461):120–124. https://doi.org/10.1126/science.aaw1313
66. Runge CA, Watson JE, Butchart SH, Hanson JO, Possingham HP, Fuller RA (2015) Protected areas and global conservation of migratory birds. Science 350(6265):1255–1258. https://doi.org/10.1126/science.aac9180
67. Schuyler Q, Hardesty BD, Wilcox C, Townsend K (2014) Global analysis of anthropogenic debris ingestion by sea turtles. Conserv Biol 28(1):129–139. https://doi.org/10.1111/cobi.12126
68. Shackelford N, Paterno GB, Winkler DE, Erickson TE, Leger EA, Svejcar LN, Breed MF, Faist AM, Harrison PA, Curran MF, Suding KL (2021) Drivers of seedling establishment success in dryland restoration efforts. Nat Ecol Evol 5(9):1283–1290. https://doi.org/10.1038/s41559-021-01510-3
69. Shan V, Singh SK, Haritash AK (2021) Present status, conservation, and management of wetlands in India. In: Advances in energy and environment: select proceedings of TRACE 2020. Springer, Singapore, pp 235–256
70. Sharma LK, Naik R (2024) Saline wetland ecosystems. In: Conservation of Saline Wetland ecosystems: an initiative towards UN decade of ecological restoration. Springer Nature Singapore, Singapore, pp 33–75. https://doi.org/10.1007/978-981-97-5069-6_2
71. Soliveres S, Van Der Plas F, Manning P, Prati D, Gossner MM, Renner SC, Alt F, Arndt H, Baumgartner V, Binkenstein J, Allan E (2016) Biodiversity at multiple trophic levels is needed for ecosystem multifunctionality. Nature 536(7617):456–459. https://doi.org/10.1038/nature19092
72. Sonntag E (2021) Overcoming adversity: promise of recovery of native plants in a temperate wetland community following removal of yellow flag Iris (*Iris pseudacorus L.*) using benthic barriers (Master's thesis, Royal Roads University (Canada))
73. Strassburg BB, Iribarrem A, Beyer HL, Cordeiro CL, Crouzeilles R, Jakovac CC, Braga Junqueira A, Lacerda E, Latawiec AE, Balmford A, Visconti P (2020) Global priority areas for ecosystem restoration. Nature 586(7831):724–729. https://doi.org/10.1038/s41586-020-2784-9

74. Studds CE, Kendall BE, Murray NJ, Wilson HB, Rogers DI, Clemens RS et al (2017) Rapid population decline in migratory shorebirds relying on Yellow Sea tidal mudflats as stopover sites. Nat Commun 8(1):14895. https://doi.org/10.1038/ncomms14895
75. Sutherland WJ, Atkinson PW, Butchart SH, Capaja M, Dicks LV, Fleishman E, Gaston KJ, Hails RS, Hughes AC, Le Anstey B, Thornton A (2022) A horizon scan of global biological conservation issues for 2022. Trends Ecol Evol 37(1):95–104. https://doi.org/10.1016/j.tree.2021.10.014
76. Tang Z, Wood J, Smith D, Thapa A, Aryal N (2021) A review on constructed treatment wetlands for removal of pollutants in the agricultural runoff. Sustainability 13(24):13578. https://doi.org/10.3390/su132413578
77. Trettin CC, Kolka RK, Marsh AS, Bansal S, Lilleskov EA, Megonigal P et al (2020) Wetland and hydric soils. In: Forest and rangeland soils of the United States under changing conditions: a comprehensive science synthesis, pp 99–126
78. Twomey AJ, Nunez K, Carr JA, Crooks S, Friess DA, Glamore W, Orr M, Reef R, Rogers K, Waltham NJ, Lovelock CE (2024) Planning hydrological restoration of coastal wetlands: key model considerations and solutions. Sci Total Environ 169881. https://doi.org/10.1016/j.scitotenv.2024.169881
79. Urfi AJ, Sen M, Kalam A, Meganathan T (2005) Counting birds in India: methodologies and trends. Curr Sci:1997–2003. http://www.jstor.org/stable/24111060
80. van Dam AA, Robertson HA, Prieler R, Dubey A, Finlayson CM (2024) Recognizing diversity in wetlands and farming systems to support sustainable agriculture and conserve wetlands. https://doi.org/10.31223/X5HT23
81. Venier LA, Holmes SB, Holborn GW, Mcilwrick KA, Brown G (2012) Evaluation of an automated recording device for monitoring forest birds. Wildl Soc Bull 36(1):30–39. https://doi.org/10.1002/wsb.88
82. Vera-Herrera L, Romo S, Soria J (2022) How agriculture, connectivity and water management can affect water quality of a Mediterranean coastal wetland. Agronomy 12(2):486. https://doi.org/10.3390/agronomy12020486
83. Wahlberg M, Tougaard J, Møhl B (2003) Localising bitterns Botaurus stellaris with an array of non-linked microphones. Bioacoustics 13(3):233–245. https://doi.org/10.1080/09524622.2003.9753500
84. Wimmer JD (2015) Acoustic sensing: roles and applications in monitoring avian biodiversity (Doctoral dissertation, Queensland University of Technology)
85. Yodha K, Singh D, Kour A, Yadav R, Kumar R (2024) Species diversity and abundance of birds in ber (*Zizyphus mauritiana*) crop. Asian J Conserv Biol 13(1). https://doi.org/10.53562/ajcb.83683
86. Zedler JB (2003) Wetlands at your service: reducing impacts of agriculture at the watershed scale. Front Ecol Environ 1(2):65–72. https://doi.org/10.1890/1540-9295(2003)001[0065:WAYSRI]2.0.CO;2
87. Zedler JB, Kercher S (2005) Wetland resources: status, trends, ecosystem services, and restorability. Annu Rev Environ Resour 30(1):39–74. https://doi.org/10.1146/annurev.energy.30.050504.144248

# Chapter 7
# Wetland Restoration and Management Practices to Sustain Aquatic Bird Populations

**Deepak Vats, Geeta Rani, Nikita Sharma, Ravikant Verma, and Mayur Mukut Murlidhar Sharma**

**Abstract** 'Wetland specialists' and 'Generalist' aquatic birds rely on the wetlands for food, shelter, breeding or perching, completely or partially. Multinational governments are continually making efforts and signing treaties to save the deteriorating wetlands to sustain wetland-dependent birds. Major wetlands, including Ambuja Wetland in West Bengal and Perur-Sundakamuthur Lake in Coimbatore, are experiencing a decline in aquatic bird species because of wetland degradation. Ramsar Convention is an intergovernmental treaty that establishes various principles, management strategies and guidelines for wetlands. The Ministry of Environment, Forest and Climate Change (MoEFCC), in India, has initiated various campaigns and missions like the Amrit Dharohar Initiative and Mission Sahbhagita to engage local communities and authorities for the sustainable use of wetlands. There should be regional and national capacity-building programmes to strengthen wetland managers, people for their livelihood and other stakeholders. These programmes will help to save wetlands from degradation and thus help in protecting aquatic bird species.

**Keywords** Aquatic Birds · Management Practices · Ramsar Convention · Restoration · Wetland

D. Vats (✉) · G. Rani
Department of Biochemistry, AIIMS, New Delhi, India
e-mail: vatsdeepak887@aiims.edu

N. Sharma
Department of Zoology, Shivaji College, University of Delhi, New Delhi, India

R. Verma
Department of Zoology, Chaudhary Charan Singh Haryana Agricultural University, Hisar, Haryana, India

M. M. M. Sharma
Department of Surgery, University of Illinois, Chicago, IL, USA



R. Kumar et al. (eds.), *Ecology, Conservation of Aquatic Birds*, Wetlands: Ecology, Conservation and Management 15, https://doi.org/10.1007/978-3-032-17125-2_7

## 7.1 Introduction

Wetlands are the lands that hold water and become an ecosystem by providing food, habitat, and other ecological services to the life forms. Water is the primary factor in wetlands; however, soil, vegetation, and animal life are other factors that create favourable conditions for diverse species [13]. Wetlands are called 'kidneys of the landscape' because of their crucial role in hydrological and chemical cycles. 'Nature's Supermarket' is also another term that can be used for wetlands because of the services provided by wetlands to the inhabiting species [18]. Wetlands can be found on the edges of rivers, along vast coastlines, and inside dense forests. Wetlands sustain vast biodiversity, regulate climate and provide livelihood to the locals. Birds are classified as either "Wetland Specialists", which are entirely reliant on wetlands or "Generalists", which use wetlands opportunistically for food, shelter, breeding or perching [4].

The framework for the sustainable use of wetlands, their resources and conservation is provided by the Ramsar Convention. This is a multinational treaty adopted in 1971 in Ramsar, Iran. This intergovernmental initiative has almost 90% of United Nations (UN) members from all the world's geographic regions acceding to become 'Contracting Parties'. The 173 contracting parties, including the recent addition of the Kingdom of Saudi Arabia, have designated 2536 sites as Ramsar Sites, covering an area of 257,989,130 hectares. India, a major biogeographical realm, has designated 89 Ramsar Sites to strengthen wetlands management [23].

Avian diversity, including various species like the great blue heron, teals, coots, moorhens, and the wood duck, relies on wetlands for their survival. In a Haryana-based study that covered water pond sites in different villages, 115 bird species of 18 orders and 46 families were documented [9]. Another study in Koonthankulam village pond, Tirunelveli district, Tamil Nadu, by Selvam [24] recorded 90 avian species of 21 orders and 42 families. Bharathapuzha River Basin, the second largest west-flowing river in Kerala, Western Ghats, holds 262 bird species [20]. Sixty-nine wetland birds of 20 different families in the six rural ponds of District Kurukshetra, Haryana, were documented [14]. These studies underscore the importance of wetlands as a favourite habitat for birds. However, agricultural practices such as drainage and irrigation have degraded or destroyed wetland habitats, resulting in declining bird populations [12, 25]. So, effective wetland management strategies are critical to restore wetlands, which can help to sustain bird populations. These habitats can be recreated and made functional by applying practical strategies that integrate ecological principles and adaptive practices. This chapter will cover the wetland degradation factors, potential threats to aquatic birds, and the major principles and adaptive management practices to restore wetlands to sustain aquatic birds.

## 7.2 Wetland Degradation: Key Threats to Aquatic Birds

Fragmentation of the hydrological regime and habitat loss because of anthropogenic and non-anthropogenic human activities are the leading causes of the decline in the population of aquatic birds. Drainage and utilisation, particularly reclamation for aquaculture and agriculture, are the main factors contributing to wetland deterioration [15, 16, 26]. Wetlands are also being adversely affected by irrigation projects and dams. Various regions across the globe are expanding and intensifying agriculture to cope with the higher demand for food. Crop and grazing lands have increased from 27.2% to 46.5% of the world's total land area over the past 100 years [5]. But this expansion in the crop and grazing land has come at a significant ecological cost, particularly affecting wetlands' ecological health. According to the Review of Ramsar Information Sheet (RIS) data 2015, around 50% of wetlands of international importance have been negatively affected due to the expansion of agriculture-related practices.

Natural wetlands have declined by 35% during the period from 1970 to 2015 because of rapid land conversion, while human-made wetlands such as paddy fields and reservoirs have increased by 233% [3]. The primary cause of loss of natural wetlands is agricultural development, which leads to increased extraction and drainage of water, as well as increased usage of fertilisers and pesticides [7, 22]. documented the loss of 3.4 million $km^2$ of inland wetlands over the past three centuries, and 21% of global wetland area has been converted to croplands or human land uses between 1700 and 2020 AD. This Wetland loss expanded rapidly from the mid-twentieth century, and major loss was found in the United States, Europe, and China. Aquatic birds are more vulnerable to changes in climate when compared to other avian species [10]. Recent climate change is impacting both terrestrial and wetland ecosystems in intricate ways. Climate change causes fluctuation in water levels, which affects wetland salinity, thus regulating wetland communities and influencing waterbird habitat use. Waterbirds are subjected to greater water stress because of shifts toward more saline and fewer freshwater wetlands. These changes degrade vital continental migration corridors and crucial breeding grounds [8].

**Case Studies**

(i) **Ambuja Wetland, West Bengal, India** [1]

Rapid urbanisation is one of the major threat factors that disturbs the natural structure and function of wetland ecosystems, which has a detrimental impact on the population of aquatic birds. A study was carried out in a rapidly deteriorating wetland, the Ambuja Wetland, to assess the changes in avifaunal congregation pattern and their diversity. The Ambuja Wetland (23.54°N, 87.31°E) is located in the heavily populated and industrial city of Durgapur, West Bengal. This wetland shares the flora and fauna of both the Gangetic Plains and the Chotanagpur Plateau. It was observed that the wetland harboured 73 bird species from 34 families, including two summer migrants, 10 winter migrants, and the remaining were resident species. Out of the recorded 73 bird species, 20 species were waterbird species and 53 were non-waterbird species. A significant decrease in waterbird numbers was observed from 2013 to 2014 and from 2016 to 2017 ($p = 0.049$). The population of Lesser Whistling Duck (*Dendrocygna javanica*), a dominant species, declined significantly, from approximately 3000 individuals in 2013 to less than 500 by 2017. They found an inverse relationship between the vegetation and aquatic avifauna species in the wetland. Besides the declining *Dendrocygna javanica* population, three waterbird species -Cotton Teal (*Nettapus coromandelianus*), Purple Swamphen (*Porphyrio porphyrio*), and Pheasant-tailed Jacana (*Hydrophasianus chirurgus*)—completely disappeared from the Ambuja Wetland by the end of the study period. In addition, seven other waterbird species have declined, indicating the severe impact of habitat loss on local aquatic bird diversity. Despite the decline in waterbird population, habitat changes in the Ambuja wetland appeared to benefit some non-waterbird species as populations of these non-waterbirds—Dusky Warbler (*Phylloscopus fuscatus*), Rose-ringed Parakeet (*Psittacula krameri*), Yellow-eyed Babbler (*Chrysomma sinense*), and Ashy Prinia (*Prinia socialis*), increased significantly from 2013 to 2017. The bird species richness was highest in 2014–2015, but a gradual reduction was observed over the years because of rapid habitat degradation and illegal bird hunting. Satellite data showed a significant increase in vegetation cover from 47.6% in 2013 to 88.4% in 2017. And this increase in vegetation cover was positively correlated with a rise in non-waterbird-to-waterbird ratios. They observed that the lake was heavily polluted with household waste, thermoplastic and plastic waste, and other solid waste. Chemicals released from this waste are likely to release nutrients and thus promote excessive plant growth that reduces open water areas in the wetland. This study strongly emphasises the urgent need for a comprehensive management plan, including cessation of bird hunting, waste dumping, dredging of the lake parts for the restoration of deeper water areas, and macrophyte control, along with public awareness of restoration and sustainable management of the degraded wetland.

(ii) **Perur-Sundakamuthur Lake, Coimbatore, Tamil Nadu, India, May 2022** [21]

Urban development constantly threatens the habitats at Perur-Sundakamuthur Lake (Perur Lake), an urban wetland, as shown by a long-term (2014–2020) bird monitoring project, negatively impacting its bird communities. Perur Lake (10.97°N, 76.93°E) is located southwest of Coimbatore city, Tamil Nadu. In the study, they recorded 125 bird species in Perur Lake and categorised them into 49 resident species (e.g. *Anas poecilorhyncha*), 15 local migrants (e.g. *Pelecanus philippensis*), 21 winter migrants (e.g. *Tringa glareola*), 0 passage migrants and 40 uncertain. They documented the potential threats of wetland degradation and changes in water level and their impact on the diversity and population of aquatic bird species, both before and after the threat event. Before each count, the water level was visually estimated. The level of water in the wetland also affects the population of waterbirds. The highest population of 29 out of 49 recorded resident aquatic birds (e.g. dabbling ducks, storks, and shorebirds) was found at medium water level (water was present in the whole lake with exposed periphery shoreline), with about 80% of shorebirds. For the comparison of the trend in the decrease in biodiversity and population of birds in wetland with the national and global data they selected 12 representative aquatic birds—Indian Spot-billed Duck, Lesser Whistling Duck, Garganey, Little Grebe, Little Ringed Plover, Eurasian Coot, Common Sandpiper, Black-winged Stilt, Asian Openbill, Little Egret, Indian Pond Heron, Wood sandpiper. In the Perur Lake, the total number of all these 12 representative aquatic birds declined in a range of 43% (Common Sandpiper) to 100% (Garganey), but this trend was not reflected in national or global data. Previously, it has been recorded that local factors have a more severe effect than regional and global factors. Therefore, local human activities can be the primary factor responsible for the decrease in the population and diversity of aquatic birds, supporting the view that local factors have a more substantial impact on bird populations than regional or global factors [6]. Perur Lake was once considered a designated bird sanctuary because of its higher species richness of birds. Since it is an urban wetland, sand mining, road expansion, and commercial fishing are some causes of its deterioration, and thus decline its waterbird populations. According to [11], waterbirds typically avoid heavily disturbed areas and prefer to select roosting and foraging sites that have the least disruption and maximum accessibility to food.

Although there is a clear correlation between these recorded degrading factors and the decline in bird numbers, some other multiple factors beyond the study area have also contributed collectively. Human interference can be managed by creating buffer zones around the lake, which can increase natural purification and thus improve the quality of the wetland's ecosystem. However, failure to implement these measures promptly has led to the overall deterioration of the wetland ecosystem and reduced bird populations because of habitat loss. So, wetland degradation is a major threat to aquatic birds. The various anthropogenic factors, such as increased agricultural practices, urbanisation, sand mining, hunting, and water pollution, are deteriorating the wetlands and affecting the aquatic birds.

## 7.3 Principles of Wetland Restoration for Bird Conservation

Bird Conservation, specifically aquatic bird conservation, depends on wetland restoration. Wetland restoration lies in regaining degraded wetlands' functions related to the natural habitat and ecosystem for aquatic and migratory birds. It begins with understanding the basic causes of quality degradation, like changes in the wetland hydrology, water and air pollution, or invasion of alien species. A full recreation of the wetland may take time, so restoring the degraded wetland could prevent the harmful effects of degradation. According to the Ramsar Convention on Wetlands, the seven best practices in wetland restoration include restoring multiple benefits, developing a restoration plan, involving the community, addressing the causes of degradation, restoring native species, cleaning the degraded area and creating specific spaces for people to access the wetland i.e. structural access to the wetland [2]. To achieve this bigger goal of wetland restoration, a national programme should be formulated to address these seven practices. The programme's objectives should recognise that wetlands perform multiple functions; only then can the aim of bird conservation be achieved.

The well-defined standards linked to the programme's objectives should measure the wetland restoration achievement. The number of breeding pairs of avian species, other wildlife forms, and the percentage of area covered by specific plant species can be among the few measurable outcomes of these standards. Another key process to achieve this goal is to monitor the program regularly. The regular interval evaluation will help recognise the limitations and challenges so that necessary adjustments can be made.

**So, the key principles for wetland restoration for bird conservation are:**

1. Identifying and establishing aims and objectives to restore wetlands can attract local and migratory bird species.
2. Applying the catchment-level planning approach, which involves managing the land from which water flows into a river or any other water body, will help manage the water resources. This integrated approach will focus on the interaction between land use, human activities, and water flows, restoring wetland hydrology.
3. The local people, communities, and other stakeholders must be involved throughout the entire restoration process. Local authorities can run several programmes to foster community participation and support. This will help rebuild the native vegetation and habitat complexity for several bird species.
4. Incorporation of traditional and religious ecological knowledge in restoration efforts to gather support from the indigenous people.
5. Plan for a long time, and use adaptive management where modifications can be made periodically after monitoring the projects.
6. Practical constraints, technical limitations, and socioeconomic factors should be considered during project planning and implementation to ensure their feasibility.

## 7.4 Adaptive Management Strategies

Adaptive Management Strategies are a crucial part of the restoration programme. Management and monitoring help track progress and refine strategies so that wetlands remain functional and resilient and continuously provide ecosystem services to the avian species.

(i) **Revised National Plan for Conservation of Aquatic Resources guidelines [19]:**

India has a rich diversity of wetlands and is committed to wetland conservation, as reflected in Article 51-A, fundamental duties (g), Constitution of India, which urges every Indian citizen to save and improve the natural environment and to show kindness towards living organisms. Since the establishment of the Ministry of Environment, Forest and Climate Change (MoEFCC) in 1985, it has prioritised wetland conservation. In September 1982, India's accession to the Ramsar Convention reinforced this commitment. In 1986, the MoEFCC launched a National Wetlands Conservation Programme (NWCP) to support state wetland management by providing a policy framework and financial assistance. The National Lake Conservation Plan (NLCP) was introduced in 2001 to combat pollution in urban and peri-urban lakes by interception, redirection and elimination of pollution that enters lakes. Considering the need for a common regulatory framework to conserve and sustainably manage wetlands, urban and semi-urban lakes, in February 2013, the Union Cabinet made a decision to combine the two programmes in one plan called the "National Plan for Conservation of Aquatic Ecosystems (NPCA)".

**Some essential points of the NPCA are:**

1. **Programme coverage:** The wetland covers the areas of fen, marsh, peatland, flowing or stagnant water that can be fresh, brackish, salty, or marine. Wetlands can be artificial or natural, temporary or permanent. Marine water areas, with a depth of less than six meters at low tide, can also be defined as wetlands. The Scheme for Conservation and Management of Mangroves and Coral Reefs will continue to guide the sustainable management and conservation of the coral reefs and mangrove wetlands.
2. **Goal and Purpose:** The primary goal of NPCA is to create and maintain healthy wetlands that support human well-being and also to maintain wetland biodiversity. The primary purpose of NPCA is to integrate the wide range of wetland biodiversity and its ecological contributions into developmental schemes at all levels.
3. **Objectives:** NPCA offers a scientific and integrated framework to conserve and sustainably manage the wetlands in India. The main objectives for this plan include the development of policy guidelines, the formation of a national inventory and information and decision support system, promotion and support of integrated management and managers and stakeholders to improve wetlands' management.
4. **Strategy:** NPCA encourages states and UTs to lead wetland conservation efforts, with the MoEFCC, which supports this by setting policies, connecting development sectors, enhancing research-management coordination, and building stakeholder capacity. Some strategies to integrate its vision are:

   (a) **Integrated Management Plan (IMP):** Investment for the wetland conservation should be based on an integrated management plan which ensures sustainable management of ecosystem services and biodiversity. The Integrated Management Plan for wetlands is developed by State/UT authorities with the help of domain expertise by following NPCA guidelines. Based on the uniqueness of each wetland, NPCA guidelines focus on the comprehensive assessment of ecological, hydrological, socio-economic, and institutional aspects of each wetland to set clear objectives and actions in IMP. Guidelines also recommend identifying the linkages of stakeholders with the wetlands and their participation in designing the plan and its implementation. IMP will help to integrate wetlands into broader sectoral and regional development strategies within catchment and coastal zones. Inadequate IMP leads to insufficient wetland degradation management. Assessment of the current state of wetlands, identification of causes of degradation, identification of affected areas and extent of degradation, and implementation of targeted restoration measures are essential for effective conservation and management.

(b) **Framework Management Plan:** This plan is an interim and supportive step towards the development of the IMP. It will be prepared based on currently available information, recommend non-regrettable actions, and help identify activities for the IMP preparation. It also ensures the commitment from the States/UTs to make and implement IMP within a set timeframe.

(c) **Funds**: Financial management is necessary to implement the wetland restoration plans. The primary source for the funding is public and private developmental sector programmes that impact wetland functioning and support integrated management. The NPCA will provide core funding to initiate and support the integrated management of high-priority wetlands. Different agencies at local, district, and state levels, in coordination with State/UT Wetlands Authorities, must ensure adequate funding is allocated for the completion of the IMP or FMP.

(d) **Cross-Sectoral Governance:** State/UT authorities must establish a local authority for executing policies for the adequate management of wetland ecosystems. These authorities will have representation from all relevant sectors. It also ensures support from the concerned governments, urban local bodies, NGOs, political, technical, and administrative leadership. The Wetlands Conservation and Management Rules—2017, established State Wetlands Authorities for the cross-sectoral governance and stakeholder involvement.

## 7.5 Financial Assistance from Central Government

According to the NPCA guidelines, State Governments and Union Territory administrations can request financial support for the preparation and implementation of the IMP or FMP for wetlands. The MoEFCC will fund eligible activities based on the policies of the Central Government to the states/UTs/ Special/ North-Eastern states. Only those activities that lack any other funding sources and are listed as core or non-core of NPCA are eligible to receive financial support. Core activities will receive 75% of the total allocated budget, and the remaining 25% will be given for non-core activities.

## 7.6 Institutional Arrangements

MoEFCC coordinates NPCA at the National level through a designated division of the Ministry, by providing a national policy framework, financial support, detailed guidelines and technical assistance, research funding, capacity development, periodic monitoring of management plans, national wetland inventory development and outreach on wetlands.

(a) In accordance with Wetlands Conservation and Management Rules (2017), the National Wetlands Committee (NWC) was established to serve as the NPCA's nodal advisory body.
(b) Technical Appraisal Committee (TAC), composed of individual or institutional domain experts, shall be constituted by the MoEFCC to assess comprehensiveness of FMP to complete IMP, adequacy of study, estimated cost and time, and feasibility of actions. It will also recommend a budget and timeline for implementing FMP/IMP. The TAC may also conduct periodic site visits and hold review meetings to assess the progress and to give suggestions for improvements.

II. State Governments and UT Administrations hold primary responsibility for wetland conservation and management, with State/UT Wetlands Authorities—established under the Wetlands Rules, 2017- will be the nodal agency for NPCA implementation.

## 7.7 Strategy for the Implementation of NPCA

(i) NPCA aims to establish a healthy network of conserved and sustainably managed wetlands. The implementation strategies to ensure the sustainable management and conservation of wetland's ecosystem services and biodiversity are:
    (a) State governments/UT administrations must proactively participate to ensure the selection of a representative set of wetlands for management, with a special focus on the wetlands that lie beyond the protected areas and are under threat due to human activities.
    (b) For effective wetland management, the Ministry will provide handholding support through national and regional capacity-building programmes to strengthen the capacity of wetland managers and other stakeholders.
    (c) Biodiversity conservation and ecological balance measures may include the protection and conservation of endangered or threatened wetland species (both plant and animal) and management of the natural environment to preserve the ecological integrity of the wetland.

(d) Tourism infrastructure development must comply with existing regulations and align with the wise use of wetlands to ensure that the ecological integrity of wetlands remains intact.
(e) Integration of ecosystem services with other livelihoods to enhance the sustainability and well-being of wetland-dependent communities, emphasising the socio-economic benefits along with ecosystem conservation.
(f) Promoting the wetlands as long-term carbon sinks and recognising their role in climate change mitigation through managing healthy wetland ecosystems.

(ii) **National Wetlands Atlas**: It covers the spatial distribution, importance, and need for sustained management and conservation of wetlands in India. Data on the number, biogeography, health condition, and biodiversity are collected and analysed. Based on the accurate and updated data, necessary management strategies are made and implemented to address wetland issues. As per India's Wetland Atlas, wetlands cover 4.63% of the total land area [17].
(iii) **Launch of Amrit Dharohar Initiative**: MoEFCC, Government of India, launched the Amrit Dharohar Initiative in June 2023 to promote the unique conservation value of Ramsar sites in India. This initiative is a capacity-building approach to support the local livelihoods by encouraging sustainable and optimal use of wetlands, ecotourism, and biodiversity. The scheme aims to transition strategically from tourism to nature tourism at ecologically sensitive wetlands. The major pilot sites are Sultanpur National Park, Haryana and Chilika Lake, Odisha.
(iv) **Launch of Mission Sahbhagita:** Mission Sahbhagita was launched in 2022 by the MoEFCC, India, to encourage community involvement and ownership in wetland conservation and management. It empowers local people, communities and stakeholders to lead the conservation and management efforts. Under this mission, the 'Save Wetlands' campaign was launched in 2023 to educate and raise awareness among the people to promote wetland conservation. The main goals under this mission are conservation of species, cultural heritage, climate change mitigation and adaptation, flood and drought mitigation, and water and food security.

India, famous for its environmental and eco-cultural heritage, has run various communication and outreach programmes to educate, raise awareness, and involve people in protecting, conserving, and managing wetlands. Thus, protecting the wetlands will also help sustain avian species.

## 7.8 Future Prospects and Challenges

An increased human population requires a larger area for agriculture and aquaculture, which reduces the wetland area. So, the challenges in wetland restoration start with identifying suitable sites for restoration. If the identification is done, then site-specific constraints, which can be ecological, technical, social or economic, are other challenges to implement the planned projects. The priorities of the local communities, stakeholders, and indigenous peoples are different. So, engaging the locals and stakeholders might be challenging when starting and implementing the projects. But, the restoration efforts could also lead to unintended and unwanted side effects, like forming favourable mosquito breeding sites, flooding, etc. This could lead to the spread of various diseases locally. This challenge should also be carefully addressed.

To address these challenges, behavioural change programs, expansion of capacity-building programs, and fostering local community involvement could be seen as future directions. The regional cooperation network should be expanded, and practical day-to-day guidance for local wetland managers should be developed. Integration of birds and wetland conservation into the broader climate and water conservation projects should be done under the umbrella programmes.

## 7.9 Conclusion

Birds, especially aquatic avian fauna, depend on wetlands. Wetlands are the most favourable feeding, breeding or perching sites for birds. But wetland degradation is an alarming challenge for these species. Wetland restoration, protection, and conservation programmes are needed to foster participation from the local people and stakeholders. Capacity-building programs and initiatives must be implemented and monitored periodically to strengthen the policies. The effective implementation of umbrella programmes that can integrate birds and wetland conservation can enhance the aims and objectives of aquatic bird population conservation.

## References

1. Adhurya S, Adhurya S, Roy S (2019) Rapid degradation of wetlands and its impact on avifauna: a case study from Ambuja Wetland, West Bengal, India. Indian BIRDS 15(2):43–48
2. Convention on Wetlands (2002) Principles and guidelines for wetland restoration, 8th Meeting of the Conference of the Contracting Parties to the Convention on Wetlands (Ramsar, Iran, 1971) Valencia, Spain
3. Darrah SE, Shennan-Farpón Y, Loh J, Davidson NC, Finlayson CM, Gardner RC, Walpole MJ (2019) Improvements to the Wetland Extent Trends (WET) index as a tool for monitoring natural and human-made wetlands. Ecol Indic 99:294–298. https://doi.org/10.1016/j.ecolind.2018.12.032
4. Delany S, Scott D, Helmink ATF (2006) Waterbird population estimates. Wetlands International
5. Dudley N, Alexander S (2017) Global land outlook, 1st edn. UNCCD, Bonn
6. Evans KL, Newson SE, Gaston KJ (2009) Habitat influences on urban avian assemblages. Ibis 151(1):19–39. https://doi.org/10.1111/j.1474-919X.2008.00898.x

7. Fluet-Chouinard E, Stocker BD, Zhang Z, Malhotra A, Melton JR, Poulter B, Kaplan JO, Goldewijk KK, Siebert S, Minayeva T, Hugelius G, Joosten H, Barthelmes A, Prigent C, Aires F, Hoyt AM, Davidson N, Finlayson CM, Lehner B, Jackson RB, McIntyre PB (2023) Extensive global wetland loss over the past three centuries. Nature 614(7947):281–286. https://doi.org/10.1038/s41586-022-05572-6
8. Haig SM, Murphy SP, Matthews JH, Arismendi I, Safeeq M (2019) Climate-altered wetlands challenge Waterbird use and migratory connectivity in Arid landscapes. Sci Rep 9(1):4666. https://doi.org/10.1038/s41598-019-41135-y
9. Jangra L, Verma R, Vats D (2024) Assessing Avian diversity and the impact of air pollution on ecological communities in Western Haryana, India. IJE. https://doi.org/10.55362/IJE/2024/4441
10. Jordán DP (2017) Waterbirds in a changing world: effects of climate, habitat and conservation policy on european waterbirds. Acad dissertat Univ Helsinki. pp 146
11. Khan TN, Sinha A, Hazra P (2016) Population trends and community composition of migratory waterbirds in three emerging wetlands of global significance in southwestern Bengal, India. J Threat Taxa 8(3):8541. https://doi.org/10.11609/jott.2652.8.3.8541-8555
12. Kiran SD, Kour A, Priya DV, Kumar R (2022) Community composition and status of avian diversity at Campus and Agricultural landscapes of Chaudhary Charan Singh Haryana Agricultural University, Hisar (Haryana). J Appl Nat Sci 14(4):1130–1140. https://doi.org/10.31018/jans.v14i4.3784
13. Kotze DC (1996) How wet is a wetland?: an introduction to understanding wetland hydrology, soils and landforms. Grassland Science Department, Univ. Natal
14. Kumar P, Rai D, Gupta SK (2016) Wetland bird assemblage in rural ponds of Kurukshetra, India. Waterbirds 39(1):86–98. https://doi.org/10.1675/063.039.0111
15. Kumar R, Yadav R, Gupta RK, Pal A, Yodha K, Kumar A (2024) Antioxidant enzyme activities and markers of oxidative stress in the life cycle of different earthworm species. Indian J Exp Biol 62(07):578–583
16. Kumar R, Yadav R, Kumar Gupta R, Pooja (2022) A sustainable way for fish health management by replacement of chemical and drugs by earthworm. In: Environmental degradation in Asia: land degradation, environmental contamination, and human activities. Springer International Publishing, Cham, pp 329–352
17. Ministry of Environment, Forest and Climate Change (2025) Ramsar National Report to COP15
18. Mitsch WJ, Bernal B, Hernandez ME (2015) Ecosystem services of wetlands. Int J Biodivers Sci Ecosyst Serv Manag 11(1):1–4. https://doi.org/10.1080/21513732.2015.1006250
19. NPCA, Ministry of Environment, Forest and Climate Change (2024) National Plan for Conservation of Aquatic Ecosystems (NPCA) Guidelines
20. Anoop Raj PN, Velankar AD, Pramod P (2023) Diversity and distribution of birds in the Bharathapuzha river basin, Kerala, India. J Threat Taxa 15(11):24169–24183. https://doi.org/10.11609/jott.8573.15.11.24169-24183
21. Parameswaran G, Sivashankar R (2022) Rapid decline of waterbirds in urban wetlands: a case study from Perur-Sundakamuthur Lake, Coimbatore, Tamil Nadu, India
22. Ramsar Convention on Wetlands (2021) Wetlands and agriculture: impacts of farming practices and pathways to sustainability
23. Ramsar: The Convention on Wetlands (2025) The List of Wetlands of International Importance
24. Muralikrishnan S, Shanmugam E, Nagendran NA, Pandiaraja D (2023) Diversity and abundance of aquatic birds in Koonthankulam village pond, Tamil Nadu, India. J Threat Taxa 15(6): 23297–23306. https://doi.org/10.11609/jott.6612.15.6.23297-23306
25. Vats D, Verma R, Dahiya T (2023) Ornithology for enhancing agriculture production in a sustainable way. In: Multidisciplinary approach: enhanced agriculture production in a sustainable way, 1st edn. Department of Agriculture Entomology Shri Vaishnav Institute of Agriculture, Shri Vaishnav Vidyapeeth Vishwavidyalaya, Indore, pp 99–108
26. Yadav R, Gupta RK, Kumar R, Kaur T, Moun R (2022) Biochemical response of earthworm, *Eisenia fetida* to heavy metals toxicity. J Appl Nat Sci 14(3):990–998

# Chapter 8
# Potential Threats to Water Bird Population: Habitat Deterioration and Urbanization

**Amit Kour, Dharambir Singh, Kiran Yodha, and Renu Yadav**

**Abstract** Water birds are vital pointers of aquatic ecosystem health playing pivotal place in maintaining ecosystem functions, but they face raising pitfalls due to combined impacts of niche declination and urban expansion. The fast pace of urbanization has exacerbated niche fragmentation, rising pollution levels and degraded natural ecosystems as well as aids new stressors like artificial lighting, noise pollution, and changed hydrological patterns. The addition of such stressors with binary pressure also interferes with ecological processes supporting ecosystem services and biodiversity and in order to understand the pitfalls that urbanization and niche deterioration pose to water bird populations it is imperative to synthesize current knowledge, identify exploration gaps, develop a comprehensive framework for conservation strategies, and punctuate their synergistic impacts on water bird populations with highlights of effective conservation measures by addressing related issues and shaping policies with the help of scientific and community based research amalgamations. The development of a holistic perspective on how these forces interact, enabling researchers and policymakers to prioritize interventions and allocate coffers effectively seeks to bridge that gap, furnishing a precious resource for stakeholders, including ecologists, civic itineraries, and original communities. By synthesizing scientific evidences allowing the collation of fractured data, it is necessary to raise mindfulness about the long-term consequences of inactivity and emphasize the significance of espousing integrative approaches to alleviate these pitfalls and to inspire cooperative interventions to balance experimental requirements with biodiversity conservation, icing the survival of water bird populations and the ecosystems they inhabit.

**Keywords** Decision-making · Habitat deterioration · Threats · Urbanization · Water birds · Wetlands

A. Kour (✉) · D. Singh · K. Yodha · R. Yadav
Department of Zoology, Chaudhary Charan Singh Haryana Agricultural University, Hisar, Haryana, India



R. Kumar et al. (eds.), *Ecology, Conservation of Aquatic Birds*, Wetlands: Ecology, Conservation and Management 15, https://doi.org/10.1007/978-3-032-17125-2_8

## 8.1 Introduction

Water birds are a different group of avian species distributed across the globe, inhabiting marshes, lakes, mangroves, and littoral regions acting as vital factors of submarine ecosystems by serving as pointers of ecological health [6, 41]. Water birds play a critical part in ecosystem maintaining the health and functionality of ecosystems performing ecological functions that are integral to the stability of submarine and swamp surroundings while significantly contributing to global biodiversity [111, 120]. Encyclopedically, water birds represent a significant proportion of avian diversity, encompassing families similar as Anatidae (ducks, geese, and swans), Ardeidae (herons and egrets), Scolopacidae (sandpipers), and Phalacrocoracidae (Cormorants), among others [16, 61, 107]. One of their primary role is nutrient cycling, wherein, water birds transport nutrients across ecosystems through their feces, also known as stool, rich in nitrogen and phosphorus, thereby, enhancing the productivity of wetlands, promoting the growth of submarine foliage and algae that form the foundation of food webs [15, 35, 125]. Water birds are essential seed dispersers, abetting in the rejuvenescence of communities by transporting seeds over long distances during their rustling and migrant conditioning fostering foliage diversity, which in turn supports a wider range of wildlife [88]. As bio-indicators, water birds reflect the health of wetland ecosystems as population trends, actions, and distribution patterns frequently provide early warnings about changes in water quality, pollution situations, and niche declination as a decline in water bird populations may indicate issues similar as inordinate fungicide use, water pollution, or reduced prey vacuity [3, 32]. Monitoring these species can help identify and alleviate pitfalls to wetland ecosystems, making them inestimable for conservation sweats. Water birds also play a pivotal part in supporting trophic relations by acting as both bloodsuckers and prey, thereby help to regulate populations of fish, amphibians, insects, and other pets, maintaining ecological balance [68, 113]. Also, water birds' rustling actions, similar as stirring up sediments or stamping foliage, produce microhabitats for lower organisms, further perfecting biodiversity [14]. This ecotourism promotes community-grounded conservation sweats while raising mindfulness about the significance of conserving these species and their territories. Incipiently, water birds contribute to global ecosystem connectivity through their migrant geste numerous species travel thousands of kilometers annually, connecting ecosystems across mainland [26, 88]. Water birds also help to connect the subaqueous and terrestrial food chain as many bigger bloodsuckers feed on them just as raptors and rapacious mammals do [69]. Water has a strong influence on culture and economics.

This chapter is aimed to collect recent information, find areas that require further exploration and support conservation tactics. Emerging need of research studies intending to highlight the urgent need for efficient conservation measures is required by examining these related hazards in order to punctuate their synergistic effects on water bird populations. From a scientific perspective, knowing these pitfalls are crucial for directing the creation of policies and specialized operational procedures to limit this binary pressure as it also interferes with ecological processes that support ecosystem services and biodiversity. This book chapter serves to call for actions from international conservation communities by bringing together disparate data and creating a comprehensive understanding of the ways in which these forces interact to help policymakers and experimenters with prioritized interventions and funds allocation. This book chapter seeks to bridge that gap, furnishing a precious resource for stakeholders, including ecologists, civic itineraries, and original communities. By synthesizing scientific shoes, the book chapter aims to raise mindfulness about the long- term consequences of inactivity and emphasize the significance of espousing integrative approaches to alleviate these pitfalls.

## 8.2 Water Bird Population Trend

While Wetlands International and BirdLife International have established, a significant number of water birds species are experiencing population reduction [25, 109]. This trend is particularly pronounced in Asia and Africa, where rapid-fire urbanization, agrarian expansion, and swamp declination have redounded in substantial niche loss [7, 122]. For illustration, the East Asian- Australasian Flyway, one of the world's major migrant routes for water catcalls, has seen a dramatic decline in populations of shorebirds similar as the ladle-billed Sandpiper (*Calidris pygmaea*), which is critically risked. These declines punctuate the need for immediate conservation action to cover critical territories along migrant routes. Regionally, the status of water birds populations varies significantly [29, 102, 103]. In Europe and North America, where expansive swamp protection programs and monitoring programs are in place, certain populations have shown signs of recovery [83, 97] for example case, conservation sweats have led to the stabilization of species like the Common Crane (*Grus grus*) and the Great Egret (*Ardea alba*). Still, indeed in these regions, localized pitfalls, including niche fragmentation and climate change, pose challenges to water birds conservation. In discrepancy, in developing regions similar as South Asia and Sub-Saharan Africa, the lack of strict swamp operation programs and adding anthropogenic pressures have aggravated water birds population declines [30, 112]. Iconic washes like India's Keoladeo National Park and Nigeria's Hadejia- Nguru washes have witnessed significant reductions in bird diversity due to water birth and niche encroachment [4, 17, 76]. The way water birds travel around the world suggests that countries must collaborate to protect them. During their migrations, many watersheds welcome fish as they cross several countries and rest, eat and reproduce. For example, birds such as the Bar-headed Goose (*Anser indicus*) and the Black-tagged Godwit (*Limosa limosa*) navigate using the washes in rivers to calculate their landmark. A change in the declination of any important place can influence many people. In some places, the issue is not as bad, as water birds have been helped by various conservation programs. Ramsar spots help to reserve areas where water catcalls can happen. Now, with the recovered and restored Chilika Lake in India, a Ramsar site, there are more water birds because operating swamps. Although water calls affect people everywhere, by conserving them it is possible to improve ecosystems and biodiversity. This work requires varying efforts such as rebuilding niche habitats, engaging with other countries and engaging communities [59, 73, 96].

## 8.3 Role of Habitat Deterioration as a Threat to Water Bird Population

Degraded habitats lose their ability to support the plant, animal and microbial communities that rely and inhabit specific habitats [55, 90]. Human actions, changes in the environment or both, are often the reasons for shrinking of habitats with significant effect on biodiversity and throws off the balance in different kinds of ecosystems all over the globe [105]. Many threatened water bird species, global biodiversity and ecosystem health are concerned by decline in their habitats, since they need the wetlands, mangroves, floodplains and similar aquatic ecosystems to survive affecting species at every level in food webs due to their ecosystem needs [62, 83, 117]. Improving habitat in combination with habitat restoration, intelligent land-use planning, pollution control and actions to prevent climate change is important to care for habitats well, as this seeks to protect water birds and ensure human society benefits from the many services provided [45, 66].

## 8.4 Causes of Habitat Deterioration

(a) **Water bird Habitat Deterioration due to Pollution:** Water bird wildlife is threatened most significantly when their habitat becomes polluted by industrial effluents, pesticides or plastic soups, pesticides and plastic trash pollute water (Fig. 8.1), in turn water birds are affected by less food, fewer successful breeding chances and an increase in death tolls [56, 57, 72, 79, 118, 119]. Troubling water pollution results when industrial effluents which are often discharged without insufficient treatment contain heavy metals, poisonous chemicals and too many nutrients upset the balances in lakes and rivers and balance among species in water systems flow into wetlands, rivers or reservoirs effluents Furthermore,

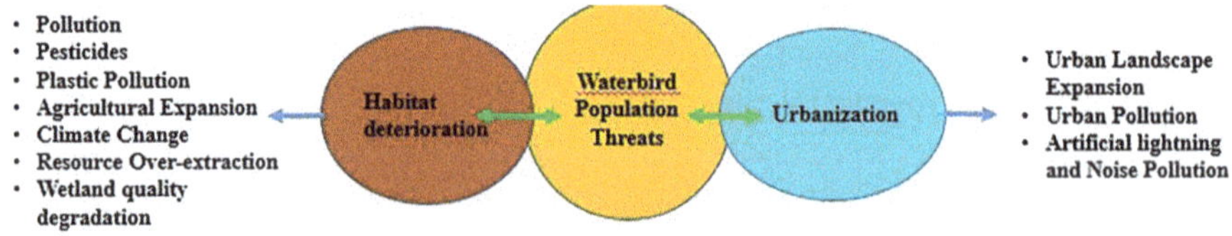


**Fig. 8.1** Causes of Habitat deterioration and Urbanization as threat to water bird depletion

Birds such as herons, cormorants and pelicans that eat tropical fish are particularly at risk from water contamination by heavy metals mercury, lead and cadmium that accumulate in the food chain by bio-magnification, gather in food chains causing problematic neurological functioning, reproductive anomalies and even fatality leading to population decline [53, 99, 108]. Besides, too many nutrients pour into rivers and seas from industries causes eutrophication and the blooms of algae resulting lower oxygen levels in the water disturbing the waterbody ecosystem causing harsh conditions for inhabiting fish species to thrive Fish die-offs often happen due to deoxygenation leading to fish die-offs removing the main food source for water birds and disturbing all parts of the ecosystem which in turn leads to water bird decline [2, 82].

(b) **Pesticides**: Many water birds have suffered due to the widespread use of pesticides in farming as rain or runoff can introduce organophosphates and neonicotinoid pesticides into water systems, contaminating lakes, rivers and seas [58, 77, 95]. Water bird that takes in contaminated food or water soon develops acute or chronic poisoning affecting bird eggshells, resulting in fewer successfully reared raptors who live mainly off aquatic food as well as loss of aquatic insects through pesticides also removes a significant food supply for water birds, so they either need to leave polluted lakes or end up with smaller populations [38, 80, 87]. Apart from that, being continuously exposed to pesticides leads water birds more vulnerable to different infections.

(c) **Plastic pollution:** Plastic pollution including micro-plastics has become a serious problem for water birds worldwide as they can mistake plastic waste in wetlands, along coastlines and in freshwater ecosystems for food and consume them which may develop blockages and starve because of lack of nutrition. Many seabirds, including gulls, terns and pelicans, have been found swallowing plastic which can end in their death. Most marine animals, like birds, are struck by plastic pollution in an additional way—when they become trapped in discarded netting, ropes or similar waste, either preventing movement, feeding or leading to drowning. As a result, micro-plastic particles that small aquatic creatures eat can build up in the food chain and endanger water birds, since the chemicals in the plastics may change their ability to reproduce [39, 47, 60, 92, 104, 110].

(d) **Agricultural Expansion and Wetland Conversion:** Among the major causes of habitat loss for water birds are the speedy growth of agriculture and the large-scale conversion, loss, fragmentation and degradation of wetlands which water birds need to breed, feed and migrate into other areas causing serious changes in the environment leading to water bird populations experiencing negative consequences which straight away impact water bird numbers, types and the overall health of the whole ecosystem [26, 27, 75]. Water birds are at risk because the way lands are being developed for farming and using up wetland areas leads to their loss of suitable habitats. Sustainable methodologies are needed to promote sustainable farming, guarding wetlands and help restore damaged habitats [40]. When agriculture and wetland conservation are both important, the water birds can still exist and essential ecosystems will not be disturbed [54].

(e) **Climate change:** Climate change is having major effects on many parts of the world by causing unusual rainfall and higher temperatures and floating ecosystems like wetlands, floodplains and mangroves—relied on by water birds—are easily affected by climate changes with key aspects of climate change—rising temperatures and changes in rain patterns—are messing with these habitats, leading to problems for the birds and their inhabiting areas [44, 70]. As a consequence of these changes, aquatic environments lose their desired traits which leads to water bird population decline [100]. Preserving wetlands, preventing habitat loss and combating climate change are needed to ensure water birds are safe and well. Addressing causes of climate change and building resilient aquatic habitats helps water birds remain safe and prevent biodiversity from decreasing [36].

(f) **Over-extraction of water resources:** Water birds are losing their habitat because too much water is taken away for farming, factories and cities. Water birds depend on water bird habitats that are well supplied with water for their normal functions. If too much water is removed from these areas, the balanced patterns needed for water birds are lost and this often leads to both diversity and abundance declines [42].

(g) **Degradation of Wetland Quality and Ecosystem Services:** Wetlands are world's most diverse ecosystems acting as an important habitat for birds that have evolved to live in water and are essential for keeping the atmosphere in balance by provision of services like carbon sequestration, water purification, control floods offering critical role in ecosystem [55]. The loss of wetlands, wetland quality and the ecosystem services due to human activity, pollution, and weather trade causes significant impacts on the ability of these ecosystems to support water bird populations [89] by alteration in their ability to survive, reproduce, and grow their populations. The loss of breeding grounds, the destruction of food sources, the fragmentation of habitats, and the increase in disturbances paves a huge effect on the number and diversity of water birds [48] creating a feedback loop which renders these environments more extreme thereby making it more challenging for aquatic birds to survive. To solve these problems, it is important to stress how important it is to protect, restore, and manage wetlands in a way that is good for the environment as safeguarding these areas not only helps birds, but also makes sure that important ecosystem services that are important for both biodiversity and human health continue to be accessible [11].

## 8.5 Role of Urbanization as Water Bird Population Threat

The rapid pace of growing population as well as development of cities and towns attracts people to move from rural to urban areas for a number of reasons, but two of the most important are population growth and migration [64, 111]. People move to cities to get better employment opportunities, schools, healthcare, and living conditions in general. Urbanization, although promoting economic development and technical advancement, also results in significant transformations in the environment, social structures, and cultural contexts (Fig. 8.1). Urbanization is a complicated and multi-faceted problem that has both good and bad repercussions [19].

(a) **Expansion of Urban Landscapes:** The increase in urbanization has generally led to the destruction of several natural habitats and in particular wetlands, important refuge ecosystems for a variety of species, including water birds. One of the most serious impact of urbanization, on environment, has resulted in wetlands conversion for the purpose of urban infrastructure. The draining or filling of wetlands to accommodate roads, buildings and other development projects is a common byproduct of expanding cities [8]. Wetlands perform all sorts of services from water purification, flood protection to ecosystem for various species. Wetlands are important feeding, nesting, and roosting areas for water birds. With the disturbance of these habitats, birds lose the space for breeding and feeding which affects their life cycles and results in reduced population [74]. Over and above the outright loss of wetland habitat, urbanization also results in the expansion of impervious surfaces (e.g., roads, parking lots and roofs) which increases the rate of runoff. These ground surfaces also obstruct the long-existing rain water seepage into earth, thereby changing the water regime of this habitat [63]. When rainwater can't seep into the ground, it flows across surfaces and into water bodies, carrying pollutants such as oil, heavy metals and sediments. This shift in water dynamics results in modified hydrological regimes, which have important consequences for water bird populations. For example, water birds depend on even water levels in wetlands for feeding and breeding. Urbanization alters the dynamics of these levels by higher run-off and by water bodies being more susceptible to flooding or desiccation [43]. Changes in water levels can be destructive for nests, for food availability and for birds that need 'the right conditions' in order to survive. Moreover, polluted water in urbanized water systems results in poor water quality that is harmful to the survival of both plant and animal organisms. Because water birds are one of the most vulnerable group of populations to these changes, the extending of urban landscape is very likely to exert significant long-term effects on water bird survival and biodiversity in these areas [9]. These factors of wetland encroachment and expanding impervious surfaces must therefore be crucial in the decline in habitats used by water birds, thereby indicating the intricate relationship between urbanization and the health of natural ecosystems.

(b) **Urban Pollution:** Urban pollution, in particular the discharge of untreated sewage and industrial waste, poses excessive danger to the fitness of aquatic ecosystems and the wildlife that depend on them, specially water bird populations. Towns generate massive quantities of wastewater, much of which incorporates dangerous pathogens, chemical substances, and natural count number in many urban areas, inadequate sewage treatment systems bring about untreated or in part treated sewage being launched at once into rivers, lakes, and wetlands [52]. Eutrophication can purpose the speedy boom of algae, which consumes big quantities of oxygen as it decomposes, main to hypoxic conditions thus suffocating fish and different aquatic organisms, which, in turn, reduces the meals supply for water birds [91]. Further, high nutrient masses encourage the boom of harmful bacteria, viruses, and pathogens which could further damage water birds both without delay through infection or indirectly by means of reducing the best and availability of their food sources. Industrial discharge is some other main pollutant in urban areas, with factories and plant life liberating diverse toxic chemicals consisting of heavy metals, solvents, petrochemicals, and different dangerous substances into close by water bodies [46]. Untreated sewage and business discharge contribute to water pollutants that notably impacts water hen populations. Pollutants no longer only reduce the quality of the water and the food supply but additionally pose direct fitness risks to the birds themselves [114]. The combined impact of city pollutants exacerbates the demanding situations faced by means of water birds, making the protection of easy, wholesome aquatic habitats essential challenge for both flora and fauna conservation and the sustainability of ecosystems.

(c) **Artificial Lighting and Noise Pollution:** Synthetic lighting fixtures and noise pollutants are developing issues in urban environments, both of that have tremendous outcomes on bird conduct, especially for water bird populations by interfering with birds' natural patterns of activity, migration, and conversation, main to disruptions of their life cycles and habitat use [34]. Artificial lights in urban areas alters natural mild cycles, regularly known as light pollution [115]. Birds, which includes water birds, depend upon herbal mild cues for crucial behaviors such as migration, feeding, and nesting. For migratory species, artificial mild can disorient birds, main them to fly off direction or become trapped in illuminated areas, growing their hazard of collisions with homes or energy traces. During the breeding season, immoderate light can disturb water birds' nesting behaviors. Many species of water birds are particularly sensitive to light tiers and might abandon their nests or delay egg-laying if synthetic lighting interferes with their reproductive cycles. The alteration of circadian rhythms because of mild pollution can also result in disruptions in foraging styles, as water birds can be pressured to feed all through unusual hours or in much less most desirable conditions, affecting their survival and reproductive success [31]. Noise pollution, by and large generated by means of human activities which include visitors, production, and industrial methods, additionally has a profound impact on fowl behavior. Water birds, like many other species, use vocalizations for communication, consisting of territory protection, mating calls, and alerts to predators. Chronic noise pollutants can mask these vocalizations, making it tough for birds to communicate correctly especially, noise can also disrupt mating behaviors, making it tougher for water birds to attract mates or establish territories moreover, the strain resulting from non-stop noise can bring about decreased reproductive fulfillment, as the birds may also expend more power on averting noise resources or can be too harassed to take care of their young correctly [123]. For water birds, which can be already facing pressures from habitat loss and pollution, the additional stressors of synthetic lighting fixtures and noise can appreciably compromise their survival. Migratory water birds may modify their migration routes or postpone their departure, leading to timing mismatches with meals availability or breeding web sites. Additionally, immoderate noise and mild could make in any other case appropriate habitats uninhabitable, forcing water birds to abandon these areas in search of quieter, darker areas, which might not be without difficulty to be had in urbanized landscapes. Water bird populations diminishes especially in places where cities are built close to important wetland and coastal habitats [1] affecting the behavior and survival of water birds while disturbing and interfering important behaviors like migration, breeding, and communication, which has negative effects on population balance and reproduction. To protect water birds and their habitats, it is important to deal with the sources of pollution in urban land as a primary necessity for safeguarding aquatic species [21].

## 8.6 Synergistic Effects of Habitat Deterioration and Urbanization on Water Bird Population

The combined impacts of urbanization and habitat degradation on water bird populations are highly synergistic leading to increased risks in a manner that may be more severe than each cause when considered separately.

I. **Synergistic Effects of Habitat Deterioration and Urbanization**

(a) **Habitat Fragmentation and Loss**: One of the most common effects of urban growth is the direct change of wetlands, lakes, and marshes into land that is used for residential occupancy, businesses, and farming. Pollution via introduction of invasive species, environmental interchange and their combined effects is lowering the quality of wetland habitats leading habitat destruction of waterbirds resulting in more isolated, and less than ideal places to spawn, feed and resting place causing genetic bottlenecks and a loss of resilience, both of which are detrimental to the species' survival [117].

(b) **Pollution as a Dual Threat**: Development of cities and side-by-side increase of pollutants, heavy metals, plastics, and untreated wastewater in wetlands causes plant cover decrease or algal blooms outburst because of eutrophication and their combination with habitat degradation leads to water bird inhabiting environment toxication as a result of which water birds are not able to thrive thus limiting their survival. The flood of nutrients that is induced by urban effluent can have a negative impact on the populations of fish and invertebrates that are necessary for the survival of water birds like herons and storks [111].

(c) **Climate Stress and Urban Heat Islands**: It is possible for the heat island effect of the city to combine with the degradation of habitat that is brought on by changes in the weather, such as the decline of wetland areas as temperature swings in metropolitan settings can have a negative impact on migratory indicators, reproductive seasons, and the availability of agricultural products because water birds that are dependent on seasonal wetlands may miss out on opportunities to reproduce or graze as a consequence along with change in the migration timeline [50].

II. **Population-Level Impacts**

(a) **Decline in Breeding Success**: Increasing human activity, noise pollution, and mild pollution are all factors that contribute to the disruption of nesting environments that are commonly caused by urbanization. The success rate of breeding is lowered when the factors of habitat degradation, decreased availability of food, and invading predators are taken into consideration. However, nocturnal animals like as bitterns may get disoriented due to night disruptions induced by synthetic lights. This is the case despite the fact that damaged wetlands provide only a minimum amount of protection against predators [105].

(b) **Altered Migration and Foraging Patterns**: A community of wetland ecosystems that are located along migration pathways is necessary for aquatic species. Because of the frequent interruption of those stopping places caused by urban development, birds are forced to travel greater distances across ecosystems that have been damaged. This is mostly owing to the fact that insufficient amounts of electricity are available [44]. The presence of developed areas in close proximity to main stopover wetlands, such as the Chilika Lagoon in India, might make it more difficult for migrating shorebirds to access feeding grounds [93].

(c) **Increased Mortality Rates**: The burden of identifying suitable assets in degraded ecosystems, in conjunction with the collision with urban infrastructure (such as powerlines and residences), leads in improved mortality estimates as a result of the destruction of wetland areas in India and China, as well as the expansion of urban areas into the areas where Siberian Cranes spend their winters, the population of these birds is in danger of reaching extinction [117].

III. **Ecosystem Consequences**

(a) **Disruption of Trophic Interactions**: The lack of water bird populations can lead to imbalances in aquatic habitats, which can have negative consequences like loss of water birds with cascading consequences, such as changes in the structure of vegetation or outbreaks of algae for which water birds are typically responsible for regulating the populations of fish and invertebrates living in the water [27].

(b) **Reduced Seed Dispersal and Nutrient Cycling**: Water birds play an important role in the process of seed distribution and nutrient cycling that occurs within wetland habitats, however, degradation of their habitat, which is further aggravated by urbanization, therefore has an effect on the functioning of the atmosphere, which in turn has a negative influence on their existence [88].

## 8.7 Conservation Efforts and Mitigation Strategies

(i) **Policy and Community Engagement in Water bird Conservation Efforts for Strengthening Wetland and Biodiversity Policies**: In order to ensure the well-being of aquatic avian populations, it is of the utmost importance to be in possession of solid legislation that encourages the preservation of biodiversity and wetland areas. Waterbirds are reliant on wetland habitats, the strengthening of legislation that protect them is a means of protecting them from the negative effects of urban growth, pollution, and deterioration by enforcing laws such as the Ramsar Convention on Wetlands and the CBD to protect both the wildlife that thrive in wetlands along with existing conditions. It is possible to protect wetlands by limiting the amount of land that is reclaimed for farming, and reducing the number of times that unsustainable farming practices are used as well as to improve water quality, pollution reduction by building sensitive zones near wetlands [109]. India passed the Wetland Conservation and Management Rules in 2017 to protect both water birds and wetlands. Land restoration activities should be encouraged by governments, and support should be provided to manage wetlands for the conservation of wetland areas as part of flood prevention and control measures as well as support from the community contributes to their overall success (Fig. 8.2). The existence of wetland buffers and green spaces is beneficial to the conservation of water bird populations in areas that are experiencing rapid urbanization. The rehabilitation of the environment, the reduction of the climate in urban areas, and the formation of connections between the many components of the ecosystem are all outcomes of these acts.

(ii) **Role of Local Communities and Citizen Science in Water bird Monitoring**: Through their participation in the upkeep and repair of wetland regions, local people make a great contribution to the conservation of waters birds as the participation of communities in conservation efforts not only assures that conservation methods will not fail, but it also helps citizens develop a feeling

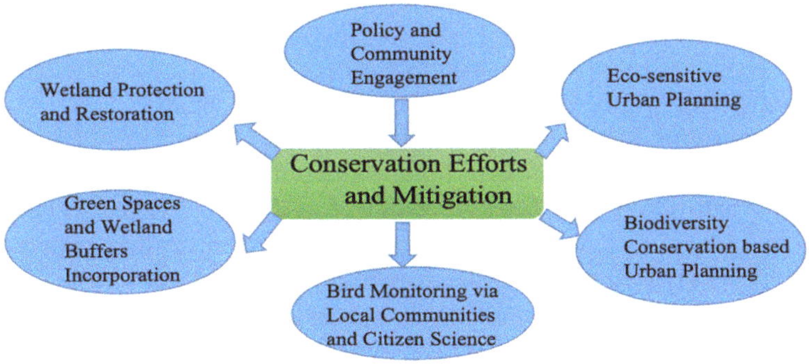


**Fig. 8.2** Conservation efforts and mitigation strategies for water bird conservation

of responsibility as people from a wide range of nations, particularly those who support themselves via activities that are dependent on wetland ecosystems, such as farming or fishing, are making concerted efforts to safeguard their own health. There is a reduction in the risk of human-wildlife conflicts occurring when citizens of the region participate in conservation projects [18]. The rehabilitation of mangrove wetlands in the Niger Delta, Nigeria, for example, has been made possible by the efforts of the local community as these efforts have proved beneficial to the local populace as well as a range of species, such as the African Fish Eagle and the White-faced Whistling Duck. Citizen scientists in India are responsible for determining the arrival and departure of migratory birds in Chilika Lake thus, helps to ensure the lake's resources are preserved and managed over the long term. In addition to monitoring, members of the community also take action to support policies that are designed to conserve resources through the engagement of local inhabitants that are able to improve their understanding of wetland systems and motivated to support and promote the implementation of protection measures for water birds and their inhabiting environments as during the planning process, conservationists can have the assurance that the policies they adopt are not only scientifically sound but also in accordance with the customs and practices of the region. People who live in close proximity to coastal wetland regions in Australia have worked along with conservation groups to reduce the negative effects of human activity, which include the blocking of reproductive areas during crucial times of the year through the implementation of this program, waterbirds such as the Eastern Curlew have been able to survive. Educating individuals and providing them with the skills to participate in conservation programs allows communities to participate in these initiatives and provide support to take action for an extended length of time when they are exposed to programs that educate the community about the necessity of maintaining nature. Creating community conservation oriented programs, seminars, and workshops in conjunction with a variety of groups may help local communities better preserve and conserve water avian habitats ensuring necessary incorporation of new legislation regarding wetland and biodiversity with the participation of the community as it provide equal importance to include the local people in the different conservation operations. The conjugation of policymakers, local communities, and conservationists will be help to conserve the habitat of water birds, and these species will be able to continue existing in the future.

(iii) **Incorporating Green Spaces and Wetland Buffers in Urban Landscapes for Water bird Conservation**: The expansion of cities leads to the disintegration of natural ecosystems, the separation of various groups and for the purpose of ensuring that water birds have access to the resources they require, town planners can incorporate wetland areas and other green spaces into their designs. The incorporation of green regions and wetland zones to preserve and maintain the aquatic species with the support of green spaces and buffers, urban areas have the potential to reduce the damage that is caused by urban growth, to maintain ecosystems, and to merge municipal projects and environment in a pleasant and harmonious manner by providing food sourcing opportunities in settings such as urban parks, gardens, and artificial wetlands as water birds are given with new habitats and routes for migration and food search by replicating the qualities of wetland environments in urban settings by incorporating elements like as tiny ponds, native flora, and undeveloped spaces for the construction of nests [12]. Irregular birds like egrets, kingfishers, and herons are drawn to Singapore Wetland Park, which is a city that integrates wetland elements and attracts these species. The requirements of water fowl are met by these essential venues, which also contribute to an improvement in the quality of life in urban areas by providing a wide range of activities and a gorgeous setting. Wetland buffers, on the other hand, are vegetated sections that are strategically placed around wetland areas in order to shield them from the effects of urban expansion providing the role as shields that capture pollutants, reduces the volume of discharge, and mitigate the impact of human interference. It is possible to find wetland buffers in the Pallikaranai Marsh Reserve Forest in Chennai providing a habitat for the Painted Stork and the Black-winged Stilt, while also protecting the marsh from being taken over by land grabs. Additionally, the presence of green areas and wetland buffers within the habitat is a crucial element in the maintenance of high water quality ensuring the survival of a large number of aquatic species. The deployment of these measures also contributes to the reduction of urban heat islands (UHIs), which have the potential to worsen wetland difficulties by rising temperatures and altering water levels because UHIs have the ability to cause wetland concerns to become more severe. The municipal waterways of Amsterdam, which are kept in good condition, are home to a variety of water birds, including swans and coots. Herons and ibises, for example, are able to migrate in Nairobi despite the high amount of urbanization that exists because of buffer zones and lush places in close proximity to the river.

(iv) **Eco-Sensitive Urban Planning to Minimize Wetland Encroachment for Water bird Conservation**: Environmentally friendly urban planning aims to protect wetland areas while also promoting development and environment side-by-side to ensure eco-sensitive urban design. Some of the ways that eco-sensitive urban design tries to solve these problems are by protecting and restoring wetlands, conserving biodiversity, and encouraging sustainable development as well as using zoning laws and regulations to protect wetland areas as one of the most important parts of environment-friendly urban design as a result of which city planners can stop the growth of urban lands while saving wetland areas by designating them as protected or environment-sensitive zones. The Ramsar Convention protects the East Kolkata Wetlands in India from the bad effects of urban growth. The wetland regions not only function as a system for the treatment of wastewater, but they also provide a habitat for a wide variety of aquatic bird species, such as the Purple Swamphen and the Lesser Whistling Duck. Creating buffer zones and verdant belts is the most efficient way to stop people from entering wetland regions once they have already been established. The use of both physical and biological barriers helps to mitigate, to a limited degree, the effects of urbanization that have spread from neighboring regions into these zones. These consequences have been brought about by the expansion of urbanization. In order to fulfill their requirements for breeding and their activities of searching for food, water birds are dependent on verdant boundaries for their continuing survival. By establishing buffer zones surrounding the Bellanwila-Attidiya Wetlands in Sri Lanka, urban expansion has been effectively limited, and the Black Bittern and Yellow Bittern species have been protected. This is a direct result of the development of these buffer zones. Both the preservation of these natural regions and the improvement of living conditions for urban dwellers are attainable goals that may be accomplished via the intentional incorporation of wetland areas into urban planning frameworks. By recognizing the ecological worth, economic benefits, and social relevance of wetland areas in the context of water purification, recreational activities, and flood management, this approach acknowledges the triple value that wetland areas possess. The cities like Singapore, have become a very good example of how you can utilize wetlands for urban purposes. The Bishan-Ang Mo Kio Park creation of a restored wetland here is the home for birds like egrets and kingfishers and also serves the humans aesthetically and recreationally. Sustainable urban drainage systems (SUDS) are very important for eco-sensitive planning. SUDS includes built wetlands, retention ponds, and bioswales, all of which serve double purposes of rainwater management and providing the bird habitats. SUDS by imitating the natural hydrological processes help prevent flooding and water pollution besides improving the quality of urban wetlands [78]. The use of such systems in the urban planning of Melbourne, Australia, has alleviated the situation caused by urban runoff and thus has made conditions more comfortable for water birds such as herons and cormorants. Participation of community members and the creation of awareness are the main factors of success in eco-sensitive urban planning. Locals often have such a personal relationship with wetlands that they can go to the extent of being their stewards. When community members are directly involved in the conservation of wetlands through providing support to the various eco-sensitive planning teams, it becomes a win-win situation and the success of the transition can be much longer and sustainable. They can be used as a case in point among others, the Loktak Lake region in Manipur, India, where the community-led wetland restoration projects have proven to be the most powerful force in preserving this Ramsar site not only that but supporting species like the Black-necked Grebe and the globally threatened Baer's Pochard. Dealing with the effects of urban heat islands (UHIs) is also among the main goals of eco-sensitive urban planning. UHIs make it worse for wetlands to suffer from degradation by changing the temperature and the hydrological cycle. These strategies include reforesting riparian zones, planting native species, and designing urban layouts that give precedence to green spaces, such as parks and gardens, among others; all these strategies are geared towards mitigating the effects of UHI [49].

(v) **Urban Planning for Biodiversity Conservation as a Strategy for Water bird Conservation**: Urban planning that includes biodiversity conservation is very important in reducing negative effects of urbanization on water bird populations. Nature is the source of life, and its habitats keep the water birds fed and safe. As cities grow bigger and bigger, they take over the space that is natural for animals. In particular, the wetlands get destroyed, hence causing a loss of habitat, fragmentation, and a downhill trend in their quality. The urban planning, integrating the ecological principles, can become a suitable system for the water birds if it supports them with the capacity of habitat preservation and restoration, thus informing elimination of the threats, repopulation of the species, and peaceful co-existence of the city-urban environments and wildlife [85]. One of the most powerful features of urban planning, that makes it very supportive for water birds, is green infrastructure. The elements of green infrastructure like urban wetlands, green roofs, bioswales, and rain gardens are able to perform the ecological functions of nature which are essential for water birds. For instance, the urban wetlands were constructed as a part of the city planning in Singapore and also in Bangkok. They are also very good shelters for water birds because of their shallow water which gives access to food and resting places for birds such as waders, herons, and other water bird species. These places ensure that the urban areas can be a part of the biodiversity conservation and not be the cause of it. Zoning and land-use planning can also be regarded as the main instruments with which urban planning can embrace water bird conservation. It is also very important to designate some spaces as buffer zones around the wetlands so that human beings will not disturb the birds while they are in the wetlands [98]. Cities like Pune have taken steps to apply zoning regulations to uphold the important wetlands that are the main source of life for the migratory water birds. Land-use plans that give the highest priority to the conservation of critical habitats are the ones that will be at the highest level of assurance that the destruction of the habitat will be kept at the minimum possible level if any. Another factor that plays a significant role in the planning of the cities is the regaining of the connections by means of the ecological corridors. The fragmentation caused by the urban development can isolate water bird populations, which in turn means less genetic diversity and less access to resources. Building ecological corridors or flyways through urban areas lets water birds travel between habitats without danger. To give an example, the incorporation of green corridors in the city planning of Curitiba, Brazil, ensures that various bird species, including water birds, can still move after urban development has not cut their migratory paths. Water-sensitive urban design (WSUD) is a sustainable method of urban planning that considers the importance of water and biodiversity. WSUD is such an approach that if we follow it, with features such as constructed wetlands, retention basins, and permeable surfaces, it leads to the reduction of urban runoff and pollution, which is an important factor for the water birds to be a threat that is eliminated. The water bodies not polluted and therefore clean attract water birds which in turn provide the birds with food and shelter and also raise the aesthetic and eco-value of the cities [84].

(vi) **Role and Use of Constructed Wetlands in Biodiversity Conservation for Water birds**: Constructed wetlands are man-made habitats designed to imitate natural wetlands in terms of both function and structure. At first, the artificial wetlands were designed for such purposes as water treatment, flood control, and wastewater management. However, these systems are now indispensable as the biodiversity conservation efforts continue to intensify as very important to water bird population providing the bird new habitats in places where the natural wetlands have been lost or greatly disturbed. One of the most important pluses of constructed wetlands is represented by their functions as sanctuaries for water birds in urbanized and industrialized regions. They create these habitats that have the same ecological conditions of natural wetlands, thus they give water birds critical resources such as nesting sites, roosting areas, and food sources that are plentiful [121]. An example of this is the constructed wetlands that are located in urban parks and wastewater treatment plants in Europe and Australia. These wetlands are the main rest and food sites for migratory birds like Black-tailed Godwits, Northern Shovelers, and various wader species. In addition, these constructed wetlands are highly efficient in providing countermeasures to the damage suffered by habitats also, in such regions that witness unimpeded urbanization and an increase in agricultural activities, these wetlands can give back to nature the areas that are lost to development [22]. Constructed wetlands acting along migratory flyways are environmentally friendly solutions to the issue of water birds' rest and food. It is a similar approach to reuse the available water for one species in place of another, thus balancing the climate of the ecosystem. The constructed wetlands are made exemplified by the one in the People's Republic of China along the East Asian-Australasian Flyway, and this is a list of wetlands in China along the East Asian-Australasian Flyway that support over a million migratory birds, of which several species are endangered [102, 103].The role of the constructed wetlands is not limited to habitat provision only. In fact, they also act as natural purifiers of the water as they rid the water of pollutants, absorb the necessary nutrients, and get rid of harmful substances through the natural process of decomposition. This means that the cleaner water leads to better and more water species to be used as food for the water birds. For example, the Nakivubo Swamp in Uganda is a partly natural and partly man-made wetland clearing waste water thus enabling different species like herons and jacanas to find a habitat [51]. Additionally, the carbon stock of water bodies is currently limited because the carbon stock is lost in the process of desiccation. Around the globe, different places face different problems due to climate change but in the process, wetlands also mitigate this however by regulating water levels and thus cooling the local temperatures, they help the ecosystems become more stable and protected against hotter or drier periods/floods, some of which are becoming more frequent in the water birds' habitats. Furthermore, in that part of the world, where it is also getting hotter and dryer, land reclamation becomes more and more frequent, so the wetlands created artificially are very important as they can provide a permanent water source and at the same time prevent further degradation of the natural wetlands. Additionally, these systems are good platforms to share and spread conservation ideas and eco-tourism. These areas can bring in the funds they need to support the local communities and plans to protect the birds. Artificial wetlands are an extremely effective means for water bird conservation and also help in the protection of biodiversity. They not only act as the primary source of habitat for animals, but also work as a vehicle for habitat restoration, improving ecosystem services, and increasing climate resilience. Since natural wetlands are under great pressure and their condition is deteriorating gradually, artificial wetlands' role in water bird population sustenance and ecological balance only becomes more necessary. If these ecosystems can be merged into a bigger conservation strategy, then the future will become more sustainable for the water birds and the ecosystems which depend on them.

(vii) **Role of Wetland Protection and Restoration in Conserving Water bird Populations**: The protection and restoration of wetlands are the main strategies for water bird conservation that not only provide habitats but also tackle the problems caused by urbanization, pollution, and climate change. Wetland protection means implementing different measures such as legal tools, programs for community involvement, and reinforcement of regulations to prevent further degradation of wetland ecosystems. Protected wetlands, for example, national parks, wildlife sanctuaries, and Ramsar sites, are places where many water bird species can find safety [33]. The case of India's Bharatpur Keoladeo National Park being recognized as a Ramsar site, it has become a very important place of shelter for the Siberian Crane and Painted Stork, species which are in danger of extinction globally. To be more specific, the policy regulations that are strongly enforced such as those concerning the management of land use, the prevention of water pollution, and the control of the spread of invasive species are really important for the habitat conditions to remain suitable for the water birds [20, 101]. On the other hand, wetland restoration refers to choosing the wetlands that have been destroyed or have lost their natural functions, and bringing them up to a good ecological state again. There are different activities which involve the reestablishing of natural water movement, eliminating of invasive species, the replanting of indigenous vegetation, and the control of pollution. Restoration projects have turned out to be particularly successful in restoring the ecological balance of wetlands and in improving the carrying capacity of water birds. The utmost example is the restoration of Chilika Lagoon in India, where dredging the silted channels facilitated water exchange and decreased the salinity. This measure triggered a tremendous rise in the population of migratory water birds, such as flamingos and pelicans, thus turning the lagoon into a biodiversity hotspot [67, 81]. In addition, wetland protection and restoration projects also represent the alleviation of urbanization and climate change. Wetlands function as natural buffers, absorbing pollutants, regulating water cycles, and mitigating urban heat island effects. Water birds will gain access to these clean and safe places because protecting these habitats will guarantee it to them, even if the surrounding landscapes are facing the pressure of urban areas. Moreover, restored wetlands act as an alternative home for water birds which have been forced out of their original habitats that have been degraded or destroyed. As an illustration, human-made wetlands in cities, similar to the ones that have been established in Bangkok, Thailand, function as habitats for birds such as herons and egrets, while at the same time, they recycle water and remove pollutants from the environment. The factor that influences the greatest the progress of these ventures is local people involvement and awareness. Local communities can be deeply committed to conservation through participation, thus, if they are empowered to decide and participate in decision-making, the success will be long-term [94]. In Assam, India, there are various programmes where the villagers are deeply committed to the protection of the nesting sites of the Greater Adjutant Storks and they show how conservation contributions of communities can complement the efforts of wetland restoration. Promoting eco-tourism as an economic incentive is the way it will mobilize local participation and provide both financial and other supports to conservation projects [10, 24]. On the other hand, the conservation and restoration of wetlands enrich the intention of biodiversity that is the main goal of the conservation in general. Wetlands are indispensable not just for water birds but they also provide for a countless number of other species and in this way, they increase the resilience of ecosystems and promote ecological balance. Hence, by rehabilitating and conserving these habitats, biodiversity protection activities create a domino effect that pulls along the whole ecosystems to the good, thus being able to guarantee not only the survival of birds but also the innumerable species that cohabit wetland environments [86]. Wetland protection and restoration are fundamentally important instruments in the maintenance of water bird populations. Such initiatives (Fig. 8.2) stem from the need to prevent habitat loss, they also help reduce the impact of human activities and furthermore they assist in the preparation for climate change. In addition to providing habitats for water birds, protecting wetlands means ecosystems that are sources of life for both wildlife and human communities become healthy and can be sustained.

## 8.8 Future Research Directions in Water Bird Conservation

Futuristic water birds preservation through the implementation of conservation plans defined by policies, ensures appropriate wetlands management, and habitat restoration. Combination of modern techniques with more conventional tactics, encourages researchers to become active in the advances technological pursuits enhancing the safety of protected areas beneficial to conservation efforts via pollution reduction, climatic change control, and the reduction of disturbances caused by anthropogenic activities. The advancement of technology in the areas of remote sensing, geographic information systems (GIS), and ecological modeling will make it much simpler for researchers to monitor and make more informed decisions for the purpose of protecting water birds and the inhabiting environments essential to strictly enforce environmental regulations and collaborate with individuals from all regions of the world.

(i) **Need for Comprehensive Studies on Urbanization Impacts on Water birds**: The global expansion of urbanization and its consequences on wildlife has become more and more apparent however, additional research is required to determine the ways in which urbanization is influencing water birds and the inhabiting environments imperative to investigate the ways in which urban growth leads to the fragmentation of habitats, pollution, increase in urban heat, and the movement of humans into regions occupied by wildlife [37]. There has been research conducted on certain aspects of this subject; however, the majority of studies are still primarily focused on a limited number of regions or are only conducted for a brief period of time. It is important for researchers to adopt a more holistic perspective and investigate the ways in which urbanization impacts water birds at various moments in time. For instance, researchers could investigate the impact that the expansion of cities from rural, suburban, and urban areas to industrial zones has on birds that are adapted to living in water. Additionally, research that compares urban wetlands to non-urban wetlands would provide crucial insights into the responses of water birds to the development that is caused by humans to better understand the ways in which urbanization impacts wildlife to devise individualized solutions to mitigate the negative consequences of urbanization and enhance the quality of life for aquatic birds residing in urban areas [70, 116].

(ii) **Long-Term Monitoring of Habitat Quality and Bird Population Trends**: Several studies follows populations over time are absolutely necessary for gaining an understanding of aquatic bird populations by performing continual bird observation and study, as well as checking the state of vegetation, monitoring water levels, and evaluating contaminants, scientists are able to spot changes in the environment [66]. Long-term approach enables researchers to document changes in water bird populations through long-term monitoring, quick and timely identification of population reduction, and assessment of effectiveness conservation initiatives conducting a more accurate assessment of the degree to which water birds are able to withstand a variety of environmental and anthropogenic disturbances by using specialized monitoring technologies in conjunction with abundant historical data to track the effects of urban growth, land reclamation, and climate change on wetland areas [5]. Lesser Flamingo and Black-headed Ibis migrations can be tracked over a number of years to better understand the factors that lead to the observed fluctuations in these two species' populations. Based on the data collected, conservation efforts can be adjusted as needed to improve the results of agricultural and conservation efforts [65].

(iii) **Role of Technology in Advancing Research**: A tremendous growth in the pursuit of ecological study and the conservation of animals with technological advances with the assistance of satellite photographs and several other cutting-edge monitoring tools, wetland ecosystems and populations of water birds are being monitored in great detail. Scientists are able to identify locations where wetlands are being damaged, identify regions where changes in land use are occurring, and determine the extent to which urban areas are expanding, monitoring changes in water quality and identifying instances of pollution in wetland areas that are difficult to access or located at a significant distance. e-Bird, a citizen science website revolutionizing bird monitoring by enabling users to immediately record sightings in addition to the use of satellite photos paves way for previously unexplored areas indicating that contributors give essential information on the presence of various bird species over the whole world, as well as their mobility and the number of individuals providing a deeper understanding of water birds and contribute to the efforts that are being made to monitor them on a broad basis [23].

In order to explore the connections between habitat and the actions of water birds, it is possible to use the data of e-Bird in conjunction with additional environmental records with advancement of ornithology facilitated by utilization of drones for aerial research and specialized equipment for bioacoustics research can quantify the niche and habitat use by specific avian species [13, 111]. The use of drones allows bird populations monitoring, the acquisition of high-quality photographs of wetland areas, and the monitoring of nests without causing the birds to be disturbed. It is possible for scientists to employ acoustic sensors to capture the noises of birds enabling them to observe birds that are active at night and to keep track of population numbers in areas where people have an effect on nature [124]. Artificial intelligence and machine learning are being used in analysis of large amounts of data collected through online citizen science platforms, research questionnaires, and satellite photos [28, 71, 106] helping the scientists to identify species, track their movements, spot trends, and forecast future population changes using historical data by utilizing state-of-the-art technology and conducting long-term water bird observations to better understand the impacts of urban development on water birds as the most promising direction for water bird conservation research with technological developments for easier data collection and management of water bird habitats paving the way for important and successful administrative decisions [28, 71, 106].

## 8.9 Conclusion

This chapter emphasizes the vital connection between urbanization, habitat degradation, and water bird population survival. Several research findings shows that aquatic bird species are seriously threatened by the confluence of habitat loss, pollution, urban heat, and human activity, particularly in areas that are rapidly becoming more urbanized leading to wetlands being invaded, fragmented, and damaged as a result of urbanization affecting water bird population. Recognizing the pressing need to address these issues is crucial as aquatic avifauna populations are important indicators of the health of wetland ecosystems, and their decline is a sign of wider environmental deterioration. A significant number of species are currently endangered as a result of the rapid expansion of urban areas due to the rise in buildings, reduction in amount of land suitable for human habitation, and the harmful air pollution. The emphasis only on urban growth may negatively affect aquatic marine life and the environment urging the need to address conservation measures via scientific and community science work together to find solutions to these issues. Enhancing their habitats, enacting sensible legislation, raising the standard of living in cities, and involving the community can help to protect aquatic birds by creating habitat corridors, wetland areas, and green spaces in urban areas leading to preservation and restoration of vital ecosystems. The preservation of aquatic species can also be positively impacted by ongoing monitoring of locations, the use of satellites to monitor such areas, and the coordination of local and global activities where wetland areas have been destroyed due to the substantial increase in settlements, endangering water birds to ensure their survival and safeguard the inhabiting ecosystem by developing a thorough plan that takes into account both environmental and economic factors.

## References

1. Adams CA, Blumenthal A, Fernández-Juricic E, Bayne E, St Clair CC (2019) Effect of anthropogenic light on bird movement, habitat selection, and distribution: a systematic map protocol. Environ Evid 8(1):1–16
2. Adhurya S, Das S, Ray S (2020) Guanotrophication by waterbirds in freshwater lakes: a review on ecosystem perspective. In: Mathematical analysis and applications in modeling: ICMAAM 2018, Kolkata, India, January 9–12, pp 253–269
3. Aghababyan K (2024) Birds as potential bioindicators for terrestrial ecosystems. Int J Life Sci Res Arch 6(01):001–022
4. Akani GC, Amuzie CC, Alawa GN, Nioking A, Belema R (2022) Factors militating against biodiversity conservation in The Niger Delta, Nigeria: the way out. In: Biodiversity in Africa: potentials, threats and conservation. Springer Nature Singapore, Singapore, pp 573–600
5. Amano T, Székely T, Sandel B, Nagy S, Mundkur T, Langendoen T, Sutherland WJ (2018) Successful conservation of global waterbird populations depends on effective governance. Nature 553(7687):199–202
6. Amat JA, Green AJ (2010) Waterbirds as bioindicators of environmental conditions. In: Conservation monitoring in freshwater habitats: a practical guide and case studies, Dordrecht: Springer Netherlands:45–52.
7. Andrade R, Bateman HL, Franklin J, Allen D (2018) Waterbird community composition, abundance, and diversity along an urban gradient. Landsc Urban Plan 170:103–111
8. Aouadi A, Samraoui F, Touati L, Nedjah R, Souiki L, Samraoui B (2021) Close to the madding crowd: waterbird responses to land use conversion in and around a Mediterranean urban wetland. Wetlands 41:1–12
9. Banville MJ, Bateman HL, Earl SR, Warren PS (2017) Decadal declines in bird abundance and diversity in urban riparian zones. Landsc Urban Plan 159:48–61
10. Bhatta LD, Chaudhary S, Pandit A, Baral H, Das PJ, Stork NE (2016) Ecosystem service changes and livelihood impacts in the Maguri-Motapung Wetlands of Assam, India. Land 5(2):15

11. Brandis KJ, Bino G, Spencer JA, Ramp D, Kingsford RT (2018) Decline in colonial waterbird breeding highlights loss of Ramsar wetland function. Biol Conserv 225:22–30
12. Callaghan CT, Major RE, Lyons MB, Martin JM, Kingsford RT (2018) The effects of local and landscape habitat attributes on bird diversity in urban greenspaces. Ecosphere 9(7):e02347
13. Cao L, Meng F, Zhao Q (2021) Understanding effects of large-scale development on bird migration and habitats through cutting edge avian monitoring techniques. Bull Chin Acad Sci (Chin Ver) 36(4):436–447
14. Caseiro-Silva F, Faria FA, Barreto CT, Fernandez CN, Bugoni L (2023) Colonial waterbirds provide persistent subsidies to swamp forests along an estuarine Island food chain. Oecologia 202(1):113–127
15. Catry T, Lourenço PM, Lopes RJ, Carneiro C, Alves JA, Costa J, Granadeiro JP (2016) Structure and functioning of intertidal food webs along an avian flyway: a comparative approach using stable isotopes. Funct Ecol 30(3):468–478
16. Cerda-Peña C, Rau JR (2023) The importance of wetland habitat area for waterbird species-richness. Ibis 165(3):739–752
17. Chakraborty SK, Chakraborty SK (2021) Diversity and conservation of wildlife associated with rivers: an eco-ethological analysis. In: Riverine ecology volume 2: biodiversity conservation, conflicts and resolution, Springer, pp 287–441
18. Chandler M, See L, Copas K, Bonde AM, López BC, Danielsen F, Turak E (2017) Contribution of citizen science towards international biodiversity monitoring. Biol Conserv 213: 280–294
19. Chen S, Zhang Y, Xu C, Cao L, Huang ZY, Li C, de Boer WF (2022) Neighbourhood threats: landscape context and anthropogenic changes can trigger waterbird population collapse. Landsc Ecol 37(12):3141–3158
20. Cherkaoui SI, Selmi S, Amhaouch Z, Hanane S (2018) Assessment of the effectiveness of wetland protection in improving waterbird diversity in a Moroccan wetland system. Environ Monit Assess 190(12):699
21. Cianchetti-Benedetti M, Becciu P, Massa B, Dell'Omo G (2018) Conflicts between touristic recreational activities and breeding shearwaters: short-term effect of artificial light and sound on chick weight. Eur J Wildl Res 64(2):19
22. Čížková H, Kučera T, Poulin B, Květ J (2023) Ecological basis of ecosystem services and management of wetlands dominated by common reed (Phragmites australis): European perspective. Diversity 15(5):629
23. De Barbosa KV, Develey PF, Ribeiro MC, Jahn AE (2021) The contribution of citizen science to research on migratory and urban birds in Brazil. Ornithol Res 29:1–11
24. Doley DM, Barman P (2025) The role of biodiversity communication in mobilizing public participation in conservation: a case study on the greater adjutant stork conservation campaign in Assam, Northeast India. J Nat Conserv 84:126847
25. Donald PF, Fishpool LD, Ajagbe A, Bennun LA, Bunting G, Burfield IJ, Wege DC (2019) Important Bird and Biodiversity Areas (IBAs): the development and characteristics of a global inventory of key sites for biodiversity. Bird Conserv Int 29(2):177–198
26. Donnelly JP, King SL, Silverman NL, Collins DP, Carrera-Gonzalez EM, Lafón-Terrazas A, Moore JN (2020) Climate and human water use diminish wetland networks supporting continental waterbird migration. Glob Chang Biol 26(4):2042–2059
27. Donnelly JP, Moore JN, Casazza ML, Coons SP (2022) Functional wetland loss drives emerging risks to waterbird migration networks. Front Ecol Evol 10:844278
28. Dronova I, Kislik C, Dinh Z, Kelly M (2021) A review of unoccupied aerial vehicle use in wetland applications: emerging opportunities in approach, technology, and data. Drones 5(2): 45
29. Duan H, Yu X (2022) Linking landscape characteristics to shorebird habitat quality changes in a key stopover site along the East Asian–Australasian Flyway migratory route. Ecol Indic 144: 109490

30. Dube T, Dube T, Marambanyika T (2023) A review of wetland vulnerability assessment and monitoring in semi-arid environments of sub-Saharan Africa. Phys Chem Earth, Parts A/B/C 132:103473
31. Dutta H (2017) Insights into the impacts of four current environmental problems on flying birds. Energy Ecol Environ 2:329–349
32. Egwumah FA, Egwumah PO, Edet DI (2017) Paramount roles of wild birds as bioindicators of contamination. Int J Avian Wildlife Biol 2(6):194–199
33. Fan J, Wang X, Wu W, Chen W, Ma Q, Ma Z (2021) Function of restored wetlands for waterbird conservation in the Yellow Sea coast. Sci Total Environ 756:144061
34. Ferraro DM, Le MLT, Francis CD (2020) Combined effect of anthropogenic noise and artificial night lighting negatively affect Western Bluebird chick development. Condor 122(4):duaa037
35. Fujita M, Kameda KO, Şekercioğlu ÇH, Wenny DG, Whelan CJ (2016) Nutrient dynamics and nutrient cycling by birds. In: Why birds matter: Avian ecological function: nutrient dynamics and nutrient cycling by birds, The University of Chicago Press, pp 271–291
36. Gaget E, Galewski T, Jiguet F, Le Viol I (2018) Waterbird communities adjust to climate warming according to conservation policy and species protection status. Biol Conserv 227: 205–212
37. Gao M, Fang S, Deitch MJ, Hu Y, Zhang D, Wan Z, Gebremicael TG (2023) Prioritizating Birds' habitats for conservation to mitigate urbanization impacts using field survey-based integrated models in the Yangtze River Estuary. Land 12(12):2115
38. Gibbs JP, Rouhani S, Shams L (2017) Scale-dependence in polychlorinated biphenyl (PCB) exposure effects on waterbird habitat occupancy. Ecotoxicology 26:762–771
39. Gil-Delgado JA, Guijarro D, Gosálvez RU, López-Iborra GM, Ponz A, Velasco A (2017) Presence of plastic particles in waterbirds faeces collected in Spanish lakes. Environ Pollut 220:732–736
40. Grande JM, Orozco-Valor PM, Liébana MS, Sarasola JH (2018) Birds of prey in agricultural landscapes: the role of agriculture expansion and intensification. In: Birds of prey: biology and conservation in the XXI century, pp 197–228
41. Green AJ, Elmberg J (2014) Ecosystem services provided by waterbirds. Biol Rev 89(1): 105–122
42. Grimes ES, Kneer ML, Berkowitz JF (2024) Military activity and wetland-dependent wildlife: a warfare ecology perspective. Integr Environ Assess Manag 20(6):2153–2161
43. Haaksma LR (2022) Surviving urbanisation and agricultural intensification: a review of the threats for farmland and urban birds in human-modified landscapes in The Netherlands
44. Haig SM, Murphy SP, Matthews JH, Arismendi I, Safeeq M (2019) Climate-altered wetlands challenge waterbird use and migratory connectivity in arid landscapes. Sci Rep 9(1):4666
45. Hamza F, Abdi AM, Chokri MA, Yacoubi L, Dlensi H, Hammouda N, Hanane S (2024) Landscape context and wetland attributes influence wintering waterbirds in important bird and biodiversity areas: implications for conservation and management. Landsc Ecol 39(8):151
46. Holbech LH, Cobbinah CC (2021) Pollution or protection-what early survey data shows on rapid waterbird utilisation of a newly established sewage treatment plant in urban Ghana, West Africa. Wetlands 41(8):110
47. Holland ER, Mallory ML, Shutler D (2016) Plastics and other anthropogenic debris in freshwater birds from Canada. Sci Total Environ 571:251–258
48. Imdad K, Rihan M, Sahana M, Parween S, Ahmed R, Costache R, Tripathi R (2023) Wetland health, water quality, and resident perceptions of declining ecosystem services: a case study of Mount Abu, Rajasthan, India. Environ Sci Pollut Res 30(55):116617–116643
49. Irfeey AMM, Chau HW, Sumaiya MMF, Wai CY, Muttil N, Jamei E (2023) Sustainable mitigation strategies for urban heat Island effects in urban areas. Sustainability 15(14):10767
50. Jandaghian Z, Colombo A (2024) The role of water bodies in climate regulation: insights from recent studies on urban Heat Island mitigation. Buildings 14(9):2945

51. Kansiime F, Nalubega M (2020) Socio-economics of the Nakivubo swamp. In: Wastewater treatment by a natural wetland: the Nakivubo Swamp, Uganda. CRC Press, pp 265–281
52. Kekkonen J (2017) Pollutants in urbanized areas: direct and indirect effects on bird populations. In: Ecology and conservation of birds in urban environments, Springer, Heidelberg, pp 227–250
53. Khan BN, Ashfaq Y, Hussain N, Atique U, Aziz T, Alharbi M, Alasmari AF (2023) Elucidating the effects of heavy metals contamination on vital organ of fish and migratory birds found at fresh water ecosystem. Heliyon 9(11):e20968
54. Kiran SD, Kour A, Priya DV, Kumar R (2022b) Community composition and status of avian diversity at Campus and Agricultural landscapes of Chaudhary Charan Singh Haryana Agricultural University, Hisar (Haryana). J Appl Nat Sci 14(4):1130–1140. https://doi.org/10.31018/jans.v14i4.3784
55. Kiran SD, Kour A, Priya DV, Kumar R, Punia N, Sunita (2022a) Different strategies adopted by birds to sustain ecosystem: a review. Pharm Innov SP-11(9):412–422
56. Kolawole AS, Iyiola AO (2023) Environmental pollution: threats, impact on biodiversity, and protection strategies. In: Sustainable utilization and conservation of Africa's biological resources and environment. Springer Nature Singapore, Singapore, pp 377–409
57. Kumar R, Singh D, Yadav R (2025) Toxic effects of chlorantraniliprole and fipronil on *Eisenia fetida*: from growth and reproductive parameters to avoidance behavior. Environ Eng Manag J 24(7):1499–1508. Accepted manuscript
58. Kumar R, Singh D (2022) Insecticide-tolerant bacterial population in *Eisenia fetida*'s Gut and Vermicast exposed to Chlorantraniliprole and Fipronil. Appl Biol Res 24(3):273–279
59. Kumar R, Finlayson CM, Pattnaik AK (2020) Ecological characterization of Chilika: defining strategies and management needs for wise use. In: Ecology, conservation, and restoration of Chilika Lagoon, India, Springer International Publishing: pp 23–61
60. Kumar R, Yadav R, Gupta RK, Pal A, Yodha K, Kumar A (2024) Antioxidant enzyme activities and markers of oxidative stress in the life cycle of different Earthworm species. Indian J Exp Biol 62(07):578–583
61. Lee AT, Geary C, Wright DR, Dean WRJ (2019) Vulnerability of birds to contaminated water sources in the Karoo region of South Africa. Ostrich 90(4):397–406
62. Lehikoinen A, Rintala J, Lammi E, Pöysä H (2016) Habitat-specific population trajectories in boreal waterbirds: alarming trends and bioindicators for wetlands. Anim Conserv 19(1):88–95
63. Leung F, Doherty PD, Liu M, Metcalfe K, Godley B, Lee SY (2024) Rise and fall of an avian oasis: tracking the impacts of land use change in a key coastal wetland in the world's largest megalopolis. Sci Total Environ 906:167231
64. Li Y, Mao D, Wang Z, Wang X, Tan X, Jia M, Ren C (2021) Identifying variable changes in wetlands and their anthropogenic threats bordering the Yellow Sea for water bird conservation. Global Ecol Conserv 27:e01613
65. Liu Y, Ji X, Zhou L (2024) Assessment of Waterbird habitat importance and identification of conservation gaps in Anhui Province. Animals 14(7):1004
66. Ma Z, Choi CY, Gan X, Li J, Liu Y, Melville DS, Zhang Z (2023) Achievements, challenges, and recommendations for waterbird conservation in China's coastal wetlands. Avian Res 14:100123
67. Manivanan R (2017) Coastal wetland restoration: concepts, methodology, and application areas along the Indian coast. In: Coastal wetlands: alteration and remediation, pp 437—454, Springer, switzerland
68. Mariyappan M, Rajendran M, Velu S, Johnson AD, Dinesh GK, Solaimuthu K, Sankar M (2023) Ecological role and ecosystem services of birds: a review. Int J Environ Climate Change 13(6):76–87
69. Martín-Vélez V, Mohring B, van Leeuwen CH, Shamoun-Baranes J, Thaxter CB, Baert JM et al (2020) Functional connectivity network between terrestrial and aquatic habitats by a generalist waterbird, and implications for biovectoring. Sci Total Environ 705:135886

70. Matchett EL, Fleskes JP (2017) Projected impacts of climate, urbanization, water management, and wetland restoration on waterbird habitat in California's Central Valley. PLoS One 12(1):e0169780
71. Medvedev A, Telnova N, Alekseenko N, Koshkarev A, Kuznetchenko P, Asmaryan S, Narykov A (2020) UAV-derived data application for environmental monitoring of the coastal area of Lake Sevan, Armenia with a changing water level. Remote Sens 12(22):3821
72. Meng W, He M, Hu B, Mo X, Li H, Liu B, Wang Z (2017) Status of wetlands in China: a review of extent, degradation, issues and recommendations for improvement. Ocean Coast Manag 146:50–59
73. Mishra M, Acharyya T, Chand P, Santos CAG, da Silva RM, Dos Santos CAC, Kar D (2022) Response of long-to short-term tidal inlet morphodynamics on the ecological ramification of Chilika lake, the tropical Ramsar wetland in India. Sci Total Environ 807:150769
74. Mohring B, Henry PY, Jiguet F, Malher F, Angelier F (2021) Investigating temporal and spatial correlates of the sharp decline of an urban exploiter bird in a large European city. Urban Ecosyst 24:501–513
75. Morissette JL, Bayne EM, Kardynal KJ, Hobson KA (2019) Regional variation in responses of wetland-associated bird communities to conversion of boreal forest to agriculture. Avian Conserv Ecol 14(1):12
76. Muhammad SI, Junior HH, Ringim AS, Muhammad IL, Onoja J (2022) Waterbird population estimates in Hadejia-Nguru wetlands: analysis of a five-year monitoring program. Wetlands 42(1):12
77. Nam HK, Kim M, Choi G, Jang D, Choi SH, Cho KJ, Kim MH (2016) Influence of pesticide use on distribution of waterbirds in rice fields at mid-western part of South Korea. Korean J Environ Biol 34(4):361–364
78. Nguyen TT, Ngo HH, Guo W, Wang XC, Ren N, Li G, Liang H (2019) Implementation of a specific urban water management-Sponge City. Sci Total Environ 652:147–162
79. Ogidi OI, Akpan UM (2022) Aquatic biodiversity loss: impacts of pollution and anthropogenic activities and strategies for conservation. In: Biodiversity in Africa: potentials, threats and conservation. Springer Nature Singapore, Singapore, pp 421–448
80. Pandiyan J, Poiyamozhi A, Mahboob S, Al-Ghanim KA, Al-Misned F, Ahmed Z, Govindarajan M (2022) Assessment of the toxic effects of heavy metals on waterbirds and their prey species in freshwater habitats. Toxics 10(11):641
81. Pattnaik AK (2024) Reversing the degeneration of an ecological hotspot through an ecosystem approach: a case study of Chilka Lake. In: Regenerative ecosystems in the Anthropocene: a transdisciplinary ecosystemic framework for regenerativeness. Springer Nature Switzerland, Cham, pp 129–150
82. Pavón-Jordán D, Santangeli A, Lehikoinen A (2017) Effects of flyway-wide weather conditions and breeding habitat on the breeding abundance of migratory boreal waterbirds. J Avian Biol 48(7):988–996
83. Pearce-Higgins JW, Brown DJ, Douglas DJ, Alves JA, Bellio M, Bocher P, Verkuil YI (2017) A global threats overview for Numeniini populations: synthesising expert knowledge for a group of declining migratory birds. Bird Conserv Int 27(1):6–34
84. Pille L, Säume I (2021) The water-sensitive city meets biodiversity: habitat services of rain water management measures in highly urbanized landscapes. Ecol Soc 26(2):23
85. Plummer KE, Gillings S, Siriwardena GM (2020) Evaluating the potential for bird-habitat models to support biodiversity-friendly urban planning. J Appl Ecol 57(10):1902–1914
86. Ponzio KJ, Osborne TZ, Davies GT, LePage B, Sundareshwar PV, Miller SJ, Sandvik C (2019) Building resiliency to climate change through wetland management and restoration. In: Wetlands: ecosystem services, restoration and wise use. Springer International Publishing, Cham, pp 255–309
87. Puthur S, Anoopkumar AN, Rebello S, Aneesh EM, Sindhu R, Binod P, Pandey A (2020) Toxic effects of pesticides on avifauna inhabiting wetlands. In: Sustainable agriculture reviews

47: pesticide occurrence, analysis and remediation Vol. 1 biological systems. Springer International Publishing, Cham, pp 335–349
88. Qiu J, Zhang Y, Ma J (2024) Wetland habitats supporting waterbird diversity: conservation perspective on biodiversity-ecosystem functioning relationship. J Environ Manag 357:120663
89. Rahimi L, Malekmohammadi B, Yavari AR (2020) Assessing and modeling the impacts of wetland land cover changes on water provision and habitat quality ecosystem services. Nat Resour Res 29(6):3701–3718
90. Rajashekara S, Venkatesha MG (2018) Impact of urban threats and disturbance on the survival of waterbird communities in Wetlands of Bengaluru City, India. In: Proceedings of the zoological society, vol 71, No. 4. Springer India, New Delhi, pp 336–351
91. Rehman S, Tiwary NK, Urfi AJ (2021) Conservation monitoring of a polluted urban river: an occupancy modeling study of birds in the Yamuna of Delhi. Urban Ecosyst 24(6):1399–1411
92. Reynolds C, Ryan PG (2018) Micro-plastic ingestion by waterbirds from contaminated wetlands in South Africa. Mar Pollut Bull 126:330–333
93. Sarkar D, Maji N (2022) Status and threats of wetland change in land use pattern and planning: impact of land use patterns and urbanization. In: Handbook of research on monitoring and evaluating the ecological health of wetlands. IGI Global Scientific Publishing, pp 106–127
94. Scholte SS, Todorova M, Van Teeffelen AJ, Verburg PH (2016) Public support for wetland restoration: what is the link with ecosystem service values? Wetlands 36:467–481
95. Sievers M, Hale R, Parris KM, Swearer SE (2018) Impacts of human-induced environmental change in wetlands on aquatic animals. Biol Rev 93(1):529–554
96. Singhar AS, Gundimeda H (2024) Ranking sustainable management alternatives for Chilika wetland ecosystem, in India, using decision support system. In: Environment, development and sustainability, Springer, switzerland, pp 1–32
97. Smith PA, Smith AC, Andres B, Francis CM, Harrington B, Friis C, Brown S (2023) Accelerating declines of North America's shorebirds signal the need for urgent conservation action. Ornithol Appl 125(2), 1-14
98. Snep RP, Kooijmans JL, Kwak RG, Foppen RP, Parsons H, Awasthy M, van Heezik YM (2016) Urban bird conservation: presenting stakeholder-specific arguments for the development of bird-friendly cities. Urban Ecosyst 19:1535–1550
99. Sonone SS, Jadhav S, Sankhla MS, Kumar R (2020) Water contamination by heavy metals and their toxic effect on aquaculture and human health through food chain. Lett Appl NanoBioSci 10(2):2148–2166
100. Steen VA, Skagen SK, Melcher CP (2016) Implications of climate change for wetland-dependent birds in the Prairie Pothole Region. Wetlands 36(Suppl 2):445–459
101. Szabo JK, Mundkur T (2017) Conserving wetlands for migratory waterbirds in South Asia. In: Wetland science: perspectives from South Asia, Springer, New Delhi, pp 105–127
102. Szabo JK, Battley PF, Buchanan KL, Rogers DI (2016a) What does the future hold for shorebirds in the East Asian–Australasian flyway? Emu 116(2):95–99
103. Szabo JK, Choi CY, Clemens RS, Hansen B (2016b) Conservation without borders–solutions to declines of migratory shorebirds in the East Asian–Australasian Flyway. Emu 116(2): 215–221
104. Tariq A, Qadir A, Ahmad SR (2022) Consequences of plastic trash on behavior and ecology of birds. In: Microplastic pollution: environmental occurrence and treatment techno*logies*. Springer International Publishing, Cham, pp 347–368
105. Taylor CM, Stutchbury BJ (2016) Effects of breeding versus winter habitat loss and fragmentation on the population dynamics of a migratory songbird. Ecol Appl 26(2):424–437
106. Tripathi RN, Ramachandran A, Tripathi V, Badola R, Hussain SA (2024) Optimizing riparian habitat conservation: a spatial approach using aerial and space technologies. IEEE J Select Top Appl Earth Observ Remote Sens 17:16932
107. Tryjanowski P, Jankowiak Ł, Czechowski P, Dulisz B, Golawski A, Grzywaczewski G, Wuczyński A (2022) Summer water sources for temperate birds: use, importance, and threats. Eur Zool J 89(1):913–926

108. Varagiya D, Jethva B, Pandya D (2022) Feather heavy metal contamination in various species of waterbirds from Asia: a review. Environ Monit Assess 194(1):26
109. Waliczky Z, Fishpool LD, Butchart SH, Thomas D, Heath MF, Hazin C, Allinson TS (2019) Important Bird and Biodiversity Areas (IBAs): their impact on conservation policy, advocacy and action. Bird Conserv Int 29(2):199–215
110. Wang L, Nabi G, Yin L, Wang Y, Li S, Hao Z, Li D (2021) Birds and plastic pollution: recent advances. Avian Res 12:1–9
111. Wang X, Kuang F, Tan K, Ma Z (2018) Population trends, threats, and conservation recommendations for waterbirds in China. Avian Res 9:1–13
112. Wantzen KM, Alves CBM, Badiane SD, Bala R, Blettler M, Callisto M, Zingraff-Hamed A (2019) Urban stream and wetland restoration in the Global South—a DPSIR analysis. Sustainability 11(18):4975
113. Whelan CJ, Tomback DF, Kelly D, Johnson MD (2016) Trophic interaction networks and ecosystem services. In: Why birds matter: avian ecological function and ecosystem services, The University of Chicago Press, Chicago, pp 49–72
114. Whiteside M, Herndon JM (2018) Aerosolized coal fly ash: a previously unrecognized primary factor in the catastrophic global demise of bird populations and species. Asian J Biol 6(4):1–13
115. Wilson AA, Ditmer MA, Barber JR, Carter NH, Miller ET, Tyrrell LP, Francis CD (2021) Artificial night light and anthropogenic noise interact to influence bird abundance over a continental scale. Glob Chang Biol 27(17):3987–4004
116. Xie S, Marzluff JM, Su Y, Wang Y, Meng N, Wu T, Ouyang Z (2022) The role of urban waterbodies in maintaining bird species diversity within built area of Beijing. Sci Total Environ 806:150430
117. Xu X, Xie Y, Qi K, Luo Z, Wang X (2018) Detecting the response of bird communities and biodiversity to habitat loss and fragmentation due to urbanization. Sci Total Environ 624: 1561–1576
118. Yadav R, Gupta RK, Kumar R, Kaur T, Moun R (2022a) Biochemical response of earthworm, *Eisenia fetida* to heavy metals toxicity. J Appl Nat Sci 14(3):990–998
119. Yadav R, Gupta RK, Kumar R, Kaur T (2022b) Assessment of toxicity of lead and nickel on the biochemical and immunological parameters of earthworm, *Eudrilus eugeniae*. Agric Sci Dig 44(5):922–929
120. Yodha K, Singh D, Kour A, Yadav R, Kumar R (2024) Species diversity and abundance of birds in ber (*Zizyphus mauritiana*) crop. Asian J Conserv Biol 13(1) https://doi.org/10.53562/ajcb.83683
121. Zhang C, Wen L, Wang Y, Liu C, Zhou Y, Lei G (2020) Can constructed wetlands be wildlife refuges? A review of their potential biodiversity conservation value. Sustainability 12(4):1442
122. Zhang L, Ouyang Z (2019) Focusing on rapid urbanization areas can control the rapid loss of migratory water bird habitats in China. Glob Ecol Conserv 20:e00801
123. Zhao X, Zhang M, Che X, Zou F (2020) Blue light attracts nocturnally migrating birds. Condor 122(2):duaa002
124. Zhao Y, Sheppard S, Sun Z, Hao Z, Jin J, Bai Z, Wang C (2022) Soundscapes of urban parks: an innovative approach for ecosystem monitoring and adaptive management. Urban For Urban Green 71:127555
125. Zou K, Thébault E, Lacroix G, Barot S (2016) Interactions between the green and brown food web determine ecosystem functioning. Funct Ecol 30(8):1454–1465

# Chapter 9
# Evaluation of Diminishing Seabird Abundance Trends: Implications for Monitoring

**Sheetal, Savita Sihag, Parveen Kumar Nimbrayan, and Nisha Kumari**

**Abstract** The number of seabirds is steadily declining worldwide, clearly showing that these birds have become very susceptible due to natural and human burdens. The main problems are degradation of their essential habitats, ocean pollution, overfishing, weather changes due to climate change, and the introduction alien (invasive) species into their areas. Especially the wetlands, where these birds depend for breeding and food, are deteriorating rapidly. This not only affects their breeding, but it also impacts the entire ecosystem. Seabirds are considered an indicator of the ecological health of marine and coastal ecosystems by exchanging nutrients between the sea and land. Their numbers and health are deeply linked to the condition of these places. Therefore, integrated conservation strategies are needed. To prevent the current situation from recurring, scientific, local monitoring programs must identify the underlying causes and mechanisms of change in small populations. Protecting the birds is not enough, but protecting their habitats and implementing a comprehensive biodiversity policy is also essential. This chapter emphasizes that scientific monitoring, well-planned interventions, and adaptive approaches can protect the future of seabirds and the entire ecosystems on which our lives and theirs depend.

**Keywords** Biodiversity · Conservation Policy · Monitoring · Population Decline · Seabird

Sheetal (✉) · S. Sihag
Department of Environmental Science & Engineering, Guru Jambheshwar University of Science & Technology, Hisar, Haryana, India

P. K. Nimbrayan
Department of Agricultural Economics, CCS Haryana Agricultural University, College of Agriculture, Bawal, Rewari, Haryana, India

N. Kumari
Department of Commerce, Gurugram University, Gurugram, Haryana, India

 
R. Kumar et al. (eds.), *Ecology, Conservation of Aquatic Birds*, Wetlands: Ecology, Conservation and Management 15, https://doi.org/10.1007/978-3-032-17125-2_9

# 9.1 Introduction

## 9.1.1 Sea Birds

Seabirds are considered the most visible sentinels of the ocean's health, linking marine and terrestrial ecosystems and serving as indicators of environmental changes. Seabirds are birds that spend most of their lives in ocean environments and get most of their food from marine sources — small fish, crustaceans like crabs, molluscs like snails, and even sea insects [90]. These birds belong to different scientific classification groups but depend on the sea for their livelihood. Some species live entirely in the sea and come to land only during breeding time, while some occasionally visit coastal or marine areas for feeding [34, 52]. Seabirds are found in the world's oceans—from polar regions to tropical seas—and have adapted to significant variations in climate, seasonal changes, and marine productivity [106]. By scientific taxonomy, seabirds include: Suliformes, Charadriiformes (such as auks, gulls, and skuas—but not waders and other families), Sphenisciformes (penguins), Procellariiformes (tube-nosed birds such as albatrosses, petrels, and shearwaters), Pelecaniformes (pelicans) [55]. Beyond their physical adaptations, seabirds play essential roles in marine ecosystems. These include long-distance transport of nutrients from the sea to land, maintaining the balance of prey species populations, and serving as bioindicators for fisheries resource management [129].

Even though these birds come from different families, the similar challenges of marine life have caused them to develop many convergent features in size, physiology, and behavior. The biggest challenge for aquatic life is osmoregulation, i.e., maintaining the right balance of salt and water in the body [92]. Many seabirds get the water they need from food, but drinking seawater becomes essential when hunting. The salt in seawater is very high, which can be dangerous for the body. That is why they have special glands (supraorbital salt glands) above the eyes, creating a very dense salty solution and expelling it from the nostrils [57]. In most of the other vertebrates, salt regulation is performed by the kidneys, but in seabirds, this is the primary function of these glands [124]. In recent decades, populations of many seabird species have declined dramatically due to overfishing, plastic ingestion, oil pollution, and habitat changes associated with climate change.

## 9.1.2 Importance of Sea Birds in Marine Ecosystem

Seabirds can serve as significant indicators for environmental conditions and contribute to the maintenance and revival of life in extremely harsh environments such as cold, dry, and volcanic regions. The hydrosphere covers about 70% of the Earth's surface, making the marine realm the largest and most important body of water. It includes a variety of natural habitats such as salt marshes, tidal estuaries, lagoons, mangroves, coral reefs, and deep-sea areas [8]. They serve as suitable homes (i.e.,

**Table 9.1** List of seabird species that various ornithologists have identified

| Family | Scientific Name | Common Name | Ecological Importance | References |
|---|---|---|---|---|
| Alcidae | *Pinguinus impennis* | Great auk | The extinct was an essential predator in North Atlantic marine food webs | [71] |
| Diomedeidae | *Diomedea exulans* | Wandering albatross | Largest wingspan seabird, key long-distance nutrient transporter, and top predator in open oceans | [59] |
| Fregatidae | *Fregata andrewsi* | Christmas frigatebird | Aerial predators are impacting flying fish and small seabird populations | [58] |
| Phalacrocoracidae | *Phalacrocorax aristotelis* | European shag | Coastal predators contribute to fish population control | [38] |
| Rhynchopidae | *Rhynchops niger* | Black skimmer | Feeds on surface fish and invertebrates, an indicator of coastal health | [94] |
| Spheniscidae | *Spheniscus mendiculus* | Galapagos penguin | Endemic species are important benthic predators in marine food webs | [91] |
| Sternidae | *Sterna hirundo* | Common tern | Surface feeder controlling small fish populations is an indicator of water quality | [20] |
| Sulidae | *Sula leucogaster* | Brown booby | Mid-level predator affecting fish populations, nutrient recycler via guano | [2] |
| Stercorariidae | *Catharacta antarctica* | Brown skua | Aggressive predator and scavenger, influences seabird colony dynamics | [78] |
| Laridae | *Anous minutus* | Black noddy | Coastal and island species contributing to marine-terrestrial nutrient cycling | [65] |

places to live, food, shelter, and breeding grounds) for a vast variety of vertebrate animal species (i.e., fish, birds, reptiles, and mammals) and invertebrate species (i.e., millions of species). However, only a tiny portion of the world's marine areas are under protection, such as 1.2% of the world's marine area is protected within exclusive economic zones (EEZ), while 4.3% of the continental shelf and 0.9% of offshore waters [24]. The most productive ecology for seabird species is found in marine areas, which offer a variety of habitats that are abundant in food supplies, attracting a variety of seabird species that can be used all year round. Effective identification of key feeding and breeding sites is essential so these areas can be declared marine protected and managed sustainably. Such efforts are necessary to ensure the successful breeding of seabirds and increase their population worldwide. Table 9.1 gives a detailed description of the various seabird species classified by leading ornithologists, showing their environmental importance and distribution patterns in the marine realm.

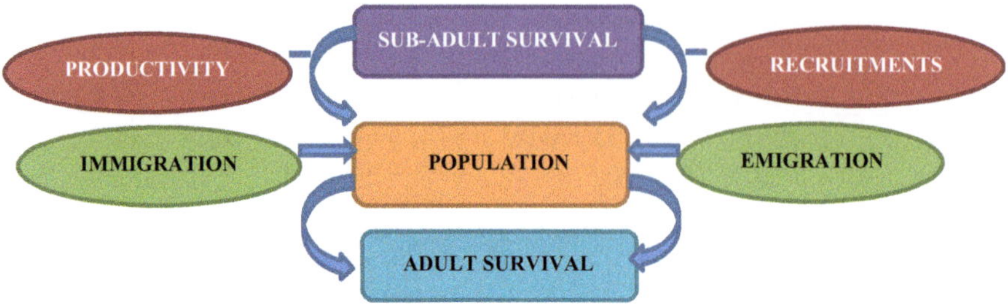


**Fig. 9.1** Primary determinant that controls seabird population and community characteristics in the marine environment

Black, white, or black-and-white seabirds provide information about environmental conditions, food availability, and the coastal or land areas they live in. They are bio-indicators that help us ascertain the environment's quality and sustainability [33]. Some sea birds, such as boobies, gulls, terns, and alcids, often live in large clusters and form colonies in one place. Some species are small in number in these colonies, while millions of birds are found together in some areas, such as Wilson's Storm-petrel or the Sooty Shearwater [32]. Some other species, such as the Chatham Island Petrel and the Chinese Crested Tern, prefer to live alone, which is rarely seen [21].

Seabirds frequently favour upland (erosional) environments for nesting and lounging, and marine nearshore (depositional) regions for foraging [67]. By observing the number of sea birds, the mixture of species and their behaviour, we can estimate the life of the sea, its usefulness, and the dangers present there [101]. To make conservation successful, it is essential to keep observing different things related to seabirds [100]. These include their travel, where they are found in which season, what they eat, how and where they make their nests, and their physical health [130]. Many things affect their population, like how many eggs are given, how many fish or other food are available, how many babies are born, and how many birds have come or gone. All these can change every year or after some time (Fig. 9.1) [117, 150].

The information needed to assess the variables causing population changes across several marine habitats is provided by monitoring the seabirds' parameters in those environments. Monitoring also aids in the conservation and improved management of seabird species that are endangered or threatened. Till now, there is no complete information on the life and behaviour of birds in the sea. Therefore, we cannot know properly how well these birds reflect the changes in the marine environment [77]. On the other hand, not enough research has been done on seabirds' long-term populations, their small geographic range, diverse climates (microhabitats and microclimates), and their interactions. The ecological responsibilities of seabird species concerning microhabitat and habitat disturbances- that is, what would happen to the species when their habitat is changed—are poorly understood. Would there be a rise or fall in the number of seabirds? Or would they go to places that aren't as good for breeding and foraging? Seabirds have adapted to various environments, ranging from the Antarctic to the North Pole. They rely on the marine environment, including coastal areas (such as mangroves, mudflats, estuaries, and islands), either directly or indirectly, to carry out a variety of tasks for their survival and existence, including inhabiting, foraging, perching, loafing, roosting, and breeding [114].

Seabirds spend most of their time in the ocean—sometimes for a few hours, sometimes for months or even years. These birds fly in the air, often hovering over the sea so that they can find food. When they find suitable food, like fish or small marine creatures, they catch them and take them to their colony—even if they must travel a long distance. Pelicans, cormorants, gulls, terns, and skimmers are among the seabirds frequently seen near the coast and in estuaries [97]. Similarly, boobies, petrels, and albatrosses are always found offshore. The fact that all seabirds graze in saline water is one of their most common traits. For nesting and foraging purposes, seabirds frequently use a variety of environments.

### 9.1.3 Purpose and Scope

Through trophic and non-trophic processes, seabirds can directly affect trophic status, biodiversity, environmental contamination, and food webs, among other ways, to shape coastal ecological processes and services. Nowadays, the number of sea birds is decreasing rapidly. The main reason for this is many human activities. When foreign animals like rats and cats are brought into the nesting areas of birds, they destroy their eggs. Birds get trapped and die in fishing nets, and food is also reduced due to humans catching too many fish. On the other hand, increased pollution and oil cleaning also harm them. Sharp weather changes, frequent storms, and climate change have also become a big problem for them. Because of all these reasons, many seabird species are now at risk—some are on the verge of extinction or have disappeared from their original habitat. These birds are essential to our marine environment, so protecting them is our responsibility. Everyone should know the importance of seabirds and their safety, but saving them is difficult until the general public understands why these birds are essential. Saving sea birds is not the job of just one or two organizations or governments; instead, all of us should come together and contribute towards their safety and protection so that the coming birds can also see them.

### 9.1.4 Seabirds as Bioindicators in Marine Ecosystem

Seabirds are useful bioindicators of marine environments. They are easier to spot and recognise, vocalise when they occur, and react to environmental changes more rapidly [98, 113, 116]. Seabirds have previously been employed as bioindicators of environmental pollution in aquatic systems [26], pollution in the Antarctic ecosystem [9], pollution [87], oil spills [120], wetland ecosystem health [109], climate change, primary productivity [115], and pollution in the Antarctic ecosystem [1]. Some examples of seabirds important as environmental indicators include the common term (*Sterna hirundo*), which is used to monitor pollution levels and detect changes in marine food webs [155]; the wandering albatross (*Diomedea exulans*), an indicator of climate change and long-distance transport of oceanic nutrients; the Galápagos penguin (*Spheniscus mendiculus*) [48], whose populations are helpful in assessing the effects of climate change and pollution; the black skimmer (*Rynchops niger*), which is a key indicator of estuarine and coastal water quality; the European shag (*Phalacrocorax aristotelis*) [132], which indicates the status of regional pollutants and fish resources; the brown skua (*Catharacta antarctica*), which means pollution in the Antarctic ecosystem; and the black noddy (*Anous minutus*), which plays a vital role in the cycling of nutrients between the sea and land through its feces (guano) in island and coastal environments [95]. Selecting the right seabird species and researching the location is vital. Some species (such as Great Cormorants and Black-legged Kittiwakes) become agitated when people are around and may kill eggs or chicks [141]. In contrast, other species (like Northern Gannet) can withstand a lot of disturbance. Those sea birds should be chosen as bioindicators that have some special properties [110]. For example, those that easily absorb pollutants in their body and yet are not much affected by poisonous substances. Those birds eat the same food that is part of the food web we have studied, or which are often found in significant habitats, and whose breeding, social structure, and life cycle are already well understood [11]. These species are sensitive to environmental changes and play essential roles in the transfer of nutrients, regulation of food webs, and assessment of habitat quality; for this reason, they are considered sentinels of the health of the marine ecosystem. It's possible that seabirds can exhibit a range of adaptations to take advantage of aquatic resources and may exhibit distinct habitat preferences, which can be used to assess the integrity of marine ecosystems [80]. While seabird community characteristics emphasise environmental conditions and productivity, variable or abiotic factors may reveal the current state of the specific location. The cause-and-effect relationship is essential for conserving and safeguarding a particular place. Since seabirds often build their nests in large pools, they are easy to count and monitor. Therefore, research on these birds or their community information can be used as bioindicators to identify any hazard or pollutant in the sea [126]. The foundation for interpretation to look at the contamination levels and patterns of harmful effects is provided by comprehensive data on seabird ecology [99].

### 9.1.5 Sensitivity of Seabirds to Environmental Changes

Seabirds, which are exposed to hundreds of artificial pollutants, can be significantly impacted by environmental contamination. More than fifty species of seabirds, especially penguins, cormorants, and small petrels, are thought to be at risk from chemical or light pollution [42, 81]. Given the rise in chemical production worldwide [82, 154], the release of plastic into the environment [18, 105, 156, 157], and the construction of offshore or coastal structures and related artificial lighting [37], the effects are probably going to get worse.

Due to their lengthy lifespans and position at the top of marine food webs, seabirds exhibit extremely high concentrations of some chemical pollutants that build up in their tissues and multiply along food chains. These contaminants can therefore have significant sub-lethal impacts on seabirds, influencing their behaviour and physiology and ultimately their ability to reproduce, survive, and maintain a population [95]. Exposure to contaminants can impact a bird's metabolism by disrupting its hormones [14, 93] or decrease hatching success by preventing the bird from maintaining energetically demanding incubation behaviour [139]. According to research on Herring Gulls (*Larus argentatus*), neurotoxic pollutants may affect a young bird's ability to learn and think [23]. This could impact the bird's ability to adapt its behaviour and, consequently, its ability to survive. Beyond these indirect effects, thousands of seabirds die every year as a direct result of acute environmental pollution after oil spills caused by unauthorized or unintentional releases from shipping, oil and gas extraction [29, 151, 153], which may have significant demographic effects [148].

Seabirds are increasingly at risk from plastic pollution, as evidenced by reports of plastic ingestion in various species worldwide [5, 79, 112]. Eating plastic has many effects on sea birds. This can cause plastic to get stuck in their stomach, which causes difficulty and trouble in digesting the food [123]. Secondly, plastic floating in the sea also carries toxic substances with it. When this plastic enters the body of birds, the chemicals and other pollutants found in it also reach their body [138, 159]. However, till now, its direct effect on the entire population of birds has been seen only in some places. It has been seen in many reports that many seabirds, especially seabirds that live in seawater, petrel and storm-petrel, die after colliding with ships, oil platforms, and other large structures in the sea. Birds also die near the seashore due to buildings and road lights. Artificial light has become a different kind of pollution for them, which often attracts them towards itself and becomes the cause of collision [118]. The likelihood of a crash can be significantly impacted by environmental factors, including the moon phase and the weather, most collisions occur on darker nights, in fog, or under rainy circumstances [96]. Birds migrating along coastal areas may also have their ability to navigate altered by artificial lights [146]. It has been observed from experience and even in the open field that excessive light (light pollution) affects the sleeping behaviour of birds. This can disrupt their sleep, negatively impacting their body and behaviour [44, 136]. It is not yet fully understood whether artificial light harms seabirds' abilities. To reduce the effects of light pollution, island residents, shipping companies, and ship crews are changing the type of lights or their direction by installing shields or even turning off the lights when necessary, such as when chicks are urinating or at night on shores. However, investigations into the extent to which light pollution from ocean shipping and offshore structures affects seabirds are still incomplete [96].

### 9.1.6 Factors Affecting Seabird Abundance

A multitude of factors impact seabird abundance, such as:

- **Environmental factors:** Seabird abundance may be affected by coastal upwelling, ocean depth, and distance to the shore [128].
- **Prey abundance:** Seabird demography may be impacted by the quantity of prey, such as forage fish [127]
- **Fishing:** When paired with offshore renewable energy sources like wind, tidal, and wave power, fishing can hurt seabirds [64, 83].
- **Breeding success:** Several variables, including the availability of waste, might influence breeding success [13].
- **Seasonality:** The quantity of seabirds can be influenced by the season [4].
- **Water mass boundaries:** Seabird abundance may be impacted by water mass boundaries [135].
- **Physical processes:** The ecology and evolution of seabirds can be impacted by physical processes in the water [142].

The complex interplay between environmental factors and vital rates causes seabird populations to vary. Many definite factors (which sometimes also include human activities, such as the arrival of new species) affect the life of sea birds [102, 160]. Apart from these, the number of populations also has an effect—sometimes this effect is good (for example, when the number of birds is less, they help each other, this is called Allee effect) and sometimes bad (when the flock of birds becomes very large, the competition for food increases) [121]. Due to all these things, essential rates like survival of birds, birth of their babies, arrival of new birds in the group, and movement to other places can change [140]. Therefore, each element has a crucial role in controlling seabird populations. Even though many of these characteristics have undergone independent analysis in the literature, little is known about their relative relevance, and it is likely to fluctuate with time. Like most species, seabird populations are directly impacted by food availability, habitat, and predation [16]. Changes may indirectly influence these environmental parameters [43, 103, 161]. Over a comparatively long lifespan, seabirds have developed a life history strategy based on iteroparity, or repeated fecundity, and modifications that lower adult survival tend to have the most significant impact on the increase of the seabird population. However, populations may also be impacted by substantial shifts in productivity such as, total breeding failures. According to recent research, recruitment and breeding success significantly affect population variations in several species [47, 74]. Dispersal is an essential factor in seabird population changes. Young birds or small birds move from their birth area to another place (natal dispersal), and adult birds move to a new place to build a nest (breeding dispersal) [25]. These movements have a significant impact on the growth and decrease of their population, and if the birds are unable to move to another place, they may be at risk of extinction [72]. The discontinuous distribution of colonies in space and the vulnerability of isolated populations with limited connectivity are highlighted by the metapopulation paradigm in conjunction with accurate dispersal rate estimation [56].

### 9.1.7 Case Studies on Diminishing Seabird Populations

According to our study, most seabird species in Britain and Ireland will see their numbers decline due to climate change. Our future predictions can help us understand which species are at risk and which need to be paid attention to urgently. However, there is always some uncertainty in these circumstances, which can vary from species to species. Knowing how seabird species respond to climate change and reducing misconceptions about future estimates can help conservation [19, 62]. Therefore, the Seabird Monitoring Program should include more species and sites in its surveys every year, so that we can better understand all changes and take timely action on them. It was found that the maximum temperature during the breeding season is often inversely correlated with seabird occurrence in Britain and Ireland. Other connections showed significant variations in sensitivity between functional categories due to species-specific variations in climate and oceanography. To assist in determining which facets of seabird ecology should be the focus of conservation initiatives aimed at mitigating the effects of climate change, future research can delve deeper into these links [36].

Previously thought to be decreasing, the number of breeding pairs of flesh-footed shearwaters on Lord Howe Island seems to have steadied or even grown between 2009 and 2017. According to earlier estimates, road mortality is high. Nonetheless, the risk of road fatalities has decreased, most likely due to island-wide awareness-raising initiatives and the corresponding backing from the local populace. Although there is still some evidence, it is generally poor, that particular demographic indicators, like breeding success or burrow density, fall close to possible human impacts, such as roads, on the colonies. These findings for the people of Lord Howe Island imply that the hazards that have been previously reported are lessening. Previously, significant habitat loss was no longer happening. Concurrently, the island's flesh-footed shearwater colony area is currently growing. It appears that road fatalities have decreased. Previously reported to be a significant bycatch in longline fisheries, the number has diminished due to current fishery management initiatives [152].

In the world's monitored seabirds (i.e., populations observed at least five times, representing around 19% of the worldwide seabird population), our research showed a long-term and significant population drop of 69.7% between 1950 and 2010. Given the size and apparent representativeness of the sample, this diminishing tendency might reflect the worldwide seabird population trend. Families with widely distributed pelagic species showed the most significant reductions, indicating that pan-global populations may be more vulnerable than coastal populations with limited ranges [108].

On the small atoll of Surprise Island, New Caledonia (southwestern tropical Pacific), we monitored the number of breeding pairs of seven seabird species for an extended period. The island's marine wildlife was examined every 4 years before and 4 years following the 2005 rodent eradication, and we carried out several one-time surveys from around 10 years before and 15 years following eradication. In an isolated setting, we aimed to ascertain various seabird species' reactions to removing invasive rats. With variations in lag time and sensitivity, three species responded favourably to eradication (population size increased by two to ten times). Over 4 years after eradication, immigration increased the number of breeding pairs for two species (impact sizes = 0.49–0.95 and 0.35–0.52). One species responded more slowly than the others (at least 5 years); breeding pairs grew for more than 10 years following elimination [111]. Three critical factors need to be taken into account to better understand how island species react to the removal of invasive non-native species. First, monitoring communities before eradication is essential to understand ecosystem functioning through at least primary species interactions, draw attention to the detrimental effects of invasive non-native species, and minimise unanticipated cascade effects after eradication (e.g., increased abundance of other invasive species in response to reduced predation [27].

### *9.1.8 Declining Trends in Seabird Population*

Marine bird populations have decreased due to habitat degradation brought on by water pollutants. Some sea birds are in trouble, as many are threatened, critically endangered, or completely extinct [70]. The main reasons for this are loss of their natural habitats, excessive hunting, pollution, and climate change. Seabirds make their lives difficult when their forests and oceans are not protected, and many species are lost forever [145]. These birds face many problems, like weather changes while washing food, some species (like diving petrels) swallow salty water while drowning, human pollution is increasing, and they also have to compete for food with fishing companies that catch fish. All these problems together reduce the population of seabirds or make their lives even more difficult [60].

The graph shows that from 1950 to 2025, the species and number of seabirds have declined steadily [78, 108, 144] (Fig. 9.2). 1950 was taken as the baseline, from which the numbers gradually declined. First, a decline of around 17% was observed from 1950 to 1970, which probably reflects the industrial impact and greater use of marine resources [163]. Between 1970 and 1990, this decline increased rapidly to around 40.5%. During this time, the increase in fish catching rates reduced the food availability of seabirds. New insects (invasive species) such as rats, cats, and stoats, which attack the birds' eggs and chicks, have increased their reproduction [31]. The sharpest decline was observed from 1990 to 2010, reaching about 69.7%. During this time, pollution such as plastic waste, chemical substances, and changes in the natural environment, such as climate change, also had an effect [40].

Near the shore or in slightly deep water, seabirds are most at risk from fishing nets or pressures. Fishing nets often kill many species, such as auk, guillemot, razorbill, and puffin [54]. They also face food shortages as many fish are caught. Oil spills,

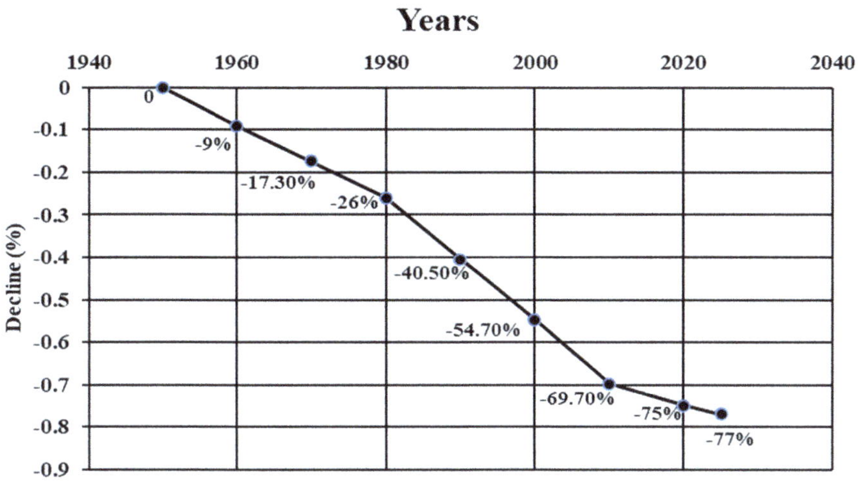


**Fig. 9.2** Year-wise global decline in seabird species diversity and population from 1950 to 2025. (**References**: [78, 108, 144])

water pollution, predation, attacks by other animals such as rats or cats, human movement, climate change, and introducing exotic species to the breeding site are also significant problems. Changes in the natural environment of the ocean also particularly affect food facilities, making it difficult for seabirds to survive [6]. These are essential driving variables that affect seabird population community parameters directly or indirectly; for example, particular species become threatened or endangered and are at risk of extinction. It has been reported that within the past 10 years, the number of people living in some coastal areas has increased by up to 40% [17]. Oil spills seriously threaten seabirds, which can potentially kill many seabird species. Compared to other creatures, seabirds are the most noticeable and vulnerable to marine oil spills [73]. This may be because they spend much of their lives at sea, and their populations are dispersed and concentrated in offshore habitats and coastal locations that frequently experience oil spill issues. In the event of an oil leak, their chances of surviving are extremely low [7, 147].

### *9.1.9 Global Pattern of Seabird Population Declines*

Invasive alien species (IAS), bycatch, and climate change are the primary causes of the decrease of many seabird species, according to the most recent worldwide evaluation of hazards to seabirds conducted in 2012. The majority of threatened seabirds (about 70%), especially those globally vulnerable, are threatened by many factors. Although bycatch has a larger impact in terms of severity (pace of decline) and scope (percentage of the population impacted), IAS affects the most significant number of species overall (almost 50%) [3]. Seabirds are the most endangered group of birds, and the situation has worsened more quickly in recent years. Ten percent are Near Threatened, and twenty-eight percent are threatened worldwide, with five percent falling into the highest category of Critically Endangered. Particularly concerning are the 64 species whose populations are declining along with their limited distribution. Compared to coastal species, pelagic species are over-represented; the species designated under the Agreement on the Conservation of Albatross and Petrels have suffered the most [84]. Ten main pressures have contributed to declines. These include overfishing, mining, oil spills, marine debris pollution, and incidental bycatch in longline, gillnet, and trawl fisheries. Human disturbance, infrastructure, commercial and residential development, hunting, trapping, invading alien species, and troublesome native species (such as those that have become super-abundant) have all contributed to reductions in land. Seabirds, both on land and at sea, are impacted by climate change and extreme weather [63].

## 9.2 Key Threats to Seabird Populations

Before we know the main threats to the seabird populations, we must remember that these species have a special ecological niche in the marine and terrestrial ecosystems. Their life cycles depend on healthy ecosystems, making them highly responsive to modifications in both systems. As globally distributed indicators, the health of seabirds offers essential information about the general comfort of oceanic environments. Seabirds are vulnerable to major environmental threats from human activities and loss of their natural habitat. These threats can cause bird populations to decline, disrupt breeding (egg-laying), and alter their migration routes [161]. The highlights in Fig. 9.2 show that seabirds are extremely sensitive to these changes, so their conservation and management of the entire marine ecosystem is necessary. In recent decades, however, various environmental concerns largely stemming from human activities have led to severe declines in seabird populations worldwide. Understanding the specific factors responsible for these declines is crucial for creating conservation strategies to protect seabirds and their ecosystems (Fig. 9.3).

### *9.2.1 Overfishing and Competition for Food*

If you look out across the world's coasts and islands today, it's clear that seabirds are struggling like never before. To locate enough food, these birds have travelled great distances for thousands of years, perfecting their routines to correspond with the tides' rise and fall. However, those journeys have become more of a desperate search [30]. A significant reason for this strain is large-scale overfishing. Massive fishing fleets with cutting-edge equipment capture massive marine life at a rate that the ocean cannot readily recover [61, 107]. For people, they end up as ingredients in fish meal, oils, or make their way to markets. But for seabirds, these fish are life itself:

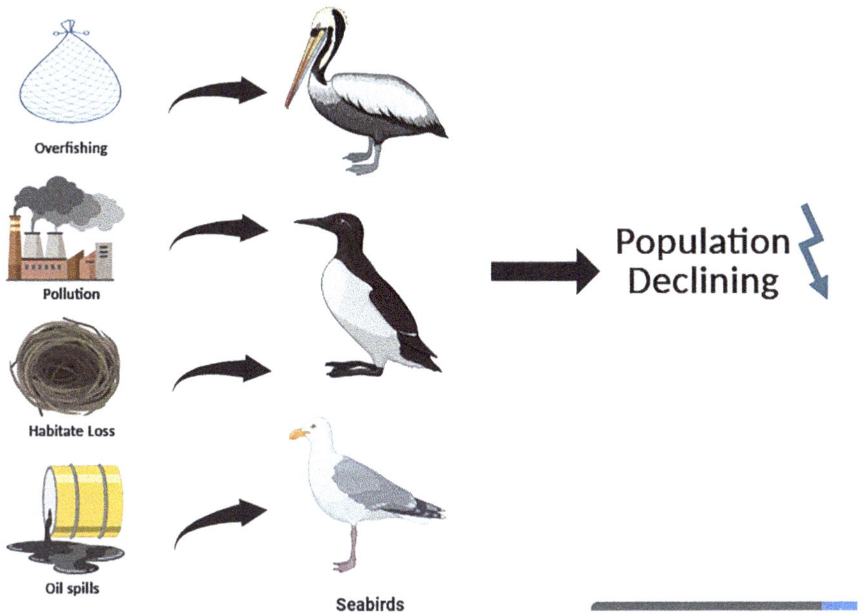


**Fig. 9.3** Environmental threats causing the decline of the Seabird population decline

they keep adults healthy and strong, and more importantly, provide the food that chicks depend on to survive their first weeks [41]. Beyond food, additional pressures are piling onto seabird populations. The location of prey species is changing due to changing water temperatures brought on by climate change [45]. If there is any clear result from current studies, seabirds bear one of the highest costs when industrial fishing reduces these tiny fish supplies to all-time lows. Addressing these challenges requires more innovative fisheries management, global cooperation, and a closer look at how we manage the ocean's shared resources.

### *9.2.2 Habitat Loss and Degradation*

It is easy to notice how much has changed the world's beaches that were once known for an abundance of birds. The growth of industrialization and human activities profoundly impacts the lifestyle and population of seabirds [35]. Very few places are left for them to live; buildings are often constructed there, or it is surrounded (fenced). Pollution from runoff and nearby industry affects food and water resources. Invasive plants might crowd out the native vegetation birds need for cover [85]. Over the years, seabirds have lost many of their safe nesting places. As people build more along the coasts, birds often nest in areas that are not ideal. This makes it harder for their chicks to survive and for colonies to stay healthy. As a result, we now see fewer seabirds in some places, and old nesting spots are sometimes left empty.

### 9.2.3 Loss in Wetland Ecosystem

Wetlands, marshes, and seashore land are the most important and valuable places for seabirds. These are the areas where these birds can comfortably build nests and find food. However, nowadays these wetlands are decreasing rapidly, because man has caused the spread of cities, excessive water extraction, dirty water from factories, and the pollution of farms [88]. The immaculate and natural areas were now either dirty or foreign plants had entered them, causing harm to the real birds [164]. Where the natural form of wetlands has been formed, the number of species is also higher, and different types of birds are found, which are considered different not only in number, but also in their work (like catching fish, eating birds). As the wetlands dry up or become polluted, the species of birds decrease, and the whole system gradually becomes weak. Seabirds are not just residents but also mean a lot to their environment. By bringing their waste (guano) from the sea and leaving it on islands or wetlands, they make the soil fertile, providing essential nutrients for the good growth of plants [162]. Guano benefits plants, insects, and fish, which works like a chain in the ecosystem [131]. If, for some reason, the number of seabirds decreases, those nutrients cannot reach the land, and gradually these areas become less productive and weaker. As soon as this link is broken, such areas start facing more problems, be it drought, pollution, or other natural hazards.

### 9.2.4 Invasive Species

On many remote islands, seabirds used to nest safely where there were no land predators. But after people brought animals like rats and cats, things changed quickly. These new predators often find and eat eggs, chicks, and sometimes adult birds. In some places, just a few years of this can wipe out entire seabird colonies, especially for species that nest on the ground or in burrows [122]. Removing these animals from islands helps seabirds rebound, sometimes seeing more surviving populations [134].

### 9.2.5 Pollution (Plastics, Oil, Chemicals)

Seabirds often swallow pieces of plastic floating in the sea, mistaking them for food. This plastic can enter their stomach and obstruct their digestive process, which causes them to feel hungry and sometimes even leads to their death [119]. Toxic substances found inside plastic also cause harm to the bodies of seabirds, which can harm both their health and breeding [15]. Oil spills are uncommon, but they are cumbersome to seabirds when they happen. The oil gets on their feathers, making them incapable of retaining water and warmth [53]. The seabirds shiver in the cold, find breathing difficult, and sometimes even die. Apart from this, heavy metals and dangerous pollutants mixed in the sea keep accumulating in the bodies of seabirds. Due to this, their offspring are born fewer, they face problems in reproduction, and they become more susceptible to diseases [158]. Today, the situation is such that plastic and chemicals are visible even in remote islands and snowy places, where it is difficult for humans to live. All this proves that this problem is not limited to just one place, but is a threat to the sea birds of the entire world [66].

### *9.2.6 Climate Change*

One of the most pervasive and complex issues facing marine communities, including seabirds, is likely climate change. Many changes are being observed in the ocean environment in the coming time, such as increasing water temperatures, strong winds, and changing tides, which all affect seabirds. Rising sea levels could submerge their nesting sites. Storms and bad weather have also started increasing, diseases are spreading, and, in some places, snow is also melting. All these changes are already affecting the lives of seabirds in different parts of the world.

Long-term bottom-up mechanisms impacting food resources and productivity are anticipated to affect climate change significantly. Sea ice phenology governs primary production at the poles, with ramifications for the entire trophic chain. There are beginning to be bottom-up effects on arctic seabird groups as sea ice shrinks and spring melting timing changes [86]. When the water temperature in the sea becomes higher than usual and remains so for a long time, it is called a "marine heatwave". In the past few years, the number of such hot days has increased, possibly changing the marine life as well. When water becomes too hot, fish and other food either move to the depths, or shift to cold places, or their nutrition gets reduced [51]. Such a lack of prey can lead to low reproductive outputs in seabirds by delaying breeding or inducing nest abandonment and directly causing illness. When the seawater in the north-eastern part of the Pacific Ocean became very hot, the movement of black-legged kites was observed using GPS. It was found that not only during these hot days, but even when the water became normal again, these birds started going much farther than before to wash their food. This suggests that the effect of a temperature change is not just for a few days, but persists for a long time [104]. Most birds affected by climate change also face other major sea and land problems—such as being trapped in fishing nets (bycatch) or being threatened by exotic predators. Although humans cannot immediately control climate change entirely, these other problems can be prevented. Reducing these pressures through bycatch regulation, establishing safe zones that support biodiversity, and maintaining or eradicating alien species may help seabird populations become more resilient to the adverse effects of climate change. By this point, many high-priority areas have been cleared of alien species; the largest attempt to date was to eradicate rats and mice from South Georgia, covering an area of 3500 $km^2$ [89]. Various bycatch mitigation strategies have also been successful in several fisheries, though efforts to reduce gillnet bycatch have been less successful [22]. Many species of marine birds travel far and wide and are found in aquatic areas of many different countries. Therefore, the work of one country is not enough for its protection. Successful conservation can only be achieved when other countries and fisheries organizations work together to save the diversity of marine life [10]. The direction of the ocean's waves and currents is changing, so the location of small fish and other prey or their numbers are decreasing [76]. For this reason, when seabirds return to their old nesting places, they often check whether the place is filled with water or the land has disappeared completely [69]. Rain and winds have also started coming more intensely, harming searbirds [28]. The seawater becomes warmer, making it difficult to get fish or other food, and many birds have to jump long distances for their chicks, which increases the risk of them getting tired and falling. When everything is becoming abnormal due to climate change, other threats—such as bycatch, pollution, or habitat loss—become even more dangerous, because the birds no longer have many ways to survive [39, 137].

### *9.2.7 Monitoring Approaches and Current Challenges*

Counting seabirds and monitoring their condition is very important for their protection. Earlier, colonies were counted in the old ways, like standing on land or from a boat, or an airplane [68]. But often the place is so remote or complex that it becomes difficult to reach. The same person cannot go to all the places every time, and because of this, information often remains incomplete, or there is a difference in the bird count according to different people [149]. Nowadays, technology has given us many new facilities. With methods like satellite tracking and bio-logging, we can now follow the entire journey of birds, their day-long behaviour, and understand their numbers, without disturbing them [46]. Tools like drones and acoustic sensors also take pictures of colonies that are difficult to reach, or their sounds tell us how many birds are there [49]. In this way, we are also learning about the previously hidden things from us, like which routes a bird takes to migrate or whether its number is increasing or decreasing. Still, monitoring has many challenges. When it is difficult to reach the place or the situation changes very quickly, it is difficult to get the correct data. Due to a lack of money, people, or time, monitoring is often incomplete. Therefore, the same method does not work everywhere; rather, old and new techniques must be combined and chosen according to the situation.

### *9.2.8 Conservation Strategies: Evidence and Best Practices*

Seabirds are in decline, but there are simple and effective ways to save them. First, restoring old seabird colonies in places where they have been lost is a powerful solution. This involves several things, such as moving small seabirds to a safe location or relocating them to new areas using sounds or decoys [75]. It has been observed many times that if a predatory animal, like a rat or a cat, is removed from an island, then in some years, birds return there and new nostrils also start emerging. Another critical step is to improve the habitat. When original plants are replanted or unknown habitats are cleared, the entire area becomes favorable for birds again. If wetlands have dried up, reviving them by replenishing them with water also helps protect them [143]. These measures yield good results when the local community is involved—such as school children, villagers, or NGOs working together for sanitation, tree plantation, or monitoring [12]. As new technology is used—drones, camera traps, or satellite trackers—it lets us know how many seabirds are returning, whether breeding is taking place, and how healthy the colony is [50]. Today's research shows that conservation is successful only when we look at nature conservation as a cohesive, continuous process—security, improvement, people's participation, and new methods all work together to create safe places for birds.

## 9.3 Future Directions and Recommendations

Nowadays, the number of sea birds is gradually decreasing. Where earlier large flocks of birds were seen on seashores, cliffs, or islands, now very few birds are left. There are many reasons for this decline, such as their living area being cut off, becoming more vulnerable to fish, pollution, and increasing human activity. It has often been seen that urban development and the spread of factories are destroying areas where birds used to dominate. Fishing companies also take in many small fish and sea creatures, without which the birds cannot feed their chicks. Stains from plastic and oil are also a danger to the birds, while the weather and climate are also changing due to climate change. This has also created problems like increased storms, sea water, and food shortages. Amidst many issues, counting the birds in the old ways is not easy. That place has often become distant, or the birds have left their old place and gone to another place. Today, with new technology like drones, GPS trackers, or sound recording devices, we can know the number and condition of birds without disturbing them. This helps us to know which areas are more affected or which species are more vulnerable. Therefore, monitoring is not a one-day job but a continuous process to catch the problems before they occur. Moving forward, it is essential to have integrated rules for both the sea and wetlands. New fishing techniques should be used, such as changing nets and monitoring birds in unexplored areas. Locals should also be involved, as they are often the first to feel the changes. Nowadays, forest diseases also pose a threat, so it is essential to adopt new methods of restoration and monitoring. If we improve our monitoring and conservation strategies, use the latest tools, and involve people, bird counts can increase again, and their lives will remain the same—even as the world changes rapidly.

## 9.4 Conclusion

Seabird populations are declining in many parts of the world, and this trend is a clear warning sign about the health of our marine and coastal environments. The loss of these birds is not happening by chance—it is deeply linked with various human activities and environmental changes. Wetland areas where seabirds nest and feed are shrinking, often because of land development and pollution. Polluted waters, with plastics and chemicals, have made food scarce or unsafe for many species. Invasive animals like rats and cats threaten eggs and chicks, and the ever-shifting climate brings new challenges, such as stronger storms, rising sea levels, and changes in ocean temperatures. These combined pressures have made it difficult for seabirds to find safe places to breed, raise young ones, and get enough food.

These changes in seabird numbers are not just a problem for the birds. When seabird populations fall, it often reveals deeper issues in their ecosystems. Seabirds are closely connected to the overall health of the ocean and the coastal lands—they help move nutrients and are sensitive to changes in the food chain. If they start disappearing, it suggests that other life forms might also be at risk, and the balance of nature is being disturbed. Keeping track of seabirds is, therefore, more than just a research project—it's a meaningful way to check the well-being of our shared environment. Tried-and-tested practices, like colony surveys or counting nests, are valuable. At the same time, new tools, such as drones for aerial photos or listening devices to record their sounds, make it easier to spot changes or find hidden issues. The best results come from a mix of traditional methods and modern technology, repeated regularly over many years.

What is urgently needed is a commitment to ongoing and flexible monitoring efforts. Conditions in the natural world are constantly changing, but recent shifts have happened so quickly that long-term observation is essential to notice trends early and respond in time. Knowledge from these monitoring programs helps scientists, conservationists, and communities make wise choices—whether restoring habitats, limiting hunting or fishing, stopping pollution, or controlling invasive species. In summary, paying attention to seabird trends is vital for the birds and the entire ecosystem. If we act wisely and stay alert to their needs, we can help stop their decline and protect the natural balance of coasts and oceans for future generations. Taking care of seabirds is, in the end, about guarding the health of our whole planet.

## References

1. Ainley DG, Morandini V, Salas L, Nur N, Rotella J, Barton K et al (2024) Response of indicator species to changes in food web and ocean dynamics of the Ross Sea, Antarctica. Antarct Sci 36(5):290–318
2. Almalki M (2024) Nesting Distribution and Conservation of the Brown Booby (Sula leucogaster) on the Red Sea Coast of Saudi Arabia
3. Antonio V, Claire AR (2024) Seabirds of Europe: current status, main threats and way forward
4. Appoo J, Bunbury N, Letori J, Hector A, Gendron A, Graham NA et al (2024) Seabird presence and seasonality influence nutrient dynamics of atoll habitats. Biotropica 56(4): e13354
5. Baak JE, Provencher JF, Mallory ML (2020) Plastic ingestion by four seabird species in the Canadian Arctic: comparisons across species and time. Mar Pollut Bull 158:111386
6. Baker GB, Gales R, Hamilton S, Wilkinson V (2002) Albatrosses and petrels in Australia: a review of their conservation and management. Emu-Austral Ornithol 102(1):71–97
7. Balseiro A, Espí A, Marquez I, Perez V, Ferreras MC, Marín JG, Prieto JM (2005) Pathological features in marine birds affected by the Prestige's oil spill in the north of Spain. J Wildl Dis 41(2):371–378
8. Barbier EB, Hacker SD, Kennedy C, Koch EW, Stier AC, Silliman BR (2023) Marine ecosystems: a global perspective. Annu Rev Mar Sci 15:1–27. https://doi.org/10.1146/annurev-marine-032122-105632
9. Bargagli R, Rota E (2024) Environmental contamination and climate change in Antarctic ecosystems: an updated overview. Environ Sci Adv 3(4):543–560
10. Beal M, Dias MP, Phillips RA, Oppel S, Hazin C, Pearmain EJ et al (2021) Global political responsibility for the conservation of albatrosses and large petrels. Sci Adv 7(10):eabd7225
11. Beauchamp G (2022) Flocking in birds is associated with diet, foraging substrate, timing of activity, and life history. Behav Ecol Sociobiol 76(6):74
12. Belcher JS et al (2024) Community-based habitat restoration supports seabird recovery and ecosystem resilience. Auk Ornithol Adv 96(1):art7
13. Bialas JT, Dylewski Ł, Dylik A, Janiszewski T, Kaługa I, Królak T et al (2021) Impact of land cover and landfills on the breeding effect and nest occupancy of the white stork in Poland. Sci Rep 11(1):7279
14. Blevin P, Tartu S, Ellis HI, Chastel O, Bustamante P, Parenteau C et al (2017) Contaminants and energy expenditure in an Arctic seabird: Organochlorine pesticides and perfluoroalkyl substances are associated with metabolic rate in a contrasted manner. Environ Res 157:118–126
15. Bond AL, Lavers JL (2023) Persistent threats of marine plastic and chemical pollution to seabirds worldwide. Sci Total Environ 858:159925. https://doi.org/10.1016/j.scitotenv.2022.159925
16. Bonnaud E, Bourgeois K, Vidal E, Legrand J, Le Corre M (2009) How can the Yelkouan shearwater survive feral cat predation? A meta-population structure as a solution? Popul Ecol 51:261–270
17. Booth BP, Bio RP, Federation CN (2001) Southern Vancouver Island marine waters and seabird islands important bird areas conservation plan. Canadian Nature Federation, Bird Studies Canada, Prince George
18. Borrelle SB, Ringma J, Law KL, Monnahan CC, Lebreton L, McGivern A et al (2020) Predicted growth in plastic waste exceeds efforts to mitigate plastic pollution. Science 369(6510):1515–1518
19. Bowgen KM, Kettel EF, Butchart SHM, Carr JA, Foden WB, Magin G et al (2022) Conservation interventions can benefit species impacted by climate change. Biol Conserv 269: 109524

20. Bracey AM, Etterson MA, Strand FC, Matteson SW, Niemi GJ, Cuthbert FJ, Hoffman JC (2020) Foraging ecology differentiates life stages and mercury exposure in common terns (Sterna hirundo). Integr Environ Assess Manag 17(2):398–410
21. Brooke MDL (2002) Seabird systematics and distribution: a review of current knowledge. In: Biology of marine birds. Taylor and Francis Group, pp 57–85
22. Bull LS (2007) Reducing seabird bycatch in longline, trawl and gillnet fisheries. Fish Fish 8(1):31–56
23. Burger J, Gochfeld M (2005) Effects of lead on learning in herring gulls: an avian wildlife model for neurobehavioral deficits. Neurotoxicology 26(4):615–624
24. Butchart SH, Walpole M, Collen B, Van Strien A, Scharlemann JP, Almond RE et al (2010) Global biodiversity: indicators of recent declines. Science 328(5982):1164–1168
25. Cam E, Oro D, Pradel R, Jimenez J (2004) Assessment of hypotheses about dispersal in a long-lived seabird using multistate capture–recapture models. J Anim Ecol 73(4):723–736
26. Carrasco L, Jiménez-Mora E, Utrilla MJ, Pizarro IT, Reglero MM, Rico-San Román L, Martin-Maldonado B (2025) Birds as bioindicators: revealing the widespread impact of microplastics. Birds 6(1):10
27. Caut S, Angulo E, Courchamp F (2009) Avoiding surprise effects on Surprise Island: alien species control in a multitrophic level perspective. Biol Invasions 11:1689–1703
28. Chastel O et al (2022) Food shortage and reproductive success in seabirds: a changing relationship under climate change? Sci Total Environ 845:157333. https://doi.org/10.1016/j.scitotenv.2022.157333
29. Chilvers BL, Morgan KJ, White BJ (2021) Sources and reporting of oil spills and impacts on wildlife 1970–2018. Environ Sci Pollut Res 28(1):754–762
30. Clay TA, Oppel S, Phillips RA, Matthiopoulos J (2023) Global patterns in seabird movement strategies reveal key drivers of long-distance foraging. Curr Biol 33(2):289–298.e4. https://doi.org/10.1016/j.cub.2022.11.032
31. Cooper DJ, Schweiger EW, Shaw JR, Westbrook CJ, Kaczynski K, Abouelezz H et al (2025) Rapid riparian ecosystem decline in Rocky Mountain National Park. Conserv Biol 39:e70053
32. Coulson JC, White S (2022) Coloniality in seabirds: ecology, evolution, and behaviour. Mar Ornithol 50(2):123–134
33. Couper-Johnston R (2000) El Niño: the weather phenomenon that changed the world. Hodder & Stoughton
34. Croxall JP, Butchart SH, Lascelles BEN, Stattersfield AJ, Sullivan BEN, Symes A, Taylor PHIL (2012) Seabird conservation status, threats and priority actions: a global assessment. Bird Conserv Int 22(1):1–34. https://doi.org/10.1017/S0959270912000020
35. Da Silva DS, Couto GS et al (2024) The impact of habitat loss and disturbance on coastal seabird communities. Curr Res Environ Sustain 6:100183. https://www.sciencedirect.com/science/article/pii/S2053716625000064
36. Davies JG, Humphreys EM, Evans T, Howells RJ, Hara-Murray RO, Pearce-Higgins JW (2023) Seabird abundances projected to decline in response to climate change in Britain and Ireland. Mar Ecol Prog Ser 725:121–140
37. Davies W, Van Alstine J, Lovett JC (2016) 'Frame Conflicts' in natural resource use: exploring framings around Arctic offshore petroleum using Q-methodology. Environ Policy Gov 26(6): 482–497
38. Dehnhard N, Mattisson J, Tarroux A, Anker-Nilssen T, Lorentsen SH, Christensen-Dalsgaard S (2022) Predicting foraging habitat of European shags-a multi-year and multi-colony tracking approach to identify important areas for marine conservation. Front Mar Sci 9:852033
39. Descamps S et al (2022) Climate impacts on seabirds, a global review. Nat Rev Earth Environ 3(8):642–658. https://www.nature.com/articles/s43017-022-00321-3
40. Dhaka V, Singh S, Anil AG, Sunil Kumar Naik TS, Garg S, Samuel J et al (2022) Occurrence, toxicity and remediation of polyethylene terephthalate plastics. A review. Environ Chem Lett 20(3):1777–1800

41. Dias MP et al (2023) Seabird population responses to fisheries management: global insights. Sci Adv 9(8):eade7654
42. Dias MP, Martin R, Pearmain EJ, Burfield IJ, Small C, Phillips RA et al (2019) Threats to seabirds: a global assessment. Biol Conserv 237:525–537
43. Doherty PF Jr, Schreiber EA, Nichols JD, Hines JE, Link WA, Schenk GA, Schreiber RW (2004) Testing life history predictions in a long-lived seabird: a population matrix approach with improved parameter estimation. Oikos 105(3):606–618
44. Dominoni D, Quetting M, Partecke J (2013) Artificial light at night advances avian reproductive physiology. Proc R Soc B Biol Sci 280(1756):20123017
45. Dunn DC et al (2022) Trajectories of climate-driven marine ecosystem change point to critical management needs. Nat Commun 13:1202
46. Edney AT, Wood MJ (2021) Modern tools for monitoring bird behaviour and populations: progress and challenges. Mar Policy 131:104699. https://www.sciencedirect.com/science/article/pii/S0964569123003587
47. Ezard THG, Becker PH, Coulson T (2006) The contributions of age and sex to variation in common tern population growth rate. J Anim Ecol 75(6):1379–1386
48. Fernández CE, Luna-Jorquera G, Encinas VG, Lancelloti AA, Lantadilla C, Aguilar-Pulido R et al (2024) Seabirds as biovectors in the transport of plastic debris across ecosystem borders: a case study from the Humboldt Current Upwelling System. Sci Total Environ 952:175938
49. Fischer S, Edwards AC, Garnett ST, Whiteside TG, Weber P (2023) Drones and sound recorders increase the number of bird species identified: a combined surveys approach. Eco Inform 74:101988
50. Francis RJ, Kingsford RT, Brandis KJ (2022) Using drones and citizen science counts to track colonial waterbird breeding, an indicator for ecosystem health on the Chobe River, Botswana. Glob Ecol Conserv 38:e02231
51. Frolicher TL, Fischer EM, Gruber N (2018) Marine heatwaves under global warming. Nature 560(7718):360–364
52. Furness RW, Monaghan P (1987) Seabird life styles. In: Seabird ecology. Springer US, Boston, MA, pp 3–22
53. García-Barón I, González-Solís J, Jiménez S (2021) The underestimated effects of oil spills on seabird mortality. Mar Pollut Bull 167:112320. https://doi.org/10.1016/j.marpolbul.2021.112320
54. Garthe S (1997) Influence of hydrography, fishing activity, and colony location on summer seabird distribution in the south-eastern North Sea. ICES J Mar Sci 54(4):566–577
55. Gatt MC, Granadeiro JP, Catry P (2022) An introduction to seabirds and their study. In: Volume 1: seabird biodiversity and human activities. CRC Press, pp 3–18
56. Genovart M, Oro D, Juste J, Bertorelle G (2007) What genetics tell us about the conservation of the critically endangered Balearic Shearwater? Biol Conserv 137(2):283–293
57. Gerstberger R, Gray DA (1993) Fine structure, innervation and functional control of Avian Salt Glands. Int Rev Cytol 144:129–215. https://doi.org/10.1016/S0074-7696(08)61515-6
58. Giambalvo G, Albores-Barajas YV, Soldatini C, Hernandez MPR, Rattenborg NC (2022) Environmental drivers of sex-biased foraging behavior in Magnificent Frigatebird in Baja California Sur, Mexico. Avian ConservEcol 17(2):25
59. Gillies N, Thorley J, Weimerskirch H, Jenouvrier S, Barbraud C, Delord K, Patrick SC (2024) Plastic behaviour buffers climate variability in the Wandering Albatross. Ecol Evol 14(12):e70631
60. Green B, Brothers N (1989) Water and sodium turnover and estimated food consumption rates in free-living fairy prions (Pachyptila turtur) and common diving petrels (Pelecanoides urinatrix). Physiol Zool 62(3):702–715
61. Grémillet D et al (2022) Industrial fishing, forage fish depletion, and seabird population declines. Nat Ecol Evol 6:354–362

62. Hakkinen H, Petrovan SO, Taylor NG, Sutherland WJ, Pettorelli N (2023) Seabirds in the North-East Atlantic: climate change vulnerability and potential conservation actions. Open Book Publishers
63. Harris MP (2022) Coastal development and its effect on seabird breeding habitats: past, present and future perspectives. Environ Conserv 49(3):245–256
64. Harwood AJ, King S (2022) Seabirds and marine renewable energy sources. In: Volume 1: seabird biodiversity and human activities. CRC Press, pp 106–125
65. Healing S, Benkwitt CE, Dunn RE, Graham NA (2024) Seabird-vectored pelagic nutrients integrated into temperate intertidal rocky shores. Front Mar Sci 11:1343966
66. Herzke D et al (2023) Plastics and persistent chemicals reach seabirds in remote polar regions. Environ Sci Technol Lett 10(2):148–155. https://doi.org/10.1021/acs.estlett.2c00887
67. Heubeck MARTIN, Mellor RM, Harvey PV, Mainwood AR, Riddington R (1999) Estimating the population size and rate of decline of Kittiwakes Rissa tridactyla breeding in Shetland, 1981–97. Bird Study 46(1):48–61
68. Hodgson JC, Mott R, Baylis SM, Pham TT, Wotherspoon S, Kilpatrick AD, Raja Segaran R, Reid I, Terauds A (2018) Drones count wildlife more accurately and precisely than humans. Methods Ecol Evol 9(5):1160–1167. https://doi.org/10.1111/2041-210X.12974
69. Hughes BB et al (2023) Increasing risk of extreme events threatens seabird breeding on low-lying islands. Glob Chang Biol 29:4032–4044. https://onlinelibrary.wiley.com/doi/10.1111/gcb.16751
70. Huppop O, Wurm S (2000) Effects of winter fishery activities on resting numbers, food and body condition of large gulls Larus argentatus and L. marinus in the south-eastern North Sea. Mar Ecol Prog Ser 194:241–247
71. Iglésias SP, Fournier-Sowinski J (2024) An account of the natural history and exploitation of the great auk (Pinguinus impennis) in 'Histoire des pesches', an illustrated eighteenth-century manuscript. Arch Nat Hist 51(2):234–252
72. Inchausti P, Weimerskirch H (2002) Dispersal and metapopulation dynamics of an oceanic seabird, the wandering albatross, and its consequences for its response to longline fisheries. J Anim Ecol 71:765–770
73. Irons DB, Kendall SJ, Erickson WP, McDonald LL, Lance BK (2000) Nine years after the Exxon Valdez oil spill: effects on marine bird populations in Prince William Sound, Alaska. Condor 102(4):723–737
74. Jenouvrier S, Barbraud C, Cazelles B, Weimerskirch H (2005) Modelling population dynamics of seabirds: importance of the effects of climate fluctuations on breeding proportions. Oikos 108(3):511–522
75. Kappes MA, Jones HP (2022) Translocation and social attraction in seabird restoration: current practices and recent advances. Glob Ecol Conserv 33:e02015
76. Keogan K et al (2018) Global phenological insensitivity to shifting ocean temperatures among seabirds. Nat Clim Chang 8:313–318. https://www.nature.com/articles/s41558-018-0115-z
77. Kiran SD, Kour A, Priya DV, Kumar R (2022) Community composition and status of avian diversity at Campus and Agricultural landscapes of Chaudhary Charan Singh Haryana Agricultural University, Hisar (Haryana). J Appl Nat Sci 14(4):1130–1140. https://doi.org/10.31018/jans.v14i4.3784
78. Komarowska K, Fudala K, Dziembowski M, Hagge A, Bialik RJ (2025) Population status of sympatrically breeding Skuas (Catharacta spp.) at Admiralty Bay, King George Island, Antarctica: a case report for 2020–2024. Biology 14(3):305
79. Kuhn S, Van Franeker JA (2020) Quantitative overview of marine debris ingested by marine megafauna. Mar Pollut Bull 151:110858
80. Kuletz KJ, Ferguson SH, Frederiksen M, Gallagher CP, Hauser DD, Hop H et al (2024) A review of climate change impacts on migration patterns of marine vertebrates in Arctic and Subarctic ecosystems. Front Environ Sci 12:1434549

81. Kumar R, Singh D, Yadav R (2025) Toxic effects of chlorantraniliprole and fipronil on *Eisenia fetida*: from growth and reproductive parameters to avoidance behavior. Environ Eng Manag J 24(7):1499–1508. Accepted manuscript
82. Kumar R, Singh D (2022) Insecticide-tolerant bacterial population in *Eisenia fetida*'s Gut and Vermicast exposed to Chlorantraniliprole and Fipronil. Appl Biol Res 24(3):273–279
83. Kumar R, Yadav R, Kumar Gupta R, Pooja (2022) A sustainable way for fish health management by replacement of chemical and drugs by earthworm. In: Environmental degradation in Asia: land degradation, environmental contamination, and human activities. Springer International Publishing, Cham, pp 329–352
84. Lascelles B, Rice J, Sato M, Tarzia M, Wanless RM (2016) Seabirds. United Nations Report
85. Lavers JL, Bond AL (2022) Seabird population decline linked to habitat disturbance and reduced breeding success. Ecol Evol 12(8):e9201. https://doi.org/10.1002/ece3.9201
86. Lecomte N, Gremillet D, Tarroux A et al (2023) Climate-driven changes in Arctic sea ice impact seabird populations through bottom-up food web effects. Glob Chang Biol 29(4): 1234–1248. https://doi.org/10.1111/gcb.16270
87. Lopes CS, Laranjeiro MI, Lavers JL, Finger A, Provencher J (2022) Seabirds as indicators of metal and plastic pollution. In: Volume 1: seabird biodiversity and human activities, pp 169–188
88. Ma Z, Melville DS, Liu J, Chen J (2014) Global wetlands loss and species decline. Bioscience 64(10):939–948
89. Martin AR, Richardson MG (2019) Rodent eradication scaled up: clearing rats and mice from South Georgia. Oryx 53(1):27–35
90. Martínez-Curci NS, García GO, Marbán L, Simonetti P, Zalba SM (2021) Shorebirds and seabirds' ecology and conservation. In: The Bahía Blanca Estuary: ecology and biodiversity. Springer International Publishing, Cham, pp 327–358
91. McMullen K (2023) The Galápagos penguin as the "canary in the coal mine" for microplastics research in the Galápagos Marine Reserve & plastic pollution perceptions in Ecuadorian mangrove communities (Doctoral dissertation, University of British Columbia)
92. McNamara JC, Freire CA (2022) Strategies of invertebrate osmoregulation: an evolutionary blueprint for transmuting into fresh water from the sea. Integr Comp Biol 62(2):376–387
93. Melnes M, Gabrielsen GW, Herzke D, Sagerup K, Jenssen BM (2017) Dissimilar effects of organohalogenated compounds on thyroid hormones in glaucous gulls. Environ Res 158:350–357
94. Mendez LX, Hedrick TL (2024) Wind gradient exploitation during foraging flights by black skimmers (Rynchops Niger). J Exp Biol 227(16):jeb246855
95. Mills WF, Bustamante P, McGill RA, Anderson OR, Bearhop S, Cherel Y et al (2020a) Mercury exposure in an endangered seabird: long-term changes and relationships with trophic ecology and breeding success. Proc R Soc B 287(1941):20202683
96. Montevecchi WA (2006) Influences of artificial light on marine birds. In: Ecological consequences of artificial night lighting, pp 94–113
97. Nelson B (1980) Seabirds: their biology and ecology. A & W Publishers, New York, p 1979
98. Nisha A, Rani S, Kumar S, Kumar R, Kumar A (2025) Recent technologies and strategies for bioremediation of toxic and hazardous waste. In: Dubey KK, Shah MP, Yadav A (eds) Challenges and sustainable solutions in bioremediation. CRC Press, pp 1–16. https://doi.org/10.1201/9781003407317-1
99. Noyes PD, Miranda D, Oliveira de Carvalho G, Perfetti-Bolaño A, Guida Y, Barbosa Machado Torres F et al (2025) Climate change drives persistent organic pollutant dynamics in marine environments. Commun Earth Environ 6(1):363
100. Oro D, Furness RW (2002) Influences of food availability and predation on survival of kittiwakes. Ecology 83(9):2516–2528
101. Oro D, Pradel R, Lebreton JD (1999) Food availability and nest predation influence life history traits in Audouin's gull, Larus audouinii. Oecologia 118:438–445

102. Oro D, Tavecchia G, Pradel R (2006) The role of recruitment in long-lived bird populations: the importance of a reliable estimation. J Ornithol 147(suppl 1):36–37
103. Oro D, Torres R, Rodríguez C, Drummond H (2010) Climatic influence on demographic parameters of a tropical seabird varies with age and sex. Ecology 91(4):1205–1214
104. Osborne OE, Hara PD, Whelan S, Zandbergen P, Hatch SA, Elliott KH (2020) Breeding seabirds increase foraging range in response to an extreme marine heatwave. Mar Ecol Prog Ser 646:161–173
105. Ostle C, Thompson RC, Broughton D, Gregory L, Wootton M, Johns DG (2019) The rise in ocean plastics evidenced from a 60-year time series. Nat Commun 10(1):1622
106. Paiva VH (2022) Seabirds as indicators of oceanographic changes. In: Volume 1: seabird biodiversity and human activities. CRC Press, pp 149–168
107. Paiva VH et al (2024) Seabird population resilience linked to regional fisheries management choices. Conserv Lett 17(2):e12922
108. Paleczny M, Hammill E, Karpouzi V, Pauly D (2015) Population trend of the world's monitored seabirds, 1950-2010. PLoS One 10(6):e0129342
109. Panda BP, Parida SP, Pradhan A, Mohanta YK, Parthipan P, Narayan M, Sarma H (2025) Identifying ecological risk zones using spatial prioritization of heavy metal pollution and bioaccumulation in birds. Environ Sci Pollut Res 32:1–22
110. Peschko V, Mendel B, Mercker M, Dierschke J, Garthe S (2021) Northern gannets (Morus bassanus) are strongly affected by operating offshore wind farms during the breeding season. J Environ Manag 279:111509
111. Philippe-Lesaffre M, Thibault M, Caut S, Bourgeois K, Berr T, Ravache A et al (2023) Recovery of insular seabird populations years after rodent eradication. Conserv Biol 37(3): e14042
112. Phillips RA, Wood AG (2020) Variation in live-capture rates of albatrosses and petrels in fisheries, post-release survival and implications for management. Biol Conserv 247:108641
113. Piatt JF, Parrish JK, Renner HM, Schoen SK, Jones TT, Arimitsu ML et al (2020) Extreme mortality and reproductive failure of common murres resulting from the northeast Pacific marine heatwave of 2014-2016. PLoS One 15(1):e0226087
114. Rajpar MN, Ozdemir I, Zakaria M, Sheryar S, Rab A (2018) Seabirds as bioindicators of marine ecosystems. In: Seabirds. IntechOpen, pp 47–65
115. Ramírez F, Vicente-Sastre D, Afan I, Igual JM, Oro D, Forero MG (2021) Stable isotopes in seabirds reflect changes in marine productivity patterns. Mar Ecol Prog Ser 662:169–180
116. Rani S, Sharma S, Singh D, Sheoran R, Kumar R (2025) Improved production of anti-cancer compounds from medicinal plants through molecular and biotechnological mutagenesis strategies. In: Chen JT (ed) Biotechnology, multiple omics, and precision breeding in medicinal plants. CRC Press, pp 287–297
117. Raymond B, Shaffer SA, Sokolov S, Woehler EJ, Costa DP, Einoder L et al (2010) Shearwater foraging in the Southern Ocean: the roles of prey availability and winds. PLoS One 5(6): e10960
118. Rodriguez A, Arcos JM, Bretagnolle V, Dias MP, Holmes ND, Louzao M et al (2019) Future directions in conservation research on petrels and shearwaters. Front Mar Sci 6:94
119. Roman L, Hardesty BD, Hindell MA, Wilcox C (2019) A global assessment of the impact of marine debris on seabirds: plastic ingestion and entanglement. Sci Total Environ 646:85–91. https://doi.org/10.1016/j.scitotenv.2018.07.320
120. Rooney RC, Daniel J, Mallory M, Hedd A, Ives J, Gilchrist G et al (2023) Fuzzy cognitive mapping as a tool to assess the relative cumulative effects of environmental stressors on an Arctic seabird population to identify conservation action and research priorities. Ecol Solut Evid 4(2):e12241
121. Ruiz X, González-Solís JACOB, Oro D, Jover L (1998) Body size variation in Audouin's Gull Larus audouinii; a density-dependent effect? Ibis 140(3):431–438

122. Russell JC, Towns DR, Anderson SH, Bellingham PJ (2020) Predator eradication and Island conservation: what can we learn from successful campaigns? Conserv Sci Pract 2(9):e256. https://doi.org/10.1111/csp2.256
123. Ryan PG, Jackson S (1987) The lifespan of ingested plastic particles in seabirds and their effect on digestive efficiency. Mar Pollut Bull 18(5):217–219
124. Sabater González M (2022) Ophthalmology of Gruiformes and Aequorlitonithes: Flighted Seabirds & Relatives. In: Wild and exotic animal ophthalmology: volume 1: invertebrates, fishes, amphibians, reptiles, and birds. Springer International Publishing, Cham, pp 505–536
125. Samuel PO, Edo GI, Oloni GO, Ugbune U, Ezekiel GO, Essaghah AEA, Agbo JJ (2023) Effects of chemical contaminants on the ecology and evolution of organisms a review. Chem Ecol 39(10):1071–1107
126. Sangwan S, Kumar M, Lamba R, Singh S, Kumari A, Wati L (2024) Bioindicators: natural biotic sensors of environmental pollution and ecological disturbance. In: Environmental nexus approach. CRC Press, pp 311–337
127. Saraux C, Sydeman WJ, Piatt JF, Anker-Nilssen T, Hentati-Sundberg J, Bertrand S et al (2021) Seabird-induced natural mortality of forage fish varies with fish abundance: evidence from five ecosystems. Fish Fish 22(2):262–279
128. Serratosa J, Hyrenbach KD, Miranda-Urbina D, Portflitt-Toro M, Luna N, Luna-Jorquera G (2020) Environmental drivers of seabird at-sea distribution in the Eastern South Pacific Ocean: assemblage composition across a longitudinal productivity gradient. Front Mar Sci 6:838
129. Setufe SB, Bilikoni J, Ofori E, Ogwu MC (2024) Fish as a sustainable biomonitoring tool in aquatic environments. In: Biomonitoring of pollutants in the global south. Springer, Singapore, p 421
130. Shaffer SA, Weimerskirch H, Scott D, Pinaud D, Thompson DR, Sagar PM et al (2009) Spatiotemporal habitat use by breeding sooty shearwaters Puffinus griseus. Mar Ecol Prog Ser 391:209–220
131. Signa G, Mazzola A, Vizzini S (2021) Seabird influence on ecological processes in coastal marine ecosystems: an overlooked role? A critical review. Estuar Coast Shelf Sci 250:107164
132. Sobrino-Monteliu M, Navarro A, Rodríguez B, Tejera G, Herrera A, Rodríguez A (2025) Seabird biomonitoring indicates similar plastic pollution throughout the Canary Current. Mar Pollut Bull 211:117424
133. Sonne C, Bustnes JO, Herzke D, Jaspers VL, Covaci A, Halley DJ et al (2010) Relationships between organohalogen contaminants and blood plasma clinical–chemical parameters in chicks of three raptor species from Northern Norway. Ecotoxicol Environ Saf 73(1):7–17
134. Spatz DR, Zilli F, Newton KM, Holmes ND, Butchart SHM, Tershy BR (2023) Island restoration can benefit seabird populations and ecosystem health. Conserv Lett 16(1): e12936. https://doi.org/10.1111/conl.12936
135. Stempniewicz L, Weydmann-Zwolicka A, Strzelewicz A, Goc M, Głuchowska M, Kidawa D et al (2021) Advection of Atlantic water masses influences seabird community foraging in a high-Arctic fjord. Prog Oceanogr 193:102549
136. Svechkina A, Portnov BA, Trop T (2020) The impact of artificial light at night on human and ecosystem health: a systematic literature review. Landsc Ecol 35:1725–1742
137. Sydeman WJ, Piatt JF, Thompson SA (2021) Climate change and marine top predators: seabirds and their prey. Front Ecol Evol 9:621119. https://www.frontiersin.org/articles/10.3389/fevo.2021.621119/full
138. Tanaka K, van Franeker JA, Deguchi T, Takada H (2019) Piece-by-piece analysis of additives and manufacturing byproducts in plastics ingested by seabirds: implication for risk of exposure to seabirds. Mar Pollut Bull 145:36–41
139. Tartu S, Gabrielsen GW, Blévin P, Ellis H, Bustnes JO, Herzke D, Chastel O (2014) Endocrine and fitness correlates of long-chain perfluorinated carboxylates exposure in Arctic breeding black-legged kittiwakes. Environ Sci Technol 48(22):13504–13510
140. Tavecchia G, Pradel R, Genovart M, Oro D (2007) Density-dependent parameters and demographic equilibrium in open populations. Oikos 116(9):1481–1492

141. Taylor M (2022) How birds live together: colonies and communities in the Avian World. Princeton University Press, Princeton
142. Thorne LH, Clay TA, Phillips RA, Silvers LG, Wakefield ED (2023) Effects of wind on the movement, behavior, energetics, and life history of seabirds. Mar Ecol Prog Ser 723:73–117
143. Tozer DC, Hvenegaard GT, Bouvier L (2023) Wetland restoration benefits biodiversity and waterbird populations in temperate regions. Ecol Sol Evid 4(2):e12242. https://doi.org/10.1002/2688-8319.12242
144. Tremlett CJ, Cleasby IR, Bolton M, Wilson LJ (2024) Declines in UK breeding populations of seabird species of conservation concern following the outbreak of high pathogenicity avian influenza (HPAI) in 2021–2022. Bird Study 71(4):293–310
145. Tuck GN, Polacheck T, Bulman CM (2003) Spatio-temporal trends of longline fishing effort in the Southern Ocean and implications for seabird bycatch. Biol Conserv 114(1):1–27
146. Van Doren BM, Horton KG, Dokter AM, Klinck H, Elbin SB, Farnsworth A (2017) High-intensity urban light installation dramatically alters nocturnal bird migration. Proc Natl Acad Sci 114(42):11175–11180
147. Velando A, Munilla I, Lopez-Alonso M, Freire J, Perez C (2010) EROD activity and stable isotopes in seabirds to disentangle marine food web contamination after the Prestige oil spill. Environ Pollut 158(5):1275–1280
148. Votier SC, Hatchwell BJ, Beckerman A, McCleery RH, Hunter FM, Pellatt J et al (2005) Oil pollution and climate have wide-scale impacts on seabird demographics. Ecol Lett 8(11): 1157–1164
149. Wakefield ED, Cleasby IR, Bearhop S, Bodey TW, Davies R, Miller PI, Newton J (2015) Long-term individual foraging site fidelity—why some gannets don't change their spots. Ecology 96(12):3058–3074. https://doi.org/10.1890/14-1405.1
150. Weimerskirch H, Louzao M, de Grissac S, Delord K (2012) Changes in wind pattern alter albatross distribution and life-history traits. Science 335(6065):211–214
151. Wiese FK, Robertson GJ (2004) Assessing seabird mortality from chronic oil discharges at sea. J Wildl Manag 68(3):627–638
152. Wilcox C, Carlile N, Hardesty BD, Reid T (2021) Assessing multiple threats to seabird populations using flesh-footed shearwaters Ardenna carneipes on Lord Howe Island, Australia as case study. Sci Rep 11(1):7196
153. Wilhelm SI, Robertson GJ, Ryan PC, Schneider DC (2007) Comparing an estimate of seabirds at risk to a mortality estimate from the November 2004 Terra Nova FPSO oil spill. Mar Pollut Bull 54(5):537–544
154. Wilson MP, Schwarzman MR (2009) Toward a new US chemicals policy: rebuilding the foundation to advance new science, green chemistry, and environmental health. Environ Health Perspect 117(8):1202–1209
155. Yacoubi L, Savoca D, El Zrelli RB, Gopalan J, Nazal M, Lin YJ et al (2024) Trace element levels in the muscles of three tern species (Aves: Laridae) from the western Arabian Gulf: environmental assessment and implications for conservation. Environ Monit Assess 196(3): 235
156. Yadav R, Gupta RK, Kumar R, Kaur T, Moun R (2022a) Biochemical response of earthworm, *Eisenia fetida* to heavy metals toxicity. J Appl Nat Sci 14(3):990–998
157. Yadav R, Gupta RK, Kumar R, Kaur T (2022b) Assessment of toxicity of lead and nickel on the biochemical and immunological parameters of earthworm, Eudrilus eugeniae. Agric Sci Dig 44(5):922–929
158. Yamashita R et al (2020) Accumulation and health risks of heavy metals and organochlorines in seabird populations. Environ Pollut 263:114428. https://doi.org/10.1016/j.envpol.2020.114428
159. Yamashita R, Hiki N, Kashiwada F, Takada H, Mizukawa K, Hardesty BD et al (2021) Plastic additives and legacy persistent organic pollutants in the preen gland oil of seabirds sampled across the globe. Environ Monitor Contamin Res 1:97–112. http://www.gadflypetrel.ning.com

160. Yodha K, Singh D, Kour A, Yadav R, Kumar R (2024) Species diversity and abundance of birds in ber (*Zizyphus mauritiana*) crop. Asian J Conserv Biol 13(1). https://doi.org/10.53562/ajcb.83683
161. Yodha K, Singh D, Kour A, Yaday R, Dahiya P, Kumar R (2023) Estimation of predatory bird damage to winter guava crop. Indian J Ecol 50(5(SI)):1787–1793. https://doi.org/10.55362/IJE/2023/4138
162. Zedler JB, Kercher S (2005) Wetland resources: status, trends, ecosystem services, and restorability. Annu Rev Environ Resour 30:39–74
163. Zeller D, Ansell M, Andreoli V, Heidrich K (2023) Trends in Indian Ocean marine fisheries since 1950: synthesis of reconstructed catch and effort data. Mar Freshw Res 74(4):301–319
164. Zhao J, Chen L, Wang Y (2024) Loss of wetland habitats and declining waterbird diversity in Asia. Avian Res 15:100227. https://www.sciopen.com/article/10.1016/j.avrs.2025.100227

# Chapter 10
# Effect of Anthropogenic Activities on Nesting Ecology of Colonial Seabirds: A Conservative Implication

**Vibhu Vaid and Nisha Vashishat**

**Abstract** Anthropogenic activities have profound impact on terrestrial, marine and freshwater ecosystems worldwide. Among these, seabirds are considered the most conspicuous in the marine ecosystems. In addition, they are highly sensitive birds due to their late ages at first breeding with lower reproductive output making them vulnerable to threats with slow recovery rates. Therefore, the present chapter was planned to elaborate different anthropogenic activities such as tourism, anthropogenic nesting materials, environment pollution, artificial night lightening, anthropogenic noise, ground and aerial-based surveys, which can affect nesting ecology of colonial seabirds. In addition, the impact of human activities on seabirds depends on their species, age, diet, life history, gender and overall health. Furthermore, different human disturbances can cause physiological, behavioural and developmental alterations and sometimes even death. The chapter ultimately underscores the importance of collaborative approach across policy, industry and society to conserve seabird species by developing and implementing laws to conserve their habitat, control human disturbance, revegetation of indigenous plants used by them for nesting materials and removal of debris from their nesting sites.

**Keywords** Anthropogenic activities · Anthropogenic materials · Pollution · Human disturbance · Seabirds

## 10.1 Introduction

Contact of public and wildlife is crucial for creating awareness for the conservation of biodiversity, however, extensive anthropogenic-related disturbances including habitat loss, degradation and pollution can have negative impacts on wildlife including bird diversity such as behavioural and physiological alterations in them [22, 137, 139]. With increase in the anthropogenic disturbances, the Earth is entering Anthropocene era [119, 226]. Among different birds, seabirds, spending a significant

V. Vaid (✉) · N. Vashishat
Department of Zoology, Punjab Agricultural University, Ludhiana, Punjab, India

 
R. Kumar et al. (eds.), *Ecology, Conservation of Aquatic Birds*, Wetlands: Ecology, Conservation and Management 15, https://doi.org/10.1007/978-3-032-17125-2_10

amount of their life in marine or coastal systems, are among the most sensitive group of birds due to late ages at first breeding with lower reproductive output making them susceptible to threats with slow recovery rates [35]. They also support numerous cavity and burrow nesting species, and act as vital bioindicator for evaluating overall health of marine ecosystem [50, 244].

Likewise, they play an essential role in ecosystem functioning processes including shaping community, transportation of nutrients and trophic regulation [86, 155, 164] by acting as predators in food web, fertilizing vegetation through their urine and faeces, disturbing soil by digging burrows and constructing nests [5, 114, 217], and contributing to the influx of allochthonous materials [131]. They also serve as monitors of impacts induced by human activities at terrestrial breeding colonies as well as ocean foraging grounds because they typically construct nest in denser coastal areas that coincide with popular tourist attractions [19, 21].

Among seabirds, responses to human disturbance depend the type of species, where some seabirds adapt to visitor disturbances, yet other species can experience decline in breeding success, increase in nestling mortality, physiological changes, internal stress and behavioural alterations including fleeing behaviour [192]. For instance, Brandt's cormorants amplified their behaviour of vigilance in the presence of anthropogenic noises [34], whereas, *Pygoscelis adeliae* (Adelie penguins) did not showed response [51]. Likewise, *Morus bassanus* (northern gannets) displayed avoidance behaviour in sites with offshore wind turbines whereas *Phalacrocorax carbo* (great cormorants) did not evade sites with wind turbines and utilised stationary structures as roosting places [59]. Similarly, Schaefer and Colombelli-Négrel [176] observed that *Eudyptula minor* (little penguins) located in highly disturbed colonies had more intense physiological and behavioural responses compared to lesser disturbed areas.

Similarly, anthropogenic disturbance responses also upon gender. For instance, an investigation conducted by Ellenberg et al. [62] showed that female *Megaduptes antipodes* (yellow eyed penguins) exhibited longer recovery periods and lesser heart rates in the presence of human beings compared to males. This might be because males have higher testosterone, less stressed and more defensive than female penguins [107]. On another hand, females are more vulnerable to disturbances at nest because they are already under pressure to lay eggs and store body reserves [83]. Moreover, other human activities including drone-based surveys, anthropogenic materials, environmental pollution and artificial lightening could impact nesting ecology of colonial seabirds; therefore, the present literature was designed to describe different anthropogenic activities that can impact nesting ecology of colonial seabirds.

## 10.2 Anthropogenic Activities and Their Impacts on Seabirds

The negative effects of human disturbances to marine bird species, especially pollution and tourism are widely documented [45, 82, 120, 122, 146, 209]. The negative of human activities on seabirds include temporary nest abandonment, rise in the risk of predation, nest desertion and decline in breeding success [60, 209]. Moreover, human disturbances can alter temporal and spatial foraging patterns of seabirds in coastal environments [49]. Therefore, different human activities that can impact nesting ecology of seabirds are shown in Fig. 10.1.

### *10.2.1 Environmental Pollution*

Globally, infrastructural development, industrial revolution and agricultural expansion has led to exponential rise in environmental pollution such as water, light, air, oil spill and noise pollution [118, 157, 184, 240, 241]. Avian species act as pest control agents, ecosystem engineers, pollinators, seed dispersers and serve as valuable environmental health bioindicators [232] and their decline indicates broader ecological disturbances which can also disrupt human health, as shown by vulture decline in Kenya and India [158]. Among these, seabirds act as useful sentinels of

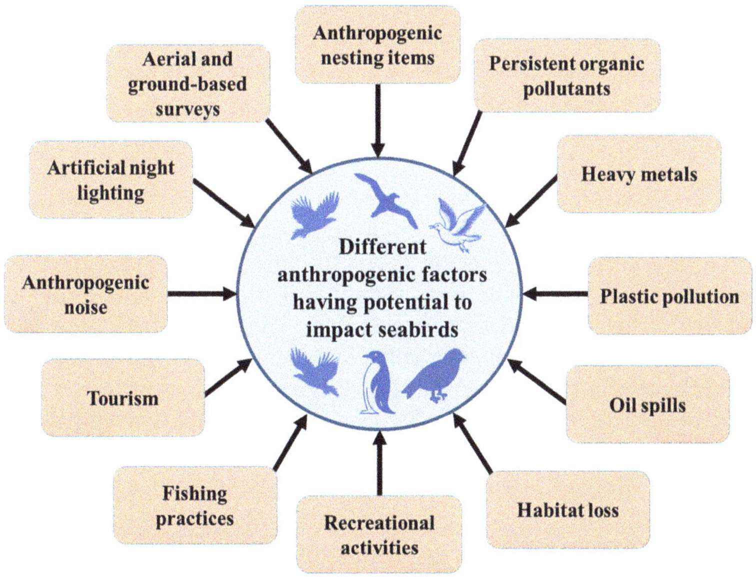


**Fig. 10.1** Different anthropogenic factors that can impact seabird species

marine and coastal marine pollution [73]. They have different responses to environmental pollution due to diverse behaviours, physiology, life histories and diet [33].

#### 10.2.1.1 Oil Spills

Petroleum, a key global energy source, also cause the degradation and pollution of marine environments worldwide [111]. Acute oil spills profoundly challenge global environmental sustainability, ecosystem functioning, impact biodiversity and health of human beings [100, 157, 242] indirectly (habitat degradation) or directly (ingestion, physical contact, absorption and inhalation) [44, 93]. Among these, seabirds are typically affected by oil spills due to the plumage's oil fouling which further damages their capability to fly, thermoregulate, dive, float or swim resulting in starvation, hypothermia, drowning and mortality [215]. Their survival rate further depends on the condition of plumage, duration of oil exposure and degree of oiling in sea birds such as *E. minor* [82].

Troisi et al. [214] observed that the seabirds exposed to oil cannot dive and hence forage for food. Culik et al. [52] observed significant decrease in the body mass, increase metabolic rate and death in oiled penguins. Oiled plumage of penguins' results in rise in the metabolic rate which usually occurs due to increase in heat loss and reduction in thermal insulation [110]. Furthermore, a minute amount of oil on the penguins' plumage led to waterlogging, thus, affecting their waterproofing, thermal insulation and buoyancy which further causes dehydration and hypothermia [82]. In addition, petroleum products also cause chronic toxic effects effecting their survival and reproduction [110]. Additionally, penguins exposed to hydrocarbon pollution can also show mortality due to narcotic-like intoxication or poisoning [169]. Vanstreels et al. [223] reported that 39 oil spills in South Africa and Nambia affected 83,224 different seabirds especially *Morus capensis* (Cape gannets) (8.5%) and *Spheniscus demersus* (African penguins) (91%). Moreover, they discussed different sources of oil spills such as ship to ship transfers (14%), tankers (38%), cargo carriers (43%) and others (46%). Similarly, Barreto et al. [14] observed that penguins died from oil spill pollution showed lung and hepatic congestion.

The negative impacts of oil spills on diversity, richness and functional traits of bird species in the Niger Delta region were also found by Iniunam et al. [100]. The diversity and richness of birds were significantly lesser at impacted regions in comparison to reference sites which suggested loss of ecological functions including pest control, seed dispersal and pollination. In addition, the oil spilled-sites showed predominance of generalist species whereas it showed lesser abundance of specialist species. The study also highlighted significant decline in functional diversity in oil spilled sites as shown by Rao's entropy and functional richness. Likewise, Troisi [213] noticed polycyclic aromatic hydrocarbons (PAH) metabolites in the liver of oiled seabird species from 0.04 to 0.97 μg/g. Taniguchi et al. [204] observed that the distribution of PAHs in the fat tissue of penguins in Antarctica was similar to the petroleum derivatives in Diesel Fuel. These PAHs in diesel can cause immunotoxic conditions by disturbing the cell-mediated response of bird species to immunogens.

Its accumulation in urine and bile by seabirds is due to monooxygenase enzyme system, which is having ability to convert PAHs into water soluble metabolites [14]. Hydrocarbons affect reproductive performance of seabirds like penguins by reducing their egg production, fertility as well as hatching rates and embryotoxicity [190].

#### 10.2.1.2 Plastic Waste

Plastic waste is a ubiquitous environmental issue and ingestion of plastic by seabird species is a matter of rising concern worldwide [11]. It enters sea through off shore dumping, plastic industries, fishing nets, objects left behind by beach goers, waste disposal from drainage and river systems [123, 235]. In addition, plastic particles such as microplastics (MPs) on the ocean surface severely impact seabirds through ingestion or entanglement because they intentionally select these particles due to their resemblance with prey [150, 219]. MPs refer to small (0.1–5000 μm in diameter) pellets, flakes, fibres, granules, fragments, spheroids or beads formed by plastic degradation [61], and were initially informed in marine bodies in early 1970s [210]. Entanglement of plastic cause wounds, upsurge in the chances of being preyed upon, drowning and decrease in efficiency of predation suffocation [225]. Similarly, physical ingestion of plastic further causes detrimental effects in seabirds such as choking, feeling of false satiation, reduction in nutrient uptake, damage of digestive tracts, ulcers or even mortality [117, 128, 198]. Furthermore, guano deposition and regurgitation by birds also spread pollutants to other places acting as vectors of plastic debris derived from marine waters in other water bodies and terrestrial ecosystems [89]. Likewise, the number of plastic pieces in stomach of different seabirds including petrels, albatrosses, phalaropes and storm-petrels show negatively correlation with body fat, a measure of energy reserves [8, 159].

Numerous seabirds ingest plastic particles, however, procellariiforms suffer the most because it is difficult for them to regurgitate solid substances such as plastics because they have small gizzards and a narrowing between proventriculus and gizzard [10, 159]. Since 1960s, MPs have been studied in digestive systems of roughly 78% seabirds [15] and by 2050, more than 99% of over 300 seabirds are expected to ingest MPs [234]. From 1983–2023, around 398, 75, 40 and 3013 MPs were found in pellets, guano, pouch and stomachs, respectively about thirteen seabirds. On an average, 7.2 and 1.1 MPs were noticed in Arctic region and Antarctica, respectively. The literature also recorded that polyethylene (PE) was predominant MP followed by polypropene and polystyrene [205]. Baak et al. [11] examined 38 different studies reporting plastic ingestion by seabirds in the Arctic region. The researchers examined 51 seabird species and observed that over half were having plastic ingestion. Apart from dietary sources, birds can also gather MPs from air as they have efficient respiratory system. To confirm air borne MPs in birds, Tokunaga et al. [211] investigated lungs of about 22 wild birds (Japan) and observed the presence of 6 MPs in 3 birds ranging from 28.0 to 70.5 μm.

#### 10.2.1.3 Heavy Metals (HMs)

HMs refer to soluble natural elements available in trace levels (<10 ppm) in environment with specific weight more than 5 g/cm$^3$ [2]. HMs including selenium (Se), magnesium (Mg), nickel (Ni), copper (Cu), zinc (Zn), manganese (Mn), cobalt (Co), iron (Fe) and molybdenum (Mo) are essential for numerous biochemical and physiological functions of biota such as maintenance of skeletal system and regulation of acid-base balance. In addition, they act as constituents of enzymes, hormones as well as proteins and their inadequate levels can lead to different health issues [109, 240–242]. In contrast, non-essential HMs are toxic, have no role in body functions and can adversely impact biota as well as environment including water, atmosphere and soil [28, 211, 218, 220]. Moreover, different HMs including lead (Pb), mercury (Hg), chromium (Cr), antimony (Sb), thallium (Ti), cadmium (Cd), Zn and Cu can cause cardiovascular toxicity, hepatotoxicity, skin damage, nephrotoxicity and neurotoxicity [147].

Their prolonged exposure is perilous to aquatic and terrestrial systems due to their bioaccumulation, non-biodegradability and toxic nature [206]. Furthermore, the overexploitation of HM ores for different natural and anthropogenic activities (construction, metallurgical industry or mining) has led to tremendous accumulation of HMs in the marine ecosystems [182]. Due to their direct connect with water, fishes accumulate HMs through various pathways (gills, body surface and digestive tract) which further leads to HM bioaccumulation and biomagnification in in top predators such as coastal seabirds through drinking or fish consumption [3, 48, 243]. HMs have been found in cormorants, albatross and penguins worldwide, even in remote areas like Antarctic [40, 105]. Moreover, HMs also enter marine food webs through various mechanisms such as via their uptake by phytoplankton and microorganisms or adsorption onto dissolved organic matter or suspended particles [32]. Once prey is ingested, HMs assimilate in their intestine and are carried from lower to higher trophic levels [201]. Moreover, guano deposition by penguins could transfer HMs from marine bodies to the terrestrial ecosystems. Chu et al. [45] observed that the levels of HMs such as Cd, Pb, Cu, Hg and Zn increased Ardley Island after contamination with penguin guano. Likewise, Evans et al. [67] reported higher concentrations of different HMs *viz.* Cu, arsenic (As) and Cd were in orinthogenic soils from penguin rookeries in the Antarctic Peninsula and South Shetland Islands.

#### 10.2.1.4 Persistent Organic Pollutants (POPs

POPs which are highly resistant to degradation such as organochlorine pesticides (OCPs), polybrominated diphenyl ethers (PBDEs) and polychlorinated biphenyl ethers (PCBs), act a serious threat to different marine predators at trophic levels including seabirds despite of being banned in many countries for decades. They act as endocrine disruptors and impair reproductive and immune system [84]. They impact their development, physiology, reproduction and survival depending on different factors such as age, gender, species and the nature of chemical compound [85, 92, 118, 123]. Moreover, they transmit contaminants to their offsprings [95] via eggs leading to reduction in their size, embryo mortality, decline in egg hatchability, changes in composition and thickness [7]. For example, mothers of different species such as *P. adeliae* [202], *L. argentatus* (Herring gulls) and *Sterna paradisaea* (Arctic terns) [130] transfer 4, 24 and 45% of PCBs to their eggs, respectively. In addition, Tanabe et al. [202] noticed the positive relation between egg size and PCBs.

Campioni et al. [38] examined the annual alterations in PAH, OCP, PCB and PBDE their impacts on breeding success of *Pterodroma cahow* (Bermuda petrel) during 2019 and 2022. The results indicated that 33 pollutants were noticed in the blood of Bermuda petrels. Moreover, only 9 samples were ≥50% and out of these, PCBs were most abundant in blood of petrels in both studied years. The study also shown that females with more contaminants laid eggs with lesser possibility of hatching. Wurster and Wingate [238] also observed decline in breeding success with rise in DDTs in Bermuda petrel eggs. Miljeteig et al. [146] recorded the impact of various contaminants *viz.* OCPs, PCBs, Hg, perfluorinated alkyl substances and brominated flame retardants (BFRs) on α-tocopherol (vitamin E), retinol (vitamin A) and egg thickness of *Pagophila eburnea* (ivory gull). The results shown no relation between Hg and perfluorinated alkyl substances. However, 7–17% egg thinning was observed by the researchers. Colabuono et al. [46] assessed OCPs and PCBs in the liver, muscle along with subcutaneous fat of total 100 birds belonging to 8 species of Procellariiformes in southern Brazil. The results indicated that organochlorine compounds were present in all the analysed samples. The scientists observed that dichlorodiphenyltrichloroethane (DDTs) and PCBs were predominant in tissues of studies species.

Along with this, summary of other recent investigations showing the accumulation of multiple stressors including plastic, HMs, POPs and anthropogenic materials in nests or bodies of seabirds are depicted in the Table 10.1.

### *10.2.2 Artificial Night Lighting*

Human beings utilise artificial night lights either for convenience, safety or productivity typically in coastal areas (lighthouses, urban settlements, light induced fisheries, port facilities and other infrastructures) [148]. Slight changes in the levels of light, hardly distinguishable to human eye, can significantly affect ecological processes and functioning of organisms i.e. their behaviours and physiology [174] which further alters species abundance and impacts biodiversity [17, 174]. Due to higher radiance and cost efficiency, light-emitting diode (LED) lights are frequently used for lighting [47, 143].

Among different organisms, seabirds are extremely sensitive to short wavelength LED lights, typically those in the violet-blue region of the visible spectrum [39]. Additionally, night lights distract nocturnal seabirds [167] and expose them

**Table 10.1** Recent studies showing different anthropogenic stressors found in seabirds or their nests

| Sr. No. | Stressors studied | Species studied | Observations | References |
|---|---|---|---|---|
| 1. | Plastic debris and anthropogenic material | *L. michahellis* | 88% nests were having anthropogenic material account for 3% weight of nest items<br>Plastic represented 78% of recorded anthropogenic material and 71% of weight of anthropogenic material | [90] |
| 2. | MPs and macroplastics (>5 mm in size) | *L. michahellis* chicks | The stomach contents displayed the occurrence of MPs in all the studied samples, however, 19.64% showed macroplastics in their stomach content | [140] |
| 3. | 12 plastic additives [including 9 Benzotriazole UV-Stabilizers (BZT-UV)] | *L. gulls* | 96% gulls had around 1 BZT-UV at detectable liver levels<br>Among different BZT-UVs, UV328 (76%) and UVP (91.4%) were the most abundant | [179] |
| 4. | 29 *Per*- and polyfluoroalkyl substances | *P. papua* (Gentoo penguins) | Adults accumulated more NaDONA, PFOA and PFPeS than chicks<br>The compounds 4:2 FTS, PFOS, PFDoDA, PFEESA, PFHxA, 6:2 FTS and PFBS were only found in excreta of penguins showing their inability to absorb into species | [40] |
| 5. | As, Cu, Zn, Pb, Hg, vanadium (V), potassium (K), calcium (Ca), sodium (Na), Ni, Co, barium (Ba), Cd, Cr, Fe, Mg, strontium (Sr) and Mn | *Onychoprion anaethetus* (bridled tern), *T. bergii* (greater crested tern) and *S. repressa* (white-cheeked tern) | 15 out of 19 elements studied were found in muscles of terns<br>Co, Ni, Cd, V and Cr were found below the detection limit<br>Different individuals displayed high variability in the level of different HMs | [239] |
| 6. | Mn, Mg, Se, Cu, Ni, Ca, Cd, V, Cr, Zn, Pb, Co and Fe | *L. fuscus* | HMs in feather samples were found in the order:<br>Mg > Ca > Fe > Zn > Cr > Mn > Ni > Cu > Pb > Co > Cd > Se > V | [154] |
| 7. | Plastic debris and anthropogenic material | *L. michahellis* | 92% pellets contained at least 1 anthropogenic item whereas 79% contained at least 1 plastic debris | [4] |
| 8. | POPs and Hg | *U. lomvia* | The blood showed the presence of methylmercury followed by PBDEs, PCBs, hexachlorobenzene and *p,p*'dichlorodiphenyltrichloroethylene | [66] |

to risks of collision with vehicles, surfaces or structures as well as predation [58, 144]. They can be attracted to areas with lights to increase their vision or easy location of prey [144]. A study by Rodríguez et al. [168] on Phillip Island in Victoria, Australia showed that *E. minor* prefers illuminated paths while returning to their nests at night to improve predator detection and their navigation. However, environmental factors including precipitation, cloud cover, moon phase or closeness to nesting sites can influence the preference of lights by seabirds [43, 47, 233]. Iasiello and Colombelli-Négrel [99] examined the impact of LED floodlight at night on behaviours *viz.* group size, vigilance arrival duration and number of groups of little penguins while returning to their colony on Lipson Island in South Australia. The scientists found that in the presence of light lesser number of groups arrived earlier. Furthermore, the scientists observed effect of light on vigilance behaviour of penguins. In Longyearbyen, different seabirds decreased with increase in the intensity of artificial light, however, piscivorous Black Guillemot showed adaption to the presence of light [132].

### *10.2.3 Anthropogenic Noise*

Seabirds depend on their senses to perform daily activities such as navigation, acquisition of resources and avoidance of threats [72, 163]. Anthropogenic noise is a globally pervasive ecological concern, produced from coastal developments (marine industries, transportation, infrastructure and buildings) can induce stress response in birds [104, 161], where their attention is grasped by stimulus, which alters their behaviours such as breeding, foraging and communication [18, 126, 186]. Senses of seabirds are extremely adapted to their surrounding environment and make them sensitive to even sensory pollution [23, 149].

Additionally, long term fitness in seabirds may occur due to short term exposure to anthropogenic noise, which in turn declines the population or lead to rejection of disturbed areas [17, 160]. Artificial light can affect photoperiod, physiological traits and foraging success [75, 199]. Moreover, the rise in light has been recognised as a threat for seabird species such as *Puffinus puffinus* (Manx Shearwater) and petrels [200]. Since the last century, the rise in the light intensity has been assessed to grow by 6% ever year [96].

Artificial lights also cause them to wrongly select poor quality sites over better substitutes and hence act as an ecological trap [177, 188]. It may also avert them from migrating south to their wintering sites leading to higher energy demands during the harsh nights in Arctic region [166, 191]. In illuminated area in Longyearbyen, seabird diversity clearly decreases with increasing levels of artificial light. The only species that appears to be most adapted to or capable of utilizing artificial light sources is the piscivorous Black Guillemot [132]. Furthermore, Iasiello and Colombelli-Négrel [98] examined the physiological and behavioural response of *E. minor* to anthropogenic noises from coastal developments. The scientists recorded increase in vigilance time in the presence of noise. Repetitive occurrence of construction noises also reduce time spent for self-care behaviours including preening or resting [41, 68] and divert them predation threat [34]. Chatwin et al. [42] observed alert response at shorter distances in breeding colonies of *Cepphus columba* (pigeon guillemots), *Haematopus bachmani* (black oyster-catchers), *Phalacrocorax auritus* (double-crested), *P. pelagicus* (pelagic cormorants) and *L. glaucescens* (glaucous-winged gulls) on Vancouver Island (Canada).

### *10.2.4 Anthropogenic Materials: Their Characterization, Nest Composition and Impacts*

Anthropogenic materials refer to human-made solid objects having different types such as HMs, plastic, fibre, robe, rubber and paper [12]. Due to incorrect disposal, many of these items enter into land and water bodies [65, 216] leading to their interaction with wildlife [9, 212]. Numerous studies reported these materials in the digestive system of various turtles, mammals (Sirenia, Mysticeti and Otariidae) and seabirds (Pelecaniformes, Procellariiformes and Charadriiformes) [116, 207] which led to behavioural and development alterations [88]. The incidence of human items in the nests of various water birds has been found to range from 4–80% [16, 236] which are connected to distance from pollution source and reusing of nest [208]. Some species including *Ciconia Ciconia* (white stork) reutilise the same nests whereas spoonbills do not [102]. Moreover, collecting items from con-specific nests including older nests is an infamous behaviour of Pelecaniformes [87, 153].

Nests usually comprise four sections i.e. lining materials to protect eggs and to create a suitable microenvironment within nest, supporting structural layer, outer decorative layer and attachment which includes materials required for the fixation of nest to the substrate [91]. Moreover, these nest characteristics have a key role in defining breeding success of birds [138]. The incorporation of anthropogenic items in nests, both for lining and structural purposes, are linked with the availability of anthropogenic items near nesting sites [6] along with the absence and scarcity of natural materials near nesting area [165]. Furthermore, these materials aid in reducing the time and energy required to search natural materials compared to durable, readily available and light weight debris such as plastic foils or strings [6]. Lee et al. [129] observed 71% anthropogenic material in nesting sites of *Platalea minor* (black-faced spoonbills) breeding on rocky islands lacking vegetation, which was halved in the presence of natural items.

However, several researchers have anticipated that the utilisation of these materials may be favoured over natural nesting objects by different species [29, 101] due to various reasons:

(i) **Chemical composition:** Cigarette butts waste hold considerable quantities of nicotine and various components, which act as ecto-parasite repellents despite of their long-term impacts [197] favouring their selection by *Haemorhous mexicanus* (house finch) [196] to increasing breeding success [195].

(ii) **Resemblance:** The presence of anthropogenic items in nests is also associated with their similarity with natural items. For example, *Anous stolidus* (brown noddy) mostly accumulates rope and similar substances to build their nests, probably due to their resemblance natural materials like twigs [173].

(iii) **Durability:** The usage of plastic debris by *Lanius excubitor* (great grey shrike) [6] and artificial plants by *Fulica atra* (common coot) [94] play a vital role in strengthening nests, protecting nestlings and eggs.

(iv) **Mate attraction:** For example, the decoration of bowers with plants, flowers and debris i.e. straws, bottle tops by bowerbirds to attract females for having better reproductive output [26]. Similarly, *Milvus migrans* (black kites) utilise plastic debris in their nests to indicate better quality of its nest [183].

Among different anthropogenic items, plastic debris are considered the most common contaminants [31] and are classified based on their purpose into user (thread and sheet like plastics) and industrial plastic debris (nurdles) [162, 222]. In 1966, Kenyon and Kridler primarily documented hard items such as plastics in the gastrointestinal tract of juvenile Laysan albatrosses (Kenyon and Kridler [108]). Similarly, anthropogenic debris also include fishing materials, personal hygiene products, construction or demolition remains, food packaging items and textile [74, 152]. Domínguez-Hernández et al. [60] observed the proportion of anthropogenic items in 48 left-out nests of *Larus michahellis atlantis* (yellow-legged gull) from 5 hardly accessible remote areas in biosphere reserves of the Canary Islands in Spain. The results indicated that 81.3% nests contained debris, out of which 34.7, 33.6 and 19.6% were plastic, HM and paper items, respectively. Based on origin of waste, the proportion of items were in the order: food packaging materials > personal hygiene stuffs > fabric. Additionally, Fourier-transform infrared spectroscopy (FTIR) indicated polyester as the predominant polymer followed by PE and rayon.

Ferreira et al. [70] also observed at least 1 anthropogenic material typically plastic of white-transparent colour and sheet-like shape in 39% of total studied nests (439) of *P. leucorodia* (Eurasian spoonbill) monitored in the Camargue wetland, Southern France during the breeding seasons of 2021 and 2022. The quantity of anthropogenic nest items showed negative correlation with natural vegetation. Furthermore, higher amount of natural materials were noticed during the egg phase in comparison to nestling phase. Battisti [16] also observed different anthropogenic debris such as 31% plastic, 27% glass and 16% pieces of paper in 307 roosting and nesting sites situated in the western region of the Italian San Pietro Island. Jagiello et al. [102] also observed anthropogenic materials in 50 and 42% broods of *Ciconia Ciconia* (white stork) during egg and chick stage respectively. Moreover, the researchers observed that 38% anthropogenic materials were plastic strings followed by foil, fabric, paper, different plastic items and other materials. They also recorded positive correlation between age of bird and debris present in the nest structure. Delgado et al. [57] also found 1.3 anthropogenic litter per nest typically plastic, textile and ropes in nests of Ulia. Likewise, Lopes et al. [133] observed that plastic was the most abundant material followed by textile and paper in nests studied in the Atlantic coast of Portugal. Similarly, Holland et al. [97] also found 11.1% plastic and other anthropogenic objects in 350 birds of 18 species (marine common eider and freshwater birds) from Canada.

Nevertheless, they also have several negative impacts on different sea animals, including marine birds, chiefly through entanglement or ingestion [116, 134, 172] (Bourdages et al. [27]). Additionally, the type and number of anthropogenic items could affect their functional performances such as thermoregulation, camouflage and structural integrity which further impact their fitness related traits including fledging and hatching success [55, 208]. Similarly, the loss of container of ship in the Wadden Sea caused the contamination of the breeding area of spoonbill with plastics, though, no macroplastics (>20–100 mm) were noticed in the birds' regurgitates [71]. In addition, numerous studies showed that different chemical components in anthropogenic objects such as polychlorinated biphenyls impact animal physiology [69, 116]. For example, studies showed that biphenyl exposure to birds lead to alterations in hormonal levels, increase in risk of disease, reproductive disorders and even death [56, 124, 203].

### *10.2.5 Aerial and Ground-Based Surveys*

Economical, accurate and consistent surveys of wildlife act as fundamental elements to support wildlife science and management. In response to disturbances, nesting birds show increase in agonistic, escape or vigilant behaviour [13, 229]. Traditional censusing methods for colonial birds and distribution of nesting included aircraft and ground-based survey and remote sensing techniques (satellite imagery) [36]. Aerial surveys have been conventionally utilised to study spatial distribution [136], fine-scale foraging behaviours [103], operational sex ratios [178], count individuals [171] and nest survival estimates [125]. Ground-based surveys necessitate close proximity of scientists to nests [30], however, aerial surveys are considered to quickly cover more regions but can be dangerous, luxurious and difficult [156].

Similarly, *Leucophaeus pipixcan* (Franklin's gulls) [142] and swifts [145] revealed behavioural disturbances when the drone is flying below 50 m. However, Western Grebes [125] and *Leucogeranus leucogeranus* (Siberian cranes) [230] displayed no or lower disturbance irrespective of altitude. Moreover, drone aerial surveys are lesser invasive substitute to aircraft and ground surveys but lead to increase in heart rate by colonial nesting seabirds such as wandering albatrosses *Diomedea exulans*, Giant petrels *Macronectes halli* and Adélie penguins [54, 77, 127]. Moreover, they caused less disturbance compared to ground surveys in case of *L. fuscus* (Lesser black-backed gull) [171]. In addition, there size can also influence animal behaviour, with larger drones displaying more reactivity [151].

Recent techniques such aerial drones [136] are used to gather large amounts of information, and can be flown in unreachable sites [141]. Despite of their advantages, the study of colonial birds elevated the concerns about their effects on nesting birds [231]. So far, studies on the impacts of drones on bird species has focused chiefly on behavioural changes of colonial birds [64]. Weimerskirch et al. [229] examined the impact of drone passage on heart rate responses in chick and adult *Aptenodytes patagonicus* (King penguins). Likewise, scientists study the 'distance hypothesis' which showed animals exhibit fewer responses to drones at higher elevations [229]. The scientists observed that heart rates of both size groups amplified in response to the presence of drones about 2–50 m. Similarly, Brisson-Curadeau et al. [30] recorded rise in avoidance behaviour and hence predation potential on eggs of *Uria lomvia* (thick-billed murres) in the presence of experiment drone flights.

Among different drones, the noise of multirotor drone more in comparison to fuel powered and fixed wing drone [81]. Numerous studies also shown that birds respond more to drones displaying flight patterns alike avian predators including target-oriented approach angles and vertical flight patterns [151, 170]. Scarton and Valle [175] observed non-significant different in ground and drone survey of *Chroicocephalus ridibundus* (black-headed gulls) during breeding seasons from 2017 to 2019 on marsh islands, Lagoon of Venice (Italy). Valle and Scarton [221] also examined non-significant difference between drone and ground-based surveys of terns [*Gelochelidon nilotica* (gull-billed terns) and *Thalasseus sandvicensis* (sandwich terns] and gulls [*L. melanocephalus* (Mediterranean gulls) and *C. genei* (slender-billed gulls) breeding in Adriatic wetlands, Italy. Likewise, Geldart et al. [77] found the effect of 30 m above quadcopter drone on the heart rates of incubating female *Somateria mollissima* (common eider ducks) for 17 days. The findings suggested that planned drone surveys avoid physiological impacts among eiders. Besides, planned drone-based aerial surveys could minimise disturbance on nesting bird surveys.

### 10.2.6 Presence of Human (Tourism/Recreational Visits)

Bird watching is a fast growing, wildlife and nature-based tourism which has both negative and positive impacts [53, 115]. It can raise awareness to protect birds and generate funding for their conservation [25, 181], however, it also disturbs wildlife at species, individual and population level [194]. Besides, birds adapted to trouble by human tourist may become braver rising their susceptibility to predation [76, 193]. Globally, ecotourism places and their popularity has amplified since the introduction of term in 1980s [228, 237]. Seabirds are also susceptible to anthropogenic disturbance in coastal regions [112, 185].

The recreational visits are usually less invasive than the presence of scientific researchers, but have comparatively greater effects [78] because they are unregulated and involve more people for longer durations. Few studies showed elevated heart rates [63] and corticosterone stress responses in penguins in the presence of tourists, however, few reports displayed no alterations in breeding success [245]. Earlier researches have presumed that visual contact is essential to alter the behaviour of animals so, nest-burrowing animals are less likely to be affected by human disturbances (Lunn et al. [135]). However, Ellenberg et al. [63] showed visitors passing nearby without any visual contact can also elevate the heart rate of Humboldt penguins. Watson et al. [227] observed higher nestling mortality in cavity nesting seabird *Hydrobates pelagicus* (European storm petrels) followed by higher human visitation. Similarly, rise in heart rate of *P. adeliae* was observed without any behavioural modification when human beings were present within 15 m [79]. Besides, repeated tourist pressure discouraged breeding of mature Brandt's cormorants in their actual breeding locations [34].

Similarly, Thompson [209] observed that *T. elegans* (elegant terns) abandoned their nesting places after a single short-term contact. Furthermore, Walker et al. [224] observed habituation to human trouble in adults of Magellanic penguin, however, not in chicks. Likewise, other direct or indirect human disturbances that can impact seabirds include habitat loss and fishing practices. Kavelaars et al. [106] observed negative consequences of habitat loss on foraging behaviour and development of offspring of colonial breeding seabird *L. fuscus* using GPS-tracking. Besides, recently relocated gulls visited distant foraging sites for foraging in comparison to residents to avoid areas in home range of resident colonies where they settled. Similarly, the population of penguins of Falkland Islands reduced by about 84% during the 1980s and 1990s due to commercial fishing around Falklands [20].

## 10.3 Conservation Implications

In light of the ongoing threats posed by human disturbances, there is a necessity to implement multiple management techniques and sustainable practices to mitigate these impacts. Governments should implement strict policies for improvement and restoration of nature and marine net gain [180] including managing the use of artificial lighting and limiting anthropogenic noise. Restoration of indigenous plants and removal of litter in nesting and foraging locations of seabirds can provide food and nesting items to them, limiting their dependence on anthropogenic nesting items. The conservation of seabirds also requires the development strict marine reserves with ban of all human activities including resource utilisation [1, 80, 113]. Static constructions such as wind turbines can affect seabirds only at particular site whereas contaminants such as HMs, plastic, oil spills and POPs freely flow in aquatic and terrestrial ecosystems, emphasising the need to manage them at different spatial scales [24]. Few methods to avoid negative impacts of different pollutants include limiting usage of microbeads in laundry, cosmetics and other industries, regulating release of pesticides and HMs from their sources in water bodies, encouraging utilisation of sustainable materials and evolving new strategies to eradicate POPs, HMs and plastic debris from marine ecosystems. Numerous studies [180, 187, 189] documented the rise in breeding success in different species upon fisheries closure suggesting its potential to offset anthropogenic impacts.

## 10.4 Future Perspectives

Future studies should focus on the continuous and long-term impacts of anthropogenic materials and activities on the colonial seabirds. In addition, advanced technologies can be used to map and monitor vulnerable nesting zones for further conservation.

## 10.5 Conclusion

The chapter shown that anthropogenic activities including HM and POP pollution, anthropogenic nesting materials, tourism, recreational activities, research based-surveys, habitat loss, fishing, anthropogenic noise and artificial lighting can negatively impact seabird populations. These human disturbances can alter their physiology, behaviour, decline in breeding success, increase in the predation risk, temporary nest abandonment and sometimes even death. Furthermore, the chapter emphasized multiple management techniques to control the impact of human disturbances on seabirds.

## References

1. Agardy T, Di Sciara GN, Christie P (2011) Mind the gap: addressing the shortcomings of marine protected areas through large scale marine spatial planning. Mar Policy 35(2):226–232
2. Alengebawy A, Abdelkhalek ST, Qureshi SR, Wang MQ (2021) Heavy metals and pesticides toxicity in agricultural soil and plants: ecological risks and human health implications. Toxics 9(3):42
3. Ali H, Khan E (2019) Trophic transfer, bioaccumulation, and biomagnification of non-essential hazardous heavy metals and metalloids in food chains/webs—concepts and implications for wildlife and human health. Hum Ecol Risk Assess Int J 25(6):1353–1376
4. Almeida FN, Leray C, Souc C, Scotto S, Selmi S, Hammouda A, Ramos R, Ter Halle A, McCoy KD, Vittecoq M (2024) Among-colony variation in plastic ingestion by yellow-legged gulls (*Larus michahellis*) across the western Mediterranean basin. Mar Pollut Bull 204:116508
5. Anderson WB, Mulder CP, Ellis JC (2017) Seabird Island ecology. eLS, pp 1–8
6. Antczak M, Hromada M, Czechowski P, Tabor J, Zabłocki P, Grzybek J, Tryjanowski P (2010) A new material for old solutions—the case of plastic string used in Great Grey Shrike nests. acta ethologica 13:87–91
7. Assersohn K, Marshall AF, Morland F, Brekke P, Hemmings N (2021) Why do eggs fail? Causes of hatching failure in threatened populations and consequences for conservation. Anim Conserv 24(4):540–551
8. Auman HJ, Ludwig JP, Giesy JP, Colborn THEO (1997) Plastic ingestion by Laysan albatross chicks on Sand Island, Midway Atoll, in 1994 and 1995. Albatross biology and conservation 239244
9. Azevedo-Santos VM, Brito MF, Manoel PS, Perroca JF, Rodrigues-Filho JL, Paschoal LR, Gonçalves GR, Wolf MR, Blettler MC, Andrade MC, Nobile AB (2021) Plastic pollution: a focus on freshwater biodiversity. Ambio 50(7):1313–1324

10. Azzarello MY, Van Vleet ES (1987) Marine birds and plastic pollution. Mar Ecol Prog Ser 37(2/3):295–303
11. Baak JE, Linnebjerg JF, Barry T, Gavrilo MV, Mallory ML, Price C, Provencher JF (2020) Plastic ingestion by seabirds in the circumpolar Arctic: a review. Environ Rev 28(4):506–516
12. Barnes DK, Galgani F, Thompson RC, Barlaz M (2009) Accumulation and fragmentation of plastic debris in global environments. Philos Trans R Soc B Biol Sci 364(1526):1985–1998
13. Barr JR, Green MC, DeMaso SJ, Hardy TB (2020) Drone surveys do not increase colony-wide flight behaviour at waterbird nesting sites, but sensitivity varies among species. Sci Rep 10(1): 3781
14. Barreto IS, de Albergaria-Barbosa ACR, Patire VF, de Jesus Silva M, Baldassin P, Taniguchi S, Montone RC, Gallo H, Maranho A, Bícego MC (2020) Bioavailability of polycyclic aromatic hydrocarbons to penguins on the coast of southeastern Brazil. Mar Pollut Bull 157:111306
15. Basto MN, Nicastro KR, Tavares AI, McQuaid CD, Casero M, Azevedo F, Zardi GI (2019) Plastic ingestion in aquatic birds in Portugal. Mar Pollut Bull 138:19–24
16. Battisti C (2020) Heterogeneous composition of anthropogenic litter recorded in nests of Yellow-legged gull (*Larus michahellis*) from a small Mediterranean Island. Mar Pollut Bull 150:110682
17. Bennie J, Davies TW, Cruse D, Bell F, Gaston KJ (2018) Artificial light at night alters grassland vegetation species composition and phenology. J Appl Ecol 55(1):442–450
18. Bevan E, Whiting S, Tucker T, Guinea M, Raith A, Douglas R (2018) Measuring behavioral responses of sea turtles, saltwater crocodiles, and crested terns to drone disturbance to define ethical operating thresholds. PLoS One 13(3):e0194460
19. Bianchini K, Mallory ML, Braune BM, Muir DC, Provencher JF (2022) Why do we monitor? Using seabird eggs to track trends in Arctic environmental contamination. Environ Rev 30(2): 245–267
20. Bingham M (2002) The decline of Falkland Islands penguins in the presence of a commercial fishing industry. Rev Chil Hist Nat 75(4):805–818
21. BirdLife International (2008) State of the world's birds: indicators for our changing world. Birdlife International
22. Bisht S, Bargali SS, Bargali K, Rawat GS, Rawat YS, Fartyal A (2022) Influence of anthropogenic activities on forest carbon stocks—a case study from Gori Valley, Western Himalaya. Sustainability 14(24):16918
23. Blumstein DT, Whitaker J, Kennen J, Bryant GA (2017) Do birds differentiate between white noise and deterministic chaos? Ethology 123(12):966–973
24. Boersma PD, Parrish JK (1999) Limiting abuse: marine protected areas, a limited solution. Ecol Econ 31(2):287–304
25. Borges de Lima I, Green RJ (2017) Introduction: wildlife tourism management and phenomena: a web of complex conceptual, theoretical and practical issues. In: Wildlife tourism, environmental learning and ethical encounters: ecological and conservation aspects. Springer International Publishing, Cham, pp 1–17
26. Borgia G (1985) Bower quality, number of decorations and mating success of male satin bowerbirds (*Ptilonorhynchus violaceus*): an experimental analysis. Anim Behav 33(1): 266–271
27. Bourdages MP, Provencher JF, Baak JE, Mallory ML, Vermaire JC (2021) Breeding seabirds as vectors of microplastics from sea to land: evidence from colonies in Arctic Canada. Sci Total Environ 764:142808
28. Briffa J (2020) Heavy metals in Maltese agricultural soil (Master's thesis, University of Malta)
29. Briggs KB, Deeming DC, Mainwaring MC (2023) Plastic is a widely used and selectively chosen nesting material for pied flycatchers (*Ficedula hypoleuca*) in rural woodland habitats. Sci Total Environ 854:158660
30. Brisson-Curadeau É, Bird D, Burke C, Fifield DA, Pace P, Sherley RB, Elliott KH (2017) Seabird species vary in behavioural response to drone census. Sci Rep 7(1):17884

31. Browne MA, Underwood AJ, Chapman MG, Williams R, Thompson RC, van Franeker JA (2015) Linking effects of anthropogenic debris to ecological impacts. Proc R Soc B Biol Sci 282(1807):20142929
32. Budko DF, Martynova DM (2019) Trace element uptake assessment in the planktonic biofiltration system. Oceanology 59:27–36
33. Burger J, Gochfeld M (2001) Effects of chemicals and pollution on seabirds. In: Biology of marine birds, pp 503–544. CRC press, USA
34. Buxton RT, Galvan R, McKenna MF, White CL, Seher V (2017) Visitor noise at a nesting colony alters the behavior of a coastal seabird. Mar Ecol Prog Ser 570:233–246
35. Buxton RT, Jones C, Moller H, Towns DR (2014) Drivers of seabird population recovery on New Zealand islands after predator eradication. Conserv Biol 28(2):333–344
36. Callaghan CT, Brandis KJ, Lyons MB, Ryall S, Kingsford RT (2018) A comment on the limitations of UAVS in wildlife research–the example of colonial nesting waterbirds. J Avian Biol 49(9):e01825
37. Calle P, Alvarado O, Monserrate L, Cevallos JM, Calle N, Alava JJ (2015) Mercury accumulation in sediments and seabird feathers from the Antarctic Peninsula. Mar Pollut Bull 91(2): 410–417
38. Campioni L, Oró-Nolla B, Granadeiro JP, Silva MC, Madeiros J, Gjerdrum C, Lacorte S (2024) Exposure of an endangered seabird species to persistent organic pollutants: assessing levels in blood and link with reproductive parameters. Sci Total Environ 930:172814
39. Capuska GEM, Huynen L, Lambert D, Raubenheimer D (2011) UVS is rare in seabirds. Vis Res 51(12):1333–1337
40. Celis JE, Espejo W, Groffen T, Bervoets L, Padilha J, Mello FV, Sandoval M, Chiang G (2025) Per-and polyfluoroalkylated substances (PFAS) in the feathers and excreta of Gentoo penguins (*Pygoscelis papua*) from the Antarctic Peninsula. Sci Total Environ 959:178333
41. Chan AAYH, Blumstein DT (2011) Attention, noise, and implications for wildlife conservation and management. Appl Anim Behav Sci 131(1–2):1–7
42. Chatwin TA, Joy R, Burger AE (2013) Set-back distances to protect nesting and roosting seabirds off Vancouver Island from boat disturbance. Waterbirds 36(1):43–52
43. Chiaradia A, McBride J, Murray T, Dann P (2007) Effect of fog on the arrival time of little penguins *Eudyptula minor*: a clue for visual orientation? J Ornithol 148:229–233
44. Chilvers BL, Morgan KJ, White BJ (2021) Sources and reporting of oil spills and impacts on wildlife 1970–2018. Environ Sci Pollut Res 28(1):754–762
45. Chu Z, Yang Z, Wang Y, Sun L, Yang W, Yang L, Gao Y (2019) Assessment of heavy metal contamination from penguins and anthropogenic activities on Fildes Peninsula and Ardley Island, Antarctic. Sci Total Environ 646:951–957
46. Colabuono FI, Taniguchi S, Montone RC (2012) Organochlorine contaminants in albatrosses and petrels during migration in South Atlantic Ocean. Chemosphere 86(7):701–708
47. Commonwealth of Australia (2020) National light pollution guidelines for wildlife: including marine turtles, seabirds and migratory shorebirds. https://environment.gov.au/biodiversity/publications/national-light-pollution-guidelines wildlife
48. Correia AAS, Matos MP, Gomes AR, Rasteiro MG (2023) Soils contaminated by heavy metals-study of the addition of carbon nanotubes and montmorillonite. https://doi.org/10.53243/ICEG2023-429
49. Costa L, Range DF, Zalmon I (2023) Effect of touristic activities on seabirds' habitat selection on sandy beaches. Oceanol Hydrobiol Stud 52(3):287–298
50. Croxall JP, Butchart SH, Lascelles BEN, Stattersfield AJ, Sullivan BEN, Symes A, Taylor PHIL (2012) Seabird conservation status, threats and priority actions: a global assessment. Bird Conserv Int 22(1):1–34
51. Culik B, Adelung D, Woakes AJ (1990) The effect of disturbance on the heart rate and behaviour of Adélie penguins (*Pygoscelis adeliae*) during the breeding season. In: Antarctic ecosystems: ecological change and conservation. Springer, Berlin Heidelberg, pp 177–182

52. Culik BM, Wilson RP, Woakes AT, Sanudo FW (1991) Oil pollution of Antarctic penguins: effects on energy metabolism and physiology. Mar Pollut Bull 22(8):388–391
53. De Salvo M, Cucuzza G, Ientile R, Signorello G (2020) Does recreation specialization affect birders' travel intention? Hum Dimens Wildl 25(6):560–574
54. De Villiers M, Bause M, Giese M, Fourie A (2006) Hardly hard-hearted: heart rate responses of incubating Northern Giant Petrels (*Macronectes halli*) to human disturbance on sub-Antarctic Marion Island. Polar Biol 29:717–720
55. Deeming DC, Reynolds SJ (eds) (2015) Nests, eggs, and incubation: new ideas about avian reproduction. Oxford University Press, Oxford
56. DeLeon S, Webster MS, DeVoogd TJ, Dhondt AA (2020) Developmental polychlorinated biphenyl exposure influences adult zebra finch reproductive behaviour. PLoS One 15(3): e0230283
57. Delgado S, Zorrozua N, Arizaga J (2020) Marginal presence of plastic in nests of yellow–legged gulls (*Larus michahellis*) in the southeastern Bay of Biscay. Anim Biodivers Conserv 43(2):191–195
58. Deppe L, Rowley O, Rowe LK, Shi N, Gooday OLIVER, Goldstien SJ (2017) Investigation of fallout events in Hutton's shearwaters (*Puffinus huttoni*) associated with artificial lighting. Notornis 64(4):181–191
59. Dierschke V, Furness RW, Garthe S (2016) Seabirds and offshore wind farms in European waters: avoidance and attraction. Biol Conserv 202:59–68
60. Domínguez-Hernández C, Villanova-Solano C, Álvarez-Méndez SJ, Pestano M, Tejera G, Arias Á, Díaz-Peña FJ, Hernández-Borges J, Hernández-Sánchez C (2024) Anthropogenic debris pollution in yellow-legged gull (*Larus michahellis atlantis*) nests in biosphere reserves of the Canary Islands. Sci Total Environ 949:175209
61. EFSA Panel on Contaminants in the Food Chain (CONTAM) (2016) Presence of microplastics and nanoplastics in food, with particular focus on seafood. EFSA J 14(6):e04501
62. Ellenberg U, Mattern T, Seddon PJ (2009) Habituation potential of yellow-eyed penguins depends on sex, character and previous experience with humans. Anim Behav 77(2):289–296
63. Ellenberg U, Mattern T, Seddon PJ, Jorquera GL (2006) Physiological and reproductive consequences of human disturbance in Humboldt penguins: the need for species-specific visitor management. Biol Conserv 133(1):95–106
64. Ellis-Felege SN, Stechmann T, Hervey S, Felege CJ, Rockwell RF, Barnas AF (2021) Nesting common eiders (*Somateria mollissima*) show little behavioral response to fixed-wing drone surveys. Drone Syst Appl 10(1):1–14
65. Eriksen M, Lebreton LC, Carson HS, Thiel M, Moore CJ, Borerro JC, Galgani F, Ryan PG, Reisser J (2014) Plastic pollution in the world's oceans: more than 5 trillion plastic pieces weighing over 250,000 tons afloat at sea. PLoS One 9(12):e111913
66. Esparza I, Elliott KH, Choy ES, Braune BM, Letcher RJ, Patterson A, Fernie KJ (2022) Mercury, legacy and emerging POPs, and endocrine-behavioural linkages: implications of Arctic change in a diving seabird. Environ Res 212:113190
67. Evans C, Brandsma J, Pond DW, Venables HJ, Meredith MP, Witte HJ, Stammerjohn S, Wilson WH, Clarke A, Brussaard CP (2017) Drivers of interannual variability in virioplankton abundance at the coastal western Antarctic peninsula and the potential effects of climate change. Environ Microbiol 19(2):740–755
68. Fanning L, Larsen H, Taylor PS (2020) A preliminary study investigating the impact of musical concerts on the behavior of captive Fiordland penguins (*Eudyptes pachyrhynchus*) and collared peccaries (*Pecari tajacu*). Animals 10(11):2035
69. Faure F, Demars C, Wieser O, Kunz M, De Alencastro LF (2015) Plastic pollution in Swiss surface waters: nature and concentrations, interaction with pollutants. Environ Chem 12(5): 582–591
70. Ferreira HR, Hadden AC, Champagnon J, Lok T, Vittecoq M, Alves JA (2025) Presence and potential impact of anthropogenic nesting materials on a colonial breeding waterbird. Sci Total Environ 964:178588

71. Foekema E, van der Molen J, Asjes A, Bijleveld A, Brasseur SMJM, Camphuysen K, Van Franeker JA, Holthuijsen S, Kentie R, Kühn S, Leopold M (2021) Ecologische effecten van het incident met de MSC Zoe op het Nederlandse Waddengebied, met focus op microplastics. NIOZ-rapport 03
72. Friesen MR, Beggs JR, Gaskett AC (2017) Sensory-based conservation of seabirds: a review of management strategies and animal behaviours that facilitate success. Biol Rev 92(3): 1769–1784
73. Furness RW, Camphuysen K (1997) Seabirds as monitors of the marine environment. ICES J Mar Sci 54(4):726–737
74. Galimany E, Navarro J, Martino I, Aymí R, Cermeño P, Montalvo T (2023) Gulls as potential sentinels for urban litter: combining nest and GPS-tracking information. Environ Monit Assess 195(4):521
75. Gaston KJ, Bennie J, Davies TW, Hopkins J (2013) The ecological impacts of nighttime light pollution: a mechanistic appraisal. Biol Rev 88(4):912–927
76. Geffroy B, Samia DS, Bessa E, Blumstein DT (2015) How nature-based tourism might increase prey vulnerability to predators. Trends Ecol Evol 30(12):755–765
77. Geldart EA, Barnas AF, Semeniuk CA, Gilchrist HG, Harris CM, Love OP (2022) A colonial-nesting seabird shows no heart-rate response to drone-based population surveys. Sci Rep 12(1):18804
78. Giese M (1996) Effects of human activity on Adelie penguin *Pygoscelis adeliae* breeding success. Biol Conserv 75(2):157–164
79. Giese M (1998) Guidelines for people approaching breeding groups of Adélie penguins (*Pygoscelis adeliae*). Polar Rec 34(191):287–292
80. Gilman E (2002) Guidelines for coastal and marine site-planning and examples of planning and management intervention tools. Ocean Coast Manag 45(6–7):377–404
81. Goebel ME, Perryman WL, Hinke JT, Krause DJ, Hann NA, Gardner S, LeRoi DJ (2015) A small unmanned aerial system for estimating abundance and size of Antarctic predators. Polar Biol 38:619–630
82. Goldsworthy SD, Gales RP, Giese M, Brothers N (2000) Effects of the Iron Baron oil spill on little penguins (*Eudyptula minor*). I. Estimates of mortality. Wildl Res 27(6):559–571
83. Goutte A, Angelier F, Chastel CC, Trouvé C, Moe B, Bech C, Gabrielsen GW, Chastel O (2010) Stress and the timing of breeding: glucocorticoid-luteinizing hormones relationships in an arctic seabird. Gen Comp Endocrinol 169(1):108–116
84. Goutte A, Barbraud C, Herzke D, Bustamante P, Angelier F, Tartu S, Clément-Chastel C, Moe B, Bech C, Gabrielsen GW, Bustnes JO (2015) Survival rate and breeding outputs in a high Arctic seabird exposed to legacy persistent organic pollutants and mercury. Environ Pollut 200:1–9
85. Goutte A, Barbraud C, Meillère A, Carravieri A, Bustamante P, Labadie P, Budzinski H, Delord K, Cherel Y, Weimerskirch H, Chastel O (2014) Demographic consequences of heavy metals and persistent organic pollutants in a vulnerable long-lived bird, the wandering albatross. Proc R Soc B Biol Sci 281(1787):20133313
86. Graham NA, Wilson SK, Carr P, Hoey AS, Jennings S, MacNeil MA (2018) Seabirds enhance coral reef productivity and functioning in the absence of invasive rats. Nature 559(7713): 250–253
87. Grant ML, Lavers JL, Stuckenbrock S, Sharp PB, Bond AL (2018) The use of anthropogenic marine debris as a nesting material by brown boobies (*Sula leucogaster*). Mar Pollut Bull 137: 96–103
88. Gregory MR (2009) Environmental implications of plastic debris in marine settings—entanglement, ingestion, smothering, hangers-on, hitch-hiking and alien invasions. Philos Trans R Soc B Biol Sci 364(1526):2013–2025
89. Hammer S, Nager RG, Johnson PCD, Furness RW, Provencher JF (2016) Plastic debris in great skua (*Stercorarius skua*) pellets corresponds to seabird prey species. Mar Pollut Bull 103(1–2):206–210

90. Hammouda A, Youssef FB, McCoy KD, Selmi S (2025) Use of anthropogenic debris as nesting materials in a south-Mediterranean yellow-legged gull (*Larus michahellis*) breeding colony in relation to their environmental availability. Environ Monit Assess 197(4):1–12
91. Hansell M (2000) Bird nests and construction behaviour. Cambridge University Press, Cambridge
92. Hao Y, Zheng S, Wang P, Sun H, Matsiko J, Li W, Li Y, Zhang Q, Jiang G (2021) Ecotoxicology of persistent organic pollutants in birds. Environ Sci Proc Impacts 23(3): 400–416
93. Henkel LA, Ziccardi MH (2018) Life and death: how should we respond to oiled wildlife? J Fish Wildl Manag 9(1):296–301
94. Hiemstra AF, Gravendeel B, Schilthuizen M (2021) Birds using artificial plants as nesting material. Behaviour 159(2):193–205
95. Hitchcock DJ, Andersen T, Varpe Ø, Borgå K (2019) Effects of maternal reproductive investment on sex-specific pollutant accumulation in seabirds: a meta-analysis. Environ Sci Technol 53(13):7821–7829
96. Hölker F, Wolter C, Perkin EK, Tockner K (2010) Light pollution as a biodiversity threat. Trends Ecol Evol 25(12):681–682
97. Holland ER, Mallory ML, Shutler D (2016) Plastics and other anthropogenic debris in freshwater birds from Canada. Sci Total Environ 571:251–258
98. Iasiello L, Colombelli-Négrel D (2023) Noisy neighbours: effects of construction noises on nesting seabirds. Mar Freshw Res 74(7):573–585
99. Iasiello L, Colombelli-Négrel D (2025) Bright and early: artificial light affects arrival time, but not group size or vigilance in Little Penguins (*Eudyptula minor*). Ibis 167(1):248–262
100. Iniunam IA, Tende T, Chaskda AA (2025) Assessing the ecological consequences of crude oil spills on bird communities in an Afrotropical landscape. Environ Sustain Indicat 26:100600
101. Jagiello Z, Reynolds SJ, Nagy J, Mainwaring MC, Ibanez-Alamo JD (2023) Why do some bird species incorporate more anthropogenic materials into their nests than others? Philos Trans R Soc B 378(1884):20220156
102. Jagiello ZA, Dylewski Ł, Winiarska D, Zolnierowicz KM, Tobolka M (2018) Factors determining the occurrence of anthropogenic materials in nests of the white stork Ciconia ciconia. Environ Sci Pollut Res 25:14726–14733
103. Jagielski PM, Dey CJ, Gilchrist HG, Richardson ES, Semeniuk CA (2021) Polar bear foraging on common eider eggs: estimating the energetic consequences of a climate-mediated behavioural shift. Anim Behav 171:63–75
104. Jerem P, Mathews F (2021) Trends and knowledge gaps in field research investigating effects of anthropogenic noise. Conserv Biol 35(1):115–129
105. Jimenez-Uzcategui G, Vinueza RL, Urbina AS, Egas DA, Garcia C, Cotin J, Sevilla C (2017) Lead and cadmium levels in Galapagos penguin *Spheniscus mendiculus*, flightless cormorant *Phalacrocorax harrisi*, and waved albatross *Phoebastria irrorata*. Mar Ornithol 45:159–163
106. Kavelaars MM, Baert JM, Stienen EW, Shamoun-Baranes J, Lens L, Müller W (2020) Breeding habitat loss reveals limited foraging flexibility and increases foraging effort in a colonial breeding seabird. Move Ecol 8:1–11
107. Kazama K, Niizuma Y, Sakamoto KQ, Watanuki Y (2011) Factors affecting individual variation in nest-defense intensity in colonially breeding Black-tailed Gulls (*Larus crassirostris*). Can J Zool 89(10):938–944
108. Kenyon KW, Kridler E (1969) Laysan albatrosses swallow indigestible matter. The Auk 86(2): 339–343
109. Khalef RN, Hassan AI, Saleh HM (2022) Heavy metal's environmental impact. In: Environmental impact and remediation of heavy metals. IntechOpen, United Kingdom
110. King MD, Elliott JE, Williams TD (2021) Effects of petroleum exposure on birds: a review. Sci Total Environ 755:142834
111. Kingston PF (2002) Long-term environmental impact of oil spills. Spill Sci Technol Bullet 7(1–2):53–61

112. Kiran SD, Kour A, Priya DV, Kumar R (2022b) Community composition and status of avian diversity at Campus and Agricultural landscapes of Chaudhary Charan Singh Haryana Agricultural University, Hisar (Haryana). J Appl Nat Sci 14(4):1130–1140. https://doi.org/10.31018/jans.v14i4.3784
113. Kiran SD, Kour A, Priya DV, Kumar R, Punia N, Sunita (2022a) Different strategies adopted by birds to sustain ecosystem: a review. Pharm Innov SP-11(9):412–422
114. Kiran KN, Sindhu S, Gawaria B, Ansari A, Dhariwal A, Sharma P, Kumar R (2025) GWAS approaches in medicinal plants for enhanced secondary metabolite production. In: Chen JT (ed) Biotechnology, multiple omics, and precision breeding in medicinal plants, pp 119–134. CRC Press, USA
115. Kronenberg J (2016) Birdwatchers' wonderland? Prospects for the development of birdwatching tourism in Poland. J Ecotour 15(1):78–94
116. Kühn S, Van Franeker JA (2020) Quantitative overview of marine debris ingested by marine megafauna. Mar Pollut Bull 151:110858
117. Kumar R, Singh D, Yadav R (2025) Toxic effects of chlorantraniliprole and fipronil on *Eisenia fetida*: from growth and reproductive parameters to avoidance behavior. Environ Eng Manag J 24(7):1499–1508. Accepted manuscript
118. Kumar R, Singh D (2022) Insecticide-tolerant bacterial population in *Eisenia fetida*'s Gut and Vermicast exposed to Chlorantraniliprole and Fipronil. Appl Biol Res 24(3):273–279
119. Kumar R, Gupta R, Dahiya T, Yadav R, Saifi R, Yodha K, Kumari B, Singh P, Shefali. (2022a) Surveillance of *Heteropneustes fossilis* (Stinging Catfish) diseases and health management practices in Haryana. Appl Ecol Environ Sci 10(4):261–267
120. Kumar R, Gupta RK, Yadav R, Saifi R, Yodha K, Pooja. (2022b) *Eisenia fetida* as protein source for growth enhancement of *Heteropneustes fossilis*. Egypt J Aquat Biol Fish 26(2):577–588
121. Kumar R, Yadav R, Gupta RK, Pal A, Yodha K, Kumar A (2024) Antioxidant enzyme activities and markers of oxidative stress in the life cycle of different Earthworm species. Indian J Exp Biol 62(07):578–583
122. Kumar R, Yadav R, Kumar Gupta R, Pooja (2022c) A sustainable way for fish health management by replacement of chemical and drugs by earthworm. In: Environmental degradation in Asia: land degradation, environmental contamination, and human activities. Springer International Publishing, Cham, pp 329–352
123. Kumari K (ed) (2021) Persistent organic pollutants: gaps in management and associated challenges. CRC Press, USA
124. Kuo DT, Rattner BA, Marteinson SC, Letcher R, Fernie KJ, Treu G, Deutsch M, Johnson MS, Deglin S, Embry M (2022) A critical review of bioaccumulation and biotransformation of organic chemicals in birds. Rev Environ Contam Toxicol 260(1):6
125. Lachman D, Conway C, Vierling K, Matthews T (2020) Drones provide a better method to find nests and estimate nest survival for colonial waterbirds: a demonstration with Western Grebes. Wetl Ecol Manag 28(5):837–845
126. Lai S, Loke LH, Hilton MJ, Bouma TJ, Todd PA (2015) The effects of urbanisation on coastal habitats and the potential for ecological engineering: a Singapore case study. Ocean Coast Manag 103:78–85
127. Larsen GD, Varga HF, Patterson-Fraser DL, Johnston DW, Cimino MA (2024) Drone-based monitoring and geomorphology of southern giant petrel nests near Palmer Station, western Antarctic Peninsula. Polar Biol 47(5):459–474
128. Lavers JL, Sharp PB, Stuckenbrock S, Bond AL (2020) Entrapment in plastic debris endangers hermit crabs. J Hazard Mater 387:121703
129. Lee K, Jang YC, Hong S, Lee J, Kwon IK (2015) Plastic marine debris used as nesting materials of the endangered species black-faced spoonbill *Platalea minor* decreases by conservation activities. J Korean Soc Marine Environ Energy 18(1):45–49

130. Lemmetyinen R, Rantamäki P, Karlin A (1982) Levels of DDT and PCB's in different stages of life cycle of the Arctic tern *Sterna paradisaea* and the herring gull *Larus argentatus*. Chemosphere 11(10):1059–1068
131. Linhares BDA, Bugoni L (2023) Seabirds subsidize terrestrial food webs and coral reefs in a tropical rat-invaded archipelago. Ecol Appl 33(2):e2733
132. Lopes LE, Fernandes AM, Medeiros MC, Marini MÂ (2016) A classification scheme for avian diet types. J Field Ornithol 87(3):309–322
133. Lopes CS, De Faria JP, Paiva VH, Ramos JA (2020) Characterization of anthropogenic materials on yellow-legged gull (Larus michahellis) nests breeding in natural and urban sites along the coast of Portugal. Environ Sci Pollut Res 27(29):36954–36969
134. Luna N, Ory N, Miranda-Urbina D, Morales N, Hinojosa IA, Zavalaga C, Pacheco AS, Gallardo C, Portflitt-Toro M, Luna-Jorquera G, Álvarez-Varas R (2018) Impacts of marine plastic pollution from continental coasts to subtropical gyres—fish, seabirds, and other vertebrates in the SE Pacific
135. Lunn NJ, Stirling I, Andriashek D, Richardson E (2004) Selection of maternity dens by female polar bears in western Hudson Bay, Canada and the effects of human disturbance. Polar Biol 27(6):350–356
136. Lyons MB, Brandis KJ, Murray NJ, Wilshire JH, McCann JA, Kingsford RT, Callaghan CT (2019) Monitoring large and complex wildlife aggregations with drones. Methods Ecol Evol 10(7):1024–1035
137. Magle S, Zhu J, Crooks KR (2005) Behavioral responses to repeated human intrusion by black-tailed prairie dogs (*Cynomys ludovicianus*). J Mammal 86(3):524–530
138. Mainwaring MC, Hartley IR, Lambrechts MM, Deeming DC (2014) The design and function of birds' nests. Ecol Evol 4(20):3909–3928
139. Manjeet AA, Kumar R, Saifi R (2022) Temporal and spatial impact of lockdown during COVID-19 on air quality index in Haryana, India. Sci Rep 12:20046. https://doi.org/10.1038/s41598-022-20885-2
140. Max A, Martín-Vélez V, Navarro J, Borrell A, Montalvo T, Garcia-Garin O (2025) Characterization of plastic ingestion in urban gull chicks and its implications for their use as pollution sentinels in coastal cities. Mar Pollut Bull 211:117409
141. McClelland GT, Bond AL, Sardana A, Glass T (2016) Rapid population estimate of a surface-nesting seabird on a remote Island using a low-cost unmanned aerial vehicle. Mar Ornithol 44: 215–220
142. McKellar AE, Shephard NG, Chabot D (2021) Dual visible-thermal camera approach facilitates drone surveys of colonial marshbirds. Remote Sens Ecol Conserv 7(2):214–226
143. McNaughton EJ, Beggs JR, Gaston KJ, Jones DN, Stanley MC (2021) Retrofitting streetlights with LEDs has limited impacts on urban wildlife. Biol Conserv 254:108944
144. Merkel FR, Johansen KL (2011) Light-induced bird strikes on vessels in Southwest Greenland. Mar Pollut Bull 62(11):2330–2336
145. Mesquita GP, Rodríguez-Teijeiro JD, Wich SA, Mulero-Pázmány M (2021) Measuring disturbance at swift breeding colonies due to the visual aspects of a drone: a quasi-experiment study. Curr Zool 67(2):157–163. Oxford University Press, Oxford, England
146. Miljeteig C, Gabrielsen GW, Strøm H, Gavrilo MV, Lie E, Jenssen BM (2012) Eggshell thinning and decreased concentrations of vitamin E are associated with contaminants in eggs of ivory gulls. Sci Total Environ 431:92–99
147. Mitra S, Chakraborty AJ, Tareq AM, Emran TB, Nainu F, Khusro A, Idris AM, Khandaker MU, Osman H, Alhumaydhi FA, Simal-Gandara J (2022) Impact of heavy metals on the environment and human health: novel therapeutic insights to counter the toxicity. J King Saud Univ Sci 34(3):101865
148. Montevecchi WA (2006) Influences of artificial light on marine birds. In: Ecological consequences of artificial night lighting, pp 94–113

149. Mooney TA, Smith A, Hansen KA, Larsen ON, Wahlberg M, Rasmussen M (2019, July) Birds of a feather: hearing and potential noise impacts in puffins (*Fratercula arctica*). In: Proceedings of meetings on acoustics, vol 37, No. 1. AIP Publishing
150. Moser ML, Lee DS (1992) A fourteen-year survey of plastic ingestion by western North Atlantic seabirds. Colonial Waterbirds 15:83–94
151. Mulero-Pázmány M, Jenni-Eiermann S, Strebel N, Sattler T, Negro JJ, Tablado Z (2017) Unmanned aircraft systems as a new source of disturbance for wildlife: a systematic review. PLoS One 12(6):e0178448
152. Multisanti CR, Merola C, Perugini M, Aliko V, Faggio C (2022) Sentinel species selection for monitoring microplastic pollution: a review on one health approach. Ecol Indic 145:109587
153. Nelson JB (2006) Pelicans, cormorants, and their relatives: Pelecanidae, Sulidae, Phalacrocoracidae, Anhingidae, Fregatidae, Phaethontidae. OUP, Oxford
154. Nisanth HP (2024) Investigating lesser black-backed gulls (*Larus fuscus*) Scavenging fishery bycatch as sentinels of heavy metal pollution in Kerala. Rec Zool Surv India 124:513–522
155. Nisha A, Rani S, Kumar S, Kumar R, Kumar A (2025) Recent technologies and strategies for bioremediation of toxic and hazardous waste. In: Dubey KK, Shah MP, Yadav A (eds) Challenges and sustainable solutions in bioremediation. CRC Press, pp 1–16. https://doi.org/10.1201/9781003407317-1
156. Nowak MM, Dziób K, Bogawski P (2018) Unmanned Aerial Vehicles (UAVs) in environmental biology: a review. Eur J Ecol 4(2):56–74
157. O'Hara PD, Morandin LA (2010) Effects of sheens associated with offshore oil and gas development on the feather microstructure of pelagic seabirds. Mar Pollut Bull 60(5):672–678
158. Ogada DL, Keesing F, Virani MZ (2012) Dropping dead: causes and consequences of vulture population declines worldwide. Ann N Y Acad Sci 1249(1):57–71
159. Pierce KE, Harris RJ, Larned LS, Pokras MA (2004) Obstruction and starvation associated with plastic ingestion in a Northern Gannet *Morus bassanus* and a Greater Shearwater *Puffinus gravis*. Mar Ornithol 32:187–189
160. Pink M, Abrahams MV (2018) In shallow water ecosytems the abiotic environment is more important than prey abundance for foraging terns. Environ Biol Fish 101:355–362
161. Potvin DA (2017) Coping with a changing soundscape: avoidance, adjustments and adaptations. Anim Cogn 20(1):9–18
162. Provencher JF, Bond AL, Avery-Gomm S, Borrelle SB, Rebolledo ELB, Hammer S, Kühn S, Lavers JL, Mallory ML, Trevail A, Van Franeker JA (2017) Quantifying ingested debris in marine megafauna: a review and recommendations for standardization. Anal Methods 9(9): 1454–1469
163. Rajpar MN, Ozdemir I, Zakaria M, Sheryar S, Rab A (2018) Seabirds as bioindicators of marine ecosystems. In: Seabirds, pp 47–65. IntechOpen, Croatia
164. Rani S, Sharma S, Singh D, Sheoran R, Kumar R (2025) Improved production of anti-cancer compounds from medicinal plants through molecular and biotechnological mutagenesis strategies. In: Chen JT (ed) Biotechnology, multiple omics, and precision breeding in medicinal plants, pp 287–297. CRC Press, USA
165. Reynolds SJ, Ibáñez-Álamo JD, Sumasgutner P, Mainwaring MC (2019) Urbanisation and nest building in birds: a review of threats and opportunities. J Ornith 160:841–860. [online]
166. Robertson BA, Rehage JS, Sih A (2013) Ecological novelty and the emergence of evolutionary traps. Trends Ecol Evol 28(9):552–560
167. Rodríguez A, Dann P, Chiaradia A (2017) Reducing light-induced mortality of seabirds: high pressure sodium lights decrease the fatal attraction of shearwaters. J Nat Conserv 39:68–72
168. Rodríguez A, Holmberg R, Dann P, Chiaradia A (2018) Penguin colony attendance under artificial lights for ecotourism. J Exp Zool Part A Ecol Integ Physiol 329(8–9):457–464
169. Roslee AFA, Zakaria NN, Convey P, Zulkharnain A, Lee GLY, Gomez-Fuentes C, Ahmad SA (2020) Statistical optimisation of growth conditions and diesel degradation by the Antarctic bacterium, *Rhodococcus* sp. strain AQ5–07. Extremophiles 24:277–291

170. Rümmler MC, Mustafa O, Maercker J, Peter HU, Esefeld J (2016) Measuring the influence of unmanned aerial vehicles on Adélie penguins. Polar Biol 39:1329–1334
171. Rush GP, Clarke LE, Stone M, Wood MJ (2018) Can drones count gulls? Minimal disturbance and semiautomated image processing with an unmanned aerial vehicle for colony-nesting seabirds. Ecol Evol 8(24):12322–12334
172. Ryan PG (2018) Entanglement of birds in plastics and other synthetic materials. Mar Pollut Bull 135:159–164
173. Ryan PG (2020) Using photographs to record plastic in seabird nests. Mar Pollut Bull 156: 111262
174. Sanders D, Frago E, Kehoe R, Patterson C, Gaston KJ (2021) A meta-analysis of biological impacts of artificial light at night. Nat Ecol Evol 5(1):74–81
175. Scarton F, Valle RG (2022) Comparison of drone vs. ground survey monitoring of hatching success in the black-headed gull (*Chroicocephalus ridibundus*). Ornithol Res 30(4):271–280
176. Schaefer R, Colombelli-Négrel D (2021) Behavioural and heart rate responses to stressors in two populations of little penguins that differ in levels of human disturbance and predation risk. Ibis 163(3):858–874
177. Schlaepfer MA, Runge MC, Sherman PW (2002) Ecological and evolutionary traps. Trends Ecol Evol 17(10):474–480
178. Schofield G, Katselidis KA, Lilley MK, Reina RD, Hays GC (2017) Detecting elusive aspects of wildlife ecology using drones: new insights on the mating dynamics and operational sex ratios of sea turtles. Funct Ecol 31(12):2310–2319
179. Schutten K, Morrill A, Lu Z, Chandrashekar A, Cunningham JT, Robertson GJ, Mallory ML, Jardine CM, Provencher JF (2025) Accumulation of benzotriazole UV-stabilizers in relation to ingested plastics and associated health metrics in Larus gulls feeding at a landfill in Atlantic Canada. J Hazard Mater 488:137498
180. Searle KR, Regan CE, Perrow MR, Butler A, Rindorf A, Harris MP, Newell MA, Wanless S, Daunt F (2023) Effects of a fishery closure and prey abundance on seabird diet and breeding success: implications for strategic fisheries management and seabird conservation. Biol Conserv 281:109990
181. Sekercioglu CH (2002) Impacts of birdwatching on human and avian communities. Environ Conserv 29(3):282–289
182. Sen IS, Peucker-Ehrenbrink B (2012) Anthropogenic disturbance of element cycles at the Earth's surface. Environ Sci Technol 46(16):8601–8609
183. Sergio F, Blas J, Blanco G, Tanferna A, López L, Lemus JA, Hiraldo F (2011) Raptor nest decorations are a reliable threat against conspecifics. Science 331(6015):327–330
184. Shahi C, Bargali SS, Bargali K, Vibhuti (2023) Dry matter dynamics and $CO_2$ mitigation in the herb layer of Central Himalayan agroecosystems along an altitudinal gradient, India. Trop Ecol 64(1):180–192
185. Shannon G, Larson CL, Reed SE, Crooks KR, Angeloni LM (2017) Ecological consequences of ecotourism for wildlife populations and communities. In: Ecotourism's promise and peril: a biological evaluation, pp 29–46. Springer International Publishing, Switzerland
186. Shannon G, McKenna MF, Angeloni LM, Crooks KR, Fristrup KM, Brown E, Warner KA, Nelson MD, White C, Briggs J, McFarland S (2016) A synthesis of two decades of research documenting the effects of noise on wildlife. Biol Rev 91(4):982–1005
187. Sherley RB, Barham BJ, Barham PJ, Campbell KJ, Crawford RJ, Grigg J, Horswill C, McInnes A, Morris TL, Pichegru L, Steinfurth A (2018) Bayesian inference reveals positive but subtle effects of experimental fishery closures on marine predator demographics. Proc R Soc B Biol Sci 285(1871):20172443
188. Sherley RB, Ludynia K, Dyer BM, Lamont T, Makhado AB, Roux JP, Scales KL, Underhill LG, Votier SC (2017) Metapopulation tracking juvenile penguins reveals an ecosystem-wide ecological trap. Curr Biol 27(4):563–568

189. Sherley RB, Winker H, Altwegg R, van der Lingen CD, Votier SC, Crawford RJ (2015) Bottom-up effects of a no-take zone on endangered penguin demographics. Biol Lett 11(7): 20150237
190. Sievwright KA, Battley PF, McConnell H, Armstrong DP, Morgan KJ (2019) Survival rates of oil-rehabilitated and non-rehabilitated little penguins after the C/V Rena oil spill, New Zealand. Mar Pollut Bull 146:317–325
191. Sih A, Ferrari MC, Harris DJ (2011) Evolution and behavioural responses to human-induced rapid environmental change. Evol Appl 4(2):367–387
192. Soldatini C, Albores-Barajas YV, Tagliavia M, Massa B, Fusani L, Canoine V (2015) Effects of human disturbance on cave-nesting seabirds: the case of the storm petrel. Conserv Physiol 3(1):cov041
193. Stensland S, Aas Ø, Dybsand HNH, Haugen TO (2021) A Nordic perspective on wildlife tourism experiences. In: Nordic perspectives on nature-based tourism, pp 263–280. Edward Elgar Publishing, United Kingdom
194. Steven R, Castley JG (2013) Tourism as a threat to critically endangered and endangered birds: global patterns and trends in conservation hotspots. Biodivers Conserv 22:1063–1082
195. Suárez-Rodríguez M, Garcia CM (2014) There is no such a thing as a free cigarette; lining nests with discarded butts brings short-term benefits, but causes toxic damage. J Evol Biol 27(12):2719–2726
196. Suárez-Rodríguez M, Garcia CM (2017) An experimental demonstration that house finches add cigarette butts in response to ectoparasites. J Avian Biol 48(10):1316–1321
197. Suárez-Rodríguez M, López-Rull I, Macías Garcia C (2013) Incorporation of cigarette butts into nests reduces nest ectoparasite load in urban birds: new ingredients for an old recipe? Biol Lett 9(1):20120931
198. Susanti NKY, Mardiastuti A, Wardiatno Y (2020) Microplastics and the impact of plastic on wildlife: a literature review. In: IOP conference series: earth and environmental science, vol 528, No. 1. IOP Publishing, p 012013
199. Syposz M, Gonçalves F, Carty M, Hoppitt W, Manco F (2018) Factors influencing Manx Shearwater grounding on the west coast of Scotland. Ibis 160(4):846–854
200. Syposz M, Padget O, Willis J, Van Doren BM, Gillies N, Fayet AL, Wood MJ, Alejo A, Guilford T (2021) Avoidance of different durations, colours and intensities of artificial light by adult seabirds. Sci Rep 11(1):18941
201. Szynkowska MI, Pawlaczyk A, Maćkiewicz E (2018) Bioaccumulation and biomagnification of trace elements in the environment. In: Recent advances in trace elements, pp 251–276. Wiley, New Jersey
202. Tanabe S, Subramanian AN, Hidaka H, Tatsukawa R (1986) Transfer rates and pattern of PCB isomers and congeners and p, p′-DDE from mother to egg in adelie penguin (*Pygoscelis adeliae*). Chemosphere 15(3):343–351
203. Tanaka K, Takada H, Yamashita R, Mizukawa K, Fukuwaka MA, Watanuki Y (2015) Facilitated leaching of additive-derived PBDEs from plastic by seabirds' stomach oil and accumulation in tissues. Environ Sci Technol 49(19):11799–11807
204. Taniguchi S, Montone RC, Bícego MC, Colabuono FI, Weber RR, Sericano JL (2009) Chlorinated pesticides, polychlorinated biphenyls and polycyclic aromatic hydrocarbons in the fat tissue of seabirds from King George Island, Antarctica. Mar Pollut Bull 58(1):129–133
205. Taurozzi D, Scalici M (2024) Seabirds from the poles: microplastics pollution sentinels. Front Mar Sci 11:1343617
206. Tchounwou PB, Yedjou CG, Patlolla AK, Sutton DJ (2012) Heavy metal toxicity and the environment. In: Molecular, clinical and environmental toxicology: volume 3: environmental toxicology, pp 133–164. Springer, Basel, Switzerland
207. Thiel M, Bravo M, Hinojosa IA, Luna G, Miranda L, Núñez P, Pacheco AS, Vásquez N (2011) Anthropogenic litter in the SE Pacific: an overview of the problem and possible solutions. Revista de Gestão Costeira Integrada-J Integ Coast Zone Manag 11(1):115–134

208. Thompson DL, Ovenden TS, Pennycott T, Nager RG (2020) The prevalence and source of plastic incorporated into nests of five seabird species on a small offshore Island. Mar Pollut Bull 154:111076
209. Thompson J (2021) A drone crash caused thousands of elegant terns to abandon their nests. Audubon Magazine 11
210. Thompson RC, Olsen Y, Mitchell RP, Davis A, Rowland SJ, John AW, McGonigle D, Russell AE (2004) Lost at sea: where is all the plastic? Science 304(5672):838–838
211. Tokunaga Y, Okochi H, Tani Y, Niida Y, Tachibana T, Saigawa K, Katayama K, Moriguchi S, Kato T, Hayama SI (2023) Airborne microplastics detected in the lungs of wild birds in Japan. Chemosphere 321:138032
212. Townsend AK, Barker CM (2014) Plastic and the nest entanglement of urban and agricultural crows. PLoS One 9(1):e88006
213. Troisi GM (2013) T-cell responses in oiled guillemots and swans in a rehabilitation setting. Arch Environ Contam Toxicol 65:142–148
214. Troisi GM, Bexton S, Robinson I (2006) Polyaromatic hydrocarbon and PAH metabolite burdens in oiled common guillemots (*Uria aalge*) stranded on the east coast of England (2001– 2002). Environ Sci Technol 40(24):7938–7943
215. Tseng FS, Ziccardi M (2019) Care of oiled wildlife. In: Medical management of wildlife species: a guide for practitioners, pp 75–84. Wiley-Blackwell, USA
216. UNEP (2016) Marine plastic debris and microplastics-global lessons and research to inspire action and guide policy change. United Nations Environment Programme, Nairobi
217. Urmila GRK, Sharma S, Kumar R (2025) Earthworms: the environmentally preferred bio-engineers for managing heavy metals. In: Dey S, Bhattacharya S (eds) Biotechnological interventions in the removal of emerging pollutants. Springer Nature Singapore, Singapore, pp 365–384
218. Vaid V, Hundal SS (2019) Light microscopic studies to evaluate fish scales as non-invasive indicators of heavy metal–contaminated waters. Environ Monit Assess 191:1–11
219. Vaid V, Vashishat N (2025) Impact of microplastics on aquatic organisms. In: Hundal SS, Singh J (eds) Trends in life sciences. Rubicon Publications, London, pp 187–219
220. Vaid V, Hundal SS, Vashishat N (2024) Ultrastructural studies to evaluate fish scales as indicators of heavy metals in *Labeo rohita*. J Soil Salinit Water Qual 16(1):118–125
221. Valle RG, Scarton F (2021) Drone-conducted counts as a tool for the rapid assessment of productivity of Sandwich Terns (*Thalasseus sandvicensis*). J Ornithol 162(2):621–628
222. Van Franeker JA, Meijboom A (2007) Fulmar Littter EcoQO Monitoring in the Netherlands 1982–2005 in relation to EU Directive 2000/59/EC on Port Reception Facilities. Wageningen IMARES Report, (C019-07)
223. Vanstreels RE, Parsons NJ, Sherley RB, Stander N, Strauss V, Kemper J, Waller L, Barham BJ, Ludynia K (2023) Factors determining the number of seabirds impacted by oil spills and the success of their rehabilitation: lessons learned from Namibia and South Africa. Mar Pollut Bull 188:114708
224. Walker BG, Dee Boersma P, Wingfield JC (2006) Habituation of adult Magellanic penguins to human visitation as expressed through behavior and corticosterone secretion. Conserv Biol 20(1):146–154
225. Wang L, Nabi G, Yin L, Wang Y, Li S, Hao Z, Li D (2021) Birds and plastic pollution: recent advances. Avian Res 12:1–9
226. Waters CN, Zalasiewicz J, Summerhayes C, Barnosky AD, Poirier C, Gałuszka A, Cearreta A, Edgeworth M, Ellis EC, Ellis M, Jeandel C (2016) The Anthropocene is functionally and stratigraphically distinct from the Holocene. Science 351(6269):aad2622
227. Watson H, Bolton M, Monaghan P (2014) Out of sight but not out of harm's way: human disturbance reduces reproductive success of a cavity-nesting seabird. Biol Conserv 174:127–133
228. Weaver DB, Lawton LJ (2007) Twenty years on: the state of contemporary ecotourism research. Tour Manag 28(5):1168–1179

229. Weimerskirch H, Prudor A, Schull Q (2018) Flights of drones over sub-Antarctic seabirds show species-and status-specific behavioural and physiological responses. Polar Biol 41:259–266
230. Wen L, Du D, Zhu P, Hu Q, Wang Q, Bo L, Lyu S (2021) Detection, tracking, and counting meets drones in crowds: a benchmark. In: Proceedings of the IEEE/CVF conference on computer vision and pattern recognition, pp 7812–7821
231. Weston MA, O'Brien C, Kostoglou KN, Symonds MR (2020) Escape responses of terrestrial and aquatic birds to drones: towards a code of practice to minimize disturbance. J Appl Ecol 57(4):777–785
232. Whelan CJ, Wenny DG, Marquis RJ (2008) Ecosystem services provided by birds. Ann N Y Acad Sci 1134(1):25–60
233. Wiese FK, Montevecchi WA, Davoren GK, Huettmann F, Diamond AW, Linke J (2001) Seabirds at risk around offshore oil platforms in the North-west Atlantic. Mar Pollut Bull 42(12):1285–1290
234. Wilcox C, Van Sebille E, Hardesty BD (2015) Threat of plastic pollution to seabirds is global, pervasive, and increasing. Proc Natl Acad Sci 112(38):11899–11904
235. Williams AT, Simmons SL (1997) Estuarine litter at the river/beach interface in the Bristol Channel, United Kingdom. J Coast Res 13:1159–1165
236. Witteveen M, Brown M, Ryan PG (2017) Anthropogenic debris in the nests of kelp gulls in South Africa. Mar Pollut Bull 114(2):699–704
237. Wondirad A (2019) Does ecotourism contribute to sustainable destination development, or is it just a marketing hoax? Analyzing twenty-five years contested journey of ecotourism through a meta-analysis of tourism journal publications. Asia Pacific J Tour Res 24(11):1047–1065
238. Wurster CF Jr, Wingate DB (1968) DDT residues and declining reproduction in the Bermuda Petrel. Science 159(3818):979–981
239. Yacoubi L, Savoca D, El Zrelli RB, Gopalan J, Nazal M, Lin YJ, Maccotta A, Hamza F, Bhuyan MS, Arculeo M, Rabaoui LJ (2024) Trace element levels in the muscles of three tern species (Aves: Laridae) from the western Arabian Gulf: environmental assessment and implications for conservation. Environ Monit Assess 196(3):235
240. Yadav R, Gupta RK, Kumar R, Kaur T, Moun R (2022a) Biochemical response of earthworm, *Eisenia fetida* to heavy metals toxicity. J Appl Nat Sci 14(3):990–998
241. Yadav R, Gupta RK, Kumar R, Kaur T (2022b) Assessment of toxicity of lead and nickel on the biochemical and immunological parameters of earthworm, Eudrilus eugeniae. Agric Sci Digest 44(5):922–929
242. Yadav R, Kumar R, Gupta RK, Kaur T, Yodha K, Kour A, Kaur S, Rajput A (2023) Heavy metal toxicity in earthworms and its environmental implications: a review. Environ Adv 12: 100374
243. Ye J, Li J, Wang P, Ning Y, Liu J, Yu Q, Bi X (2022) Inputs and sources of Pb and other metals in urban area in the post leaded gasoline era. Environ Pollut 306:119389
244. Yodha K, Singh D, Kour A, Yadav R, Kumar R (2024) Species diversity and abundance of birds in ber (*Zizyphus mauritiana*) crop. Asian J Conserv Biol 13(1). https://doi.org/10.53562/ajcb.83683
245. Yorio P, Boersma PD (1992) The effects of human disturbance on Magellanic penguin *Spheniscus magellanicus* behaviour and breeding success. Bird Conserv Int 2(3):161–173

# Chapter 11
# Challenges to Waterbird Conservation in the Indian Himalayan Region: The Impacts of Climate Change and Urbanization

**Santosh Sharma and Manjula Karunarathne**

**Abstract** The Indian Himalayan Region (IHR) is a Biodiversity Hotspot and a critical habitat for a diverse range of waterbird species. However, escalating challenges such as climate change and urbanization are threatening these sensitive ecosystems. The current study explores the challenges of waterbird conservation in IHR. It begins by examining the effects of climate change on waterbird habitats, focusing on altered hydrological patterns, shrinking wetlands and the shift in distribution of species. It then analyzes how rapid urbanization, including infrastructure development, pollution and land-use changes, exacerbates habitat degradation and fragmentation. To address these challenges, the current study advocates for integrated conservation strategies that incorporate habitat restoration, sustainable development practices, and community engagement. Finally, it highlights critical policy and research gaps, emphasizing the need for robust climate-adaptive frameworks, long-term monitoring programs, and cross sectoral collaboration to safeguard waterbird populations in the IHR. This comprehensive overview aims to provide a foundation for targeted conservation actions in the face of growing environmental and anthropogenic pressures.

**Keywords** Indian Himalayan Region · Central Asian Flyway · Migration · Waterbirds · Wetland birds

S. Sharma
Department of Management, The Institute of Chartered Financial Analysts of India (ICFAI) University Sikkim, Gangtok, India
e-mail: santosh2019shar@gmail.com

M. Karunarathne (✉)
Department of Geography, Faculty of Humanities and Social Sciences, University of Ruhuna, Wellamadama, Matara, Sri Lanka
e-mail: msmanjula@geo.ruh.ac.lk

 
R. Kumar et al. (eds.), *Ecology, Conservation of Aquatic Birds*, Wetlands: Ecology, Conservation and Management 15, https://doi.org/10.1007/978-3-032-17125-2_11

## 11.1 Introduction

Waterbirds exert a pivotal influence on the nutrient cycling and trophic dynamics within aquatic ecosystems, maintaining balanced vegetation growth and serving as a vital food source for apex predators. Consequently, their populations contribute significantly to the livelihoods of communities that depend on these habitats [60, 90, 111]. Typically, the waterbirds/wetland dependent birds not only play a crucial ecological role as bio indicators of health of wetlands (rivers, marshes, lakes and peat lands) but also reflect the overall environmental condition of the ecosystems they inhabit [96, 154, 197].

Wetlands as aquatic ecosystems are home for large diversity of wildlife including plants, mammals, fish, amphibians, insects, resident and migratory birds species these habitats are productive due to high nutritional content [7, 12, 29, 64, 66, 144, 202]. Migratory waterbirds facilitate transcontinental, hemispheric, cultural, and societal connections through their periodic movements, given their ecological reliance on wetlands situated between disparate sites [91]. Significant wetland loss has occurred globally over the past century, with approximately half of these ecosystems disappearing as a consequence of human-induced activities, primarily including land conversion due to habitat destruction and degradation resulting from water extraction, agricultural and urban development, water pollution stemming from domestic and industrial activities, overhunting, and climate change effects [41, 87, 113, 207]. As a consequence, waterbirds have been in continuous decline since the 1975's [55, 59, 61, 71, 112, 124, 162, 187, 212]. Consequently, global conservation efforts for water birds have been significantly increased, prompted by the steep decline in waterbird biodiversity [199]. According to the Millennium Ecosystem Assessment Report [129], in Asia, freshwater fish species (20%), mammal species (37%) and bird species (21%) are either extinct or globally threatened [129]. In particular, around half of migratory waterbird species East Asia's are suffering population declines, owing to the loss and degradation of wetland habitats due to urban expansion (SoWB 2022).

The Indian subcontinent is home to at least 370 species of migratory birds, which travel along three different flyways: the Central Asian Flyway (CAF), the East Asian-Australian Flyway (EAAF) that includes regions of eastern India, and the Asian East African Flyway (AEAF) that encompasses pTamang arts of western India. Of these species, 310 primarily rely on wetlands habitats (MoEF 2018). Majority of India's wintering shorebirds migrate along the CAF, with some also utilizing the EAAF to a lesser extent (SOIB 2023). The EAAF serves as a crucial migratory corridor for over 500 bird species across 22 countries, with its wetlands supporting approximately 50 million migratory birds (SOWB 2022). However, these migratory avifauna face significant threats due to agricultural area expansion, unsustainable development and climate change which jeopardize their ecosystem health and the consequently the species which depend on them. The wetlands of India, situated at the core of the CAF, play a critical role in sustaining migrating waterbird populations, with nearly 71 percent of migratory waterbirds using India as a stopover site [98–101]. India stands out as a major destination for migratory birds in South Asia [103, 104, 106]. This is largely due to its unique biogeographical conditions, diverse physical landscapes and climatic variation. Additionally, the presence of numerous wetlands, including coastal areas and inland lakes and reservoirs, contributes to this significance [95]. In India, there is a variation on the number of waterbirds and wetland dependent birds based on the studies conducted by different ecologists and ornithologists from time to time. In some literature, only 18% of avifauna are reported to be strictly waterbirds while in another literatures approximately 23% (310 of 1340) of the bird species found in India are known to be dependent on wetlands [98–101]. According to other studies, India is home to 245 species of waterbirds, classified into 7 orders and 21 families. Among them, 128 species are migratory [94, 98–101]. Study by BirdLife International, reported that the wetlands of India serve as habitats for 202 waterbird species of which 165 are migratory. In spite of the disparity in the number of waterbird species reported from India, it has been reported two decades before that the important feature of Indian wetlands was that many natural lakes and rivers were progressively being degraded [166]. Furthermore, during the last decade, hunting and associated disturbances resulted in the loss of wetlands (32%), human settlements (22%), fishing (19%) and through drainage (23%) from agriculture [176]. Agricultural expansion, excessive cattle grazing, fishing, and illegal hunting of waterbirds are among the major threats to wetlands in India, significantly impacting their ecological balance and biodiversity. Previous studies have reported a decline in coastal bird populations at key sites in India, likely due to habitat loss, harmful industrial developments, unregulated agricultural practices, human disturbances, and climatic changes [15]. Intensification of agriculture, overgrazing and construction of development projects have been the significant threat for Important Bird Areas (IBAs) in India [153]. Persistent and multifaceted anthropogenic threats, encompassing poaching, habitat fragmentation, and infrastructure development, as well as disruptions to feeding and breeding habitats, continue to put considerable pressure on waterbird species (Gopi et al. 2014) [132, 153]. Furthermore, significant gaps in protected area coverage for migratory species have been identified in India, leaving these species vulnerable to various threats [165]. During the 2022 Asian Waterbird Census, pollution from the improper disposal of solid and domestic waste, as well as agricultural runoff and habitat encroachment, posed significant threats to the habitats of waterbirds in India. Furthermore, irregularities in water levels, increased tourism, and the spread of invasive species like water hyacinth were observed to be impacting the wetland ecosystems. 47% decline in wetland specialists 2000–2018 in India (SOWB 2022). Illegal sand mining in rivers and plastic pollution are the additional dangers for water birds such as the India Skimmer, Black-bellied tern, Pranticoles, River Lapwings and great stone plovers [152]. A study conducted between 2006 and 2015 in the Central Asian Flyway region found that 53% of waterbird species exhibited a declining population trend, with nearly one in four species observed in Indian wetlands experiencing a decrease in population (WISA 2020). Notable declines have been observed in the population trends for various waterbird species which have been encountered in India (Gopi et al. 2014) [94].

The Indian Himalayan Region (IHR) is a distinctive geographical and geological entity characterised by a multifaceted social, cultural and environmental mosaic which is still evolving and thereby not having stabilised from geomorphological and geological considerations [135]. Owing to the IHR's unique geography and climatic conditions diversity of habitats—ranging from high-altitude wetlands to river valleys supports diversity of waterbirds and wetland dependent bird species including migratory species that depend on this area as a key stopover or breeding ground [33]. According to Price et al. [149], 80%(783 species) bird species found in the subcontinent of India are found in IHR. Recent studies have concluded that the Himalaya constitutes 987 avian species out of which 807 (81.8%) terrestrial species and 180 (18.2%) are water bird species [4]. However, climate change and rapid urbanization have posed significant threats to the waterbird populations in the IHR over the past few decades and therefore conservation efforts in the region have faced several challenges due to environmental changes, particularly climate change and urbanization [139]. Rising temperatures, shifting precipitation patterns, habitat fragmentation, and increased human settlement have led to substantial changes in wetland availability and quality, which are critical for waterbird survival [178]. The potential impact of these changes in due to climate change and urbanization are vast ranging from disrupting migratory patterns, declining species richness, and altering breeding success. As these environmental stressors continue to intensify, waterbird conservation in the region requires urgent attention to mitigate biodiversity loss [173]. However, studies dealing with waterbird and wetland dependent species in the IHR has received little attention [136] (Bhattacharjee and Bargali 2003; Gandhi and Singh 1995) [9, 14, 22, 89, 92, 98–101, 190, 191]. In the light of above, this book chapter, mainly cover (1) analyse the impacts of climate change and urbanization on waterbird of Indian Himalayan Region (2) discuss policy and research gaps in waterbird conservation and suggest integrated conservation strategies for waterbird conservation in Indian Himalayan Region. Through these objectives, this study aims to enhance the understanding of waterbird conservation challenges in the IHR, particularly in the context of climate change and urbanization. In addition, it highlights the need for a comprehensive conservation strategy that combines ecological, social, and policy-based solutions.

## 11.2 Indian Himalayan Region Study Area

The Indian Himalayan Region (IHR), stretching over 2500 kilometers across ten states and union territories from Jammu and Kashmir to Arunachal Pradesh in the northwest and in the northeast respectively, is a unique geographical and ecological entity. The IHR spans across 11 states and 2 union territories. These states and UTs are: Arunachal Pradesh, Tripura, Mizoram, Himachal Pradesh, Jammu & Uttarakhand, Kashmir (UT), Ladakh (UT), Sikkim, West Bengal (Partially—Darjeeling region), Nagaland, Manipur, Meghalaya, Assam (Partially—Hill districts), (Fig. 11.1). It constitutes about 16.2% of total land area of India and is home to

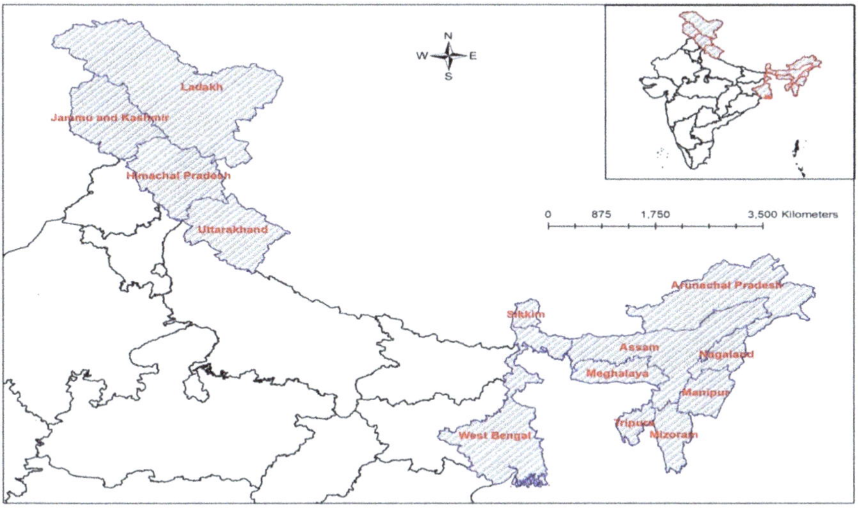


**Fig. 11.1** States of Indian Himalayan Region

diverse ecosystems, cultures, and climatic zones, making it a vital study area for researchers across multiple disciplines. The IHR is characterized by towering mountain ranges, deep river valleys, glaciers, and alpine meadows. It harbors the origin of major rivers like the Ganga, Yamuna, and Brahmaputra, which sustain millions of people downstream. The region also lies within one of the Global Biodiversity Hotspots, the Himalayan Biodiversity Hotspot, hosting endemic flora and fauna such as the red panda, snow leopard and medicinal plants. The region's population includes various indigenous communities with distinct languages, traditions, and livelihood practices (including agro-pastoralism, shifting cultivation, and spiritual traditions tied to the environment) adapted to living at high-altitude. Due to having fragile ecosystems, the IHR presents opportunities for research in climate change, ecology, hydrology, impact of urbanization and sustainable development. Its vulnerability to climate change, earthquakes, glacial retreat, and land degradation makes it a critical area for studying environmental challenges and resilience strategies.

To identify key challenges of waterbird conservation and knowledge gaps in Indian Himalayan Region (IHR). This chapter is based on studies relating to 'waterbird conservation' 'impact of climate change on water birds' "impact of urbanization on water birds', 'challenges to waterbird conservation' in IHR. Additionally, field surveys were conducted across some of the prominent wetlands of IHR, covering western and eastern Himalayas to validate the current study.

## 11.3 Wetlands of the Indian Himalayan Region

The distribution and extent of wetlands vary significantly across 11 states and 2 UT's of the Indian Himalayan Region (IHR) each with unique wetland ecosystems, topography, climate, and anthropogenic activities that support biodiversity, regulate hydrology and provide livelihoods to local communities. Out of 210,636 sq.km geographic area, there are 4632 wetlands occupying 285,539 ha (1.36%) of IHR [133]. Some internationally important Wetlands/regions of IHR are in Tripura, Meghalaya, Assam, Arunachal Pradesh, Nagaland, Manipur, Mizoram, Darjeeling, Jalpaiguri, Coochbehar, Sikkim, Ladakh and Hokar Sar wetland [76]. The wetlands of Indian Himalayan can be grouped into four categories (Table 11.1) [147].

## 11.4 Waterbirds and Wetland-Dependent Bird Species in the Indian Himalayan Region

The IHR supports a diverse range of waterbirds and wetland-dependent bird species, reflecting the ecological importance of the wetlands which provides habitats for foraging, nesting, and as stopover sites during migration. The diversity, abundance, richness and distribution of waterbird species vary across the different states due to differences in wetland habitat availability, climate characteristics, and anthropogenic pressures (Table 11.2) (Bassi 2014). The Himalayan Region acts as a critical stopover and breeding ground for waterbirds which are migratory. For instance, wetlands in Ladakh, India, are a key stopover site for migratory birds travelling between the Palearctic and the Indian Subcontinent and these wetlands provides favourable wintering habitats for a globally threatened species such as the Black-necked Crane *Grus nigricollis* [81, 114].

The Foothills within the Western IHR, specifically encompassing Corbett Landscape, Haridwar, Dehradun serves as a suitable habitat for a diverse array of waterbird species, encompassing both long-distance and short-distance migrants. Research has consistently demonstrated high diversity of waterbird species among disparate regions of Uttarakhand. Narang [136], Gandhi and Singh (1995), Tak et al.

**Table 11.1** Categories of wetlands of Indian Himalayan Region

| Sr. No. | Categories | Wetlands |
|---|---|---|
| 1 | Ladakh and Zanskar | Noorichan, Chushul, Hanlay marshes Tso Morari, Chantau and Pangong Tso |
| 2 | Kashmir Valley | Dal, Malgam, Haukersar and Kranchu, Wular, Anchar, Haigam lakes |
| 3 | Central Himalayas | Naukuchital, Bhimtal and Nainital |
| 4 | Eastern Himalayas | Numerous wetlands in Arunachal Pradesh, Nagaland and Manipur, Sikkim, Meghalaya, Assam, Beels in the Barak valley and Brahmaputra |

**Table 11.2** Number of wetlands in the different states of Indian Himalayan Region

| Sr.No. | State | No. of wetlands |
|---|---|---|
| 1 | Jammu & Kashmir | 3651 |
| 2 | Himachal Pradesh | 641 |
| 3 | Uttarakhand | 994 |
| 4 | Sikkim | 553 |
| 5 | West Bengal | 147,826 |
| 6 | Assam | 11,178 |
| 7 | Arunachal Pradesh | 2653 |
| 8 | Nagaland | 421 |
| 9 | Manipur | 708 |
| 10 | Mizoram | 234 |
| 11 | Tripura | 3415 |
| 12 | Meghalaya | 426 |

[191], Tak and Sati [190], Kumar et al. [98–101], Bhattacharjee and Bargali (2003) and Dhakate et al. [43] studied different wetlands of Corbett landscape, Kaushik et al. [89] studied at wetlands of Dehradun and very few studies [9, 14, 22, 92] focused on the wetlands of Uttarakhand, Western Himalayas. Some of the studies on the water birds of IHR have provided certain key insights (Table 11.3).

## 11.5 Threats to Wetland Ecosystems in Indian Himalayan Region

In IHR, degradation of many wetland habitats as a result of climate change and human activities have been established and which leads to habitat loss, fragmentation [53, 208, 209]. These impacts are compounded by lack of effective enforcement of legislation and policy gaps which ultimately leads to species decline (Fig. 11.2).

## 11.6 Climate Change Impacts on Water Birds/Wetland Dependent Birds in IHR

Birds are widely used as diversity surrogates for studying bird globally to estimate overall biodiversity patterns, ecosystem health, and conservation priorities [8, 13, 34, 137, 210]. Birds are also considered as the indicators and the best vertebrate models to study the impacts of habitat and climate change owing to high sensitivity [206]. Globally, it has been concluded that the climate change impacts the bird species negatively [37, 50, 160, 205]. Nevertheless, research on the impact of climate change on waterbirds and wetland-dependent birds in the IHR is limited, remains poorly quantified but growing [44]. Several studies have begun to shed light on how climate change and related factors are affecting avian populations in this

**Table 11.3** Key insights on waterbird diversity and distribution in the Indian Himalayan Region (Including Bhutan and Central Himalayas-Nepal)

| | Region | Diversity Insights | Key Species | References |
|---|---|---|---|---|
| Indian Himalayan Region | Eastern Himalayas | High diversity of water-dependent birds, with 70% of habitats located less than 2000 m altitude together with major drainages. | White-capped Water Redstart *Phoenicurus leucocephalus* and Blue Whistling Thrush *Myophonus caeruleus* | [193] |
| | Bhutan | Habitat quality and fragmentation play a considerable role in structuring waterbird communities. | | [45] |
| | Central Himalayas (Nepal) | Species richness peaks at mid-elevations (2600–3000 m asl), with insectivores dominating at higher elevations. | – | Katuwal et al. (2016), [139] |
| | Trans-Himalayas (Ladakh) | Hosts 38 waterbird species with high conservation value wetlands like Statapuk Tso. | Black-necked Crane *Grus nigricollis* | [81, 114] |
| | Northwestern Himalayas | Supports 677 bird species, with 54 threatened species, including 23 Vulnerable, 22 Near Threatened, 6 Endangered, and 3 Critically Endangeredspecies. | River Tern *Sterna aurantia,* Woolly-necked Stork *Ciconia episcopus,* River Lapwing *Vanellus duvaucelii* and Painted Stork *Mycteria leucocephala* | Yousuf et al. (2023), [92]. |

ecologically sensitive area. However, there is the evidence of the impact of climate change on birds [121].

River birds, are outstanding indicators of the anthropogenic impacts on riverine/riparian habitats in the Himalayan region [179, 180]. In the IHR, water dependent bird species such as White-capped Water Redstart *Phoenicurus leucocephalus* (Fig. 11.3) and Blue Whistling Thrush *Myophonus caeruleus* (Fig. 11.4) are the most common indicators of healthy riverine ecosystems and are regarded as model species for distribution studies [118] (Tamang et al. 2024). These water birds use the riverine habitats by sharing with other water bird species such as Plumbeous Water Redstart *Phoenicurus fuliginosus* (Fig. 11.5), Brown Dipper *Cinclus pallasii* (Fig. 11.6), Forktails *Enicurus* and Wagtails *Motacilla* (Fig. 11.7) [30, 179, 180].

According to the 2023 updated Checklist by Clement's Cornell's Lab of Ornithology, 12% of the 10,906 recorded species are globally threatened [77]. The main

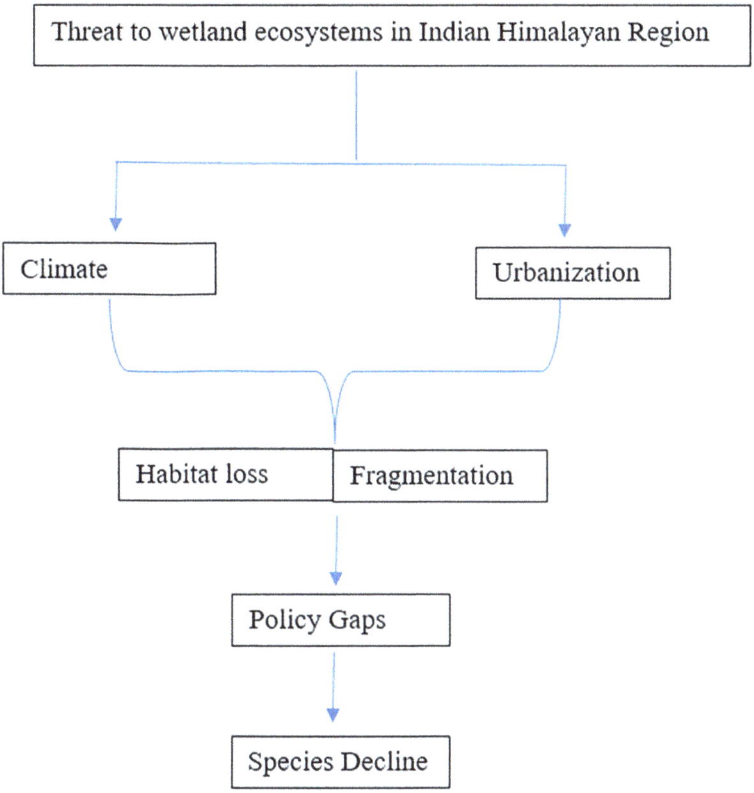


**Fig. 11.2** Flowchart illustrating the threat to waterbird conservation in Indian Himalayan Region

causes of the threat are habitat loss, increasing levels of pollution [19, 25], and climate change [167, 195]. Migratory and aquatic birds, as well as restricted-range species, are particularly vulnerable [213]. However, there is a paucity of detailed information on the migratory patterns and behaviour of waterbirds in the Indian subcontinent [171].

The IHR is considered as one of the most susceptible regions to changes which occurs globally such as global warming and climate change. Prior research indicates an accelerated warming trend in the Himalayas, outpacing the global mean temperature increase of 0.74 °C during the past century, and forecasts suggest a continued escalation, with concomitant degradation of associated natural ecosystems [5, 48, 75, 93]. Amidst the Himalayan region, accelerating thermal ascension outpacing global averages juxtaposed with accelerated land-use transformations ostensibly driven by expanding demographic requirements poses considerable threats to native biodiversity [32]. Climate change is purportedly exerting a profound impact on all terrestrial ecosystems, agricultural systems, and human infrastructure reliant on mountainous regions, including tourism sectors and associated built environments [83]. Climate change exacerbates the extinction risk among species with ranges

**Fig. 11.3** White-capped Water Redstart (*Phoenicurus leucocephalus*)

**Fig. 11.4** Blue Whistling Thrush (*Myophonus caeruleus*)

confined to narrow geographic or climatic zones, as well as disrupts existing community structures [17]. Mountain ecosystem is facing considerable threats of species loss and extinction from several factors such as fragmentation of landscape and habitat degradation mainly caused by anthropogenic reason and climate change [134, 141, 177] (Fig. 11.2). A temperature increase of 3–4 °C could eliminate 85% of the world's wetlands.

**Fig. 11.5** Plumbeous Water Redstart (*Phoenicurus fuliginosus*)

**Fig. 11.6** Brown Dipper (*Cinclus pallasii*)

According to the 2023 updated Clement's Checklist by Cornell's Lab of Ornithology, 12% of the 10,906 recorded species are globally threatened [77]. The main causes of the threat are habitat loss, increasing levels of pollution [19, 25], and climate change [167, 195]. Migratory and aquatic birds, as well as restricted-range species, are particularly vulnerable [213]. However, there is a dearth of in-depth information of migratory patterns off waterbirds in the Indian subcontinent [171].

**Fig. 11.7** Wagtails *Motacilla*

The IHR remains one of the most sensitive ecosystems to global climate change. Previous studies have indicated that warming in the Himalayas has exceeded the global average of 0.74 °C over the past century and is projected to continue rising, leading to further degradation of natural ecosystems [5, 48, 75, 93]. Rising temperatures, which are increasing at a rate faster than the global average, along with rapid land-use changes driven by the demands of a growing population, pose significant threats to biodiversity in the Himalayas [32]. Climate change is believed to be affecting all natural ecosystems, agricultural ecosystems, tourism and mountain infrastructure [83]. Climate change poses significant the risk of extinction for species with narrow geographic or climatic distributions and disrupts existing biological communities [17]. Mountain ecosystem is facing considerable threats of species loss and extinction from several factors such as habitat degradation and fragmentation of landscape elements caused by anthropogenic reason and climate change [134, 141, 177] (Fig. 11.2).

The aquatic ecosystems of IHR such as lakes and rivers are enormously sensitive to the climate change impacts which have a direct impact on diverse species of flora and fauna (Fig. 11.8, 11.9, and 11.10) [107].

With the increase in temperature and consequent melting of glaciers, the water availability is reduced on wetlands and freshwater habitats on which specific water-dependent animals are dependent [11]. Most high-altitude glacial lakes and associated wetland areas, including marshes, serve as crucial habitats for various wildlife species, particularly migratory waterbirds and cold-water fish populations [76]. Changes in bird populations is one of the key indicators of the climate change impacts [37]. Globally, climate change has affected population of many bird species and therefore many waterbird species are in decline since the last few decades [121, 174, 175, 203]. Several studies have indicated that many species of flora and

**Fig. 11.8** Changu Lake, East Sikkim

**Fig. 11.9** Gurudongmar Lake, North Sikkim

fauna together with birds are shifting their ranges towards the pole or to higher altitudes in tropical mountains as a result of climate change [146]. Numerous bird species are expanding their geographical distributions further north (Thomas and Lennon 1999), while certain migratory birds are arriving earlier at their breeding sites [120]. Additionally, some species are initiating breeding activities sooner [38, 51] or delaying their migration to wintering grounds (Gatter 1992). Migratory birds are especially susceptible to changes in the environment, including climate change, as they are highly sensitive to continuous alterations that may have negative

**Fig. 11.10** Ganga River, Rishikesh, Uttarakhand

correlations [168]. Furthermore, these migratory birds use separate habitats often geographically distant, as their wintering and breeding grounds and changes in any one of the sites during their life cycle may have serious impacts on their survival and population dynamics. Climate-induced habitat changes are expected to impact the staging sites, stopover ecology and fueling processes of migratory birds. Rising temperatures and habitat changes in breeding grounds (in Central Asia and Siberia) impact the arrival and stopover patterns of migratory waterbirds. Studies have highlighted that climate change significantly influences waterbird populations through altering habitat structure and food availability in the IHR. For instance, Sharma et al. [173] reported that temperature increases and fluctuating rainfall patterns are impacting high-altitude wetlands, which serve as breeding sites for migratory birds. These climatic changes often lead to reduced wetland areas and altered food availability, both of which can negatively impact waterbird species diversity and abundance [96]. Socolar et al. [182] noted changes in migratory timings of waterbirds in the Himalayan flyway, potentially leading to mismatched availability of food and nesting sites. Additionally, changing temperature gradients due to global warming have caused shifts in migration timings, disrupting the ecological balance in habitats [178]. According to Down to Earth [47], Black-necked Cranes (*Grus nigricollis*) modify their incubation behavior in response to weather conditions and the thermal requirements of their eggs. In response to climate change, shift in altitudinal ranges have been reported in previous studies by wetland dependent bird species such as Snow Pigeon *Columba leuconota* Ibisbill *Ibidorhyncha struthersii.* Similarly, delays in breeding activities, including habitat selection, nest building, egg-laying, and the emergence of hatchlings, have also been observed in the White-capped Water Redstart *Chaimarrornis leucocephalus* [3]. In the same way, climate change has significantly impacted breeding activities in the higher elevation areas of Sikkim, resulting in reproductive failure among water birds

**Fig. 11.11** Ruddy Shelduck (*Tadorna ferruginea*)

**Fig. 11.12** Common Redshank (*Tringa tetanus)*

and wetland-dependent species such as Black-necked Crane *Grus nigricollis*; Ruddy Shelduck *Tadorna ferruginea* (Fig. 11.11), Common Redshank *Tringa tetanus* (Fig. 11.12), and Ibisbill *Ibidorhyncha struthersii* [6, 56–58]. Bird species have a critical water surface temperature threshold, beyond which prey availability declines, leading to increased chick mortality [204].

Driven by climate change, increased temperatures and reduced precipitation leads to habitat shrinkage and wetland degradation affecting the distribution and abundance of waterbirds in IHR [110, 128] (Xu et al. 2023). In fact, climate change adversely affects wetland dependent water birds in the IHR by altering thermal structures, freshwater species composition, and water quality, leading to mismatches in migration timing and food availability, ultimately affecting their survival and ecosystem dynamics [2]. Furthermore, increasing temperatures may lead to the expansion of invasive plant species such as *Ageratina adenophora* and *Lantana camara* in wetlands, which can outcompete native vegetation that provides essential resources for waterbirds [27]. This invasion further diminishes suitable habitat for waterbirds, affecting their reproductive success and survival rates [33]. The limited capacity of species to adapt quickly to such rapid environmental changes exacerbates these issues, putting various species at higher risk of local extinction [139]. Mountain birds may perish with the alteration of climate pattern and rise in temperature, as they are confined to increasingly smaller patches of habitat [121].

Shifting weather patterns, glacier retreat, and changes in water availability alter the habitats of waterbirds. Sharma et al. [172] found that rising temperatures in the Himalayan region influence the timing of migration and breeding of waterbirds such as the Common Teal *Anas crecca*. Bassi et al. [20] reported that drying wetlands in Himachal Pradesh due to reduced glacial melt and precipitation have affected migratory bird species. Furthermore, the increased variation in precipitation patterns and river flows due to climate change is anticipated to have adverse effects on riverine bird communities [150] (Royan et al. 2015), especially in the drier habitats of the Western Himalaya. For instance, the Hokersar wetland, Western Himalaya has experienced a reduction from 18.75 $km^2$ in 1969 to 13 $km^2$ in 2008, severely impacting the marshy lands that serve as critical habitats for migratory birds [161].

Research suggests that deeper water bodies support a higher density of waterbirds, particularly ducks, as they prefer such habitats for foraging [72] (Rajpar et al. 2011). In order to sustain optimal ecological health and functionality, wetlands require an adequate supply of high-quality water, delivered at appropriate times and in a suitable pattern [155]. Due to the impact of climate change, the glaciers of IHR are prone to Glacial Lake Outburst Flood (GLOF) [125]. The GLOF disaster of 3–4 Oct. 2023 in Sikkim Himalaya resulted in a huge siltation (upto a depth of 3–4 meter) in many places on the bank of Teesta river by eroding ~270 million $m^3$ of sediment [69, 211]. Silt deposition in rivers can have several negative impacts on water and wetland-dependent birds such as loss of feeding grounds, disruption of breeding sites, alteration of wetland ecosystems, decline in water quality, loss of biodiversity (Fig. 11.13). Bank-nesting species are particularly vulnerable to riparian habitat loss, nest flooding, flow variability during critical periods of their annual cycles, such as breeding [35, 159, 186].

**Fig. 11.13** Rapid urbanization and siltation along river valleys

## 11.7 Impact of Urbanization on Waterbirds in Indian Himalayan Region

Urbanization in the IHR has accelerated in recent years, leading to increased human encroachment on wetlands, pollution, and habitat fragmentation [97]. Evidence suggests that habitat modification influences beta diversity patterns in the Western Himalaya, whereas similar effects have not been observed in the Eastern Himalayas; previous studies have documented the sensitivity of bird species in this region to anthropogenic influences [1, 126, 179, 180]. The expansion of urban areas results in habitat degradation, making it challenging for waterbirds to find suitable nesting sites and sufficient food resources. In addition to habitat loss, urbanization increases water pollution levels through runoff from urban and agricultural areas, introducing contaminants that can harm waterbird species [102–104]. Studies also indicate that urban expansion leads to disturbances such as noise and light pollution, which can disrupt the behavior of sensitive waterbird species and also affects their migratory patterns and reproductive cycles [178]. The abundance of waterbird species is significantly influenced by proximity to human settlements and the nearest roads [115]. A recent study has identified contamination of microplastic in fish fauna from the Ramsar Wetland site of Loktak, the world's only floating lake, located in the Eastern Himalayas, IHR [28]. In another study, microplastic have been observed in Markanda river, Western Himalaya [105]. This is the manifestation of the adverse impact of urbanization on the aquatic ecosystem due to ineffective solid waste management practices prevalent in the IHR (Fig. 11.14).

The wetland of IHR is subject to encroachment, wetland draining, agricultural expansion, and infrastructure development (e.g., roads, hydropower projects) threaten critical waterbird habitats [20, 63, 161]. Chandan et al. (2008) observed that rapid urbanization and agricultural intensification [106] around wetlands in the

**Fig. 11.14** Solid Waste dumping near water bodies

Ladakh region are major causes of habitat degradation, impacting waterbird populations like Bar-headed Geese *Anser indicus*. Kumar et al. (2021) highlighted wetland loss in Uttarakhand due to human activities, which poses a significant threat to species such as Ruddy Shelduck *Tadorna ferruginea*. Loss of wetland poses a significant threat to ecology of an area which can potentially lead to various undesirable consequences [117, 147].

Birds are highly sensitive to alterations in habitat structure caused by human land use, both at the local and landscape levels [108, 196]. Although riverine habitats make up only a small part of mountainous landscapes, they provide unique ecological niches for bird species from nearby forested areas [138, 179, 180, 188]. Many water birds are habitat specialists that depend on aquatic and riparian resources, relying on these narrow, linear habitats for their survival [30]. Variations in river topography, including factors such as depth of river, width of river and flow, determines community structure processes through multiple ecological pathways [164]. Regulation of river flow is one of the most harmful human-induced impacts on riverine ecosystems [52], a concern that is especially pronounced in developing countries across the Indian subcontinent [68]. Dams and hydropower projects disrupt wetland ecosystems by altering natural flow regimes and fragmenting habitats; in particular, they modify downstream channel dynamics by creating deeper, wider pools that are generally unsuitable for obligate waterbird species (Fig. 11.15) (Sinha 2019). Typically, riparian habitats at low-elevation zones, along with their associated species, face intense pressure from dam construction, industrial development, mining activities, and the expansion of road networks in IHR (Pandit et al. 2012). Furthermore, fluctuations in flow conditions—particularly pronounced changes in water velocity and depth resulting from impoundment—can affect the availability of foraging habitats, potentially leading to shifts in the composition and structure of riverbird communities [39]. Kattel et al. (2018) highlighted the adverse impacts of hydropower dams on waterbirds in the Teesta Basin, Sikkim, particularly noting the

**Fig. 11.15** Hydel Power projects

reduction in habitat connectivity for migratory species such as the Great Cormorant *Phalacrocorax carbo*. Anthropogenic activities, including road construction and hydropower development, frequently infringe upon essential habitats, leading to ecosystem fragmentation and disruptions in waterbird populations [173]. With approximately 300 dams spread across 90% of the Indian Himalayan valleys, there is a high likelihood of significant negative impacts on biodiversity, including the potential for species extinctions [140]. Among riverine birds as well, the proposal of over 160 dams in the Eastern Himalaya is expected to bring about irreversible alterations to riverine systems, resulting in significant negative impacts on riverine bird communities—patterns that have already been observed in other regions [1, 198]. The Eastern Himalaya harbors the highest diversity of riverine bird species anywhere in the world [30, 184]. In the Sikkim Himalaya, a region within the Eastern Himalaya, six hydroelectric projects have been proposed along the Teesta River, two of which—Teesta Stage-III and Stage-V—have already been constructed and are currently operational [192]. Additionally, a range of developmental activities—such as the construction of roads, bridges, and industrial facilities including pharmaceutical factories—are taking place across various riverine ecosystems in the region. Collectively, these interventions are driving large-scale environmental changes and pose substantial threats to key biodiversity areas and riverine ecosystems in the Eastern Himalayas.

Riverine forests in the region are under considerable threat from habitat loss, destructive fishing practices, and sand and boulder mining—all of which are likely to negatively impact riverine bird communities [119, 126, 185] (Figs. 11.16 and 11.17). Agricultural runoff, untreated sewage, and plastic waste degrade water quality, leading to eutrophication and loss of biodiversity. Pant et al. [142] analyzed the eutrophication of Wular Lake, Jammu and Kashmir, and its negative impact on fish and invertebrate populations, which are crucial for waterbirds. A recent study conducted between 2014 and 2018 along river stretches of the Western Himalaya, spanning an altitudinal range of 300 m to 3100 m a.s.l., reported a 5%–10% decline

**Fig. 11.16** Illegal Mining (Western Himalaya)

**Fig. 11.17** Illegal Mining (Eastern Himalaya)

in the overall population sizes of five bird species: Brown Dipper *Cinclus pallasii*, Plumbeous Water Redstart *Phoenicurus fuliginosus,* Common Kingfisher *Alcedo atthis,* White-throated Kingfisher *Halcyon smyrnensis* and White Wagtail *Motacilla alba* [181].

Tourism in IHR has experienced rapid, largely unregulated, and demand-driven growth since tourism generates lot of local employment particularly for traders, craftsmen, transporters, hoteliers and to those who have the knowledge of the local places and would like to work as guides. During the last three decades, tourism led development in terms of roads, housing, administrative and defence posts have rapidly transformed the landscape of IHR [24, 36, 157, 163, 170]. IHR has been a home to several stunning lakes that are popular among tourists for their natural beauty and cultural significance (Pangong Tso –Ladakh; Tso Moriri- Ladakh; Tsomgo Lake—Sikkim; Gurudongmar Lake—Sikkim; Chandratal Lake-Himachal

**Fig. 11.18** Encroachment of wetland habitats

Pradesh; Prashar Lake-Himachal Pradesh; Dal Lake-Jammu and Kashmir; Wular Lake—Jammu and Kashmir; Nainital Lake–Uttarakhand; Bhimtal Lake-Uttarakhand; Gurudongmar Lake—Sikkim (Fig. 11.9); Hemkund Lake (Uttarakhand). The diverse landscapes of these lakes—ranging from high-altitude deserts to lush valleys—make them must-visit destinations. The infrastructural development in the IHR along river valleys and at high altitude lakes for the defence, tourism and trade maximizes the negative consequences during the sensitive bird migration and breeding periods by causing loss of wetlands and water pollution and so the infrastructural development prompted by urbanization and tourism are threats to wetland/water birds avifauna of IHR (Blaise Humbert 2017) [86]. Previous studies have identified Hemis National Park (Central Ladakh), the wetlands of the Changthang (Tsokar, Tsomoriri and Pangong) as two regions especially impacted by environmental degradation due to tourism [62]. Landfill and construction of roads and commercial, residential and defence buildings are major causes of wetland loss and consequently to wetland birds [49, 127] as the wetlands are of critical importance as bird breeding, staging and wintering grounds. In many wetlands, the birds breeding, staging and wintering areas lie adjacent to major roads, trekking routes, campsites and other infrastructure [26, 73, 183]. Unregulated tourism, camping, and boating disturb nesting and feeding waterbirds. Kala (2014) highlighted how unregulated tourism in high-altitude wetlands like Pangong Tso affects the breeding success of Black-necked Cranes *Grus nigricollis*. Mishra et al. observed that tourism in Nainital Lake led to pollution and habitat disturbance, reducing waterbird diversity. The low-elevation zones of Sikkim in the Eastern Himalaya faced various anthropogenic threats, including sand and boulder mining, encroachment, deforestation, and habitat fragmentation. These disturbances may significantly impact critical stopover sites for Pallas's Gull *Ichthyaetus ichthyaetus* and other migratory species [171] (Fig. 11.18).

The introduction of invasive aquatic plants and fish disrupts native ecosystems, affecting food availability and nesting sites for waterbirds. Rawat and Adhikari [158] documented that invasive species like *Eichhornia crassipes* (water hyacinth) in wetlands of Uttarakhand degrade water quality, reducing the suitability of these habitats for birds such as the Eurasian Wigeon (*Mareca penelope*).

## 11.8 Threats, Conservation Challenges and Implications

Mountain ecosystems, including the Indian Himalayan Region, are exceptionally diverse and globally significant as centers of biological diversity. This ecological richness makes the region particularly important for biodiversity conservation [78]. Waterbird diversity is closely linked to the variety of habitats and resources available in wetlands—such as food, shelter, nesting, and breeding sites. This diversity plays a vital role in shaping ecosystem functions, including nutrient cycling and energy flow within wetland environments [67, 201]. Nevertheless, due to human activities degradation of wetlands is accompanied by a sharp decline in waterbird diversity [151]. The deteriorating condition of wetlands in IHR presents major challenges in sustaining a network of wetland habitats capable of supporting the life-cycle requirements of migrating waterbirds. According to [88], wetlands in the Himalayas are rapidly becoming degraded and in some areas as many as 30% of lakes and marshes have disappeared because of over-exploitation and climate change. Despite the national legislation on wetland regulation, most wetlands remain overlooked in the policy-making process [20]. Migratory waterbirds are often hunted for food or sport in remote areas [189]. Bhat et al. (2020) reported poaching of migratory bird Northern Pintail *Anas acuta* in wetlands of Kashmir Valley, despite legal protections. Limited local awareness and poor enforcement of conservation policies lead to continued degradation of wetlands. The concept of wise use of wetlands has been frequently advocated; however, a significant gap persists between theoretical frameworks and practical implementation (Bhandari 2009). According to the Ministry of Environment, Forest and Climate Change (MoEF 2018), 29 sites across India—including 20 major wetlands and nine wetland clusters—have been designated as critical bottleneck sites for migratory waterbirds.

The high-altitude wetlands in the Indian Himalayan Region (IHR) serve as vital habitats for migratory birds, providing both breeding and wintering grounds for a diverse range of species. Notable among these are Green Sandpiper *Tringa ochropus*, Common Merganser *Mergus merganser*, Brown-headed Gull *Larus brunnicephalus,* Ruddy Shelduck *Tadorna ferruginea* (Fig. 11.11), Bar-headed Goose *Anser indicus* (Fig. 11.19c), Black-necked Grebe *Podiceps nigricollis* and Great Crested Grebe *Podiceps cristatus* which are known to breed in these habitats. However, the high-altitude ecosystem is extremely fragile and equally vulnerable to anthropogenic disturbances, much like other sensitive habitats [42]. Therefore, immediate conservation efforts are necessary to protect these wetlands from human activities, particularly tourism, which can negatively impact the natural

**Fig. 11.19** Some of the threatened and migratory waterbirds/wetland associated birds of Indian Himalayan Region ((**a**) River lapwing (*Vanellus duvaucelli*), NT; (**b**) Black-Necked Stork *Ephippiorhynchus-asiaticus,* NT; (**c**) Bar-headed Goose *Anser indicus*; (**d**) Mallard *Anas platyrhynchos* (**e**) Painted Stork *Mycteria leucocephala* NT F. Pallas Gull *Ichthyaetus ichthyaetus*

habitat (Mazumdar et al. 2017) [74, 116]. As on today, there are 137 Protected Areas (PA's) representing the total area of 47, 208 sq.km. in the Indian Himalayan Region [145]. Protected areas play a crucial role in helping waterbirds adapt to climate change by preserving critical habitats, ensuring resource availability, and providing safe breeding and stopover sites [46, 54]. Nevertheless, the existing network of Protected Areas may be inadequate for safeguarding bird diversity in a fragmented tropical hotspot, particularly under varying climate change scenarios [194]. Most importantly, waterbirds outside the protected area network in the IHR remain unprotected. Existing legislation, such as the Wetland (Conservation and Management) Rules, 2017, falls short in adequately addressing the unique requirements of high-altitude wetlands in the Indian Himalayan Region (IHR), irrespective of their size, even though they are intended to be notified under these rules. This limitation arises because protected area-based management approaches often overlook the critical need to integrate wetlands into broader water and land-use planning frameworks within the surrounding landscape [94]. Despite the establishment of a network of Ramsar sites, several key wetland sites essential for the sustenance of waterbirds remain unprotected by any legislation. Consequently, wetland degradation and loss continue to persist [40]. Illegal encroachments, unregulated tourism and pollution have been affecting wetland habitats in IHR. There is an urgent need to expand protected areas and improve the implementation of wetland conservation rules. Policies do not effectively integrate local communities in wetland and bird

**Table 11.4** Ramsar Sites in Indian Himalayan Region

| Sr. No. | State | No. of Ramsar Sites | Name of Ramsar Sites | References |
|---|---|---|---|---|
| 1 | Himachal Pradesh | 3 | Renuka Lake<br>Chandra Taal<br>Pong Dam Lake | [10] |
| 2 | Jammu & Kashmir | 5 | Wular Lake<br>Hokera Wetland<br>Shallbugh Wetland Conservation Reserve<br>Hygam Wetland Conservation Reserve<br>Surinsar-Mansar Lakes | [10, 16, 79] |
| 3 | Ladakh | 2 | Tsomoriri<br>Tso Kar Wetland Complex | [10, 70] |
| 4 | Sikkim | 1 | Khecheopalri Wetland | |
| 5 | Uttarakhand | 1 | Asan Conservation Reserve | [10, 130] |
| 6 | Manipur | 1 | Loktak Lake | [85, 143] |
| 7 | Mizoram | 1 | Palak Lake | RSIS (2022) |
| 8 | Tripura | 1 | Rudrasagar Lake | [18] |
| 9 | Assam | 1 | Deepor Beel | [174, 175] |

conservation efforts. Secondly, across the extensive 595,000 km$^2$ area of IHR, the designation of only 16 Ramsar Sites fails to adequately reflect the ornithological significance of its wetlands (Table 11.4). Ghosh and Sinha (2018) stressed the gap between policy and implementation in Ramsar-listed wetlands of the Himalayan region. By addressing these challenges through science-based policy and inclusive conservation efforts, the Indian Himalayan region's waterbird populations can be better protected. Policies that focus on sustainable development and climate resilience are essential for the preservation of these critical habitats [173].

Climate change impacts on invertebrate populations can affect bird populations, and the lack of extensive long-term monitoring data for invertebrates hampers understanding these effects, complicating efforts to monitor bird populations effectively [80]. Information on the impact of climate change on waterbirds in the Indian Himalayan Region (IHR) remains limited. Therefore, there is a pressing need for expanded ornithological research addressing various aspects of bird life across this subcontinent. Such studies are essential to better understand how climate change influences the population dynamics, distribution, and survival of individual avifaunal species within the IHR. Long term monitoring schemes and databases on population and nests of several bird species can help to identify several new and unforeseen impacts of climate change. There is insufficient research on species distribution, population trends, and ecology of many wetland-dependent birds in the region. More studies are needed on how changing precipitation patterns and glacial retreat affect waterbird habitats. Few standardized long-term monitoring programs exist for waterbirds in the IHR. Research on the impact of human activities, including agriculture, hydropower projects, and tourism, on wetland birds is lacking. Another critical area requiring attention is the potential harmful impact of invasive and alien species on the survival of native flora and fauna, particularly under changing environmental conditions.

Birds dependent on wetland ecosystems are monitored globally as indicators of wetland ecosystems: however, in India the waterbirds and wetlands dependent birds are not censused across the network of rivers ecosystems in IHR (Kingsford and Porter 2009; Brandis et al. 2018; Amano et al. 2018) [181]. Monitoring programs that track the climate change impacts on waterbird populations are also necessary to inform conservation practices and examine the effectiveness of management strategies [139]. In the Indian Himalayan Region (IHR), riparian zones are subjected to a range of pressures, from global influences such as climate change and altered hydrological dynamics to localized land-use changes that impact both the riparian areas and their associated river catchments [119, 177, 179, 180]. Therefore, integrated wetland management plans should be developed/formulated based on a comprehensive diagnostic assessment of their ecological, hydrological, socioeconomic, and institutional characteristics, as well as the factors driving change within the surrounding landscape. Effective conservation of waterbird species in the IHR requires integrated strategies that address both climate change and urbanization (Fig. 11.1). Investigation on how the climate change affects wetland hydrology and bird populations and implementing strict regulations on tourism and industrial activities near key wetlands is crucial in sustaining waterbird populations in IHR. Conservationists should emphasize the importance of establishing protected areas, restoring degraded wetlands, and implementing measures to control invasive species [86]. Enhancing habitat protection through community-driven wetland management is a viable option (Kumar et al. 2021). Kumar et al. (2011) emphasized the need for community involvement in managing the Harike wetland to improve waterbird conservation. Besides, monitoring the climate change impacts using citizen science and remote sensing data can help in tracking the waterbird populations [172]. Finally, there is also dire need to strengthen policies against illegal hunting and regulate tourism activities (Bhat et al. 2020).

Tourism and biodiversity conservation can successfully coexist only when robust monitoring programs are implemented, enabling management authorities to promptly identify and mitigate any negative impacts [17]. Increased collaboration between governmental and non-governmental organizations can enhance the effectiveness of waterbird conservation initiatives in the region. Identifying declining migratory bird species and understanding the key drivers of their population decline is crucial. By integrating policy reforms, scientific research, and active public engagement, it is possible to mitigate the effects of climate change and reduce the impact of urbanization on waterbirds in the Indian Himalayan Region (IHR). Educational programs that raise awareness about the ecological significance of waterbirds and wetlands can play a crucial role in building public support for conservation efforts [96]. The Ministry of Environment, Forests, and Climate Change (MoEFCC) has formulated the National Action Plan for the Conservation of Migratory Birds and Their Habitats along the Central Asian Flyway for the period

2018–2023 (F.No. 4-37/2017 WL (Part-2), dated 30 August 2018). This plan is organized into six core components: 1. species conservation 2. habitat conservation and sustainable management 3. capacity development 4. communication and outreach 5. research and knowledge base development and 6. international cooperation. Its long-term objective is to mitigate population declines and safeguard critical habitats for migratory birds. In the short term, the plan aims to halt population declines and ensure stable or increasing trends for waterbirds and wetland birds by 2027 (MoEF 2018). Notwithstanding, adequate attention needs to be accorded to the migratory, threatened, rare and endangered waterbirds and wetland birds of IHR (Fig. 11.19a–e).

The recognition of Sikkim's Khecheopalri Wetland as a Ramsar site on World Wetland Day (February 2, 2025) offers a valuable example for identifying and nominating other important Himalayan wetlands. Such recognition is crucial for protecting fragile high-altitude wetlands and the bird species that depend on them. It should also encourage better regional planning and the inclusion of wetlands in broader development frameworks across the Indian Himalayan Region (IHR), enhancing the enforcement of conservation policies in sensitive ecological zones. Although states in the IHR have existing Wetland Conservation Programmes, more precise and site-specific strategies are needed—particularly for wetlands that are essential for waterbird breeding and feeding. Current conservation plans often do not directly address the unique requirements of these birds. Furthermore, integrating traditional ecological knowledge could significantly enrich waterbird conservation by offering local insights into their behavior, habitats, and management techniques.

There is a pressing need for more detailed and comprehensive data on the distribution, population status, and habitat preferences of waterbird species in IHR. Although general biodiversity research is ongoing, targeted studies addressing the specific threats to waterbirds—such as habitat degradation, pollution, and the impacts of climate change—remain limited. It is recommended to develop species-specific conservation plans that prioritize habitat preservation, effective management of breeding areas, and strategies to address key threats. Additionally, enhancing wetland management through regular water quality assessments, pollution control measures, and the promotion of sustainable usage practices is essential. By addressing these policy and research gaps, the IHR can improve its waterbird conservation efforts and ensure the long-term survival of these valuable species.

## 11.9 Future Prospects

The ongoing and projected impacts of climate change and urbanization on waterbird populations and wetland ecosystems in the Indian Himalayan Region (IHR) highlight an urgent need for forward-looking, evidence-based conservation planning. Future research should prioritize long-term ecological monitoring to track population dynamics, migration shifts, and breeding patterns of waterbirds under changing climatic conditions. There is also a critical need to strengthen predictive modeling capabilities to anticipate future wetland habitat changes and the impact of such changes for vulnerable waterbird species. Policy development must keep pace with scientific insights. This includes integrating wetland and biodiversity conservation into broader land-use planning and climate adaptation policies. A focus on transboundary cooperation, given the migratory nature of many waterbird species, will also be crucial in developing coordinated conservation strategies across political and ecological boundaries. Ultimately, the sustainability of waterbird populations in the IHR will depend on proactive and adaptive management, robust scientific inquiry, and inclusive governance frameworks. Ensuring future resilience of these ecosystems will require not only ecological restoration and policy innovation, but also long-term investment in capacity-building and environmental education at the community level.

## 11.10 Conclusion

Understanding the impacts of climate change and urbanization on biodiversity has been a central focus of research in the field of biodiversity conservation. This review summarized the effects of these two major drivers—climate change and urbanization—on waterbirds and bird species dependent on wetland ecosystems. The current study revealed that climate change is significantly altering waterbird habitats by degrading wetland, causing habitat shifts of waterbird species and also disrupting migratory and breeding patterns of waterbird species. The findings of the present study demonstrate that climate change is a critical and escalating threat to the ecological stability of waterbird habitats, waterbirds and wetland-dependent birds in the IHR, increasing temperatures, alteration in precipitation patterns, habitat degradation, and shifting migration timings collectively impact water bird populations, particularly migratory species and those with restricted ranges. Wetland ecosystems, which serve as crucial habitats for many waterbird species, are increasingly vulnerable due to glacial retreat, siltation, invasive species expansion, and reduced water availability. These environmental changes disrupt breeding cycles, food availability, and ecological balance, leading to potential declines in bird populations. Furthermore, rapid urbanization in IHR has led to habitat fragmentation, pollution and human disturbance thereby reducing the viability of habitats for nesting and feeding. The impact of urbanization underscores the urgent need for sustainable urban planning and wetland protection policies to mitigate habitat loss and degradation.

Given the vital role of waterbirds as bio-indicators of wetland ecosystem health, further research is needed to have a better understanding on climate-induced shifts in waterbird specues populations and to develop conservation strategies tailored to mitigate these impacts. Based on the findings, integrated conservation strategies are proposed to resolve the dual challenges of climate change and urbanization to waterbird species by habitat restoration, community-based conservation also monitoring and adaptive management. These strategies emphasize a holistic approach that combines ecological, social, and governance aspects to ensure the sustainability of waterbirds in the IHR. While both drivers independently and synergistically exacerbate habitat loss, integrated strategies that incorporate habitat restoration, community involvement, and adaptive management are essential to mitigate these threats. Policymakers and researchers must address existing gaps in action to ensure the protection of waterbird/wetland dependent bird species of IHR for maintaining ecosystem health and ensuring the resilience of these fragile habitats in the face of rapid environmental change. Strengthening conservation efforts, restoring degraded wetlands, and implementing climate-resilient policies will be essential to safeguarding the diversity these species in the face of ongoing climate change and rapid pace of urbanization.

## References

1. Abreu TL, Berg SB, de Faria IP, Gomes LP, Marinho-Filho JS, Colli GR (2020) River dams and the stability of bird communities: a hierarchical Bayesian analysis in a tropical hydroelectric power plant. J Appl Ecol 57(6):1124–1136
2. Achal G, Harinder Y, Vijai K (2015) Impacts of climate change on hydrology and biodiversity of Himalayan Region in India. Glo J Res Anal 4(5):209–210
3. Acharya BK, Chettri B (2012) Effect of climate change on birds, herpetofauna and butterflies in Sikkim Himalaya: a preliminary investigation. In: Climate change in Sikkim: patterns, impacts and initiatives. Information and Public Relations Department, Gangtok, pp 141–160
4. Acharya BK, Thakuri B, Dewan S, Jins VJ, Chettri N (2024) Birds of the Himalaya Biodiversity Hotspot: diversity, distribution, and conservation. In: Biodiversity hotspot of the Himalaya. Apple Academic Press, pp 293–316
5. Adler C, Wester P, Bhatt I, Huggel C, Insarov GE, Morecroft MD, Muccione V, Prakash A (2022) Crosschapter paper 5: mountains. In: Climate change 2022: impacts, adaptation and vulnerability. In: Pörtner H-O, Roberts DC, Tignor M, Poloczanska ES, Mintenbeck K, Alegría A, Craig M, Langsdorf S, Löschke S, Möller V, Okem A, Rama B (eds) Contribution of working group II to the sixth assessment report of the intergovernmental panel on climate change. Cambridge University Press, Cambridge, pp 2273–2318
6. Ali S (1962) The birds of Sikkim. Oxford University Press, Delhi
7. Ali A, Khan MSH, Altaf M (2018) Winter survey of birds at district of the Badin, Pakistan. JWildl Ecol 2:11–22
8. Arya SK, Gopi GV (2021) Influence of season and habitat of birds in the mid-elevation village ecosystem of Kailash Sacred Landscape-India. Environ Challeng 5:100317
9. Arya AK, Bhatt D, Singh A, Saini V, Verma P, Rathi R, Bhatnagar P (2019) Diversity and status of migratory and resident wetland birds in Haridwar, Uttarakhand, India. J Appl Nat Sci 11(3):732–737
10. Arya AK, Joshi KK, Bachheti A (2020) A review on distribution and importance of wetlands in the perspective of India. J Appl Nat Sci 12(4):710–720
11. Arya D, Chandra N, Kumar R, Upadhayay ML, Mishra AP (2025) Climate change impact on Himalayan biodiversity. Springer
12. Ashraf S, Riaz A, Muhammad N (2019) Assessments of avian diversity of uchhali lake. Pak J Wild Ecol 3:8–15
13. Asmare BD, Derebe Y, Tamer M (2023) Diversity and relative abundance of bird species in the two habitat types of Dokima forest Awi zone, Ethiopia. PLoS ONE 18(3): e0281595. https://doi.org/10.1371/journal.pone.0281595

14. Bachheti A, Arya AK, Tyagi S, Deepti. (2023) An assessment of Wetland and water-dependent Avifaunal diversity in selected Wetlands of Uttarakhand, Western Himalayas, India. Biol Bull 50(4):729–735
15. Balachandran S (2012) Avian diversity in coastal wetlands of India and their conservation needs. Uttar Pradesh State Biodiversity Board, pp 155–163
16. Banerjee D, Bharti D, Kumar S, Mitra A, Joshi R, Gupta D (2023) Faunal composition of Ramsar wetlands from India: An analysis. Rec Zool Surv India 123:01–15
17. Baral HS, Bhandari BB (2011) Importance of high altitude wetlands for protection of avian diversity in the Hindu Kush Himalayas. Initiation 4:96–102
18. Barman D, Mandal SC, Bhattacharjee P, Datta SP (2013) Status of Rudrasagar Lake (Ramsar Site) in Tripura, India. Environ Ecol 31(3):1320–1325
19. Barton MG, Handerson I, Border JA, Siriwardena G (2023) A review of the impacts of air pollution on terrestrial birds. Sci Total Environ 873:162136
20. Bassi N, Kumar MD, Sharma A, Pardha-Saradhi P (2014) Status of wetlands in India: a review of extent, ecosystem benefits, threats and management strategies. J Hydrol Region Stud 2:1–19
21. Bhat H, Mahapatra DM, Boominathan M, Rao S (2010) Avian diversity of Ladakh wetlands H. in Lake 2010: Wetlands, Biodiversity and Climate Change
22. Bhatt D, Sethi VK, Sharma S, Kumar A, Saini V, Singh A (2015) Waterbirds of selected wetlands of Uttarakhand. In: Waterbirds of India, ENVIS WII, vol 16, pp 140–159
23. Bhattacharjee A, Bargali HS (2012) Status and trend of waterbirds in Wetlands around Corbett. The Corbett Foundation, India. Technical report submitted to the Uttarakhand State Forest Department, India
24. Bhattacharya S (2019) Environmental crisis in the Eastern Himalayan landscapes in India. Consilience 21:66–85
25. Bhowmick SR (2021) Biodiversity assessment of bird species as bioindicators and the impact of air pollution on the ecological communities. Int J Pure Appl Zool 9(2):18–25
26. Bishop MA, Tsamchu D, Li FX (2012) Number and distribution of Black-necked Cranes wintering in Shigatse prefecture, Tibet. Chinese Birds 3:191–198
27. Biswas T, Subramanian S, Niyogi R, Mylliemngap W, Mitra S, Das DK, Sarkar MS (2025) Conservation challenges and adaptation strategies for Indian Himalayan biodiversity in a changing climate. In: Climate change impact on Himalayan biodiversity. Springer Nature Switzerland, Cham, pp 179–224
28. Borah P, Kshiar N, Reang D, Nath AJ, Baruah KK (2024) Incidence of microplastic contamination in fishes of the Ramsar Wetland, Loktak–The world's only floating lake from the Indian Himalayan region. J Environ Manag 358:120928
29. Buckton S (2007) Managing wetlands for sustainable livelihoods at Koshi Tappu. Danphe 16: 12–13
30. Buckton ST, Ormerod SJ (2002) Global patterns of diversity among the specialist birds of riverine landscapes. Freshw Biol 47:709–695. https://doi.org/10.1046/j.1365-2427.2002.00891.x
31. Buckton ST, Ormerod SJ (2008) Niche segregation of Himalayan river birds. J Field Ornithol 79(2):176–185
32. Chakraborty A, Sah S, Sachdeva K, Joshi PK (2018) Vulnerability of forests in the Himalayan region to climate change impacts and anthropogenic disturbances: a systematic review. Reg Environ Chang 18:1783–1799. https://doi.org/10.1007/s10113-018-1309-7
33. Chatterjee A, Gopi GV, Sreenivasan R (2017) Assessing the impact of climate change on the avifauna of the Indian Himalayan Region. J Ornithol Environ Sci 12(3):45–54
34. Chettri A, Sharma K, Dewan D, Acharya BK (2018) Bird diversity of tea plantations in Darjeeling Hills, Eastern Himalaya, India. Biodiversitas 19(3):1066–1073
35. Chiu MC, Kuo MH, Hong SY, Sun YH (2013) Impact of extreme flooding on the annual survival of a riparian predator the Brown Dipper *Cinclus pallasii*. Ibis 155(2):377–383
36. Cole V, Sinclair AJ (2002) Measuring the ecological footprint of a Himalayan tourist center. Mt Res Dev 22(2):132–141

37. Crick HQ (2004) The impact of climate change on birds. Ibis 146:48–56
38. Crick HQ, Sparks TH (1999) Climate change related to egg-laying trends. Nature 399(6735): 423–423
39. Cumming GS, Paxton M, King J, Beuster H (2012) Foraging guild membership explains variation in waterbird responses to the hydrological regime of an arid-region flood-pulse river in Namibia. Freshw Biol 57(6):1202–1213
40. Dar SA, Bhat SU, Rashid I (2021) The status of current knowledge, distribution, and conservation challenges of wetland ecosystems in Kashmir Himalaya, India. In: Wetlands conservation: current challenges and future strategies, pp 175–200
41. Davidson N (2018) Wetland losses and the status of wetland-dependent species. In: The wetland book
42. Denniston D (1993) Saving the Himalayas. Worldwatch. 1 November 1993
43. Dhakate PM, Patil TA, Bhartari R (2008) Wetland birds of Corbett Tiger Reserve Landscape. In: Sengupta M, Dalwani R (eds) Proceedings of Taal 2007: The 12th World Lake Conference, Jaipur, India, 29th October-2nd November 2007
44. Dimri AP, Allen S, Huggel C, Mal S, Ballesteros-Cánovas JA, Rohrer M et al (2021) Climate change, cryosphere and impacts in the Indian Himalayan Region. Curr Sci 120(5):774–790
45. Dorji U (2022) Waterbird assemblage along Punatsangchhu River, Punakha and Wangdue Phodrang, Bhutan. J Threat Taxa 14(6):21179–21189. https://doi.org/10.11609/jott.7681.14.6.21179-21189
46. Down to Earth (DTE) (2020) Protected areas can help water birds adapt to climate change: study Down to Earth 22 Oct. Center for Science and Environment, Faridabad, New Delhi
47. Down to Earth DTE (2023) Climate change severely affecting biodiversity in Hindu Kush Himalayas: ICIMOD. https://www.downtoearth.org.in/news/climate-change/climate-change severely-affecting-biodiversity-in-hindu-kush-himalayas-icimod-90155. Accessed 12 April 2024
48. Du MY, Kawashima S, Yonemura S, Zhang XZ, Chen SB (2004) Mutual influence between human activities and climate change in the Tibetan Plateau during recent years. Glob Planet Chang 41:241–249
49. Dugan P (ed) (2005) Guide to Wetlands. Philip's Octopus Publishing, Londong
50. Dunn PO, Møller AP (eds) (2019) Effects of climate change on birds. Oxford University Press, Oxford
51. Dunn PO, Winkler DW (1999) Climate change has affected the breeding date of tree swallows throughout North America. Proc R Soc Lond Ser B Biol Sci 266(1437):2487–2490
52. Dynesius M, Nilsson C (1994) Fragmentation and flow regulation of river systems in the northern third of the world. Science 266(5186):753–762
53. Erwin KL (2009) Wetlands and global climate change: the role of wetland restoration in a changing world. Wetl Ecol Manag 17(1):71–84
54. Gaget E, Johnston A, Pavón-Jordán D, Lehikoinen AS, Sandercock BK, Soultan A, Brommer JE (2022) Protected area characteristics that help waterbirds respond to climate warming. Conserv Biol 36(4):e13877
55. Galewski T, Devictor V (2016) When common birds became rare: historical records shed light onlong-term responses of bird communities to global change in the largest wetland of France. PLoSONE 11:e0165542
56. Ganguli-Lachungpa U (1990) Brahminy Duck Tadorna ferruginea (Pallas) breeding in Sikkim. J Bombay Nat Hist Soc 87:290
57. Ganguli-Lachungpa U (1998) Attempted breeding of the Blacknecked Crane Grus nigricollis przevalski in north Sikkim. J Bombay Nat History Soc 95(2):341
58. Ganguli-Lachungpa U, Islam MZ, Rahamani AR (2007) Important bird areas of Sikkim: priority sites for conservation. Department of Forest Environment and Wildlife Management, Gangtok

59. Gardner R, Finlayson CM (2018) Global Wetland Outlook - State of the world's wetlands and their service to people 2018. Ramsar Convention Secretariat. Gopi, G. V, S, Gland, Switzerland
60. Garg J (2015) Wetland assessment, monitoring and management in India using geospatial techniques. J Environ Manag 148:112–123
61. Gaston AJ (1975) Methods for estimating bird populations. J Bombay Nat Hist Soc 72:271–283
62. Geneletti D, Dawa D (2009) Environmental impact assessment of mountain tourism in developing regions: A study in Ladakh, Indian Himalaya. Environ Impact Assess Rev 29(4): 229–242
63. Ghosh K, Chakraborty P (2021) Assessing the benefits, threats and conservation of reservoir-based wetlands in the Eastern Himalayan River Basin. In: Wetlands conservation: current challenges and future strategies, pp 140–161
64. Gibbs JP (1993) Importance of small wetlands for the persistence of local populations of wetland-associated animals. Wetlands 13:25–31
65. Gopal B, Chatterjee A, Gautam P (2008) Sacred waters of the India Himalaya. WWF-India, New Delhi
66. Greb SF, DiMichele WA, Gastaldo RA (2006) Evolution and importance of wetlands in earthhistory. https://doi.org/10.1130/2006.2399(01)
67. Green AJ, Elmberg J (2014) Ecosystem services provided by waterbirds. Biol Rev 89(1): 105–122. https://doi.org/10.1111/brv.12045
68. Grumbine RE, Pandit MK (2013) Threats from India's Himalaya dams. Science 339(6115): 36–37. https://doi.org/10.1126/science.1227211
69. Gupta A, Chattaraj R (2024) The challenges faced in maintaining the Sevoke– Rangpo Section of NH– 10. In: Proceedings of the 228th Mid– Term Council Meeting of IRC, Siliguri, West Bengal, India, pp 14–15
70. Gupta SK, Shukla DP (2016) Assessment of land use/land cover dynamics of Tso Moriri Lake, a Ramsar site in India. Environ Monit Assess 188(12):700
71. Hardy AR, Stanley PI, Greeing SPW (1987) Birds as indicator of the intensity of use of agricultural pesticide in UK. In: Diamond AW, Falion FN (eds) The value of birds, vol 6. Tech Publ, pp 119–121
72. Hattori A, Mae S (2001) Habitat use and diversity of waterbirds in a coastal lagoon around Lake Biwa, Japan. Ecol Res 16(3):543–553
73. Humbert-Droz B (2011) Trends in waterfowl populations and development of toruism in high –altitude wetlands of Ladakh, north-western India. Casarca 114:184–203
74. Hussaina SA, Mahara N, Tuboia C, Badolaa R (2018) Wetlands of the Indian Himalayas: status and conservation initiatives. In: Wetlands in the Himalaya: securing services for livelihoods, 33. ii: Distribution, description, and conservation. Springer, pp 369–381. https://doi.org/10.1007/978-94-007-4001-3_197
75. IPCC (2007) Climate Change (2007): the physical sciences basis. In: Summary for policy makers. IPCC: 21. IPCC, Geneva
76. Iqbal A, Shang Z (2019) Wetlands as a carbon sink: Insight into the Himalayan Region. In: Carbon management for promoting local livelihood in the Hindu Kush Himalayan (HKH) Region. Springer International Publishing, Cham, pp 125–144
77. IUCN (2024) The IUCN Red list of threatened species. Version 2024-1. https://www.iucnredlist.org. Accessed on 15th June 2024
78. Ives JD (2004) Himalayan perceptions: environmental change and the well-being of mountain peoples. Himal J Sci 2(3):17–19
79. Jamal S, Ahmad WS, Ajmal U, Aaquib M, Ashif Ali M, Babor Ali M, Ahmed S (2022) An integrated approach for determining the anthropogenic stress responsible for degradation of a Ramsar Site–Wular Lake in Kashmir, India. Mar Geod 45(4):407–434
80. James W, Pearce-Higgins, Roger M (2022) 4. Declines in invertebrates and birds – could they be linked by climate change? Bird Study. https://doi.org/10.1080/00063657.2022.2157373

81. Jamwal PS, Shrotriya S, Takpa J (2020) The pattern of waterbird diversity of the trans-Himalayan wetlands in Changthang Wildlife Sanctuary, Ladakh, India. J Threat Taxa 12(1): 15129–15139
82. Jessop CJR, Melville DS (2017) Rapid population decline in migratory shorebirds relying on yellow sea tidal mudflats as stopover sites. Nat Commun 8:14895. https://doi.org/10.1038/ncomms14895
83. Jianchu X, Shrestha A, Vaidya R, Eriksson M, Hewitt K (2007) The Melting Himalayas. Regional challenges and local impacts of climate change on mountain ecosystems and livelihoods. International Center for Integration and Mountain Development, Kathmandu
84. Jiaxuan L (2023) 1. Response of the biological components to climate change in the Wetland ecosystem of the Qinghai-Tibet Plateau. Highlights Sci Eng Technol. https://doi.org/10.54097/hset.v69i.11904
85. Joshi A, Solanki V (2019) Degradation and conservation of Indian Ramsar wetlands: a study of Loktak Lake and Deepor Beel. Our Heritage 67(5):71–90
86. Joshi S, Thapa G, Singh D (2018) Conserving high-altitude wetlands in the face of urbanization and climate change: A case study from the Indian Himalayas. Mt Res Dev 38(1):50–59
87. Joshi K, Kumar D, Arya AK, Bachheti A (2021) An assessment of Water bird species and associated water bird composition in the Haiderpur Wetland of Hastinapur Wildlife sanctuary Uttar Pradesh, India. Asian J Conserv Biol 10(1):141–145
88. Kanwal KS, Samal PK, Lodhi MS, Kuniyal J (2013) Climate change and high-altitude wetlands of Arunachal Pradesh. Curr Sci 105(8):1037–1038
89. Kaushik TK, Gupta RC (2013) Understanding and analyzing the coordinates of diversity of wetland birds of Asan Barrage near Paonta Sahib, northern India. Our Nat 11:192–200
90. Kingsford RT, Jenkins K, Porter J (2004) Imposed hydrological stability on lakes in arid Australia andeffects on waterbirds. Ecology 85:2478–2492
91. Kirby JS, Stattersfield AJ, Butchart SHM, Evans MI, Grimmett RFA et al (2008) Key conservation issues for migratory land- and waterbird species on the world' s major flyways. Bird Conserv Int 18:S49S73. https://doi.org/10.1017/S0959270908000439
92. Kukreti M, Arya AK (2022) A survey of waterbirds of Khoh and Kohlu rivers of Lansdowne forest division (western Himalayas), Uttarakhand, India. Asian J Conserv Biol 11(2):266
93. Kulkarni A, Patwardhan S, Kumar KK et al (2013) Projected climate change in the Hindu Kush-Himalayan region by using the high-resolution regional climate model PRECIS. Mt Res Dev 33:142–151
94. Kumar R (2019) Wetlands and waterbirds in Central Asian Flyway: an overview of status, management and conservation priorities for India. Special Issue Environ J Govern 18:111
95. Kumar A, Alam I (2023) Migration of birds and their flyways in India. Rec Zool Surv India 123:25–35
96. Kumar R, Gupta N (2019) Climate change and its impact on waterbirds in India: challenges and mitigation strategies. Biodivers J 5(2):134–146
97. Kumar R, Singh D (2022) Insecticide-tolerant bacterial population in *Eisenia fetida*'s Gut and Vermicast exposed to Chlorantraniliprole and Fipronil. Appl Biol Res 24(3):273–279
98. Kumar A, Sati JP, Tak PC et al (2005) Handbook on Indian Wetland Birds and their conservation: i-xxvi; 1–468. Published by the Director, Zoological Survey of India
99. Kumar et al (2005) list 128 species of migratory waterbirds in India. This forms 71 per cent of total migratory waterbird species reported in CAF by CMS, i.e. 182 species (CMS 2005)
100. Kumar A, Sati JP, Tak PC, Alfred JRB (2005) Handbook on Indian Wetland Birds and their conservation. Zoological Survey of India, Kolkata
101. Kumar A, Sati JP, Tak PC, Alfred JRB (2005) Handbook on Indian waterbirds and their conservation, vol 472. Zoological Survey of India, Kolkata
102. Kumar R, Gupta R, Dahiya T, Yadav R, Saifi R, Yodha K, Kumari B, Singh P, Shefali. (2022) Surveillance of *Heteropneustes fossilis* (Stinging Catfish) diseases and health management practices in Haryana. Appl Ecol Environ Sci 10(4):261–267

103. Kumar R, Gupta RK, Yadav R, Saifi R, Yodha K, Pooja (2022) *Eisenia fetida* as protein source for growth enhancement of *Heteropneustes fossilis*. Egypt J Aquat Biol Fish 26(2):577–588
104. Kumar R, Yadav R, Kumar Gupta R, Pooja (2022) A sustainable way for fish health management by replacement of chemical and drugs by earthworm. In: Environmental degradation in Asia: land degradation, environmental contamination, and human activities. Springer International Publishing, Cham, pp 329–352
105. Kumar S, Behera D, Ajay K, Karthick B, Dharia C, Anoop A (2024) Microplastics and heavy metal contamination along a land-use gradient in a Himalayan foothill river: Prevalence and controlling factors. J Contam Hydrol 266:104411
106. Kumar R, Singh D, Yadav R (2025) Toxic effects of chlorantraniliprole and fipronil on *Eisenia fetida*: from growth and reproductive parameters to avoidance behavior. Environ Eng Manag J 24(7):1499–1508. Accepted manuscript
107. Lacoul P, Freedman B (2005) Relationships between aquatic plants and environmental factors along a steep Himalayan altitudinal gradient. Aquat Bot 84(1):3–16
108. Larsen S, Sorace A, Mancini L (2010) Riparian bird communities as indicators of human impacts along Mediterranean streams. Environ Manag 45:261–273
109. Lei X, Le Y, Cai L, Qing Z, Shengling Z, Yongbing Y, Shansi L, Zhaxijie, Li ZWD (2009) Status of waterbirds in Asia: results of the Asian waterbird census: 1987–2007. Wetlands International
110. Li, Guangchun L (2023) Impacts of environmental factors on over-wintering Aquatic Bird Communities in Yamzho Yumco Lake, China. Sustainability. https://doi.org/10.3390/su16010254
111. Lorenzón RE, Beltzer AH, Olguin PF, León EJ, Sovrano LV, Antoniazzi CE, Ronchi-Virgolini AL (2019) Temporal variation of bird assemblages in dynamic fluvial wetlands: seasonality and influence of water level and habitat availability. Revista de Biología Trop 67: 1131–1145
112. Ma Z, Wang Y, Gan X, Li B, Cai Y, Chen J (2009) Waterbird population changes in the wetlands at Chongming Dongtan in the Yangtze River estuary, China. Environ Manag 43: 1187–1200
113. Ma Z et al (2014) Rethinking China's new great wall. Science 346:912–914. https://doi.org/10.1126/science.1257258
114. Mahar N, Habib B, Hussain SA, Shawl T, Takpa J (2022) Imperiled prancing crane: population status and breeding performance of black-necked crane Grus nigricollis in trans-Himalayan Ladakh region. In: Proceedings of the Zoological Society, vol 75, No. 2. Springer India, New Delhi, pp 181–189
115. Mahar N, Habib B, Hussain SA, Shawl T, Takpa J (2023) Influence of anthropogenic factors on the waterbirds in Trans-Himalayan wetlands. Global Ecol Conserv 46:e02567
116. Maheswaran G, Alam I (2017) Waterbirds of Arunachal Pradesh with special reference to high altitude rivers and wetlands. Rec Zool Surv India 117:315–325
117. Maikhuri RK, Rawat LS, Negi VS, Phondani P, Bahuguna A, Chamoli KP, Farooquee N (2009) Impact of climate change and coping strategies in Nanda Devi Biosphere Reserve Central Himalayas, India. In: Sharma E (ed) Proceedings of the international mountain biodiversity conference Kathmandu, 16–18 November 2008. International Centre for Integrated Mountain Development, Kathmandu, Nepal, pp 138–148
118. Manel S, Dias JM, Ormerod SJ (1999) Comparing discriminant analysis, neural networks and logistic regression for predicting species distributions: a case study with a Himalayan river bird. Ecol Model 120:337–347. https://doi.org/10.1016/S0304-3800(99)00113-1. [DOI] [Google Scholar][Ref list]
119. Manel S, Buckton ST, Ormerod SJ (2000) Testing large-scale hypotheses using surveys: the effects of land use on the habitats, invertebrates and birds of Himalayan rivers. J Appl Ecol 37(5):756–770
120. Mason CF (1995) Long-term trends in the arrival dates of spring migrants. Bird Study 42(3): 182–189

121. Mazumdar S, Saha GK (2012) Birds in a warming world: a review. In: Saha GK (ed) Climate change: man and environmnet. Daya Publishing House, New Delhi, pp 263–276
122. Mazumdar K, Maheswari A, Dutta PK, Borah PJ (2011) High altitude wetlands of western Arunachal Pradesh: new breeding ground for Ruddy Shelduck (Tadorna ferruginea). ZOO'S PRINT 26(8):9–10
123. McAllister DE, Craig JF, Davidson N, Delany S, Seddon M (2001) Biodiversity impacts of large dams. Background paper 1
124. Mckinney ML (2002) Urbanization, biodiversity, conservation. Bioscience 52:883–890
125. Mehta M (2024) Existing and potential changes in Himalayan Glaciers. In: Climate change perspective. In natural hazards and risk mitigation: natural hazards in Himalaya and Risk Mitigation. Springer Nature Singapore, Singapore, pp 149–172
126. Menzies RK, Rao M, Naniwadekar R (2021) Assessing the status of the critically endangered White-bellied Heron Ardea insignis in north-east India. Bird Conserv Int 31(2):255–267
127. Mercer DC (1990) Recreating and wetlands:impacts, conflict and policy issues. In: Williams M (ed) Wetlands: a threatened landscape. Blackwell, Oxford, UK/Cambridge, MA, pp 267–295
128. Michael Intal Magcamit (2022) 3. Identification of faunal indicators of climate in a Himalayan transect to assess climate change. https://doi.org/10.21203/rs.3.rs-2157865/v1
129. Millennium Ecosystem Assessment (MEA) (2005) Ecosystems and human well-being: Wetlands and water synthesis. World Resources Institute, Washington, DC
130. Mishra AP, Chandra N, Srivastava A, Singh SK (2023) An updated checklist of the vascular flora of Asan Conservation Reserve, Uttarakhand, India. Indian J For 46(1):12–21
131. Møller AP, Rubolini D, Lehikoinen E (2008) Populations of migratory bird species that did not show a phenological response to climate change are declining. PNAS 105(42):16195–16200
132. Mundkur T (2017) Conserving birds and their habitats along the Central Asian Flyway. Hornbill, (April–June): 5–11
133. Murthy TVR, Patel JG, Panigrahy S, Parihar JS (eds) (2013) National Wetland Atlas: Wetlands of International Importance under Ramsar Convention, SAC/EPSA/ABHG/NWIA/ATLAS/38/2013. Space Applications Centre (ISRO), Ahmedabad, India, p 230
134. Myers N, Mittermeier RA, Mittermeier CG, da Fonseca GAB, Kent J (2000) Biodiversity hotspots for conservation priorities. Nature 403:853–858
135. Nandy SN, Dhyani PP, Samal PK (2006) Resource information database of the Indian Himalaya. Environmental Information System on Himalayan Ecology, GP Institute of Himalayan Environment and Development, Almora, pp 27–30
136. Narang ML (1990) Migratory waterbirds of Dhalipur Lake near Dehradun (U.P.). Indian J For 131:63–64
137. Neupane J, Khanal L, Gyawli B, Chalise MK (2020) Elevation pattern and seasonality of avian diversity in Kaligandaki river basin, Central Himalaya. J Threat Taxa 12(14):16927–16943
138. Palmer GC, Bennett AF (2006) Riparian zones provide for distinct bird assemblages in forest mosaics of south-east Australia. Biol Conserv 130(3):447–457
139. Pandey R, Bhattacharya S, Mohanty M (2020) Conservation of waterbirds in the Indian Himalayan Region: Addressing climate and anthropogenic stressors. Ecohydrol Conserv 7(4):29–41
140. Pandit MK, Grumbine RE (2012) Potential effects of ongoing and proposed hydropower development on terrestrial biological diversity in the Indian Himalaya. Conserv Biol 26:1071–1061. https://doi.org/10.1111/j.1523-1739.2012.01918.x
141. Pandit MK, Sodhi NS, Koh LP, Bhaskar A, Brook BW (2007) Unreported yet massive deforestation driving loss of endemic biodiversity in Indian Himalaya. Biodivers Conserv 16:153–163
142. Pant RR, Singh UK, Sinha B (2017) Pollution and eutrophication of Wular Lake: an ecological and conservation perspective. Ecotoxicol Environ Saf 142:210–218. https://doi.org/10.1016/j.ecoenv.2017.04.030

143. Paonam J, Chatterjee S (2022) Threat perception of stakeholders: case study of Loktak Lake, a Ramsar Site under Montreux Record in North East India. Wetlands 42(8):114
144. Paracuellos M (2006) How can habitat selection affect the use of a wetland complex by waterbirds? Biodivers Conserv 15(14):4569–4582
145. Parimala B, Nuzhat FA (2022) An overview on the status of Wetland Avian biodiversity in India. Appl Aquat Terrest Eco-Biol 57:114
146. Parmesan C, Yohe G (2003) A globally coherent fingerprint of climate change impacts across natural systems. Nature 421(6918):37–42
147. Prasad SN, Ramachandra TV, Ahalya N, Sengupta T, Kumar A, Tiwari AK, Vijayan VS, Vijayan L (2002) Conservation of wetlands in India: a review. Trop Ecol 43(1):173–186
148. Prem K, Chhetri K, Gurung T, Namgyal L, Bijoy C (2021) High altitude wetland migratory birds in the Sikkim Himalaya: a future conservation perspective. J Threat Taxa 13(2): 17832–17836. https://doi.org/10.11609/JOTT.5610.13.2.17832-17836
149. Price T, Zee J, Jamdar K, Jamdar N (2003) Bird species diversity along the Himalaya: a comparison of Himachal Pradesh with Kashmir. J Bombay Nat Hist Soc 100(2&3):394â
150. Prudhomme C, Giuntoli I, Robinson EL, Clark DB, Arnell NW, Dankers R et al (2014) Hydrological droughts in the 21st century, hotspots and uncertainties from a global multimodel ensemble experiment. Proc Natl Acad Sci 111(9):3262–3267
151. Qiu J, Zhang Y, Ma J (2024) Wetland habitats supporting waterbird diversity: conservation perspective on biodiversity-ecosystem functioning relationship. J Environ Manag 357:120663
152. Rahmani AR (2025) Flights cut short. Down to Earth, pp 132–134
153. Rahmani AR, Islam MZ, Kasambe RM (2016) Important Bird and Biodiversity Areas in India, 2nd edn. Bombay Natural History Society, Indian Bird Conservation Network, Royal Society for Protection of Birds, and BirdLife International (UK), Mumbai
154. Rajpar MN, Zakaria M (2011) Effects of habitat characteristics on water bird distribution and richness in wetland ecosystem of Malaysia. J Wildlife Parks 18:107–122
155. Ramsar (2010) Water-related guidance: an Integrated Framework for the Convention's water-related guidance, 4th edn. Ramsar Convention Secretariat, Gland, Switzerland
156. Ramsar Sites Information Service (2022) Pala Wetland. https://rsis.ramsar.org/ris/2484
157. Rasool M, Johari GK (2021) The process of development and landscape change in South Asia: an overview of transformation of Himalayan environment. Ilkogretim Online 20(3): 3089–3310
158. Rawat S, Adhikari BS (2005) Invasive alien flora and its impact on the wetland ecosystems of Uttarakhand, India. Curr Sci 89(10):1772–1773
159. Roche EA, Gratto-Trevor CL, Goossen JP, White CL (2012) Flooding affects dispersal decisions in piping plovers (*Charadrius melodus*) in prairie Canada. Auk 129(2):296–306
160. Rodenhouse NL, Matthews SN, McFarland KP, Lambert JD, Iverson LR, Prasad A et al (2008) Potential effects of climate change on birds of the Northeast. Mitig Adapt Strateg Glob Chang 13:517–540
161. Romshoo SA, Rashid I (2014) Assessing the impacts of changing land cover and climate on Hokersar wetland in Indian Himalayas. Arab J Geosci 7:143–160
162. Rosenberg KV, Dokter AM, Blancher PJ, Sauer JR, Smith AC, Smith PA, Stanton JC, Panjabi A, Helft L, Parr M (2019) Decline of the North American avifauna. Science 366: 120–124
163. Rout PC (2024) Scoping sustainable tourism transformation in the Indian Himalayan Region. IGI Global, pp 17–30. https://doi.org/10.4018/979-8-3693-6260-0.ch002
164. Royan A (2015) The influence of river flow on the distribution and community organisation of river birds (Doctoral dissertation, University of Birmingham)
165. Runge CA, Watson JEM, Butchart SHM, Hanson JO, Possingham HP, Fuller RA (2015) Protected areas and global conservation of migratory birds. Science 350(6265):1255–1258
166. Sanjit L, Bhatt D, Sharma RK (2005) Habitat Heterogeneity of Loktak Lake. Curr Sci 88(7): 1027–1028

167. Sekercioglu CH (2012) Bird functional diversity and ecosystem services in tropical forests, agroforests and agricultural areas. J Ornithol 153(S1):153–161
168. Seri NA, Rahman AA (2021) Impact of climate change on migratory birds in Asia. Pertanika J Sci Technol 29(4):2937
169. Shakil A, Romshoo, Irfan R (2014) 5. Assessing the impacts of changing land cover and climate on Hokersar wetland in Indian Himalayas. Arab J Geosci. https://doi.org/10.1007/S12517-012-0761-9
170. Sharma P (1998) Sustainable tourism in the Hindu Kush-Himalayas:issues and approaches. In: East P, Luger K, Inmann K (eds) Sustainability in mountain tourism. Book Faith India and Insbruck-Vienna: Studienverlag, Delhi, pp 47–69
171. Sharma S, Bhatt D (2016) Migratory Pallas' s Gull *Larus ichthyaetus* (Pallas, 1773): a new record from Sikkim, the eastern Himalaya, India. J Threat Taxa 8(12):9449–9453
172. Sharma E, Chettri N, Tse-ring K, Shrestha AB, Jing F, Mool P, Eriksson M (2015) Climate change impacts and vulnerability in the Eastern Himalayas. Sustainability 7(1):654–679. https://doi.org/10.3390/su7010654
173. Sharma N, Sondhi S, Bhatt S (2021) Waterbird conservation in the Indian Himalayan Region: addressing the impacts of climate change and urban expansion. Asian J Ecol 9(3):99–112
174. Sharma P, Sarkar R, Deka JP, Koley S, Saha B (2024) Assessing water quality of Deepor Beel, Assam, NE India, using water quality index: a case of Ramsar wetland. Arab J Geosci 17(1):20
175. Sharma S, Arya AK, Bachheti A (2024) The impact of climate change in the Eastern Himalayan State of Sikkim, India: a review. In: Climate crisis and sustainable solutions: strategies for adaptation, mitigation and sustainable development, pp 265–283
176. Shintre S (2012) Ecological & environmetnal reporting in India, 1st edn. Sakal Papers Pvt. Ltd., Pune, India
177. Shrestha UB, Gautam S, Bawa KS (2012) Widespread climate change in the Himalayas and associated changes in local ecosystems. PLoS One 7(5):e36741
178. Sikder I, Bisht G, Varma K (2018) Climate change effects on migratory patterns of birds in the Indian Himalayan region. Ecol Sci Conserv 15(4):89–102
179. Sinha A, Hariharan H, Adhikari BS, Krishnamurthy R (2019) Bird diversity along riverine areas in the Bhagirathi Valley, Uttarakhand, India. Biodivers Data J 7:e3158
180. Sinha A, Chatterjee N, Ormerod SJ, Adhikari BS, Krishnamurthy R (2019) River birds as potential indicators of local-and catchment-scale influences on Himalayan River Ecosystems. Ecosystems and People 15(1):90–101
181. Sinha A, Chatterjee N, Ormerod SJ, Krishnamurthy R (2025) Population variability and apparent recent decline of river birds in the Indian Himalaya. Biotropica 57(2):e70017
182. Socolar JB, Robinson SK, Wilcove DS (2022) Climate change and migratory waterbirds in the Indian Himalayan region: a global review. Biol Conserv 273:109612. https://doi.org/10.1016/j.biocon.2022.109612
183. Song H, Zhang Y, Gao H, Guo Y, Li S (2014) Plateau wetlands, an indispensible habitat for the black-necked crane (Grus nigricollis)-a review. Wetlands 34:629–639
184. Srinivasan U, Tamma K, Ramakrishnan U (2014) Past climate and species ecology drive nested species richness patterns along an east-west axis in the H imalaya. Glob Ecol Biogeogr 23(1):52–60
185. Srivastava S, Singh TP, Singh H, Kushwaha SPS, Roy PS (2002) Assessment of large-scale deforestation in Sonitpur district of Assam. Curr Sci 82:1479–1484
186. Strasevicius D, Jonsonn M, Nyholm NEI, Malmqvist B (2013) Reduced breeding success of pied flycatchers Ficedula hypoleuca along regulated rivers. Ibis 155(2):348–356
187. Studds CE, Kendall BE, Murray NJ, Wilson HB, Rogers DI, Clemens RS et al (2017) Rapid population decline in migratory shorebirds relying on Yellow Sea tidal mudflats as stopover sites. Nat Commun 8(1):14895
188. Sullivan SMP, Watzin MC, Keeton WS (2007) A riverscape perspective on habitat associations among riverine bird assemblages in the Lake Champlain Basin, USA. Landsc Ecol 22: 1169–1186

189. Szabo JK, Mundkur T (2017) Conserving wetlands for migratory waterbirds in South Asia. In: Wetland science: perspectives from South Asia, pp 105–127
190. Tak PC, Sati JP (2003) AVES. in Fauna of Asan wetland, wetland ecosystem series. Zool Survey India 5:31–38
191. Tak PC, Sati JP, Kumar A (1998) Waterfowl potential of Asan Reservoir, Dehradun Valley, India. Zoologie 5:111–132
192. Tamang KL, Mohapatra S (2022) Teesta hydro-electricity projects and livelihood security in the Sikkim Himalaya: issues and challenges. IJRAH 2:54–48
193. Tamang R, Jins VJ, Dewan S, Chaudhry S, Rawat S, Acharya BK (2023) Ecological niche modelling of two water-dependant birds informs the conservation needs of riverine ecosystems outside protected area network in the Eastern Himalaya, India. PLoS One 18(11):e0294056
194. Tonetti V, Bocalini F, Schunck F, Vancine MH, Butti M, Ribeiro M, Pizo M (2024) The Protected Areas network may be insufficient to protect bird diversity in a fragmented tropical hotspot under different climate scenarios. Perspect Ecol Conserv 22(1):63–71
195. Trautmann S (2018) Climate change impacts on bird species. In: Tietze D (ed) Bird species. Fascinating life sciences. Springer, Cham, pp 217–234
196. Tscharntke T, Klein AM, Kruess A, Steffan-Dewenter I, Thies C (2005) Landscape perspectives on agricultural intensification and biodiversity–ecosystem service management. Ecol Lett 8(8):857–874
197. Ullah I, Xue-Ying S, Qing-Ming W, Wen-You D, Khan TU, Rajpar MN, Suliman M (2024) Evaluation of water birds population trends, threats, and conservation status in selected Wetlands of Pakistan
198. Vagholikar N, Das PJ (2010) Damming the Northeast. Kalpavriksh, Aaranyak and ActionaAid India. Kalpavriksh, Pune
199. Wang X, Kuang F, Tan K, Ma Z (2018) Population trends, threats, and conservation recommendations for waterbirds in China. Avian Res 9:1–13. https://doi.org/10.1186/s40657-018-
200. Wetlands International South Asia (2020) Asian Waterbird census: results for coordinated January counts for India 2006–2015. Wetlands International, New Delhi, India
201. Whelan CJ, Wenny DG, Marquis RJ (2008) Ecosystem services provided by birds. Ann N Y Acad Sci 1134(1):25–60. https://doi.org/10.1196/annals.1439.003
202. Whittaker RH, Likens GE (1973) Primary production: the biosphere and man. Hum Ecol 1: 357–369
203. Wormworth J, Mallon K (2006) Bird species and climate change. World Wide Fund for Nature Australia, Sydney, Australia. https://link.springer.com/chapter/10.1007/978-81-322-3715-0_6
204. Wormworth J, Sekercioglu CH (2011) Winged sentinels: birds and climate change. Cambridge University Press, Cambridge
205. Wu W, Xu H, Wu J, Cao M (2012) The impact of climate change on birds: a review. Biodivers Sci 20(1):108
206. Wu TY, Walther BA, Chen YH, Lin RS, Lee PF (2014) Reassessment of the conservation status and protected area coverage of Taiwanese birds: how distribution modelling can help species conservation. Bird Conserv Int 24:238–223. https://doi.org/10.1017/S0959270913000336
207. Xu W, Fan X, Ma J, Pimm SL, Kong L, Zeng Y, Li X, Xiao Y, Zheng H, Liu J (2019) Hidden loss of wetlands in China. Curr Biol 29:3065–3071. e3062. https://doi.org/10.1016/j.cub.2019.07.053
208. Yadav R, Gupta RK, Kumar R, Kaur T, Moun R (2022) Biochemical response of earthworm, *Eisenia fetida* to heavy metals toxicity. J Appl Nat Sci 14(3):990–998
209. Yadav R, Gupta RK, Kumar R, Kaur T (2022) Assessment of toxicity of lead and nickel on the biochemical and immunological parameters of earthworm, *Eudrilus eugeniae*. Agric Sci Dig 44(5):922–929
210. Yodha K, Singh D, Kour A, Yadav R, Kumar R (2024) Species diversity and abundance of birds in ber (*Zizyphus mauritiana*) crop. Asian J Conserv Biol 13(1). https://doi.org/10.53562/ajcb.83683

211. Zhang T, Wang W, An B (2025) A massive lateral moraine collapse triggered the 2023 South Lhonak Lake outburst flood, Sikkim Himalayas. Landslides 22(2):299–311
212. Zou Y, Pan A, Zhang BH, Zhang H, Yao P-Y, Liu Y, Gao X, Xie DL (2017) Impacts of microhabitat changes on wintering waterbird populations. Sci Rep 7:13934
213. Zurell D, Graham CH, Gallien L, Thuiller W, Zimmermann N (2018) Long distance migratory birds threatened by multiple independent risks from global change. Nat Clim Chang 8:992–996

# Chapter 12
# Persistent Organic Pollutants and Avian Toxicology: Mechanisms and Biomagnification

**Kapish Samota, Amit Kumar, Sharda Kalra, and Asha Poonia**

**Abstract** Persistent organic pollutants, because of their persistent nature and ability to accumulate in living organisms, pose considerable risks to avian populations across the globe after multiple way entry into ecosystems via home, agricultural, and industrial sources. The article explains absorption, distribution, metabolism and excretion of POPs in birds, and highlights the biochemical and molecular means by which POPs interfere with immunological, reproductive, neurological and endocrine systems. The main mechanisms underlying immunotoxic effects are the interference with signalling pathways and cytokine generation. Biomagnification, the mechanism by which POP concentrations escalate at every trophic level and primarily effect predatory bird species such as raptors and seabirds is also examined. Such biomagnification exacerbates the toxic effects, leading to severe health effects like organ damage, immunosuppression, infertility, population declines, altered reproductive success, behavioural changes, and mortality in birds. High levels of PCBs and dioxins have been found in the tissues of albatrosses, gulls, and other seabirds. These avian toxicology studies also serve as crucial bioindicators of environmental health, providing insights into the broader ecological impacts of POPs.

**Keywords** Avian toxicology · Biomagnification · DDT · PCB · Persistent organic pollutants

## 12.1 Introduction to Persistent Organic Pollutants (POPs)

In recent years, environmental science and public health have become increasingly concerned with the persistent behaviour of toxic compounds in the biosphere. In response to the increase in scientific and public concern over this new type of pollutant, policies to control them have also increased on a global scale. Among persistent chemicals, a special role is played by Persistent Organic Pollutants

K. Samota · A. Kumar · S. Kalra · A. Poonia (✉)
Department of Zoology, Faculty of Life Science, Chaudhary Bansi Lal University, Bhiwani, Haryana, India
e-mail: asha.poonia@cblu.ac.in



R. Kumar et al. (eds.), *Ecology, Conservation of Aquatic Birds*, Wetlands: Ecology, Conservation and Management 15, https://doi.org/10.1007/978-3-032-17125-2_12

(POPs), which, due to their resistance to degradation, can accumulate transboundary across the globe and have a substantial period of environmental residence [1]. In their initial development, Persistent Organic Pollutants (POPs) exhibit five significant characteristics, which encompass: (1) harmful to and a target for plants, animals, and humans; (2) highly conserved in human and animal fat body tissues; (3) resistance to physical, chemical, and organic weathering; (4) bioaccumulated, biomagnified, and bioconcentrated as they develop toward the larger and top trophic levels; and (5) ability to facilitate international environmental transport through their associated components [2–4]. POPs include various families of chemicals, such as polychlorinated biphenyls (PCBs), organochlorinated pesticides, and brominated flame retardants [5].

### 12.1.1 Sources of POPs

POPs are various chemicals that have become widespread in air, water, soil, and living organisms due to human activities. Main sources of POPs are-

1. **Industrial Processes**: POPs primarily originate from industrial activities, wherein chemicals are released into the air, water, and soil through manufacturing and production processes [6]. Waste disposal significantly impacts POP release from chemical, metal production, coal conversion, pesticide creation, oil and gas extraction, and textiles. These activities emit substances like phthalates, dioxins, insecticides, and solvents [3, 7].
2. **Agricultural Practices**: The exposure of pesticides and fertilizers in agriculture leads to POPs accumulation as these substances can leach into the soil and surrounding water systems [8]. Developing countries may see a sharper rise in POP concentrations due to pesticide inefficiency against heavy pest populations, leading to more frequent applications [9].
3. **Waste Disposal**: Waste disposal contributes to environmental pollution from persistent organic pollutants (POPs). Common methods include combustion, landfilling, recycling, composting, and gasification [10]. Improper waste dumping, especially hazardous waste, releases POPs into the environment [11]. The incineration of products containing organic chemicals results in the release of POPs, impacting both air quality and the surrounding environment [12].

### 12.1.2 Classification of POPs

The POPs classification based upon the environment contamination sources, on the basis of source and environmental release is well explained in the Fig. 12.1.

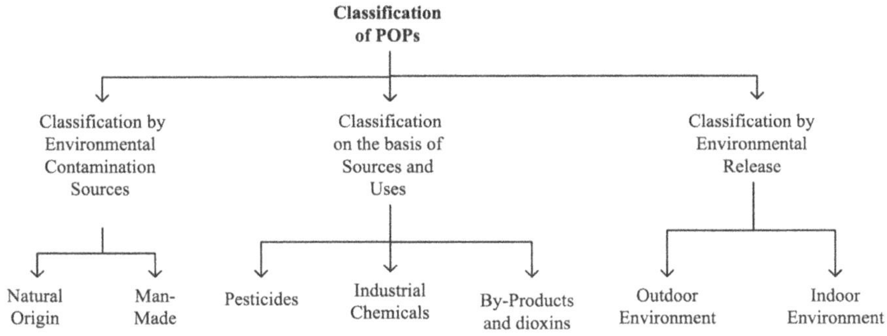


**Fig. 12.1** Classification of POPs

#### 12.1.2.1 Classification by Environmental Contamination Sources

**Natural Origin**- Substances linked to naturally occurring processes and phenomena. Example: Naturally occurring dioxins generated by wildfires [13].

**Man-Made**—Substances formed during the manufacturing and environmental transport of other POPs. Polychlorinated biphenyls (PCBs) created during industrial processes [13].

#### 12.1.2.2 Classification by Sources and Uses

**Pesticides**—Widely produced chemicals aimed at protecting crops and structures from pests. Example: Glyphosate used for weed control in agriculture [14].

**Industrial Chemicals**—Widely used industrial chemicals such as PCBs, chlordane, and dieldrin are identified as POPs [14]. These play crucial roles in various industrial applications, functioning as catalysts, hardeners, or colour additives [15].

**By-Products and Dioxins**—Waste which are unintentionally produced in various industrial processes. Dioxins, a significant type of by-product, form during combustion and chemical reactions, particularly in metal production and waste incineration [16].

#### 12.1.2.3 Classification by Environmental Release

**Outdoor Environment**—Accidental releases, resuspension from the earth's surface, and agricultural pesticide applications like a chemical spill during pesticide spraying [17].

**Indoor Environment**—Emissions from household products, activities, and building materials like formaldehyde released from pressed wood products [18].

Once released into the environment, POPs show a high persistence due to slow degradation and are transported throughout the world atmosphere. However, in some places on Earth, very high POP concentrations have been measured, indicating regional associated long-range atmospheric and water transports [2, 3]. Global distribution of pesticides is controlled by geographical and climatic factors; however, some areas in the world are subject to higher contamination levels than others [19]. Typically, these regions are contaminated due to large-scale agricultural activity or industrial production [20]. Countries in the southern hemisphere and typically Asian countries are heavily affected by electronic waste imports, which are openly incinerated in some instances, and hence increased levels of Polybrominated diphenyl ethers are commonly detected in air worldwide [21].

Numerous toxicological studies on organochlorine and organophosphate insecticides have been published, showing the potential effects of these toxicants on birds and mammals. In the case of organochlorine insecticides, the adverse impacts that these compounds induced on birds and fish were the major causes for the implementation of a worldwide ban on the agricultural use of these compounds [22]. The fact that biologists studied the effects of organochlorine compounds on birds was one of the main reasons for the implementation of the Stockholm Convention on Persistent Organic Pollutants, as high levels of several DDT-related compounds and other organochlorine compounds were such a hazard to humans that a worldwide ban was warranted [23].

## 12.2 Mechanisms of Toxicity in Avian Species

### *12.2.1 Absorption of POPs in Birds*

The absorption of POPs in birds compromise a complex process which is provoked by various chemical and biological factors. The absorption process is mainly controlled by how effectively these chemicals can move through cellular membranes via passive diffusion. Molecular weight and lipophilicity of POPs are the primary elements which influence the absorption of these pollutants [24]. Furthermore, the presence of specific transport proteins can either assist or hamper the absorption of these pollutants [25]. Birds frequently take POPs from their dietary intake, recurrently coming from the aquatic prey that is contaminated with high levels of POPs [26]. Birds can accumulate POPs through several different routes including consumption of contaminated food and water, inhalation of polluted air and dermal absorption from the contaminated environments in which they reside [27].

### 12.2.2 Distribution of POPs in Birds

POPs are stored in the lipid-rich tissues of organisms. In Birds also the POPs are accumulated into the fat [28]. POPs have high lipophilicity which influence their accumulation in birds' tissues [29]. After intestinal absorption POPs can enter the bloodstream and get distributed to various organs based on blood flow [30]. They accumulate primarily in the liver where they can get metabolized, and the unmetabolized waste mostly excreted out in feces or some exhaled out through air.

### 12.2.3 Metabolism of POPs in Birds

The metabolism of POPs in birds generally occurs through biotransformation in the liver, which is categorized into two phases [29].

**Phase I:** Functionalization- This primary phase involves modifying POPs by adding or removing functional groups, such as hydroxyl (-OH) or carboxyl (-COOH) groups which is regulated by the enzymes such as cytochrome P450 monooxygenases (CYP450s). Predominant reactions in this phase include hydroxylation, where a hydroxyl group is added; dechlorination, where chlorine atom is removed from certain POPs like polychlorinated biphenyls (PCBs); and epoxidation, where a cyclic ether is formed [31].

**Phase II:** Conjugation- In the next phase conjugation reactions occurs. In this hydrophilic molecules including glutathione and glucuronic acid are attached to the functionalized POPs produced in Phase I. The main purpose of these reactions is to reinforce the water solubility of the generated compounds, thereby assisting their excretion from the body. Main enzymes responsible for this phase includes glutathione-S-transferase and UDP-glucuronosyltransferase [30]. Despite these metabolic processes, many POPs exhibit significant resistance to complete degradation due to their inherent chemical stability. As a result, they may undergo only partial biotransformation and can accumulate in various tissues within avian species, posing a risk of toxicity and ecological impact [6, 32].

### 12.2.4 Excretion of POPs in Birds

A significant portion of conjugated POPs in birds is eliminated via bile, which transports these compounds into the digestive tract and facilitates their removal through feces [33]. Female birds may transfer POPs to their eggs, thereby increasing the risk to the developing embryos and subsequent hatchlings [34]. Certain POPs can also be excreted through preening oils or during feather molting, adding another pathway for elimination [35].

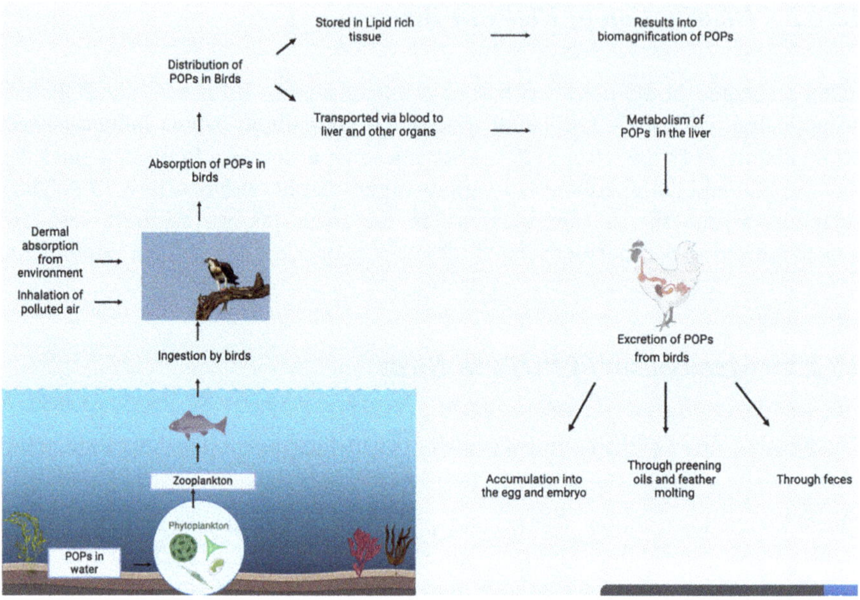


**Fig. 12.2** Mechanisms of toxicity in avian species showing ingestion, absorption, distribution and excretion of POPs

Physiological barriers in birds, such as the gastrointestinal tract, liver, blood-brain barrier (BBB), egg membranes, immune system, and fat storage, serve to protect against harmful substances [36]. However, POPs bypass these defenses due to their lipophilic and chemically stable nature [25]. They are efficiently absorbed in the gut, resist enzymatic detoxification in the liver, and diffuse through the Blood brain barrier and egg membranes, often mimicking natural hormones [37]. POPs accumulate in fat stores, evading immediate excretion, but are released during fat mobilization, increasing toxicity [38]. The tissue distribution of pollutants has shown that they mainly accumulate in lipid-rich tissues such as the liver, fat, and egg yolks [33]. Additionally, they suppress immune responses, leaving birds vulnerable to chronic exposure and adverse effects. In Fig. 12.2 the mechanism of toxicity in avian species showing ingestion, absorption, distribution and excretion of POPs is explained through the example of a food chain which starts from the aquatic phytoplankton and ends at excretion of POPs through birds.

## 12.3 Molecular Mechanisms of POP Toxicity

POPs demonstrate detrimental effects in avian species by means of endocrine disruption, immunotoxicity, and neurotoxicity, thereby influencing essential physiological systems [39]. Acting as endocrine disruptors, POPs can mimic or obstruct vital hormones such as estrogens and thyroid hormones, hinder hormone synthesis and metabolism, and interact with receptors including the aryl hydrocarbon receptor [40]. The effect of these interaction can be seen in reproductive dysfunction, growth abnormalities, and modifications in behaviour [41]. The POPs interfere with the B and T cell activities which results into the induction of oxidative stress, alternations in epigenetic regulations and increased vulnerability to infections and compromised vaccine efficacy [42]. POPs exhibit neurotoxic effect by disrupting neurotransmitter pathways, inhibiting calcium signalling and obstructing thyroid hormone-mediated neural development [43]. These interferences may lead to cognitive impairments, behavioural changes, and developmental delays [39]. By combined actions, these molecular mechanisms remarkably impact avian health, reproductive success and overall survival which leads to population decline and causing ecological imbalances.

## 12.4 Biomagnification of POPs in Avian Food Chains

In avian food chains POPs accumulate and increase in concentration at each trophic level causing the biomagnification of POPs [44]. POPs persist within the tissues of various organisms due to their lipophilic nature and resistance to degradation [45]. Primary producers, such as algae, absorb these POPs from their surroundings, facilitating their transfer to zooplankton and subsequently to fish, with concentrations intensifying as they ascend the food chain. Avian species, especially those that are predatory or piscivorous, ingest prey that harbours these contaminants, resulting in substantial accumulation within their own bodies [46]. Factors including lipid content, feeding behaviors, and limited detoxification capabilities further intensify this issue, rendering apex predators, such as raptors and fish-eating birds, particularly susceptible to elevated burdens of POPs [47]. Through Fig. 12.3 the process of biomagnification along with the sources of POPs is well explained.

## 12.5 Case Study: The Impact of DDT and PCBs on Avian Predators

DDT, which has been extensively utilized as a pesticide, along with its metabolites, critically affected avian raptor populations by interfering with calcium metabolism [48]. This disruption led to the formation of eggs with diminished shell thickness, thereby increasing the incidence of egg breakage [49]. Notable species such as the bald eagle and the peregrine falcon underwent significant declines in their populations until the prohibition of DDT allowed for their subsequent recovery [50]. In a similar vein, polychlorinated biphenyls (PCBs) exerted detrimental effects on avian species through mechanisms of neurotoxicity and disruption of endocrine

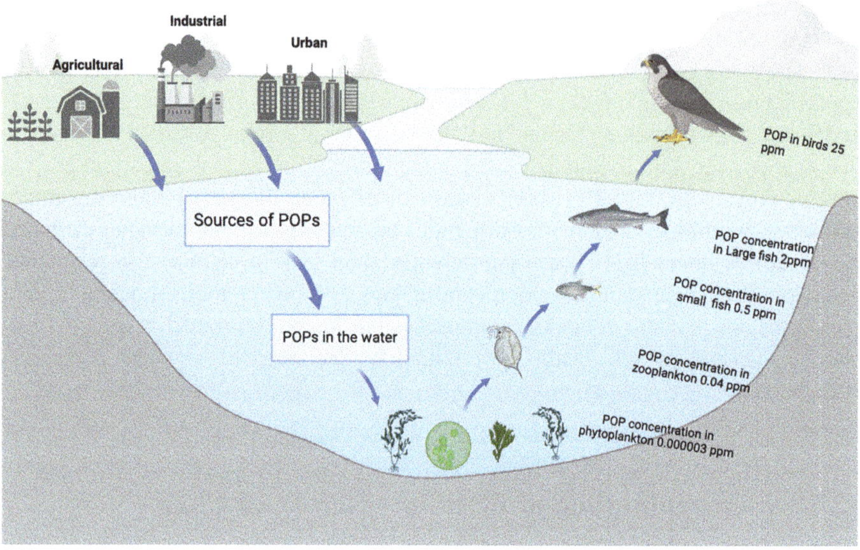


**Fig. 12.3** Process of biomagnification in birds along with the sources of POPs

functions, ultimately compromising reproduction, immune response, and behavioural patterns [51, 52]. Within ecosystems such as the Great Lakes, piscivorous birds, including herring gulls, displayed physical deformities, cognitive impairments, and failures in reproductive success attributable to exposure to PCBs [53].

The Table 12.1 clearly shows the effect studies on the avian species along with the location and the effect of chemical analysed.

## 12.6 Reproductive and Developmental Toxicity

Persistent Organic Pollutants (POPs) have a profound adverse effect on the reproductive health of avian species, fundamentally disrupting hormonal pathways and inflicting direct harm to reproductive organs [77]. Compounds such as DDT and its metabolite DDE function by either mimicking or obstructing the activity of natural hormones, which results in hormonal imbalances that interfere with oviposition behaviours and diminish fertility rates [78]. A particularly significant consequence of DDE exposure is the thinning of eggshells, attributable to its inhibition of calcium deposition, culminating in eggs that are more susceptible to breakage during the incubation period [79]. Exposure to POPs has an detrimental impact on male fertility characterized by diminished sperm quality which results into mortality of embryo and reduction in clutch size [80]. Together, these factors impair reproductive success and cause a critical risk to avian populations.

**Table 12.1** The major findings by the researchers on effect of PCB and DDT in Avian species

| Chemical analysed in the Study | Species | Stage at which evaluation was done | Effect of the chemical analysed | Location | References |
|---|---|---|---|---|---|
| PCB | Tree swallows (*Tachycineta bicolor*) | Adult, egg and chicks | Reproductive success impaired | Hudson River of New York | [51] |
| DDT | Robin | Adult | Population decreased | America | [54] |
| DDT | Peregrine | Adult | Population decreased | Southern England | [55] |
| DDT &PCB | Peregrine falcons (*F. Peregrinus*) | Egg | Egg shell thining | Colorado and northern New Mexico | [56] |
| DDT & PCB | Sharp- shinned hawks (*Accipiter striatus*), Cooper's hawks (A. *cooperii*), northern goshawks (*A. gentilis*) | Egg | Egg shell thickness | South- Central Ontario | [57] |
| PCB | Tree swallows (*Tachycineta bicolor*) | Adult | Abnormal Nest building behaviour due to endocrine distruption | Hudson River of New York | [58] |
| PCB | Tree swallows (*Tachycineta bicolor*) | Adult | Different plumage color and reproductive success | Hudson River of New York | [59] |
| DDT | Great white pelican (*Pelecanus onocrotalus*), Marabou stork (*Leptoptilos crumeniferus*), African sacred ibis (*Threskiornis aethiopicus*), hamerkop (*Scopus umbretta*) | Adult | 26.7% of birds contained high DDE levels above thresholds which leads to reproductive failure | Lake Ziway, Ethiopia | [60] |
| DDT | American robins (*Turdus migratorius*) | Egg | Egg shell thining | Okanagan valley, British Columbia, Canada | [61] |
| DDT | Peregrine falcon (*Falco peregrinus*) | Adult | Population decline due to compromised immune system | An analysis of the study involved all over the world | [48] |

(continued)

**Table 12.1** (continued)

| Chemical analysed in the Study | Species | Stage at which evaluation was done | Effect of the chemical analysed | Location | References |
|---|---|---|---|---|---|
| | | | and reduced reproductive success | | |
| PCB & DDT | Marabou stork (*Leptopilos crumeniferus*), African sacred ibis (*Threskiornis aethiopicus*), Hamerkop (*Scopus umbretta*) | Adult | Concentration of these POPs were seen elevated in these. Due to this elevation the 26.7% birds showed exceed-ing reproductive failure. | Lake Hawassa, Ehtiopion Rift valley lakes found in Sidama region, Whtiopia | [62] |
| PCB & DDT | Egret (*Ardea intermedia*), Red-footed booby (*Sula sula*), Pea-cock grouper (*Cephalopholis argus*), Goldspot seabream (*Gnathodentex aureolineatus*), Japan surgeon-fish (*Acanthurus japonicus*), Stri-ated surgeonfish (*Ctenochaetus straitus*) | Adult | Reproductive failure in the species. | Qilianyu Island | [63] |
| DDT & PCB | Black- tailed gull, Greenshank, common gull, black- headed gull | Adult | Accumulation of DDT and PCB in the resident spe-cies is more then the migratory species which cause harm to the birds | Nakdong River estuary in Korea | [64] |
| PCB | Great lake colo-nial waterbirds | Egg | The eggshell thining through the disruption of calcium metalbolism | North Ameri-can Great Lakes | [65] |
| PCB | Tree swallow (*T. bicolor*) | Egg & nestlings | Effect on the reproductive suc-cess of the birds | New Bedford harbor superfund site (NBH, | [66] |

(continued)

**Table 12.1** (continued)

| Chemical analysed in the Study | Species | Stage at which evaluation was done | Effect of the chemical analysed | Location | References |
|---|---|---|---|---|---|
| | | | | Massachisetts, USA) | |
| PCB | Common terns | Adult | Effect on the reproductive success and reduced hatching success. | Lake Michigan | [67] |
| PCB | Bermuda petrel (*Pterodroma cahow*) | Adult | Reduced egg hatching success. | Island Nature reserve, Bermuda | [68] |
| PCB & DDT | Great gorned owl (*Bubo virginianus*) | Adult & egg | Accumulation of the POPs in the liver,kidney, blood plasma and gall blader | Kalamazoo River superfund site, Michigan, USA | [69] |
| PCB | *Phalacrocorax carbo*, *Podiceps cristatus*, *Ardea cinerea*, *Egretta garzetta*, *Nycticorax*, *Alburnus alburns*, *Rutilus rutilus*, *Lepomis gibbosus*, *Carassius auratus*. | Egg | Bioaccumulation of PCBs in the seabirds lead to the detrimental effect and *Phalacrocorax carbo* can be considered as the biomonitor of target PCBs. | Lake kerkini. Northern Greece | [70] |
| PCB & DDT | Six species of wild aquatic birds | Egg | The some proportion of the population is affected due to these | North China | [71] |
| PCB & DDT | Seabirds species (*Calonectris shearwaters*) | Adult | The accumulation of PCB & DDT in these birds make them a suitable biomonitoring organism in marine environment | Atlantic Ocean & Mediterranean Sea | [72] |
| PCB & DDT | Seven aquatic bird species | Egg | Level of PCB was seen much higher in the eggs of the aquatic birds | Lake Paijanne | [73] |

(continued)

**Table 12.1** (continued)

| Chemical analysed in the Study | Species | Stage at which evaluation was done | Effect of the chemical analysed | Location | References |
|---|---|---|---|---|---|
| PCB & DDT | 2 species of auatic birds (Grey heron- *Ardea cinerea*, Great crested grebe- *Podiceps cristatus*), 5 species of terrestrial birds (Barn owl- *Tyto alba*, long-eared owl- *Asio otus*), common buzzard- *Buteo buteo*, kestrel- *Falco tinnunculus*, sparrowhawk- *Accipiter nisus*, | Adults- liver and muscle | Different concentration of POPS in the birds accumulated in the liver and muscles. | Belgium | [74] |
| PCB & DDT | Pied kingfisher (*Ceryle rudis*), Cattle egret (*Bubulcus ibis*), Grey heron (*Ardea cinera*), House sparrow (*Passer domesticus*) | Egg | Eggshell thinning due to the PCB and DDT | Thohoyandou, South Africa | [75] |
| PCB & DDT | Great crested Grebe (*Podiceps cristatus*), Mallard(*Anas platyrhynchos*) | Egg | The much higher level of POPs in the fish eating great egret. | Lake Maggiore, Italy | [76] |

### *12.6.1 Developmental Toxicity in Hatchlings*

In newly hatched birds the teratogenicity generally stems from the transfer of POPs from the mother through the egg [81]. These harmful chemicals hinder with the functioning of thyroid hormones which are essential for proper brain development [82]. Therefore, newly hatched birds often suffer from neurodevelopmental disorder and show abnormality in their motor skills. Along with this, they have a compromised immune system that makes them increasingly prone to infections and show stunted growth due to interruption in their metabolism [83]. They often struggles with foraging and avoiding predators which place their survival in the wild at risk. POPs exposure can lead to physical abnormalities such as deformities in their beaks and limbs, which further reduce the fitness of these affected birds [34].

## 12.7 Population-Level Effects of POP Exposure

The effect of POPs on avian populations is intense which leads to a significant drop in their reproductive success and reduced survival rates [84]. POPs exposure can result into reduction in the numbers of various bird species resulting into a serious threat to biodiversity and disturb ecosystem equilibrium [85]. With time these consequences have the potential to reshape community dynamics and initiate a chain reaction that affects entire ecosystems which will ultimately compromise the health and stability of natural environments.

### *12.7.1 Case Studies on Species Affected by POPs*

Seabirds encompassing species like albatrosses and gulls, are highly susceptible to POPs, generally due to their dependence on polluted marine habitats [86]. Their lengthy lifespans combined with a slow reproductive rate lead to significant accumulation of POPs through their diet, which consists mainly of fish and squid [87]. This build-up results in adverse repercussion such as reproductive failures, decreased survival rate and various developmental challenges [88]. The transfer of these pollutants from mother to offspring further intensifies these adverse effects. In the Great Lakes region, gulls like herring gulls exhibit alarming levels of PCB contamination, which has been linked to egg deformities, compromised development, and diminished immune functionality [89]. Other seabird species like brown pelicans, have experienced population declines due to eggshell thinning caused by DDT exposure [90]. Arctic seabirds such as northern fulmars also accumulate POPs in their bodies, which are brought to the polar regions, negatively affecting their reproductive success and the well-being [91]. The thorough repercussions of POPs on the reproduction, development and survival of seabirds underscore the intense threats that these pollutants pose to their populations.

## 12.8 Ecological and Evolutionary Implications

Information about the health of ecosystems and the dynamics of evolutionary processes can be uncovered by investigating the avian toxicology in the context of POPs. Birds act as bioindicators which provide a clear reflection of the overall environmental impacts that POPs exert on food webs and biodiversity [92–94]. Prolonged exposure to POPs can negatively affect reproductive success, compromise immune function, and alter behaviour, which may lead to declines in bird populations and disturb the equilibrium of ecosystems. As this detrimental pressure persists, it may gradually shape the evolution of resistance mechanisms or create selective pressures within avian species [95–98]. However, these adaptations usually take a lot of generations to really come together and often don't quite keep up with how quickly the environment is changing.

## 12.9 Future Directions and Mitigation Strategies

Future studies should aim to explore the prolonged impacts of POP exposure across molecular, population, and ecosystem scales. It's super important to expand monitoring to cover areas and species that haven't been looked at much. To tackle this issue, we need to step up efforts around global regulations like the Stockholm Convention, while also pushing for safer chemical alternatives and strategies to reduce POPs right from the source. When it comes to conservation, we should focus on protecting habitats, improving pollution control, and raising public awareness to help bird populations and keep our ecosystems healthy.

## 12.10 Conclusion

POPs cause serious menace to bird species, ecosystems and global biodiversity. POPs solubility into lipids, permanence and resistance to breakdown cause global contamination of the environment thus causing biomagnification across trophic levels. Birds is an important bioindicators illustrating the substantial results of POPs, which includes reproductive issues, developmental toxicity, compromised immune system and declining populations. The molecular mechanisms responsible for POP toxicity such as hormonal disruption, neurotoxic effects and oxidative stress highlight the heterogenous risks caused by these substances. Species such as albatrosses, gulls and raptors highlights the urgent need to address this matter. Strategies for mitigating these effects should involve stricter implementation of global agreements like the Stockholm Convention and promoting international cooperation to control and reduce POPs emissions. Future studies need to focus on the influence of POPs at both molecular and population levels as well as on discovering new methods for the remediation and reduction of POP exposure in bird habitats. By adopting a comprehensive approach that merges scientific research, conservation efforts and policy execution, it is possible to perfect the protection of avian species and promote the health of ecosystems globally.

## References


1. Travnikov O, Batrakova N, Gusev A, Ilyin I, Kleimenov M, Rozovskaya O, Shatalov V, Strijkina I, Aas W, Breivik K, Nizzetto PB (2020) Assessment of transboundary pollution by toxic substances: heavy metals and POPs. EMEP Status Rep 2:2020
2. Jones KC (2021) Persistent organic pollutants (POPs) and related chemicals in the global environment: some personal reflections. Environ Sci Technol 55(14):9400–9412
3. Li L, Chen C, Li D, Breivik K, Abbasi G, Li YF (2023) What do we know about the production and release of persistent organic pollutants in the global environment? Environ Sci Adv 2(1): 55–68
4. Mahire S, Tiwana AS, Khan A, Nalawade PM, Bandekar G, Trehan N, Mukkannawar U, Kaur S, Pandit V, Kamble PN (2023) Accumulation and effects of persistent organic pollutants and biogeographical solutions: appraisal of global environment. Arab J Geosci 16(10):570
5. Mohapatra P, Nøklebye E, Arora T, Basu AR (2023) International regulatory frameworks and best practices in management of POPs. In: Managing persistent organic pollutants in India: case studies on Vapi and Surat, Gujarat. Springer Nature Switzerland, Cham, pp 127–151
6. Akhtar ABT, Naseem S, Yasar A, Naseem Z (2021) Persistent organic pollutants (POPs): sources, types, impacts, and their remediation. In: Environmental pollution and remediation. Springer, Singapore, pp 213–246
7. Shen J, Yang L, Liu G, Zhao X, Zheng M (2021) Occurrence, profiles, and control of unintentional POPs in the steelmaking industry: a review. Sci Total Environ 773:145692
8. Ghosh A, Manna MC, Jha S, Singh AK, Misra S, Srivastava RC, Srivastava PP, Laik R, Bhattacharyya R, Prasad SS, Singh SP (2022) Impact of soil-water contaminants on tropical agriculture, animal and societal environment. Adv Agron 176:209–274
9. Townson H (1992) Public health impact of pesticides used in agriculture. World Health Organization, Geneva. 1990. 128 pp. Price Sw. fr. 21.00. ISBN 92-4-156139-4
10. Nanda S, Berruti F (2021) Municipal solid waste management and landfilling technologies: a review. Environ Chem Lett 19(2):1433–1456
11. Siddiqua A, Hahladakis JN, Al-Attiya WAK (2022) An overview of the environmental pollution and health effects associated with waste landfilling and open dumping. Environ Sci Pollut Res 29(39):58514–58536
12. Di Maria F, Mastrantonio M, Uccelli R (2021) The life cycle approach for assessing the impact of municipal solid waste incineration on the environment and on human health. Sci Total Environ 776:145785
13. Weber R, Iino F, Imagawa T, Takeuchi M, Sakurai T, Sadakata M (2001) Formation of PCDF, PCDD, PCB, and PCN in de novo synthesis from PAH: mechanistic aspects and correlation to fluidized bed incinerators. Chemosphere 44(6):1429–1438
14. Clapp J (2021) Explaining growing glyphosate use: the political economy of herbicide-dependent agriculture. Glob Environ Chang 67:102239
15. Mrozik W, Rajaeifar MA, Heidrich O, Christensen P (2021) Environmental impacts, pollution sources and pathways of spent lithium-ion batteries. Energy Environ Sci 14(12):6099–6121
16. He H, Lu S, Peng Y, Tang M, Zhan M, Lu S, Xu L, Zhong W, Xu L (2022) Emission characteristics of dioxins during iron ore Co-sintering with municipal solid waste incinerator fly ash in a sintering pot. Chemosphere 287:131884
17. Briffa J, Sinagra E, Blundell R (2020) Heavy metal pollution in the environment and their toxicological effects on humans. Heliyon 6(9):e04691
18. Zhou X, Yan Z, Zhou X, Wang C, Liu H, Zhou H (2022) RETRACTED: an assessment of volatile organic compounds pollutant emissions from wood materials: a review. Chemosphere 308:136460
19. Tang FH, Lenzen M, McBratney A, Maggi F (2021) Risk of pesticide pollution at the global scale. Nat Geosci 14(4):206–210
20. Zahoor I, Mushtaq A (2023) Water pollution from agricultural activities: a critical global review. Int J Chem Biochem Sci 23(1):164–176

21. Cai K, Song Q, Yuan W, Ruan J, Duan H, Li Y, Li J (2020) Human exposure to PBDEs in e-waste areas: a review. Environ Pollut 267:115634
22. Keswani C, Dilnashin H, Birla H, Roy P, Tyagi RK, Singh D, Rajput VD, Minkina T, Singh SP (2022) Global footprints of organochlorine pesticides: a pan-global survey. Environ Geochem Health 44:1–29
23. Zaynab M, Fatima M, Sharif Y, Sughra K, Sajid M, Khan KA, Sneharani AH, Li S (2021) Health and environmental effects of silent killers Organochlorine pesticides and polychlorinated biphenyl. J King Saud Univ Sci 33(6):101511
24. Gaur N, Narasimhulu K, PydiSetty Y (2018) Recent advances in the bio-remediation of persistent organic pollutants and its effect on environment. J Clean Prod 198:1602–1631
25. Karnwal A, Martolia S, Dohroo A, Al-Tawaha ARMS, Malik T (2024) Exploring bioremediation strategies for heavy metals and POPs pollution: the role of microbes, plants, and nanotechnology. Front Environ Sci 12:1397850
26. Borgå K, McKinney MA, Routti H, Fernie KJ, Giebichenstein J, Hallanger I, Muir DC (2022) The influence of global climate change on accumulation and toxicity of persistent organic pollutants and chemicals of emerging concern in Arctic food webs. Environ Sci: Processes Impacts 24(10):1544–1576
27. Zheng XB, Luo XJ, Zheng J, Zeng YH, Mai BX (2015) Contaminant sources, gastrointestinal absorption, and tissue distribution of organohalogenated pollutants in chicken from an e-waste site. Sci Total Environ 505:1003–1010
28. Gray F, Derous D, Bize P (2024) Is minimally-invasive sampling the future of persistent organic pollutant (POP) research in birds? A meta-analysis on tissue comparisons. Chemosphere 362: 142591
29. Katagi T, Fujisawa T (2021) Acute toxicity and metabolism of pesticides in birds. J Pestic Sci 46(4):305–321
30. Nagar N, Saxena H, Pathak A, Mishra A, Poluri KM (2023) A review on structural mechanisms of protein-persistent organic pollutant (POP) interactions. Chemosphere 332:138877
31. Nasra S, Kumar A (2022) Metabolism of environmental pollutants with specific reference to pesticides, endocrine disrupters, and mutagenic pollutants. In: Environmental toxicology and ecosystem. CRC Press, pp 165–177
32. Muñoz-Arnanz J, Cortés-Avizanda A, Donázar-Aramendía I, Arrondo E, Ceballos O, Colomer-Vidal P, Jiménez B, Donázar JA (2024) Levels of persistent organic pollutants (POPs) and the role of anthropic subsidies in the diet of avian scavengers tracked by stable isotopes. Environ Pollut 343:123188
33. Hakk H, Pfaff C, Lupton SJ, Singh A (2021) Absorption, distribution, metabolism, and excretion of three [14C] PBDE congeners in laying hens and transfer to eggs. Xenobiotica 51(3):335–344
34. Coppock RW, Dziwenka MM (2022) Reproductive and developmental toxicity in avian species. In: Reproductive and developmental toxicology. Elsevier, pp 1461–1486
35. Pacyna-Kuchta AD (2023) What should we know when choosing feather, blood, egg or preen oil as biological samples for contaminants detection? A non-lethal approach to bird sampling for PCBs, OCPs, PBDEs and PFASs. Crit Rev Environ Sci Technol 53(5):625–649
36. Wolfe MF, Schwarzbach S, Sulaiman RA (1998) Effects of mercury on wildlife: a comprehensive review. Environ Toxicol Chem Int J 17(2):146–160
37. Lushchak VI, Matviishyn TM, Husak VV, Storey JM, Storey KB (2018) Pesticide toxicity: a mechanistic approach. EXCLI J 17:1101
38. Maurya PK, Malik DS (2016) Bioaccumulation of xenobiotics compound of pesticides in riverine system and its control technique: a critical review. J Ind Pollut Control 32:580–594
39. Grace J, Duran E, Ottinger MA, Maness T (2024) Sublethal effects of early-life exposure to common and emerging contaminants in birds. Curr Res Toxicol 7:100190
40. Kadhim SJ, Sahib MA, Khlaif SM, AL-Azzawi MK, Zubaidi NA (2024) Evaluating the impact of endocrine-disrupting chemicals on fertility in wildlife: from amphibians to mammals. J Med Genet Clin Biol 1:216–228

41. Amir S, Shah STA, Mamoulakis C, Docea AO, Kalantzi OI, Zachariou A, Calina D, Carvalho F, Sofikitis N, Makrigiannakis A, Tsatsakis A (2021) Endocrine disruptors acting on estrogen and androgen pathways cause reproductive disorders through multiple mechanisms: a review. Int J Environ Res Public Health 18:1464
42. Scharf P, Aschner M, Farsky S (2024) Neutrophils in toxicology: a forgotten field. J Toxicol Environ Health Part B:1–32
43. Latchney SE, Majewska AK (2021) Persistent organic pollutants at the synapse: shared phenotypes and converging mechanisms of developmental neurotoxicity. Dev Neurobiol 81: 623–652
44. Fremlin KM, Elliott JE, Green DJ, Drouillard KG, Harner T, Eng A, Gobas FA (2020) Trophic magnification of legacy persistent organic pollutants in an urban terrestrial food web. Sci Total Environ 714:136746
45. La Merrill M, Emond C, Kim MJ, Antignac JP, Le Bizec B, Clément K, Birnbaum LS, Barouki R (2013) Toxicological function of adipose tissue: focus on persistent organic pollutants. Environ Health Perspect 121:162–169
46. Mansfield I, Reynolds SJ, Lynch I, Matthews TJ, Sadler JP (2024) Birds as bioindicators of plastic pollution in terrestrial and freshwater environments: a 30-year review. Environ Pollut 348:123790
47. Kean EF, Shore RF, Scholey G, Strachan R, Chadwick EA (2021) Persistent pollutants exceed toxic thresholds in a freshwater top predator decades after legislative control. Environ Pollut 272:116415
48. Castagna F, Montano L, Lombardi R, Pagano A, Gigliotti A, Bava R, Lupia C, Costagliola A, Giordano A, Palma E, Britti D (2024) Understanding environmental contamination through the lens of the peregrine falcon (*Falco peregrinus*). Environments 11:264
49. De Solla SR, King LE, Gilroy ÈA (2023) Environmental exposure to non-steroidal anti-inflammatory drugs and potential contribution to eggshell thinning in birds. Environ Int 171: 107638
50. Walker A (2024) Using population modelling to understand the dynamics of the Sussex peregrine falcon (*Falco Peregrinus*) population. Master's thesis, Canterbury Christ Church University (United Kingdom)
51. McCarty JP, Secord AL (1999) Reproductive ecology of tree swallows (*Tachycineta bicolor*) with high levels of polychlorinated biphenyl contamination. Environ Toxicol Chem Int J 18: 1433–1439
52. Carro T (2012) The effects of polychlorinated biphenyls (PCBs) on avian cardiac development. PhD thesis, University of Maryland, College Park
53. Hornbuckle KC, Carlson DL, Swackhamer DL, Baker JE, Eisenreich SJ (2006) Polychlorinated biphenyls in the Great Lakes. Persist Organ Pollut Great Lakes 965:13–70
54. Carson R (1962) Silent Spring. Houghton Mifflin Company. ISBN-10:0618249060
55. Ratcliffe DA (1963) The status of the Peregrine in Great Britain. Bird Study 10:56–90
56. Enderson JH, Craig GR, Burnham WA, Berger DD (1982) Eggshell thinning and organochlorine residues in Rocky Mountain (USA) peregrines, *Falco peregrinus* and their prey. Canadian Field Nat 96:255–264
57. Elliott JE, Martin PA (1994) Chlorinated hydrocarbons and shell thinning in eggs of (*Accipiter*) hawks in Ontario, 1986–1989. Environ Pollut 86:189–200
58. McCarty JP, Secord AL (1999) Nest-building behavior in PCB-contaminated tree swallows. Auk 116:55–63
59. McCarty JP, Secord AL (2000) Possible effects of PCB contamination on female plumage color and reproductive success in Hudson River tree swallows. Auk 117:987–995
60. Shiferaw A, Yosef M, Deribe E, Eklo OM (2022) Organochlorine pesticides and polychlorinated biphenyls in carnivorous waterbird species from Lake Ziway, Ethiopia. SN Appl Sci 4. Available from: https://doi.org/10.1007/s42452-022-05215-5

61. Robert K, Elliott JE, Fremlin KM, Gauthier L, Drouillard KG, Bishop CA (2021) Continuing persistence and biomagnification of DDT and metabolites in northern temperate fruit orchard avian food chains. Environ Toxicol Chem 40:3379–3391. Available from: https://doi.org/10.1002/ETC.5220
62. Ayele S, Mamo Y, Deribe E, Eklo OM (2022) Organochlorine pesticides and polychlorinated biphenyls in carnivorous waterbird and fish species from Lake Hawassa, Ethiopia. SN Appl Sci 4(10):285
63. Wang Q, Xie C, Long C, Yang W, Wang Y, Xu W et al (2022) Bioaccumulation and biomagnification of polychlorinated biphenyls and dichlorodiphenyltrichloroethane in biota from Qilianyu Island, South China Sea. Toxics 10(6):324
64. Choi JW, Matsuda M, Kawano M, Min BY, Wakimoto T (2001) Accumulation profiles of persistent organochlorines in waterbirds from an estuary in Korea. Arch Environ Contam Toxicol 41(3):353–363
65. Giesy JP, Ludwig JP, Tillitt DE (1994) Dioxins, dibenzofurans, PCBs and colonial, fish-eating water birds. In: Dioxins and health. Springer US, Boston, pp 249–307
66. Jayaraman S, Nacci DE, Champlin DM, Pruell RJ, Rocha KJ, Custer CM et al (2009) PCBs and DDE in tree swallow (Tachycineta bicolor) eggs and nestlings from an estuarine PCB superfund site, New Bedford Harbor, MA, USA. Environ Sci Technol 43(21):8387–8392
67. Ward MP, Jablonski C, Semel B, Soucek D (2010) The biological pathway and effect of PCBs on common terns in Lake Michigan. Ecotoxicology 19(8):1513–1522
68. Campioni L, Oró-Nolla B, Granadeiro JP, Silva MC, Madeiros J, Gjerdrum C, Lacorte S (2024) Exposure of an endangered seabird species to persistent organic pollutants: assessing levels in blood and link with reproductive parameters. Sci Total Environ 930:172814
69. Strause KD, Zwiernik MJ, Im SH, Bradley PW, Moseley PP, Kay DP et al (2007) Risk assessment of great horned owls (Bubo virginianus) exposed to polychlorinated biphenyls and DDT along the Kalamazoo River, Michigan, USA. Environ Toxicol Chem 26(7):1386–1398
70. Antoniadou V, Konstantinou IK, Goutner V, Sakellarides TM, Albanis TA, Bintoudi E (2007) PCB levels and accumulation patterns in waterbird eggs and in their prey at Lake Kerkini, a north-eastern Mediterranean wetland of international importance. Arch Environ Contam Toxicol 53(2):249–260
71. Gao F, Luo XJ, Yang ZF, Wang XM, Mai BX (2009) Brominated flame retardants, polychlorinated biphenyls, and organochlorine pesticides in bird eggs from the Yellow River Delta, North China. Environ Sci Technol 43(18):6956–6962
72. Roscales JL, Munoz-Arnanz J, Gonzalez-Solis J, Jimenez B (2010) Geographical PCB and DDT patterns in shearwaters (Calonectris sp.) breeding across the NE Atlantic and the Mediterranean archipelagos. Environ Sci Technol 44(7):2328–2334
73. Paasivirta J, Särkkä J, Pellinen J, Humppi T (1981) Biocides in eggs of aquatic birds. Completion of a food chain enrichment study for DDT, PCB and Hg. Chemosphere 10(7):787–794
74. Jaspers VLB, Covaci A, Voorspoels S, Dauwe T, Eens M, Schepens P (2006) Brominated flame retardants and organochlorine pollutants in aquatic and terrestrial predatory birds of Belgium: levels, patterns, tissue distribution and condition factors. Environ Pollut 139(2):340–352
75. Bouwman H, Viljoen IM, Quinn LP, Polder A (2013) Halogenated pollutants in terrestrial and aquatic bird eggs: converging patterns of pollutant profiles, and impacts and risks from high levels. Environ Res 126:240–253
76. Galassi S, Saino N, Melone G, Croce V (2002) DDT homologues and PCBs in eggs of great crested grebe (Podiceps cristatus) and mallard (Anas platyrhynchos) from Lake Maggiore (Italy). Ecotoxicol Environ Saf 53(1):163–169
77. Lewis PJ, Lashko A, Chiaradia A, Allinson G, Shimeta J, Emmerson L (2022) New and legacy persistent organic pollutants (POPs) in breeding seabirds from the East Antarctic. Environ Pollut 309:119734
78. Tyler C, Jobling S, Sumpter JP (1998) Endocrine disruption in wildlife: a critical review of the evidence. Crit Rev Toxicol 28:319–361

79. Grasman KA, Scanlon PF, Fox GA (1998) Reproductive and physiological effects of environmental contaminants in fish-eating birds of the Great Lakes: a review of historical trends. Environ Monit Assess 53:117–145
80. Mineau P (2005) A review and analysis of study endpoints relevant to the assessment of "long term" pesticide toxicity in avian and mammalian wildlife. Ecotoxicology 14:775–799
81. Ricolfi L, Taylor MD, Yang Y, Lagisz M, Nakagawa S (2024) Maternal transfer of per- and polyfluoroalkyl substances (PFAS) in wild birds: a systematic review and meta-analysis. Chemosphere 142346. https://doi.org/10.1016/j.chemosphere.2023.142346
82. Mughal BB, Fini JB, Demeneix BA (2018) Thyroid-disrupting chemicals and brain development: an update. Endocr Connect 7:R160–R186
83. Lavoie ET, Grasman KA (2007) Effects of in ovo exposure to PCBs 126 and 77 on mortality, deformities and post-hatch immune function in chickens. J Toxic Environ Health A 70:547–558
84. Vos JG, Dybing E, Greim HA, Ladefoged O, Lambré C, Tarazona JV, Brandt I, Vethaak AD (2000) Health effects of endocrine-disrupting chemicals on wildlife, with special reference to the European situation. Crit Rev Toxicol 30:71–133
85. Giesy JP, Feyk LA, Jones PD, Kannan K, Sanderson T (2003) Review of the effects of endocrine-disrupting chemicals in birds. Pure Appl Chem 75:2287–2303
86. Burger J, Gochfeld M (2001) Effects of chemicals and pollution on seabirds. In: Biology of marine birds. CRC Press, pp 503–544
87. Tanabe S, Watanabe M, Minh TB, Kunisue T, Nakanishi S, Ono H, Tanaka H (2004) PCDDs, PCDFs, and coplanar PCBs in albatross from the North Pacific and Southern Oceans: levels, patterns, and toxicological implications. Environ Sci Technol 38:403–413
88. Elliott JE (2005) Chlorinated hydrocarbon contaminants and stable isotope ratios in pelagic seabirds from the North Pacific Ocean. Arch Environ Contam Toxicol 49:89–96
89. Grasman KA (1995) Immunological and hematological biomarkers for contaminants in fish-eating birds of the Great Lakes
90. Bouwman H, Yohannes YB, Nakayama SMM, Motohira K, Ishizuka M, Humphries MS, Van der Schyff V, Du Preez M, Dinkelmann A, Ikenaka Y (2019) Evidence of impacts from DDT in pelican, cormorant, stork, and egret eggs from KwaZulu-Natal, South Africa. Chemosphere 225:647–658
91. Vander Pol SS, Vander Pol SS, Pugh RS, Becker PR (2005) Chemicals in Arctic seabirds: I. Annotated bibliography, US Department of Commerce, National Institute of Standards and Technology
92. Saha A, Ghosh RK, Jesna PK, Choudhury PP (2021) Bioindicators of pesticide contaminations. In: Sustainable agriculture reviews 48: pesticide occurrence, analysis and remediation, vol 2. Springer, pp 185–231
93. Kiran SD, Kour A, Priya DV, Kumar R (2022) Community composition and status of avian diversity at Campus and Agricultural landscapes of Chaudhary Charan Singh Haryana Agricultural University, Hisar (Haryana). J Appl Nat Sci 14(4):1130–1140. https://doi.org/10.31018/jans.v14i4.3784
94. Yodha K, Singh D, Kour A, Yadav R, Kumar R (2024) Species diversity and abundance of birds in ber (*Zizyphus mauritiana*) crop. Asian J Conserv Biol 13(1). https://doi.org/10.53562/ajcb.83683
95. Kumar R, Singh D (2022) Insecticide-tolerant bacterial population in *Eisenia fetida*'s Gut and Vermicast exposed to Chlorantraniliprole and Fipronil. App Biol Res 24(3):273–279
96. Yadav R, Gupta RK, Kumar R, Kaur T, Moun R (2022) Biochemical response of earthworm, *Eisenia fetida* to heavy metals toxicity. J Appl Nat Sci 14(3):990–998
97. Richard FJ, Gigauri M, Bellini G, Rojas O, Runde A (2021) Warning on nine pollutants and their effects on avian communities. Glob Ecol Conserv 32:e01898
98. Wang Y, Chen J (2024) Genetic adaptation in avian species to rapid environmental changes. Int J Mol Evol Biodivers 14:197

# Chapter 13
# Accomplishment, Issues and Strategies for Water Bird Conservation in Coastal Wetland

**Pooja Kadyan, Renu Gill, Mohit Deswal, Sudhir Kumar Kataria, Parmesh Kumar, and Monika Pilania**

**Abstract** Coastal wetlands are rich in life and provide vital homes for many waterbirds, including some that travel long distances and are rare worldwide. Here, we discuss the successes, existing problems, and future approaches in safeguarding waterbirds in fragile ecosystems. Over the years, great work has been done thanks to setting up protected areas, international agreements, and community efforts. Still, there are many threats to these birds such as reducing habitats, pollution, rising climate, human effect, and the arrival of new invasive species. Because of these impacts, various species have declined and wetlands vital to the environment have deteriorated. Following strategies like habitat restoration, responsible land use, long-term ecological monitoring, including the public, and working with other nations are suggested in this chapter to address these problems. It is important that research, policies, and community efforts cooperate to protect waterbirds and coastal wetlands for many years ahead. The discovery aims to guide action in conservation and join efforts on a local, national, and global basis.

**Keywords** Biodiversity · Climate change · Coastal wetlands · Conservation strategies · Migratory waterbirds · Wetland degradation

P. Kadyan · R. Gill · M. Deswal · S. K. Kataria
Department of Zoology, Maharshi Dayanand University, Rohtak, Haryana, India
e-mail: sudhir.zoology24@mdurohtak.ac.in

P. Kumar
Department of Zoology, Institute of Integrated and Honors Studies, Kurukshetra University, Kurukshetra, Haryana, India

M. Pilania (✉)
Health and Life Sciences, Conestoga College, Kitchener, ON, Canada
e-mail: Monika5090@conestogac.on.ca

 
R. Kumar et al. (eds.), *Ecology, Conservation of Aquatic Birds*, Wetlands: Ecology, Conservation and Management 15, https://doi.org/10.1007/978-3-032-17125-2_13

## 13.1 Introduction

Wetlands near the seashore are special habitats where land meets water, distinguished by their special water systems and wide-ranging flora [41]. They are important for carbon recycling, diversity in the environment, and reducing climate change [46]. They play an important role in nature and provide homes for various different animals and plants. How many and what kinds of waterbirds are found in a wetland is considered a good sign of whether the wetland is healthy and functioning properly [33, 51]. Because of what people are doing, coastal wetlands are becoming smaller with each passing day. According to Davidson [10], over 50% of wetlands have vanished or degraded in the last century, and over 35% have been destroyed since 1970 [55]. This has negatively affected the fauna that inhabits in wetlands, and has especially harmed those waterbird species that rely heavily on wetland habitats for their survival. According to Wetlands International [76], 38% of the world's waterbird populations are in decline, making most of them endangered and the subject of conservation efforts. Some of these species have even emerged as the flagship species for biodiversity protection. As a result, the condition of waterbirds largely mirrors the conservation and management status of coastal wetlands. Recently, there has been growing recognition of the importance of these coastal wetlands being among the planet's most productive ecosystems offering a multitude of ecological, economic, and social advantages [4].

### *13.1.1 Characteristics of Coastal Wetlands*

**Tidal Influence** Coastal wetlands are typically influenced by tidal cycles, meaning they are regularly flooded and exposed to the changing water levels from the sea [8]. The wetlands near the coast usually contain water that is a blend of saltwater and freshwater. Salinity in the water changes due to tides, freshwater influx, and nearness to the ocean, as research shows [62].

Within these wetlands are plants capable of enduring both high and low salt levels. Due to Mangroves, salt grasses, and halophytic plants, coastal wetlands are common [42].

Coastal wetlands are home to many different kinds of animals and plants. Habitat islands are crucial for many animals and bird species to breed, find food, and have shelter [5].

Coastal wetlands have several useful functions, such as blocking floods, capturing carbon, restricting erosion, cleaning the water, and supporting a wide range of animals, especially farmed fish [44].

### *13.1.2 Types of Coastal Wetlands*

Coastal wetlands appear when the coastline is influenced by both land and water. All types of coastal wetlands offer services like carbon storage, protection against waves, and cleaner water for the environment. They serve as habitats to a variety of lifeforms, acting as breeding and nesting grounds for many migratory bird species [44]. They play crucial role in maintaining ecological balance and enhance ecotourism as well. They can be categorized into several types depending upon their hydrology and vegetation (as mentioned in Table 13.1).

## 13.2 Coastal Wetlands as a Habitat

Wetlands are varied biotic communities that combine upland and aquatic traits, offering various habitats and supporting a variety of invertebrate and vertebrate organisms [41]. Here, we concentrate on these systems as bird habitats, viewing them as suppliers of the resources necessary for birds to thrive and procreate. Coastal wetlands are abundant with food sources such as plants, animals, and invertebrates [18]. Birds use these resources via foraging in wetland soils, water columns, and submerged and emergent vegetation. Vegetarian bird species eat marsh plants' fruits, tubers, and leaves, while others feed on the various invertebrate communities that thrive in these habitats. For millions of migratory birds annually, these wetlands offer essential feeding, migration, and breeding habitats [31]. Many species of birds get protected and find a place to nest thanks to the interaction of marsh plants and water. Birds can live securely in coastal wetlands, as they are rich in plants and include various kinds of landforms. Because the park is well-protected, birds can safely build their nests and rest there [38]. Thanks to the different conditions around and varying water levels, many types of wildlife can be seen in coastal wetlands [63]. It is possible that the reasons are that, compared to other creatures, birds in wetlands are more easily seen, currently studied a lot and carefully managed, resulting in deeper understanding of their habits in wetlands. Coastal wetlands provide refuge for waterbirds by helping them breed, obtain food and relax while migrating. Waterbirds rely on Patagonia to the Carpathians for a habitat, where many migratory birds can live [84].

**Table 13.1** Types of coastal wetlands along with their location and characteristic features

| S. no. | Type | Location | Characteristics | Reference |
|---|---|---|---|---|
| 1. | Salt marshes | Found in temperate and subtropical regions along coastlines. | Dominated by salt-tolerated grasses, shrubs, and herbs.<br>Periodically flooded by tides.<br>High productivity and carbon storage capacity. | Adam et al. [1] |
| 2. | Freshwater marshes | Found in areas with consistent supply of freshwater such as floodplains, riverbanks, or lakeshores. | Dominated by herbaceous plants such as grasses, reeds with minimal presence of trees.<br>Soils are typically waterlogged and rich in organic content.<br>Act as natural filters, removing pollutants and sediments from runoff. | Mitsch et al. [44] |
| 3. | Mangrove forests | Found in tropical and subtropical regions along coasts and estuaries. | Dominated by mangrove trees and shrubs with specialized prop roots.<br>Adapted to saline and brackish water conditions.<br>Act as natural barriers against storm surges and coastal erosion. | Alongi et al. [2] |
| 4. | Seagrass beds | Found in shallow, clear, and sheltered coastal waters. | Dominated by submerged, flowering plants like eelgrass and turtle grass.<br>Stabilize sediments and improve water quality by trapping particles.<br>Act as carbon sinks, storing "blue carbon". | Orth et al. [50] |
| 5. | Tidal flats (mudflats and sandflats) | Found in sheltered coastal areas, such as bays, estuaries, and lagoons. | Composed of fine sediment (mud or sand) deposited by tides.<br>Periodically exposed and submerged by tides.<br>Support diverse invertebrate communities (e.g., worms, crabs). | Davidson et al. [11] |
| 6. | Coral reefs (shallow marine wetlands) | Found in warm, shallow tropical waters. | Built by colonies of coral polyps and associated organisms.<br>Protect coastlines from wave action and erosion.<br>Support immense biodiversity, including fishes, molluscs, and crustaceans. | Spalding et al. [65] |
| 7. | Peatlands (coastal bog and fen wetlands) | Found in cool coastal areas with freshwater or brackish water. | Composed of waterlogged, organic soils rich in decayed plant material (peat).<br>Often dominated by mosses and sedges.<br>Serve as carbon storage and provide habitat to wildlife. | Mitch et al. [44] |
| 8. | Estuaries | Found at the mouths of rivers where freshwater mixes with saltwater. | Include a mix of tidal marshes, mudflats, and seagrass beds.<br>Highly productive ecosystems with diverse plants and animals.<br>Provide natural flood protection and water filtration. | Goss-Custard et al. [23] |

### 13.2.1 Importance of Coastal Wetlands as Waterbird Habitats

(i) *Breeding and Nesting Grounds*: Waterbirds prefer to lay their eggs and raise their young. Features that include shallow water and plenty of vegetation, along with little human disturbance, are what make these habitats needed by young animals. Protection and suitable places for nesting are frequently found by birds during key life stages through the links between plants and water found in marshes [38].

- Salt marshes are found close to the coast and are regularly filled with saltwater from the sea. They exist in the zone located between land and sea, including grasses, sedges, and shrubs that are used to salt-tolerant plants. Several ground-nesting birds like shell duck, teal, sarus crane, and egrets choose this place to nest, thanks to the dense vegetation by the tidal flats [27].
- Mangrove Forests: Living in the coastal regions of tropical and subtropical regions, mangrove ecosystems create great nesting and protection for Mangrove Cuckoos, Spoonbill, Cormorants, and Herons, which all use the branches for both nesting and shelter [38].

(ii) *Feeding and Foraging Areas*: The land and sea around wetlands is full of food for animals and birds, including fish, insects, and seaweed. Birds take advantage of these resources by feeding on emergent and submerged vegetation, water columns, and wetland soils. Shallow, nutrient-rich waters support abundant populations of invertebrates like crabs, molluscs, and worms, which are a primary food source for many waterbird species. Vegetarian bird species consume fruits, tubers, and leaves of wetland plants [31, 80].

- *Mudflats*: Mudflats are rich in benthic organisms (living in or on the seafloor), including worms, molluscs, and crustaceans, which act as the base of the food web. These tidal areas are highly productive feeding grounds for migratory waders like *Shorebirds*, *Sandpipers*, and *Godwits*, which forage for small invertebrates exposed during low tides [23].
- *Seagrass Beds*: The flowering plants known as seagrass, which thrive in shallow coastal waters, create seagrass beds, which are underwater ecosystems. Seagrass meadows provide important foraging areas for species like *Ducks*, *Geese*, and *Swans*, which feed on submerged plants and small fish [50]

(iii) *Migration Stopovers*: Along birds' journeys, coastal wetlands are valuable places to find food and rest for many migratory birds. Wetlands are popular due to their advantageous placement, allowing many forms of life to establish safe havens for themselves from predators and tough weather [47]. migrating birds use the wetlands during migration to find food and safety from other animals [49].

They follow migratory flyways when traveling from where they breed to where they spend winters and back [49]. Many times, mountains, ocean coasts, and large seas help birds discover vital pathways during their trips. Many migratory birds stop and feed in wetlands along the East Atlantic Flyway, Pacific Flyway, and African-Eurasian Flyway. Millions of migratory birds use coastal wetlands like the Wadden Sea and the Yellow Sea as important stopovers, including birds such as the Bar-tailed Godwit and Dunlin as reported by Deboelpaep et al. [12]

(iv) *Protection from Predators and Storms*: Salt marshes and other wetlands along the coast provide a safe haven for waterbirds against threats and weather changes. Nesting birds are protected from threats on land and weather by the mangrove trees and grasses along the salt marsh [63]. Because of this protective environment, both resident and migratory birds have places where they can rest and build nests.

- Because of their root and branch systems, mangrove trees offer a safe haven for Herons and Egrets against predators [66].
- Coastal wetlands serve to absorb strong winds and rising seas from storms and hurricanes, also helping to control the erosion of the coastline. Without this protection, the environments for waterbirds to breed and nest would be unstable [45].

(v) *Rich Biodiversity*: Coastal wetlands are some of the most productive ecosystems around the globe. Besides waterbirds, the variety in this ecosystem contains many other species, including fishes, shellfish, amphibians and invertebrates [84]. If wetlands are healthy, waterbirds can find various types of food and shelter from different elements. They give the food that waterbirds need such as fish or some small, invertebrate prey [54].

### *13.2.2 Key Waterbird Species in Coastal Wetlands*

- You can find many waterbirds in the significant wetlands along the coast. The main waterbird species found here depend on the climate, the area's habitat (like salt marshes, mangroves and tidal flats) and geographic location [84].
- In the case of Shorebirds and Waders, coastal wetlands such as tidal flats and mudflats, are essential since they hunt for invertebrates at low tide. These birds are found in Sandpipers, Plovers and Curlews [7].
- Ducks and Geese: These animals benefit from coastal wetlands with abundant seagrass, shallow lakes and marshes in which to feed. Places like this are used by species such as the Northern Pintail, American Black Duck and Brant Goose for feeding along their migration route [3].
- Great Egret, Little Blue Heron and Reddish Egret are some of the herons and egrets you often see hunting for fish and other living things in the coastal wetlands. In particular, mangrove wetlands are perfect for these birds to make their nests [37].

- Pelicans and Cormorants: Many pelicans and cormorants populate coastal wetlands, including marshes and mangrove swamps, since they need these places for feeding and finding partners for breeding [61].
- Swans: Many Wetlands offer safe breeding grounds and food sources for the Tundra Swan and Mute Swan which build their nests in gently sloping areas in the wet zone [3].

Dense tropical and wetland coastal vegetation is where species called Oystercatchers, Kingfishers and Skimmers make their homes [44]. Different habitats see variations in animals ranging from one area to another according to the environment. Asian Dowitchers and Black-faced Spoonbills are critical to the coastal wetland birds of Asia. African Sacred Ibis and Lesser Flamingo are the most frequent birds seen in Africa. Of the notable bird species in Australia are the Pied Oystercatcher and the Royal Spoonbill [11]. For many birds, coastal wetlands are useful areas where they stopped during their annual migrations [49].

## 13.3 Accomplishments in Waterbird Conservation

Waterbirds use coastal wetlands to search for food, reproduce and build their nests. The conservation of waterbirds in coastal wetlands must happen by adopting treaties from around the world, laws at the national level and rules at the local community level. Cross-border co-operation, use of data collected by scientists and involving local people support the success of these efforts. Since climate change and human actions continue to affect coastal wetlands, developing new policies is essential to keep them suitable for birds. During the past several decades, important progress has been reached in waterbird conservation through Legislation and the development of new policies. A number of nations enforce national laws and on-site restrictions to safeguard vital wetland areas. Changes in waterbird conservation for coastal wetlands have come about as a result of new laws and policies on habitat loss, global climate changes, pollution and actions of people (see Table 13.2). Examples of such steps include developing local, national and global laws, as well as managing various regions.

### *13.3.1 International Frameworks and Agreements*

- Waterbirds use coastal wetlands to search for food, reproduce and build their nests. The conservation of waterbirds in coastal wetlands must happen by adopting treaties from around the world, laws at the national level and rules at the local community level. Cross-border co-operation, use of data collected by scientists and involving local people support the success of these efforts. Since climate change and human actions continue to affect coastal wetlands, developing new policies is essential to keep them suitable for birds.

**Table 13.2** Measures undertaken globally in waterbird conservation

| S. no. | Accomplishment steps | Measures taken globally |
|---|---|---|
| 1. | International framework and agreements | Ramsar Convention on Wetlands (1971)<br>Convention on migratory species (CMS)<br>World Heritage Convention (1972)<br>Convention on Biological Diversity (CBD) |
| 2. | National Legislation | Endangered Species Act, USA (1973)<br>Migratory Bird Treaty Act, USA (1918)<br>Habitat Conservation Plans<br>Wildlife Protection Acts |
| 3. | Regional and local policies | Coastal zone management plans<br>Marine Protected Areas (MPAs)<br>National Wetland Policies |
| 4. | Sustainable management practices | Ecological Restoration Programs<br>Incentives for conservation<br>Monitoring and data collection |
| 5. | Newly emerging policies | Climate change adaptation<br>Pollution control<br>Conflict resolutions |

- During the past several decades, important progress has been reached in waterbird conservation through Legislation and the development of new policies. A number of nations enforce national laws and on-site restrictions to safeguard vital wetland areas. Changes in waterbird conservation for coastal wetlands have come about as a result of new laws and policies on habitat loss, global climate changes, pollution and actions of people (see Table 13.2). Examples of such steps include developing local, national and global laws, as well as managing various regions.
- If a coastal wetland is important for biodiversity, it may be recognized as a World Heritage Site which both legally protects it and gives it attention from countries all over the globe. It can also mean safeguarding the habitats and birds found in wetlands [72].
- The CBD tries to ensure that benefits from genetic resources are supplied justly, to support their sustainable use and to protect different life forms. Under CBD guidelines, many wetlands along the coast that hold large groups of waterbirds are designated as protected areas [71].

### *13.3.2 National Legislation*

- The Endangered Species Act (from 1973 in the USA) aims to protect threatened or endangered waterbirds. Along with others, the Piping Plover and California Least Tern are species whose normal habitats are found in coastal wetlands [14].
- Migratory Bird Treaty Act [43]: The law is in place to protect migratory birds and their habitats in the U.S. Through the act, waterbirds are protected in coastal areas by laws against killing or destroying their nests [43].

- In many cases, Habitat Conservation Plans are created with local landowners and other people involved to ensure protection of coastal wetlands for waterbirds and sustainable development. Many of these plans often include coastal wetlands that serve as habitats for waterbirds [70].
- Many countries have put laws in place that are dedicated to protecting all kinds of wildlife, including birds in coastal wetlands. Certain laws aim to provide safety for some animals and oversee fishing, hunting and building in delicate regions [24].

### *13.3.3 Regional and Local Policies*

- Managing the Coastal Zone: Many nations near the ocean have created plans that include both conservation and sustainable development. Many plans for wetlands give waterbirds priority, by managing activities such as tourism, construction and farming that might harm these areas [48].
- Marine Protected Areas (MPAs): Many wetlands along coasts are included in ocean conservation effort networks. Waterbirds have a place to rest, feed and raise their young in MPAs. MPAs may limit the impact of fishing, pollutants and other activities by humans on waterbirds [59].
- Countries like the UK, Australia and Canada have policies in place that protect the wetlands where waterbirds can be found. Waterbird-protection policies usually involve surveying and understanding their populations, preserving vital habitats and supporting efforts for conservation among stakeholders [57].

### *13.3.4 Sustainable Management Practices and Policy Tools*

- To help waterbirds, both governments and NGOs work on programs that restore damaged coastal wetlands. Restoration may include planting new vegetation, managing water levels and removing illegitimate species [81]. Legislation and support through different funds often assist these actions at a national and worldwide level [15].
- Financial benefits, as well as tax breaks, can motivate people involved in farming and land management to carry out actions for waterbird protection. One can maintain wetlands, keep invading species in check and prevent harmful construction [53].
- The development of policies may include starting programs to oversee waterbirds, their habitats and any dangers they face. Complete data is necessary to check if efforts to conserve species are effective. The AEWA sets up the Waterbird Population Estimates program and countries also use national systems such as State of the Environment reports [30].

### 13.3.5 Emerging Policies

- As climate change increases, many coastal wetland areas are dealing with higher sea waters and different patterns of water through the land. As a result, new policies are being formed to help waterbirds by adapting to climate change by choosing managed retreat or developing artificial wetlands [77].
- Pollution Control: Waterbirds are often threatened by pollution from runoff and plastics. Policies that help reduce pollution in coastal wetlands are necessary for protecting waterbirds. Stricter controls are set on chemicals, discarding waste and dealing with oil spills.
- Reaching a resolution between local needs, industry and waterbirds is sometimes tricky. Therefore, nowadays, making policies involves several parties to arrive at solutions where useful conservation activities do no harm to local people [74].

## 13.4 Issues Related to Waterbird Conservation in Coastal Wetlands

Various natural factors as well as actions by humans such as the loss of habitats, its division, climate change, disturbance, invasive species and using resources too much, create difficulties for waterbirds in wetlands near the shore (Fig. 13.1). They are connected and, at times, make the situation worse for conservationists who must approach solving them in several different ways [10]. For planning to be effective, scientists, lawyers, community groups and those involved in sustainable development must all work together [22, 79]. Resolving these problems calls for

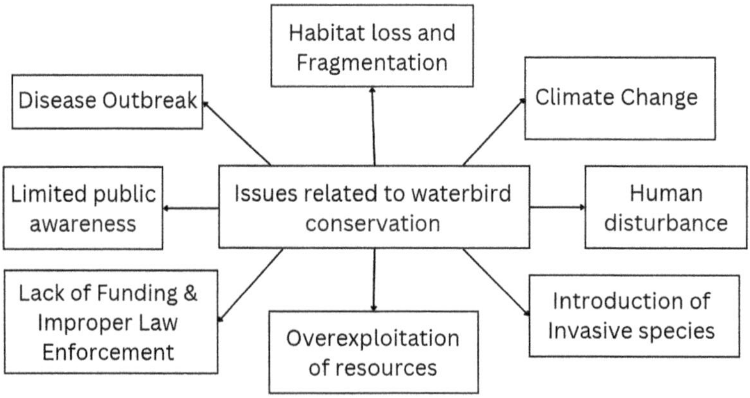


**Fig. 13.1** Issues related to waterbird conservation

stability in actions, flexible approaches and international efforts to assist those species that travel outside one country.

*Some of the fundamental issues linked to waterbird conservation in coastal wetlands*:

### 13.4.1 Habitat Loss

Urbanisation, building new roads and structures and tourism put coastal wetlands at great risk. The construction of structures such as buildings, roads and resorts may interrupt or damage regions where waterbirds live. Creating new communities often changes the habitat of waterbirds, encouraging them to leave their usual breeding and feeding places [10, 35, 78]. Making wetlands into farm or grazing land or into just one type of crop, brings about considerable loss of wetland habitats. Waterbirds have fewer areas to feed in when coastal wetlands are used for farming and this also results in more polluted water [40].

### 13.4.2 Breaking Up of Habitats

Due to the construction of roads, dams and levees, wetlands are often divided into smaller sections, making it challenging for waterbirds to travel or breed and reducing their food supply. Because many wetland areas are not connected, it is hard for waterbirds to change habitats when faced with threats to the environment [21, 36]. Waterbirds often use coastal wetlands when moving on their migration paths. Still, if stopover sites are harmed or lost along these migration routes, it can influence the populations. For example, projects such as dams may cut off birds' migration paths to locations where they look for food or a place to breed [34, 68].

### 13.4.3 Climate Change

When sea levels rise due to climate change, it is more difficult for waterbirds to find a habitat in many low-lying coastal wetlands. If their coastal habitat is eliminated, organisms that depend on a certain tidal level could move away or become less abundant [16]. Global warming's impact on wetlands can disrupt their ecosystems by causing different rainfall, flooding and water levels [20]. If too much freshwater or saltwater is present, certain birds that require particular wetland conditions for survival may not thrive [44]. With the effects of climate change, there are now more damaging storms that can harm the environment, destroy wetlands, take away the shore and erase places where sea creatures nest [16]. Storms also interrupt the ability of waterbirds, especially those nesting in undisturbed places, to find food and breed successfully [28].

### 13.4.4 Human Disturbance

Some human actions such as boating and fishing can interfere with waterbirds in coastal wetlands, mainly during their mating season and while migrating. Any disturbance in their environment can force parents to leave their nest, affect the way they feed their young, cause more stress and influence how successfully the young are able to reproduce and survive [75]. Despite being illegal in most places, many people still participate in both hunting and poaching where large numbers of waterbirds are thought to exist. Waterbirds can be taken for their meat, their feathers or because they are sold illegally. Poaching often reduces the numbers of species that are at risk which may result in problems for nearby habitats [13]. These wetlands are being damaged by pollution brought by industrial waste, agricultural wastes, oil spills and plastics. All these effects can contaminate the water, harm different types of food and have negative results for the health of waterbirds [17]. Waterbirds are mostly affected by toxic substances from heavy metals, pesticides and fertilizers that build up in the food they eat [6].

### 13.4.5 The Practice of Using Too Many Resources

Many coastal areas with wetlands are important for waterbirds who eat fish. When the population of fish decreases due to overfishing, waterbirds could face food shortages, leading to problems with their health, reproduction and survival [75]. The act of removing shellfish, peat and water from wetlands can be harmful to their ecosystems. Usually, these actions result in harming the quality of habitats and sometimes the waterbirds themselves [52].

### 13.4.6 Invasive Species

Common reed (Phragmites australis) may reach coastal wetlands and transform the habitat, thus leading to fewer waterbirds. These plants frequently come to dominate in the area, reduce the variety of plants and animals there and alter the amount of water in the wetlands, so that waterbirds face limited opportunities for food and nesting [82]. When rats, foxes or predatory fishes are present in an area, they can seriously affect waterbirds and their eggs or hatchlings. They may significantly affect reproduction by eating eggs, chicks or grown birds [9].

### *13.4.7 Disease Outbreak*

Birds living on water bodies can easily catch diseases such as avian influenza or botulism and the infection can quickly spread as the birds tend to live close together. If there is an outbreak, it can result in the death of many waterbirds in coastal wetlands where waterbirds gather to breed or rest during their travels [64].

### *13.4.8 Conservation Funding and Enough Political Motivation*

Restoring, securing and watching over coastal wetlands and waterbirds costs a lot of money. Even so, it is common for conservation programs to depend on fluctuating funds from donors or the government [26]. In a lot of places, despite the existence of laws, they are often not enforced. If illegal activities such as poaching or harming natural waterbird habitats are not monitored and prevented, waterbirds will remain threatened. There are also regions where politicians fail to prioritize important policies [26].

### *13.4.9 Lack of Public Awareness and Stakeholder Engagement*

People may not have enough understanding of the coastal wetlands and how waterbirds contribute to these areas. When people are not aware of the problem, it may cause less importance to be put on conservation and not providing enough support for conservation [69]. Farming, tourism and development can arise in competition with protecting waterbirds. For these priorities to balance, all stakeholders must get involved, gain new knowledge and develop measures that ensure both the environment's protection and the development of society. If neighbors and key members of the community are not consulted when making decisions, the efforts to protect nature may be rejected or inefficient [21].

## 13.5 Ways to Care for Waterbirds in Coastal Wetlands

Waterbird conservation depends on a plan that merges efforts to protect places where they live, manage environments in a sustainable manner, restore habitats, research and back it up with solid laws (Fig. 13.2). The success of a project can be sustained through greater involvement from stakeholders and more awareness among the public. Since new threats to wetlands and waterbirds keep arising, it will be

**Fig. 13.2** Strategies for waterbird conservation and wetland management

necessary for people and organizations to cooperate and find effective solutions. Any strategy should take into account all the dangers to waterbirds and review how actions might influence coastal wetlands on all levels. Attention to both conservation and sustainability will allow these ecosystems to be protected for the generations to come.

There are a number of strategies that have been put into place around the globe to protect waterbirds in coastal wetlands.

### *13.5.1 Conservation and Identification of Habits*

National Parks, Nature Reserves and Marine Protected Areas play an important role in helping to conserve waterbirds. This makes it possible for animals to live without interference from humans. When a wetland is given the Ramsar Site (Wetlands of

International Importance) designation, it automatically receives legal protection and recognition on a global scale. The purpose of such sites is to ensure conservation, give support to wise use of resources and promote worldwide agreements on migratory species [54]. Establishing waterways between isolated wetlands is crucial for waterbirds to move, find their food and connect with others. Avoiding gaps in connectivity allows many species to maintain their populations, mainly species that stop over in certain places [21].

### *13.5.2 Habitat Restoration*

To benefit waterbirds, you can restore coastal wetlands by planting new vegetation and modifying their water sources. To insert suitable conditions for feeding and nesting, wetlands should be planted with indigenous wetland species, have their waterways restored to their natural route and invasive plants should be removed [40]. Because of the rise in sea levels and things people do, coastal wetlands are more likely to erode and their home habitat decrease. If salt marshes or mangroves are restored, there can be fewer losses of habitat as the shorelines are less likely to be damaged by storms [44].

### *13.5.3 The Goal of Sustainable Management Is to Limit Damage to the Environment*

Because of ICZM, the management of coasts includes the environment, the community and economic activities. Its strategies consist of safe conservation of nature, responsible use of natural resources, controlling pollution and educating about it to the public. As part of protecting waterbirds, ICZM ensures that development remains sustainable and water species are not harmed by typical human activities [32]. When sustainable farming and fishing are preferred near coastlines, waterbirds' habitats and access to food are less at risk [83]. To keep wetlands clean and safe, we must control pollution from runoff, industries and sewage. Proper nutrient management, control of soil erosion and handling of wastewater help ensure the necessary water quality for waterbirds that depend on fish, other invertebrates and plants for food [6].

### *13.5.4 Saving the Areas Migratory Species Use*

They use certain places along their way to feed and rest from their travels. Maintaining these habitats allows waterbirds to easily start their long journey to other regions. This goal can be accomplished by setting aside protected areas, restoring habitat and avoiding disrupting the area during the main migration period

[49]. Many waterbirds fly across countries which means nations have to collaborate to protect them. The AEWA of CMS helps to bring countries together to ensure the conservation of waterbirds and their resting spots along migratory routes [67].

### 13.5.5 It Is Important to Study the Effects and to Monitor Progress

It is important to keep track of waterbirds to identify their changes in number, location and number of offspring. Techniques such as flying over and imaging through satellites, as well as banding birds, aids in tracking how wild birds behave and if they are healthy or not. It can help in deciding what and how to conserve and direct actions taken to manage the wetlands. Checking the vegetation, water and wetland habitats allows us to assess the suitability of coastal wetlands for waterbirds. Using the outcomes of monitoring, adaptive management helps take action in time when any habitat is being threatened. We must learn how climate change affects waterbirds and their habitats to better conserve them in the future. Strategies for conserving wildlife are adjusted using information on sea levels, changing rainfall patterns and differences in animal behavior [19].

### 13.5.6 It Is Important to Involve Stakeholders and Increase Their Awareness

If communities are involved in waterbird conservation, the efforts will pay off over the years. Collaborating with local inhabitants in care for wetlands and ecological tourism involves them in looking after and understanding their wetlands better. Besides, community outreach programs can help locals gain more information about waterbird conservation [58]. By raising knowledge about the waterbirds and their living areas in schools and the media, we gain widespread support for them. Involving the public can also help people comply with conservation laws [69]. A successful approach to conservation depends on the cooperation of governments, NGOs, the private sector and local citizens. They advocate for including waterbird conservation in national laws and budgets and encourage organizations to improve their sustainability habits [58].

### 13.5.7 Rules and Regulations

The right laws can set guidelines for protecting waterbirds. For coastal wetlands and waterbirds to be properly protected, policies should merge environmental interests with ways to maintain a developing economy. This is necessary to help stop illegal hunting, damage to habitats and poaching. Proper enforcement depends on having enough resources, trained individuals and crowds supporting it.

### *13.5.8 Approaches to Handling Climate Change*

Considering climate change, efforts should be made to increase the protective capacity of coastal wetlands against rising oceans and powerful storms. Commentators suggest restoring mangroves and salt marshes that support waterbirds and protect the area from erosion [20]. In locations where this is appropriate, moving communities away from the coast will help coastal wetlands move deeper into land because of rising sea levels. One must first clarify which wetland areas need special protection and look into possible ways to relocate or restore any destroyed wetland habitat [77].

### *13.5.9 How Can Conflict in a Region Be Resolved and Development Made Sustainable?*

It is important to find a balance between conservation and developing coastal areas that grow swiftly. Enacting laws that zone areas, do environmental studies and motivate actions that preserve wetlands is one solution. It can aid in raising money for saving waterbirds and drawing people to care about their protection. Still, it should be managed with caution to keep wetland habitats unharmed. Setting rules for visitors such as not visiting important bird nesting areas and being careful when birdwatching, is necessary [59].

## 13.6 Funding Agencies for Waterbird Conservation in Coastal Wetland Ecosystem

Some organizations provide funds for waterbird conservation in areas surrounding the ocean [29]:

- In this program, states, U.S. territories and those along the Great Lakes can receive grants of up to $1 million to help preserve wetlands.
- Mohamed bin Zayed Species Conservation Fund helps projects aimed at conservation of birds and some of these initiatives focus on waterbirds.
- Leonardo DiCaprio Foundation: The foundation works to care for vulnerable wildlife, as well as balance and restore ecosystems at risk.

- The Skoll Foundation: It helps social entrepreneurs and innovators who contribute to the preservation of coastal wetlands and waterbirds.
- The Oak Foundation: They assist communities that address climate change, marine life and help wildlife, including those involving waterbirds.
- The Pentair Foundation: This foundation supports initiatives that provide sustainable access to water and address community needs, including projects focused on waterbird conservation.
- Patagonia International Grants Programme: This program supports grassroots organizations working on environmental campaigns, including those focused on waterbird conservation.

## 13.7 Significance of Waterbird Conservation in Coastal Wetland Ecosystem

Coastal wetlands, including estuaries, salt marshes, mangroves, and mudflats, provide vital services to both wildlife and humans, and waterbirds are key players in maintaining the health and balance of the nature as mentioned in Fig. 13.3. Waterbird conservation in coastal wetland ecosystems is crucial not only for the birds themselves but also for maintaining the health of the entire ecosystems. By protecting these birds, we safeguard biodiversity, ecosystem services, and resilience

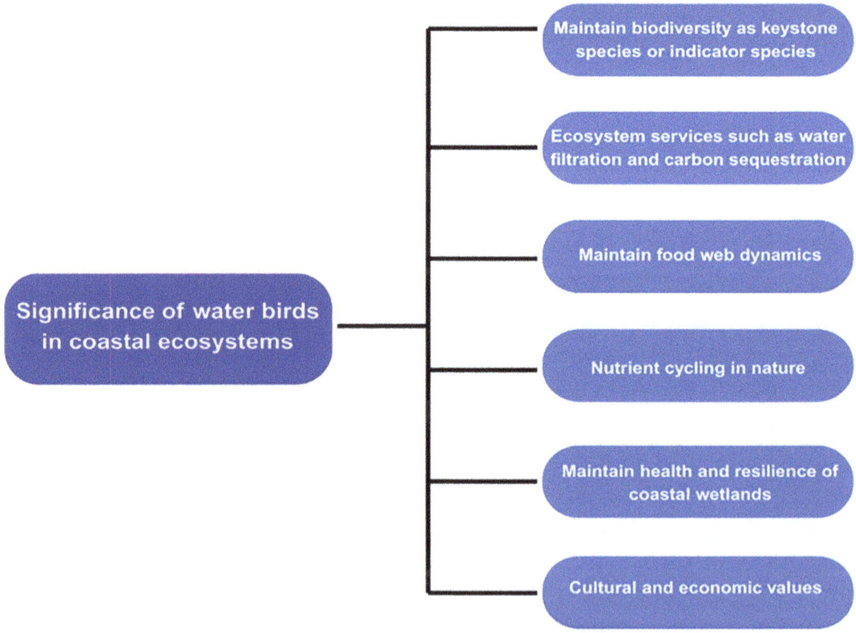


**Fig. 13.3** Flowchart showing significance of waterbirds to coastal ecosystems

to environmental changes. Coastal wetlands are dynamic and vital ecosystems, and waterbirds are essential to their functioning. Conserving these birds helps ensure that coastal wetlands continue to provide critical services to wildlife, humans, and the planet.

### 13.7.1 Biodiversity Maintenance

In coastal areas, species like waterbirds, waders, shorebirds, ducks and seabirds are usually called keystone species. The way they behave and exist in the area can shape the environment around them. In addition, many waterbirds consume various invertebrates, fish and plants from the water which helps control their populations and supports healthy growth of nearby plants as well as the quality of the water [5]. Waterbirds provide information about the quality of the wetland in which they live. Monitoring species living in the wetlands is helpful for tracking their health and detecting any signs of pollution, degraded habitats and similar dangers quickly [38].

### 13.7.2 Ecosystem Services

Many ecosystem services that rely on waterbirds are found in coastal wetlands [44].

- Wetlands act as filters for pollutants in water, making the water healthier. They keep the ecosystem balanced by eating algae, insects in the water and small fish.
- Carbon Sequestration: Coastal wetlands hold and store significant amounts of carbon that used to be in the air. Ecosystems nurtured by species of waterbirds help deal with climate change.
- Flood Control: Wetlands can protect areas from floods by absorbing most of the water. By finding food and adopting their feeding/nesting behaviors, waterbirds preserve the basic features and usefulness of wetlands.

### 13.7.3 Food Web Dynamics

Waterbirds play a key role in the food web found in coastal wetlands. They are responsible for keeping invertebrates, small fish and plants in balance by feeding on them. Some animals also eat invasive species that could upset the ecosystem's balance [3]. In addition, other animals eat waterbirds and their movements between different habitats help increase diversity across broad areas of the world [49].

### 13.7.4 Nutrient Cycling

The presence of waterbirds helps to maintain nutrient levels in coastal wetlands. Because guano is abundant in nitrogen and phosphorus, it helps plants and algae thrive in wetlands. Nutrients make it possible for the wetland ecosystem to produce enough food for many other animals. As waterbirds eat, they also help move plants seeds around and promote the development of aquatic plants in the area [25].

### 13.7.5 Help for Migratory Species

These wetlands on the coast are vital places for migratory waterbirds to rest while on journeys from their breeding grounds to their winter homes. They support food, shelter and resting areas. Maintaining the habitats of migratory species is necessary as it helps them survive and forms an important pattern in the ecosystem and its genetics [12]. Among the dangers facing these migratory waterbirds are changes in their habitat, impacts from climate change and human activities. To help the birds, it is important to maintain and protect these wetlands [47].

### 13.7.6 Coastal Wetland Health and Resilience

Waterbirds help ensure the health and strength of coastal wetlands. Stirring mud with their beaks on mudflats or shallow water, waterbirds support plant growth and preserve the wetland's role in nature. Climate change, like rising sea levels and severe weather, is unlikely to have a big impact on strong and healthy wetlands. Because they aid ecology, waterbirds indirectly help ecosystems withstand changes in their environment [39].

### 13.7.7 The Benefits of Traditions

Observing birds, especially the many waterbirds, is the reason many people are drawn to coastal wetlands for ecotourism. When we protect waterbirds, the tourism industry that relies on these places benefits the local community economically. In various countries, people hold waterbirds in high regard for their traditions and religions. Looking after cultural heritage and enhancing people's bond with the environment are some benefits of conserving these sites [5].

### 13.7.8 Threat Mitigation

- A variety of things such as pollution, loss of habitat, new invasive species and changes in the climate, impact coastal wetlands. Working on waterbird conservation addresses many of these matters because preserving these species often helps to save the habitats where they live [10]. For instance:
  - By cleaning water sources, we help the birds and nature.
  - Conserving breeding and migration areas helps waterbirds.
  - By restoring damaged wetlands used by waterbirds, the entire ecosystem can adapt better to any changes in the environment.

### 13.7.9 International Collaboration

Due to their migration across borders, many waterbirds face challenges that are important to people everywhere. Often, conservation of wetlands along coastlines depends on countries working together, as shown by the Ramsar Convention on Wetlands, the Migratory Birds Treaty and the Agreement on the Conservation of African-Eurasian Migratory Waterbirds (AEWA) (Fig. 13.3). Thanks to these agreements, countries work together to save coastal wetlands and waterbirds and the areas where birds migrate remain protected [60].

## 13.8 Future Prospects

Observing birds, especially the many waterbirds, is the reason many people are drawn to coastal wetlands for ecotourism. That will create economic to local communities. There is a need to develop new technologies for the conservation of aquatic birds. In various countries, people hold waterbirds in high regard for their traditions and religions. Due to their migration across borders, many waterbirds face challenges that are important to people everywhere. There is also a need for more policies for local communities so that they can maintain their traditional and ethical aspects through aquatic birds.

## 13.9 Conclusion

Birds living in coastal wetlands need to be protected, as their conservation benefits the entire ecosystem as well. If we care for these birds, biodiversity, their role in nature and resistance to changes in the environment are guaranteed. Balanced wetlands on the coast depend greatly on waterbirds. Yet, several factors are now placing these ecosystems at great risk which is resulting in fewer species and some of them facing danger of extinction. The main conservation problems include the destruction of habitats, global warming and interaction with people. With a lot of waterbirds migrating and using wetlands around the world, governments from different nations need to cooperate. Treaties related to migratory birds and the Ramsar Convention assist countries in working together to save animals and their habitats. A strategy that considers ecological, social and economic factors should be in place to help waterbirds in coastal wetlands. Combining environmental protection for these areas, research on them, engaging the local community and making effective policies ensures the conservation of waterbirds that live there. If these waterbirds are conserved, the coasts will help continue to supply important resources to wildlife, humans and our planet.

## References


1. Adam P (1990) Saltmarsh ecology. Cambridge University Press, Cambridge
2. Alongi DM (2009) The energetics of mangrove forests. Springer, Dordrecht
3. Baldassarre G (2014) Ducks, geese, and swans of North America. Johns Hopkins University Press, Baltimore
4. Barbier EB (2019) The value of coastal wetland ecosystem services. In: Coastal wetlands. Elsevier, pp 947–964
5. Barbier EB, Hacker SD, Kennedy C, Koch EW, Stier AC, Silliman BR (2011) The value of estuarine and coastal ecosystem services. Ecol Monogr 81(2):169–193
6. Burger J, Gochfeld M (2004) Metals in albatross feathers from midway atoll: influence of species, age, and nest location. Environ Res 98(1):125–133
7. Colwell MA (2010) Shorebird ecology, conservation, and management. University of California Press, Berkeley
8. Costanza R, Pérez-Maqueo O, Martinez ML, Sutton P, Anderson SJ, Mulder K (2008) The value of coastal wetlands for hurricane protection. Ambio 37(4):241–248
9. Cuthbert RJ, Hilton GM (2004) Introduced house mice (*Mus musculus*) on Gough Island: a major threat to seabird conservation. Biol Conserv 117(5):483–489
10. Davidson NC (2014) How much wetland has the world lost? Long-term and recent trends in global wetland area. Mar Freshw Res 65(10):934–941
11. Davidson NC, Stroud DA (2006) African–Eurasian flyways and wetland conservation. Conserv Biol 20(2):318–326
12. Deboelpaep E, Partoens L, Koedam N, Vanschoenwinkel B (2022) Highway (s) overhead: strong differences in wetland connectivity and protected status challenge waterbird migration along the four palearctic-Afrotropical flyways. Divers Distrib 28(5):1067–1080
13. Donald PF, Anderson GQA (2013) The global impact of illegal killing on birds. Bird Conserv Int 24(1):1–11
14. Endangered Species Act of 1973, 16 U.S.C. §§ 1531–1544 (1973)
15. Environmentalist Foundation of India (n.d.) Our projects. Environmentalist Foundation of India. Retrieved February 11, 2025
16. Erwin RM (2009) Coastal habitats and waterbirds: how rising sea levels will affect coastal species. Condor 111(4):744–757
17. Finkelstein ME, Keitt BS, Croll DA, Tershy BR, Jarman WM, Smith DR (2006) Albatross species demonstrate regional differences in North Pacific marine contamination. Ecol Appl 16(2):678–686

18. Fretwell JD, Williams JS, Redman PJ (eds) (1996) National water summary on wetland resources, vol 2425. US Government Printing Office, Washington
19. Galbraith H, Jones R, Park R, Clough J, Herrod-Julius S, Harrington B, Page G (2002) Global climate change and sea level rise: potential losses of intertidal habitat for shorebirds. Auk 119(2):465–482
20. Galbraith H, DesRochers DW, Brown S, Reed JM (2014) Predicting vulnerabilities of north American shorebirds to climate change. PLoS One 9(9):e108899
21. Gardner RC, Barchiesi S, Beltrame C, Finlayson CM, Galewski T, Harrison I, Paganini M, Perennou C, Pritchard DE, Rosenqvist A, Walpole M (2015) State of the world's wetlands and their services to people: a compilation of recent analyses. Ramsar Convention Secretariat, Gland
22. Gopal B, Junk WJ (2000) Biodiversity in wetlands: assessment, function, and conservation. Backhuys Publishers, Leiden
23. Goss-Custard JD et al (1993) The carrying capacity of estuarine habitats for shorebirds. Biol Conserv 65(2):135–148
24. Government of India (1972) The wildlife (protection) act, 1972 (no. 53 of 1972). Ministry of Law and Justice
25. Green AJ, Elmberg J (2014) Ecosystem services provided by waterbirds. Biol Rev 89(1): 105–122
26. Green RE, Sutherland WJ (2021) Addressing funding and political challenges in waterbird conservation: a global perspective. Conserv Biol 35(3):654–662
27. Greenberg R, Maldonado JE, Droege S, McDonald MV (2006) Tidal marshes: a global perspective on the evolution and conservation of their terrestrial vertebrates. Bioscience 56(8):675–685
28. Iwamura T, Possingham HP, Chadès I, Minton C, Murray NJ, Rogers DI, Treml EA, Fuller RA (2013) Migratory connectivity magnifies the consequences of habitat loss from sea-level rise for shorebird populations. Proc R Soc B Biol Sci 280(1761):20130325
29. Johnson RD, Miller EA (2020) Waterbird conservation and funding strategies in coastal wetlands. Wetland Ecology Press
30. Kabra K, Xiong A, Li W, Luo M, Lu W, Garcia R, Vijay D, Yu J, Tang M, Yu T, Arnold H, Vallery A, Gibbons R, Barman A (2022) Deep object detection for waterbird monitoring using aerial imagery. arXiv preprint arXiv:2210.04868
31. Kathiresan K, Bingham BL (2001) Biology of mangroves and mangrove ecosystems. Adv Mar Biol 40:81–251
32. Kay R, Alder J (2005) Coastal planning and management: integrated coastal zone management revisited. Ocean Coast Manag 48(7–8):493–516
33. Kiran, Singh D, Kour A, Priya, Delu V, Kumar R (2022) Community composition and status of avian diversity at campus and agricultural landscapes of Chaudhary Charan Singh Haryana Agricultural University, Hisar (Haryana). J Appl Nat Sci 14(4):1130–1140. https://doi.org/10.31018/jans.v14i4.3784
34. Kirby JS, Stattersfield AJ, Butchart SHM, Evans MI, Grimmett RFA, Jones VR, O'Sullivan J, Tucker GM, Newton I (2008) Key conservation issues for migratory land- and waterbird species on the world's major flyways. Bird Conserv Int 18(S1):S49–S73
35. Kumar R, Singh D (2022) Insecticide-tolerant bacterial population in *Eisenia fetida*'s gut and vermicast exposed to chlorantraniliprole and fipronil. Appl Biol Res 24(3):273–279
36. Kumar R, Singh D, Yadav R (2025) Toxic effects of chlorantraniliprole and fipronil on *Eisenia fetida*: from growth and reproductive parameters to avoidance behavior. Environ Eng Manag J 24(7):1499–1508. Accepted manuscript
37. Kushlan JA, Hancock J (2005) The herons. Oxford University Press, Oxford
38. Luther DA, Greenberg R (2009) Mangroves: a global perspective on the evolution and conservation of their terrestrial vertebrates. Bioscience 59(7):602–612
39. Ma Z, Cai Y, Li B, Chen J (2010) Managing wetland habitats for waterbirds: an international perspective. Wetlands 30(1):15–27

40. Ma Z, Melville DS, Liu J, Chen Y, Yang H, Ren W, Zhang Z, Piersma T, Li B (2014) Rethinking China's new great wall: massive seawall construction in coastal wetlands threatens biodiversity. Science 346(6212):912–914
41. Maynard LA, Wilcox DA (1996) Coastal wetlands. Environment Canada, Guelph
42. McLeod E, Chmura GL, Bouillon S, Salm R, Björk M, Duarte CM, Lovelock CE, Schlesinger WH, Silliman BR (2011) A blueprint for blue carbon: toward an improved understanding of the role of vegetated coastal habitats in sequestering CO2. Front Ecol Environ 9(10):552–560
43. Migratory Bird Treaty Act of 1918, 16 U.S.C. §§ 703–712 (1918)
44. Mitsch WJ, Gosselink JG (2015) Wetlands, 5th edn. Wiley, Hoboken
45. Möller I, Spencer T, French JR, Leggett DJ, Dixon M (1999) Wave transformation over salt marshes: a field and numerical modelling study. Estuar Coast Shelf Sci 49(3):411–426
46. Mustafa G, Hussain S, Liu Y, Ali I, Liu J, Bano H (2024) Microbiology of wetlands and the carbon cycle in coastal wetland mediated by microorganisms. Sci Total Environ 954:175734
47. Nagelkerken I et al (2008) The habitat function of mangroves for terrestrial and marine fauna: a review. Aquat Bot 89(2):155–185
48. National Oceanic and Atmospheric Administration (2020) Coastal zone management program strategic plan 2020–2025. U.S. Department of Commerce
49. Niles LJ et al (2009) Status of the red knot (*Calidris canutus rufa*) in the Western hemisphere. Stud Avian Biol 36:1–185
50. Orth RJ, Carruthers TJB, Dennison WC et al (2006) A global crisis for seagrass ecosystems. Bioscience 56(12):987–996
51. Péron G, Crochet P-A, Dunn P, Johnson AR, Lefebvre G (2013) Waterbird communities as indicators of ecological condition in mediterranean wetlands: new metrics for conservation. Ecol Indic 34:490–498
52. Piersma T, Baker AJ (2000) Life history characteristics and the conservation of migratory shorebirds. Behav Conserv 9:105–124
53. Point Blue Conservation Science (2018) Financial incentives create critical waterbird habitat in extreme drought. ScienceDaily, Encinitas
54. Ramsar Convention on Wetlands (1971) Protecting wetlands of international importance for biodiversity and human well-being
55. Ramsar Convention on Wetlands (2018) Global wetland outlook: state of the world's wetlands and their services to people 2018. Ramsar Convention Secretariat, Gland
56. Ramsar Convention Secretariat (1971) The convention on wetlands of international importance especially as Waterfowl Habitat. Retrieved from https://www.ramsar.org
57. Ramsar Convention Secretariat (2016) The Ramsar convention and national wetland policies: a framework for wetland conservation and wise use
58. Reed MS (2008) Stakeholder participation for environmental management: a literature review. Biol Conserv 141(10):2417–2431
59. Roberts CM, Hawkins JP (2000) Fully-protected marine reserves: a guide. Environ Conserv 27(2):103–125
60. Runge CA, Watson JEM, Butchart SHM, Hanson JO, Possingham HP, Fuller RA (2015) Protected areas and global conservation of migratory birds. Science 350(6265):1255–1258
61. Schreiber EA, Burger J (eds) (2002) Biology of marine birds. CRC Press, Boca Raton
62. Shepard CC, Crain CM, Beck MW (2011) The protective role of coastal marshes: a systematic review and meta-analysis. PLoS One 6(11):e27374
63. Shriver WG et al (2007) Population trends of breeding salt marsh birds: evidence from the Gulf of Maine. Wetlands 27(3):717–729
64. Smith JA, Brown PR (2022) Avian influenza outbreaks in migratory waterbirds: patterns and implications. J Wildl Dis 58(4):789–805
65. Spalding M, Ravilious C, Green EP (2001) World atlas of coral reefs. UNEP World Conservation Monitoring Centre, Berkley
66. Spalding M, Kainuma M, Collins L (2010) World atlas of mangroves. Earthscan, London

67. Stroud DA, Davidson NC, West RJ (2006) Waterbird conservation and international collaboration: an overview of flyway-scale agreements. Bird Conser Int 16(2):147–157
68. Studds CE, Kendall BE, Murray NJ, Wilson HB, Rogers DI, Clemens RS, Gosbell K, Hassell CJ, Jessop R, Melville DS, Milton DA, Minton CDT, Possingham HP, Riegen AC, Straw P, Woehler EJ, Fuller RA (2017) Rapid population decline in migratory shorebirds relying on Yellow Sea tidal mudflats as stopover sites. Nat Commun 8:14895
69. Taylor JP, Carter RL (2021) Overcoming barriers to stakeholder engagement in wildlife conservation: the case of waterbirds. J Environ Manag 288:112450
70. U.S. Fish and Wildlife Service (2020) Habitat conservation plan for the northern spotted owl (Report No. FWS-2020-HCP). U.S. Department of the Interior
71. United Nations (1992) Convention on biological diversity. Retrieved from https://www.cbd.int
72. United Nations Educational, Scientific and Cultural Organization (UNESCO) (1972) Convention concerning the protection of the world cultural and natural heritage. Retrieved from https://whc.unesco.org/en/conventiontext
73. United Nations Environment Programme (1979) Convention on the conservation of migratory species of wild animals (Bonn Convention). Retrieved from https://www.cms.int
74. Vaithianathan K (2020) Coexistence and conflict: case study on colonial waterbirds in southern India. Palgrave Macmillan, Singapore
75. Weston MA, Elgar MA (2005) Disturbance to brood-rearing hooded plover *Thinornis rubricollis*: responses and consequences. Bird Conserv Int 15(2):193–209
76. Wetlands International (2012) Waterbird population estimates, 5th edn. Wetlands International. https://www.wetlands.org/publications/waterbird-population-estimates-fifth-edition/
77. Wetlands International (2019) Climate change adaptation and waterbirds: a more integrated approach is needed. Wetlands International, Wageningen
78. Yadav R, Gupta RK, Kumar R, Kaur T, Moun R (2022) Biochemical response of earthworm, *Eisenia fetida* to heavy metals toxicity. J Appl Nat Sci 14(3):990–998
79. Yodha K, Singh D, Kour A, Delu V, Kumar R (2022) Community composition and status of avian diversity at campus and agricultural landscapes of Chaudhary Charan Singh Haryana Agricultural University, Hisar (Haryana). J Appl Nat Sci 14(4):1130
80. Yodha K, Singh D, Kour A, Yaday R, Dahiya P, Kumar R (2023) Estimation of predatory bird damage to winter guava crop. Indian J Ecol 50(5 (SI)):1787–1793. https://doi.org/10.55362/IJE/2023/4138
81. Yodha K, Singh D, Kour A, Yadav R, Kumar R (2024) Species diversity and abundance of birds in ber (*Zizyphus mauritiana*) crop. Asian J Conserv Biol 13(1):11725–11731. https://doi.org/10.53562/ajcb.83683
82. Zedler JB, Kercher S (2004) Causes and consequences of invasive plants in wetlands: opportunities, opportunists, and outcomes. Crit Rev Plant Sci 23(5):431–452
83. Zhao M, Cong P, Barter M, Fox AD (2012) The changing abundance and distribution of greater white-fronted geese *Anser albifrons* in the Yangtze River floodplain: impacts of recent hydrological changes. Bird Conserv Int 22(2):135–143
84. Zöckler C et al (2010) Wader studies and conservation of spoon-billed sandpipers in the Gulf of Mottama, Myanmar, Wader Study Group. Bulletin 117(2):10–113

# Chapter 14
# Aquatic Bird Conservation in a Changing Climate: Effective Enforcement Strategies

**Kiruthika Dhanapal, Baalaaji Damo, and Kamo Sende**

**Abstract** Birds, as a diverse and widely distributed class of vertebrates, serve as a vital force in preserving ecological balance and sustaining biodiversity. Among avifauna, waterbirds represent a specialized ecological group adapted to aquatic and semi-aquatic environments. Waterbirds act as keystone species in wetland ecosystems. Birds are extremely responsive to the environmental shifts, which enables them valuable gauges of ecosystem health. Understanding the factors that influence bird diversity is essential for biodiversity conservation and the protection of wetland ecosystems. On a global scale, climate change and land-cover alterations are among the most significant stressors affecting biodiversity. These factors often interact, intensifying the consequences of diminishing and fragmented habitats on species numbers. However, most studies do not span a sufficient timeframe to comprehensively assess the long-term impacts of climate variation on waterbirds. This Chapter seeks to bridge this critical research gap by examining the multifaceted interaction driven by climate change, habitat dynamics, and the conservation of aquatic bird species. It emphasizes the urgent need for adaptive enforcement strategies and policy reforms. By advocating for climate-resilient conservation frameworks, the chapter aims to contribute to sustainable biodiversity management and the long-term protection of waterbirds in a rapidly changing world.

**Keywords** Aquatic bird · Climate change · Laws · Conservation · Strategies

K. Dhanapal
Chennai Dr. Ambedkar Government Law College, Pudupakkam, Tamil Nadu, India

B. Damo
Madras High Court, Chennai, Tamil Nadu, India

Tamil Nadu Police Academy, Chennai, Tamil Nadu, India

K. Sende (✉)
The School of Law and Social Sciences, Robert Gordon University, Aberdeen, Scotland, UK
e-mail: k.sende@rgu.ac.uk

 
R. Kumar et al. (eds.), *Ecology, Conservation of Aquatic Birds*, Wetlands: Ecology, Conservation and Management 15, https://doi.org/10.1007/978-3-032-17125-2_14

## 14.1 Introduction

Birds, as a diverse and widely distributed class of vertebrates, serve as a vital force in preserving ecological balance and sustaining biodiversity [142]. With over 11,000 species globally, they occupy almost every habitat on Earth from high-altitude mountain ranges and dense rainforests to arid deserts, coastal regions, and wetlands. Their ecological functions are deeply interwoven into the integrity of ecosystems, while their cultural, economic, and scientific value spans human history and modern conservation strategies [7–10, 41].

Birds act as vital ecological service providers. Insectivorous species such as swallows and drongos act as natural pest controllers in agricultural settings, minimizing reliance on chemical pesticides while promoting sustainable agriculture [135]. Frugivorous birds like hornbills and barbets support seed dispersal, promoting forest regeneration and genetic diversity. Many species, such as hummingbirds and sunbirds, serve as pollinators of flowering plants, especially in tropical ecosystems. Scavenger birds, including vultures, accelerate the decomposition of animal carcasses, playing a role in limiting disease spread [141]. The catastrophic decline of vultures in India due to veterinary diclofenac contamination led to increased carcass persistence, higher populations of feral dogs, and a rise in rabies cases underscoring their critical role in ecosystem health and public hygiene [26, 91].

Birds also function as the bioindicators, their population trends, migratory shifts, breeding cycles, and phenological changes reflect underlying environmental changes, including climate anomalies, pollution, and habitat degradation [45, 53]. Their sensitivity to ecological stressors makes them valuable early warning systems for ecosystems in peril. Birds occupy a central position in trophic networks as predators, prey, and nutrient recyclers. Conservation efforts targeting bird habitats such as grasslands, forests, and wetlands often have umbrella effects, safeguarding a wide array of associated fauna and flora. Their ecological reach ensures that bird conservation strategies benefit entire ecosystems.

Birds also carry immense cultural and symbolic value. In Indian mythology and tradition, birds such as the peacock (India's national bird), eagle (Garuda), and swan (Hamsa) have spiritual and philosophical importance. Globally, birds appear in folklore, literature, and religious rituals as emblems of freedom, prophecy, and divinity [4]. Economically, birdwatching and nature tourism are growing industries. Sanctuaries like Keoladeo National Park (Rajasthan), Ranganathittu Bird Sanctuary (Karnataka), and Chilika Lake (Odisha) draw thousands of eco-tourists annually, supporting local livelihoods and incentivizing conservation through community participation [129].

The challenge of conserving aquatic birds in light of the climate change, has resulted in the evolution of a diverse body of scholarly work focused on adaptation strategies, habitat vulnerability, and policy integration. One of the seminal contributions in this field is the work of Mawdsley et al. [64], which provides a comprehensive assessment of strategies for adapting to climate change relevant to conservation of biodiversity and regulation of wildlife populations. The authors identify 16 overarching strategies, categorized under habitat and water resource conservation, along with targeted species interventions, monitoring and research, and legislative and institutional responses. These strategies underscore the need for flexible and anticipatory conservation practices that respond to dynamic climatic shifts. Importantly, the study advocates for increased protection of climatic refugia and habitat corridors, elements critical for migratory and aquatic birds whose survival hinges on seasonal movements and interconnected wetland systems [64].

Building on the theme of adaptive conservation, Pearce-Higgins and Green [93] present an exhaustive analysis of climate change effects on avian populations across global biogeographic regions. Their book emphasizes the importance of integrating climate forecasts into bird conservation policies and species action plans. They explore how the climatic variables, like shifts in temperature and rainfall patterns, influence birds phenology, distribution, and population viability. The authors argue that successful conservation amid climate change demands a twofold strategy: protecting current high-value habitats while also identifying and securing future climatic refugia. Their insights are particularly pertinent to aquatic birds, many of which rely on ephemeral wetland habitats whose hydrological characteristics are highly sensitive to changing rainfall regimes [93].

Complementing this discussion with a spatially explicit focus, Huntley et al. [49] offer a forward-looking analysis through their *Climatic Atlas of European Breeding Birds*. The atlas showcases projected future range shifts of bird species based on various climate scenarios. The authors conclude that numerous species are anticipated to relocate their breeding ranges significantly northward or to elevated altitudes, with some potentially losing significant parts of their existing habitats. Although focused on European species, the methodological framework employed has global applicability. For aquatic bird conservation in India and elsewhere, this study highlights the importance of developing dynamic reserve networks that can accommodate such range shifts, thereby reinforcing the case for long-term conservation planning [49].

In a more general yet crucial compilation, *Møller, Fiedler, and Berthold* [81] synthesize a wide array of studies in their edited volume *Effects of Climate Change on Birds*. This work delves into the biological and ecological consequences of climate change, including shifts in migration patterns, mismatches in breeding phenology, and alterations in food availability. Particularly relevant are chapters examining migratory waterbirds whose migratory cues are often linked to climatic patterns that are becoming increasingly erratic. The volume also stresses the need for multinational conservation frameworks, for instance the Convention on Migratory Species (CMS), to adjust their strategies in light of climate-induced disruptions to migratory routes and stopover sites [81].

Focusing specifically on aquatic bird habitats, Galbraith et al. [38] analyze the projected impacts of rising sea levels on intertidal zones, vital feeding and roosting areas for shorebirds. The study models habitat loss scenarios for various coastal regions and finds significant potential reductions in suitable habitats due to inundation and erosion. These findings highlight the urgency of implementing site-specific enforcement strategies, such as managed realignment and coastal buffer zoning, to offset habitat loss. Furthermore, the study argues for proactive adaptation planning that includes both physical infrastructure (e.g., dikes and artificial wetlands) and policy instruments (e.g., zoning regulations and habitat compensation schemes), thereby reinforcing the importance of integrative enforcement mechanisms in conservation practice [38].

This research adopts a purely doctrinal methodology, primarily relying on the analysis of existing literature, statutes, case laws, policy documents, international conventions, and previously published research studies. The doctrinal method, with its descriptive and analytical approach, enables a thorough examination of legal and institutional mechanisms overseeing the conservation of aquatic birds, particularly amid climate change and enforcement difficulties. The research is qualitative, analytical, and non-empirical, aiming to interpret legal provisions, evaluate enforcement mechanisms, and explore their adequacy in addressing climate-induced threats to aquatic birds. It also critically examines gaps in law and policy, providing recommendations for more effective conservation frameworks. The study focuses primarily on India, but situates the discussion within the broader international legal framework for bird conservation. The research is confined to aquatic bird species, particularly those dependent on inland wetlands, coastal areas, and estuarine ecosystems, and it analyses the legal and policy responses to their conservation in a climate change context.

## 14.2 Waterbirds and Their Ecological Significance

Among avifauna, waterbirds represent a specialized ecological group adapted to aquatic and semi-aquatic environments. They include species such as ducks, geese, swans, storks, herons, pelicans, ibises, flamingos, and shorebirds. These birds thrive in diverse habitats including wetlands, estuaries, lakes, mangroves, riverbanks, and coastal lagoons, and their ecological roles are crucial in maintaining aquatic ecosystem health.

Waterbirds act as keystone species in wetland ecosystems. Their foraging activities regulate populations of aquatic invertebrates and plants, maintaining open water areas and preventing excessive algal or weed growth [44]. Through their droppings (guano), waterbirds help transfer nutrients across ecosystems, transporting marine-derived nutrients inland during nesting or roosting. This guano input contributes significantly to soil and aquatic productivity [42]. Many waterbirds facilitate the dispersal of aquatic vegetation and seeds, promoting wetland plant propagation. This enhances wetland resilience against climate-driven changes and human disturbance. As wetlands are highly sensitive ecosystems, fluctuations in waterbird populations are often reliable indicators of ecological change such as altered hydrology, eutrophication, pollution, and the invasion of exotic species. Their presence or absence can be directly correlated with water quality and habitat viability [62].

Many waterbirds are long-distance migrants, using flyways such as the Central Asian Flyway (CAF) which spans 30 countries including India. These birds rely on a network of intact wetlands for resting, feeding, and breeding during their journeys, which can cover thousands of kilometers annually [13, 14]. India serves as a crucial stopover and wintering habitat for migratory waterbirds such as the Bar-headed Goose, Common Teal, Black-tailed Godwit, and Northern Pintail. Wetlands such as Chilika Lake, Pulicat Lake, Harike Wetland, and Point Calimere are vital for these species. Their conservation requires transboundary collaboration and the enforcement of global accords like the Ramsar Convention on Wetlands and the Central Asian Flyway (CAF) Action Plan.

In India, waterbirds support traditional livelihoods such as sustainable fishing, reed collection, salt pan activities, and handicrafts. Wetlands with high avian biodiversity are also centers for recreational tourism, birdwatching, spiritual practices, and community education. These areas provide climate resilience, flood mitigation, and water purification services further reinforcing their importance. Yet, many waterbird species are in decline due to wetland degradation, pollution, unsustainable land-use patterns and shifting climate conditions impacts. The *State of India's Birds 2023* report highlights population decreases in key species such as the Black-headed Ibis, River Tern, and Lesser Adjutant, emphasizing the need for urgently scaling up wetland protection and avian monitoring programs [114].

## 14.3 Climate Change

Birds exhibit a high degree of responsiveness to environmental changes, making them reliable indicators of ecosystem health [45]. Gaining insights into the drivers of bird diversity is essential for conserving biodiversity and safeguarding wetland ecosystems [68]. Globally, climate change and alterations in land cover represent two of the most critical stressors impacting biodiversity [51]. These drivers often interact synergistically, exacerbating the effects of habitat degradation and fragmentation on species populations, increasing the extinction risk for those already at risk, isolated populations [47].

Nevertheless, the majority of existing studies are limited in temporal scope, restricting their ability to fully assess the enduring effects of climate fluctuations on waterbirds and other biotic communities, as they tend to focus predominantly on landscape characteristics [39]. Climate change is expected to limit the ability of certain species to adapt to changes in land cover, thereby compounding existing conservation challenges [90]. Waterbirds, in particular, are known to be more sensitive to climatic fluctuations than many other avian groups, rendering them especially vulnerable to environmental changes and habitat degradation [93].

Climate change denotes sustained alterations in temperature, precipitation patterns, and other climate-related variables on a global scale. While natural phenomena like fluctuations in solar output, volcanic eruptions, and shifts in the Earth's orbital patterns have traditionally shaped the climate trends, a robust body of scientific evidence affirms that anthropogenic activities have been the predominant force driving climate change since the nineteenth century [51].

The principal driver of modern climate change is the accumulation of greenhouse gases, largely stemming resulting from the burning of fossil fuels such as coal, oil, and natural gas. These gases trap infrared radiation within the Earth's into the atmosphere, amplifying the greenhouse effect and leading to an increase in global temperatures [83]. Carbon dioxide ($CO_2$), primarily released through fossil fuel use, deforestation, and industrial processes, is the most impactful contributor to human-induced global warming. Since before the industrial revolution, atmospheric $CO_2$ levels have surged by approximately 50%, reaching 419 parts per million in 2022 [87]. Methane, a potent greenhouse gas emitted through agriculture, landfills, and fossil fuel extraction, is more than 25 times as effective effective than $CO_2$ in retaining heat over a 100-year period [34]. Likewise, nitrous oxide released from fertilizers, industrial activities, and fuel combustion possesses a global warming potential nearly 300 times greater than $CO_2$ [51]. Deforestation exacerbates climate change by reducing the planet's ability to naturally absorb $CO_2$, while industrial and agricultural practices continue to elevate greenhouse gas concentrations, thereby disturbing atmospheric equilibrium [139].

Empirical data confirm that the Earth's climate is changing at an unprecedented pace. As reported by NASA [83], the Earth's average surface temperature has risen by approximately 1.2 °C since the late 1800s, with most of this increase taking place over the last 40 years. The years 2016 and 2020 remain tied as the warmest on record. Rising ocean temperatures are among the most significant signs of climate change. Oceans absorb nearly 90% of the surplus heat caused by greenhouse gas emissions, and since 1969, the top 100 m of ocean water have warmed by about 0.33 °C [51]. Among the most significant consequences of climate change is the accelerated melting of ice sheets and glaciers. According to NASA's GRACE satellite data, Greenland lost an average of 279 billion tons of ice per year from 1993 to 2019, while Antarctica saw an annual loss of approximately 148 billion tons during the same period [83]. Glaciers in major mountain ranges including the Alps, Himalayas, Andes, and Rockies are retreating at record-breaking speeds [120]. Global sea levels have risen by around 20 centimeters over the last century, with the rate of rise in the past two decades nearly doubling that of the twentieth century [89]. In parallel, extreme weather events such as heatwaves, hurricanes, wildfires, and droughts have become increasingly frequent and intense. In the United States, there has been a notable rise in record-high temperatures since the 1950s, alongside a decline in record lows [34].

Climate change carries profound consequences for ecosystems, human communities, and the global economy. Shifts in temperature and rainfall patterns have triggered severe droughts in areas such as the Western United States, Australia, and parts of Africa [51]. Meanwhile, rising sea levels present a critical risk to coastal cities and island nations. Urban centers like Mumbai, Jakarta, and Miami are already facing repeated flooding events, highlighting the growing vulnerability of low-lying regions [128]. The agricultural sector is also vulnerable, as climate-induced changes significantly disrupt crop production and threaten global food security [36]. Beyond environmental and economic concerns, climate change presents critical public health challenges. Rising temperatures exacerbate heat-related health issues, promote the spread of vector-borne diseases such as malaria and dengue, and deteriorate air quality, heightening the risk of respiratory conditions [138].

There is strong scientific consensus that recent climate change is predominantly caused by human activities. The Intergovernmental Panel on Climate Change (IPCC), a foremost authority on climate science, has clearly affirmed that human-induced factors are the principal drivers of global warming [51]. Paleoclimate records drawn from ice cores, tree rings, and ocean sediments show that the current pace of warming is roughly ten times faster than the natural rate observed during past post-glacial periods. Additionally, Human activities are driving a rapid increase in atmospheric $CO_2$ levels at a rate approximately 50 times faster than the natural rise that occurred after the last Ice Age [83]. This compelling evidence underscores the critical urgency for coordinated global action. Addressing the most severe effects of climate change will demand significant reductions in greenhouse gas emissions, a swift transition to renewable energy sources, and the implementation of comprehensive adaptation strategies. The choices made in the coming decades will play a decisive role in shaping the future climate and determining how well humanity can withstand its wide-ranging impacts.

## 14.4 Global and Indian Coastal and Waterbird Trends

The health of aquatic bird populations is intrinsically linked to the condition of coastal and wetland habitats that function as breeding, nesting, and migratory stopovers. Assessing the geographical and ecological features of these environments, both globally and in India is crucial for developing effective climate-responsive conservation strategies. An overview of global and Indian coastal and waterbird trends is shown in Table 14.1.

### *14.4.1 Global Coastal Overview*

The global coastline stretches approximately 356,000 km, shaped by diverse coastal geomorphic processes such as erosion, accretion, wave action, and sediment dynamics [89]. Coastal regions support around 40% among people worldwide, often concentrated in mega-cities that increase the pressure on these fragile environments [121]. The Intergovernmental Panel on Climate Change [52] reports that rising global temperatures have led to an average sea-level rise of 3.3 mm per year. This accelerating trend intensifies coastal erosion, increases the frequency of flooding, and contributes to the salinization of freshwater resources in coastal areas. Over the past century, approximately 50% of the world's coastal wetlands have been lost, largely due to land reclamation, pollution, and expanding infrastructure [27, 99].

**Table 14.1** Global and Indian coastal and waterbird trends

| Details of parameters | | References |
|---|---|---|
| *Coastal length comparison* | | [80, 89] |
| *Region* | *Kilometers* | |
| Global coastline | 356,000 | |
| Indian coastline | 7516.6 | |
| *Average sea-level rise* | | |
| *Region* | *mm/year* | [50, 52] |
| Global | 3.3 | |
| India | 1.3 | |
| *Coastal erosion in India* | | |
| *Type* | *Percentage* | [80, 88] |
| Eroding | 34% | |
| Stable/accreting | 66% | |
| *Wetland loss globally* | | |
| *Type* | *Percentage* | [27, 99] |
| Global wetland loss | 64% | |
| Coastal wetlands lost (last century) | 50% | |
| *Species decline trends* | | |
| *Type* | *Percentage* | [7–10, 140] |
| Freshwater vertebrate decline Since 1970 | 70% | |
| Waterbird decline in India (20 years) | 25% (avg.) | |

## 14.4.2 Indian Coastal Profile

India's coastline, which extends for 7516.6 km, spans nine states and four Union Territories, hosting 66 coastal districts that are highly populated and economically vital [80]. The mangrove cover in India is approximately 4992 sq. km, with significant concentrations in the Sundarbans, Bhitarkanika, and Pichavaram regions [37]. The country is also home to major estuarine systems and deltas, including those formed by the Ganga-Brahmaputra, Godavari, Krishna, Mahanadi, and Cauvery rivers. As reported by the National Centre for Coastal Research (NCCR), around 34% of India's coastline is experiencing erosion, with hotspots observed in Kerala, Odisha, and West Bengal [88]. Research conducted by the Indian National Centre for Ocean Information Services (INCOIS) report an observed average sea-level increase of 1.3–2 mm per year along India's coastline [50].

## 14.4.3 Global Waterbird Statistics

There are over 870 known species of waterbirds globally, many of which engage in extensive migratory movements that span continents [130]. Migration pathways follow key flyways, including the East Asian—Australasian Flyway (EAAF), the

African–Eurasian Flyway, and the Central Asian Flyway (CAF). The CAF alone supports over 270 migratory bird species across 30 countries, connecting Arctic breeding grounds with wintering areas in South and Central Asia [23]. However, anthropogenic pressures have severely impacted waterbird populations. The Living Planet Report [140] notes a 70% decline in freshwater vertebrates, including waterbirds, since 1970. It is estimated that over 130 waterbird species are currently threatened, and 64% of a substantial portion of global wetland ecosystems has been lost since the early 1900s [99].

Agricultural intensification in regions adjacent to wetlands—often driven by global food trade and input-heavy farming systems—has led to increased habitat degradation and water stress. These land-use changes diminish the ecological integrity of wetland ecosystems and reduce their suitability for migratory and resident waterbirds. Recognizing and supporting the diversity of farming systems that coexist with wetlands is vital for ensuring long-term sustainability of both food production and bird conservation in India.

### *14.4.4 Waterbird Trends in India*

India is a critical country within the Central Asian Flyway, annually hosting over 150 migratory species and nearly 200 resident waterbird species [15]. Notable migratory species include the Bar-headed Goose, Eurasian Wigeon, Northern Pintail, and Common Teal, while Painted Stork, Black-headed Ibis, and Sarus Crane represent significant resident populations. India currently has 80 Ramsar sites, covering over 1.3 million hectares of wetland area, many of which support large numbers of waterbirds [79]. During the 2024 Asian Waterbird Census, Chilika Lake recorded over 1.1 million waterbirds representing 180 species, reaffirming its global importance [19]. The National Wetland Atlas published by ISRO's Space Applications Centre has mapped over 2,01,503 wetlands in India, yet many remain vulnerable due to encroachment, eutrophication, and unsustainable development [108]. Long-term monitoring at sites such as Keoladeo National Park and Pulicat Lake reveals a 20–30% decline in waterbird numbers at some locations over the past two decades [7–10].

## 14.5 The Impact of Climate Change on Aquatic Birds: A Complex Interaction of Environmental Factors

Climate change has profound and multifaceted effects on waterbirds, influencing their habitat, food availability, migration patterns, and reproductive success. The intricate interplay of climate variables, such as temperature fluctuations, precipitation levels, and extreme weather events, significantly impacts waterbird species and their ecosystems. Understanding these relationships is critical for biodiversity conservation and managing wetlands in response to global climate change [51].

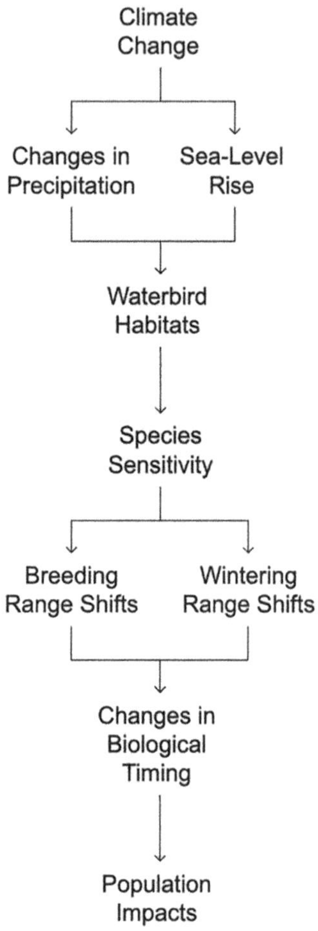


**Fig. 14.1** Effects of climate change on waterbird populations

Figure 14.1 illustrating the effects of climate change on waterbird populations. The diagram starts with "Climate Change" at the top, which leads to two branches: "Changes in Precipitation" and "Sea-Level Rise." Both branches converge into "Waterbird Habitats," which then leads to "Species Sensitivity." From there, two outcomes are shown: "Breeding Range Shifts" and "Wintering Range Shifts," both of which lead to "Changes in Biological Timing," culminating in "Population Impacts."

### *14.5.1 Influence of Precipitation on Waterbird Habitats*

Precipitation plays a vital role in shaping waterbird habitats and community structures. Variations in rainfall frequency and intensity directly affect river flows, which in turn influence estuarine conditions such as salinity levels, water clarity, and primary productivity. Increased precipitation can lead to flooding events that disrupt nesting efforts and submerge essential mudflat habitats used by waders. Conversely, these flood events can create more open water habitats favorable to swimming birds [131]. Moreover, increased river inflows can negatively impact fish populations, which constitute a vital food source for numerous waterbird species, while large flooding events shift benthic organisms from deep to shallow waters, altering the food web dynamics [120].

Extreme rainfall events also contribute to habitat instability through strong winds and increased sedimentation, which can destroy nests and disrupt breeding cycles. In contrast, reduced rainfall, coupled with higher temperatures, increases evapotranspiration rates, resulting in shorter flooding durations and decreased wetland habitat availability, particularly in coastal regions [139].

### 14.5.2 Species Sensitivity to Climate Change

Migratory waterbirds are especially susceptible to climate change due to their dependence on consistent wetland habitats along their migratory pathways. Changes in water levels affect their breeding, foraging, and wintering grounds, often forcing them to adapt to new habitats or altering migration patterns. Species with small populations, fragmented distributions, or specialized dietary requirements are especially susceptible to climate-induced habitat degradation [140].

### 14.5.3 Breeding Range Shifts

Temperature plays an essential role in determining species distribution, and many waterbird species have experienced significant shifts in their breeding ranges due to climate change. Studies have shown that rising global temperatures have driven some species to expand their breeding ranges poleward. A notable example is the Black tailed Godwit (*Limosa limosa*) in Europe has expanded its breeding range northward into previously unsuitable habitats as warmer temperatures have made these areas more viable for nesting. Similarly, the Eurasian Spoonbill (*Platalea leucorodia*) has extended its breeding range in response to changing climatic conditions in northern Europe.

### 14.5.4 Wintering Range Shifts

The wintering distribution of many waterbird species has also changed significantly due to shifts in climate conditions. Milder winters have allowed some species to overwinter at higher latitudes rather than migrating to traditional wintering grounds. A notable example is the Eurasian Wigeon (*Mareca penelope*), which has reduced its migration distance and increasingly winters in northern Europe instead of traveling further south. In North America, the Northern Pintail (*Anas acuta*) has demonstrated a similar pattern, with increasing numbers wintering in the central U.S. instead of migrating further south into Mexico.

### 14.5.5 Changes in the Timing of Biological Events

Climate change has altered the phenology (timing of biological events) of many waterbird species, with earlier breeding and migration patterns observed worldwide. For instance, the Common Cuckoo (*Cuculus canorus*) in the UK has been recorded laying eggs earlier in the season, corresponding to warmer spring temperatures. Similarly, research on the Great Crested Grebe (*Podiceps cristatus*) in Germany has shown that earlier ice melting in lakes has resulted in an advancement of their breeding season.

### 14.5.6 Mistiming of Biological Events

As species adapt their life cycles to shifting climate conditions, they may become out of sync with the timing of essential resources. For example, migratory waterbirds like the Pied Flycatcher (*Ficedula hypoleuca*) have been observed arriving earlier at their breeding sites in response to rising temperatures, potentially leading to mismatches with peak food availability. However, their primary food source, caterpillars may not be available in sufficient quantities at that time, leading to reduced reproductive success. A comparable mismatch has been documented in Arctic-breeding geese, including the Barnacle Goose (Branta leucopsis), which arrive at breeding sites before optimal vegetation growth has occurred due to earlier snowmelt, leading to lower chick survival rates.

### 14.5.7 Sea-Level Rise

The rise in sea levels presents a significant threat to coastal waterbird populations by flooding nesting sites and causing habitat loss. In the United States, the breeding colonies of the Piping Plover (*Charadrius melodus*) along the Atlantic coast are at risk due to increasing storm surges and coastal erosion [38]. In the Netherlands, the Spoonbill (*Platalea leucorodia*) s threatened by habitat degradation caused by sea-level rise and coastal habitat degradation. The degradation of salt marshes and intertidal zones has a direct impact on shorebirds that depend on these habitats for nesting and feeding.

### *14.5.8 Changes in Rainfall*

Waterbirds are particularly sensitive to changes in rainfall patterns, which directly impact wetland habitats. For instance, the Sahel region in Africa has experienced severe droughts due to reduced rainfall, leading to habitat degradation for migratory waterbirds such as the Black Stork (*Ciconia nigra*) and the White-faced Whistling Duck (*Dendrocygna viduata*). In contrast, extreme rainfall events can cause flooding, submerging nesting sites and reducing the availability of food resources. This has been observed in the Everglades, where increased rainfall variability has disrupted the nesting cycles of species like the Wood Stork (*Mycteria americana*).

### *14.5.9 Population Impacts*

Climate change affects waterbird populations by influencing survival rates and reproductive success. For example, studies on the Common Eider (*Somateria mollissima*) in the Arctic have shown that increasing temperatures lead to higher mortality rates due to reduced food availability and heat stress. In the case of the Emperor Penguin (*Aptenodytes forsteri*), melting sea ice in Antarctica threatens breeding colonies by reducing the availability of stable ice platforms necessary for chick rearing. In Australia, prolonged drought conditions have led to population declines in the Australian Pelican (*Pelecanus conspicillatus*), which relies on wetland habitats for breeding.

Climate change influences species both directly and indirectly, primarily through altered resource availability. The complex interplay between climatic conditions, food web dynamics, and habitat stability highlights the urgent necessity for holistic conservation strategies focused on addressing climate impacts on waterbird populations.

## 14.6 Conservation Policies and Legal Frameworks on Aquatic Birds at the International Level

The international community has implemented various legal frameworks and policy instruments to safeguard aquatic bird species and promote the sustainable management of their habitats. These agreements emphasize both the conservation of these birds and the importance of cross-border collaboration to effectively address shared environmental challenges, given the migratory nature of many species. Through conventions, treaties, and collaborative networks, countries across the world are working together to safeguard aquatic bird populations for current and future generations [28, 123, 124].

### 14.6.1 Ramsar Convention (1971)

The Ramsar Convention, officially titled the Convention on Wetlands of International Importance especially as Waterfowl Habitat was adopted in 1971 in Ramsar, Iran. It is recognized as the first modern global treaty dedicated to the conservation and sustainable use of natural resources, specifically wetlands. The Convention recognizes wetlands as ecosystems of immense value not only for biodiversity but also for their crucial role in water purification, flood control, climate regulation, and human livelihoods. Among its foundational principles is the recognition of wetlands as essential habitats for waterbirds, making aquatic bird conservation a central objective of the treaty [35].

A distinctive feature of the Ramsar Convention is its emphasis on the concept of "wise use" of wetlands. Rather than advocating a purely preservationist stance, the Convention promotes the sustainable utilization of wetland resources in a manner that sustains their natural ecological functions [97]. This approach is particularly relevant for aquatic birds, which rely on wetlands for breeding, nesting, feeding, and as stopover sites during migratory journeys. By advocating for balanced human use and ecological protection, the Ramsar framework ensures that conservation goals are integrated with local livelihoods and broader environmental governance.

A key component of the Ramsar Convention's approach is the identification and designation of wetlands of international importance, referred to as Ramsar Sites. These sites are selected based on scientific criteria, many of which are directly related to waterbird populations. For example, a site may qualify if it consistently supports at least 20,000 waterbirds or hosts 1% of the population of a particular species or subspecies [98]. Through this mechanism, the Convention ensures that attention and conservation efforts are directed toward wetlands that are critical for the global survival of aquatic birds.

Moreover, the Ramsar Convention actively promotes international cooperation, especially in the context of migratory waterbirds that move across national borders. It encourages contracting parties to engage in shared management of wetland systems and the protection of migratory flyways. By aligning with related frameworks like the Convention on Migratory Species (CMS) and the African—Eurasian Waterbird Agreement (AEWA), the Ramsar Convention enhances the legal and policy architecture supporting migratory bird conservation on a global scale [1, 2, 122].

In addition to legal and technical instruments, the Ramsar Convention supports education, research, and awareness-raising initiatives. The global observance of World Wetlands Day on February second commemorates the Convention's adoption and is a platform for promoting the significance of wetlands and their avian inhabitants [103]. These outreach efforts foster a global conservation ethic, engaging policymakers, researchers, and communities in collective action to preserve aquatic bird habitats.

### *14.6.2 Convention on Migratory Species (CMS or Bonn Convention 1979)*

The Convention on the Conservation of Migratory Species of Wild Animals (CMS), commonly referred to as the *Bonn Convention*, was adopted in 1979 in Bonn, Germany, and came into force in 1983. Administered by the United Nations Environment Programme (UNEP), this international treaty is distinctive in its emphasis on conserving migratory species throughout their entire migratory range [123, 124]. Migratory animals including many species of aquatic birds cross international borders and depend on various habitats in different countries. The CMS offers a legal and collaborative framework to facilitate coordinated global action for the protection of such species [13, 14].

A defining strength of the CMS is its two-tier system of protection, the treaty is organized through its Appendices, with Appendix I identifying migratory species at risk of extinction and obligating member states to provide strict protection measures, such as conserving habitats and eliminating obstacles to migration. Many endangered aquatic bird species, such as the Siberian Crane and certain storks, fall under this category [24]. Appendix II, in contrast, includes species that would benefit from international cooperation, and it encourages the development of specialized regional or global agreements tailored to those species' needs. This tiered and adaptable approach has made CMS highly effective in dealing with the diverse challenges faced by migratory aquatic birds [119].

Among the key instruments developed under the CMS is the Agreement on the Conservation of African—Eurasian Migratory Waterbirds (AEWA). AEWA focuses on 255 species of migratory waterbirds utilizing the African—Eurasian Flyway, a critical migratory corridor of global significance [1, 2]. It has enabled the implementation of species action plans, harmonized national conservation laws, and supported research and habitat protection efforts that span multiple continents [123, 124].

In addition to binding legal commitments, CMS also prioritizes research, data-sharing, and capacity-building. Through population monitoring, satellite tracking, and habitat mapping, it facilitates the identification of critical flyway sites, many of which are essential for the survival of aquatic birds during migration [12]. This science-policy interface has enhanced decision-making among member states.

Moreover, CMS supports public outreach and environmental education by recognizing migratory species, especially birds, as flagship species for wider conservation campaigns. Aquatic birds are often emblematic of broader ecological issues like wetland degradation and climate change, helping to raise awareness and mobilize political will for environmental protection.

### 14.6.3 Agreement on the Conservation of African-Eurasian Migratory Waterbirds (AEWA 1995)

The AEWA, adopted in 1995 and effective since 1999, is a landmark treaty developed under the CMS. It is among the most comprehensive multilateral agreements aimed specifically at the protection of migratory waterbirds and their habitats [1, 2]. It encompasses 119 countries across Africa, Europe, the Middle East, portions of Asia, and even extends to Canada, AEWA focuses on safeguarding species along the African-Eurasian Flyway, a complex web of migration routes that connect continents and ecosystems [122].

AEWA addresses the needs of over 250 species, including ducks, swans, cranes, storks, gulls, and terns. These birds depend on diverse wetland types such as estuaries, floodplains, coastal zones, and inland lakes which are increasingly under threat. AEWA promotes an integrated conservation approach, balancing human use of wetland resources with the imperative of ecological integrity [116].

A cornerstone of AEWA is its legally binding Action Plan, which mandates specific conservation measures for contracting parties. These include habitat protection and restoration, regulation of hunting, monitoring programs, and public engagement. The Action Plan also stresses international cooperation, urging countries to align their national efforts with flyway-level objectives and support capacity development in less-resourced regions [1, 2].

AEWA has made significant contributions to flyway conservation, by identifying key migratory sites and supporting collaborative projects such as the African-Eurasian Flyway Project. It partners with major NGOs like BirdLife International and Wetlands International to ensure that conservation is science-based and participatory [12]. These partnerships help mitigate threats like habitat loss, pollution, and climate-induced shifts in migratory patterns.

The Agreement also places strong emphasis on capacity-building and knowledge-sharing, offering workshops, technical assistance, and training sessions. This is especially crucial for countries in the Global South, where institutional and financial resources for conservation may be limited. By supporting community-based wetland management and promoting equity in conservation governance, AEWA advances both biodiversity goals and sustainable development [122].

### 14.6.4 Convention on Biological Diversity [16]

The Convention on Biological Diversity (CBD), established at the 1992 Earth Summit in Rio de Janeiro, is a landmark international treaty that provides a comprehensive framework for conserving biological diversity, promoting the sustainable use of its components, and ensuring the fair and equitable sharing of benefits arising from genetic resources [16]. While the CBD encompasses a wide range of biodiversity, terrestrial, freshwater and marine it has substantial implications for the conservation of aquatic birds, which are deeply dependent on wetland and coastal ecosystems for breeding, foraging, and rest during migration.

Unlike species-specific conventions such as AEWA or CMS, the CBD promotes an ecosystem-based approach to conservation [17]. This approach is highly relevant for aquatic birds whose survival relies on the connectivity and integrity of diverse ecosystems such as marshes, estuaries, mangroves, and floodplains. The CBD encourages parties to identify and protect key habitats, integrate biodiversity into national strategies, and reduce pressures like pollution and habitat conversion factors that critically affect aquatic bird populations [110].

A key contribution of the CBD to aquatic bird conservation has been the formulation of Strategic Plans, including the Aichi Biodiversity Targets (2011–2020) and the Post-2020 Global Biodiversity Framework adopted in 2022. For example, Target 11 of the Aichi framework sought to safeguard 17% of terrestrial and inland water ecosystems and 10% of marine areas, directly benefiting critical bird habitats [110]. These goals are complemented by the Post-2020 Framework's focus on halting biodiversity loss and restoring degraded ecosystems, including wetlands.

The CBD also promotes the use of scientific research, traditional knowledge, and multi-stakeholder engagement in decision-making. This has led to integrated wetland management, cross-sectoral policies, and community-based conservation programs that directly support both human development and bird protection [17]. Additionally, by mainstreaming biodiversity into agriculture, fisheries, and urban planning, the CBD addresses many indirect drivers of aquatic bird decline.

By leveraging financial instruments like the Global Environment Facility (GEF), the CBD enables developing countries to carry out conservation programs, often focusing on wetland restoration and waterbird habitat protection [40]. Moreover, it fosters synergies with other treaties, notably Ramsar and CMS, ensuring coordinated international efforts. By embedding aquatic bird conservation into broader biodiversity and climate agendas, the CBD plays a vital role in securing the long-term survival of these species.

### *14.6.5 Convention on International Trade in Endangered Species of Wild Fauna and Flora (CITES 1973)*

The Convention on International Trade in Endangered Species of Wild Fauna and Flora (CITES), adopted in 1973 and entering into force in 1975, plays a vital role in regulating international trade in wild species to ensure it does not jeopardize their survival [22]. Although often associated with iconic land animals, CITES is also essential in protecting aquatic birds from overexploitation through trade, particularly those targeted for ornamental use, private collections, or their feathers and eggs.

CITES organizes species into three Appendices according to their conservation status. Appendix I lists species at risk of extinction and bans their commercial trade.

Appendix II permits regulated trade to avoid unsustainable exploitation, while Appendix III applies to species protected in at least one country [22]. Aquatic birds such as the Shoebill (Balaeniceps rex), certain cranes, and spoonbills are listed under these appendices due to their vulnerability to illegal trade.

CITES operates through a permit system, requiring scientific and legal assessments before allowing export or import. This helps control and monitor trade and prevents further population declines [105]. It also provides support to enforcement bodies to detect and intercept illicit trafficking. By addressing trade-related threats, CITES complements habitat-centered treaties like CMS and Ramsar.

The Convention also fosters capacity-building and training to improve implementation, particularly in developing countries with limited enforcement capacity. It encourages international cooperation, information exchange, and legislative strengthening. Although it does not address habitat degradation directly, its focus on curbing trade-driven threats fills a critical niche in global bird conservation.

### *14.6.6 UNESCO World Heritage Convention (1972)*

Adopted in 1972, the UNESCO World Heritage Convention seeks to recognize, safeguard, and maintain cultural and natural heritage sites of outstanding universal value. While widely known for safeguarding historic monuments, it also protects ecologically significant natural sites, many of which serve as essential habitats for aquatic birds [125].

By recognizing wetlands, river deltas, lagoons, and other bird-rich ecosystems as World Heritage Sites, the Convention elevates global awareness of their ecological importance. Sites like the Sundarbans (India and Bangladesh), Danube Delta (Romania), and Okavango Delta (Botswana) are prime examples of avian biodiversity hubs under UNESCO protection [126]. These areas support breeding colonies, migratory stopovers, and feeding grounds for endangered bird species.

The international recognition associated with World Heritage status often leads to increased funding, better legal protections, and eco-tourism opportunities. This helps reduce local threats such as illegal development, pollution, and unregulated tourism [63]. In many cases, inscription catalyzes national governments to enhance management efforts and integrate conservation into broader policy.

The Convention implements monitoring tools like Periodic Reporting and the List of World Heritage in Danger to assess and respond to threats facing designated sites, which prompt corrective actions when ecological degradation is detected [125]. These tools are particularly valuable for aquatic birds, which respond rapidly to environmental changes.

By promoting community engagement and sustainable development, UNESCO encourages a shared sense of responsibility in conserving biodiversity. The Convention aligns well with other conservation efforts and promotes a global conservation ethic that supports aquatic bird protection within the broader framework of natural heritage preservation.

The conservation of aquatic birds, key ecological indicators and iconic inhabitants of the world's wetlands and coastal regions, requires a comprehensive and collaborative international approach. The global community has responded with a robust suite of legal instruments and cooperative frameworks, each addressing specific dimensions of conservation from habitat protection and migratory connectivity (e.g., Ramsar, CMS, AEWA), to ecosystem-based governance (CBD), regulation of international trade (CITES), and the recognition of ecologically critical sites (UNESCO World Heritage Convention). These mechanisms, while distinct, are complementary and together create a multilayered safety net that addresses the diverse threats facing aquatic birds today. The success of these frameworks lies not only in their legal structures but also in their implementation through science, policy, local engagement, and international cooperation, all crucial for ensuring the survival of aquatic bird species for generations to come.

As international organizations promote sustainability in wetland ecosystems, they must also critically examine how global environmental finance mechanisms—such as REDD+ and sustainability certification—affect smallholder participation. In many cases, compliance burdens and technical criteria under these schemes marginalize resource-poor producers, limiting their access to ecological incentives. A shift toward more inclusive, rights-based frameworks is needed to ensure that carbon and biodiversity markets support both ecological integrity and livelihood justice in wetland-rich regions.

## 14.7 The Role of International Organizations and Initiatives in Aquatic Bird Conservation

International organizations and initiatives serve as essential operational arms that complement and strengthen the enforcement and integration of international legal instruments like the Ramsar Convention, CBD, CMS, AEWA, and CITES. While these legal instruments provide the formal structure for international cooperation, standard-setting, and policy mandates, international organizations translate those commitments into on-the-ground action, scientific research, capacity-building, and community engagement [11]. Their work ensures that conservation strategies are data-driven, participatory, and adaptive, particularly in response to evolving challenges like climate change, habitat degradation, and human-wildlife conflict [132–134].

Moreover, by facilitating cross-border cooperation, supporting local conservationists, and influencing national and international policies, these organizations bridge the gap between global ambitions and local realities, playing a pivotal role in securing a sustainable future for aquatic birds and the ecosystems they inhabit [7].

### 14.7.1 BirdLife International

BirdLife International is one of the most prominent global conservation organizations dedicated to the protection of birds, their habitats, and global biodiversity. Functioning as an international alliance of NGOs across more than 100 countries, BirdLife plays a crucial role in aquatic bird conservation by combining scientific research, grassroots action, and policy advocacy [8]. One of its most significant contributions is the identification and surveillance and assessment of Important Bird and Biodiversity Areas (IBAs) sites deemed critical for the conservation of bird populations [30]. These IBAs often include key wetlands, coastal zones, and inland water bodies essential for aquatic birds' breeding, feeding, and migratory stopovers.

BirdLife works in partnership with the IUCN and other international organizations to deliver scientifically grounded data for the IUCN Red List, especially in evaluating the conservation status of avian species [8]. The organization also contributes expert input to the CBD and CMS, ensuring that global policy frameworks reflect current avian conservation priorities [7].

Through its extensive network, BirdLife supports community-led conservation, promotes sustainable land and water management, and raises awareness about major threats facing aquatic birds, including habitat destruction, invasive species, and climate change. Its integrated approach ensures that conservation is locally grounded but globally coordinated.

### 14.7.2 Wetlands International

Wetlands International is a leading global non-profit dedicated to the conservation and restoration of wetland ecosystems. Through evidence-based advocacy and practical interventions, it champions the sustainable management of wetlands, critical habitats for a wide range of aquatic bird species [132–134].

A cornerstone of its work is the International Waterbird Census (IWC), a long-running, global monitoring program that tracks the status and trends of waterbird populations across more than 100 countries. This data underpins decision-making under treaties such as AEWA, Ramsar, and CBD [131]. Wetlands International also engages in the development of ecosystem service models, demonstrating how healthy wetlands contribute to biodiversity, climate resilience, and sustainable livelihoods.

Working with governments, communities, and multilateral platforms, the organization ensures that wetland values are mainstreamed into national development plans, strategies for building resilience to climate change, nd minimizing risks associated with natural hazards frameworks [107]. Wetlands International emphasizes the ecological connectivity of wetland systems, thereby promoting landscape-level conservation that reflects the migratory nature of many aquatic bird species.

### *14.7.3 East Asian–Australasian Flyway Partnership (EAAFP)*

The East Asian–Australasian Flyway Partnership (EAAFP) represents a collaborative regional effort dedicated to conserving migratory waterbirds and their wetland habitats along the East Asian–Australasian Flyway through a voluntary, non-binding framework. Spanning over 22 countries from Alaska and Siberia to New Zealand and Australia, the flyway supports more than 200 species and over 50 million migratory waterbirds [31].

EAAFP engages governments, NGOs, intergovernmental bodies, site managers, and research institutions to identify, manage, and conserve Flyway Network Sites, wetlands and coastal areas critical to bird migration [32]. These sites benefit from technical support, shared data, research collaboration, and capacity-building initiatives. EAAFP also promotes community participation, awareness campaigns, and education programs, fostering public stewardship of wetland ecosystems.

The partnership addresses acute threats such as coastal reclamation, pollution, and unsustainable urban development, which are particularly severe in parts of East and Southeast Asia. By linking local action to global frameworks, EAAFP contributes substantially advancing the goals of the CMS, Ramsar Convention, CBD, and the United Nations Sustainable Development Goals.

## 14.8 India's Response and Participation in International Conventions for Aquatic Bird Conservation

India, with its vast network of wetlands, coastal zones, rivers, and inland lakes, is both a destination and a transit point for millions of migratory waterbirds. Recognizing the ecological significance of these birds and the habitats they rely upon, India has actively engaged with numerous international environmental treaties. This participation not only reflects India's commitment to global biodiversity goals but also significantly informs its domestic conservation policies.

India acceded to the Convention on Biological Diversity (CBD) in 1994, affirming its commitment to conserving biodiversity, promoting sustainable use, and ensuring fair benefit-sharing. To operationalize these objectives, the country introduced the National Biodiversity Action Plan (NBAP) in 2008, later revised in 2014, incorporating specific measures for the protection of inland and coastal wetlands, vital habitats for aquatic birds. Complementing this, the Biological Diversity Act, 2002, was enacted to fulfill India's obligations under the CBD, leading to the establishment of the National Biodiversity Authority (NBA) to coordinate conservation efforts and regulatory frameworks [67] (Biological Diversity Act 2002).

India has been a party to the Convention on International Trade in Endangered Species of Wild Fauna and Flora (CITES) since 1976. To align domestic legislation with CITES obligations, the [136], was amended to restrict the trade of listed endangered species, including various aquatic birds such as herons, cranes, and spoonbills. The CITES Management Authority of India, operating under the Ministry of Environment, Forest and Climate Change (MoEFCC), oversees the issuance of trade permits, promotes public awareness, and coordinates with enforcement agencies to combat illegal trafficking of birds and their parts (CITES, India Country Profile) [74, 75].

Demonstrating its ongoing commitment to migratory species, India became a party to the Convention on the Conservation of Migratory Species of Wild Animals (CMS) in 1983. A significant milestone was achieved when India hosted the 13th Conference of the Parties (COP13) to CMS in 2020 in Gandhinagar, Gujarat, under the theme "Migratory species connect the planet and together we welcome them home." On this occasion, India unveiled the National Action Plan for the Conservation of Migratory Birds and their Habitats along the Central Asian Flyway (CAF), a vital route for numerous aquatic migratory birds such as ducks, geese, and waders. The action plan emphasizes habitat restoration, species monitoring, and community engagement as core strategies [25, 76, 77].

India's involvement in the Ramsar Convention on Wetlands, which it joined in 1982, has resulted in the recognition of 80 wetlands as Ramsar Sites of international importance (as of 2024). Notable among these are Chilika Lake, Keoladeo National Park, and Loktak Lake, all renowned for their abundant waterbird populations. The Ramsar framework has guided the formulation of India's National Plan for Conservation of Aquatic Ecosystems (NPCA), which underscores the ecological and socioeconomic significance of wetlands and facilitates state-level conservation initiatives [78, 104].

Although India is not a formal signatory to the Agreement on the Conservation of African-Eurasian Migratory Waterbirds (AEWA), it actively engages in aligned conservation efforts through the Central Asian Flyway (CAF) framework. India's CAF National Action Plan (2021–2030) incorporates several AEWA-aligned goals, including fostering international collaboration, tracking migratory pathways, and promoting sustainable wetland management. This strategic alignment enables India to play a significant role in global initiatives for the conservation of migratory waterbirds, despite not being an AEWA party [76, 77].

In addition to biodiversity and species-specific treaties, India is a signatory to the UNESCO World Heritage Convention (1972), having ratified it in 1977. The World Heritage Convention has facilitated the recognition of ecologically vital wetlands such as Keoladeo National Park (Rajasthan), Sundarbans National Park (West Bengal), and the Western Ghats as World Heritage Sites. These designations underscore their importance for avian biodiversity and ensure enhanced global oversight, access to funding mechanisms, and strengthened conservation responsibilities under the convention [127].

India's proactive engagement with these conventions exemplifies its broader environmental diplomacy and its integration of international conservation standards into national policy. By linking treaty commitments to national legislation such as the [136], and the Biological Diversity Act, 2002, India ensures that its obligations are not merely symbolic but are operationalized through actionable frameworks. These efforts collectively strengthen the protection of aquatic birds and their habitats across the country.

## 14.9 Conservation Policies and Legal Frameworks on Aquatic Birds at the Indian National Level

India, with its diverse ecological regions from the Himalayas to the coastal wetlands and inland lakes supports a vibrant assemblage of aquatic bird species, both resident and migratory [3, 46]. Recognizing their ecological importance and vulnerability, the country has established a multi-layered legal and policy framework to protect these birds and their habitats. These frameworks are built upon constitutional mandates [43], environmental legislations including legislative measures like the [136] and the [33, 65, 66], strategic frameworks such as the National Action Plan for Conservation of Migratory Birds [73], and commitments under international accords like the Ramsar Convention and the Convention on Biological Diversity [16, 101].

### *14.9.1 Constitutional Provisions*

The Constitution of India provides a strong legal foundation for environmental protection, encompassing the preservation of biodiversity and wildlife, including aquatic bird species. Though the original text of the Constitution did not contain explicit environmental provisions, significant amendments and judicial interpretations have since embedded ecological protection as a constitutional imperative.

One of the most significant constitutional developments came with the 42nd Amendment Act, 1976, This amendment introduced Article 48A and Article 51A (g) into the Constitution. Article 48A, under the Directive Principles of State Policy, directs the State to "endeavour to protect and improve the environment and to safeguard the forests and wildlife of the country" [43]. This provision places a constitutional responsibility on the State to actively pursue wildlife conservation, encompassing the protection of aquatic birds and their habitats.

Similarly, Article 51A(g) establishes a Fundamental Duty for every individual of India "to protect and improve the natural environment including forests, lakes, rivers, and wildlife, and to have compassion for living creatures." This provision not only creates public awareness but also empowers citizens to hold the government accountable for failing to safeguard ecological interests, including wetlands and waterbird populations [29]. Article 21, which guarantees the right to life and personal liberty, has been expansively interpreted by the Supreme Court of India to encompass the right to a clean and healthy environment, as recognized in Subash Kumar v. State of Bihar [117]. This interpretation has provided a judicial basis for challenging environmental degradation that threatens habitats critical to aquatic birds.

The allocation of legislative authority between the Central and State governments under Schedule VII also plays a pivotal role. Originally part of the State List, wildlife and forests were moved to the Concurrent List by the 42nd Constitutional Amendment, empowering both the central and state governments to legislate on conservation matters [106]. This transition facilitates the uniform enforcement of national laws like the [136], which plays a vital role in protecting species, including aquatic birds.

Articles 243G and 243W of the Constitution empower Panchayati Raj Institutions (PRIs) and Urban Local Bodies (ULBs), respectively, to function as institutions of self-government with responsibilities that include environmental conservation and the management of natural resources, thereby facilitating community-based approaches to conserving aquatic ecosystems and bird habitats [112]. Overall, the Constitution of India not only empowers the government and citizens to take action for ecological conservation but also ensures that the protection of aquatic birds is embedded in the broader framework of environmental governance and justice.

### *14.9.2 Statutory Laws*

India has established an extensive legal framework for biodiversity conservation, including the protection of aquatic bird species, through key environmental legislations such as the [136], the [33], and the Biological Diversity Act, 2002, and others which collectively aim to regulate, conserve, and manage wildlife habitats, particularly wetlands and water bodies critical for aquatic bird survival.

#### 14.9.2.1 [136]

The [136] (WPA) is India's principal legislation for wildlife conservation. It provides legal protection to wild animals and birds and regulates hunting, trade, and poaching. Aquatic birds, including species like herons, storks, flamingos, pelicans, ducks, and cranes, are listed in Schedule I and Schedule IV of the Act, depending on their conservation status. Species listed in Schedule I receive the highest level of protection, with stringent penalties for offenses [136].

The WPA also Enables the central and state governments to declare protected areas such as National Parks, Wildlife Sanctuaries, and Conservation Reserves, which are crucial for aquatic bird conservation. Many wetlands, estuaries, and lakes such as Bharatpur's Keoladeo National Park are protected under this Act due to their importance as waterbird habitats [74, 75].

#### 14.9.2.2 Environment (Protection) Act, 1986

The [33] (EPA) is a Comprehensive legislation enacted as a consequence of the Bhopal Gas Tragedy. It gives wide powers to The Central Government has implemented measures to safeguard and enhance the environment, as well as to prevent, control, and mitigate environmental pollution. Under this Act, Environmental Impact Assessment (EIA) notifications have been issued which require environmental clearance for projects likely to affect wetland ecosystems [66].

Wetlands, which are critical for aquatic birds, often come under threat from developmental activities. The EPA enables regulatory interventions through EIA to assess the ecological impact of such projects and enforce mitigation measures to preserve habitats [54].

#### 14.9.2.3 Biological Diversity Act, 2002

This legislation was enacted to fulfill India's commitments under the Convention on Biological Diversity (CBD). It aims to promote the conservation of biological diversity, ensure sustainable use of its components, and facilitate fair and equitable benefit-sharing. The Act mandates the formation of Biodiversity Management Committees (BMCs) at the local level and highlights the importance of documentation through People's Biodiversity Registers (PBRs), which frequently catalog local aquatic fauna, including bird species [86].

By encouraging decentralized and participatory biodiversity governance, the Act supports habitat conservation efforts that benefit aquatic birds, particularly in community-managed wetlands and traditional water bodies.

#### 14.9.2.4 Wetlands (Conservation and Management) Rules, 2017

These rules, established under the Environmental Protection Act (EPA), create a regulatory framework for the conservation and sustainable management of wetlands across India, which are key habitats for aquatic birds. The Rules prohibit Certain activities carried out in wetlands, including waste dumping, reclamation, and construction, and require each state to set up a State Wetlands Authority for conservation oversight [71, 72].

The inclusion of Ramsar Sites wetlands of international importance within the scope of these rules has provided an additional layer of protection to globally significant aquatic bird habitats such as Chilika Lake, Loktak Lake, and Point Calimere.

#### 14.9.2.5 Forest (Conservation) Act, 1980

Though primarily aimed at preventing deforestation, this Act plays a supportive role in aquatic bird conservation by controlling the conversion of forest land for non-forest uses. Forested wetlands and mangroves, key habitats for aquatic birds receive legal protection under this Act [20].

### *14.9.3 Policy Initiatives*

India's conservation strategy for aquatic birds is reinforced through several policy initiatives that align with constitutional mandates, statutory laws, and international commitments. These policies provide the strategic vision and administrative mechanisms needed for implementing conservation programs, particularly in wetland ecosystems where aquatic birds thrive.

#### 14.9.3.1 National Wildlife Action Plan (2017–2031)

The National Wildlife Action Plan (NWAP) 2017–2031, the third such plan since independence, emphasizes landscape-level conservation, climate change adaptation, and community participation. It highlights the importance of protecting wetlands, coastal ecosystems, and aquatic habitats that support rich avifauna, including migratory and resident waterbirds.

The NWAP promotes integrated management of wetlands and water bodies with wildlife values; strengthening of protected area networks and eco-sensitive zones around key aquatic bird habitats and increased support for transboundary conservation efforts, particularly for migratory bird species protected under the Convention on Migratory Species (CMS).

#### 14.9.3.2 National Wetland Conservation Programme (NWCP)

Launched in 1987–88, the NWCP was formulated to conserve the country's most significant wetlands through management planning, community involvement, and ecological restoration. Many of the wetlands under this program like Chilika Lake (Odisha) and Harike Wetland (Punjab) are internationally recognized for hosting large populations of waterbirds [70].

The NWCP has now been incorporated into the National Plan for Conservation of Aquatic Ecosystems (NPCA), which broadens the scope to include lakes, rivers, and other aquatic ecosystems that serve as bird habitats.

#### 14.9.3.3 National Plan for Conservation of Aquatic Ecosystems (NPCA)

Introduced in 2013, The NPCA combines the goals of both the NWCP and the National Lake Conservation Plan (NLCP). Its goal is to holistically manage aquatic ecosystems, including those that are not part of designated protected areas but are vital habitats for aquatic birds. NPCA supports ecological assessment and monitoring of wetlands; restoration of degraded aquatic ecosystems and stakeholder participation, including local communities and panchayats.

#### 14.9.3.4 National Biodiversity Action Plan (NBAP)

In response to obligations pursuant to the Convention on Biological Diversity, India adopted its NBAP in 2008 (updated in 2014) to establish a strategic framework for biodiversity conservation. It includes specific actions for protecting wetlands and inland aquatic ecosystems, directly benefiting aquatic birds.

The Addendum to NBAP aligns with the Aichi Biodiversity Targets and underscores the importance of preventing habitat fragmentation and pollution in ecosystems frequented by aquatic birds [85].

#### 14.9.3.5 National Action Plan on Climate Change (NAPCC) and State Action Plans

Climate change poses significant threats to aquatic bird habitats, particularly wetlands and coastal ecosystems at risk from rising sea levels and altered hydrology. The NAPCC, especially through its National Mission for Sustaining the Himalayan Ecosystem and National Water Mission, indirectly contributes to protecting habitats of aquatic birds in sensitive regions. State Action Plans on Climate Change (SAPCCs) often include components addressing the resilience of wetlands and species dependent on them, such as migratory waterfowl.

#### 14.9.3.6 National Avian Influenza Surveillance Programme

Migratory waterfowl and other aquatic bird species serve as natural reservoirs for avian influenza viruses. The National Avian Influenza Surveillance Programme, coordinated by the Department of Animal Husbandry and the MoEFCC, monitors outbreaks and promotes safe interactions between wildlife and domestic birds, ensuring the health of both populations. This policy, though health-focused, intersects with conservation by maintaining safe waterbird populations and managing the risks of mass mortality events or culling during disease outbreaks.

## *14.9.4 Implementation Mechanisms and Conservation Schemes*

To operationalize legal and policy mandates, India has instituted several implementation mechanisms and conservation schemes focused on protecting wetlands and aquatic birds. These mechanisms span statutory protection, community engagement, monitoring programs, and international cooperation.

### 14.9.4.1 Protected Area Network

India's Protected Area Network comprising national wildlife parks, wildlife sanctuaries, conservation reserves, and communities reserves offers legal protection for many critical wetland ecosystems. These areas, administratively declared under the [136], provide refuge to numerous aquatic bird species. Notable examples include Keoladeo Ghana National Park (formerly Bharatpur Bird Sanctuary) in Rajasthan, an UNESCO World Heritage Site renowned for exceptional bird species diversity, and Nal Sarovar Bird Sanctuary in Gujarat, which supports over 250 species of aquatic birds [137].

### 14.9.4.2 Ramsar Sites and the Ramsar Convention

India is a Contracting Party as a signatory to the 1971 Ramsar Convention on Wetlands, and has officially designated 75 Ramsar sites as of 2023, covering ecologically significant wetlands such as Chilika Lake (Odisha) and Loktak Lake (Manipur). These sites provide critical habitats for migratory birds, including flamingos, pelicans, and storks [100].

### 14.9.4.3 Eco-Sensitive Zones (ESZs)

To create protective buffers around core conservation areas, the MoEFCC has established regulatory requirements for Eco-sensitive Zones (ESZs) under Section 3 of the Environment Protection Act [33]. These zones limit industrial and urban development near wetlands and bird habitats, thereby offering indirect protection to aquatic avifauna [76, 77].

### 14.9.4.4 Biodiversity Management Committees and PBRs

As per the provisions of the Biological Diversity Act, 2002, Biodiversity Management Committees (BMCs) are formed at the community level to create People's

Biodiversity Registers (PBRs). These registers record biodiversity elements, including wetlands and waterbirds, and facilitate community-based conservation [84].

#### 14.9.4.5 Wetland Mitras and Community Stewardship

India has introduced "Wetland Mitras" or Friends of Wetlands as part of a participatory conservation model. These volunteers assist in monitoring, patrolling, and raising awareness about wetland ecosystems. This approach has been successfully implemented in Ramsar sites like Sasthamkotta Lake (Kerala) and Asan Conservation Reserve (Uttarakhand) ( Wetlands International South Asia 2020).

#### 14.9.4.6 National Waterbird Census and Migratory Bird Monitoring

The Asian Waterbird Census (AWC) is a long-term monitoring program coordinated for the Indian subcontinent by Wetlands International's South Asia office, with support from Bombay Natural History Society (BNHS) and MoEFCC. Conducted annually during January, it collects population data on waterbirds and informs wetland management and species conservation planning [132–134].

## 14.10 Challenges and Gaps in Enforcement of Conservation Laws

Despite the existence of robust legislative frameworks for wildlife conservation in many jurisdictions, including the protection of aquatic bird species, there remain several challenges and enforcement gaps that undermine effective conservation outcomes. These gaps are especially evident within the framework of climate change, which introduces new stressors and complexities.

### *14.10.1 Fragmented Legal Framework*

Among the most significant challenges is the fragmented and sectoral nature of India's environmental and wildlife legislation. Multiple statutes govern various aspects of biodiversity conservation, such as forests, water bodies, pollution control, land use, and protected species. For instance, the [136], examines species protection and establishment of protected areas, while the [33], provides broader powers to regulate pollutants and set environmental standards. Wetlands, which are crucial habitats for aquatic birds, are regulated separately under the Wetlands (Conservation and Management) Rules, 2017, framed under the EP Act.

This compartmentalization leads to jurisdictional overlaps and ambiguities. For example, a wetland located within a forest area may simultaneously fall under the jurisdiction of the forest department, the pollution control board, and the State Wetlands Authority. Consequently, Coordination mechanisms are frequently inadequate and duplication of efforts, which weakens enforcement and monitoring [55]. Moreover, the absence of an integrated ecological governance framework hampers the ability to respond adaptively to climate change, which impacts ecosystems across sectors and boundaries. As climate change alters the hydrological regimes of wetlands and migratory bird patterns, a siloed legal approach fails to provide a coherent response mechanism. Scholars argue that integrated landscape-level planning and cross-sectoral collaboration are necessary for effective biodiversity conservation under climate stress [115].

### *14.10.2 Weak Institutional Capacity*

A primary obstacle to effective conservation laws, especially in India, is the limited institutional capacity of the agencies responsible for implementing these frameworks. Forest and wildlife departments in many states face shortages of trained personnel, funding, and logistical support. The State Wetland Authorities, which are entrusted with the implementation of the Wetlands Rules, 2017, often remain under-resourced and poorly staffed, limiting their ability to conduct surveys, regulate developmental encroachments, and monitor biodiversity health [21].

Enforcement agencies often lack specialized knowledge of avian biology or wetland ecology, making it difficult to tackle the multifaceted challenges presented by aquatic bird conservation. This gap is exacerbated in the impacts of climate change, where scientific input is essential to understand shifting migratory patterns, changing water regimes, and breeding behaviors. The lack of technical collaboration between enforcement agencies and scientific institutions often results in ad hoc or reactive conservation interventions rather than proactive, evidence-based strategies [96].

### *14.10.3 Lack of Real-Time Monitoring*

One of the major gaps in the effective enforcement of conservation laws concerning aquatic birds is the absence of real-time monitoring mechanisms. The application of geospatial technologies including Geographic Information Systems (GIS), satellite telemetry, drone surveillance, and remote sensing remains limited, particularly in developing countries like India. These tools are crucial for mapping migratory routes, identifying changes in habitat conditions, and monitoring anthropogenic activities in and around sensitive wetland ecosystems. Without these systems, enforcement agencies often operate reactively rather than proactively, responding to ecological threats only after damage has already occurred. For Instance, a research by Prasad et al. [94] emphasizes the importance of integrating spatial data and real-time mapping for wetland monitoring in India, noting that traditional survey methods are insufficient for the dynamic and widespread habitats of migratory birds. Similarly, the Ministry of Environment, Forests and Climate Change (MoEFCC) has acknowledged in various reports the need to improve India's ecological monitoring infrastructure but has yet to implement a comprehensive nationwide system [71, 72].

In many wetland areas, particularly across the Global South, conservation enforcement frameworks often fail to account for the structural constraints small-scale farmers face due to global commodity markets. Export-oriented agriculture and rigid contract farming schemes can restrict local producers from adopting conservation-friendly practices, even when they are ecologically aware. Such mismatches between market logic and conservation imperatives deepen local resistance to enforcement efforts. Aligning conservation with food sovereignty and participatory trade governance is crucial to achieving equitable wetland protection.

### *14.10.4 Limited Community Participation*

A critical gap in enforcement strategies is the inadequate Participatory engagement of local residents, especially those living in and around wetlands, river basins, and coastal zones. Despite the growing recognition that community-based conservation leads to more sustainable outcomes, most enforcement measures in India remain top-down, with minimal participation from indigenous groups, farmers, or fisherfolk who directly interact with aquatic bird habitats [111].

Local knowledge about bird migratory patterns, nesting sites, and habitat conditions remains underutilized in enforcement strategies. Moreover, many wetland-dependent communities are unaware of the ecological and legal significance of aquatic birds, resulting in activities such as unregulated fishing, waste dumping, and bird poaching. Climate change further exacerbates the situation by altering local livelihoods and increasing dependence on natural resources, sometimes bringing these communities into conflict with conservation laws [109].

There is also limited implementation of the Environmental Protection Act's mandate for public participation and people's biodiversity registers as mandated by India's Biological Diversity Act, 2002, especially in areas critical for waterbird conservation.

### *14.10.5 Ineffective Penalties and Prosecution*

Although wildlife crimes, including offenses against aquatic birds, are legally punishable under the [136], enforcement is often hindered by slow judicial processes, poor investigation quality, and low conviction rates. Offenses such as illegal trapping, poisoning, or habitat encroachment frequently go unpunished due to delays in lodging FIRs, lack of forensic evidence, or inadequate training of investigating officers [57]. The Schedule listing under the Wildlife Act often does not reflect updated conservation priorities. Several wetland bird species are poorly categorized or underprotected due to outdated assessments, making enforcement ineffective. Additionally, most courts are ill-equipped to deal with environmental matters, and environmental tribunals such as the National Green Tribunal (NGT) are often overburdened.

### *14.10.6 Climate-Induced Habitat Shifts*

Climate change has led to increased variability in temperature and precipitation patterns, resulting in significant habitat shifts for aquatic birds. Many species are now observed in non-traditional migratory or breeding areas due to changing hydrological cycles, drying wetlands, or flooding of nesting grounds. However, India's current protected area network and conservation legislations, like the [136], are largely site-specific and static. They do not accommodate the ecological fluidity needed to protect migratory species across newly occupied or transitional habitats.

Pearce-Higgins and Green [93] point out that conservation strategies must evolve to include climate-responsive habitat corridors and dynamic protected area designations to reflect shifting biodiversity patterns. Indian researchers like Mondreti and Rajagopalan [82] have similarly advocated for flexible policy mechanisms that can track and adapt to the real-time responses of waterbirds to climate change.

### *14.10.7 Pollution and Unregulated Development*

Aquatic habitats crucial for bird life are increasingly under threat from pollution and unsustainable development. Industrial effluents, untreated sewage, agricultural runoff, and plastics [60, 61] are polluting India's inland and coastal wetlands. At the same time, infrastructure development, tourism, and real estate pressures are causing direct habitat loss and fragmentation. Even where laws exist for example, under the [33] and Coastal Regulation Zone (CRZ) notifications, political interference and economic priorities often dilute enforcement.

Galbraith et al. [38] warn that the Degradation of intertidal ecosystems caused by pollution and sea-level increase poses a serious threat to global shorebird populations. In the Indian context, the East Kolkata Wetlands and Chilika Lake are examples where judicial intervention was necessary to curb encroachments and pollution [92]. Nevertheless, many other wetlands remain vulnerable due to weak enforcement and corruption.

### 14.10.8 Lack of Data-Driven Enforcement

Data scarcity significantly hampers the enforcement of aquatic bird conservation strategies. Reliable and current biodiversity databases covering species population trends, migratory routes, breeding behavior, and habitat conditions are crucial for designing and executing targeted conservation interventions. However, many developing nations lack the capacity or institutional infrastructure to maintain such datasets. For instance, India's National Biodiversity Authority and State Biodiversity Boards have made some progress in creating People's Biodiversity Registers, but these are often incomplete and lack specific information on aquatic avifauna [6]. Moreover, obligations under international conventions like the Ramsar Convention and the Convention on Migratory Species (CMS) require regular reporting based on scientific data, which remains a challenge due to inconsistent monitoring. Mawdsley et al. [64] argue that climate-resilient conservation depends heavily on accurate, long-term ecological datasets. Without such data, conservation efforts may misidentify priorities or overlook at-risk species until it is too late.

### 14.10.9 Cross-Border Enforcement Challenges

A significant proportion of aquatic birds in India are migratory species that traverse thousands of kilometers along Central Asian Flyways (CAF) and East Asian-Australasian Flyways (EAAF). Although India is a contracting party to multilateral agreements including the Convention on Migratory Species and Ramsar Convention, there remains a gap in translating these obligations into coordinated regional enforcement actions.

Migratory birds are vulnerable to habitat loss at any point along their routes. For example, degradation of stopover wetlands in Central Asia or South India can disrupt entire populations. While India has developed a National Action Plan for Conservation of Migratory Birds and their Habitats under CAF (2018), coordination with other flyway countries on monitoring, research, and enforcement remains limited [73]. Legal and administrative measures often fail to align migratory timelines with regulatory actions, such as seasonal wetland protection or hunting bans. There is also limited data-sharing or joint patrolling across borders, which is essential to protect long-distance migrants like the Bar-headed Goose, Black-tailed Godwit, or Eurasian Spoonbill.

### 14.10.10 Lack of Climate Adaptation in Legal Frameworks

Another critical gap is that existing wildlife and environmental laws in India are not adequately climate-responsive. The [136] and associated rules primarily focus on static protection measures like establishing protected areas or prescribing punishments for poaching. These laws do not account for dynamic ecological changes such as climate-driven shifts in migration routes or breeding patterns of waterbirds. For instance, migratory birds that once frequented particular wetlands may alter their routes or timings in response to rising temperatures or changed precipitation patterns. However, the current legal framework does not facilitate adaptive management or real-time decision-making based on such ecological shifts. As a result, conservation strategies often become outdated or misaligned with ground realities [56].

Climate change poses compound threats, such as sea-level rise affecting coastal bird habitats or changes in water salinity influencing food chains. These interconnected phenomena demand a holistic legal and policy response, which is currently lacking. The need for incorporating climate risk assessments and ecosystem resilience planning into wildlife laws has been emphasized by both Indian and global scholars [58, 64].

## 14.11 Innovative Enforcement Strategies for Aquatic Bird Protection

Traditional enforcement strategies, though critical, often fall short in addressing the dynamic challenges posed by habitat loss, pollution, and migratory shifts. Hence, innovative enforcement strategies, particularly those involving technological integration and community-based conservation have gained prominence as effective, scalable solutions to protect aquatic avifauna and their ecosystems.

Beyond traditional conservation frameworks, market-based mechanisms can create incentives for wetland protection while supporting agricultural livelihoods. For small-scale farmers in wetland-adjacent communities, certification schemes recognizing conservation-friendly practices can provide premium market access while preserving aquatic bird habitats. Nigeria's Hadejia-Nguru wetlands illustrate this potential, where sustainable rice farming practices could maintain critical habitat for migratory birds while securing food production systems. However, these approaches must address power imbalances within global trade systems. Nigerian smallholders face significant barriers to certification compliance that reflect broader North-South inequities in environmental governance [5]. These include prohibitive costs, technical requirements, and documentation demands that disadvantage resource-limited producers. Effective conservation strategies must therefore couple market incentives with capacity building, simplified compliance mechanisms, and fair cost distribution across supply chains. When thoughtfully designed, such approaches can create scenarios where communities maintain ecological knowledge that supports both food security and wetland-dependent birds, particularly in biodiversity-rich developing regions where conservation and livelihood needs frequently intersect.

### 14.11.1 Role of Technology in Monitoring and Enforcing Conservation Efforts

Technological advancements have significantly enhanced the scope and precision of conservation enforcement in recent years. Tools Like remote sensing, Geographic Information Systems (GIS), drone surveillance, satellite telemetry, and bioacoustic monitoring enable real-time monitoring of wetlands, migratory routes, nesting patterns, and habitat encroachments.

In India, initiatives like the Wetlands of India Portal, launched by the Ministry of Environment, Forest and Climate Change (MoEFCC), serve as a centralized platform using spatial data to track the health of over 2300 wetlands across the country. These data-driven platforms assist in early warning systems and habitat assessments essential for proactive conservation enforcement [76, 77].

Geospatial technologies, including remote sensing and GIS, have been effectively employed to assess wetland ecosystem degradation and model habitat suitability for threatened avian species such as the Black-necked Stork (Ephippiorhynchus asiaticus) and Indian Skimmer (*Rynchops albicollis*) [59]. Furthermore, drone-based surveillance is increasingly being used by forest departments to identify unauthorized encroachments and illegal fishing within protected wetlands such as the Chilika Lake and Keoladeo National Park [58]. Satellite telemetry has enabled ornithologists to track transboundary Migration routes of species like the Bar-headed Goose (Anser indicus) and the Common Teal (Anas crecca) support coordinated enforcement efforts under the Convention on Migratory Species (CMS) framework [48].

### 14.11.2 Community Engagement and Collaborative Conservation Efforts

Technology alone is not sufficient unless accompanied by active community participation. Local communities, particularly those living around wetlands, rivers, and coastal ecosystems, are crucial stakeholders in any enforcement strategy. Engaging them through awareness campaigns, citizen science, eco-tourism incentives, and community-based monitoring not only reduces conflicts but also leads to sustainable conservation practices.

In India, the Loktak Lake Community Conservation Program in Manipur, supported by the Loktak Development Authority, stands as a model for integrating community knowledge in managing floating phumdis and safeguarding waterbird populations like the Lesser Whistling Duck (*Dendrocygna javanica*) [113]. Similarly, the Chilika Development Authority has successfully collaborated with fisherfolk in Odisha to monitor and protect migratory birds, reducing poaching incidents through education and livelihood alternatives [18].

Citizen science initiatives, such as eBird India and the Asian Waterbird Census (AWC) coordinated by Wetlands International South Asia, have enabled volunteers to participate in waterbird population assessments. This decentralized approach empowers communities while enhancing data collection for enforcement agencies [118]. Moreover, under the National Biodiversity Action Plan (NBAP) and State Action Plans on Climate Change (SAPCCs), participatory approaches involving Gram Panchayats, Eco Development Committees (EDCs), and Joint Forest Management (JFM) groups have been institutionalized to link local governance with ecological enforcement.

## 14.12 Future Directions: Adaptive Strategies for Climate Resilience

As climate change accelerates, conservation strategies must evolve to not only preserve current habitats and populations but also to anticipate future environmental shifts. Although reducing greenhouse gas emissions remains the primary long-term goal for addressing climate change, there is an urgent need to implement adaptation strategies that enhance the resilience of species, especially vulnerable aquatic birds. Given their dependency on sensitive ecosystems such as wetlands, estuaries, and inland water bodies, adaptive management approaches must be proactive, dynamic, and evidence-based.

### *14.12.1 Site Management for Microclimatic Resilience*

Site-level interventions represent the frontline of adaptive conservation. Managing key aquatic bird habitats to maintain or enhance their suitability under changing climatic conditions can significantly reduce extinction risk. Manipulating vegetation, for example, can influence microclimates by providing shade and reducing evaporation. Similarly, water-level management through controlled damming, drainage regulation, and water-use restrictions can ensure the availability of shallow foraging areas for waders and nesting zones for breeding birds [93].

In India, the management of wetlands like Bharatpur (Keoladeo Ghana) has involved periodic water release from nearby sources to simulate seasonal flooding, crucial for the nesting of species like the Painted Stork and Black-necked Stork. Such ecological engineering interventions serve as successful examples of adaptive site management [95].

### 14.12.2 Strengthening Protected Area Networks

With increasing instances of climate-induced range shifts, static protected areas may no longer align with suitable habitats. For species to relocate and thrive in new areas, a dynamic and interconnected network of protected zones must be established especially in regions predicted to become climatically favorable [49]. The Ramsar Convention encourages this by urging member states, including India, to develop wetland site networks that are ecologically representative and resilient to climate impacts [102].

In India, strengthening wetland protection through the Wetlands (Conservation and Management) Rules, 2017, and incorporating climate projections into site selection can improve ecological connectivity essential for the survival of migratory and dispersing bird species such as the Northern Pintail and the Common Teal.

### 14.12.3 Integrating Conservation in the Wider Countryside

Beyond the borders of protected areas, the wider landscape must be made permeable to the movements of aquatic birds. This entails promoting bird-friendly land-use practices, particularly in agricultural zones adjacent to wetlands and river systems. Integrating climate-smart agri-environment schemes can help maintain ecological corridors and reduce mortality during dispersal.

Programs like the National Mission for Sustainable Agriculture (NMSA), part of India's National Action Plan on Climate Change (NAPCC) can be leveraged to incentivize the preservation of natural water bodies and minimize agrochemical runoff, thus indirectly benefiting bird habitats [69].

### 14.12.4 Minimizing Compounding Anthropogenic Impacts

Climate change often amplifies the effects of existing environmental stressors. Hence, buffering aquatic birds from non-climatic pressures is a crucial part of adaptation. This includes curbing wetland encroachment, drainage, pollution, and unregulated tourism. These measures not only help reduce habitat vulnerability but also enhance their ability to withstand climate shocks, such as extended droughts or saline intrusion from rising sea levels [38].

Preserving Loktak Lake, a designated Ramsar site in Manipur, provides a pertinent case where reducing anthropogenic pressure on phumdis (floating biomass) has helped sustain critical habitats for species like the Lesser Whistling Duck and the Northern Shoveler, even as rainfall patterns have become erratic [113].

## 14.13 Future Prospects

The future of aquatic bird conservation hinges on how effectively science, policy, and community engagement converge in the face of climate uncertainty. As wetlands continue to shrink and sea levels rise, adaptive and anticipatory conservation models must replace static management practices. Dynamic protected area networks that shift in tandem with habitat changes will become essential. Climate forecasting technologies, coupled with artificial intelligence, can play a transformative role in predicting migration disruptions, breeding failures, and population declines. Incorporating such tools into enforcement strategies could enable real-time responses to ecological stress. Transboundary cooperation will be critical, especially along flyways like the Central Asian Flyway. Nations must move beyond treaty signatories to enforce joint monitoring and shared habitat restoration efforts. Regional hubs for waterbird conservation science could facilitate such collaborations. Community-based wetland stewardship offers a promising path forward. Indigenous knowledge, when integrated with modern ecological frameworks, can guide sustainable wetland use. Programs like "Wetland Mitras" must be scaled and supported by both state and central governments. Youth-led conservation movements and citizen science initiatives also offer powerful tools to broaden public engagement. Platforms like eBird and Waterbird Census portals can crowdsource valuable data and build ecological awareness. Legal reforms are necessary to strengthen the implementation of existing laws. Fast-track tribunals, increased penalties for wetland encroachment, and legally bindinq state wetland management plans could ensure accountability. At the global level, synergizing conservation finance with social equity is key. Mechanisms like REDD+ and biodiversity credits must prioritize inclusivity and livelihood resilience alongside ecological outcomes. Ultimately, the future depends on how boldly and inclusively we act today. Aquatic birds are not just indicators of environmental health, they are testaments to our commitment to coexistence in an era of planetary change.

## 14.14 Conclusion

Aquatic birds serve as critical indicators of the health of wetland ecosystems and play vital roles in maintaining ecological balance. However, the twin threats of environmental degradation and accelerating climate change have placed these species under unprecedented stress. The paper has underscored that while India has a robust legislative and institutional framework for wildlife and wetland conservation including Legislative frameworks such as the Wildlife Protection Act of 1972, the Wetlands (Conservation and Management) Rules of 2017, along with constitutional mandates outlined in Articles 48A and 51A(g) significant gaps in enforcement and adaptability remain.

Climate-induced habitat loss, shifting migratory patterns, pollution, and the lack of dynamic legal and policy instruments have rendered many traditional conservation strategies insufficient. These challenges are compounded by fragmented laws, weak coordination among enforcement agencies, and inadequate real-time data on biodiversity. The need to integrate climate resilience into conservation is no longer optional but imperative. In this context, the paper proposed a shift towards innovative and adaptive enforcement strategies. The use of technology, including GIS, drone surveillance, and citizen-science-based monitoring, can significantly enhance surveillance and data collection. Equally important is the empowerment of local communities through participatory conservation programs that foster stewardship and reduce human-wildlife conflict.

The future of aquatic bird conservation must rest on the principles of ecological flexibility and climate responsiveness. This entails adaptive site management, the expansion of protected area networks along migratory routes, sustainable landscape-level planning, and the minimization of anthropogenic pressures. International cooperation through instruments like the Ramsar Convention, the Convention on Migratory Species (CMS), and the Convention on Biological Diversity (CBD), with India maintaining its proactive engagement.

Ultimately, ensuring a resilient and sustainable future for aquatic bird species requires a paradigm shift from reactive conservation models to proactive, climate-smart strategies rooted in ecological science, community engagement, and technological innovation. Only then can we ensure that the waterways, wetlands, and skies continue to resonate with the presence and songs of these irreplaceable species.

## References


1. AEWA (2023a) African-Eurasian migratory waterbird agreement: about AEWA. Retrieved from https://www.unep-aewa.org/about/introduction
2. AEWA (2023b) Agreement on the conservation of African-Eurasian migratory waterbirds. Retrieved from https://www.unep-aewa.org/sites/default/files/uploads/aewa_agreement_text_2023-2025_corrected%20version%20as%20of%2010%20August%202023_EN_rev.pdf
3. Ali S, Ripley SD (1987) Handbook of the birds of India and Pakistan, 2nd edn. Oxford University Press, Oxford
4. Anderson S (2003) Animals and the symbolic in world religions. Routledge, London
5. Asiyanbi A, Lund JF (2020) Policy persistence: REDD+ between stabilization and contestation. J Polit Ecol 27(1):378–400
6. Bhatt J, Patel S (2016) People's biodiversity registers and avifaunal documentation: a review of practices in India. Indian J Biodiver Manag 3(1):32–44
7. BirdLife International (2023a) State of India's birds report 2023
8. BirdLife International (2023b) State of the world's birds 2022 report
9. BirdLife International (2023c) BirdLife and global conservation policy. Retrieved from https://www.birdlife.org
10. BirdLife International (2023d) Who we are. Retrieved from https://www.birdlife.org/about-us/
11. Blihd H (2019) NGOs as implementers of international environmental conventions: between global norms and local practices. Global Environ Politics 19(2):87–106. https://doi.org/10.1162/glep_a_00505

12. Boere GC, Stroud DA (2006) The flyway concept: what it is and what it isn't. In: Boere G, Galbraith C, Stroud D (eds) Waterbirds around the world. The Stationery Office, Edinburgh, pp 40–47
13. Boere GC, Galbraith CA, Stroud DA (eds) (2006a) Waterbirds around the world. The Stationery Office, Edinburgh
14. Boere G, Galbraith C, Stroud D (eds) (2006b) Waterbirds around the world: a global overview of the conservation, management and research of the world's waterbird flyways. The Stationery Office, Edinburgh
15. Bombay Natural History Society (BNHS) (2020) Waterbird monitoring handbook. Bombay Natural History Society (BNHS), Mumbai
16. CBD (1992) Convention on biological siversity. United Nations. https://www.cbd.int/convention/
17. CBD Secretariat (2010) Global biodiversity outlook 3. Secretariat of the Convention on Biological Diversity, Montreal
18. Chilika Development Authority (2019) Annual report on wetland management and bird protection. Government of Odisha
19. Chilika Development Authority (2024) Annual bird census report
20. Chitale VS, Behera MD (2012) Forest conservation and wetland protection in India. J Environ Plan Manag 55(6):741–756
21. Chopra K, Sinha R (2019) Strengthening institutional mechanisms for wetland governance in India. J Environ Stud Policy 6(1):34–50
22. CITES (2023) What is CITES?. https://cites.org/eng/disc/what.php
23. CMS (2022) Central Asian Flyway action plan. UNEP/CMS Secretariat
24. CMS (2023) Appendices I and II of the convention on migratory species. Retrieved from https://www.cms.int/species/appendix-i-ii-cms
25. CMS Secretariat (2020) CMS COP13 India outcome documents. https://www.cms.int/en/meeting/thirteenth-meeting-conference-parties-cms
26. Cuthbert R, Parry-Jones R, Green RE, Pain DJ (2007) NSAIDs and scavenging birds: potential impacts beyond Asia's critically endangered vultures. Biol Lett 3(1):91–94
27. Davidson NC (2014) How much wetland has the world lost? Mar Freshw Res 65(10):934–941
28. Davidson NC, Fluet-Chouinard E, Finlayson CM (2018) Global extent and distribution of wetlands: trends and issues. Mar Freshw Res 69(6):620–627. https://doi.org/10.1071/MF17019
29. Divan S, Rosencranz A (2001) Environmental law and policy in India, 2nd edn. Oxford University Press, Oxford
30. Donald PF, Buchanan GM, Balmford A, Butchart SHM (2019) Important bird and biodiversity areas (IBAs): the development and characteristics of a global inventory of key sites for biodiversity. Bird Conserv Int 29(2):177–198. https://doi.org/10.1017/S0959270918000102
31. EAAFP (2023a) East Asian-Australasian flyway partnership overview. Retrieved from https://www.eaaflyway.net
32. EAAFP (2023b) Flyway network sites. Retrieved from https://www.eaaflyway.net/flyway-network-sites/
33. Environment (Protection) Act, 1986. Ministry of Environment, Forest and Climate Change
34. Environmental Protection Agency (EPA) (2023) Greenhouse gas emissions and climate change. Retrieved from epa.gov
35. Finlayson CM, Everard M, Irvine K, McInnes RJ, Middleton BA, van Dam AA, Davidson NC (2011) The Ramsar Convention and ecosystem-based approaches to the wise use and sustainable development of wetlands. J Int Wildl Law Policy 14(3–4):176–198. https://doi.org/10.1080/13880292.2011.626704
36. Food and Agriculture Organization (FAO) (2023) Climate change and agriculture. Retrieved from fao.org
37. FSI (2021) India state of forest report. Forest Survey of India, Dehradun

38. Galbraith H, Jones R, Park R, Clough J, Herrod-Julius S, Harrington B, Page G (2002) Global climate change and sea level rise: potential losses of intertidal habitat for shorebirds. Waterbirds 25(2):173–183
39. Galbraith H, Jones R, Park R, Clough J, Herrod-Julius S, Harrington B, Page G (2014) Global climate change and sea level rise: potential losses of intertidal habitat for shorebirds. Condor 104(2):451–457
40. GEF (2020) The global environment facility and biodiversity. https://www.thegef.org
41. Gill FB (2007) Ornithology. W. H. Freeman, New York
42. González-Bergonzoni I et al (2017) Transport of nutrients by migratory birds from productive to nutrient-poor lakes during stopovers. Freshw Biol 62(3):569–579
43. Government of India (1950) The constitution of India. Retrieved from https://legislative.gov.in
44. Green AJ, Elmberg J (2014) Ecosystem services provided by waterbirds. Biol Rev 89(1): 105–122
45. Gregory RD, van Strien A (2010) Wild bird indicators: using composite population trends of birds as measures of environmental health. Ornithol Sci 9(1):3–22
46. Grimmett R, Inskipp C, Inskipp T (2011) Birds of the Indian subcontinent, 2nd edn. Oxford University Press, Oxford
47. Haddad NM, Brudvig LA, Clobert J, Davies KF, Gonzalez A, Holt RD, Cook WM (2015) Habitat fragmentation and its lasting impact on earth's ecosystems. Sci Adv 1(2):e1500052
48. Higuchi H et al (2004) Satellite tracking of the migration of birds. Glob Environ Res 8(1): 75–85
49. Huntley B, Green RE, Collingham YC, Willis SG (2007) A climatic atlas of European breeding birds. Lynx Edicions, Barcelona
50. INCOIS (2021) Sea level rise studies along the Indian coastline
51. IPCC (2021a) Climate change 2021: the physical science basis. Intergovernmental Panel on Climate Change, Geneva
52. IPCC (2021b) Sixth Assessment Report (AR6). Retrieved from ipcc.ch
53. Kiran, Singh D, Kour A, Priya, Delu V, Kumar R (2022) Community composition and status of avian diversity at campus and agricultural landscapes of Chaudhary Charan Singh Haryana agricultural university, Hisar (Haryana). J Appl Nat Sci 14(4):1130–1140. https://doi.org/10.31018/jans.v14i4.3784
54. Kohli K, Menon M (2016) Environmental impact assessment in India: a handbook for legal practitioners. Centre for Policy Research-Namati Environmental Justice Program, New Delhi
55. Kohli K, Menon M (2020) Wildlife law: a practitioner's manual. Centre for Policy Research & Environmental Justice Initiative, New Delhi
56. Kumar R, Agarwal N (2021) Mainstreaming climate adaptation into wildlife Laws in India: a critical review. Indian J Environ Law 7(2):91–105
57. Kumar S, Bhupathy S (2019) Wildlife offences in India: status, gaps and the way forward. J Wildlife Legal Stud 8(2):101–115
58. Kumar S, Sharma R (2020) Drone-based surveillance for wetland protection in India: a case study from Chilika Lake. J Environ Prot 11(2):112–123
59. Kumar R, Kumar A, Singh D (2016) Using GIS and remote sensing for waterbird habitat mapping in Uttar Pradesh. Indian Forester 142(10):980–988
60. Kumar R, Gupta R, Dahiya T, Yadav R, Saifi R, Yodha K, Kumari B, Singh P, Shefali. (2022a) Surveillance of *Heteropneustes fossilis* (stinging catfish) diseases and health management practices in Haryana. Appl Ecol Environ Sci 10(4):261–267
61. Kumar R, Gupta RK, Yadav R, Saifi R, Yodha K, Pooja. (2022b) *Eisenia fetida* as protein source for growth enhancement of *Heteropneustes fossilis*. Egypt J Aquat Biol Fish 26(2): 577–588
62. Kushlan JA (2000) Heron conservation. Academic, San Diego
63. Levin J (2013) UNESCO world heritage sites and biodiversity conservation: a review. J Environ Policy Plan 15(2):223–238

64. Mawdsley JR, O'Malley R, Ojima DS (2009) A review of climate-change adaptation strategies for wildlife management and biodiversity conservation. Conserv Biol 23(5):1080–1089
65. Ministry of Environment, Forest and Climate Change (MoEFCC) (1972) The Wildlife (Protection) Act, 1972. Retrieved from https://moef.gov.in
66. Ministry of Environment, Forest and Climate Change (MoEFCC) (1986) The Environment (Protection) Act, 1986. Retrieved from https://moef.gov.in
67. Ministry of Environment, Forest and Climate Change (MoEFCC) (2014) India's fifth national report to the convention on biological diversity. https://www.cbd.int/doc/world/in/in-nr-05-en.pdf
68. Mitra A, Banerjee K, Sundaray JK (2022) Wetland conservation and management: perspectives and challenges. Springer
69. MoEFCC (2012) National Action Plan on Climate Change (NAPCC). Government of India
70. MoEFCC (2013) Guidelines for national wetland conservation programme
71. MoEFCC (2017a) Wetlands (conservation and management) rules, 2017: explanatory document. Government of India
72. MoEFCC (2017b) Wetlands (conservation and management) rules, 2017
73. MoEFCC (2018) National action plan for conservation of migratory birds and their habitats along Central Asian Flyway. Ministry of Environment, Forest and Climate Change, Government of India
74. MoEFCC (2020a) Annual report 2019–2020. https://moef.gov.in
75. MoEFCC (2020b) Guidelines for implementation of NPCA scheme
76. MoEFCC (2021a) National action plan for conservation of migratory birds and their habitats along Central Asian Flyway (2021–2030). https://moef.gov.in
77. MoEFCC (2021b) Wetlands of India portal. Government of India. https://wetlands.india.gov.in
78. MoEFCC (2022) Wetlands of India: national inventory and assessment. https://moef.gov.in
79. MoEFCC (2024) List of wetlands designated as Ramsar sites in India
80. MoES (2021) Coastal vulnerability mapping for the Indian coast
81. Møller AP, Fiedler W, Berthold P (eds) (2010) Effects of climate change on birds. Oxford University Press, Oxford
82. Mondreti S, Rajagopalan A (2019) Climate change and wetland conservation in India: emerging challenges. J Clim Change Res 5(2):101–114
83. NASA (2022) Evidence of climate change. Retrieved from climate.nasa.gov
84. National Biodiversity Authority (2022) Status of people's biodiversity registers (PBRs) in India. NBA, Chennai. Retrieved from https://nbaindia.org
85. National Biodiversity Authority (NBA). (2014). NBAP addendum 2014 to NBAP 2008
86. National Biodiversity Authority (NBA) (2021) Biodiversity management committees and PBR guidelines
87. National Oceanic and Atmospheric Administration (NOAA) (2023) Global climate change indicators. Retrieved from noaa.gov
88. NCCR (2020) Shoreline change atlas of the Indian coast. National Centre for Coastal Research
89. NOAA (2023) Shoreline data and global coastline monitoring
90. Pacifici M, Visconti P, Butchart SH, Watson JE, Cassola FM, Rondinini C (2017) Species' traits influenced their response to recent climate change. Nat Clim Chang 7(3):205–209
91. Pain DJ, Bowden CGR, Cunningham AA (2008) The race to prevent the extinction of south Asian vultures. Bird Conserv Int 18(S1):S30–S48
92. Patnaik S, Dey A (2018) Urban expansion and wetland degradation: a case study of East Kolkata wetlands. Environ Manag J 9(3):54–61
93. Pearce-Higgins JW, Green RE (2014) Birds and climate change: impacts and conservation responses. Cambridge University Press, Cambridge
94. Prasad MNV et al (2020) Ecological modelling and remote sensing for wetland monitoring in India. Environ Earth Sci 79(6):1–15

95. Rahmani AR, Islam MZ (2008) Important bird areas in India: priority sites for conservation. Bombay Natural History Society and BirdLife International, Mumbai
96. Ramanathan U (2020) Whose habitat, whose law? Ecological justice and wildlife conservation. In: Wildlife law and conservation in India: a reader. Oxford University Press, Oxford
97. Ramsar Convention Secretariat (2010) Handbook 1: wise use of wetlands, 4th edn. Ramsar Convention Secretariat, Gland
98. Ramsar Convention Secretariat (2012) The criteria for identifying wetlands of international importance. Ramsar Secretariat, Gland
99. Ramsar Convention Secretariat (2022) Global wetland outlook: special edition 2021
100. Ramsar Convention Secretariat (2023) Ramsar Sites Information Service (RSIS). Retrieved from https://rsis.ramsar.org/?q=india
101. Ramsar Secretariat (1971) The Ramsar convention on wetlands. Retrieved from https://www.ramsar.org
102. Ramsar Secretariat (2010) Managing wetlands in a changing climate. Ramsar COP10 Resolution X.24
103. Ramsar Secretariat (2023) World wetlands day. Retrieved from https://www.worldwetlandsday.org
104. Ramsar Secretariat (n.d.) Ramsar Sites Information Service (RSIS). https://rsis.ramsar.org/
105. Reeve R (2002) Policing international trade in endangered species: the CITES treaty and compliance. Routledge, London
106. Rosencranz A (2007) Constitutional environmental law in India: the importance of judicial activism. Widener Law Rev 13(1):1–22
107. Russi D, ten Brink P, Farmer A, Badura T, Coates D, Förster J, Kumar R, Davidson N (2013) The economics of ecosystems and biodiversity for water and wetlands. Institute for European Environmental Policy (IEEP), Ramsar Secretariat, London
108. SAC-ISRO (2011) National wetland atlas: India. Space Applications Centre (ISRO), Ahmedabad
109. Sarin M, Menon M (2020) Community-based wetland governance in the face of climate change. In: Climate resilient development in India. Sage, New Delhi
110. SCBD (2014) Global biodiversity outlook 4. Secretariat of the Convention on Biological Diversity, Montreal
111. Singh S (2015) Community participation in wetland management: a study of Chilika Lagoon. Indian J Environ Protect 35(9):783–789
112. Singh S (2019) Role of local governance in biodiversity conservation in India. Indian J Public Admin 65(4):1042–1055
113. Singh HR et al (2013) Community participation in the conservation of Loktak Lake, Manipur. Indian J Environ Protect 33(8):639–645
114. SoIB (2023) State of India's birds 2023: overall assessment report
115. Srinivas KR, Krishnan R (2020) Ecological governance in a changing climate: towards an integrated legal framework. Indian J Environ Law 5(2):112–130
116. Stroud DA, Davidson NC, West R, Scott DA, Haanstra L, Thorup O, Ganter B, Delany S (2004) Status of migratory waterbirds in the AEWA region. Wetlands International, Wageningen
117. Subhash Kumar v. State of Bihar, AIR 1991 SC 420
118. Sundar KSG, Subramanya S (2010) Bird monitoring in India: challenges and opportunities. Indian Birds 6(6):149–153
119. Trouwborst A (2010) Managing migratory species through networks of protected areas: important achievements and future challenges. Pace Environ Law Rev 27(2):407–435
120. U.S. Geological Survey (USGS) (2022) Glacier and ice sheet changes. Retrieved from usgs.gov
121. UNEP (2018) The weight of cities: resource requirements of future urbanization. United Nations Environment Programme, Paris

122. UNEP/AEWA (2020) AEWA strategic plan 2019–2027. Retrieved from https://www.unep-aewa.org
123. UNEP/CMS (2020a) Convention on the conservation of migratory species of wild animals: strategic plan 2020–2023. Retrieved from https://www.cms.int
124. UNEP/CMS (2020b) Convention on the conservation of migratory species of wild animals: strategic plan and achievements. Retrieved from https://www.cms.int
125. UNESCO (2023a) World heritage convention. https://whc.unesco.org/en/convention/
126. UNESCO (2023b) Natural world heritage sites. https://whc.unesco.org/en/list/
127. UNESCO World Heritage Centre (n.d.) India. https://whc.unesco.org/en/statesparties/in/
128. United Nations Environment Programme (UNEP) (2022) The state of climate action. Retrieved from unep.org
129. UNWTO (2014) Tourism and biodiversity: achieving common goals towards sustainability. World Tourism Organization, Madrid
130. Wetlands International (ed) (2020) Waterbird population estimates, 5th edn. Wetlands International, Wageningen
131. Wetlands International (2021) International waterbird census. Retrieved from https://www.wetlands.org/knowledge-base/international-waterbird-census/
132. Wetlands International (2022a) Asian waterbird census report – India. https://south-asia.wetlands.org
133. Wetlands International (2022b) Asian waterbird census: country summary for India. Retrieved from https://south-asia.wetlands.org/our-approach/healthy-wetland-nature/asian-waterbird-census/
134. Wetlands International (2022c) Who we are and what we do. Retrieved from https://www.wetlands.org
135. Whelan CJ, Wenny DG, Marquis RJ (2008) Ecosystem services provided by birds. Ann N Y Acad Sci 1134(1):25–60
136. Wildlife (Protection) Act (1972) Ministry of Environment, Forest and Climate Change. Available at: https://moef.gov.in
137. Wildlife Institute of India (2022) Database on protected areas in India. WII, Dehradun. Retrieved from https://wiienvis.nic.in/Database/Protected_Area_854.aspx
138. World Health Organization (WHO) (2023) Climate change and human health. Retrieved from who.int
139. World Meteorological Organization (WMO) (2023) State of the global climate report. Retrieved from wmo.int
140. WWF (2022) Living planet report 2022
141. Yodha K, Singh D, Kour A, Yaday R, Dahiya P, Kumar R (2023) Estimation of predatory bird damage to winter guava crop. Indian J Ecol 50(5 (SI)):1787–1793. https://doi.org/10.55362/IJE/2023/4138
142. Yodha K, Singh D, Kour A, Yadav R, Kumar R (2024) Species diversity and abundance of birds in ber (*Zizyphus mauritiana*) crop. Asian J Conserv Biol 13(1):118–126. https://doi.org/10.53562/ajcb.83683

# Chapter 15
# Conserving Aquatic Avifauna: Tackling the Threats and Ensuring Survival Through New Technology

**Neha Bhatnagar and Pankaj Sharma**

**Abstract** Aquatic avifauna or water birds, are crucial for the health of aquatic ecosystems and for sustainable development. They have specific ecological, economic and cultural roles. Aquatic avifauna are the birds that depend on water bodies for their survival. Aquatic birds, a large number of bird species inhabit and use various aquatic environments such as freshwater habitats, for example, ponds, lakes, rivers, marshes, and wetlands. The other one is coastal and marine areas like saltmarshes, mangroves, seagrass beds, estuaries, oceans, and shallow seas. This includes bird species that have adaptations for living in or near water, including specialized beaks, webbed feet, and exceptional diving or foraging abilities. Many species of aquatic avifauna are migratory, traveling along global flyways. They are a vital focus for conservation efforts as their populations serve as an indicator of big ecological changes, such as shifts in climate and increased pollution. However, human activities have led to habitat destruction, pollution, and climate change, all of these changes pose a risk to Aquatic biodiversity. Artificial Intelligence (AI) is emerging as a powerful tool to address these challenges and support conservation efforts globally. AI helps in the conservation of aquatic birds by making monitoring and analysis more efficient and accurate. It provides large amounts of data from various sources, which help conservationists to understand and to make strategies for protecting bird populations and their habitats more effectively than traditional methods alone. This paper presents a detailed knowledge about the significance of Artificial intelligence in Aquatic biodiversity conservation that leads us towards achieving the goal of ecological sustainability in future.

**Keywords** Biodiversity conservation · Avifauna · Wetland ecosystem · Climate change · Endangered species · Artificial intelligence

---

N. Bhatnagar
Department of Zoology, G.G.D.S.D. College, Palwal, Haryana, India

P. Sharma (✉)
Department of Ophthalmology and Visual Sciences, University of Illinois Chicago, Chicago, IL, USA
e-mail: sharmap@uic.edu



R. Kumar et al. (eds.), *Ecology, Conservation of Aquatic Birds*, Wetlands: Ecology, Conservation and Management 15, https://doi.org/10.1007/978-3-032-17125-2_15

## 15.1 Introduction

Aquatic avifauna refers to bird species that depend on aquatic ecosystems such as rivers, lakes, wetlands, estuaries, and oceans for feeding, nesting, or breeding. The presence of aquatic avifauna shows healthy status of their habitats as they are a good indicator of any change in ecosystem. Avifauna have a specific key role in an ecosystem as they are considered an important biotic factor, as they contribute highly in enhancing nutritive value and productivity of an ecosystem [33]. Aquatic Avifauna are the species that live in aquatic habitats like lakes, ponds, wetlands and the Ocean. Water birds function as a crucial link in an aquatic food chain and food web as various trophic levels in food chain possessed by these birds. They provide sustenance to predators like larger birds of prey and mammals while feeding on aquatic organisms themselves. Aquatic avifauna are excellent bioindicators as they are the most sensitive to environmental changes, they also help in maintaining nutrient cycle seed dispersal, pest control and also in habitat maintenance. The environmental condition of these birds is continuously deteriorating due to human activities which also affect the survival of these aquatic birds. It is essential to protect and conserve these birds in their natural habitat as their declining number adversely affects the environmental sustainability [8]. Firstly, we need to understand their feeding, breeding and migratory behaviour so that we can make an effective planning to protect these amazing animals from being extinct globally [40].

## 15.2 Ecological Importance of Aquatic Birds

Aquatic Avifauna play a crucial role in the ecosystem. Some important roles of aquatic avifauna are shown in Fig. 15.1.

### *15.2.1 As a Bioindicator*

Aquatic birds are used to evaluate the health and condition of an ecosystem since they are universally acknowledged to be excellent bioindicators. Their physiological state, behaviour, quantity, and existence can all reveal important details about the condition of the environment, especially wetlands and other aquatic ecosystems [2, 49, 56].

*Reasons behind the aquatic birds being considered as a bioindicator are as follows*:

1. *Sensitivity to Environmental Change*: Due to their severe reactivity to environmental changes, aquatic birds serve as useful "early warning systems." They react to environmental stressors such as pollution, shifting water levels, and the depletion of food supplies. As an illustration of a change in the health of the

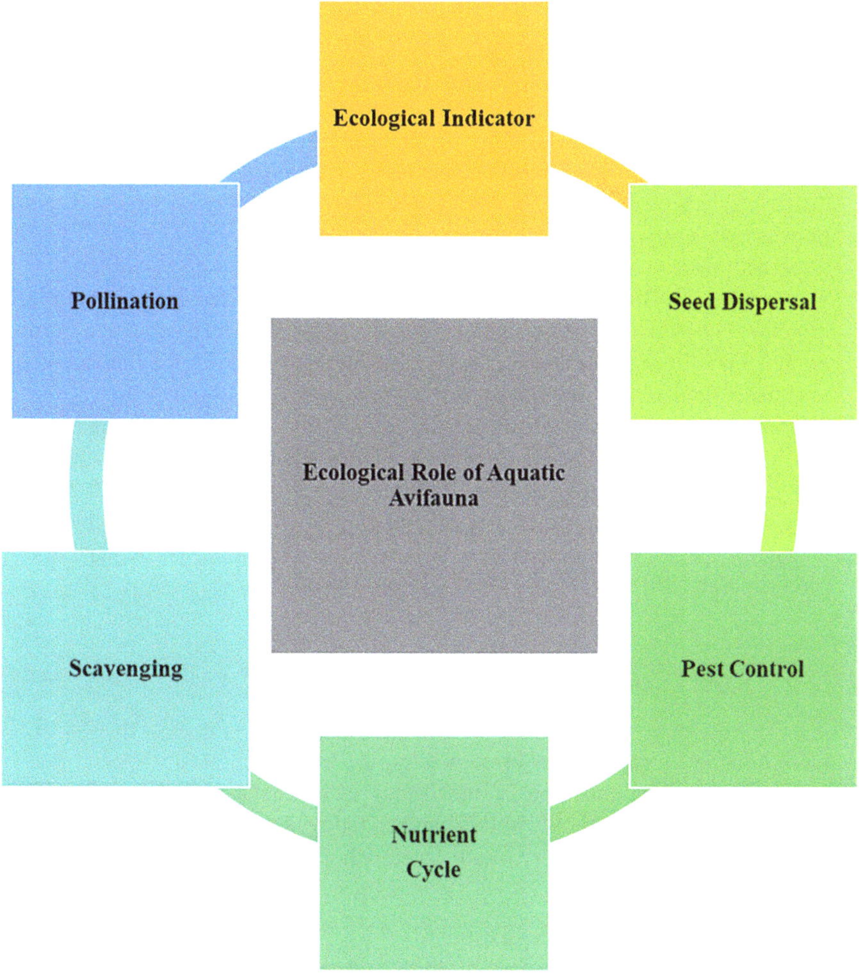


**Fig. 15.1** Beneficial attributes of aquatic avifauna

ecosystem, certain species may flourish in nutrient-rich (eutrophic) waters while others may perish. For example, Heavy metals and microplastics found in the tissues, eggs, and feathers of birds such as coots, ducks, and other seabirds have been utilized in studies to determine the degree of contamination in aquatic environments [13]. The Dipper (*Cinclus spp.*), a passerine bird that feed on aquatic invertebrates in rivers, has been used as a bioindicator of river pollution.

2. *Place in the Food Web*: A lot of aquatic birds eat earthworm, fish, insects, and other creatures as top predators. This makes them vulnerable to the bioaccumulation process, which concentrates toxins like heavy metals (like lead and mercury) and persistent organic pollutants at higher trophic levels [37, 77, 78]. The degree of contamination throughout the entire food chain can be

determined by scientists by examining the quantities of contaminants in the animals' tissues, eggs, or feathers. For example, it has been found that eutrophication (excessive nutrient enrichment) in wetlands increases the population growth of some bird species like great crested grebe due to increase in their food supply [42].

3. *Mobility and Wide Range*: Many aquatic migratory birds are able to cover large distances. This enables them to assess pollution levels and environmental conditions across large geographic areas [62]. A decrease in their number indicates an issue with their nesting areas as well as their complete migratory journey.
4. *Easy to study*: Birds are easily observable, simple to recognize, count and monitor. The statistics on Bird populations are available, which serve as a starting point for monitoring any changes in population over time [15]. So, they are more affordable than certain other organisms for environmental monitoring.
5. *Indicator of Biodiversity*: A measure of biodiversity, the variety of aquatic bird species in a region can reveal information about the ecosystem's overall health and species richness. For instance, siltation or a change in water depth may be indicated by a drop in species that depend on shallow water for foraging, but a change in species composition may be a result of the loss of particular flora or food supplies [28].

### *15.2.2 Seed Dispersal*

Aquatic birds are important for ornithochory, the process of seed dispersal that is essential for plant colonization of new areas and genetic variation [47]. Their function is especially crucial for wetland and aquatic plant species. The two primary mechanisms of endozoochory and epizoochory are responsible for this dissemination.

1. Endozoochoory: Internal Seed Dispersal

   Fruits or seeds are consumed by birds and subsequently processed by their digestive systems. Most of the time, the seeds are expelled in a different place. Examples: The seeds of aquatic plants such as pondweed, water lilies, and different types of grasses are consumed by numerous species of ducks, geese, and swans. When they fly between various bodies of water, they frequently spread these seeds across great distances.
2. Epizoochory: External Seed Dispersal

   This is the mechanism via which seeds or plant pieces adhere to the bird's exterior with the help of hooks, barbs and sticky coating and as a result travel to a new area. For instance, aquatic birds that wade through mudflats or marshes sometimes gather plant seeds from sedges and rushes, which can then be distributed as the bird flies to a new area [46].

   The importance of seed dispersal by aquatic bird are long distance dispersal, new habitat colonization and Genetic diversity. The distribution and composition of plant life in aquatic environments are shaped by this process, which is an essential ecological service [72].

### 15.2.3 Pest Control

Aquatic birds aid in the management of pests by consuming a variety of insects, rodents, and other small animals that are regarded as ecological or agricultural pests. For example,

1. *Feeding on Insect*: Numerous aquatic and semi-aquatic insects, such as mosquito larvae, dragonflies, and midges, are consumed by birds such as herons, egrets, ducks, and kingfishers [61]. This lowers the number of pests that could harm crops or spread illness.
2. *Agriculture Pest control*: Many aquatic birds search for food in rice paddies, wetlands, and flooded fields, where they feed on pests like grasshoppers, caterpillars, snails, and small rodents. For instance, Egrets frequently follow plough tractors in order to consume insects and rodents that are disturbed [58].
3. *Eating Pest Fishes and Amphibians*: Birds like cormorants and pelicans play an important role in maintaining ecological balance by feeding on invasive fish species that disrupt native aquatic ecosystems. They also help control the population of amphibians, such as frogs and tadpoles, which may overpopulate and negatively impact biodiversity.
4. *Limiting Disease-Carrying Organisms*: By consuming mosquito larvae, aquatic birds indirectly help in reducing the risk of diseases like malaria and dengue by feeding on mosquito larvae as they are the vector of many diseases [23, 34, 35].

### 15.2.4 Nutrient Cycle

Aquatic birds play a vital role in nutrient cycling within and across ecosystems. They are a crucial link in the transfer and cycling of nutrients in aquatic and adjacent terrestrial ecosystems [21]. Their primary contribution comes through guano (bird droppings), which is rich in essential nutrients like nitrogen (N), phosphorus (P), and potassium (K). There are many other ways in which they contribute to maintaining the nutrient cycle.

1. *Guano as a Natural Fertilizer*: Aquatic birds typically feed in nutrient-rich environments such as oceans, lakes, and rivers, but they often nest or roost in distant locations like islands, shorelines, or wetlands [24]. In doing so, they deposit guano, a nutrient-rich waste that acts as a natural fertilizer, significantly enhancing the nutrient content and productivity of these new habitats.
2. *Cross-Ecosystem Nutrient Transfer*: Aquatic birds consume nutrient-rich food such as fish and invertebrates in one location, and then excrete waste (guano) in another, effectively translocating essential nutrients like nitrogen and phosphorus across ecosystems. The transfer of nutrients occurs from marine to terrestrial system and within different parts of lakes and wetlands, like from deeper, nutrient-rich waters to shallower or sunlit areas [24]. This cross-ecosystem nutrient flow helps in phytoplankton growth and improves vegetation health.

3. *Enhancement of Primary Production*: Increased nutrient input from bird guano can significantly enhance primary production in aquatic environments by stimulating the growth of algae and phytoplankton, which form the base of the food web. This enrichment also leads to a boost in biodiversity and biomass of species that depend on nutrient-rich habitats, including fish, invertebrates, and aquatic plants [23].
4. *Bioturbation*: Certain foraging behaviours of aquatic birds, such as flamingos sifting through sediment can disturb the substrate and release nutrients that are trapped in the bottom layers of water bodies, making them available to primary producers [22].
5. *Carrion Consumption*: Scavenging birds such as vultures and eagles play a vital role in nutrient cycling by consuming the carcasses of dead animals [50]. This accelerates the decomposition process, preventing nutrients from being locked within decaying bodies and allowing them to be rapidly reintroduced into the ecosystem.
6. *Bottom-up Effects*: The influx of nutrients from bird guano can trigger a 'bottom-up' effect within ecosystems. By enhancing the growth of primary producers such as algae and aquatic plants, this nutrient input increases the energy available at the base of the food web. As a result, it supports a greater abundance and diversity of herbivores and higher-level consumers, potentially reshaping the entire food web structure and increasing overall biodiversity.

### 15.2.5 Scavenging

Scavenging plays a vital role in ecosystems by facilitating the removal of dead organic material, thereby preventing the spread of diseases and accelerating decomposition. This process contributes to the efficient recycling of nutrients back into the environment, supporting plant growth and overall ecosystem productivity [51]. For aquatic birds, carrion often serves as a critical food source, particularly in habitats where food availability is seasonal or unpredictable. By relying on scavenged resources, these birds can maintain energy balance and survive during periods of scarcity.

*Examples of Scavenging Aquatic Birds are as follows*:

**Gulls** Gulls, especially larger species like the herring gull, are well-known for their opportunistic scavenging behaviour. They frequently feed on discarded food scraps found in human-altered environments such as beaches, docks, and garbage dumps. In addition to human waste, gulls also scavenge on natural food sources like dead fish and other marine organisms that wash ashore [53].

**Northern Giant Petrel** The Northern Giant Petrel is a truly oceanic scavenger, well-adapted to life in the Southern Ocean. These large seabirds are notorious for their aggressive scavenging behaviour, often feeding on the carcasses of dead seals, penguins, and other marine animals [67]. They are also commonly attracted to offal and other waste discarded from fishing vessels. Their powerful beaks and strong flight capabilities enable them to cover vast distances in search of carrion, making them key players in nutrient recycling within marine ecosystems [53].

**Cormorants** They are primarily predators, but their occasional scavenging adds nuance to their ecological role. While live fish are their main food source, a cormorant's diet can change significantly based on the local availability of prey. When fish populations are low, or in environments where they have access to other food, they may supplement their diet with other aquatic animals such as crustaceans, amphibians, or even insects. The occasional scavenging on carrion is an example of this adaptability, allowing them to exploit alternative food sources during challenging times and demonstrating their survival skills in variable environments [53].

### *15.2.6 Fish Population Control*

Aquatic birds play a significant and often complex role in regulating fish populations, acting as natural predators in both wild ecosystems and aquaculture environments. By feeding on a variety of fish species, these birds contribute to maintaining ecological balance in both freshwater and marine systems. By preying on various fish species, they help maintain ecological balance in freshwater and marine environments [36, 42]. In wild habitats, this predation can prevent overpopulation of certain fish species, promoting biodiversity. In aquaculture, however, their presence can pose challenges, as birds may target farmed fish stocks. This dual role underscores the importance of a subtle understanding of their ecological functions, as well as the development of management strategies that balance wildlife conservation with the needs of the aquaculture industry.

## 15.3 Economic Importance of Aquatic Birds

### *15.3.1 Ecotourism*

Birdwatching is one of the fastest-growing niches within the ecotourism industry. Aquatic birds, with their dramatic flocking behaviour, striking plumage (such as that of flamingos and spoonbills), and distinctive behaviours, are a major attraction for birdwatchers. Avitourists are a dedicated group who frequently travel long distances specifically to observe certain bird species in their natural habitats. Compared to the average tourist, avitourists tend to spend more on travel, accommodation, professional guides, and specialized gear, making them a valuable segment of the ecotourism market [66]. The seasonal migration of aquatic birds can be the foundation for local festivals and events. These festivals, such as those celebrating the arrival of cranes, geese, or flamingos, attract thousands of visitors. This influx of tourists generates income for local businesses, from hotels and restaurants to tour operators and gift shops. This also creates jobs for local communities like professional bird guides, tour boat operators, lodge staff, and conservationists. These birds are the "flagship species" that bring attention to and help protect fragile wetland and coastal ecosystems like Flamingos, Pelicans, Migratory Waterfowl, Seabirds, albatrosses, cormorants and Penguins etc. [12].

### *15.3.2 Fishery Support*

Aquatic birds and fisheries have a complicated and multidimensional connection because they are both competitors and markers of the health of ecosystems. Depending on the particular situation and management techniques, their role may be neutral, constructive, or destructive. Aquatic birds can support fisheries by serving as bioindicators, providing an early warning system for a decline in fish stocks or water quality [16, 36]. They also play a role in nutrient cycling, helping to fertilize the water and maintain a healthy food web. In some cases, they can act as natural predators of species that compete with or harm commercial fish. Additionally, they can help with waste removal by scavenging on discarded fish from fishing vessels.

### *15.3.3 Climate Change Research*

Aquatic birds are useful markers for studies on climate change [13]. They assist researchers in monitoring the consequences of global warming by:

1. Demonstrating life cycle changes: Variations in the timing of migration or breeding indicate how animals are reacting to temperature variations.
2. Their relocation to new geographic areas reveals changes in habitat, which in turn reveals changes in ecosystems.
3. Disruptions to food webs can be shown by differences in a bird's mating cycle and the availability of its prey, which can reveal how warming impacts food chains.
4. Providing information from remote places: Sea surface temperatures and other important data can be gathered by tiny sensors mounted on birds.

### *15.3.4 Environmental Education*

Aquatic birds are excellent for environmental education because they make complex ecological concepts concrete and relatable. They serve as highly visible bioindicators, showing the health of an ecosystem through their presence or absence.

By studying them, people can easily understand the impact of human activities like pollution and habitat loss [13]. Additionally, migratory birds are a perfect example for demonstrating the local effects of climate change, as their behaviour and life cycles are directly influenced by a warming world. They provide a relatable way to discuss how human actions like pollution and habitat destruction directly impact wildlife [13].

## 15.4 Cultural Importance of Aquatic Birds

Aquatic birds have a rich cultural significance, often serving as symbols of purity, grace, and wisdom. Swans, for instance, represent grace and purity. These birds are central to numerous myths and legends. The Greek myth of the kingfisher, or "halcyon," explains the origin of calm seas during the winter, giving us the term "halcyon days." In European folklore, the white stork is famously believed to deliver babies, a myth popularized by Hans Christian Andersen [3, 80]. Aquatic birds are prominent subjects in art and literature, serving as powerful metaphors for themes of nature, time, and the human spirit.

## 15.5 Major Threats and Challenges to Aquatic Birds

Aquatic avifauna play a pivotal role in maintaining the health and balance of ecosystem. Any type of changes in the aquatic environment changes the feeding and breeding behaviour of these birds. To make a proper strategy it is necessary first to overlook the major threats to these birds. Some of these challenges are listed below:

1. *Habitat loss and degradation*: The habitat loss and degradation are the major cause of avifauna biodiversity loss and their increasing chances of extinction [79]. Habitat destruction affects the distribution and behaviour of species. We know that wetlands are the highly biodiverse, productive and vulnerable areas for avifauna. The risk of their loss is increasing day by day due to human activities like hunting, deforestation, pollution.
2. *Climate change*: Climate change has a critical role in avifauna diversity variation. Any type of disturbance in climate parameters like rainfall, temperature, cloudiness and humidity affects the bird communities [19]. Avifauna species distribution and behaviour depend on these specific climate factors. Human activities are solely responsible for these types of changes in the natural habitat of these animals.
3. *Hunting and Poaching*: Hunting negatively affects the bird population in their specific habitat. It causes a potential decline in species richness and disturbed community composition. The area with better accessibility to humans has high

hunting pressure and greater bird population depletion. It also contributes indirectly to habitat fragmentation and ecosystem changes in a particular area [55].

4. *Pollution*: As we know, avifauna work as a bioindicator and are highly sensitive towards any type of changes in the aquatic habitat. Pollution in aquatic habitats has a huge negative impact on avifauna diversity [19]. It causes decrease in the bird population due to habitat destruction and reduced reproductive success. These pollutants are heavy metals, pesticides, pharmaceuticals and plastic waste which accumulate in the ecosystem and reduce the survival efficiency of these birds.
5. *Oil spills* also have a devastating effect on aquatic birds by decreasing availability of clean feeding and nesting areas. They affect their feather by contaminating them and decreasing their waterproofing and insulating properties [20]. Oiled feathers lost buoyancy and making it difficult for these birds to fly. It also contaminates eggs and reduces hatchling survival rate, which in turn decreases the number by reducing reproductive success.
6. *Invasive species*: Introduction of invasive species in ecosystems negatively affects aquatic avifauna diversity by declining the native species diversity. They have a profound effect on the basic needs of these birds like habitat, food availability, foraging behaviour and ecosystem dynamics [14]. E.g., invasive plants like water hyacinth reduce the open water area for aquatic birds. They can increase pollution, toxicity and also spread pathogens in the water which also affect the bird population. One such example of these is Zebra mussels (*Brissena Polymorphs*) and American mink disturb the food chain by filtering out plankton.
7. *Overexploitation of Resources*: One of the main reasons of declining the birds is the overuse of aquatic resources like fishes, plants which also contribute in vulnerable status of these birds [38].

These threats create major issues for the richness, health and stability of ecosystems. So, there is a need to address these issues on an urgent basis to prevent Avifauna and the natural environment from the deadly consequences.

## 15.6 Conservation Status of Some Aquatic Birds

Scientists usually record data based on counts and observation of the behaviour of sea birds at the wetland site with the help of AI tools software and websites [11] (Fig. 15.2). This data is used by them for seeing population dynamics and also providing them conservation status. Ther are many aquatic birds that have endangered status in Red List (Table 15.1). The IUCN Red List is generally regarded as the most objective and authoritative system available for classifying species in terms of their risk of extinction [57].

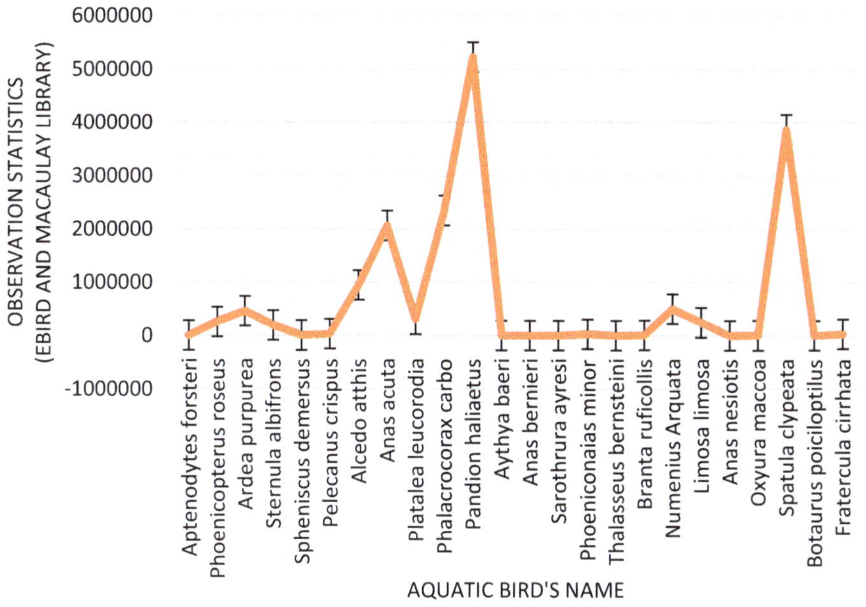


**Fig. 15.2** Statistics of aquatic bird's observation according to eBird and Macaulay library, Cornell Lab dataset

## 15.7 Conservation Strategies

There are various recent innovative methods offered by environmentalists to conserve these aquatic avifauna species. The International Union for Conservation of Nature's Red List shows that endangered sea bird species is approximately 31%. Although there has also been significant development also occurred in technologies used for conservation and assessment of sea birds. [45] (Fig. 15.3). These powerful tools are equipped with Artificial Intelligence that makes these conservation efforts more effective and productive [7]. We must adopt these recent advancements to protect these animals from extinction and also sustain the ecosystem balance (Referred from Datasets, Seabird Tracking).

1. *AI-Powered monitoring system*: An AI-powered monitoring system leverages artificial intelligence and machine learning to continuously track, analyze, and interpret data from various sources (Fig. 15.4). This provides real-time insights and can even automate responses to potential problems. It does not depend on static thresholds and manual analysis like traditional monitoring methods [10]. The main advantage of this monitoring system is proactive issue detection, reducing human error, and improving efficiency.

   Smart buoys are equipped with eDNA sensors that help in detecting genetic materials from aquatic birds. This technology aids in tracking presence of

**Table 15.1** List of some Aquatic Avian species facing danger of extinction [29]

| Common name | Scientific name | Native country | Ecosystem role | Major threats | IUCN status |
|---|---|---|---|---|---|
| Emperor Penguin | *Aptenodytes forsteri* | Antarctica | Predator of fish and krill in Antarctica | Climate change, sea ice loss | Near threatened |
| Greater Flamingo | *Phoenicopterus roseus* | Africa, Southern Europe and South Asia | Filters small organisms from brackish water; keeps wetlands healthy. | Habitat loss, Water fragmentation and Hunting | Least concern |
| Purple Heron | *Ardea purpurea* | Europe, Asia and Africa | Predator of fish, frogs, and insects in wetlands. | Habitat loss and degradation of wetland | Least Concern (declining in some regions) |
| Little Tern | *Sternula albifrons* | Africa, coastal South Asia, Australia and temperate Eurasia | Coastal predator; indicates estuary and shoreline health. | Human disturbance, habitat loss and degradation, predation, and climate change | Least Concern |
| African Penguin | *Spheniscus demersus* | South Africa, Namibia | Coastal predator; helps manage fish populations. | Oil spills, overfishing, climate change | Critically Endangered |
| Dalmatian Pelican | *Pelecanus crispus* | Eastern Europe, Central Asia | Top predator in lakes and rivers; controls fish populations. | Wetland loss, human disturbance | Near threatened (Global)<br>Least concern (Europe) |
| Common Kingfisher | *Alcedo atthis* | Europe, Asia, and North Africa | Feeds on small fish; helps maintain fish population balance in freshwater bodies. | Habitat loss and degradation, Climate change | Least Concern (locally vulnerable) |
| Northern Pintail | *Anas acuta* | Northern areas of Europe and across the Palearctic and North America | Migratory duck; important for wetland nutrient cycles. | Habitat loss, Pollution and Climate change | Least Concern (some decline noted Global)<br>Vulnerable (Europe) |
| Eurasian Spoonbill | *Platalea leucorodia* | Europe, Asia, and North Africa | Feeds on small fish and aquatic invertebrates by sweeping its bill through water. | Habitat loss and degradation, wetland drainage and pollution | Least Concern (protected in many regions) |

| | | | | | |
|---|---|---|---|---|---|
| Great Cormorant | *Phalacrocorax carbo* | Eurasia, Africa, Indian subcontinent and Australia | Excellent fisher; controls fish populations. | Habitat loss and conflict with fisheries | Least Concern (can be a conflict with fisheries) |
| Osprey | *Pandion haliaetus* | Northern Europe, Scandinavia, Russia, South America and Siberia | Feeds almost exclusively on fish; indicator of aquatic ecosystem health. | Environmental contamination, habitat loss, human disturbance, and historical pesticide exposure | Least Concern (recovering, Global) Endangered in mediterranean |
| Baer's Pochard | *Aythya baeri* | China, Russia, India | Consumer of aquatic plants, seeds, and some invertebrates, contribute to food web | Wetland loss, hunting | Critically Endangered |
| Madagascar Teal | *Anas bernieri* | Madagascar | Foraging on small invertebrates and plants in wetland ecosystem | Mangrove destruction, hunting | Endangered |
| White-winged Flufftail | *Sarothrura ayresi* | Ethiopia, South Africa | Flagship species for wetland, Bio indicator species | Wetland draining, fragmentation | Critically Endangered |
| Lesser Flamingo | *Phoeniconaias minor* | East Africa, India | Specialist filter feeders in alkaline lakes<br>Bio indicator species | Pollution, habitat degradation | Near threatened |
| Chinese Crested Tern | *Thalasseus bernsteini* | China, Taiwan | Top predator in coastal marine ecosystems | Egg collection, habitat destruction | Critically Endangered |
| Red-breasted Goose | *Branta ruficollis* | Russia, Bulgaria, Romania | Key indicator species, reflect the health of wetlands | Habitat loss, hunting, climate change | Vulnerable |
| Eurasian Curlew | *Numenius arquata* | Europe, Central Asia | Distributing chemical nutrients, disperse seeds, invertebrates and beneficial microbes which stick to their feathers, feet and bills | Agriculture, habitat loss | Near threatened |
| Black-tailed Godwit | *Limosa limosa* | Europe, Asia | Migratory shorebirds, connecting wetlands across vast distances | Wetland destruction, farming expansion | Near threatened |
| | *Anas nesiotis* | | | Invasive predators, restricted range | vulnerable |

(continued)

**Table 15.1** (continued)

| Common name | Scientific name | Native country | Ecosystem role | Major threats | IUCN status |
|---|---|---|---|---|---|
| Campbell Teal | | Campbell Island (New Zealand) | Foraging through piles of seaweed or rocks | | |
| Maccoa Duck | *Oxyura maccoa* | South Africa, Kenya, Ethiopia | Consuming benthic invertebrates, plants, and seeds in wetlands | Water pollution, habitat conversion | Endangered |
| Northern Shoveler | *Spatula clypeata* | North America, Eurasia | Feeding on aquatic invertebrates, particularly crustaceans and plankton | Habitat degradation, pollution | Least concern |
| Australasian Bittern | *Botaurus poiciloptilus* | Australia, New Zealand | Wetland predator, helping to control populations of fish, eels, frogs, and other aquatic animals | Wetland drainage, climate change | Vulnerable |
| Tufted Puffin | *Fratercula cirrhata* | North Pacific (USA, Russia, Japan) | Predators in marine ecosystems, specifically focusing on small fish and marine invertebrates | Ocean warming, prey shortages | Least concern |

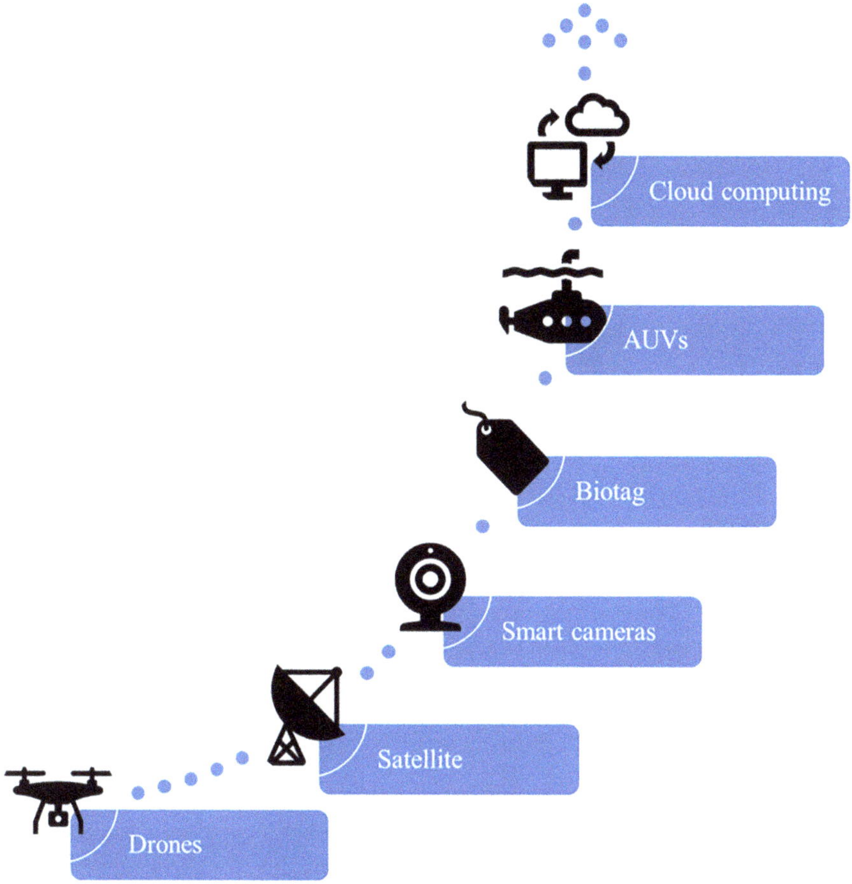


Fig. 15.3 AI tools used for biodiversity conservation

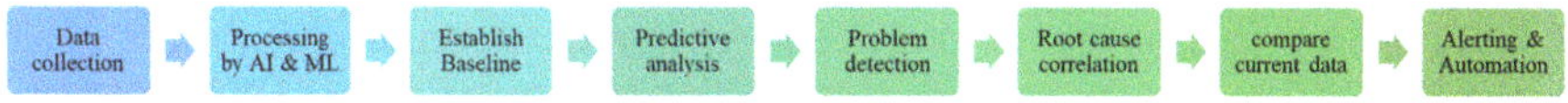


Fig. 15.4 Working of AI modal

animals and monitoring their habitat in real time. The main advantage of this technology is that while using this these devices do not cause any disturbance to other aquatic animals. It detects the presence of rare or elusive species without needing to see them.

2. *Remote Sensing and Drones for Habitat Restoration*: Remote sensing uses sensors on satellites and aircraft to collect data about the Earth's surface from a distance. Drones are wonderful tools to track these animals and also restore the health of aquatic ecosystem [59]. Drones can carry a variety of sensors beyond standard cameras, including multispectral sensors to assess plant health, thermal

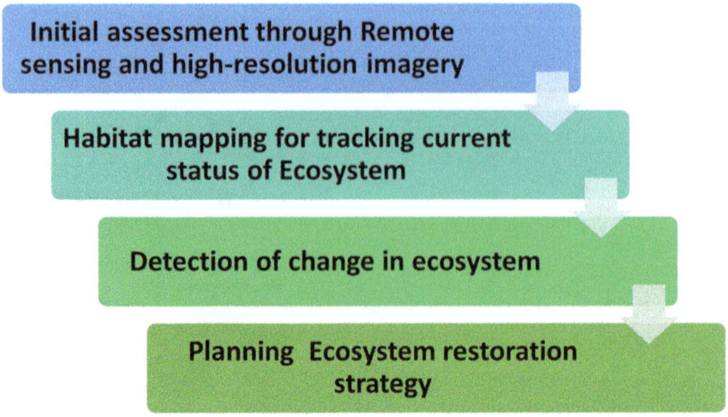


**Fig. 15.5** Working of remote sensing technology

cameras to track wildlife or identify sources of pollution, and LiDAR (Light Detection and Ranging) to create highly accurate 3D models of the terrain and vegetation. With the help of drones, illegal hunting in the conserved areas can be prevent and location and behaviour of birds also be observed. Drones also helpful in planting seeds in far-reaching areas for reforestation purpose. Along with drones, thermal cameras, and multispectral sensors are also used for monitoring wetlands and nesting sites without disturbing birds (Fig. 15.5). It is also useful in counting population of Bird Species. The role of remote sensing is as follows:

3. *Autonomous underwater vehicles (AUVs)*: These are automatically driven vehicles that help in monitoring water quality and detecting any type of physico-chemical changes in the aquatic ecosystem. In this way, they help in maintaining the bird population. It alerts when conditions become harmful to aquatic birds. AUVs are equipped with a wide array of sensors depending on their mission like sonars for creating detailed maps of the seafloor, cameras and lights for high-resolution visual surveys of marine life, environmental sensors to measure water characteristics like temperature, salinity, oxygen levels, and pH [17]. There are different types of AUVs are used for monitoring like gliders, hovering AUVs and hybrid AUVs.
4. *AI-driven predictive analysis*: Now some AI machines help to predict any type of changes in environment like Pollution, water level fluctuation and toxicity. This helps in planning proactive conservation measures [41]. It analyses large datasets from various sources (satellites, sensors, tracking). It helps identify patterns in migration, breeding success, and environmental threats to enables environmentalist in predictive modelling to anticipate changes and take proactive conservation measures (Fig. 15.6). Its working process involve:
5. *Satellite tracking and GPS Telemetry*: Recent advanced satellite technologies allow researchers and conservationists to monitor behaviour of aquatic birds like migratory pattern, reproductive behaviour, competition. This technology is used

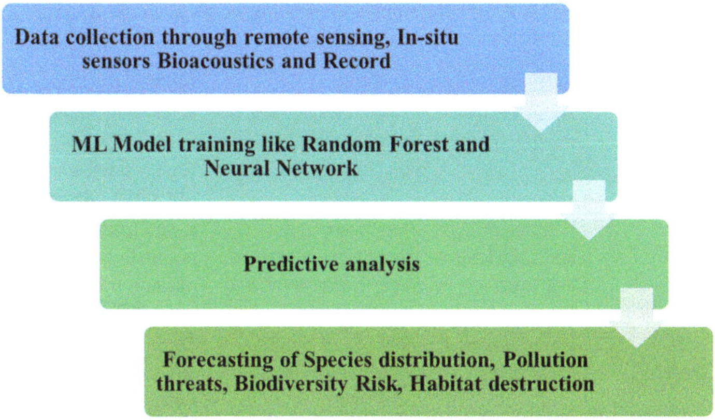


**Fig. 15.6** Predictive analysis by Artificial Intelligence

by conservationists to identify critical habitats and potential threats (e.g., pollution, habitat loss) and to protect vulnerable sites for aquatic birds. For this, lightweight GPS tags are placed on birds like geese or flamingos that record their location at set intervals. These GPS devices transmit data via satellites (like Argos or Iridium) to researchers, often in near real-time [25].

*Collars/Tags*: Lightweight, solar-powered or battery-operated.
*Satellites*: Argos, Iridium, GPS, and sometimes GSM (cellular networks).
*Software*: GIS tools, data visualization platforms (e.g., Movebank, EarthRanger).

6. *Citizen Science Apps*: It's a mobile application that collects data through a citizen science project. This technology allows the public to log sightings and upload photos (Table 15.2). These apps enable people to contribute to scientific research and conservation efforts by collecting data, making observations, and sharing findings through mobile or web platforms [73]. This application is helpful in making massive, crowd-sourced databases for tracking bird populations and distributions.
7. *AI-Powered Image and Sound Recognition*: Machine learning models that identify bird species by their calls or appearance. There are various tools and projects available that are used by scientists to record data of birds (Table 15.3). Main use of this technology automates population surveys and behavioural studies [71].
8. *Genetic and Genomic Tools*: These are used to track genetic diversity, evolutionary history, adaptive traits and health of populations. It assists in breeding programs and identifying distinct subspecies [68]. The key application of genetic tool in the identification of morphological similar species, in population

**Table 15.2** List of some Citizen Science Apps and their importance in conservation

| S. no. | Citizen science apps | Purpose | Features | Impact | Reference |
|---|---|---|---|---|---|
| 1. | iNaturalist | Biodiversity documentation | Users upload photos of plants and animals; AI helps identify species | Data contributes to global biodiversity databases like GBIF | Forti [18] |
| 2. | eBird (Cornell Lab of Ornithology) | Birdwatching and migration tracking | Users log bird sightings; data used in avian research and conservation | Helps track bird populations and migration patterns globally | Cornell Lab Dataset [11] |
| 3. | Earth Challenge | Environmental monitoring | Tracks air quality, plastic pollution, and insect populations | Supports global environmental policy and awareness | Lee et al. [39] |
| 4. | Marine Debris Tracker | Ocean and coastal cleanup | Users log litter and debris found in natural areas | Data informs cleanup and pollution prevention strategies | Merlino et al. [43] |
| 5. | Zooniverse | Crowdsourced research across disciplines | Users help classify images, transcribe data, and more | Supports projects in ecology and more areas | Smith et al. [63] |
| 6. | Wildlife Spotter (Australia) | Wildlife monitoring via camera trap images | Volunteers help identify animals in photos | Supports ecological research and species protection | Wildlife Spotter [76] |

**Table 15.3** List of some AI tools and their importance in conservation

| S. no. | Tools & project | Purpose | Impact | Reference |
|---|---|---|---|---|
| 1. | BirdNET Wildbook | Image Recognition for Aquatic Birds | Identify species from photos or videos captured by drones, camera traps, or citizen scientists<br>Monitor nesting sites<br>Detect behavioral patterns, such as feeding or mating<br>Track migration using satellite imagery and automated detection | Kahl et al. [31], Holmberg [27] |
| 2. | BirdNET-Pi Merlin Bird ID Ecoacoustics platforms | Sound Recognition for Aquatic Birds | Detect and classify bird calls using spectrogram analysis<br>Monitor presence and abundance of birds in wetlands, estuaries, and coastal zones<br>Track migration and breeding by analyzing seasonal vocal activity<br>Filter background noise to isolate bird calls in noisy environments | Martínez, Joachin [72], Nieto-Mora et al. [48] |

**Table 15.4** List of some Genetic tools and their importance in conservation

| S. no. | Genetic tools | Application | Reference |
|---|---|---|---|
| 1. | Whole Genome Sequencing (WGS) | Reveals adaptive traits, evolutionary history, and disease resistance | Hohenlohe et al. [26] |
| 2. | Transcriptomics | Studies gene expression under different environmental conditions (e.g., pollution, climate change) | Patnaik and Lee [52] |
| 3. | Epigenetics | Explores how environmental factors influence gene expression without altering DNA sequence | Balard et al. [5] |
| 4. | Metagenomics | Analyses microbial communities in aquatic bird habitats to assess ecosystem health | Wei et al. [74] |

genetics and kinship analysis (Table 15.4). Example: DNA barcoding for species ID and monitoring illegal trade or poaching of protected bird species [9].

**Technologies Used**

- PCR (Polymerase Chain Reaction)
- Next-Generation Sequencing (NGS)
- CRISPR (for functional genomics)
- Bioinformatics platforms (e.g., Galaxy, Geneious)

9. *Smart Cameras and Environmental Sensors*: Solar-powered camera systems and environmental sensors are used for studying breeding success, temperature effects, and human interference. We can get Real-time data on nest conditions and bird behaviour through this technology [7]. Example: Accelerometers, gyroscopes, and bio-loggers attached to birds [64]. Smart cameras equipped with AI and edge computing are transforming bird monitoring like Real-time species identification using computer vision models (e.g., YOLO, ResNet) and Motion-triggered recording to capture their behaviour.
10. *Geospatial Mapping & GIS (Geographic Information Systems)*: It helps conservationists to plan protected areas and analyse habitat loss over time by mapping aquatic bird habitats, migratory paths, and environmental threats. It is a powerful tool that integrates spatial data (maps, satellite imagery, GPS coordinates) with ecological and environmental datasets to support conservation planning, habitat monitoring, and species management [70].

**Tools and Techniques Used in GIS**

- *Websites: ArcGIS* (ArcGIS Geospatial Platform), *QGIS* (Open Source GIS), *InVEST* (Integrated Valuation of Ecosystem Services and Tradeoffs), *Google Earth Engine*
- *Remote sensing data*: Landsat, Sentinel, MODIS
- *GPS and telemetry*: For tracking bird movements
- *Spatial modeling*: Habitat suitability, corridor design, change detection

**Table 15.5** List of some cloud-based data platforms and their importance in conservation

| S. no. | Cloud-based data platforms | Application | References |
|---|---|---|---|
| 1. | Earth Ranger | Aggregates GPS tracking, camera trap, and sensor data to monitor wildlife, including aquatic birds | Wall [72] |
| 2. | Wildlife Insights | Identifies species from images, including waterbirds, and supports global biodiversity assessments | Ahumada [1] |
| 3. | Seabird Tracking Database | Supports marine Important Bird Area (IBA) identification and conservation planning. | Seabird Tracking Database [60] |
| 4. | Global Biodiversity Data Sources (IUCN SSC) (GBIF, eBird, BioTIME, BOLD, and others.) | Analyses microbial communities in aquatic bird habitats to assess ecosystem health | Stephenson and Stengel [65] |
| 5. | AWS + NatureServe Environmental Review Tool | Provides high-resolution species distribution data for conservation planning | Kotagi et al. [32] |

11. *Cloud-based Data Platforms*: These are helpful to store and share wild animal tracking and research data globally [54]. This technology connects the government with the researchers [69]. These platforms offer scalable, real-time, and collaborative tools for monitoring aquatic bird populations and habitats (Table 15.5).
12. *Acoustic Telemetry*: Acoustic telemetry involves attaching small transmitters (tags) to animals that emit sound pulses. These pulses are detected by underwater receivers. Through this technology, scientists can track the movements and behaviour of birds via underwater microphones (hydrophones). It is helpful in monitoring water-based birds, especially in environments where GPS is ineffective (e.g., underwater or dense canopies) [75]. Global Network also collaborates with Acoustic Telemetry like Ocean Tracking Network (OTN) supports collaborative tracking of aquatic species, including birds in coastal and marine environments & Atlantic Cooperative Telemetry Network (ACT) supports data sharing across regions to monitor tagged animals at continental scales.
13. *Portable Nesting Boxes with IoT Integration*: Nesting boxes equipped with sensors to monitor temperature, humidity, and occupancy are used for Real-time monitoring of nesting success and environmental conditions [6]. It provides a comfortable, controlled nesting environment and allows researchers to intervene when necessary. It provides insights into how birds interact with their environment, especially in controlled settings. Example IoT bird nest monitoring system by Mifra electronics.
14. *Biodegradable Tags and Sensors*: These sensors are attached to birds without leaving permanent environmental impacts. These are non-toxic and leave no

trace once the device naturally disintegrates, reducing ecological footprint [44]. These are made from natural polymers, proteins, or biodegradable plastics and include RFID, GPS, piezoelectric sensors, and environmental monitors. These are often powered by motion (piezoelectricity) or solar energy, eliminating the need for batteries.

15. *PTT (Platform Transmitter Terminal)*: This is a type of satellite-based tracking device used by researchers to monitor the movements of animals over long distances, particularly migratory birds and marine animals. Animals are fitted with a PTT device, which is often a tiny, light tag [30]. The gadget's transmitter broadcasts a signal to a group of satellites in polar orbit, including the well-known Argos satellite system [4].

These technologies are essential for sustainable conservation, resource management, and climate action. Efforts to rescue and rehabilitate these birds are critical but challenging. These technologies are revolutionizing conservation efforts making them more effective and efficient. Some organization like International Bird Rescue play a crucial role in saving the aquatic birds from extinction (Reference [5] Article, 2024).

## 15.8 Future Prospect of AI Conservation Strategies

Looking ahead, the potential for AI in this area is enormous and still mostly unrealized. Future research should concentrate on developing machine learning models that can forecast the impact of specific pollutants on bird health, applying deep learning to examine the acoustic signatures of bird calls for non-invasive monitoring, and creating more approachable AI platforms to enable citizen scientists globally. Furthermore, the ethical application of these technologies requires careful consideration. We must ensure that AI models are transparent and unbiased, that data privacy is maintained and that the technology is used as a conservation tool rather than as a substitute for the priceless labour done by human specialists on the ground. Ultimately, our capacity to utilize these potent technologies responsibly and effectively will determine the success of aquatic bird conservation as well as the ecosystems restoration. The future of these species depends s on our innovative application of human intelligence and its artificial counterparts.

## 15.9 Conclusion

In conclusion, the integration of Artificial intelligence into aquatic bird conservation marks a new and promising era for wildlife protection. This study has shown how AI-powered solutions, such as advanced image recognition for precise species identification and predictive modelling for habitat risk assessment, provide a revolutionary way to address the persistent issues that are faced by conservationists. By leveraging AI to analyse vast datasets collected by drones and other remote sensing technologies, we can monitor bird populations with unprecedented accuracy and efficiency, moving beyond manual surveys that are often resource-intensive and prone to error.

We can say that AI is not merely a supplementary tool but a vital partner in developing a more data-driven and proactive conservation strategy. AI makes it possible to go from reactive to proactive treatments by offering precise and timely insights. In light of growing concerns including pollution, habitat destruction, and climate change, this is essential for protecting aquatic bird populations.

## References

1. Ahumada JA et al (2020) Wildlife Insights: a platform to maximize the potential of camera trap and other passive sensor wildlife data. Environ Conserv 47:202003
2. Amat JA, Green AJ (2009) Waterbirds as bioindicators of environmental conditions. In: Conservation monitoring in freshwater habitats: a practical guide and case studies. Springer Netherlands, Dordrecht, pp 45–52
3. Andersen HC (1835) Fairy tales told for children. First collection. Third Booklet. Reitzel CA, Copenhagen
4. Argos Satellite System (2025) Retrieved in August 2025. https://www.argos-system.org/
5. Balard A, Baltazar-Soares M, Eizaguirre C, Heckwolf MJ (2024) An epigenetic toolbox for conservation biologists. Evol Appl 17(6):e13699
6. Basnet M (2020) Intelligent nest box using particle IoT device[Bachelor's thesis, Centria University of Applied Sciences]. Theseus. https://www.theseus.fi/handle/10024/346853
7. Berenguer Agulló A (2024) Edge computing solutions for real-time birds species analysis [Master's thesis, University of Alicante]. RUA Repository. https://rua.ua.es/handle/10045/14 5040
8. Bernard A, Rodrigues ASL, Cazalis V, Gremillet D (2021) Toward a global strategy for seabird tracking. Conserv Lett 14:20210501
9. Cassin-Sackett L, Welch AJ, Venkatraman MX, Callicrate TE, Fleischer RC (2019) The contribution of genomics to bird conservation. In: Avian genomics in ecology and evolution. Palgrave Macmillan, Cham, pp 295–330
10. Chisom ON, Biu PW, Umoh AA, Obaedo BO, Adegbite AO, Abatan A (2024) Reviewing the role of AI in environmental monitoring and conservation: a data-driven revolution for our planet. World J Adv Res Rev 21(1):161–171
11. Cornell Lab Dataset (2025) Birds of the world. Observation statistics by eBird & Macaulay Library Datasets | Seabird Tracking – Data Portal. https://birdsoftheworld.org/bow/home
12. Davis S (2010) Ecotourism and conservation: a case study of migratory waterfowl at the Salton Sea, California. J Ecotour 9(2):113–128
13. Djerboua S, Djerdali S, José G-C, Abdelkrim BS, Ali G (2022) Water birds as indicators of ecological conditions in a ramsar wetland (Sebkhet Bazer, East of Algeria). Agric Sci Dig 42(3): 341–350. https://doi.org/10.18805/ag.D-334
14. Duenas MA, Hemming DJ, Roberts A, Diaz-Soltero H (2021) The threat of invasive species to IUCN-listed critically endangered species: a systematic review. Global Ecol Conserv 26: e01476
15. Egwumah FA, Egwumah PO, Edet DI (2017) Paramount roles of wild birds as bioindicators of contamination. Int J Avian Wildlife Biol 2(1):194–200

16. Einoder LD (2009) Waterbirds as indicators of fish stocks in the Coorong, South Australia. In: Ecology and management of a large, interacting food web in a human-dominated world, pp 36–41 [Publisher/Institution, often University of Adelaide or SARDI].
17. Fish FE (2020) Advantages of aquatic animals as models for bio-inspired drones over present AUV technology. Bioinspir Biomim 15(2):025001
18. Forti LR (2023) Students as citizen scientists: project-based learning through the iNaturalist platform could provide useful biodiversity data. Biodiversity 24(1–2):76–78
19. Fox AD, Nielsen RD, Petersen IK (2019) Climate-change not only threatens bird populations but also challenges our ability to monitor them. Ibis 161(2):0019–1019
20. Fraser GS, Robertson GJ, Stenhouse IJ, Ellis JI (2022) Estimating the numbers of aquatic birds affected by oil spills: pre-planning, response, and post-incident considerations. Environ Rev 30(2):323–341
21. Fujita MS, Kameda KO (2019) Nutrient dynamics and nutrient cycling by birds. In: Sekercioglu ÇH, Wenny DG, Whelan CJ (eds) Why birds matter: avian ecological function and ecosystem services. University of Chicago Press, Chicago Scholarship Online, Chicago
22. Gonzalez SV, Johnston E, Gribben PE, Dafforn K (2019) The application of bioturbators for aquatic bioremediation: review and meta-analysis. Environ Pollut 250:426–436
23. Green AJ, Elmberg J (2014) Ecosystem services provided by waterbirds. Biol Rev 89(1): 105–122
24. Gwiazda R (1996) Contribution of water birds to nutrient loading to the ecosystem of mesotrophic reservoir. Pol J Ecol 44:289–297
25. Hart KM, Hyrenbach KD (2010) Satellite telemetry of marine megavertebrates: the coming of age of an experimental science. Endanger Species Res 10(1):9–20
26. Hohenlohe PA, Funk WC, Rajora OP (2021) Population genomics for wildlife conservation and management. Mol Ecol 30(1):62–82
27. Holmberg JA (2023) Wildbook: shared AI infrastructure for wildlife monitoring. In: Open infrastructure fund/Fondo de Infraestructura Abierta. OpenReview.net
28. Huang RM, Bass OL Jr, Pimm SL (2017) Sooty tern (Onychoprion fuscatus) survival, oil spills, shrimp fisheries, and hurricanes. PeerJ 10(5):e3287
29. IUCN (2025) Red list of threatened species. Retrieved in August 2025. https://www.iucnredlist.org/resources/summary-statistics
30. Iverson AR, Schaefer JL, Skalos SM, Hawkins CE (2023) Global positioning system (GPS) and platform transmitter terminal (PTT) tags reveal fine-scale migratory movements of small birds: a review highlights further opportunities for hypothesis-driven research. Ornithological Appl 125(3):duad014
31. Kahl S, Wood CM, Eibl M, Klinck H (2021) BirdNET: a deep learning solution for avian diversity monitoring. Eco Inform 61:101236
32. Kotagi V, Vaibhav D, Kumar S, Swamy A (2025) Empirical technique of forest conservation using machine learning and cloud computing. In: 2025 3rd International Conference on Data Science and Information System (ICDSIS). IEEE, pp 1–7
33. Kumar CS (2020) Conservation of waterbirds through conservation of wetlands report 2003 W&WT-converted
34. Kumar R, Gupta R, Dahiya T, Yadav R, Saifi R, Yodha K, Kumari B, Singh P, Shefali (2022) Surveillance of *Heteropneustes fossilis* (Stinging Catfish) diseases and health management practices in Haryana. Appl Ecol Environ Sci 10(4):261–267
35. Kumar R, Gupta RK, Yadav R, Saifi R, Yodha K, Pooja (2022) *Eisenia fetida* as protein source for growth enhancement of *Heteropneustes fossilis*. Egyptian J Aquat Biol Fish 26(2):577–588
36. Kumar R, Yadav R, Kumar Gupta R, Pooja (2022) A sustainable way for fish health management by replacement of chemical and drugs by earthworm. In: Environmental degradation in asia: land degradation, environmental contamination, and human activities. Springer, Cham, pp 329–352

37. Kumar R, Singh D, Yadav R (2025) Toxic effects of chlorantraniliprole and fipronil on *Eisenia fetida*: from growth and reproductive parameters to avoidance behavior. Environ Eng Manag J 24(7):1499–1508. Accepted manuscript
38. Kupekar S et al (2015) Aquatic and semi aquatic birds, threats and conservation of bird fauna of Ballaleshwar lake, Panvel. Dist Raigad (Maharashtra) IOSR J Environ Sci Toxicol Food Technol 9(11 (II)):29–36
39. Lee KA, Lee JR, Bell P (2020) A review of citizen science within the earth sciences: potential benefits and obstacles. Proc Geol Assoc 131(6):605–617
40. Lewison R, Oro D, Godley BJ, Underhill L, Bearhop S, Wilson RP (2012) Research priorities for seabirds: improving conservation and management in the 21st century. Endanger Species Res 17(2):93–121
41. Manjunatha GR, Raveesha HR (2025) AI for biodiversity conservation: harnessing predictive analytics and habitat monitoring to protect endangered species and ecosystems. In: AI and ecological change for sustainable development. IGI Global Scientific Publishing, Hershey, pp 33–64
42. Maznikova V, Ormerod S, Gómez-Serrano M (2024) Birds as bioindicators of river pollution and beyond: specific and general lessons from an apex predator. Ecol Indic 158:111366. https://doi.org/10.1016/j.ecolind.2023.111366
43. Merlino S, Locritani M, Guarnieri A, Delrosso D, Bianucci M, Paterni M (2023) Marine litter tracking system: a case study with open-source technology and a citizen science-based approach. Sensors 23(2):935
44. Mohsan SAH, Li Y, Sadiq M, Liang J, Khan MA (2023) Recent advances, future trends, applications and challenges of internet of underwater things (iout): a comprehensive review. J Mar Sci Eng 11(1):124
45. Morrison A, Lukaszewicz S (2023) A perspective on conservation technologies for endangered marine birds. Insights Vet Sci 7:010–014
46. Navarro Ramos MJ (2024) Waterbirds dispersing native and alien plant species: new methods, new interactions, new insights. Tesis Doctoral Inédita, Universidad de Sevilla, Sevilla
47. Nie S, Zhou L, Xu W (2022) Effect of seed traits and waterbird species on the dispersal effectiveness of wetland plants. Biology 11(5):629
48. Nieto-Mora DA, Rodriguez-Buritica S, Rodriguez-Marin P, Martínez-Vargaz JD, Isaza-Narvaez C (2023) Systematic review of machine learning methods applied to ecoacoustics and soundscape monitoring. Heliyon 9(10):e20275
49. Nisha A, Rani S, Kumar S, Kumar R, Kumar A (2025) Recent technologies and strategies for bioremediation of toxic and hazardous waste. In: Dubey KK, Shah MP, Yadav A (eds) Challenges and sustainable solutions in bioremediation. CRC Press, Boca Raton, pp 1–16. https://doi.org/10.1201/9781003407317-1
50. Orihuela-Torres A, Morales-Reyes Z, Hermoso V, Picazo F, Sánchez Fernández D, Pérez-García JM et al (2024) Carrion ecology in inland aquatic ecosystems: a systematic review. Biol Rev 99(4):1425–1443
51. Orihuela-Torres A, Pérez-García JM, Arrondo E, Pessano-Serrat T, Green AJ, Naves-Alegre L, Sebastián-González E (2025) Vertebrate scavenger assemblages and their functioning differ between artificial and natural wetlands: implications for ecosystem management. Biol Conserv 302:110929
52. Patnaik BB, Lee YS (2023) The utility of transcriptomics in the conservation of sensitive and economically important species. J Environ Biol 44(4):I–IV
53. Péron A, Oheix N (2013) Behavioral exploitation of human maritime activities by the Great Cormorant *Phalacrocorax carbo*. Mar Ornithol 41(1):1–6
54. Raghav YY, Vyas V (2024) Harnessing cloud computing and IoT for wildlife conservation. In: Integration of cloud computing and IoT: trends, case studies and applications. CRC Press, Boca Raton, p 142

55. Ramachandran R, Kumar A, Gopi Sundar KS, Bhalla RS (2017) Hunting or habitat? Drivers of waterbird abundance and community structure in agricultural wetlands of southern India. Ambio 46(5):613–620
56. Rani S, Sharma S, Singh D, Sheoran R, Kumar R (2025) Improved production of anti-cancer compounds from medicinal plants through molecular and biotechnological mutagenesis strategies. In: Chen JT (ed) Biotechnology, multiple omics, and precision breeding in medicinal plants. CRC Press, Boca Raton, pp 287–297
57. Rodríguez JP, Rodríguez-Clark KM, Keith DA, Barrow EG, Benson J, Nicholson E, Wit P (2012) IUCN Red List of ecosystems. SAPI EN S Surveys and Perspectives Integrating Environment and Society (5.2). Institut Veolia Environnement
58. Saikia P (2024) Pest management with predatory birds. Presentation on 05-01-2015 in 21 days Winter School organized by Department of Agronomy, Assam Agricultural University, Jorhat. https://doi.org/10.13140/RG.2.2.18247.51369
59. Scalici M, Perrone M, Battisti J, Benedini L, Malavasi M (2021) First perceptions of hydroperiod mapping and assessment of shallow waters in coastal landscapes by drone-based monitoring activities: a remote-sensing and GIS approach. Appl Sci 11(20):9773
60. Seabird Tracking Database (2025) Information retrieved August 2025. https://data.seabirdtracking.org/
61. Seni A, Halder J (2022) Role of predatory birds in insect pest management in agriculture. Agric Lett 3(05):2022
62. Shaffique S, Kang S-M, Ashraf MA, Umar A, Khan MS, Wajid M, Al-Ghamdi AA, Lee I-J (2024) Research progress on migratory water birds: indicators of heavy metal pollution in inland wetland resources of Punjab, Pakistan. Water 16(8):1163
63. Smith P, King ON, Pennington A, Tun W, Basham M, Jones ML et al (2023) Online citizen science with the Zooniverse for analysis of biological volumetric data. Histochem Cell Biol 160(3):253–276
64. Stenhouse IJ (2018) All bird bulletin
65. Stephenson PJ, Stengel C (2020) An inventory of biodiversity data sources for conservation monitoring. PLoS One 15(12):e0242923
66. Steven R, Jones B (2013) The importance of ecotourism for the conservation of flamingos and their wetland habitats in the Lake Bogoria National Reserve, Kenya. Tour Manag 35:172–179
67. Sutton G, Pichegru L, Botha JA, Kouzani AZ, Adams S, Bost CA, Arnould JPY (2020) Multi-predator assemblages, dive type, bathymetry and sex influence foraging success and efficiency in African penguins. PeerJ 30(8):e9380
68. Taylor HR et al (2021) Showcasing the application of genetics for the conservation management of austral birds. Emu Austral Ornithol 121(1–2):1–9
69. Upreti H, Malhotra RK (2025) Greening the future: the synergy of digital transformation and environmental sustainability. In: Industry 5.0: sustainable business practices for a bright future. Complimentary Copy, p 17
70. USGS Report (2007) Applying radar technology to migratory bird conservation and management. https://www.usgs.gov/publications/applying-radar-technology-migratory-bird-conservation-and-management-strengthening-and
71. van Merriënboer B, Dumoulin V, Hamer J, Harrell L, Burns A, Denton T (2025) Perch 2.0: the bittern lesson for bioacoustics. arXiv:2508.04665
72. Wall et al (2024) EarthRanger: an open-source platform for ecosystem monitoring, research and management. Methods Ecol Evol 15:1968–1979
73. Wang R, Ma K (2025) Trends in the application of citizen science in waterbird conservation: a bibliometric analysis. Animals 15(3):368
74. Wei F, Wu Q, Hu Y, Huang G, Nie Y, Yan L (2019) Conservation metagenomics: a new branch of conservation biology. Sci China Life Sci 62(2):168–178
75. Whoriskey (2024) Building ocean biodiversity monitoring capacity with acoustic telemetry. Oceanography 38(1):41–43

76. Wildlife Spotter (2025) SciStarter. Information retrieved August 2025. https://scistarter.org/wildlife-spotter
77. Yadav R, Gupta RK, Kumar R, Kaur T (2022) Assessment of toxicity of lead and nickel on the biochemical and immunological parameters of earthworm, Eudrilus eugeniae. Agric Sci Dig 44(5):922–929
78. Yadav R, Gupta RK, Kumar R, Kaur T, Moun R (2022) Biochemical response of earthworm, *Eisenia fetida* to heavy metals toxicity. J Appl Nat Sci 14(3):990–998
79. Zhang L, Ouyang Z (2019) Focusing on rapid urbanization areas can control the rapid loss of migratory water bird habitats in China. Global Ecol Conserv 20:e00801
80. Ziegler J (2007) The symbolism of the stork in European folklore. J Folk Res 44(2):175–194

# Chapter 16
# Role of Habitat Quality in Establishment and Growth of New Aquatic Bird Colonies

**Sarika Kumari, Nibha Singh, and Monika Pilania**

**Abstract** The development and growth of aquatic bird colonies are tied to how good the environment is which is made up of many ecological, chemical and physical factors. In this study, we look at how water presence, food sources, the structure of vegetation, the suitability of their habitat and disturbance due to humans shape the growth of aquatic bird populations. When habitats are clean, abundant in fish, not too shallow or deep and protected from people, many breeding pairs can successfully raise chicks and keep the colony alive. If habitats are degraded or fragmented, nesting efforts may fail, chicks tend to die and eventually the colony will leave. This study also investigates how interspecific competition, predators and climate changes play a role in changing colony behaviors, along with their influence from habitat characteristics. The study emphasizes the essential levels of difficult habitats needed for new colonies to develop, as shown through case studies from varied wetlands. Besides, it notes that conservation of large areas and protected corridors between habitats helps species spread and settle. Encouraging wetland recovery, obeying regulations for their protection and regularly keeping an eye on them are advised to aid aquatic birds. Awareness of these linkages is necessary to preserve wetlands and birds, as we lose more habitats and deal with climate change. The findings of this study contribute to the evidence for habitat-focused conservation of birds and stress the importance of prioritizing wetland quality conservation.

**Keywords** Anthropogenic impacts · Aquatic birds · Bird monitoring · Interspecific competition · Wetland ecology

S. Kumari (✉)
Department of Biotechnology, College – Uka Tarsadia University, Maliba Campus, Bardoli, Gujarat, India

N. Singh
Department of Endocrinology & Metabolism, AIIMS, New Delhi, India

M. Pilania
Bioinformatics Analyst, Health and Life Sciences, Conestoga College, Kitchener, ON, Canada



R. Kumar et al. (eds.), *Ecology, Conservation of Aquatic Birds*, Wetlands: Ecology, Conservation and Management 15, https://doi.org/10.1007/978-3-032-17125-2_16

## 16.1 Introduction

Freshwater and estuarine ecosystems depend on aquatic birds for their proper functioning [26]. As members of the food chain, fish, insects, and invertebrates are regulated by scavengers, predators, and herbivores in the water [30, 38]. In addition, the number and reproduction of birds indicate how well the wetlands are doing and if the environment is worrisome. A drop in waterbird numbers is regularly a first sign that pollution, changes in wetland water levels, or bigger habitats have led to the degradation of the area [73]. So, by studying aquatic bird ecology, we help preserve the species and also assist in general conservation of all living things in the ecosystem.

Among different types of behavior, numerous aquatic bird species choose colonial nesting, during which a large group nests close to each other on islands, emerging vegetation, or on the edges of water [76]. Animals living in colonies enjoy solutions such as detecting predators faster, easier group hunting, and extra opportunities to find partners. At the same time, the formation of colonies requires more space and resources, so the quality of the nesting site and environment often determines whether a colony will be successfully established and remain alive. Whenever something happens in their natural environment or due to human activities, aquatic birds can abandon bad areas and start colonies in better areas instead [10, 60].

Many features of a habitat, such as food, clean water, suitable nesting sites, levels of disturbance, and interactions with other species, significantly influence the process of colony foundation and growth [3]. Such factors play a role in the individual's ability to reproduce, how often chicks survive, and how attached they become to the location. Therefore, it is important to study how good quality habitat impacts the spread of aquatic birds for conserving and protecting wetlands well.

This chapter looks into the various ways habitat quality helps in creating and growing aquatic bird colonies. Through focused research on essential environmental factors, significant changes made by humans, and case studies from many wetlands, we work to explain the relationship between certain habitats and the growth of colonies. Moreover, we underline various conservation strategies and point out potential new studies that could protect birds as aquatic environments become at risk.

## 16.2 Key Components of Habitat Quality for Aquatic Birds

Aquatic bird colonies are formed and remain strong for many years due to strict regulations set by various habitat characteristics. Some of these aspects are the amount of food, places to build nests, proper water conditions, and threats from both animals and humans. Every one of these factors makes a difference to how well the colony reproduces, raises babies, and survives. Getting to know these requirements well is necessary to plan whether an area is appropriate for a species and how to conserve it.

### *16.2.1 Food Availability*

The availability and accessibility of food are crucial factors that determine the quality of a habitat [80]. The way aquatic birds look for food is different and may rely on fish, crustaceans, aquatic insects, molluscs, and submerged vegetation. For example, herons and cormorants need to be near water with shallow parts and many smaller fish, but dabbling ducks make use of plants and larger aquatic creatures [70, 87]. If prey fish decline due to overfishing, lake pollution, or habitat changes, seabirds may struggle to breed, their chicks might not receive enough food, and some may abandon their colony sites.

Also, during their breeding period, animals use a lot more energy to produce eggs, hatch them, and feed the young birds. If the areas where food is found are distant or have few nutrients, parents might not be able to feed their offspring, which decreases the chances of success. For that reason, having close-by areas to feed in, with little competition, is necessary for the creation and expansion of new groups of colonies.

### *16.2.2 Nesting Site Suitability*

Proper nesting materials are a key part of how suitable is habitat. Such birds prefer places with positive features for making their nests and hiding from dangerous wildlife, and can endure weather uncertainties. Types of materials turtles generally use for nesting are floating beds of *Eichhornia* or *Phragmites*, areas with dense reeds, or isolated islands and sandbars. It is significant that raised nests stop birds from getting caught by incoming flood waters [20].

Having a reliable nest will both lessen flooding danger and make the conditions inside the nest pleasant for the eggs as they develop. Examples include issues where red-necked grebes situated on floating mats group together and result in more offspring because such places provide better protection and more stability [58]. Moreover, studies show that setting the water level in restored wetlands according to nature-like conditions boosts the availability of floating nesting sites and supports the successful nesting of marsh birds [36].

### *16.2.3 Water Quality*

Prey communities and bird health are largely affected by environmental aspects such as pH, temperature, oxygen, salinity, and both natural and polluting materials in the water. A lack of DO can lead to fewer fish, and when nutrients in water grow high, there may be eutrophication that results in less oxygen and fewer things for fish to eat [57]. High levels of heavy metals, pesticides, and microplastics are possible in aquatic organisms due to biomagnification, which may lead to difficulties with reproduction and other effects [51].

Research on riverine wetlands reveals a connection between dissolved oxygen (DO) levels, chlorophyll concentration, and the diversity of water bird species, such as Eurasian coots and wild ducks. The findings also suggest that moderate nutrient levels can enhance primary productivity, which in turn supports bird populations [44]. Besides, waterbirds living in nests produce guano that influences the amount of nutrients in the sediment and water and the community of marine invertebrates [25].

### *16.2.4 Predation Pressure and Human Disturbance*

Colonies face risks from increased predation and the effects of human activities. Specifically, if nests are exposed to birds, rats, snakes, raccoons, or foxes, the success rate of nesting declines significantly unless the nests are well guarded or situated far from the ground [33, 50]. When nests are better hidden and are placed in deep water, predation becomes less common, as proven in [32]. If there is just 5–10 cm of water around the nest, it can stop some mammals from approaching [68].

Human activities, such as boating, fishing, tourism, farming, and living in noisy urban environments, can disrupt bird nesting, causing them to abandon or neglect their young. Studies have shown that managing predators and establishing buffer zones around waterbird colonies can greatly increase the likelihood of successful offspring raising [82]. When habitats that require natural disturbances are not healthy, even with many resources, this may keep a colony from forming or force them to find alternative spaces [39].

## 16.3 Site Selection and Settlement Patterns

### *16.3.1 Evaluation of Habitat Cues and Role of Past Experience*

Birds observe their surroundings and communicate with others to assess whether a location is suitable for nesting Y. For the first time, both parents and solitary individuals often explore various habitats, evaluating the vegetation, prey availability, water depth, and the level of threat from predators [32, 40]. A bird investigates and explores to gather this crucial personal information. However, relying on trial-and-error can be very time-consuming, and in unpredictable environments, the outcomes may be unfavourable [34].

Birds often depend on observing other birds and the outcomes of their breeding to help decide when it is safe to take risks. So, instead of figuring out nesting locations by themselves, they may follow the placement of birds in the group [64]. Birds generally come back to a place where they were successful in past breeding seasons [27]. This kind of fidelity in site use allows animals to eat and hide in familiar areas, so they spend less on hunting and get used to the nearby threats.

### *16.3.2 Conspecific Attraction and Social Facilitation*

Conspecific attraction is an important social factor that encourages animals to live together in existing settlements. Especially in aquatic birds such as gulls, terns, and herons, this phenomenon is very well described in scientific literature [63, 71]. The presence of animals of the same species shows that the habitat is good and is important for reproduction. Basically, when other animals are seen at a site, birds decide that it is a secure place with enough resources and no danger from predators.

Sounds and decoys have proved useful in getting seabirds to settle in alternative nesting areas, showing that seabirds rely on sound and sight for this behaviour [21]. Besides, conspecific attraction is linked to Allee effects, since when a population is denser, it provides benefits such as better self-defence, group knowledge sharing, and easier opportunities for choosing a mate [14].

### *16.3.3 Philopatry Versus Exploratory Settlement*

Aquatic birds are generally loyal and return to the same breeding sites each year. Familiar environments, access to potential mates, or previous breeding success might motivate this behaviour [1, 12]. A lot of cormorants, ibises, and pelicans prefer to raise their young near where they were born, because they tend to stay close to the same place [88].

Although most animals tend to stay in the same region, it's not a rule for all of them. Most of the time, birds consider whether staying in one place or leaving is more beneficial to them. If the area where they live becomes unsuitable because of food shortages, flooding, more threats from predators, or changes made by people, some animals may search for another area to live [17]. Common Terns (*Sterna hirundo*) were found to change their nesting location more often after breeding failed, mainly if adults noticed more successful colonies nearby [5].

By relying on both staying put and travelling, populations can monitor how habitat conditions are changing and respond well to variations in the environment.

### *16.3.4 Habitat Degradation and Colony Relocation*

Impacts on the environment may mean certain colonies have to relocate to another area. Even something as small as a shift in the water system, the plants, or how much humans are present can turn a former site unfit for use. Activities such as using heavy metals, letting ponds grow too full with nutrients, or building close to a wetland area can easily disturb local food supplies, add toxins, or take away places where animals can make their nests. With these changes, birds tend to leave their usual nesting areas and look for other places [42, 84].

In a study in Kunming Zoo, China, the night herons (*Nycticorax nycticorax*) gave up their roosts near the people's paths and moved to safer ponds where humans could not bother them [13]. To help birds adjust to their new homes, conservationists have built islands and fixed wetlands, offering successful residence for groups of birds, especially with the use of decoys and audio devices [86]. So, damage to habitats may not only decrease the population of species but also influence their whereabouts and the various species living together in the ecosystem.

## 16.4 Influence of Habitat Heterogeneity

The existence and survival of aquatic bird colonies depend a lot on the various physical and biological features across different areas in a landscape. In such habitats, animal diversity shows up because of changes in the type of plants, water levels, ground elevation, and types of microhabitats. The variety of environments allows room for more bird species to breed and live. Since there are various habitats in the colony, birds that build their nests in different places can live together. This way, less competition for food and shelter makes the colony stronger and more likely to reproduce well [4, 46, 54].

More complex habitats increase the variety of plants and animals available for food. In wetlands, a combination of open water, underwater plants, and dense vegetation along the edges creates a diverse range of aquatic animals and plants. These resources play a crucial role in the diets of aquatic birds, determining how and what they feed their young. Different bird species occupy distinct feeding areas; some search for voracious aquatic insects near the water's edge, while others dive slightly deeper to hunt for fish or search in the reeds for molluscs and amphibians. This separation in feeding habitats reduces competition among species, allowing more breeding pairs to utilise the limited resources within the same wetland [59, 86].

Along with biodiversity, diverse habitat types make a colony stronger when facing changes and obstacles in its environment [77]. If microhabitats are not all the same in certain areas, birds are able to move their nests to avoid problems caused by predators or flooding. Having mudflats flood is usually enough for ground-nesting birds to change their nesting location to nearby, above-water surfaces like hummocks and vegetation mats. Likewise, having different types of vegetation gives babies more cover from attacks by predators and helps shield them from too-high or too-low temperatures, also helping them to survive. When certain areas in the heterogeneous site are not suitable for a while, the capacity for relocation lets animals remain there for the long term [28, 35].

Experts have noticed that bird varieties and the success of colonies usually increase in wetlands with complicated features and decrease in those that are either simple or damaged. Mixes in temperate zones allow herons, ibises, stilts, and rails to co-breed; they do so by choosing unique nesting materials, all located close by. As a consequence, the bird community is well-adapted to various surroundings, and supports biodiversity, so it normally withstands disruptive factors such as invasive species, drought, or damage from human activities [55, 66]. Therefore, to boost the establishment and development of aquatic bird colonies, it is important to maintain vegetation zones, keep natural hydrological cycles, and mix areas with and without plants [48].

## 16.5 Anthropogenic Impacts on Habitat Quality

### *16.5.1 Wetland Degradation from Human Activities*

Wetland changes caused by people have become a danger to the existence and persistence of aquatic bird colonies [7]. Habitat changes and divisions are mainly caused by development in towns, excessive farming, more tourism, industrial growth, insecticides and building dams. They commonly result in wetlands being drained, alterations in water flows, breakage of nesting surfaces, and fewer available places to find food, interfering with the natural state of bird habitats [6, 43]. When wetlands are turned into cities or fields, they lose the various habitats necessary for nesting and feeding, while more humans in the area interfere with birds' activities and may result in birds not using their nests or less reproduction [6, 91].

### *16.5.2 Pollution and Contaminant Exposure*

Release of pollutants into our water systems is among the biggest dangers of human actions [47, 81]. The pesticides and fertilizers found in agricultural runoff contribute to our water, and heavy metals and plastic waste enter it through industry and urban areas. Such pollutants build up in aquatic food webs and negatively impact the health of birds as well as the levels of their food items [23, 89]. To illustrate, when mercury and lead pollute the environment, aquatic birds might experience trouble reproducing, experience brain disorders, and have weak immune systems. If polluted water enters waterways, the fish and invertebrate population drops, and this can lead to fewer food sources for birds [69, 90].

Several researchers have reported how habitat damage harms the population and health of bird colonies that live in water. As an illustration, drastic cuts in the Siberian Crane's number have occurred because of damming and increased farming in China and India that have removed crucial rest areas along its migration path [29]. Black-necked Grebes (*Podiceps nigricollis*) living in polluted wetland areas in Spain produced fewer chicks and proved to have poor reproductive success [65]. In another situation, building dams created less water flow in rivers in India's Bharatpur Wetland Sanctuary, which reduced the nesting activities of colonial waterbirds [61]. Such incidents show that small changes to their surroundings have major effects on aquatic birds and the value of habitats for their existence.

## 16.6 Climatic and Seasonal Factors Affecting Colony Growth

Aquatic bird colonies establish themselves and grow well because of the way climatic and seasonal changes impact the basic habitat factors such as access to water, plentiful food, and suitable nesting sites. Water levels in wetlands and the lives of creatures in them depend on how much and when rain falls locally. As rainfall is higher in the wet season, wetlands expands, giving birds more places to search for food and find nesting spaces in the form of high and floating plant life. As a result, colonial waterbirds get better opportunities to breed. Also, too much rain can cause flooding of nests, mainly among bottom-dwelling birds, resulting in deaths of their eggs and young birds [41]. Sometimes, wetlands become much smaller or disappear when there is a long dry period, which lowers the food supply of fish and other small animals and leads to birds being more closely bunched [56, 85].

Because of changing climates in the past several years, it is more common for continuous droughts, unexpected weather shifts, and irregular rainfall, which leads to altered wetland water, instability for aquatic birds, and changes to their natural habitat [15]. Such changes cause chick survival rates to fall since they prevent the timely synchronization of nesting and supplies of food [18]. As wetlands shrink due to rising temperatures and reduced rainfall, the ecosystem can no longer support many animals. This leaves colonies exposed to more animals that live on land and to humans who may disturb them. Many recent studies demonstrate that a long period without rain and delayed wet seasons result in smaller pelican, heron, and ibis colonies and a lower number of young birds [19, 37].

## 16.7 Role of Invasive Species and Inter-species Competition

Competition between different species and invasive species matters a lot in determining how aquatic bird colonies are built and prosper. The arrival of invasive plant species such as water hyacinth (*Eichhornia crassipes*) can ruin the structure of wetlands by covering open water, cutting down on areas where animals feed, and blocking nests, therefore preventing colonies from forming [83]. In addition, animals such as cats and rats that were introduced to the environment by humans seriously damage eggs and chicks, mostly in ground-nesting birds or small colonies, the failing many nests [79]. Finding enough nesting sites causes battles among bird species and affects the behavior of colonies. Those that are dominant or aggressive typically use the prime spots, preventing subdominant birds from nesting comfortably or forcing them to migrate, possibly slowing down the colony's advancement and its chances of successfully raising young. This means that these factors jointly change the stability, number, and reproduction of aquatic birds [31, 78].

## 16.8 Monitoring and Conservation Strategies

Sustainable management and surveys of aquatic bird colonies need a complex perspective that starts with an in-depth evaluation of the quality of the habitat. Standardized field methods are usually employed by researchers in determining the extent of vegetation cover and diversity which include plant cover and diversity, which are primary indices of nesting suitability and food supply [2, 8]. Physicochemical analyses are conducted regularly to monitor water quality parameters such as pH, dissolved oxygen, nutrient levels, and pollutant presence as indicators of the health of the aquatic ecosystem and the sustainability of prey in the bird population [52]. Moreover, first-hand observations of the avifauna through point counts, nest success surveys, and long-term banding projects will provide essential information on colony formation, breeding productivity, and diversification [62, 72].

The restoration of wetlands is critical in rectifying the habitat degradation and improving the colony conditions of birds. Restoration plans in most cases involve elimination of exotic plants like *Eichhornia crassipes* (water hyacinth), re-vegetation with local plants, control of sediments and also alteration of the water flow patterns so that they resemble the natural hydrological cycles [22]. Such measures can substantially enhance the supply of foraging and nesting habitats, raise the productivity of wetlands, and recruit recolonization by displaced or declining species of birds.

Protected areas such as bird sanctuaries, national parks and Ramsar-listed wetlands act as important refuges because they reduce anthropogenic disturbances and maintain ecological integrity. Buffer zones around wetlands can be created to minimize edge effects associated with agriculture, urban development or tourism, and to create more habitat and connectivity between wetlands to aid in bird movements. Furthermore, community-based conservation, where local stakeholders are integrated into the management and protection of habitats, has been effective in increasing awareness, decreasing unlawful disturbance, and establishing sustainable utilization of wetlands, especially in areas with scarce conservation resources [9, 53].

Technological improvement has also facilitated increased monitoring. Satellite imagery and remote sensing can enable researchers to evaluate seasonal variations, wetland dynamics, and vegetation cover on large temporal and spatial intervals [11]. Geographic Information System (GIS) tools can be used to integrate spatial data to support habitat suitability modeling, risk analysis, and conservation planning [74]. These newer tools offer scalable, cost-efficient, and repeatable ways to monitor changes to colony health and growth over time, making sure conservation efforts are adaptive and data-driven.

## 16.9 Future Perspectives

Although much progress has been made in ecological knowledge of aquatic birds, there are still a number of important research gaps in knowledge that impede the effectiveness of conservation planning. A key deficiency is long term, large scale studies which would be able to determine population trends, habitat alterations and breeding success over time especially in the tropics and sub tropics where much of the wetlands are poorly studied [67]. Most of the assessments that are done today are done on well-known species or charismatic species and smaller species or cryptic species are less assessed although they are equally sensitive to habitat degradation. Future effort ought therefore to be directed at incorporating understudied taxa as well as unknown breeding locations so as to give a much clearer image regarding the health of wetland avifauna and their habitat requirements.

One avenue of hope is the encourage interdisciplinary studies linking ornithology to hydrology, remote sensing, climatology, and spatial modeling. This type of collaboration is capable of providing more insights on the interactive effects of water regime, vegetation cover and land-use pattern changes on colony establishment and sustainability [49]. As another example, incorporating habitat suitability models (based on GIS) and hydrological models can aid in forecasting effects of proposed development plans or climate scenarios on bird populations so that more anticipatory conservation choices can be made [24].

On policy perspective, the immediate inclusion of measures to assess the bird habitat quality in land-use planning and wetland development regulation is required. Without considering the ecological impact, wetlands are often subject to agricultural, infrastructural, and urban development. Policy performances can be improved through strengthening legal protection of wetlands, enforcement of environmental impact assessment including avifaunal survey, and the harmony between national systems of conservation and international commitments like the Ramsar Convention [16, 45]. In addition, incorporation of buffer zones, and seasonal limits on human activity in and around known breeding zone can help reduce disturbances during nesting periods.

Citizen action is promising, as well. Engaging local communities in the management of the wetlands will create conservation support as well as create long term sustainability. Citizen science websites, including eBird and the Asian Waterbird Census of Wetlands International, have already proven their worth in terms of producing large-scale monitoring data on birds to fill in gaps in time and space in the scientific literature [75]. This empowers the communities to play their role in recording biodiversity and can inform local policy with evidence-based suggestions.

## 16.10 Conclusion

The quality of habitats is crucial in the formation, survival as well as success of aquatic bird colonies. Colony dynamics are determined by a host of interacting factors, including food availability and suitability of nesting sites, water quality, predation pressure, and anthropogenic effects. Settlement choices and reproduction are also affected by the habitat heterogeneity, climatic variability and inter-species interactions. As much as there is hope of conserving these species through restoration works and monitoring technology, there exists large research gaps and policy constraints. Overcoming these issues via interdisciplinary research and thinking, more robust wetland protection policy, and effective community involvement will be necessary to maintain robust and diverse populations of aquatic birds in the face of increasing environmental stressors.

## References

1. Alerstam T, Högstedt G (1982) Bird migration and reproduction about habitats for survival and breeding. Ornis Scand 13:25–37
2. Ambarlı D, Bilgin CC (2014) Effects of landscape, land use and vegetation on bird community composition and diversity in Inner Anatolian steppes. Agric Ecosyst Environ 182:37–46
3. Antoine CM, Forrest JRK (2021) Nesting habitat of ground-nesting bees: a review. Ecol Entomol 46(2):143–159. https://doi.org/10.1111/een.12986
4. Arzel C, Rönkä M, Tolvanen H, Aarras N, Kamppinen M, Vihervaara P (2015) Species diversity, abundance and brood numbers of breeding waterbirds about habitat properties in an agricultural watershed. Ann Zool Fenn 52(1–2):17–32
5. Austin OL (1945) The role of longevity in successful breeding by the common tern (Sterna hirundo). Bird-Banding 16:21–28
6. Bahati M (2012) The impact of human activities on the wetland ecosystem. https://ir.kiu.ac.ug/handle/20.500.12306/6737
7. Bai J, Cui B, Cao H, Li A, Zhang B (2013) Wetland degradation and ecological restoration. Sci World J 2013:523632
8. Balda RP (1975) Vegetation structure and breeding bird diversity. In: Proceedings of the symposium on management of forest and range habitats for nongame birds. USDA Forest Service General Technical Report WO-1, pp 59–80

9. Barakagira A, De Wit AH (2017) Community livelihood activities as key determinants for community-based conservation of wetlands in Uganda. Environ Socio-Econom Stud 5(1): 11–24. https://doi.org/10.1515/environ-2017-0002
10. Buckley PA (1978) Guidelines for the protection and management of colonially nesting waterbirds. North Atlantic Regional Office, National Park Service, Boston
11. Chen L, Jin Z, Michishita R, Cai J, Yue T, Chen B, Xu B (2014) Dynamic monitoring of wetland cover changes using time-series remote sensing imagery. Eco Inform 24:17–26
12. Clutton-Brock TH, Lukas D (2012) The evolution of social philopatry and dispersal in female mammals. Mol Ecol 21(3):472–492. https://doi.org/10.1111/j.1365-294X.2011.05232.x
13. Dai R, Li Y (2021) A new heron colony in Yunnan, China, survives amid the COVID-19 pandemic. J Heron Biol Conserv 6(3):2
14. Donahue MJ (2006) Allee effects and conspecific cueing jointly lead to conspecific attraction. Oecologia 149(1):33–43. https://doi.org/10.1007/s00442-006-0419-y
15. Donnelly JP, Moore JN, Casazza ML, Coons SP (2022) Functional wetland loss drives emerging risks to waterbird migration networks. Front Ecol Evol 10:844278
16. Dragomir A-N, Florescu I (2024) Challenges and strategies for implementing the Ramsar convention. Balancing economic development and wetland conservation. Stud Ecol Bioeth 22(4):57–71
17. Feng C, Yang C, Liang W (2019) Nest-site fidelity and breeding dispersal by common tailorbirds in a tropical forest. Avian Res 10(1):45. https://doi.org/10.1186/s40657-019-0185-2
18. Fletcher K, Howarth D, Kirby A, Dunn R, Smith A (2013) Effect of climate change on breeding phenology, clutch size and chick survival of an upland bird. Ibis 155(3):456–463. https://doi.org/10.1111/ibi.12055
19. Forti M, Monrós JS, Vera P (2021) Habitat and weather conditions effects on long-term breeding population dynamics of five species of herons (Ardeidae) and Glossy ibis (Threskiornithidae) in the Valencian Community, Spain. Limnetica 40(2):417–433
20. Frederick PC, Ogden JC (2001) Pulsed breeding of long-legged wading birds and the importance of infrequent severe drought conditions in the Florida Everglades. Wetlands 21(4): 484–491. https://doi.org/10.1672/0277-5212(2001)021[0484:PBOLLW]2.0.CO;2
21. Friesen MR, Beggs JR, Gaskett AC (2017) Sensory-based conservation of seabirds: a review of management strategies and animal behaviours that facilitate success. Biol Rev 92(3): 1769–1784. https://doi.org/10.1111/brv.12308
22. García-De-Lomas J, González R, Dana ED (2021) Removal of water hyacinth (Eichhornia crassipes) and native plant recovery in a Mediterranean permanent pool. J Aquat Plant Manag 59(2):90–93
23. Ghosh A, Manna MC, Jha S, Singh AK, Misra S, Srivastava RC, Srivastava PP, Laik R, Bhattacharyya R, Prasad SS (2022) Impact of soil-water contaminants on tropical agriculture, animal and societal environment. Adv Agron 176:209–274
24. Gottschalk T (2014) A remote sensing and GIS-based model of avian species habitat and its potential as a part of an environmental monitoring programme. Diplom de, Hamburg
25. Grant ML, Bond AL, Lavers JL (2022) The influence of seabirds on their breeding, roosting and nesting grounds: a systematic review and meta-analysis. J Anim Ecol 91(6):1266–1289. https://doi.org/10.1111/1365-2656.13699
26. Green AJ, Elmberg J (2014) Ecosystem services provided by waterbirds. Biol Rev 89(1): 105–122. https://doi.org/10.1111/brv.12045
27. Greenwood PJ, Harvey PH (1982) The natal and breeding dispersal of birds. Annu Rev Ecol Syst 13:1–21
28. Grisham BA, Godar AJ, Boal CW, Haukos DA (2016) Interactive effects between nest microclimate and nest vegetation structure confirm microclimate thresholds for Lesser Prairie-Chicken nest survival. Condor 118(4):728–746
29. Harris J, Mirande C (2013) A global overview of cranes: status, threats and conservation priorities. Chinese Birds 4(3):189–209. https://doi.org/10.5122/cbirds.2013.0025

30. Hart DR (2002) Intraguild predation, invertebrate predators, and trophic cascades in lake food webs. J Theor Biol 218(1):111–128
31. Hermann HR (2017) Dominance and aggression in humans and other animals: the great game of life. Academic
32. Hoover JP (2006) Water depth influences nest predation for a wetland-dependent bird in fragmented bottomland forests. Biol Conserv 127(1):37–45
33. James Reynolds S, Ibáñez-Álamo JD, Sumasgutner P, Mainwaring MC (2019) Urbanisation and nest building in birds: a review of threats and opportunities. J Ornithol 160(3):841–860. https://doi.org/10.1007/s10336-019-01657-8
34. Kendal RL, Coolen I, van Bergen Y, Laland KN (2005) Trade-offs in the adaptive use of social and asocial learning. Adv Study Behav 35:333–379
35. Kim S-Y, Monaghan P (2005) Effects of vegetation on nest microclimate and breeding performance of lesser black-backed gulls (Larus fuscus). J Ornithol 146(2):176–183. https://doi.org/10.1007/s10336-005-0077-6
36. Kinder K (2011) Planning by intermediaries: making cities make nature in Amsterdam. Environ Plan A Economy Space 43(10):2435–2451. https://doi.org/10.1068/a4464
37. Kingsford RT, Thomas RF (2004) Destruction of wetlands and waterbird populations by dams and irrigation on the Murrumbidgee River in Arid. Environ Manag 34(3):383–396. https://doi.org/10.1007/s00267-004-0250-3
38. Kiran, Singh D, Kour A, Priya, Delu V, Kumar R, Punia N, Sunita (2022a) Different strategies adopted by birds to sustain ecosystem: a review. Pharm Innov SP-11(9):412–422
39. Kiran Y, Singh D, Kour A, Priya, Delu V, Kumar R (2022b) Community composition and status of avian diversity at campus and agricultural landscapes of Chaudhary Charan Singh Haryana Agricultural University, Hisar (Haryana). J Appl Nat Sci 14(4):1130–1140. https://doi.org/10.31018/jans.v14i4.3784
40. Kleindorfer S, Fessl B, Hoi H (2005) Avian nest defence behaviour: assessment about predator distance and type, and nest height. Anim Behav 69(2):307–313
41. Kluge AG (1981) The life history, social organization, and parental behavior of Hyla rosenbergi Boulenger, a nest-building gladiator frog. https://deepblue.lib.umich.edu/bitstream/handle/2027.42/56404/MP160.pdf?sequence=1
42. Kolay S, Annagiri S (2017) Deterioration in nest quality triggers relocation without affecting its dynamics in an ant. Insect Soc 64(3):321–327. https://doi.org/10.1007/s00040-017-0546-y
43. Kumar R, Singh D (2022) Insecticide-tolerant bacterial population in *Eisenia fetida*'s gut and vermicast exposed to chlorantraniliprole and fipronil. Appl Biol Res 24(3):273–279
44. Liu X, Ho L, De Cock A, De Saeyer N, Pham K, Panique-Casso D, Forio MAE, Goethals PL (2025) Water quality and its influence on waterbird habitat distribution: a study along the Lieve River, Belgium. Water 17(4):595
45. López-Calatayud N, Márquez-Romance AM, Guevara-Pérez E (2022) Wetland management modeling in the frame of the Ramsar convention. J Appl Water Eng Res 10(4):296–310. https://doi.org/10.1080/23249676.2021.1982028
46. Lorenzón RE, Beltzer AH, Olguin PF, Ronchi-Virgolini AL (2016) Habitat heterogeneity drives bird species richness, nestedness and habitat selection by individual species in fluvial wetlands of the Paraná River, Argentina: bird species richness and habitat heterogeneity. Austral Ecol 41(7):829–841. https://doi.org/10.1111/aec.12375
47. Madhav S, Ahamad A, Singh AK, Kushawaha J, Chauhan JS, Sharma S, Singh P (2020) Water pollutants: sources and impact on the environment and human health. In: Pooja D, Kumar P, Singh P, Patil S (eds) Sensors in water pollutants monitoring: role of material. Springer, Singapore, pp 43–62. https://doi.org/10.1007/978-981-15-0671-0_4
48. Marnn P, Ali H, Jiang H, Liu Y, Li Z, Ahmed S, Yang T, Li Z, He C (2025) Modelling the impact of ecological restoration on waterbird diversity and habitat quality in Myanmar's Moe Yun Gyi Wetland. Biology 14(5):519
49. Mashala MJ, Dube T, Mudereri BT, Ayisi KK, Ramudzuli MR (2023) A systematic review on advancements in remote sensing for assessing and monitoring land use and land cover changes

impacts on surface water resources in semi-arid tropical environments. Remote Sens 15(16): 3926
50. Miller JR, Hobbs NT (2000) Recreational trails, human activity, and nest predation in lowland riparian areas. Landsc Urban Plan 50(4):227–236
51. Miller ME, Hamann M, Kroon FJ (2020) Bioaccumulation and biomagnification of microplastics in marine organisms: a review and meta-analysis of current data. PLoS One 15(10):e0240792
52. Mishra AP, Kumar S, Patra R, Kumar A, Sahu H, Chandra N, Pande CB, Alshehri F (2023) Physicochemical parameters of water and their implications on avifauna and habitat quality. Sustainability 15(12):9494
53. Moore AC, Kumble S (2024) Community-based conservation and restoration in coastal wetlands: a review. Wetlands 44(5):62. https://doi.org/10.1007/s13157-024-01818-3
54. Mott R, Prowse TAA, Jackson MV, Rogers DJ, O'Connor JA, Brookes JD, Cassey P (2023) Measuring habitat quality for waterbirds: a review. Ecol Evol 13(4):e9905. https://doi.org/10.1002/ece3.9905
55. Naugle DE, Johnson RR, Meeks WA, Higgins KF (1996) Colonization and growth of a mixed-species heronry in South Dakota. Colon Waterbirds, Washington, DC, pp 199–206
56. Newton I (1980) The role of food in limiting bird numbers. Ardea 55(1–2):11–30
57. Novack M, Romano L, Nascimento L, do Canterle E, Barboza CN (2021) Benthic invertebrates associated with the aquatic macrophytes in a Brazilian subtropical reservoir. Eng Sanit Ambient 26:741–748
58. Nuechterlein GL, Buitron D, Sachs JL, Hughes CR (2003) Red-necked grebes become semi-colonial when a prime nesting substrate is available. Condor 105(1):80–94
59. Paracuellos M (2006) How can habitat selection affect the use of a wetland complex by waterbirds? Biodivers Conserv 15(14):4569–4582. https://doi.org/10.1007/s10531-005-5820-z
60. Parnell JF, Ainley DG, Blokpoel H, Cain B, Custer TW, Dusi JL, Kress S, Kushlan JA, Southern WE, Stenzel LE (1988) Colonial waterbird management in North America. Colon Waterbirds, Washington, DC, pp 129–169
61. Praveen J, Jayapal R, Pittie A (2014) Notes on Indian rarities—2: waterfowl, diving waterbirds, and gulls and terns. Indian Birds 9(5 & 6):113–136
62. Ralph CJ, Sauer JR, Droege S (1995) Monitoring bird populations by point counts. Pacific Southwest Research Station, Albany
63. Reed JM, Dobson AP (1993) Behavioural constraints and conservation biology: conspecific attraction and recruitment. Trends Ecol Evol 8(7):253–256
64. Ringhofer M, Hasegawa T (2014) Social cues are preferred over resource cues for breeding-site selection in Barn Swallows. J Ornithol 155(2):531–538. https://doi.org/10.1007/s10336-013-1035-3
65. Rodríguez-Estival J, Sánchez MI, Ramo C, Varo N, Amat JA, Garrido-Fernández J, Hornero-Méndez D, Ortiz-Santaliestra ME, Taggart MA, Martinez-Haro M (2019) Exposure of black-necked grebes (Podiceps nigricollis) to metal pollution during the moulting period in the Odiel Marshes, Southwest Spain. Chemosphere 216:774–784
66. Samraoui F, Nedjah R, Boucheker A, El-Serehy HA, Samraoui B (2025) Sky high or grounded: nest site selection of herons and egrets in a mixed-species colony. Ecol Evol 15(1):e70761. https://doi.org/10.1002/ece3.70761
67. Saunders SP, Wu JX, Gow EA, Adams E, Bateman BL, Bayard T, Beilke S, Dayer AA, Fournier AM, Fox K (2021) Bridging the research-implementation gap in avian conservation with translational ecology. Condor 123(3):duab018
68. Schmidt SM, Fournier AMV, Osborn JM, Benson TJ (2023) Water depth influences survival and predator-specific patterns of nest loss in three secretive marsh bird species. Ecol Evol 13(12):e10823. https://doi.org/10.1002/ece3.10823
69. Shesterin IS (2010) Water pollution and its impact on fish and aquatic invertebrates. Interactions 1:59–69

70. Soons MB, Brochet A, Kleyheeg E, Green AJ (2016) Seed dispersal by dabbling ducks: an overlooked dispersal pathway for a broad spectrum of plant species. J Ecol 104(2):443–455. https://doi.org/10.1111/1365-2745.12531
71. Stamps JA (1988) Conspecific attraction and aggregation in territorial species. Am Nat 131(3): 329–347. https://doi.org/10.1086/284793
72. Steenhof K, Newton I (2007) Assessing nesting success and productivity. In: Raptor research and management techniques. Hancock House, Surrey, pp 181–192
73. Stolen ED, Breininger DR, Frederick PC (2004) Using waterbirds as indicators in estuarine systems: successes and perils. In: Estuarine indicators. CRC Press, Boca Raton, pp 431–444
74. Store R, Jokimäki J (2003) A GIS-based multi-scale approach to habitat suitability modelling. Ecol Model 169(1):1–15
75. Sullivan BL, Aycrigg JL, Barry JH, Bonney RE, Bruns N, Cooper CB, Damoulas T, Dhondt AA, Dietterich T, Farnsworth A (2014) The eBird enterprise: an integrated approach to development and application of citizen science. Biol Conserv 169:31–40
76. Taylor M (2022) How birds live together: colonies and communities in the Avian World. Princeton University Press, Princeton
77. Tilman D (1994) Competition and biodiversity in spatially structured habitats. Ecology 75(1): 2–16. https://doi.org/10.2307/1939377
78. Tong W (2020) Understanding bird behavior: an illustrated guide to what birds do and why. Princeton University Press, Princeton
79. Towns DR, Atkinson IAE, Daugherty CH (2006) Have the harmful effects of introduced rats on islands been exaggerated? Biol Invasions 8(4):863–891. https://doi.org/10.1007/s10530-005-0421-z
80. Underwood AJ, Chapman MG, Crowe TP (2004) Identifying and understanding ecological preferences for habitat or prey. J Exp Mar Biol Ecol 300(1–2):161–187
81. Urmila, Gupta RK, Sharma S, Kumar R (2025) Earthworms: the environmentally preferred bioengineers for managing heavy metals. In: Dey S, Bhattacharya S (eds) Biotechnological interventions in the removal of emerging pollutants. Springer Nature Singapore, Singapore, pp 365–384
82. Valente JJ, Fischer RA Jr (2011) Reducing human disturbance to waterbird communities near Corps of Engineers projects. https://erdc-library.erdc.dren.mil/bitstream/11681/8629/1/ERDC-TN-DOER-E29.pdf
83. Villamagna AM, Murphy BR (2010) Ecological and socio-economic impacts of invasive water hyacinth (Eichhornia crassipes): a review. Freshw Biol 55(2):282–298
84. Walter ST, Carloss MR, Hess TJ, Leberg PL (2013) Hurricane, habitat degradation, and land loss effects on Brown Pelican nesting colonies. J Coast Res 29(6a):187–195
85. Wang C, Xia S, Yu X, Wen L (2025) Responses to extreme drought in wintering waterbirds: a multi-species approach. Front Zool 22(1):3. https://doi.org/10.1186/s12983-025-00557-3
86. Weller MW (1999) Wetland birds: habitat resources and conservation implications. Cambridge University Press, Cambridge
87. White CR, Day N, Butler PJ, Martin GR (2007) Vision and foraging in cormorants: more like herons than hawks? PLoS One 2(7):e639
88. Winkler DW (2016) Breeding biology of birds. In: Handbook of bird biology, 3rd edn. Wiley-Blackwell, Malden, pp 407–450
89. Yadav R, Gupta RK, Kumar R, Kaur T (2022) Assessment of toxicity of Lead and nickel on the biochemical and immunological parameters of earthworm, Eudrilus eugeniae. Agric Sci Dig 44(5):922–929
90. Yadav R, Gupta RK, Kumar R, Kaur T, Moun R (2022) Biochemical response of earthworm, *Eisenia fetida* to heavy metals toxicity. J Appl Nat Sci 14(3):990–998
91. Yodha K, Singh D, Kour A, Yadav R, Kumar R (2024) Species diversity and abundance of birds in ber (*Zizyphus mauritiana*) crop. Asian J Conserv Biol 13(1):118–126. https://doi.org/10.53562/ajcb.83683

# Chapter 17
# Nest Boxes: An Effective Management Approach to Save Threatened Water Birds Species

**Farhat Jahan and Haroon Ali**

**Abstract** Globally, numerous waterbird species are experiencing significant population declines primarily due to accelerated climate change, habitat loss and escalating human activities. The use of artificial nest boxes is emerging as a viable conservation strategy for endangered aquatic and semi-aquatic bird species to address the decline of natural nesting habitats and increasing the breeding success rate. This study comprehensively reviews strategies for successful implementation, appropriate design, placement and continued maintenance of artificial nest boxes for cavity-nesting species of birds such as ducks, herons and terns within the diverse wetland ecosystems. This chapter highlights the paramount significance of meticulously selecting materials for constructing artificial nest boxes. It emphasizes the crucial role of considering factors including the appropriate size, configuration, predator deterrent measures and design to suit the needs of specific bird species. By integrating artificial nest box programs with broader conservation efforts such as damaged habitat restoration and invasive species control can ensure long-term population recovery and protection of vulnerable waterbird communities.

**Keywords** Artificial nests · Conservation strategy · Habitat loss · Nest boxes · Threatened species · Waterbirds

## 17.1 Introduction

Aquatic ecosystems, once teeming with biodiversity, are now under increasing pressure from urban development, agricultural runoff, wetland loss, invasive species, and climate fluctuations and fluctuations in water resources [62, 63]. Species that require specific nesting habitats, such as cavity-nesting birds, aquatic birds, and

F. Jahan (✉)
Department of Zoology, Hansraj College, University of Delhi, Delhi, India

H. Ali
Department of Public Health, College of Applied Medical Sciences, Khamis Mushait Campus, King Khalid University (KKU), Abha, Kingdom of Saudi Arabia

 
R. Kumar et al. (eds.), *Ecology, Conservation of Aquatic Birds*, Wetlands: Ecology, Conservation and Management 15, https://doi.org/10.1007/978-3-032-17125-2_17

semi-aquatic birds, are particularly vulnerable to these changes. In response, conservation biologists and ecologists have adopted habitat enhancement strategies for aquatic birds, with nest boxes emerging as a practical, cost-effective, and scalable solution. The aquatic bird crisis is at its peak today, with freshwater and coastal ecosystems, crucial for biodiversity, facing severe threats. Agricultural expansion, urban development, pollution, and climate change are causing habitat loss and degradation, putting many water-dependent species at risk of extinction [59, 64]. Avian water species are particularly vulnerable due to their complex life cycles which are closely linked to aquatic environments. These species including the beautiful tern, elusive bittern, attractive duck and distinctive kingfisher are vital apex predators for seed dispersers and indicators of environmental health [60]. Freshwater, aquatic and coastal ecosystems are vital for biodiversity facing unprecedented threats worldwide. Agricultural expansion, urbanization, pollution and climate change are causing habitat loss and degradation which are threatening many water-dependent species with extinction [59]. Among the most vulnerable are avian species, whose intricate life cycles are intrinsically linked to the health and accessibility of aquatic habitats. These water birds from elegant terns to charismatic ducks and distinctive kingfishers play vital roles in ecosystems as apex predators, seed dispersers and bioindicators of environmental health [60].

With the bird habitats dwindling innovative conservation methods are urgently required to reverse population declines and foster recovery. One approach is to use artificial nesting structures called nest boxes which can be enhanced in specific contexts to increase their effectiveness. This chapter examines and tell us the application of artificial nest boxes as a management strategy for endangered water bird species. It also explores their ecological rationale, practical implementation, documented achievements and future prospects. This chapter gives information of the specific artificial nest boxes required by various aquatic avian species and examine the diverse designs of artificial nest boxes that cater to their specific needs.

Here, their placement strategies, monitoring protocols, maintenance activities, challenges, limitations and considerations for effective maintenance will also be discussed. Furthermore, the present many studies illustrate their impact on the population recovery of water bird species. Innovative conservation strategies are urgently needed to reverse water bird population declines and foster recovery as natural nesting habitats for these birds are diminishing.

In this chapter explain the application of artificial nest boxes using as a management tool for threatened water bird species. It explores their ecological significance, practical implementation, documented achievements and future prospects. The specific artificial nest boxes needed for various aquatic avian species including their diverse designs, placement strategies, monitoring protocols, maintenance activities, challenges, limitations and considerations for maintenance are explained.

This chapter also presents the case studies showing their impact on their population recovery. It is necessary to discuss behavioral performance of aquatic birds, their documented successes, limitations and future prospects of using artificial nest boxes based on their ecological properties. We will discuss the designs of artificial nest boxes tailored to meet the specific requirements of different types of aquatic bird groups and those aquatic birds they inhabit them. We will address to all those important aspects related to placement and maintenance of artificial nest boxes. We will also discuss about the factors responsible for increasing population of water birds and their case studies. Aquatic avian species face escalating threats arising from habitat degradation, pollution, changing in climate and human activities. The scarcity of secure and suitable nesting habitats presents a critical challenge now. Nest boxes are a practically species-specific conservation intervention that provides secure and accessible nesting sites especially in degraded wetlands ecosystems. This chapter explores ecological benefits, design approaches, field implementations and measurable outcomes of installing artificial nest boxes in wetland environments. This conservation approach is shown to be effective in safeguarding of threatened water bird species through practical examples, tabulated form and case studies.

### *17.1.1 Aim and Objectives*

The main aim of this study to identify the ecological needs for building artificial nesting structures for threatened and endangered aquatic bird species. Here, we will examine specific case studies involving various threatened and endangered aquatic bird species by analyzing their design, present status and their management. This study will also evaluate the effectiveness of artificial nest boxes in the recovery of threatened and endangered aquatic bird species and conservation of biodiversity.

## 17.2 Artificial Nest Boxes of Aquatic Birds: Concept, History and Evolution

The idea of building artificial nests boxes has a long history, starting in Europe with the creation of bird houses for swallows and starlings. These meticulously designed nest boxes were created to cater to specific requirements of particular species of bird contributing to conservatory efforts.

Over the time, utility of the artificial nest boxes has expanded beyond birds. It has been repurposed to accommodate mammals such as otters and bats as well as reptiles like turtles that enable to support of breeding initiatives and population monitoring. It has been used for reintroducing the specific species and restoring habitats.

### *17.2.1 Growing Threats to Aquatic Birds Species*

Nest boxes have proven to be a valuable conservation strategy for safeguarding and restoration of the threatened and endangered aquatic bird species. These artificial nest boxes structures mimic natural nesting habitats provide secure and controlled environments for aquatic bird species facing challenges like habitat degradation and predation. These following points explain how artificial nest boxes help to manage and recover threatened and endangered aquatic bird species.

### *17.2.2 Supporting Nesting Habitats for Aquatic Bird Species*

Ducks, herons, gulls, kingfishers and terns are the aquatic bird species. They spend much of their lives in water bodies like wetlands, lakes and rivers. Many water bird species have developed specialized nesting strategies for examples using hollow trees, high dunes and dense bushes along riverbanks. Rapid environmental degradation, human development and climate change are reducing the availability of these natural nesting sites of water birds. Many aquatic bird species like Baer's pochard and diving ducks that depend on specific nesting sites are declining due to loss of their habitat. The Baer's pochard population has declined from 150 to 700 birds in recent years. The decline in their population is attributed to adverse environmental factors that negatively impact on their habitat as well as human activities like fishing and hunting [61].

### *17.2.3 Ecological Significance of Nesting Sites for Aquatic Bird Species*

Many aquatic bird species are site-fidelity and prefer specific nesting conditions like vegetation cover, water height and predator free areas. The destruction of these sites often causes failures of breeding and decline of populations of aquatic bird species.

### *17.2.4 Limitations of Artificial Nesting Habitats*

Our main goal of this study is to solve a major problem for many endangered and threatened water birds by offering artificial nest boxes structures and nesting areas. We also aim to raise awareness in public about the availability of suitable and secure sites for breeding of aquatic birds.

#### **17.2.4.1 Causes of Habitat Loss**

Natural nesting habitats of water birds are destroyed by various means such as draining and converting wetlands for agriculture, aquaculture and urbanization.

This conversion eliminates vast areas for nesting including emerging vegetation, mudflats and shorelines [14].

#### 17.2.4.2 Habitat Fragmentation

The remaining wetlands might be too small and isolated to support viable breeding populations. According to Fahrig [17] resource scarcity or lack of connectivity for successful nesting and roosting may be the cause.

#### 17.2.4.3 Increased Predatory Pressure

Habitat degradation reduces natural cover that make nesting birds vulnerable to predators on land and in the air. Invasive predators make more worsen this issue [40].

#### 17.2.4.4 Water Level Fluctuations

Human activities like dam construction and water extraction can disrupt water levels unpredictably and quickly. These fluctuations can cause flooding and destroy nesting sites [31].

#### 17.2.4.5 Disturbance

Human activities like recreational activities fishing and boat traffic can disturb habits of water bird nesting. This disruption can cause birds to abandon their nests and reduce their breeding success rate [12].

#### 17.2.4.6 Climate Change Impacts

Rising sea level increased storm frequency and modified vegetation patterns threaten coastal and inland gull nesting habitats [24]. Artificial nest boxes offer secure, elevated and sheltered nesting sites providing a targeted solution. These artificial nest boxes can be strategically placed in areas with limited natural nesting spots. Conservationists can reduce predation risks by creating artificial nesting habitats that provide more breeding opportunities for endangered and threatened water bird species. In our study this approach also helps to minimize negative impacts of habitat loss and disturbance.

## 17.3 Artificial Nest Boxes: An Effective Management Strategy for Preserving Threatened Waterbird Species

**Advantages of Artificial Nesting Boxes for Aquatic Birds are**:

- Provide a suitable alternative to natural nesting cavities.
- Safeguard eggs and chicks from potential threats such as predators and adverse weather conditions.
- Enhance breeding outcomes and reduce chick mortality rates.
- Facilitate conservation efforts in urban and semi-urban wetlands.

**Essential Requirements for Artificial Nesting are**:

### *17.3.1 Availability of Empty Cavities*

Artificial nesting sites must provide accessible and suitable cavities, such as hollow trees or burrows, that can accommodate aquatic birds. The necessity of empty cavities in artificial nest boxes is paramount for supporting cavity-nesting waterbird species, particularly in habitats where natural nesting sites such as tree hollows, reed clusters, or eroded banks are scarce due to deforestation, urbanization, or wetland degradation. These artificial nest cavities provide secure, insulated and predator-resistant environments that mimic natural nesting conditions that enhance breeding success and chick survival. Tailored nest cavities with appropriate size, entrance diameter and their placement enhance occupancy rates among species like Lesser Whistling Duck, Black-crowned Night Heron and Common Moorhen. Studies corroborate that cavity availability frequently restricts population size in hole-nesting birds, and artificial structures can substantially alleviate this constraint [4–8, 37, 39]. Incorporating well-designed empty cavities into conservation strategies is crucial for preserving and restoring threatened water bird populations. This study explores how natural cavities limit cavity-nesting bird populations and how artificial nest boxes can mitigate this constraint. A meta-analysis confirms that artificial nest cavities increase bird species occupancy and breeding success. The study reveals species-specific data on nesting ecology and conservation recommendations for cavity-nesting birds worldwide including India.

### *17.3.2 Thermal Regulation*

Aquatic birds require a controlled environment to regulate their body temperature. Artificial nesting sites should offer appropriate insulation and ventilation to ensure optimal thermal conditions. Designing artificial nest boxes for breeding of water birds in open, sun-exposed and humid wetlands requires careful consideration of thermal regulation. Adequate insulation ensures optimal temperature and humidity conditions conducive to egg incubation and chick development. Conversely, inadequate ventilation and shading can result in overheating, embryo mortality, adult stress, and diminished fledging success. Wood, insulated panels, semi-shaded areas and ventilation holes help to control indoor temperatures. Neglected thermal regulation leads to nest abandonment and diminished reproductive success in birds [2, 22]. A thoughtfully designed of nest box with temperature control improves occupancy and breeding success for cavity-nesting water birds. This study investigates how nest box temperature and humidity affect the hatching, fledging, incubation and reproductive timing.

### *17.3.3 Food Resources Access*

Aquatic bird species require easy access to food. Design artificial nesting sites to make it easy for water birds to find food like fish and aquatic plants. Water birds need enough food near artificial nest boxes to stay and breed successfully. The nesting birds that build their nests need easy access to food like fish, insects and plants to feed themselves and their young. If nest boxes are too far from suitable foraging grounds, energy expenditure increases and chick survival rate decreases due to infrequent and insufficient feeding. Ideal installation of artificial nest sites should be in rich wetlands, shallow waters and aquatic vegetation zones with uninterrupted feeding opportunities. These research findings establish a positive correlation between nest box occupancy and local food availability. Highlighting the significance of considering landscape-scale resource availability in conservatory efforts [19, 40, 51–56]. Ensuring adequate food near artificial nesting sites is vital for the long-term survival of water bird populations. A detailed study explaining how food available near bird nests affects their breeding success rate. Assesses the relationship between wetland restoration and food abundance and increased nesting success in water birds. The study examines the coexistence of human-managed wetlands and bird conservation, focusing on foraging resources for large waterbirds like Sarus Cranes.

### *17.3.4 Protection from Predators and Human Activities*

Aquatic birds face threats from predators, so for enhance nest success rates, it's crucial to safeguard nesting sites from predators and human interference. Artificial Waterbirds nests are often targeted by dogs, mongooses, snakes, and other predators that could harm eggs or chicks. To mitigate this risk, installing predator guards in nesting areas and elevating nests are effective strategies. Moreover, placing nests in secluded locations with minimal human activity, such as protected water bodies, less accessible islands, or reed beds, further enhances their safety. Additionally, limiting boating, fishing, and tourism activities during the breeding season allows birds to focus on nest construction and egg hatching without stress [32]. These protection measures have led to significant increases in water birds breeding success.

## 17.4 Installation of Nest Boxes: An Alternative to Lost Waterbird Species or Damaged Nesting Areas

Nest boxes serve as an alternative to lost or damaged nesting areas, contributing to the breeding success rates of aquatic bird species. By providing a safe environment, nest boxes facilitate the survival and reproduction of bird species that require cavities and nests. These include the Wood Duck (*Aix sponsa*), Swans and other aquatic birds. Nest boxes also provide a safeplaces for species that build nests in trees like herons, buzzards, hawks and spoonbills. It allowing them to continue their breeding activities. Sandpipers, cormorants, egrets and cranes also benefit from ground-nesting boxes.

### *17.4.1 Reducing Hunting and Human Disturbance*

Nest boxes for aquatic birds serve as a safeguard against predators and human interference. Placing artificial nest boxes strategically reduces threats from hunting and human disturbances especially in shared wetlands between communities, wildlife and local people. Installing nest boxes in inaccessible and low-traffic areas like isolated sandbanks, reed patches and elevated poles in restricted zones minimizes direct human interference, egg theft and accidental nest destruction.

Deploying nest boxes in protected wetlands and community-managed reserves also ensures extra monitoring and enforcement against poaching and nest disturbance. Camouflage colours and predator guards make nests harder to spot and increase survival chances. Carefully placing nesting sites and involving local communities has been shown to greatly improve breeding success for threatened water birds like the Indian Skimmer and Black-bellied Tern [4–8, 57, 66, 67]. Choosing right location for artificial nest boxes is vital to minimize human disturbances.

### *17.4.2 Raising or Isolating Ground-Based Nests*

Building nests on the ground protects them from predators. Ground-nesting water birds face threats from natural predators like snakes, monitor lizards, dogs, crows and human activities like trampling, habitat disturbance and livestock grazing. Artificial nest boxes elevated and on platforms shield nests from ground-level dangers providing safer environments for egg-laying and chick-rearing. Raised nest boxes on poles, artificial islands, or anchored floating platforms reduce predation and flooding risk, especially during the monsoon season. These structures mimic natural nesting preferences while offering enhanced protection for species like the Baer's Pochard, Sarus Crane, and Black-bellied Tern. Studies show increased nesting success when artificial nests are elevated or isolated from ground access [1, 37, 58]. Innovative nest box installation helps isolate ground-based nests, mitigating threats and supporting the conservation of vulnerable waterbird populations.

### *17.4.3 Reducing Disturbances from Recreational Activities*

Minimizing disturbances from recreational activities in proximity to water bodies where aquatic birds nest is of paramount importance. This can be achieved by controlling invasive species of aquatic bird nests, such as adjusting the size of the entrance. Engaging in recreational activities like boating, fishing, hiking, birdwatching, and picnicking can benefit public engagement with nature, but it can also disturb nesting waterbirds, especially during breeding season. Noise, vibrations, and frequent human presence can cause nest abandonment, reduced incubation efficiency, and increased chick mortality. Installing artificial nest boxes in buffer zones away from high-traffic areas is a proven strategy to reduce these disturbances. Nest boxes in restricted zones, secluded inlets, or off-limits conservation areas minimise human interference, ensuring successful bird breeding. This practice has been effective in protected wetlands like Okhla Bird Sanctuary and Bharatpur National Park where regulated visitor access and nest site zoning have positively impacted waterbird nesting [30, 45, 57]. Conservation managers can balance eco-tourism with the ecological needs of threatened avian species by carefully placing nest boxes.

## 17.5 Facilitating Monitoring of Aquatic Bird Species

The nest boxes of aquatic bird offer valuable insights for researchers and conservationists. This allows them to monitor and study these birds effectively. The artificial nest boxes offer easy access to aquatic bird breeding success. This study allows researchers to monitor population fluctuations. It provides insights in to the reproductive strategies of these birds. The nest boxes allow researchers to observe aquatic bird behavior without disturbing their natural habitats. This allows them to take targeted actions like managing parasites, addressing health of nestlings and ensuring conservation of these magnificent creatures.

### 17.5.1 *Enhancing Aquatic Bird Population Recovery*

Artificial nest boxes are vital for the recovery of aquatic bird species especially those affected by habitat loss, limited nesting sites and human activities. Many threatened water birds like the Indian Skimmer, Baer's Pochard, and Black-bellied Tern lack secure nesting sites that directly impact on their breeding success rate. Conservationists can improve reproductive success and fledgling survival by providing species-specific nest boxes that mimic natural conditions. Floating platforms, raised sandbanks, and isolated nest boxes in managed wetlands have led to measurable increases in nesting attempts and chick recruitment [4–8, 37, 58]. Combining artificial structures with site protection, community involvement, and predator control restores breeding populations and ensures long-term species persistence in disturbed and human-modified wetland ecosystem. For critically endangered water bird species, nest boxes can play a pivotal role in facilitating aquatic bird population recovery through the following mechanisms:

#### 17.5.1.1 Facilitating Breeding Opportunities

Nest boxes provide breeding habitats in areas where natural habitats are already available. In rapidly degrading wetland habitats, safe and suitable nesting sites are a major limiting factor for aquatic bird breeding. Artificial nest boxes offer crucial alternatives for waterbirds whose natural habitats are threatened by urban expansion, pollution, and water management practices. These structures replicate species-specific breeding conditions, encouraging courtship, egg-laying, and chick rearing in previously unsuitable areas. Studies show that artificial nest sites can significantly increase nesting frequency, clutch size, and hatching success, especially when combined with habitat restoration and predator control [37, 39, 58]. Nest box interventions, which enhance breeding opportunities, support reproductive resilience and contribute to the stabilization and recovery of vulnerable aquatic bird populations.

#### 17.5.1.2 Stabilizing Declining Populations

Nest boxes serve as a temporary measure to stabilise declining populations of water bird species. Habitat fragmentation, pollution, and human interference in wetlands have led to alarming population declines in aquatic bird species. Artificial nest boxes provide safe and species-specific breeding habitats replacing lost natural ones and lowering predation. These structures enhance breeding success and act as a safeguard against the population declines. Nest box programs for Sarus Cranes, Baer's Pochards and Indian Skimmers in protected wetlands have boosted breeding attempts and chick survival [58]. Sustainable methods for population decline halt and species recovery include long-term monitoring, habitat management, community awareness, and artificial nest boxes. The strategic approach enhances the resilience of water bird communities in the face of human influences.

#### 17.5.1.3 Nest Boxes Support Relocation and Reintroduction Efforts

Relocating and reintroducing artificial nest boxes for water bird species in the areas where traditional nesting sites have been destroyed by urbanization, agriculture and wetland degradation. Reintroducing nests using artificial nest boxes is a vital conservation strategy for aquatic birds. This involves placing custom-designed nests in suitable and unused sites to encourage birds for breeding in safeguard and managed environments. For example, floating nests for diving ducks and raised sand platforms for Indian Skimmers have successfully reintroduced breeding populations in restored wetlands [37, 51–56]. Gradually relocating nesting sites away from high-disturbance areas and towards protected zones can reduce mortality risks and enhance reproductive success and repopulate declining habitats for conservationists. Long-term monitoring and adaptive management ensure relocated nests meet the ecological needs of target species. It is very effective technique when combined with wetland restoration and predator control programs.

**Observations of the Table 17.1**

- The availability of nesting sites limits the reproductive success of many wetland bird species.
- Species-specific interventions, encompassing factors such as height, material, location, and proximity to water, significantly enhance adoption rates and fledgling survival.
- The utilisation of nest boxes facilitates monitoring, predator protection, and contributes to the restoration of threatened populations (Table 17.1).

**Table 17.1** Species-specific use of nest boxes

| Aquatic birds species | Ecological context and uses of nest boxes | References |
|---|---|---|
| Common Goldeneye | Cavity-nesting ducks like Golden eyes prefer elevated wooden nest boxes near freshwater lakes. A study found that managed nest box sites had better breeding success. | [42] |
| Wood Duck | Wood ducks exhibit a marked preference for nest boxes, particularly in regions characterised by the absence of natural tree cavities. This study underscores the behavioural and reproductive benefits associated with managed nest box initiatives. | [47] |
| Common Merganser | Mergansers, despite their selective nature readily accept nest boxes when natural nesting sites are limited. The research conducted in Northern European wetlands demonstrated a positive correlation between nest adoption and the availability of such structures. | |
| Little Grebe | Young Grebes utilise floating platforms or reed-supported structures for nesting. A study conducted revealed that successful nesting can be achieved in managed aquatic environments by employing low-floating structures. | |
| Great Cormorant | Cormorants utilise substantial elevated nesting platforms, frequently in colonies. This research documented the utilisation of nest boxes in coastal and island ecosystems, thereby mitigating ground disturbance. | [46] |

## 17.6 Best Practices for Artificial Nest Box Initiatives

Initiating artificial nest box programs is crucial measure to conserving aquatic avian species. To improve the effectiveness of nest box programs consider the following key factors:

### *17.6.1 The Location and Design of Artificial Nest Boxes*

Prioritize optimizing nest box placement and design to suit water birds species' specific needs considering factors like size, entrance shape and insulation. The success of artificial nest boxes for water birds depends on two key factors e.g. strategic site selection and species-specific design. The ideal locations of nest offer safety from predators, minimal human disturbance, availability to food, water and appropriate micro-climates. For example, elevated nest platforms in shallow marshes suit Sarus Cranes, while floating reed-covered structures suit diving ducks like Baer's Pochards. The design should mimic natural nesting preferences in shape, size, entrance dimensions, materials and thermal insulation. The research shows that nest boxes made of wood, PVC and recycled plastic with adequate ventilation and drainage that improve hatching success [11, 39]. When planning a location take in to account vegetation cover, predator access and flooding risk particularly in dynamic wetlands. After monitoring and adaptive management adjust nest box placement to make sure the bird have the best chance of successfully breeding and the population recovery.

### *17.6.2 Maintenance of Artificial Nest Boxes*

Regular inspection and cleaning are essential to prevent parasitic infections affecting chicks and young waterfowl, ensuring the boxes' functionality. The significance of regular nest box maintenance requires for long-term waterbird conservation. The maintenance of artificial nest boxes is paramount for ensuring their long-term efficacy in supporting waterbird conservation efforts. Neglecting proper upkeep can render nest boxes unsafe due to weathering, parasite accumulation, fungal growth or predation-induced damage. Regular maintenance practices include:

- Cleaning the boxes after the breeding season.
- Repairing or replacing damaged components.
- Inspecting for water accumulation or structural instability.
- Removing old nesting material that may harbour pathogens.

As noted by Newton [39], neglecting maintenance can lead to reduced occupancy rates and even nest abandonment. Seasonal inspections also ensure that nest boxes are oriented correctly at the right height. The insulation remains suitable for the target species. Wetland areas require special attention to prevent water logging and algal over-growth on floating or semi-submerged platforms [11]. Long-term monitoring and maintenance of the intervention require the active participation of local stakeholders.

### 17.6.3 Engage the Community in Creating Artificial Nest Boxes

Involve local communities to ensure the continued protection of water birds nest boxes. Engage with local communities to ensure their continued support and involvement for this initiative. Communities are essential for the successful installation and long-term sustainability of nest box programs for water bird conservation. Engaging resident birds particularly those living near wetlands, rivers and agricultural lands that encourages a sense of ownership and responsibility for safeguard of the nesting habitats. Community members can learn to build, install, monitor, report sightings and maintain nest boxes. It improves project efficiency and raises awareness about biodiversity and ecological balance [57]. To organize educational workshops, bird watching events and school outreach programs can further strengthen to this connectivity. Community involvement for conservation initiatives improves their effectiveness and sustainability as seen in participatory wetland management in India and Southeast Asia. Combining traditional ecological knowledge with eco-tourism and employment benefits to make long-term community support for water bird conservation through proper strategies for artificial nest boxes.

### 17.6.4 Monitoring and Research Necessitates for Developing Artificial Nest Boxes

Use of collected data to improve strategies and evaluate long-term results. Artificial nest boxes are practical, adaptable and effective ways to conserve endangered aquatic avian species by addressing habitat loss and ecological pressures. Monitoring and research are crucial for any artificial nest box program. They ensure interventions are evidence-based, adaptive, and species-appropriate. Systematic monitoring tracks occupancy rates, breeding success, predator activity, and habitat suitability over time. This data helps determine if a design, location, or material effectively supports the target water bird species [36].

Long-term research helps scientists and conservationists assess trends, identify emerging threats, and optimise nest box strategies for maximum ecological benefit. For instance, studies on species-specific nesting preferences, like Baer's Pochard's preference for floating reed nests or Indian Skimmers' preference for raised sandbanks, guide nest box structure refinement [51–56]. Advancements in technology, such as camera traps, GPS trackers, and citizen science platforms, have improved data collection and real-time monitoring. Integrating research into nest box initiatives enhances their impact and contributes to global conservation knowledge.

## 17.7 Bird Nest Boxes Key Design Considerations

Bird nest boxes should be constructed from durable materials that are non-toxic and resistant to weather conditions. The internal temperature should be maintained at a suitable level through proper insulation to ensure the survival of eggs and chicks. Entrances should be appropriately sized to prevent the entry of predators and unwanted species. Drainage and ventilation are essential to prevent water accumulation and overheating.

### *17.7.1 Design of Nest Boxes and Implementation*

Waterbirds species encounter mounting threats from habitat degradation, pollution, climate change, and human activities. A critical challenge lies in the scarcity of secure and suitable nesting habitats. Nest boxes provide a practical, species-specific, and targeted conservation intervention by offering secure and accessible nesting sites, particularly in degraded or modified wetland ecosystems. By emulating natural cavities or protected areas, nest boxes not only enhance breeding success but also play a pivotal role in restoring avian populations. When designed in accordance with the ecological requirements of specific species, nest boxes can substantially improve reproductive success, mitigate interspecies competition, and facilitate long-term population recovery.

### *17.7.2 Customising Structures to Specific Requirements of Aquatic Bird Species*

The success of artificial nest box programs hinges on meticulously designed and strategically placed structures that closely resemble to the natural nests. Simulate the natural nesting habits of the target species. A one size fits to all approach is rarely effective in achieving this goal. It's very important to highlight these key points:

#### 17.7.2.1 Requirements for Specific Species

Different water bird species have different and unique nesting habits and habitat choices. For example- some ducks and geese prefer enclosed cavities while terns and gulls prefer open scrapes on the ground or sometimes elevated platforms. Bitterns and herons often build large nests in dense vegetation or sometimes on trees. Artificial nest box designs must be tailored to specific requirements regarding size, material and entrance hole dimensions. These factors are essential for the success of water bird nest boxes. The species-specific design and thoughtful placement of the boxes are essential for their effectiveness.

**Observations of Table 17.2**

- Waterbirds like Wood Ducks and Common Goldeneyes show high adaptation to tree-mounted cavity boxes.
- Ground and platform nesters (Sarus Cranes, Little Terns) successfully use artificial platforms or rafts in wetland habitats (Table 17.2).

**Nest Box Adoption by Water birds**

This table presents real-world nest box interventions tailored for waterbird species, which rely on aquatic ecosystems for nesting and foraging. It aims to demonstrate the effectiveness of artificial nesting structures (nest boxes, rafts, platforms) in threatened or vulnerable habitats.

**Wood Ducks (*Aix sponsa*)** nest in tree holes near water in Illinois, USA. Artificial wooden nest boxes installed in 132 trees have increased their reproductive success by 68% (occupied ~90) [47].

**The Common Goldeneye (*Bucephala clangula*)** is a diving duck found in Sweden, Canada, and the UK. Nest boxes have been installed in Scandinavia and

**Table 17.2** Waterbird species—nest boxes occupancy

| Aquatic birds species name | Place of installation of nest boxes | Total no. of nest boxes installed | Nest boxes occupied by birds | Occupancy rate (%) | References |
|---|---|---|---|---|---|
| Wood Duck (*Aix sponsa*) | Illinois, USA (wetland areas) | 132 | ~90 | 68% | [47] |
| Common Goldeneye (*Bucephala clangula*) | Sweden, Canada, UK (riverbanks, lakes) | ~100 | ~75 | 75% | [20] |
| Sarus Crane (*Antigone antigone*) | Uttar Pradesh, India (flooded wetlands) | 20 | 14–16 | 70–80% | [51–56] |
| Little Tern (*Sternula albifrons*) | Andhra Pradesh, India (artificial islands, coastal wetlands) | ~50 | ~40 | 80% | [44] |

Canada, and they are occupied by about 75% of the population. These nest boxes have led to stable breeding populations and are often cited in wetland restoration literature [20].

**Sarus Cranes (*Antigone antigone*)** nest in shallow wetlands in Uttar Pradesh, India. To protect their eggs and improve chick survival, 20 elevated nest platforms have been installed. Currently, 14–16 Sarus Cranes occupy these platforms, with a 70–80% occupancy rate [51–56].

**The Little Tern (*Sternula albifrons*)** nests on sandy or gravel islands in Andhra Pradesh, India. Installing nesting platforms in shallow lagoons or coastal wetlands has increased breeding colonies and chick survival [44].

Species-specific, habitat-appropriate nest boxes and platforms can significantly improve waterbird nesting success. Whether cavity-nesting ducks, ground-nesting cranes, or floating-nesting terns, carefully designed interventions based on species ecology and preferences consistently achieve 70–80% occupancy rates, highlighting their conservation value. These approaches are particularly relevant for urban wetlands and protected areas undergoing ecological degradation, and are recommended in Integrated Wetland Management Plans.

#### 17.7.2.2 Species-Specific Applications to the Nest Boxes

Various aquatic and semi-aquatic bird species exhibit distinct nesting requirements, influenced by their behavioral patterns, physical dimensions, habitat preferences, and potential predation risks. Customising nest box designs to accommodate specific species substantially enhances nesting success and contributes to population recovery. Species-specific design is crucial for several reasons:

- It increases nest occupancy and chick survival.
- It minimizes interspecies competition.
- It ensures optimal microclimate for the egg incubation.
- It helps mimic like natural nesting conditions in degraded habitats.

#### 17.7.2.3 Examples of the Species-Specific Nest Box Designs

**Ducks and Mergansers** Wooden boxes, meticulously crafted with specially designed diameter entrance holes, can be strategically installed in trees, particularly those situated over water bodies or in proximity to wetlands. These boxes are tailored to accommodate specific bird species and may incorporate internal ramps or leads, facilitating the efficient exit of ducks.

**Terns and Gulls** Artificial platforms made of wood or gravel, often found on islands or sandbars, can be improved by adding nesting materials like sand or shell fragments.

**Kingfishers** Kingfishers are attracted to artificial burrows made up of pipes or wooden boxes with vertical tunnels near bank water.

**Herons and Bitterns** When approaching a wetland, an artificial nesting platform constructed from woven branches or any synthetic materials can be provided. This platform should be situated in dense emergent vegetation or trees close to the wetland.

**Swans** For the Swans, we suggest large artificial nesting mounds made of vegetation or stone crags for a secure and elevated nesting site.

## 17.8 Conservation of Threatened Waterbirds

Threatened Waterbirds Species like the Indian Skimmer (*Rynchops albicollis*), Baer's Pochard (*Aythya baeri*), and Black-bellied Tern (*Sterna acuticauda*) are of paramount importance due to habitat destruction, unregulated tourism, and the loss of secure nesting sites. Nest box programs, meticulously designed to cater to the specific ecological requirements of these species, emerge as a viable management strategy. Conservationists have documented a surge in nesting activity and a corresponding increase in fledgling success rates in regions where artificial nesting structures have been introduced. Notably, these observations have been particularly pronounced in protected wetlands and human-altered landscapes. This research highlights the increased breeding productivity achieved in managed wetlands through the utilization of artificial nesting aids. This study underscores the effectiveness of artificial nesting platforms in enhancing the survival and breeding success of Sarus Cranes in Indian wetlands [51–56].

**Threatened Water Bird Species: Nest Box Interventions and Conservation Rationale**

**Observations of Table 17.3**

Artificial nest-box platforms effectively address three significant threats to aquatic bird populations

**Nest Loss**: Rising water levels can destroy nesting sites.

**Human Disturbance**: Human activities and agricultural practices can disrupt bird nesting and cause stress to them.

**Habitat Degradation**: Habitat degradation reduces the availability of suitable natural nesting sites.

- These interventions not only prove to be cost-effective but also contribute to the overall well-being of the community. They are adaptable to diverse wetland restoration programs, irrespective of whether they are executed in urban or rural environments.

- To ensure the occupancy and breeding success of waterbirds, successful nest box interventions require meticulous site selection, regular monitoring, and seasonal maintenance.
- Indian Skimmer: According to BirdLife International [9] and Choudhury [13], the creation of artificial islets and raised sandbanks is recommended to emulate natural nesting conditions in open riverine habitats. Provides species-specific habitat and threat data for the Indian Skimmer, Baer's Pochard, Black-bellied Tern, and others [4–8], IUCN [27], IUCN [26]. Details habitat threats and conservation needs for the Indian Skimmer [50].
- Baer's Pochard In their book titled "Gibbons and Hirons" published in 1986, Baer's Pochard proposed the effectiveness of floating nest platforms covered with reeds as a suitable solution for diving ducks in shallow lakes and ponds. Provides insight into the species critical habitat needs and use of artificial platforms [65].
- Black-bellied Tern BirdLife International [9] indicates that the species favours low-lying sandy or muddy riverbanks. Nest trays can replicate this habitat in degraded areas. Reports on habitat loss and breeding success -recommends nest protection strategies [43].
- Sarus Crane In the 2013 film "Sundar and Kittur," it was demonstrated that artificial nesting platforms in shallow marshes can facilitate successful Sarus Crane breeding, even within human-managed wetlands. They describe nesting behavior, threats, and how artificial nest boxes platforms can help [51–56].
- Lesser Adjutant Newton [39] and Gibbons and Hirons [20] suggest constructing large, elevated nest platforms on trees or poles for this cavity-nesting species. Highlights key sites for species such as Lesser Adjutant and discusses nesting site protection Table 17.3 [25].

## *17.8.1 Material Selection for Artificial Nest Boxes*

Artificial nest boxes can be constructed from diverse materials, such as wood, plastic, and repurposed items. The selection of materials should prioritize factors like durability, weather resistance, thermal properties, and ease of maintenance. Natural materials like untreated wood are often preferred to minimize breathability and potential toxicity within the artificial nest box environment.

This Table 17.4 illustrates species-specific insights into practical nest box construction, crucial for conservation efforts in both natural and human-altered wetlands. These designs are tailored to reflect nesting habits, their body size, risk of predator and fluctuations of water level. Physical attributes like shape, size and materials used in nest boxes affect occupancy of these aquatic avian species and breeding success.

**Table 17.3** Threatened water bird species, their International Union for Conservation of Nature (IUCN) conservation status, and proposed nest box interventions

| Threatened Waterbirds Species | IUCN Status | Recommended Artificial Nest Box Intervention | Explanations of the Species | References |
|---|---|---|---|---|
| Indian Skimmer | Endangered | Raised sandbanks or islets with open visibility | In India, skimmers breed on open sandbanks in large rivers susceptible to flooding and disturbances. Artificial sand islands or elevated islets offer flood-free nesting sites and significantly improve hatching success. | [4–8, 50] |
| Baer's Pochard | Critically Endangered | Shallow floating nest platforms with reed cover | Baer's Pochard prefers still, vegetated waters for nesting. Habitat loss has degraded wetlands, so artificial floating rafts with reeds can mimic natural cover and provide safe breeding areas. | [4–8, 65] |
| Black-bellied Tern | Endangered | Low elevated nest trays along riverbanks | These terns establish their nests on low sandbars and islands within river systems. The construction of artificial low trays or stabilised sand patches serves as a protective measure for these nests, safeguarding them from the encroachment of rising water and human interference. This intervention significantly enhances the survival rates of the chicks. | [27, 43] |
| Sarus Crane | Vulnerable | Nest platforms in shallow marshes with tall grasses nearby | Sarus cranes construct ground nests in marshes and rice fields. Platforms situated in water, surrounded by grasses, mitigate nest trampling, egg predation, and human interference. Additionally, these platforms provide protection against unforeseen floods. | [51–56] |
| Lesser Adjutant | Vulnerable | Large elevated wooden platforms on tree trunks or poles | These large storks necessitate tall nesting sites in isolated wetlands or forested edges. Artificial elevated platforms replicate high canopy nest sites, particularly in areas where large trees are scarce, facilitating breeding in semi-urban or degraded environments. | [25, 66, 67] |

**Table 17.4** Nest box design included shape, size, and materials used for selected waterbird species

| Water bird species | Nest boxes shape | Nest boxes dimensions (L × W × H) | Material used | Observations and explanations | References |
|---|---|---|---|---|---|
| Indian Skimmer (*Rynchops albicollis*) | Rectangular Open platform | 80 × 60 × 10 cm | River sand base on wooden raft | Open design mimics natural sandbanks- requires sun exposure and visibility | [4–8, 50] |
| Baer's Pochard (*Aythya baeri*) | Circular Floating Raft | Ø 70 cm, height 10–15 cm | Bamboo/ wood frame with reed mat base | Floating structure aids in flood-prone areas- reed covering adds camouflage | [27, 65] |
| Black-bellied Tern (*Sterna acuticauda*) | Shallow Nest Tray | 50 × 40 × 8 cm | Plywood tray with sand-gravel mix | Stabilizes nesting on fluctuating sandbanks- prevents egg rolling | [25] |
| Sarus Crane (*Antigone antigone*) | Elevated Circular Mound | 100–120 cm, height 20 cm | Straw, mud, grass over bamboo frame | Mimics natural nesting mounds in wetlands- placed in low-human-activity zones | [51–56] |
| Lesser Adjutant (*Leptoptilos javanicus*) | Large Platform with Guard Rails | 100 × 100 × 30 cm | Wooden planks on steel/ wooden poles | Elevated design suits large-bodied birds in degraded forest areas | [66, 67]; BNHS Reports |
| Purple Heron (*Ardea purpurea*) | Clustered Mini-platforms | 60 × 60 × 15 cm | Coir rope base on bamboo scaffold | Designed to support colonial nesters over reed beds; reduces competition | [33]; IBA Reports |
| Common Moorhen (*Gallinula chloropus*) | Box with Side Entrance | 40 × 30 × 30 cm | Waterproof plywood with straw lining | Concealed design protects from avian predators in urban wetlands | [28]; Bird Census India |

### 17.8.2 Strategies for Placement of Artificial Nest Boxes

The optimal location for the placement of artificial nest boxes is crucial for their prompt and successful establishment. When choosing a right place to live, consider factors like proximity to food, water, shelter from wind and sunlight and protection from predators. The spatial arrangement of artificial nest boxes for colonial nesting species plays a crucial role in formation of colony and stability. In flood-prone regions, it must be necessary to elevate these nest boxes on platforms. Floating islands offer safe nesting sites in open water but their elevation and orientation to sunlight and wind are crucial.

A scientifically planned strategy for placing nest boxes is crucial to maximize success of water bird conservation efforts through nest box installations. The placement of nest boxes should be in harmony with the behavioral ecology, nesting preferences and habitat utilization of the target conservatory species. For example the Sarus Cranes prefer shallow wetlands with tall grasses for nesting, while Indian Skimmers prefer open sandy isolated banks [57]. Properly spacing boxes reduces territorial disputes. Orientation must consider sunlight, wind and predator avoidance. In dynamic wetland systems, elevation from water level is crucial to prevent flooding, while for colonial nesters, grouped boxes can increase attraction rates. Research also suggests that placing boxes in low human traffic areas enhances occupancy and breeding success [36]. Periodic repositioning based on feedback helps adapt to changing habitat conditions and species responses, making it crucial for conservation planning. This study demonstrates the direct correlation between nest site availability and bird population size and breeding success [39].

### 17.8.3 Protection from Natural Enemies and Human Activities: A Holistic Nest Management Approach

Maintaining a safe distance from feeding locations and refraining from any form of human interference. Predator deterrence necessitates meticulous ethical and ecological deliberations, including the implementation of predator guards around artificial nest box supports, strategic positioning in less accessible areas, and the management of specific predator populations within certain habitats. To minimize disturbance to the target species, installation and monitoring of artificial nest boxes should be conducted with minimal disruption, except during the breeding season. Careful planning of access routes to artificial nest boxes and the use of non-invasive monitoring techniques are highly important.

## 17.9 Documented Case Studies

In the United States, the Wood Duck Nest Box Program has demonstrated remarkable success in augmenting Wood Duck populations through the widespread installation of predator-proof nesting boxes. This initiative was pioneered by Belrose in 1980. Cavity-nesting birds, such as Wood Ducks (*Aix sponsa*) and Common Goldeneyes (*Bucephala clangula*), are encountering a scarcity of suitable nesting sites due to deforestation and wetland degradation. Case studies in conservation action that have been documented and proven successful. Numerous ongoing conservation initiatives globally have substantiated the efficacy of nest boxes in augmenting threatened water bird populations. Notable examples include:

**Wood Duck (*Aix sponsa)***
The widespread use of nest boxes has been pivotal in the remarkable recovery of wood duck populations in North America. Their habitats had been severely degraded due to overhunting. Subsequently, artificial nest boxes, meticulously designed and strategically positioned to provide secure nesting cavities, have demonstrated significant success in substantially elevating breeding rates [3].

**Common Goldeneye (*Bucephala clangula)***
Those areas with limited natural tree cavities, nest boxes have proven highly effective in attracting breeding Goldeneyes and boosting local populations [16].

**Common Tern (*Sterna hirundo*) and Sandwich Tern (*Thalasseus sandvicensis*)**
Artificial nesting platforms have proven effective in boosting breeding colonies of these coastal terns. These platforms work as a safeguard against flooding and land-based. Predators as exemplified by Wayne's research. Strategic placement also helpful to establish colony formation in secure locations.

**Black-Necked Grebe (*Podiceps nigricollis*)**
These birds utilize floating nest platforms to establish stable breeding grounds in lakes and other areas with fluctuating water levels. The adaptation found in grebe boosts breeding success and stabilizes their colonies [18].

**The Ferruginous Duck (*Aythya nyroca*)**
This is an endangered duck species. In specific European wetlands nest boxes are used to provide secure nesting for this species, especially in areas where natural reed beds are disturbed [21]. These examples show how artificial nest boxes can be a valuable conservation tool when used wisely and tailored to the specific needs of the targeted species.

## 17.10 Limitations and Challenges for Artificial Nest Boxes

Artificial nest boxes are a promising conservation tool, but their use for water bird species is limited by challenges. The main concern is that a box suitable for one species may not be suitable for another due to differences in size, behavior and nesting preferences [36]. Flooding, extreme temperatures and high winds can damage boxes inaccessible due to environmental factors. Unprotected boxes in vulnerable areas increase the risk of predation and human disturbance leading to nest abandonment [33].

Invasive species like rats or squirrels can occupy boxes, reducing bird nesting sites. Long-term maintenance and monitoring require sustained effort, funding, and community support. Improper design or materials can cause thermal stress or parasite buildup, affecting breeding success. However, site-specific planning, research-based designs, and stakeholder involvement can mitigate these challenges and make nest box interventions more effective [11]. Artificial nest boxes, while offering substantial benefits, also present limitations and potential challenges that warrant careful consideration:

**Species-Specificity**
Artificial Nest Boxes, while a potential solution, are not universally effective. Their efficacy is limited to species that demonstrate a propensity for utilizing artificial structures. Most bird species are very particular about where they nest and are not likely to use artificial nest boxes.

**Habitat Dependence**
Artificial nest boxes primarily focus on providing nesting habitats. However, they cannot compensate for the decline in foraging habitats, which are equally essential for waterbird survival and breeding success. Therefore, it is crucial to integrate artificial nest box programs with comprehensive habitat conservation and restoration initiatives to ensure the long-term survival and breeding success of waterbirds.

**Requirements for Maintenance**
Regular monitoring and maintenance are essential for maintaining the effectiveness of artificial nest boxes once installed. This involves cleaning the old nesting material, repairing any damage part to the nest and continuously monitoring and controlling predators. Regular maintenance can reduce the risk of disease or parasitic infestation.

**Competition potential**
Sometimes, specific species nest boxes can be occupied by more dominant species. It's crucial to limit the availability of artificial nest boxes to the target species. Reducing competition in the species can be achieved by changing the design and placement of artificial nest boxes.

**Dependence and Habituality**
There are concerns that target species might become dependent on artificial nest boxes. These structures might prevent them from nesting naturally when they're used for natural nesting spots. Long-term conservation strategies should focus on restoring natural habitats and providing artificial nest boxes.

**Efforts and Cost-Effectiveness**
A successful artificial nest box program requires significant resources to implement and maintain. It is a big investment in materials, labor and constant monitoring. Cost-effectiveness requires thorough proper planning and prioritization.

## 17.11 Artificial Nest Monitoring Protocol

The monitoring, maintenance and adaptive management of artificial nest boxes are crucial. Before and during the breeding season thoroughly inspect artificial nest boxes. Gather detailed data on occupancy, clutch size, hatching success and predation to make informed choices. Continuous monitoring and maintenance are required for maintaining the long-term effectiveness of these nest boxes. Standardized protocols are essential for the success and sustainability of artificial nest box programs for water birds. Regular monitoring of occupancy rates, breeding success, clutch size, predation events and nest box condition helps in adaptive management and timely intervention for environmental stress, disease and structural damage [36].

Species-specific monitoring protocols ensure measurable and comparable conservation outcomes across regions. These protocols may include weekly breeding season inspections, photographic documentation, and remote sensing tools like camera traps or drones to reduce human disturbance [48]. Protocols should outline maintenance schedules, nest interior cleaning guidelines, and safe chick handling. Citizen scientists and local community members can participate using simplified checklists and apps. Monitoring helps determine if location, design, or orientation needs to be changed for better outcomes. These protocols are crucial for successful long-term nest box conservation.

### *17.11.1 Artificial Nest Maintenance Activities*

Routine maintenance is vital for the long-term success and functionality of artificial nest boxes for water bird conservation. Weather, water logging, and predator interference can damage nest boxes, causing debris, parasites, or mould that deter future occupancy [36]. Following ways are used:

**Clean the boxes**: After the breeding season, thoroughly clean the boxes to remove any debris, parasites, or eggs.

**Repair structural damage**: Identify and repair any structural damage to the nest, such as broken walls or loose supports.

**Remove Debris**: Remove any debris or objects that may be obstructing the nest or hindering the movement of the birds.

**Regular Inspections**: Ideally before and after each breeding season-are essential to assess the box's condition, check for structural integrity, and ensure they are free from invasive species such as rodents or insects. The maintenance of nest boxes also involves cleaning out old nesting material, applying protective coatings to wooden surfaces and reinforcing boxes to withstand environmental stresses [34]. Platforms may need adjustments to stay accessible and safe if water levels fluctuate. In high-use wetlands, some boxes may need replacement every 2–3 years, depending on material durability. Involving local communities and trained volunteers ensures wider coverage and long-term sustainability. A record-keeping system documents occupancy rates and repair history, aiding adaptive conservation planning.

### 17.11.2 Challenges and Limitations of Artificial Nests Boxes

Artificial nests, despite their benefits, present certain ecological and logistical challenges. One such challenge is the potential attraction of predators to their scent or visual cues if placed improperly. This can result in increased predation rates. Additionally, birds may be more susceptible to trapping in artificial nests. Furthermore, inter-species competition for resources within the same environment can be heightened in the presence of artificial nests. Climate change poses another vulnerability to artificial nests, as extreme weather events can negatively impact the occupancy rates of aquatic bird species.

Installing artificial nest boxes for waterbirds, despite their conservation value, faces challenges due to species-specific requirements for design, shape, and location. What benefits one species may harm another [36]. Environmental factors like flooding, extreme temperatures, and habitat loss can weaken and reduce the occupancy of birdhouses. Additionally, poorly designed or misplaced birdhouses increase the risk of predation, especially in areas accessible to snakes, rats, or humans [39].

Improper materials can cause thermal instability, mould, or parasite accumulation. Lack of long-term monitoring and maintenance often leads to neglect. In regions with low community awareness, vandalism or disturbance may affect bird usage. Some species may reject artificial structures, requiring trials and site-specific modifications. Science-based design, regular evaluation and community involvement are crucial for effective nest box interventions in waterbird conservation.

### 17.11.3 Involvement of Community and Citizen Science in Implementation of Artificial Nests Boxes

Involving local communities in bird nest box initiatives fosters increased awareness and support for monitoring and maintenance efforts. Notable examples include school-led bird box monitoring programs in Europe and conservation initiatives facilitated through ecotourism in wetland areas. Additionally, mobile applications such as Nestwatch, developed by the Cornell Lab of Ornithology, enable real-time collection of nesting data.

Local communities and citizen scientists actively participate in the success and sustainability of waterbird nest box initiatives. Their involvement in planning, installation, and monitoring increases awareness and fosters a sense of ownership and responsibility towards bird conservation [10].

Citizen science programs allow individuals to contribute data on nest occupancy, breeding success, and environmental threats, making conservation more inclusive and scalable, especially in resource-limited settings. Training workshops, school outreach, and eco-clubs spread best practices, while mobile apps and digital platforms facilitate easy field observation reporting [15].

Local stakeholders contribute traditional ecological knowledge, aiding in suitable location selection and understanding seasonal habitat changes. Community stewardship ensures long-term maintenance and protection of nest boxes from vandalism or neglect. This fosters community engagement, transforming conservation from a top-down approach to a collaborative and impactful movement.

### *17.11.4 Future Directions and Potential of Nest Box Implementation for Waterbirds Conservation*

The utilization of artificial nest boxes as a conservation strategy for endangered water bird species presents substantial future prospects. Ongoing scientific research and innovation can further enhance their efficacy and broaden their applicability. Artificial nest boxes for waterbird conservation have great potential, especially given habitat loss, climate change, and human activities. Future research should focus on designing nest boxes that are tailored to specific conditions, using materials like thermal insulation and smart monitoring devices, and placing them based on real-time ecological data [34].

Integrating GIS and remote sensing can help identify optimal nesting sites, especially in degraded or inaccessible wetlands. Long-term ecological studies are also needed to assess the population-level impacts of nest box use, including gene flow, site fidelity, and species interactions [36]. Public-private partnerships and policy integration can scale up nest box interventions in protected areas, urban wetlands, and community-managed habitats. Collaboration between researchers, local stakeholders, and conservation NGOs will drive innovation and knowledge sharing. If guided by science, supported by communities, and embedded in conservation policies, nest boxes can become a key component of global water bird recovery programs.

### *17.11.5 Using Advanced Materials for Artificial Nest Boxes: Selection and Benefits*

Exploring the utilization of novel materials and designs that prioritize sustainability, thermal efficiency, and seamless integration with the natural environment. This may encompass bio-based materials and designs that closely emulate natural cavities. The choice of advanced materials for artificial nest boxes significantly influences their durability, thermal regulation, and occupancy success—especially in challenging aquatic habitats. While traditional wooden boxes are still in use, they are prone to rot, warping, and pest infestations, particularly in moist wetland environments. As a result, conservationists now recommend eco-friendly composite materials such as PVC-based plastic wood, recycled HDPE (high-density polyethylene), and fiber-reinforced polymers, which are resistant to moisture, lightweight, and long-lasting [34].

These materials provide better insulation, maintaining stable temperatures inside the boxes, crucial for egg incubation and chick survival. Thermally coated metals with inner linings or biofoam composites are also being explored for high-humidity zones with frequent overheating or cold spells. Non-toxic, UV-stable materials that blend with nature reduce disturbance and enhance ecological efficiency and acceptance by various species. Modular, maintenance-free materials reduce long-term costs and improve ecological efficiency.

### *17.11.6 Smart Technology Integration for Effective Nest Box Implementation*

Integrating sensors and remote monitoring systems within nest boxes to collect data on occupancy rates, foraging success, microclimate conditions, and even individual bird behavior. This information is highly valuable for optimizing management strategies. Smart technologies integrated into artificial nest box management allow real-time monitoring, data collection and conservation planning. IoT sensors, thermal cameras and motion-activated devices monitor bird nests and control the temperature, humidity and detecting predators without disturbing the birds [41].

These systems remotely monitor breeding success, hatching rates and nesting duration and eliminating need for invasive field visits. RFID tagging aids in studying behavior of water birds their migration and site fidelity. Citizen scientists and conservationists can now contribute and access data in real-time through cloud-based data platforms and mobile apps. Solar-powered smart boxes equipped with cameras and data loggers are being used in wetlands to comprehensively study responses of water bird to environmental stressors. These innovative tools enhance monitoring efficiency and facilitate the generation of evidence based conservation.

### *17.11.7 Landscape-Scale Approaches for Effective Nest Box Implementation*

Expanding artificial nest box programs to cover larger areas, ensuring habitat connectivity and considering meta-population dynamics in to account. Strategically placing artificial nest box networks can facilitate and dispersal and exchange of genes among fragmented populations. Deploying nest boxes on a large scale is essential for long-term water bird population conservation. This strategy involves strategically placing nest boxes across wetlands to promote ecological connectivity and habitat diversity [38]. Instead of isolated interventions consider installing nest boxes in a network of protected zones taking in to account factors like proximity to foraging sites, predator density, human disturbance and hydrological dynamics. GIS-based habitat modeling helps to identify nesting gaps for targeted interventions. Buffer zones, corridors and habitat restoration along with nest box placement and enhance breeding success. Collaborating with forest departments, local communities and NGOs ensures integrated habitat management supporting meta-population dynamics and facilitating dispersal and colonization especially for threatened and reintroduced species.

### *17.11.8 Citizen Science Engagement in Nest Box Implementation*

Engaging citizen scientists in building, deploying and monitoring artificial nest boxes increases public awareness and conservation engagement providing valuable data. Citizen science is a fantastic way to get people involved in collecting data, monitoring nests, maintaining them and reporting on water bird nest box projects. It reduces burden on professional researchers and fosters local stewardship [10].

To encourage local participation, training workshops, mobile reporting tools and community-led nest box campaigns can be implemented. Data from citizen-generated platforms like e Bird and Nest Watch has proven effective in understanding species distribution, nesting success and behavioral trends [49]. In rural and wetland areas, school programs, bird festivals and nature clubs encourage water bird conservation and use of nest boxes. Combine these efforts with local governance and conservation policies ensures long-term sustainability of projects related to the nest boxes.

### *17.11.9 Climate Change Adaptation in Nest Box Implementation*

The design and deployment of artificial nest boxes are crucial for increasing populations of water bird by reducing the negative effects of climate change. To achieve this, install elevated platforms in flood-prone areas and use of shaded boxes to combat extreme heat. Artificial nest boxes are a proactive adaptation strategy for vulnerable water birds as climate change disrupts wetland hydrology, temperatures and nesting cycles. Extreme weather conditions like rising temperatures, floods and unpredictable monsoons disrupt natural nesting sites. Nest boxes, those are weather-resistant and elevated offer secure breeding options that are less affected by environmental changes [29].

Heat-resistant materials, proper ventilation and placing nest boxes in sheltered areas help regulate temperature and humidity and improving reproductive success. The adjustable nesting platforms can be moved to suit changing habitats and water levels. These nests are climate-resilient and adaptive with well-thought-out placement strategies. These are safeguards for water bird populations from climate change effects.

### *17.11.10 Integrated Habitat Restoration with Installation of Artificial Nest Box*

Restoring of an active habitat involves combining different elements to create both natural and artificial nesting sites. Integrating habitat restoration with nest box programs for water birds enhances their effectiveness. Artificial nest boxes should be part of a larger ecological restoration plan that restores wetlands, native vegetation and water quality [35].

To ensure long-term habitat suitability and nesting success, conservationists should place nest boxes, rehabilitate foraging grounds, create buffer zones and reintroduce indigenous plant species. Floating nest platforms have restored riparian zones supporting species like Baer's Pochard and Black-bellied Tern [23]. Restored habitats support insect and fish populations ensuring food availability near nesting sites. These integrated approaches enhance ecosystem resilience, improve reproductive success, and sustain bird populations over time.

## 17.12 Artificial Nest Boxes: A Practical and Cost-Effective Solution- A Comprehensive Analysis of Their Impact on Bird Conservation and Future Prospects

Well-managed nest box programs have successfully reversed local bird extinctions and enhanced breeding success within protected areas. Artificial nest boxes for birds have proven instrumental as beneficial research tools. Strategies to consider include employing GIS and drone technology for optimal nest box placement and constructing climate-resilient nesting structures. Nest boxes, despite their simplicity, represent a practical, cost-effective and scalable solution in the conservation toolkit for waterbirds. They bridge the gap between shrinking habitats and the pressing need for safe nesting sites, offering renewed hope for vulnerable aquatic avifauna. Nest boxes are a tangible and effective management tool for conserving threatened waterbird species. Habitat degradation, intensified predation, and environmental change are threatening waterbird populations. Providing artificial nesting structures addresses a fundamental limiting factor for reproductive success which secure and suitable nesting sites.

We have comprehensively evaluated the multifaceted applications of nest boxes across a diverse range of waterbird taxa. These include secure nesting cavities for protected platforms for colonial nesting terns on altered coastlines which previously had fragmented forest ecosystems due to ducks. The evidence presented shows that well-designed and strategically deployed nest box programs can greatly enhance breeding parameters. These parameters include larger clutches and higher hatching rates and improved fledging success. Nest boxes offer shelter from predators, protect against extreme weather and minimize human interference. These all help declining species recover and stabilize their populations.

This chapter stresses the importance of comprehensive approach to nest box initiatives. It highlights the need to consider species-specific nesting requirements like dimensions, materials, placement and orientation to optimize occupancy. Breeding outcomes require ongoing maintenance, predator control and integrating nest box programs in to broader habitat restoration strategies. Monitoring nest box usage gives insights into breeding trends, habitat preferences and intervention effectiveness, aiding adaptive management and improvement of techniques. Public engagement and education are unique opportunities offered by nest box programs. They raise awareness about the challenges faced by water birds and significance of conserving wetlands. Engaging local communities, volunteers and citizen scientists in nest box construction, deployment, and monitoring promotes stewardship and ensures long-term conservation success.

Nest boxes are not a magical solution for all water bird problems, but they are a fantastic and budget-friendly way to protect these birds. By creating and maintaining secure habitats for reproduction they help conserve threatened populations and support their resilience and recovery. Investing in these artificial shelters ensures the future of these iconic species and the health of their wetland habitats. Nest boxes provide a tangible and hopeful solution to ensure the survival of threatened water birds for future generations as we face escalating pressures on our natural world.

## 17.13 Conclusion

Adopting artificial nest boxes are simple yet effective conservation strategy that supports endangered aquatic and semi-aquatic species in human-affected environments. These effective tools are crucial for maintaining biodiversity in the face of increasing ecosystem pressures. Bird nest boxes are a valuable and effective management strategy for conserving threatened aquatic bird species that lack suitable nesting habitats. Nest boxes, when carefully designed, strategically implemented and integrated in to conservation efforts can be very effective. They can provide essential shelter and help threatened aquatic birds to raise their chicks and greatly help their populations recovery.

The artificial nest boxes installation highlight importance of conserving aquatic bird species. Targeted interventions like nest box programs are crucial for safeguarding aquatic ecosystems from growing threats among these vulnerable and ecologically important species. To maximize the long-term benefits of this promising conservation technique, ongoing research, adaptive management and addressing the underlying causes of habitat loss and degradation are essential.

## References

1. Anderson SH, Gutzwiller KJ (1994) Habitat threats to ground-nesting birds and mitigation strategies. Conserv Biol 8(2):397–409
2. Ardia DR, Pérez JH, Clotfelter ED (2006) Experimental cooling affects the duration of incubation in Tree Swallows (Tachycineta bicolor). J Avian Biol 37(2):137–142
3. Bellrose FC, Holm DP (1994) Ecology and management of the wood duck. Stackpole Books, Mechanicsburg
4. BirdLife International (2023) Species factsheets and conservation tools for globally threatened waterbirds. www.birdlife.org
5. BirdLife International (2023) Species factsheets for threatened aquatic birds. www.birdlife.org
6. BirdLife International (2023) Species factsheets: cavity-nesting birds. Available at: www.birdlife.org
7. BirdLife International (2023) Species factsheets: Indian Skimmer & riverine birds. www.birdlife.org
8. BirdLife International (2023) Species factsheets. Retrieved from https://www.birdlife.org
9. BirdLife International (2024) Species factsheets for Indian Skimmer, Baer's Pochard, and Black-bellied Tern. Retrieved from https://www.birdlife.org
10. Bonney R, Cooper CB, Dickinson J, Kelling S, Phillips T, Rosenberg KV, Shirk J (2009) Citizen science: a developing tool for expanding science knowledge and scientific literacy. Bioscience 59(11):977–984
11. Browne SJ, Aebischer NJ, Yfantis G (2015) Nest box design and placement: effects on occupancy and breeding success of cavity-nesting birds. Bird Study 62(3):369–379
12. Carney KM, Sydeman WJ (1999) A review of human disturbance effects on nesting colonial waterbirds. Waterbirds 22(1):67–79
13. Choudhury BC (2000) Conservation of the Indian Skimmer Rynchops albicollis in the National Chambal Sanctuary. BNHS Technical Report
14. Davidson NC (2014) How much wetland has the world lost? Long-term and recent trends. Mar Freshw Res 65(1):93–101
15. Devictor V, Whittaker RJ, Beltrame C (2010) Beyond scarcity: citizen science programmes as useful tools for conservation biogeography. Divers Distrib 16(3):354–362
16. Eadie JM, Mallory ML, Lumsden HG (1995) Technical note: design and use of nest boxes for cavity-nesting ducks. Wildl Soc Bull (1973–2006) 23(4):793–803
17. Fahrig L (2003) Effects of habitat fragmentation on animal populations. Annu Rev Ecol Evol Syst 34(1):487–515
18. Fjeldså J (1977) Guide to the young of European waterfowl. Skyttegaard Boghandel
19. Gibbons DW, Boesch DF (2006) Habitat availability and nesting success of waterbirds in restored wetlands. Wetl Ecol Manag 14(2):111–121
20. Gibbons DW, Hirons GJM (1986) Artificial nest boxes for ducks: a management success. Wildfowl 37:102–107
21. Green AJ, Hughes B (2001) Nest box use by ferruginous ducks *Aythya nyroca* in Spain. Bird Study 48(3):392–395

22. Griffiths R, Daan S, Dijkstra C (1996) Environmental effects on the nest-box microclimate and consequences for avian reproductive performance. Physiol Zool 69(6):1234–1250
23. International (2022) Waterbird habitat restoration and conservation strategies. Retrieved from: www.birdlife.org
24. IPCC (2022) Climate change 2022: impacts, adaptation, and vulnerability. In: Pörtner H-O, Roberts DC, Tignor M, Poloczanska ES, Mintenbeck K, Alegría A, Craig M, Langsdorf S, Löschke S, Möller V, Okem A, Rama B (eds) Contribution of Working Group II to the sixth assessment report of the Intergovernmental Panel on Climate Change. Cambridge University Press
25. Islam MZ, Rahmani AR (2004) Important bird areas in India: priority sites for conservation. Bombay Natural History Society and BirdLife International
26. IUCN (2024) The IUCN red list of threatened species. Version 2024-1. Retrieved from https://www.iucnredlist.org
27. IUCN Red List of Threatened Species (2023) Assessments for *Rynchops albicollis*, *Aythya baeri*, *Sterna acuticauda*, *Antigone Antigone*, *Leptoptilos javanicus*. Retrieved from https://www.iucnredlist.org
28. Jha S et al (2021) Urban wetland bird nesting adaptations in North India. Urban Ecology Reports
29. Johnson DH, Niemuth ND, Estey ME (2020) Conserving birds in the face of climate change: the role of artificial nesting structures. Clim Change Biodivers Conserv 162(2):349–364
30. Kearns LJ, Fraser JD, Dinsmore JJ (1998) A comparison of nest success and site characteristics of American Oystercatchers on barrier islands in North Carolina. Wilson Bull 110(4):521–533
31. Kingsford RT, Porter JL, Salehi A (2017) Extreme waterbird die-offs in Australia: climate change and habitat degradation. Glob Chang Biol 23(12):5265–5280
32. Kiran, Singh D, Kour A, Priya, Delu V, Kumar R (2022) Community composition and status of avian diversity at Campus and Agricultural landscapes of Chaudhary Charan Singh Haryana Agricultural University, Hisar (Haryana). J Appl Nat Sci 14(4):1130–1140. https://doi.org/10.31018/jans.v14i4.3784
33. Kumar A, Bhardwaj S, Saini R (2017) Breeding ecology of Purple Heron in Delhi NCR
34. Lambrechts MM et al (2010) The design of artificial nest boxes for the study of secondary hole-nesting birds: a review of methodological inconsistencies and potential biases. Acta Ornithol 45(1):1–26
35. Ma Z, Melville DS, Liu J, Chen Y, Yang H, Ren W (2021) Restoration of coastal wetlands enhances waterbird conservation: lessons from East Asia. Conserv Biol 35(1):98–108
36. Mainwaring MC (2015) The use of artificial nest boxes by birds in conservation and research: a review. Ecol Evol 5(16):3891–3906
37. Martinez-Abraín A, Oro D, Jiménez J, Stewart G, Pullin A (2009) A systematic review of the effectiveness of artificial nest-boxes for cavity-nesting seabirds and landbirds. J Ornithol 150(4):763–769
38. McClure CJW, Anderson DL, Franke A (2017) Landscape-scale habitat modeling for the placement of artificial nests for threatened raptors. Condor 119(2):270–285
39. Newton I (1994) The role of nest sites in limiting the numbers of hole-nesting birds: a review. Biol Conserv 70(3):265–276
40. Newton I (1998) Population limitation in birds. Academic, London
41. Owen E, McDougall M, Adriaenssens B (2021) Smart conservation: integrating sensor technology and real-time monitoring in bird nest boxes. Eco Inform 62:101261
42. Pendlebury CJ et al (1997) Nest-box use and reproductive success in Common Goldeneye
43. Rahmani AR (1996) Status of black-bellied tern (Sterna acuticauda) in India. J Bombay Nat Hist Soc 93(2):225–228
44. Rahmani AR (2000) Conservation of little terns. BNHS Coastal Birds Report
45. Rodgers JA, Schwikert ST (2002) Buffer-zone distances to protect foraging and loafing waterbirds from human disturbance in Florida. Waterbirds 25(3):437–443
46. Ronconi RA et al (2010) Nest box management for Great Cormorants in coastal reserves

47. Semel B, Sherman PW (2001) Wood duck nesting success. Condor 103:607–617
48. Spotswood EN, Goodman KR, Carlisle J et al (2012) How safe is mist netting? Evaluating the risk of injury and mortality to birds. Methods Ecol Evol 3(1):29–38
49. Sullivan BL, Wood CL, Iliff MJ, Bonney RE, Fink D, Kelling S (2014) eBird: a citizen-based bird observation network in the biological sciences. Biol Conserv 169:31–40
50. Sundar KSG (2004) Conservation of Indian Skimmer (*Rynchops albicollis*) in India. Forktail 20:13–18
51. Sundar KSG, Kittur S (2013) Can ecological research help reverse declines of Sarus Cranes? J Threat Taxa 5(6):3989–3995
52. Sundar KSG, Kittur S (2013) Breeding ecology of Sarus Cranes (Grus antigone) in agricultural landscapes: implications for conservation planning. Waterbirds 36(4):459–467
53. Sundar KSG, Kittur S (2013) Can conservation breeding help the Baer's Pochard? Waterfowl Specialist Group Newsletter, IUCN
54. Sundar KSG, Kittur SM (2013) Can wetlands maintained for human use also help conserve biodiversity? Ecol Appl 23(3):541–554
55. Sundar KSG, Kittur S (2013) Sarus cranes nesting in agricultural wetlands. J Threat Taxa 5(6): 3989–3995
56. Sundar KSG, Kittur S (2013) Can wetlands maintained for human use also help conserve biodiversity? A case study of Sarus Crane in India. Biol Conserv 167:272–279
57. Sundar KSG, Subramanya S (2010) Bird use of wetlands in agricultural landscapes: insights for conservation planning in South Asia. Biodivers Conserv 19(7):1863–1883
58. Sundar KSG, Kittur S, Sharma D (2021) Artificial nesting platforms improve breeding success of Sarus Cranes in human-dominated landscapes. Ecol Solut Evid 2(1):e12037
59. Vié J-C, Hilton-Taylor C, Stuart SN (eds) (2009) Wildlife in a changing world: an analysis of the 2008 IUCN Red List of Threatened Species. IUCN
60. Whelan CJ, Wenny DG, Marquis RJ (2008) Ecosystem services provided by birds. Ann N Y Acad Sci 1134(1):25–60
61. Wu Y, Zhang W, Yong F, Zhou D, Cui P (2020) Waterbirds' coastal habitat in danger. Science 369(6506):928–929
62. Yadav R, Gupta RK, Kumar R, Kaur T (2022) Assessment of toxicity of lead and nickel on the biochemical and immunological parameters of earthworm, Eudrilus eugeniae. Agric Sci Dig 44(5):922–929
63. Yadav R, Gupta RK, Kumar R, Kaur T, Moun R (2022) Biochemical response of earthworm, *Eisenia fetida* to heavy metals toxicity. J Appl Nat Sci 14(3):990–998
64. Yodha K, Singh D, Kour A, Yadav R, Kumar R (2024) Species diversity and abundance of birds in ber (*Zizyphus mauritiana*) crop. Asian J Conserv Biol 13(1). https://doi.org/10.53562/ajcb.83683
65. Zhang L, Cong H, Zhang ZW, Lei G (2019) Status and conservation of Baer's Pochard Aythya baeri in China and throughout its range. Wildfowl 69:160–179
66. Zöckler C, Htin Hla T, Clark N (2018) Conservation of the critically endangered Spoon-billed Sandpiper: artificial nest protection and community-based monitoring. Wader Study 125(2): 120–129
67. Zöckler C, Htin Hla T, Hein HT (2018) Conservation needs of Lesser Adjutants and other large waterbirds in Southeast Asia. Wetlands International Technical Report

# Chapter 18
# Nestling Dreams: Unveiling the Breeding and Conservation Mysteries of Aquatic Birds

**Manjinder Kaur, Mohit Pun, and Nisha Vashishat**

**Abstract** This chapter delves into the reproductive habits and conservation issues facing aquatic birds, focusing on species found in freshwater and coastal areas. It covers a range of breeding behaviors, such as how birds choose their nesting sites, perform courtship rituals, and care for their young, from incubation to the fledgling stage. The text emphasizes how critical these early life stages are for a bird's long-term survival. The discussion also highlights the major threats to these birds' breeding success and population health. It examines environmental pressures like habitat loss, climate change, water pollution, and human activity. Using examples from different parts of the world, the chapter shows how some species are able to cope with these challenges while others struggle, leading to changes in their breeding cycles and survival rates. Special attention is given to species that rely on specific nesting habitats and are at risk from altered water flow, human expansion, and fragmented ecosystems. Finally, the chapter evaluates current conservation efforts, including restoring wetlands, using artificial nesting platforms, and implementing policies to protect habitats. It also stresses the importance of community involvement, citizen science projects, and incorporating traditional knowledge into modern conservation strategies. By showcasing successful projects, the chapter argues for the need for flexible, multi-faceted approaches to protect breeding habitats and ensure the long-term survival of aquatic bird populations.

**Keywords** Breeding behaviors · Environmental pressures · Artificial nesting platforms and aquatic bird populations

M. Kaur · M. Pun (✉) · N. Vashishat
Department of Zoology, College of Basic Sciences and Humanities, Punjab Agricultural University, Ludhiana, Punjab, India



R. Kumar et al. (eds.), *Ecology, Conservation of Aquatic Birds*, Wetlands: Ecology, Conservation and Management 15, https://doi.org/10.1007/978-3-032-17125-2_18

## 18.1 Introduction

**Diversity of Aquatic Birds: From Shore to Open Ocean**

Freshwater habitats such as lakes, rivers, marshes, and ponds support a wide array of bird species, each shaped by the ecological demands of their environment. Ducks, herons, grebes, bitterns, and rails are among the most prominent freshwater species that have evolved to survive in these often seasonal and dynamic ecosystems. Many of these birds demonstrate a strong relationship between form and function. For example, the heron's long neck and dagger-like bill allow it to spear fish and amphibians with precision, a strategy that demands both patience and agility [46]. Dabbling ducks, such as mallards, feed at the water's surface by tipping forward to graze on aquatic vegetation and invertebrates, while diving ducks like pochards submerge to forage on deeper substrates. Freshwater birds also rely heavily on aquatic vegetation for cover and nesting material, making them highly sensitive to shifts in plant communities caused by pollution, invasive species, or fluctuating water levels.

In contrast, marine birds inhabit coastal regions, rocky shorelines, islands, and the expansive open ocean. These birds must contend with harsher conditions, including salt exposure, strong winds, and often vast distances between feeding and nesting grounds. As a result, marine birds such as gulls, puffins, petrels, and albatrosses have evolved exceptional adaptations for salt tolerance and efficient long-range travel.

One of the most distinctive marine bird adaptations is the presence of salt glands, which filter out excess sodium and allow the birds to drink seawater without harm [76]. Their wings are often long and narrow, as seen in the albatross, enabling them to glide effortlessly for hours or even days over open water with minimal energy expenditure [25]. Puffins and murres, meanwhile, exhibit dense plumage and compact bodies that allow for agile swimming beneath the waves in pursuit of fish.

Wading birds are specialists in the shallow margins of aquatic environments—zones rich in invertebrates and small fish. Species like stilts, avocets, plovers, and sandpipers have evolved elongated legs and fine-tipped beaks that enable them to navigate mudflats, tidal pools, and marsh edges. Their feeding methods vary: some probe into wet soil with sensitive bills that detect hidden prey, while others use rapid pecking to snatch food from the surface [40].

Their behavior often synchronizes with tidal cycles, and many are long-distance migrants, traveling thousands of kilometers between breeding and wintering grounds. These journeys depend on the availability of key stopover sites, making habitat preservation across continents crucial for their survival.

Swimmers, including many species of ducks, geese, and swans, have evolved buoyant bodies and strong, webbed feet that make them natural paddlers. Their foraging strategies vary widely—from surface feeding to grazing on submerged vegetation and even filter feeding in some cases. For example, shovelers use their wide, spatula-shaped bills to strain tiny organisms from the water.

These birds often form social groups and can be found in large flocks during migration or winter, where communal behavior can aid in predator detection and

**Fig. 18.1** Red-throated Diver / Red-throated Loon (*Gavia stellata*) and Black-winged Stilt (*Himantopus himantopus*) [7]

foraging efficiency. Many species also exhibit strong site fidelity, returning to the same breeding or wintering areas year after year.

Among the most specialized aquatic birds are the divers—species that have adapted to hunt underwater with impressive efficiency. Birds such as loons, cormorants, auks, and penguins (in the southern hemisphere) possess dense bones that reduce buoyancy, making diving easier. Their bodies are torpedo-shaped, and their feet are set far back on the body, aiding in powerful propulsion through water. Cormorants, for example, are adept underwater hunters, using sharp eyesight and swift movements to catch fish [55]. Unlike many other aquatic birds, their feathers are less waterproof, which helps reduce buoyancy but requires them to dry out frequently—hence their iconic wing-spreading behavior on rocks and docks. Loons, meanwhile, are streamlined and nearly silent underwater, using stealth and speed to outmaneuver their prey. Red-throated Diver / Red-throated Loon (*Gavia stellata*) bird has a slender body, sharp bill, and a distinct rusty-red throat patch during the breeding season. It is mostly grey and white with red eyes. It is found in freshwater lakes, ponds, and coastal waters, especially in Arctic and sub-Arctic regions. It is an expert diver, feeding mainly on fish and aquatic invertebrates, using its speed underwater to catch prey (Fig. 18.1)

## 18.2 Iconic Species and Their Global Distribution

Aquatic birds are distributed across every continent and climate zone, displaying a remarkable range of adaptations that reflect the environments in which they live. They play vital ecological roles and, in many cultures, carry symbolic significance as representations of the richness of the natural world. Highlighting a few globally recognized species illustrates both the ecological breadth of aquatic bird life and the diversity of habitats they occupy.

- **Flamingos:** Recognized for their pink plumage and graceful, curved necks, flamingos serve as symbols of wetland biodiversity. Inhabiting saline or alkaline lakes across Africa, South America, and parts of Asia, they possess a specialized filter-feeding mechanism. Using lamellae—comb-like structures inside their beaks—they extract algae, diatoms, and small crustaceans from the water [30]. Their coloration derives from carotenoid pigments in their diet, making plumage brightness a visual indicator of ecosystem health.

- **Emperor Penguins:** The largest and heaviest penguin species, emperor penguins are a prime example of survival in extreme environments. Endemic to Antarctica, they endure severe winters by incubating their eggs on their feet beneath a brood pouch while forming dense huddles to conserve heat. Their thick plumage, substantial fat reserves, and cooperative social behaviors are critical adaptations for surviving subzero conditions [86].
- **Great Egrets:** Found across freshwater and brackish wetlands in the Americas, Europe, Africa, Asia, and Australia, great egrets are distinguished by their tall stature and striking white plumage. With sharp yellow bills and deliberate hunting movements, they are highly effective fish predators. Their elegance and wide distribution have also established them as a flagship species in wetland conservation campaigns [51].

These emblematic species demonstrate how aquatic birds have conquered environments ranging from the tropics to the poles. Their global distribution not only highlights their adaptability but also their exposure to a wide array of human-induced threats—habitat loss, climate change, pollution, and overfishing. A Consequently, conserving these species is not just about protecting individual birds but safeguarding the intricate ecosystems they represent [6].

## 18.3 Habitats and Adaptations: Thriving in Wetlands, Rivers, and Seas

Aquatic birds have evolved alongside their watery habitats, developing remarkable behavioral and physical traits to exploit diverse and sometimes extreme environments. From the lush fertility of wetlands to the icy shores of the Arctic and even concrete-lined urban canals, these birds are masters of adaptation.

### (a) **The Role of Wetlands in Bird Reproduction**

Wetlands are among the most productive ecosystems on Earth—and they are absolutely vital for the life cycles of countless aquatic bird species. These areas provide abundant nesting material, dense cover for concealment, and rich feeding grounds full of insects, fish, amphibians, and aquatic plants.

Birds such as **bitterns**, **moorhens**, and **rails** often rely on the tall vegetation of marshes to hide their nests from predators. **Storks**, **herons**, and **ibises**, on the other hand, prefer elevated nest sites, such as trees or shrubs standing in shallow flooded zones [83]. Seasonal wetland flooding not only creates a surge in available resources, but it also acts as a biological signal that triggers courtship and nesting behaviors.

Unfortunately, wetlands are disappearing at an alarming rate due to urban expansion, agriculture, drainage, and pollution. This loss is one of the greatest threats facing aquatic birds today, as it directly affects their breeding success and long-term population viability [61, 88].

(b) **Extreme Breeding Grounds: Arctic, Desert, and Urban Waterways**

Some aquatic birds have taken adaptation to an entirely new level by thriving in the most unexpected corners of the world.

- **Arctic tundras**, despite their short growing seasons, serve as critical breeding grounds for migratory species such as **red-necked phalaropes** and **snow geese**. These birds take advantage of the intense summer productivity, feeding on abundant insect life and raising chicks in a matter of weeks [73].
- In **desert regions**, species like the **Egyptian goose** have adapted to isolated wetlands or oases. These birds nest in crevices, cliffs, and even abandoned burrows, relying on their ability to travel long distances in search of water and forage.
- Surprisingly, **urban waterways** have become unconventional sanctuaries for birds such as **coots**, **grebes**, and **mallards**. These birds utilize artificial lakes, drainage ditches, and canals, and may even nest in man-made structures like docks, bridge pilings, or floating platforms. Although these urban refuges offer some protection from predators, they come with challenges such as pollution, human disturbance, and altered food webs [60].

These wide-ranging habitats demonstrate the incredible flexibility and resilience of aquatic birds. However, thriving in extreme or artificial environments often means living on the edge of ecological stability.

(c) **Unique Adaptations: Waterproof Feathers, Webbed Feet, and More**

Life in aquatic environments requires a host of specialized traits, and aquatic birds have developed a fascinating toolkit to meet these demands:

- **Waterproof Feathers**: Birds maintain the integrity of their feathers through meticulous preening, spreading oils secreted from the uropygial gland. This coating helps repel water, maintains insulation, and preserves buoyancy in species like ducks and swans [35].
- **Webbed and Lobed Feet**: These provide efficient propulsion in water. While most swimmers like ducks have fully webbed feet, species such as grebes and coots have lobed toes that offer flexibility for both swimming and walking on soft substrates.
- **Dense Bones and Hydrodynamic Bodies**: In contrast to most birds, which have light skeletons for flight, diving birds like loons and cormorants have denser bones that reduce buoyancy, making underwater hunting easier. Their streamlined shapes reduce drag and increase agility beneath the surface.

**Table 18.1** Comparative table of aquatic bird adaptations

| Category | Examples | Key adaptations |
|---|---|---|
| Freshwater birds | Ducks, Herons, Grebes | Specialized beak shapes; feeding behaviors tailored to diets; patient hunting techniques |
| Marine birds | Gulls, Puffins, Albatrosses | Salt glands to excrete salt; streamlined bodies; long wingspans for efficient flight |
| Waders | Stilts, Sandpipers | Long legs and slender beaks for foraging in shallow waters; probing substrate for invertebrates |
| Swimmers | Ducks, Geese | Webbed feet for propulsion; buoyant bodies; feeding habits from dabbling to grazing |
| Divers | Loons, Cormorants, Auks | Streamlined bodies; strong muscles for diving; hydrodynamic bodies to minimize drag |
| Iconic species | Flamingos, Emperor Penguins, Great Egrets | Specialized feeding beaks; adaptations to extreme cold; elegant plumage for various habitats |

- **Specialized Feeding Tools**: Some birds have evolved extraordinary tools for capturing prey. **Pelicans** use their large, expandable throat pouches to scoop up fish. **Skimmers** have lower mandibles longer than their upper ones, allowing them to skim the water's surface for small fish and crustaceans.
- **Advanced Physiology**: Many divers, including penguins and auks, can regulate oxygen use during dives, redirecting blood flow to essential organs. Some species also possess adaptations for underwater vision, enabling them to spot and capture prey in murky conditions [78, 49, 50] (Table 18.1).

## 18.4 Breeding Strategies Across Species

Aquatic birds exhibit a remarkable array of breeding systems, reflecting the diversity of their environments and life histories. These strategies—ranging from strict monogamy to complex polyandry and cooperative breeding—demonstrate how reproductive success is shaped by ecological pressures, resource availability, and evolutionary trade-offs.

## 18.5 Cooperative Breeding: Shared Effort in Challenging Environments

Cooperative breeding, where individuals other than the breeding pair assist in raising young, is relatively rare but has evolved in response to specific ecological conditions. This system typically emerges in environments where territory, food, or nesting sites are limited, and group cooperation can enhance the chances of offspring survival.

Though more common in terrestrial birds, some aquatic and semi-aquatic species—particularly those in tropical or arid zones—have been documented engaging in cooperative behaviors. For example, **African jacanas (Actophilornis africanus)** and certain species of **rails** and **moorhens** may involve subadult offspring or non-breeding individuals in chick-feeding or nest defense. Helpers can increase the reproductive success of the breeding pair, while also gaining experience that improves their own future breeding prospects.

In some cases, cooperative breeding may also reflect kin selection, where individuals help raise related offspring, indirectly passing on shared genetic material.

## 18.6 Nesting Colonies and Social Structure

Breeding strategy is often closely tied to nesting behavior. Many aquatic birds, including **cormorants**, **pelicans**, and **terns**, breed in dense colonies, where hundreds or even thousands of pairs nest in close proximity. While this creates intense competition for space and increases the risk of disease or predation, it also offers advantages such as predator dilution, social information exchange (e.g., locating feeding grounds), and easier mate finding.

Colonial nesting species may show varied pair bonds—from monogamous pairings within the colony to opportunistic mating. In contrast, solitary nesters such as **bitterns** or **grebes** often rely on cryptic behavior and hidden nests to reduce predation risk, which typically necessitates strong pair coordination and high parental investment (Table 18.2).

**Seasonal vs. Year-Round Breeding Cycles**

**Table 18.2** Breeding systems in aquatic birds

| Breeding system | Description | Representative species |
|---|---|---|
| Monogamy | Pair bond between one male and one female for at least one breeding season. | Swans, Geese, Loons |
| Polygamy | Individuals have multiple mates during a breeding season. | Red-necked Phalarope, Pheasant-tailed Jacana |
| Cooperative breeding | Additional individuals assist a breeding pair in raising offspring. | Some species of Kingfishers, Acorn Woodpeckers |

## 18.7 Timing of Reproduction and Disruption by Climate Change

The reproductive cycles of aquatic birds are closely aligned with environmental cues, balancing physiological readiness with ecological opportunity. Most species are seasonal breeders, timing reproduction to coincide with periods of peak food availability and favorable weather. This synchronization ensures that chicks hatch during periods of resource abundance, such as insect emergence, fish spawning, or aquatic vegetation growth, thereby improving survival rates.

## 18.8 Colonial Nesting: Safety in Numbers

- Collective defense against predators through mobbing and shared vigilance.
- Synchronization of breeding, which overwhelms predators with a temporary surplus of eggs or chicks.
- Information sharing on foraging sites, observed in brown pelicans (*Pelecanus occidentalis*) and common murres (*Uria aalge*).

In South Asia, little terns (*Sternula albifrons*) and small pratincoles (*Glareola lactea*) establish communal nests on riverine sandbanks.

### *18.8.1 Solitary Nesting: Stealth and Seclusion*

- Reduced intraspecific competition for food and space.
- Lower risk of disease transmission common in dense colonies.
- Decreased attraction of predators due to less noise and activity.

Research on Adélie penguins (*Pygoscelis adeliae*) at Cape Crozier shows that while most individuals nest in large colonies, some select peripheral or isolated sites. These outliers may benefit from reduced aggression and trampling, though they face greater exposure to predators and environmental stress.

## 18.9 Environmental Influence on Nesting Strategy

Whether a species adopts colonial or solitary nesting often depends on habitat stability, food distribution, and nest site availability. For example, semipalmated sandpipers (*Calidris pusilla*), usually scattered nesters in Arctic tundra, sometimes aggregate more tightly when suitable sites are limited. However, such clustering may increase predation pressure and nest failure, as recent observations suggest [77].

**Table 18.3** Nesting strategies comparison

| Nesting strategy | Advantages | Disadvantages |
|---|---|---|
| Colonial nesting | Enhanced protection through group vigilance; increased mate availability. | Higher competition for resources; increased risk of disease transmission. |
| Solitary nesting | Reduced competition; lower risk of disease spread. | Less protection from predators; limited mate availability. |

**Fig. 18.2** Black-headed Ibis (*Threskiornis melanocephalus*) and Wood Sandpiper (*Tringa glareola*) [71]

Black-headed Ibis (*Threskiornis melanocephalus*) is a medium-to-large wading bird with a bare black head and neck, white body, and a long down-curved black bill. In the photo, iridescent sheen on wings may be visible due to lighting. Commonly found in marshes, wetlands, paddy fields, and lakes across South and Southeast Asia. Wood Sandpiper (*Tringa glareola*) is a medium-sized wader with a slender build, brown speckled upperparts, white underparts, and a straight, thin bill. Legs are usually greenish. It is found in freshwater wetlands, marshes, flooded fields, and river edges, especially during migration.

Species may also exhibit flexible strategies in response to habitat changes. In human-altered environments, some birds have shifted to semi-colonial behavior—utilizing bridges, rooftops, or artificial islands for nesting (Table 18.3 and Fig. 18.2).

## 18.10 Mating Rituals and Pair Bonds: Love on the Water

Mating rituals and the formation of pair bonds are among the most captivating and complex behaviors observed in aquatic birds. Far from being mere spectacles, these rituals play essential roles in ensuring reproductive success, particularly in environments where timing, cooperation, and synchronization are crucial. In species that require biparental care, the strength of a pair bond can directly affect chick survival, making courtship rituals a vital precursor to breeding [38].

## 18.11 Elaborate Displays: Dances, Calls, and Gifts

Aquatic bird courtship often involves highly specialized and ritualized displays designed to communicate fitness, readiness to mate, and compatibility. These behaviors can be visual, auditory, tactile, or even involve material offerings, and are shaped by both sexual selection and species-specific ecological constraints [42]. (see Table 18.4).

## 18.12 Climate and Courtship Timing

- **Phenological shifts** (changes in seasonal timing due to climate change) are causing mismatches between peak food availability and courtship behavior. If birds arrive at breeding grounds too early or late, mating may be delayed or skipped altogether [13].
- **Unseasonal storms or heatwaves** can interfere with display rituals. For example, excessive wind can prevent aerial courtship displays in **terns**, while flooding can destroy early nesting sites [62].

## 18.13 Habitat Quality and Pair Formation

- Research in *Ecology Letters* [19] shows that habitat degradation reduces courtship frequency and quality. In **wading birds** like herons and spoonbills, males in poorer territories perform fewer displays and are less likely to attract mates.
- In **variable environments**, birds may opt for **shorter-term bonds** to cope with the uncertainty, increasing the frequency of mate switching between seasons or even within a single breeding period [32].

**Table 18.4** Courtship behaviours in selected aquatic bird species and their functional significance in mate bonding and reproductive success

| Species | Courtship behavior | Function | Reference |
|---|---|---|---|
| Great Crested Grebe | "Weed dance" with synchronized head-shaking | Bond formation and mate assessment | Kear [46] |
| Western Grebe | "Rushing" across water surface | Demonstrates stamina and coordination | Storer and Nuechterlein [79] |
| Swans and Geese | Trumpeting duets | Bond reinforcement | Jones and Hunter [45] |
| Black-headed Gull | Male offers food (fish, insects) | Stimulates ovulation and shows provisioning ability | Bolund et al. [12] |
| Cormorants | Presentation of nesting materials (sticks, seaweed) | Signals nest-building readiness and commitment | Nelson [65] |

## 18.14 Environmental Stress and Pair 'Divorce'

- Aquatic birds are generally known for high pair fidelity, particularly species like **mute swans (*Cygnus olor*)** and **albatrosses**. However, recent studies suggest that **climate-induced stress**—such as food scarcity, rising temperatures, and habitat loss—is leading to increased rates of pair separation [33].
- A 2022 study in *Nature Climate Change* on **black-browed albatrosses** found a significant rise in "divorces" in warmer years, hypothesized to result from **failed breeding attempts** tied to environmental stress [33] (Table 18.5).

## 18.15 Role of Long-Term Bonds in Breeding Success

Long-term monogamous pair bonds offer numerous reproductive advantages in aquatic bird species. These partnerships, often spanning several breeding seasons or a lifetime, are key to breeding success in species that require high parental investment.

## 18.16 Benefits of Long-Term Pair Bonds

- **Improved Coordination**: Pairs that have bred together previously are more efficient in nest building, incubation shifts, and chick provisioning. This coordination minimizes the time eggs or chicks are left unguarded [11].
- **Increased Breeding Success**: Research on **black swans** and **great cormorants** indicates that stable pairs have **higher chick survival rates** and **greater fledging success** compared to new or unstable pairings [43].
- **Social Learning and Familiarity**: Experienced mates can **learn each other's signals and behaviors**, reducing conflict and stress during critical stages of breeding [22] (Table 18.6).

**Table 18.5** Environmental influences on mating behavior

| Factor | Effect on courtship and pair bonds | Example species/ study | Reference |
|---|---|---|---|
| Phenological shifts | Mating delay or failure due to food mismatch | Arctic and temperate breeders | Both et al. [13] |
| Storms and heatwaves | Disruption of aerial or aquatic displays, nesting failures | Terns, colonial seabirds | Møller et al. [62] |
| Habitat degradation | Reduced display frequency, lower pair formation success | Herons, spoonbills | Callard et al. [19] |
| Environmental stress | Increased rate of pair 'divorces' in long-term bonded species | Black-browed albatross | Fayet et al. [33] |

**Table 18.6** Benefits of long-term pair bonds

| Benefit | Description | Species examples | Reference |
|---|---|---|---|
| Improved coordination | Efficient incubation, nest defense, and feeding | Black swans, great cormorants | Black et al. [11]; Johnsgard [43] |
| Higher chick survival | Stable pairs tend to have higher fledging rates | Black swans | Johnsgard [43] |
| Social learning | Familiar mates communicate better, reducing breeding stress | Many monogamous aquatic birds | Choudhury [22] |

**Table 18.7** Flexibility of pair bonds in different species

| Species type | Mating system | Flexibility level | Adaptation potential | Reference |
|---|---|---|---|---|
| Mute Swan | Obligate monogamy | Low (breeding delay after loss) | Vulnerable to partner loss | Lifjeld et al. [52] |
| Jacanas, Phalaropes | Polyandry/ flexible | High (multiple mates) | High resilience in fluctuating habitats | Emlen and Wrege [31] |
| Albatross | Strong monogamy | Moderate | Recent increase in pair divorces | Fayet et al. [33] |

## 18.17 Pair Stability vs. Flexibility

- In **obligate monogamous species**, loss of a partner can delay or cancel a breeding season. In contrast, **facultative monogamists** may adapt by quickly re-pairing [52].
- **Flexible mating systems**, like those seen in **jacanas** and some **phalaropes**, allow for multiple mating partners and can buffer against environmental unpredictability [31] (Table 18.7).

## 18.18 Environmental Pressures on Bond Stability

- Long-term pair bonds are particularly vulnerable in **rapidly changing environments**. Habitat degradation, rising water levels, and food web disruptions can erode the benefits of pair stability by making previous breeding strategies less effective.
- As a result, some species are showing signs of **evolutionary shifts** toward more **opportunistic mating systems** under environmental stress [4].

## 18.19 Nest Building: Crafting Safe Havens for the Next Generation

Nest building is one of the most critical phases of the reproductive cycle in aquatic birds. A well-constructed nest not only provides a safe environment for egg incubation and chick development but also reflects evolutionary adaptations to specific habitats, predator threats, and climatic conditions. Aquatic birds have developed an extraordinary variety of nesting strategies that reflect their ecological diversity and ingenuity [37].

## 18.20 Diverse Nesting Habitats and Materials

Aquatic birds utilize a wide range of nesting habitats, often dictated by species-specific behavior, environmental context, and the need to minimize predation and maximize reproductive success.

### *18.20.1 Types of Nesting Sites*

- **Ground Nests**: Many waterfowl, such as ducks (*Anas* spp.) and geese (*Branta* spp.), build nests on the ground, often in tall grass, reed beds, or near water bodies. These nests are usually shallow depressions lined with vegetation and down feathers for insulation [46].
- **Floating Nests**: Species like **grebes** (*Podicipedidae*) and some **rails** construct floating nests anchored to emergent vegetation. These nests rise and fall with water levels, protecting the eggs from minor flooding. However, they are vulnerable to storms and strong currents [79].
- **Tree and Shrub Nests**: Certain **herons**, **egrets**, and **cormorants** build large stick nests in trees or bushes, often forming noisy colonies called **rookeries**. These elevated sites reduce predation by terrestrial animals and offer good vantage points [81].
- **Cliffside and Ledge Nests**: Seabirds such as **murres**, **kittiwakes**, and **guillemots** often nest on narrow cliff ledges, where the risk of mammalian predators is minimal. Some, like murres, do not build conventional nests but rely on the minimal movement of their pear-shaped eggs to prevent them from rolling off ledges [8].
- **Burrows and Crevices**: **Puffins** and some **storm petrels** use natural cavities or dig burrows in soft soil, offering protection from aerial and ground predators and harsh weather [14] (Table 18.8).

**Table 18.8** Nesting site types and associated bird species

| Nesting site type | Description | Examples of bird species |
|---|---|---|
| Ground nests | Shallow depressions lined with vegetation, typically built near water bodies. | Ducks (*Anas* spp.), Geese (*Branta* spp.) |
| Floating nests | Anchored to vegetation, rise and fall with water levels. | Grebes (*Podicipedidae*), Rails |
| Tree and shrub nests | Large stick nests in elevated vegetation, often colonial. | Herons, Egrets, Cormorants |
| Cliffside and ledge nests | Nests or eggs laid directly on narrow cliff ledges. | Kittiwakes, Guillemots, Murres |
| Burrows and crevices | Dug into soil or found in rock cavities. | Puffins, Storm Petrels |

**Table 18.9** Common nesting materials and their functional roles

| Material | Function | Birds commonly using it |
|---|---|---|
| Grasses and Reeds | Structure and insulation | Waterfowl, Gulls |
| Feathers/Down | Thermal insulation for eggs | Ducks, Swans |
| Sticks and Twigs | Structural support for tree nests | Herons, Egrets |
| Mud | Elevates nests in saline or muddy environments | Flamingos |
| Seaweed and Guano | Reinforcement in coastal nests | Cormorants |
| Plastic and Trash | Opportunistic use, sometimes harmful | Urban-nesting species |

## *18.20.2 Nest Materials and Construction Techniques*

Aquatic birds are opportunistic and innovative in their choice of nesting materials:

- **Vegetation**: Common materials include grasses, reeds, twigs, moss, and aquatic plants. These are used for structural support and lining [37].
- **Feathers and Down**: Birds like ducks pluck down from their own breasts to create soft, warm linings that help insulate the eggs [46].
- **Mud and Seaweed**: Some birds, like **flamingos**, build cone-shaped mud nests that elevate the eggs above saline or muddy ground. **Cormorants** use seaweed, sticks, and guano to reinforce their nests [65].
- **Anthropogenic Materials**: In human-impacted areas, birds may incorporate synthetic materials such as plastic strips, fishing line, or cloth, which sometimes increase the risk of entanglement or pollution [80].

Nest design often reflects a **balance between protection, accessibility, thermal regulation**, and **camouflage**. The use of certain materials may also play a role in parasite defense or signaling to mates [57] (Table 18.9).

**Parental Roles in Nest Preparation**
The roles of male and female aquatic birds in nest construction are diverse and often species-specific, ranging from shared responsibilities to tasks assigned by sex. Nest building can also serve as part of the **courtship process**, where construction effort or skill influences mate selection [23].

## *18.20.3 Biparental Nest Building*

In many species, **both parents participate** in constructing and maintaining the nest:

- **Great crested grebes** engage in cooperative nest building, collecting aquatic vegetation and reinforcing their floating nests throughout incubation [79].
- In **black-headed gulls**, males often bring nesting materials while females focus on arranging them. This cooperation ensures that the nest is quickly and efficiently completed [17].

## *18.20.4 Male-Dominated Construction*

In some species, **males take the lead**, sometimes even building multiple "practice" nests to attract females:

- **Marbled ducks** (*Marmaronetta angustirostris*) and some **jacanas** are known for male-dominated nest building, especially in species where males assume greater incubation duties [1, 31].
- **Weaver birds**, although not aquatic, present a useful comparison: males build elaborate nests, and females choose based on nest quality—a behavior observed to a lesser extent in some waterbirds [24].

## *18.20.5 Female-Led Construction*

In certain species, females take on the primary building role, especially when **males are occupied with territory defense or courtship displays**:

- **Mallards** (*Anas platyrhynchos*) typically see the female solely responsible for nest construction, including lining the nest with her own down feathers [46].
- **Phalaropes**, in which polyandry is common, show a role reversal, where the female lays the eggs and leaves, while the male builds and incubates [66].

### *18.20.6 Ongoing Maintenance and Vigilance*

Nest building doesn't end with the initial construction. Many aquatic birds continue to **reinforce, adjust, or repair** nests throughout the incubation and chick-rearing periods:

- Birds nesting in tidal zones must frequently adjust nest height or structure to avoid flooding [16].
- In **colonial nesting species**, competition for space and materials often leads to nest theft or modification, demanding regular upkeep [87] (Table 18.10).

## 18.21 Nesting as an Evolutionary Adaptation

The architecture, placement, and materials of nests are the results of long evolutionary processes aimed at improving reproductive success:

- **Predation Pressure**: Camouflaged nests in reeds or high in trees reduce exposure to predators [37].
- **Microclimate Regulation**: Nest location and structure influence internal temperature and humidity, critical for embryonic development [57].
- **Flood Avoidance**: Floating or elevated nests buffer against seasonal or unpredictable water level changes [23].

Studies such as *Mainwaring* et al. *[56], Journal of Avian Biology,* have emphasized the **plasticity and adaptability** of nest-building behaviors in aquatic birds, suggesting that these behaviors may shift in response to **climate change, habitat disturbance**, and **urban expansion** (Table 18.11).

**Table 18.10** Parental roles in nest construction

| Parental role distribution | Typical pattern | Example species |
|---|---|---|
| Bi-parental | Both sexes build or maintain nests together | Great crested grebes, Black-headed gulls |
| Male-dominated | Male constructs nests to attract mates or incubate alone | Marbled ducks, Jacanas |
| Female-led | Female builds and maintains the nest | Mallards |
| Role reversal | Female lays eggs, male builds/maintains and incubates | Phalaropes |

**Table 18.11** Adaptive significance of nest design

| Adaptation goal | Nest design feature | Species/examples |
|---|---|---|
| Predator avoidance | Camouflaged nests in vegetation or high trees | Herons, Reed-nesting ducks |
| Climate regulation | Use of insulating materials; location in shaded/warm areas | Ground-nesting ducks, Puffins |
| Flood avoidance | Floating nests, elevated mud mounds | Grebes, Flamingos |
| Space competition in colonies | Continuous nest maintenance and defense | Cormorants, Gulls |

## 18.22 Egg Incubation and Parental Roles

**Egg Characteristics and Incubation Strategies**

The diversity of aquatic bird eggs reflects a complex interplay between **evolutionary history**, **ecological demands**, and **predation pressure**. Key features like **egg shape**, **coloration**, and **surface texture** are fine-tuned adaptations to specific nesting environments [29].

### *18.22.1 Egg Shape and Functionality*

- **Pyriform (pear-shaped) eggs**, like those laid by cliff-nesting species such as **common murres (*Uria aalge*)**, are tapered and roll in tight circles, preventing them from falling off narrow ledges [9].
- **Elliptical or spherical eggs** are more common in ground-nesting or platform-nesting species, offering compactness and efficient heat distribution during incubation [53].

Recent biomechanical research ([10], *Biological Journal of the Linnean Society*) found that egg shape in aquatic birds correlates not just with nest location but also with **female pelvic anatomy and flight efficiency**, showing how reproductive traits are influenced by overall body plan [10].

### *18.22.2 Egg Coloration and Camouflage*

- Eggshell coloration is often a defense mechanism. Ground-nesting birds like **plovers** and **terns** lay **cryptically speckled eggs** that blend seamlessly with sand, pebbles, or mud [36].
- In contrast, **colonial nesters** such as **cormorants** and **herons** often lay **white or pale blue eggs**, since the risk of predation is lower due to large group defense and the need for visibility in shaded environments [48].

**Table 18.12** Egg shape, nest type, and functional significance

| Egg shape | Nest type | Function/adaptation | Example species |
|---|---|---|---|
| Pyriform (pear-shaped) | Cliff ledges | Rolls in tight circles to avoid falling | Common murres (*Uria aalge*) |
| Elliptical | Ground/platform nests | Even heat distribution and efficient space usage | Ducks, Loons |
| Spherical | Ground or burrow nests | Maximal volume for heat retention | Puffins, Petrels |

**Table 18.13** Egg coloration and nest environment

| Egg coloration | Environmental context | Function | Example species |
|---|---|---|---|
| Cryptic speckled | Ground nests on mud, sand, or pebbles | Camouflage against predators | Plovers, Terns |
| White or pale blue | Colonial nests in shaded areas | Visibility for parents; less predation pressure | Cormorants, Herons |
| Thickly pigmented (dark) | Exposed, sunlit environments | Protection from UV radiation and overheating | Stilts, Open-ground nesters |

A 2023 study in *Journal of Avian Ecology* confirmed that **eggs exposed to direct sunlight in open nests had thicker, darker pigmentation**, possibly to shield developing embryos from harmful UV radiation [63] (Tables 18.12 and 18.13).

### *18.22.3 Incubation Strategies: Who Sits on the Nest?*

Aquatic birds display a remarkable variety of incubation strategies:

- **Biparental incubation** is common in species like **loons** and **grebes**, where both sexes alternate shifts at the nest [79].
- **Uniparental incubation**, often by the female, is seen in many ducks (e.g., **mallards**) where the male departs soon after mating [46].
- **Male-only incubation** is a hallmark of polyandrous species like the **northern jacana (*Jacana spinosa*)**, where females mate with multiple males who then assume all incubation and chick-rearing responsibilities [31].

Incubation periods range from **10–11 days in small species** (e.g., **least sandpiper**) to **35–40 days in large seabirds** (e.g., **albatrosses**), with variation influenced by egg size, environmental temperature, and parental attentiveness [28].

**Thermoregulation and Environmental Challenges**
Maintaining an optimal and stable egg temperature is essential for healthy embryonic development. Most bird embryos develop best within a narrow range of **36–38 °C**, and even minor deviations can significantly affect survival and hatching success [82].

### 18.22.4 Thermal Management Techniques

- **Brood patches**: Most incubating birds develop a featherless area on their abdomen with enhanced blood flow, allowing efficient heat transfer to the eggs [84].
- **Turning the eggs**: Birds frequently rotate their eggs (up to 12 times per hour) to ensure even warming and prevent embryo adhesion to the shell membrane [28].
- **Shading and fanning**: In hot climates, birds like **black-winged stilts** and **ibises** use their wings to shade eggs or wet their feathers to cool them through evaporative heat loss [27].

### 18.22.5 Adaptation to Cold Climates

In colder regions, aquatic birds such as **eiders** and **gulls** face the challenge of preventing heat loss:

- **Dense nesting** (as in Arctic tern colonies) helps maintain ambient nest warmth [39].
- **Use of insulating materials** such as down, seaweed, and grasses helps retain heat [37].
- **Incubation constancy** (minimal breaks from the nest) is crucial. Female common eiders (*Somateria mollissima*) may fast for the entire incubation period (up to 28 days) to ensure their eggs remain warm in icy conditions [67].

## 18.23 Impacts of Climate Change on Incubation Success

Modern environmental changes, especially global warming, are directly threatening successful incubation in aquatic birds. Rising temperatures can lead to overheating in exposed nests, while altered precipitation patterns increase flooding risks in low-lying nesting areas. Additionally, shifts in prey availability may force parents to leave nests more frequently, disrupting optimal incubation conditions [56, 68].

### 18.23.1 Rising Temperatures and Heat Stress

In many coastal and wetland regions, **ambient temperatures have increased by 1.5–2 °C over the past 30 years** [41]. This has led to:

- **Egg overheating** in open or exposed nests.
- **Increased embryo mortality**, particularly in shallow or ground-level nests without adequate shading.
- **Reduced incubation constancy**, as birds are forced to leave the nest more frequently to cool down or forage further due to shifting food availability.

### *18.23.2 Extreme Weather Events*

Climate-induced increases in **storm frequency, heatwaves, and unseasonal rainfall** affect incubation through:

- **Nest flooding**, especially in wetland species like **moorhens** and **bitterns**.
- **Nest desertion** during prolonged heat or drought, particularly in temperate-zone breeders.
- **Delayed or failed hatching** due to sudden cold snaps after egg-laying, a growing problem in temperate regions.

### *18.23.3 Case Study: Pied Avocets in Europe*

A 2024 study in *Global Change Biology* tracked **pied avocet (*Recurvirostra avosetta*) colonies in the UK and France** and found that rising spring temperatures advanced the breeding season by 10–14 days. However, early-season nests had lower hatching success due to **mismatched food availability**, suggesting that phenological shifts aren't always adaptive.

### *18.23.4 The Role of Human Activity*

In addition to climate, **human disturbance** near nesting sites also affects incubation:

- **Tourism** and **boating** increase flushing behavior in colonial nesters.
- **Light pollution** affects the circadian rhythms of some nocturnal incubators like **night-herons**.
- **Habitat alteration** can lead to microclimate changes (e.g., removal of vegetation leads to greater exposure and higher egg temperatures).

Mitigation strategies such as **buffer zones**, **artificial shade**, and **controlled visitor access** have been shown to improve nesting success in protected areas.

## 18.24 Chick Development: From Hatchlings to Flight

### *18.24.1 Precocial vs. Altricial Chicks*

Aquatic birds exhibit a wide range of chick development strategies, which are broadly categorized into **precocial** and **altricial** types—each adapted to different ecological pressures and reproductive strategies.

#### 18.24.1.1 Precocial Chicks: Born Ready

Precocial chicks, like those of **ducks**, **geese**, **shorebirds**, and many **waders**, hatch with open eyes, down-covered bodies, and the ability to walk or swim within hours. These chicks are relatively independent but still rely on parents for guidance, warmth, and protection, especially from predators.

- **Advantages of Precociality**:
  - Immediate mobility reduces time spent at vulnerable nest sites.
  - Better suited for ground-nesting or open-area breeders.
  - Enables access to food sources in aquatic or semi-aquatic habitats quickly after hatching [54].
- **Notable Species**:
  - **Northern Lapwing (*Vanellus vanellus*)** chicks forage for insects soon after hatching.
  - **Mallards (*Anas platyrhynchos*)** lead their ducklings to water within 24 hours of hatching.
- **Recent Research**: A 2023 study published in *Journal of Avian Biology* found that precocial chicks from marshland habitats with dense vegetation had a **35% higher survival rate** in their first week compared to those in more exposed habitats, due to better concealment and faster self-feeding ability [21].

#### 18.24.1.2 Altricial Chicks: Nurtured to Maturity

Altricial chicks, found in species like **herons**, **terns**, **grebes**, and **cormorants**, hatch in a helpless state—naked, blind, and completely dependent on their parents for warmth, protection, and food.

- **Advantages of Altriciality**:
  - Allows parents to allocate more energy to chick development post-hatching.
  - Suited to species nesting in **colonies**, **trees**, or **cliffs** where food delivery is efficient.
  - Higher brain and sensory development occurs after hatching, allowing for enhanced learning [2].
- **Notable Species**:
  - **Great Crested Grebes (*Podiceps cristatus*)** carry altricial chicks on their backs while continuing to feed and warm them.
  - **Black-crowned Night Herons (*Nycticorax nycticorax*)** tend to their chicks with precise food delivery in densely populated nesting colonies.

- **Recent Findings**: A 2024 paper in *Ecology and Evolution* reported that altricial species in warmer climates had increased chick mortality due to **heat stress**, especially in colonial nesters where **thermal load is compounded** by proximity to other nests [59].

### *18.24.2 Feeding Strategies: Nourishing the Young*

Aquatic birds employ diverse feeding techniques adapted to their ecological niche and chick developmental type:

- **Regurgitation**: Common among altricial birds like gulls and herons; parents partially digest food and regurgitate it for their chicks.
- **Direct Leading to Food**: Precocial chicks are often led by parents to foraging areas and taught what to eat (e.g., sandpipers, moorhens).
- **Provisioning with Live Prey**: Species like terns and kingfishers deliver live fish or insects directly to chicks [47].
- **Parental Investment**:
  - In biparental species, both parents may share feeding duties (e.g., grebes).
  - In some species like **jacanas**, males alone care for and feed the chicks, a rare form of sex-role reversal.

### *18.24.3 Fledging and First Flight: Taking to the Skies*

**Fledging** is the critical milestone where chicks leave the nest and become capable of flight or independent swimming:

- **Timing**:
  - Precocial chicks may take weeks to fledge depending on their mobility needs and predator exposure.
  - Altricial chicks typically fledge after significant wing and muscle development, often **3–6 weeks post-hatching**, depending on species size and growth rate [75].

- **Behavioral Training**:
  - Chicks often undergo a learning phase where they mimic adult foraging, flying, or diving behavior.
  - Studies show that **parental guidance during fledging** increases post-natal survival by up to 40% in species like **common loons** and **terns** [3] (Table 18.14).

**Table 18.14** Fledging timelines in aquatic birds

| Species | Chick type | Fledging time (approx.) | Notes on development |
|---|---|---|---|
| Mallard (*Anas platyrhynchos*) | Precocial | 45–60 days | Early mobility, gradual wing development |
| Great Crested Grebe | Altricial | 60–70 days | Carried on back; late independence |
| Common Tern | Altricial | 21–28 days | High parental care in colonies |
| Northern Lapwing | Precocial | 35–40 days | Early foraging but delayed flight |
| Black-crowned Night Heron | Altricial | 30–40 days | Strong post-fledging parental support |

**Table 18.15** Environmental threats to chick development

| Threat | Impact on development | Affected species |
|---|---|---|
| Habitat Loss | Fewer safe nesting and foraging sites | Moorhens, Stilts, Bitterns |
| Predation (invasive species) | Chick mortality, especially ground nesters | Lapwings, Gulls, Sandpipers |
| Heatwaves | Dehydration, overheating in nests | Herons, Cormorants (colonial nesters) |
| Pollution (plastic ingestion) | Indirect chick ingestion via parental feeding | Petrels, Gulls |

## 18.25 Challenges during Development

Chick development is increasingly challenged by modern environmental pressures:

- **Habitat Degradation**: Loss of safe nesting and foraging grounds due to human encroachment [34].
- **Predation**: Increased exposure to introduced predators (e.g., rats, foxes, feral cats).
- **Climate Extremes**: Heatwaves and heavy rains disrupt both incubation and early development phases [41].
- **Pollution**: Ingestion of plastic by parent birds is transferred to chicks during feeding (notably in gulls and petrels) [72] (Table 18.15).

## 18.26 Adaptive Behaviors for Survival

Aquatic bird species demonstrate resilience through a variety of adaptations:

- **Synchronous Hatching**: Ensures all chicks are at the same development stage, which enhances brood mobility and feeding efficiency [69].
- **Back-riding**: Used by grebes and some duck species to transport young away from danger.

## 18.27 Conservation Mysteries of Aquatic Birds

Key habitats remain underprotected or suffer from inadequate management. Frameworks like the Ramsar Convention and BirdLife International's Important Bird Areas (IBAs) identify critical sites, but enforcement often falls short. The East Asian–Australasian Flyway exemplifies this issue, where ongoing habitat destruction along the Yellow Sea coast jeopardizes millions of shorebirds [64]. These challenges underscore the urgent need for transboundary conservation and adaptive, ecosystem-wide strategies.

## 18.28 Major Threats to Aquatic Birds

Aquatic birds face a multitude of anthropogenic threats:

- Habitat Loss: Wetland drainage for agriculture, urbanization, and industry has severely fragmented essential breeding and foraging habitats. Coastal systems, including mangroves and salt marshes, are disappearing at alarming rates [26].
- Pollution: Aquatic birds are exposed to toxic substances such as heavy metals, persistent organic pollutants (POPs), and microplastics. Oil spills compromise plumage integrity, leading to hypothermia and drowning [18].
- Climate Change: Altered precipitation patterns, rising sea levels, and increased temperatures disrupt migratory patterns and breeding phenology. Arctic-breeding species are especially vulnerable to habitat shrinkage [70].
- Human Disturbance: Recreational activities, boating, and habitat encroachment often lead to nest abandonment. Invasive species also degrade habitat and prey on eggs and chicks [20, 44].
- Overexploitation: In some regions, hunting, egg collection, and illegal wildlife trade further stress already-declining populations.

## 18.29 Conservation Strategies for Aquatic Birds

**Full Life-Cycle Conservation**

Protecting aquatic birds requires safeguarding their habitats across all stages of their life cycle. Species like the Red Knot (*Calidris canutus rufa*), which migrates across continents, depend on a network of sites that must be conserved through international collaboration. Tracking technologies have been instrumental in identifying critical stopover points, such as Delaware Bay, prompting targeted conservation efforts [74].

**Habitat Protection and Restoration**
Wetlands, estuaries, and mangroves are essential for aquatic birds but face rapid degradation. Restoration strategies—like reestablishing natural hydrology, replanting vegetation, and removing physical barriers—have yielded positive outcomes in regions like the Danube Delta and Cheorwon Plain [5].

**Integrated Coastal Zone Management (ICZM)**
This approach coordinates multiple stakeholders—fisheries, tourism, urban development—under a unified framework. In Peru, ICZM has minimized disturbances to breeding colonies of the endangered Peruvian Tern (*Sternula lorata*) [5].

**Community-Based Conservation**
Local stewardship is crucial, especially where governmental reach is limited. In Bangladesh, fisher communities help protect wetlands and species like the Pallas's Fish Eagle (*Haliaeetus leucoryphus*). Indigenous efforts in Papua New Guinea support mangrove conservation for species such as the Pacific Golden Plover (*Pluvialis fulva*) [15].

**Pollution Control**
Fish-eating birds in the Great Lakes continue to suffer from legacy and emerging pollutants. Regulatory bans on PCBs and DDT have helped, but new threats—such as PFAS chemicals—require renewed research and policy [85].

**Climate Resilience and Dynamic Reserve Design**
As habitats shift with climate change, conservationists are identifying refugia—stable zones resilient to climatic fluctuations. Managed retreat strategies in the Netherlands, where farmland is being returned to tidal wetlands, support bird populations while offering flood mitigation [58].

## 18.30 Future Prospects

1. Long-term Monitoring Programs—Establishing systematic population and breeding success monitoring is essential for understanding the impacts of climate change, habitat degradation, and conservation interventions on aquatic bird populations.
2. Restoration of Critical Habitats—Large-scale wetland restoration and improved connectivity between fragmented aquatic ecosystems should be prioritized to ensure long-term habitat viability and resilience.
3. Policy and Legal Frameworks—Strengthening environmental legislation, enhancing the enforcement of international treaties, and aligning national conservation priorities are necessary to safeguard freshwater and coastal ecosystems.
4. Community Engagement—Scaling up citizen science, ecotourism-based conservation models, and education initiatives empowers local communities to play an active role in aquatic bird conservation.

## 18.31 Conclusion

5. Aquatic birds are vital indicators of ecosystem health, and their reproductive ecology reflects the stability and integrity of freshwater and coastal habitats.
6. Breeding success is under growing threat from anthropogenic pressures—including habitat destruction, pollution, and climate change—that jeopardize long-term population survival.
7. Case studies across different regions reveal both the resilience and vulnerability of aquatic birds, emphasizing the importance of context-specific and adaptive conservation strategies.
8. Effective conservation requires a multi-dimensional approach, combining wetland restoration, provision of artificial nesting structures, and robust protective legislation.
9. The involvement of local communities, contributions from citizen science, and the integration of traditional ecological knowledge are critical to achieving lasting conservation outcomes.
10. By combining scientific innovation with participatory management, it is possible to safeguard breeding habitats and ensure the persistence of aquatic bird populations for future generations.

## References

1. Alonso JA, Alonso JC (1982) Parental care in the marbled teal (Marmaronetta angustirostris). Ardeola 29(1):49–58
2. Andrews RJ, Singh D, Li H (2024) Altricial versus precocial development and the role of thermal stress in avian colonies. Ecol Evol 14(1):e10328. https://doi.org/10.1002/ece3.10328
3. Avian Conservation & Ecology (2023) Parental guidance and fledging success in aquatic birds. Avian Conserv Ecol 18(2):12
4. Avian Ecology Reports (2023) Behavioral shifts in aquatic bird mating systems under climate stress. Avian Ecol Rep 12(3):221–234
5. BirdLife International (2019–2020) State of the world's birds reports
6. BirdLife International (2022) State of the world's birds 2022: taking the pulse of the planet. BirdLife International, Cambridge, UK
7. BirdLife International (2024) Species factsheets: Gavia stellata and Himantopus himantopus. BirdLife International, Cambridge, UK
8. Birkhead TR (2011) Bird sense: what it's like to be a bird. Bloomsbury
9. Birkhead TR (2016) The Most perfect thing: inside (and outside) a Bird's egg. Bloomsbury Publishing
10. Birkhead TR, Thompson JE, Biggins JD (2022) Egg shape, size, and female body shape in aquatic birds: an evolutionary and biomechanical perspective. Biol J Linn Soc 136(2):210–223. https://doi.org/10.1093/biolinnean/blab143
11. Black JM, Prop J, Larsson K (2007) Wild goose dilemmas: population consequences of individual decisions in barnacle geese. Branta Press
12. Bolund E, Schielzeth H, Forstmeier W (2009) Compensatory investment in egg size in response to partner attractiveness in the zebra finch. Ethology 115(6):597–606. https://doi.org/10.1111/j.1439-0310.2009.01642.x

13. Both C, Bouwhuis S, Lessells CM, Visser ME (2006) Climate change and population declines in a long-distance migratory bird. Nature 441(7089):81–83. https://doi.org/10.1038/nature04539
14. Brooke M (2004) Albatrosses and petrels across the world. Oxford University Press
15. Brown C et al (2017) Community-based conservation and traditional ecological knowledge. Ecol Soc 22(2):13
16. Burger J (1979) Resource partitioning: Nest site selection in gulls. Am Nat 113(4):593–597. https://doi.org/10.1086/283418
17. Burger J, Gochfeld M (1996) Family Laridae (Gulls). In: del Hoyo J, Elliott A, Sargatal J (eds) Handbook of the birds of the world, vol 3. Lynx Edicions, pp 572–623
18. Burger J, Gochfeld M (2004) Marine birds as sentinels of environmental pollution. EcoHealth 1(3):263–274
19. Callard D, Lopez G, Miskell J (2023) Habitat degradation and courtship frequency in aquatic waders. Ecol Lett 26(9):1789–1802. https://doi.org/10.1111/ele.14127
20. Carney KM, Sydeman WJ (1999) A review of human disturbance effects on nesting colonial waterbirds. Waterbirds 22(1):68–79
21. Chen Y, Patel A, Kumari R (2023) Marshland concealment and survival rates in precocial chicks. J Avian Biol 54(9):e02942
22. Choudhury S (1995) Divorce in birds: a review of the hypotheses. Anim Behav 50(2):413–429. https://doi.org/10.1016/0003-3472(95)80199-0
23. Collias NE, Collias EC (1984) Nest building and bird behavior. Princeton University Press
24. Collias NE, Victoria JK (1978) Nest and mate selection in the village weaverbird (Ploceus cucullatus). Anim Behav 26:286–289. https://doi.org/10.1016/0003-3472(78)90029-9
25. Croxall JP, Butchart SH, Lascelles B, Stattersfield AJ, Sullivan B, Symes A, Taylor P (2012) Seabird conservation status, threats and priority actions: a global assessment. Bird Conserv Int 22(1):1–34
26. Davidson NC (2014) How much wetland has the world lost? Mar Freshw Res 65(10):934–941
27. Dawson WR, Bennett AF (2009) Thermal physiology. In: Scanes CG (ed) Sturkie's avian physiology, 6th edn. Academic, pp 869–888
28. Deeming DC (2002) Avian incubation: behaviour, environment, and evolution. Oxford University Press
29. Deeming DC, Ferguson MWJ (eds) (1991) Egg incubation: its effects on embryonic development in birds and reptiles. Cambridge University Press
30. del Hoyo, J., Elliott, A., & Sargatal, J. (Eds.). (1992). Handbook of the birds of the world, Vol. 1: Ostrich to ducks. Lynx Edicions
31. Emlen ST, Wrege PH (2004) The evolution of cooperative breeding in birds. In: Koenig WD, Dickinson JL (eds) Ecology and evolution of cooperative breeding in birds. Cambridge University Press, pp 301–337
32. Ens BJ, Safriel UN, Harris MP (1996) Divorce in birds: when do pairs break up? In: Black JM (ed) Partnerships in birds: the study of monogamy. Oxford University Press, pp 344–401
33. Fayet AL, Hansen ES, Oswald SA, Trathan PN (2022) Climate change drives increase in divorce rate of a long-lived seabird. Nat Clim Chang 12(9):847–851. https://doi.org/10.1038/s41558-022-01441-4
34. Fernández R, Liu Q (2023) Anthropogenic threats to chick development in wetland birds. Wetl Ecol Manag 31(4):401–416
35. Gill FB (2007) Ornithology, 3rd edn. W. H. Freeman and Company
36. Hanley D, Cassey P, Doucet SM (2015) Eggshell pigmentation is functionally linked to gas exchange and microbial infection. Funct Ecol 29(2):209–218
37. Hansell M (2000) Bird nests and construction behaviour. Cambridge University Press
38. Harrison XA, York JE, Meade J, Hatchwell BJ (2021) Synchronization of pair behavior improves reproductive outcomes in socially monogamous birds. Behav Ecol 32(1):32–41. https://doi.org/10.1093/beheco/araa086

39. Hatch JJ (2002) Arctic tern (Sterna paradisaea). In: Poole A, Gill F (eds) The birds of North America online (No. 707). Cornell Lab of Ornithology
40. Hayman P, Marchant J, Prater T (1986) Shorebirds: an identification guide to the waders of the world. Houghton Mifflin
41. IPCC (2023) Sixth assessment report: climate change 2023. Intergovernmental Panel on Climate Change
42. Jenni-Eiermann S, Jenni L (2018) Coping with stress: hormonal responses and their behavioral impacts in birds. Ornithol Sci 17(1):1–13. https://doi.org/10.2326/osj.17.1
43. Johnsgard PA (2010) The global diversity of ducks, geese and swans. University of Nebraska Press
44. Jones HP et al (2008) Severity of the effects of invasive rats on seabirds. Conserv Biol 22(1): 16–26
45. Jones IL, Hunter FM (1999) Experimental evidence for mutual intersexual selection and preferred mating in a socially monogamous seabird. Behav Ecol 10(4):451–457. https://doi.org/10.1093/beheco/10.4.451
46. Kear J (2005) Ducks, geese and swans, vol. 1 & 2. Oxford University Press
47. Keller DL, Donlan MJ (2022) Foraging adaptations in aquatic birds: from chick provisioning to prey selection. Aquat Ornithol 37(2):145–159
48. Kilner R (2006) The evolution of egg colour and patterning in birds. Biol Rev 81(3):383–406
49. Kiran AQ (2022a) Physiological and sensory adaptations of aquatic birds for underwater foraging. J Avian Biol 53:e02841
50. Kiran AQ (2022b) Adaptive mechanisms in diving birds under low-visibility aquatic environments. Int J Zool Animal Biol 5(4):000389
51. Kushlan JA, Hancock JA (2005) The herons. Oxford University Press
52. Lifjeld JT, Dunn PO, Whittingham LA (2005) Promiscuous mating systems. In: Bird DM, Bildstein KL (eds) Raptor research and management techniques. Hancock House, pp 125–142
53. Lovell PG, Martin GR, Wilson KJ (2013) The association between egg shape and cliff-nesting in birds. Ibis 155(2):243–248
54. Lovich JE, Torres AM, Batchelder RH (2023) The ecological benefits of precocial development in wetland avifauna. Avian Ecol Rep 11(3):88–95
55. Lovvorn JR, Jones DR (1994) Biomechanics of underwater swimming in diving birds. Biol Rev 69(4):611–641
56. Mainwaring MC, Deeming DC, Hartley IR (2022) Avian nesting behavior in a changing world: plasticity and adaptation. J Avian Biol 53(9):e02987. https://doi.org/10.1111/jav.02987
57. Mainwaring MC, Hartley IR, Lambrechts MM, Deeming DC (2014) The design and function of birds' nests. Ecol Evol 4(20):3909–3928. https://doi.org/10.1002/ece3.1054
58. Martin TG, Richardson DM (2019) Dynamic conservation strategies for climate-adaptive wetlands. Nat Ecol Evol 3(9):1321–1327
59. Martínez P, Huerta L, Zhao T (2024) Heat stress and chick mortality in colonial aquatic birds. Ecol Evol 14(2):e10345. https://doi.org/10.1002/ece3.10345
60. Marzluff JM, Bowman R, Donnelly R (eds) (2001) Avian ecology and conservation in an urbanizing world. Springer
61. Mitsch WJ, Gosselink JG (2015) Wetlands, 5th edn. Wiley
62. Møller AP, Rubolini D, Lehikoinen E (2010) Populations of migratory bird species that did not show a phenological response to climate change are declining. Proc Natl Acad Sci 107(12): 6022–6026. https://doi.org/10.1073/pnas.0910246107
63. Morales S, Barret RT, Hanley D (2023) Pigment density of avian eggs correlates with exposure to sunlight. J Avian Ecol 94(1):45–56
64. Murray NJ et al (2014) Shorebird decline in the Yellow Sea. Biol Conserv 170:1–9
65. Nelson JB (2005) Pelicans, cormorants, and their relatives: the Pelecaniformes. Oxford University Press
66. Oring LW (1982) Avian polyandry. In: Farner DS, King JR, Parkes KC (eds) Avian biology, vol 6. Academic, pp 1–92

67. Parker H, Holm H (1990) Patterns of nutrient and energy expenditure in female common eiders nesting in the High Arctic. Auk 107(4):660–668
68. Peach WJ, Kuhlmann J, Sherley RB (2019) Climate change and seabird breeding failure. Nat Clim Chang 9:602–606. https://doi.org/10.1038/s41558-019-0526-x
69. Pearce JM, Gómez S (2022) Hatching synchrony in aquatic birds: evolutionary function and survival outcomes. Avian Biol Res 15(1):12–21
70. Piersma T, Lindström Å (2004) Migrating shorebirds and climate change. Glob Chang Biol 10(1):1–13
71. Rasmussen PC, Anderton JC (2012) Birds of South Asia: The Ripley Guide. Smithsonian Institution & Lynx Edicions
72. Reynolds CL, Bharucha N, Koenig A (2024) Parental ingestion and chick contamination by plastics in seabirds. Mar Ornithol 52(1):17–26
73. Rubega MA, Schamel D, Tracy DM (2000) Red-necked phalarope (Phalaropus lobatus), version 2.0. In: Poole AF, Gill FB (eds) Birds of the world. Cornell Lab of Ornithology
74. Runge MC et al (2015) Full annual cycle modeling. Ecol Lett 18(4):336–347
75. Sato K, Nishikawa M, Wilson RP (2022) Flight readiness and fledging behavior in aquatic birds. Journal of Ornithological Science 33(2):67–79
76. Schreiber EA, Burger J (2002) Biology of marine birds. CRC Press
77. Smith JT, Michel NL, Morrison RIG (2023) Nesting density and reproductive success in semipalmated sandpipers (*Calidris pusilla*) in Arctic Canada. Auk 140(2):234–245. https://doi.org/10.1093/auk/ukad004
78. Stettenheim PR (2000) The integumentary morphology of modern birds—an overview. Am Zool 40(4):461–477
79. Storer RW, Nuechterlein GL (1992) Western Grebe. In: Poole A, Gill F (eds) The birds of North America (No. 26). The Academy of Natural Sciences; The American Ornithologists' Union
80. Townsend AK, Barker CM (2014) Plastic and the nesting behavior of aquatic birds. Environ Pollut 187:249–254. https://doi.org/10.1016/j.envpol.2013.12.021
81. Voisin C (1991) The herons of Europe. T & A D Poyser
82. Webb DR (1987) Thermal tolerance of avian embryos: a review. Condor 89(4):874–898
83. Weller MW (1999) Wetland birds: habitat resources and conservation implications. Cambridge University Press
84. Whittow GC (2000) Sturkie's avian physiology, 5th edn. Academic
85. Wilcox C et al (2015) Threat of plastic pollution to seabirds. Proc Natl Acad Sci 112(38): 11899–11904
86. Williams TD (1995) The penguins: Spheniscidae. Oxford University Press
87. Wittenberger JF, Hunt GL (1985) The adaptive significance of coloniality in birds. In: Farner DS, King JR, Parkes KC (eds) Avian biology, vol 8. Academic, pp 1–78
88. Yodha K, Singh D, Kour A, Yadav R, Kumar R (2024) Species diversity and abundance of birds in ber (*Zizyphus mauritiana*) crop. Asian J Conserv Biol 13(1). https://doi.org/10.53562/ajcb.83683

# Chapter 19
# Genomic Tools for Aquatic Bird Ecology and Conservation—A Comprehensive Overview

**Varsha Malik, Ravikant Verma, Shweta Sharma, Raja Kumar, Disha Agarwal, Navdeep Godara, and Kunal Narang**

**Abstract** Genomic facilities have revolutionized avian conservation, particularly for water bird species, over the last couple of years. Researchers can now investigate ecological trends, adaptive response and evolutionary paths of species which were not researched as deeply before using such facilities and newer techniques. Along with increasing threats of climate change to which wetland and coastal environments are susceptible, habitat loss and anthropogenic disturbance, genomics provides important insight into aquatic bird reaction to these stresses. This chapter deals with how genetic diversity, migratory behavior and climate stress resistance are being addressed through the implementation of novel genomic tools in the form of whole-genome sequencing, eDNA analysis and machine learning algorithms. Researchers are now discovering more about the population structure and even the potential future response of these birds to an environmental change. Genomic data can be utilized as an effective guide in recommending well-intentioned conservation measures like the safeguarding of migratory routes and breeding grounds when complemented by ecological data and spatial modeling. This chapter will demonstrate that genomics not only expands our scientific understanding but also offers specific advice on how to maintain aquatic bird populations' diversity.

V. Malik · R. Verma
Department of Zoology, Chaudhary Charan Singh Haryana Agricultural University, Hisar, India

S. Sharma (✉)
Department of Zoology, Chaudhary Charan Singh Haryana Agricultural University, Hisar, India

Department of Pharmaceutical Sciences, University of California, Irvine, Irvine, USA
e-mail: Shwetas4@uci.edu

R. Kumar · K. Narang
Department of Molecular Biology and Biotechnology, Chaudhary Charan Singh Haryana Agricultural University, Hisar, India

D. Agarwal
Department of Agricultural Communication, GBPUAT, Pantnagar, India

N. Godara
School of Plant and Environmental Sciences, Virginia Tech, Blacksburg, VA, USA

 
R. Kumar et al. (eds.), *Ecology, Conservation of Aquatic Birds*, Wetlands: Ecology, Conservation and Management 15, https://doi.org/10.1007/978-3-032-17125-2_19

**Keywords** Anthropogenic · Aquatic Birds · Conservation · Diversity · Genetic Diversity · Genomics

## 19.1 Introduction

An extremely fascinating order of the class Aves, water birds have developed a sequence of adaptations that make them successful in a sequence of demanding aquatic environments, from wetlands in freshwater to large marine ones. Birds' unique features like their feathers, beaked jaws, possessing high metabolic rates and flight adapted skeletons make them capable of inhabiting nearly every kind of terrestrial and aquatic biome [84].

Marine birds have a number of common adaptive features which allow them to survive in aquatic habitats, such as webbed feet, body streamlined shape, waterproof feathers and beak adaptation. Webbed feet, for example, in ducks and swans, are an important adaptation for propulsion, directional control and stability in the aquatic environment [67, 143]. Much like human-made paddling devices are effective in creating thrust without energy loss, mallards' webbed feet are effective in producing thrust using very little energy [57]. Their torpedo-shaped body, which is found in birds such as cormorants and herons, reduces the resistance of water, thereby enhancing the swimming speed and saving energy during diving [53, 114]. This type of hydrodynamic shape significantly enhances their efficiency underwater [112].

Another essential feature is waterproof feathers, which are kept in place by preen gland secretions. Waterproofing aids in buoyancy, insulation, and slippery swimming in water [92]. Waterproofing of the feathers has been quantitatively investigated by Srinivasan et al. [131], who determined the microstructure of the feathers and contact angles with water to determine the efficacy. Lastly, aquatic birds exhibit specialized beak morphologies that are closely linked to their feeding strategies. Whether used for filtering tiny organisms or catching fish, these beaks enhance resource use and support niche specialization within aquatic ecosystems [93]. Aquatic birds have different beak types suited to how they feed. Strainer beaks help filter plankton, fish-catching beaks hold slippery fish, scooping beaks collect fish and remove water, probing beaks reach into mud for invertebrates and spear-shaped beaks stab or grab prey quickly. Each type matches their food needs and habitat [41, 64, 94, 97, 123].

### *19.1.1 Beneficial Attributes of Aquatic Birds in Ecosystems*

Aquatic birds are not just graceful creatures of the water but indispensable components of healthy ecosystems, especially those surrounding wetlands, lakes, rivers and coastal regions. These birds play multiple roles needed for maintaining the balance and health of their habitats. Their intricate interactions with the environment through

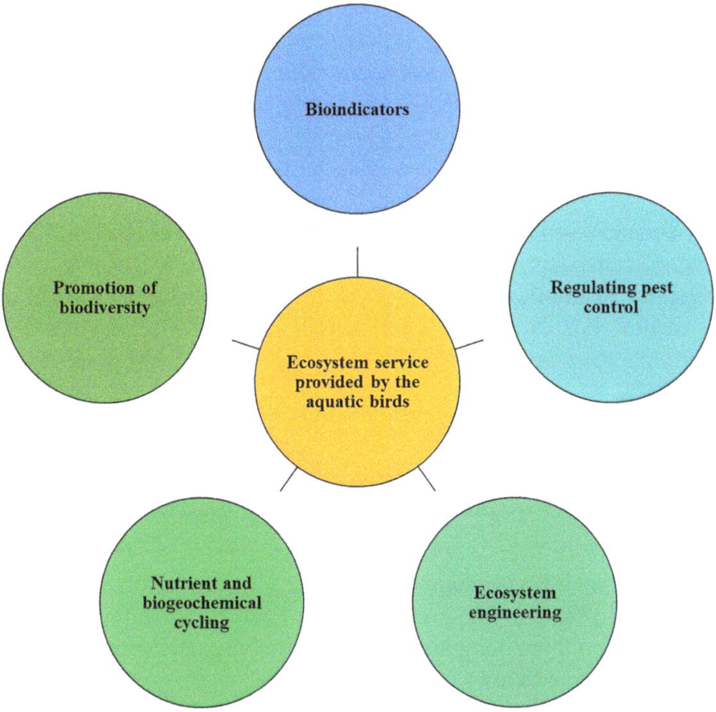


**Fig. 19.1** Importance of aquatic birds in ecosystems. (Source: Author)

feeding, migration, or nesting creates a ripple effect that benefits numerous species and sustains ecological processes. Aquatic birds contribute to the regeneration and maintenance of wetlands, the earth's most biodiverse and productive ecosystem. These also play a key role in most ecosystems for seed dispersal and pest controland offer different ecosystem services. These services directly or indirectly benefit human well-being [44, 147]. The functional role and ecosystem service provided by the aquatic birds are shown in Fig. 19.1.

- **Bioindicators:** Aquatic birds serve as cost-effective bioindicators, especially in remote or protected areas where intensive monitoring is unfeasible. Their population dynamics often correlate with ecosystem health, such as submerged macrophyte density, making them valuable proxies for biodiversity assessment [39, 96].
- **Nutrient and Biogeochemical Cycling:** Waterbirds influence nutrient fluxes through guano deposition and cross-ecosystem movement. In nutrient-limited systems, such as Arctic tundras, guanotrophication enhances productivity. In Hungary's soda pans, geese contribute over 50% of carbon and 70% of phosphorus inputs, fueling crustacean production. Feeding across terrestrial and aquatic zones, birds like gulls and swans facilitate cross-boundary nutrient transport.

- **Natural Pest Control:** Aquatic birds act as biological pest regulators by preying on agricultural pests like mosquito larvae and snails. For instance, introducing domestic ducks reduced *Pomaceacanaliculata* densities by over 80% in rice fields. Ducks have also been shown to suppress invasive slugs like *Arion vulgaris* in Europe [81].
- **Ecosystem Engineering:** Large wading birds, such as swans and flamingoes, modify habitats through bioturbation, distributing sediments while feeding. This activity enhances soil aeration nutrient cycling and promotes benthic diversity by creating microhabitats.
- **Biodiversity Promotion:** Waterbirds contribute to broader ecosystem diversity across scales. For example, in the Aleutian Islands, the decline of *Branta hutchinsiileucopareia* led to reduced plant diversity due to the absence of grazing pressure [82].

### *19.1.2 Challenges in the Conservation of Aquatic Birds*

Conserving aquatic birds is becoming more difficult due to a mix of environmental and anthropogenic factors. These challenges are not isolated; they often overlap, making the situation more complex and demanding urgent, coordinated responses. The major issues are as follows:

- **Habitat Loss and Degradation:** Industry, urbanisation and agriculture are causing an increasing loss of wetlands and coastal areas that are vital to waterbirds. Numerous population declines and the endangerment of important species have resulted from this degradation [99].
- **Invasive Species:** These species reduce the suitability of habitat for waterbirds and upset ecological balance by outcompeting native fauna for food and nesting locations.
- **Pollution:** From plastics to pesticides, pollutants build up in aquatic environments and endanger waterbirds through ingestion-related mortality, illness and poisoning, particularly in migratory species[145, 146].
- **Climate Change:** Variations in sea level, temperature & precipitation, migration and food availability,which affect breeding, lowers reproductive success and increases mortality.
- **Pathogens and Diseases:** Waterbird populations are seriously threatened by emerging diseases like West Nile virus and avian influenza, which are exacerbated by international bird migration [63].

These are interrelated problems and their solution requires an evident, multi-faceted approach involving international cooperation, sustainable management, and habitat protection. To save water bird species in the long term, it's first necessary to know how these threats interact with each other and impact them.

### *19.1.3 Genomic Approaches in Aquatic Bird Conservation and Evolution*

Aquatic birds renowned for holding a crucial position in the balance of ecosystems under severe extinction threat due to several factors. Genomic tools of the day enable us to look beyond skin-deep conservation with such methods as landscape genomics and molecular markers that enable scientists to follow gene movement along migratory routes, identify genetic isolation due to habitat breaks and measure overall genetic wellness. This is significant since increased genetic diversity gives birds increased opportunities for survival under changing circumstances, and compromised diversity indicates more risks of extinction [107].

In captive breeding, genomics has taken over traditional pedigree-based methods by offering more accurate insights into kinship and genetic variationespecially in social birds. DNA barcoding and STR markers also strengthen conservation efforts by tracing species origins and combating illegal wildlife trade [98, 127].

On the evolutionary side, population genetics helps us understand how genetic variation is distributed and maintained across fragmented habitats. High-throughput genomic instruments such as NGS (Next-Generation Sequencing) and RAD-seq are useful for detecting small DNA changes in the type of SNPs (single nucleotide polymorphisms). These are extremely helpful to describe gene flow between populations, bird adaptation to different habitats and population evolution over time. For example, studies of Greenland, Canada, and Norway Eider ducks utilized such tools to determine population structure and migration [116, 130]. SNP arrays also offer more convenient and less costly access to screen large cohorts and determine adaptations for local environmental pressures.

Accompanying this, DNA barcoding with mitochondrial and nuclear loci (COI, 16S, ITS) is especially applied in species and bottleneck population identification necessary in the monitoring of migratory or cryptic birds [6, 54]. RNA-seq is also a requirement utilized in determining the birds' adaptation against stressors like cold and salinity. Cold-tolerance genes involved in thermoregulation and metabolism, for instance, have been identified in S. mollissima by such a process [78, 140].

After gene expression analysis, phylogenomics provides evolutionary data essential to the explanation of divergence of the principal aquatic bird clades such as penguins, ducks and waders [52, 148]. Genomic data has provided the description of characters under ecological and behavioral pressures such as beak shape variation, foraging mode and reproductive separation [109, 113, 142]. Besides this, it has revealed the genetic underpinning of significant adaptations such as penguin flightlessness [20], salt gland activity in seabirds [29] and protein expression in preen glands for water repellency in waterbirds [114]. Such discoveries enrich our knowledge of aquatic bird evolution and inform conservation policy directly.

### *19.1.4 Gene Flow, Hybridization and Evolutionary Dynamics*

Hybridization and gene flow are fundamental processes that control the pattern of aquatic bird evolution. Hybridization of closely related species within areas of overlap introduces genetic variation, which promotes adaptability in a new environment [50]. Migration provides a mechanism for long-distance gene transfer that forms the backbone of genetic coherence and minimizes probability of isolation among different waterfowl species [102]. At the same time, genetic research demonstrates how environmental or behavioral differentiation leads to speciation via changes in functionally significant genes [15, 83].

From a conservation perspective, genomics enables precise monitoring of genetic health. It helps identify populations with low diversityoften more vulnerable to extinctionand supports the recognition of Evolutionarily Significant Units (ESUs) with unique adaptive value [24, 89]. These insights are especially relevant under climate change, where understanding genomic signatures linked to resilience can inform future-focused conservation efforts [139].

Aquatic birds also exhibit co-evolutionary patterns with their ecosystems. Genomic analyses reveal adaptive changes in genes involved in predator–prey dynamics, suggesting ongoing evolutionary feedback shaped by ecological interactions [28]. Such findings offer a deeper view into the functional basis of adaptation and biodiversity persistence.

### *19.1.5 Genomic Signatures of Adaptation to Aquatic Environments*

Genomic signatures refer to specific DNA patterns that enable organisms like aquatic birds to survive in their unique environments [22]. Comparative genomics has revealed aquatic birds' physiological, morphological and behavioural adaptations for thriving in aquatic habitats. Birds like penguins, loons and cormorants have adapted to diving, osmoregulation and underwater vision challenges. Key genomic signatures in aquatic birds include:

**Positive Selection on Specific Genes** Some genes evolve under environmental pressures to provide survival advantages. Notable adaptations and their genomic signatures are listed in Table 19.1 [115].

- **Convergent Evolution**: Independent lineages of aquatic birds, like penguins and loons, develop similar adaptations in response to shared ecological pressures. For example, amino acid changes in the vision genes PDE6G and PDE6A illustrate convergent evolution. Another example is the loss of olfactory receptor genes, as aquatic birds rely more on hearing and vision [11].

**Table 19.1** Genomic signatures of adaptations in aquatic birds

| Adaptation | Genomic Signature | Gene examples | Advantage |
|---|---|---|---|
| Diving and Hypoxia | Positive selection in hypoxia-related genes | EPAS1, SLC4A1 | Cope up with hypoxic conditions during diving. |
| Thermoregulation | Lipid metabolism gene adaptation | MC5R | Accumulation of fat layer (blubber) for insulation through regulation of lipid metabolism and storage of energy essential for long-duration swimming. |
| Underwater Vision | Selection of phototransduction genes | PDE6A, PDE6G | Enhancement of vision in underwater dim conditions. |
| Saltwater Tolerance | Selection of water transport genes | AQP (aquaporins) | Osmoregulation |

- **Regulatory Evolution (Non-Coding Genomic Changes)**: Changes in non-coding regions like promoters or silencers affect gene expression without altering protein structure. For instance, under low oxygen conditions, the overexpression of EPAS1 helps maintain metabolic activity during diving

Genomic tools, including NGS, SNP genotyping, transcriptomics and DNA barcoding, have revolutionized our understanding of aquatic bird biology. These technologies enable the assessment of genetic diversity, gene flow and adaptive traits crucial for survival in aquatic environments. This genetic knowledge supports conservation by informing efforts to protect aquatic bird populations from habitat loss and climate change threats.

**Molecular Markers in Waterbird Species**

The scientific study of aquatic birds is challenging because of their migratory habits and intricate ecological relationships. Conventional methods fall short of giving cues on species recognition, genetic variation, and evolutionary relationships [10]. Molecular markers, on the other hand, provide accurate and reproducible reactions, which enable one to investigate genetic variation, evolutionary lineages, and environmental influences on populations [7]. Such markers, which are recoverable non-invasively from feather or eDNA sources, have transformed research by offering comprehensive analysis without necessarily having to physically handle the birds and hence minimizing stress on the birds [12, 136]. With the issue of climate change and habitat destruction, molecular markers are central to genetic health monitoring, conservation, and migration guideline provision [101] as shown in Table 19.2 and Fig. 19.2.

**Table 19.2** Various types of molecular markers are used for aquatic bird species identification

| Marker type | Regions | Applications | References |
|---|---|---|---|
| mtDNA Markers | Cytochrome b, D-loop, COI, 16S/12S rRNA | Species ID, phylogenetics, population structure, migration tracking | [48, 60, 103] |
| Microsatellites (STRs) | Short tandem repeats in nuclear DNA | Genetic diversity, kinship, population structure | [105] |
| SNPs | Genome-wide single base-pair changes | High-res genotyping, adaptation studies, conservation genomics | [126] |
| Intron Markers | RAG1, RAG2, β-fibrinogen intron 7, Myoglobin intron 2 | Deep phylogeny, divergence in diving & aquatic species | [8, 128] |
| Z-linked Markers | ACO1I9 (Aconitase 1 intron 9) | Sex-biased dispersal, species complex resolution | [106] |

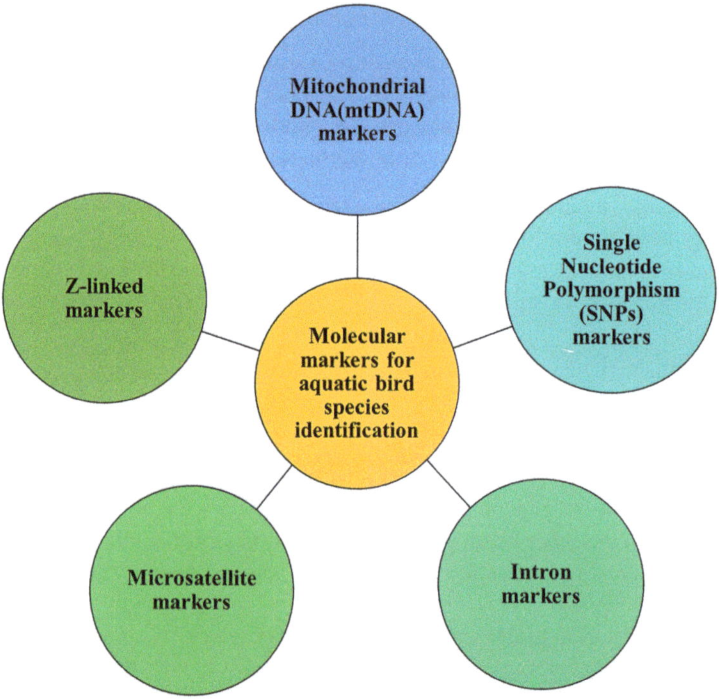


**Fig. 19.2** Various types of molecular markers used for aquatic bird species identification. (Source: Author)

## 19.2 DNA Barcoding in Aquatic Bird Identification

DNA barcoding is a key molecular tool for identifying aquatic bird species, especially in cases where visual traits overlap or hybridization occurs. It enables rapid, accurate species recognition even from feathers, droppings, or degraded tissue supporting taxonomy, migration studies and conservation [48, 60].

### *19.2.1 What is DNA Barcoding?*

DNA barcoding is a molecular method that identifies species using a short, standardized DNA sequence most often a fragment of the mitochondrial *cytochrome c oxidase I* (COI) gene. Like a supermarket barcode, this sequence serves as a unique identifier for species, helping classify even cryptic or juvenile birds where morphology fails [48, 108].

The COI gene, standard in the Barcode of Life Data System (BOLD), is highly effective for distinguishing aquatic bird species such as *Sterna*, *Larus*, *Anas crecca* and*Podiceps* [59, 60]. Procedure for identification of aquatic bird by using DNA barcoding is mentioned in Fig. 19.3.

### *19.2.2 Importance of DNA Barcoding in Aquatic Birds*

Aquatic birds often occur in remote or ecologically complex habitats where visual identification is limited by overlapping morphology, especially during non-breeding or juvenile stages. DNA barcoding offers a reliable, standardized method for distinguishing cryptic species such as *Larus* and *Calidris*, where conventional methods fail [49]. It has proven effective in identifying hybrids like *Anas platyrhynchos* × *A. rubripes*, providing insights into gene flow and species boundaries. By analysing feathers, droppings, or water samples, barcoding aids in tracking

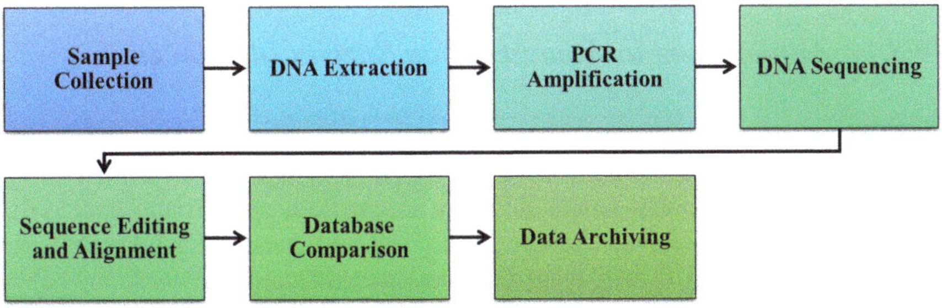


**Fig. 19.3** Workflow of DNA barcoding used for aquatic bird species identification. (Source: Author)

biodiversity and migration routes, even across vast wetland networks. Its integration with environmental DNA (eDNA) further enables non-invasive monitoring in inaccessible regions, while also supporting conservation by detecting endangered or illegally traded species.

**Case Studies: Seabird Diet Analysis (North Atlantic)**
DNA barcoding of regurgitated pellets and faeces from puffins (*Fraterculaarctica*) and shearwaters (*Puffinus* spp.) provided high-resolution dietary data, revealing temporal and spatial foraging shifts. This study advanced the understanding of marine trophic interactions and ecosystem health under climate and fisheries pressures [18].

**Limitations**
Despite its advantages, DNA barcoding has limitations. Incomplete reference databases can reduce accuracy, especially for understudied groups. Hybridization may cause mitochondrial introgression, leading to misidentification. Nuclear pseudogenes can also be mistakenly amplified. Additionally, barcoding alone may not resolve recently diverged species or subspecies without nuclear markers.

## 19.3 Microsatellites and Single Nucleotide Polymorphisms (SNPs)

Microsatellites or short tandem repeats (STRs) are also highly polymorphic DNA sequences of 1–6 base pair length that have been efficiently used to investigate species identity, population structure, and genetic diversity of aquatic birds [27]. Being of high mutation rate due to replication errors, they generate diversity in repeat counts among individuals, and therefore they play an important role in detecting fine-scale genetic differences [43]. As a result of codominant inheritance, microsatellites enable precise discrimination between homozygous and heterozygous individuals [19, 33]. Procedure for identification of aquatic bird by microsatellite is enlisted in Fig. 19.4.

### *19.3.1 Applications in Aquatic Avian Species Identification*

Microsatellites are very powerful tools in avian genetics with high resolving capacity for parentage and species identification and for population analysis. Microsatellites have been used to quantify genetic diversity among Podiceps cristatus and Sterna hirundo species [73], establish parentage in Chroicocephalus ridibundus [68] and distinguish between two closely related species such as Antarctic seabirds and Fulica atra [76, 100]. These markers are also applied in conservation planning via the demarcation of unique genetic units, as described in efforts at Himantopus

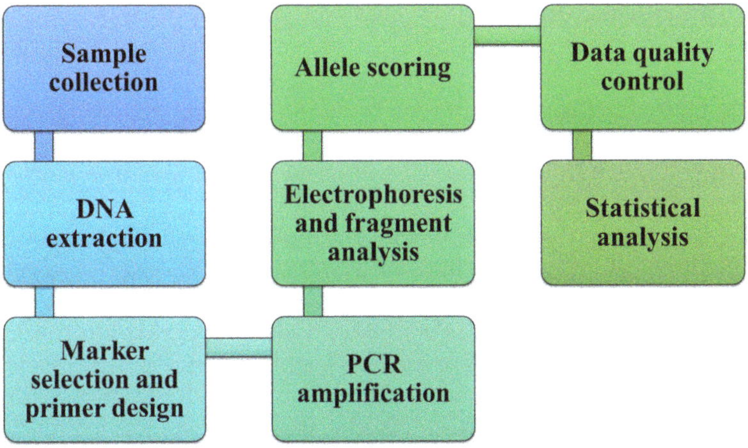


**Fig. 19.4** Microsatellite analysis workflow utilized to identify aquatic bird species. (Source: Author)

mexicanus knudseni and Chlamydotis undulata [62, 110]. But they are also in requirement for species-specific development and subject to genotyping anomalies such as null alleles. But with the advancement in high-throughput sequencing and SNPs tagging, they are becoming more effective and viable.

### *19.3.2 Single Nucleotide Polymorphisms (SNPs)*

Microsatellites used to dominate much of avian population genetics, but their liability in the guise of stutter artifacts, allelic dropout and elevated mutation rates frequently complicates large-scale analyses, particularly on deteriorated samples [51]. SNPs have proven to be a less messy, more effective substitute, though. These single base-pair substitutions occur universally throughout the genome and produce solid, high-throughput data in vast numbers [21]. With stable heredity, low rates of errors and easier analysis, SNPs are easily integrated into contemporary genomic research. Methods such as whole-genome sequencing (WGS), RAD-seq and RNA-seq enable scientists to ascertain thousands of SNPs even in non-model aquatic birds [91].

### *19.3.3 Applications in Aquatic Birds*

SNPs have revolutionized aquatic bird genetics. On the population side, they've identified morphologically non-obvious genetic divisions between puffins and common terns that helped inform targeted conservation at local scales [61, 73]. For hybrid identification, particularly in Anas spp. ducks, SNPs surpass traditional markers with more precise tracking of gene flow in areas impacted by human-induced range overlaps [65, 66].

In evolutionary research, SNPs provide fine-scale details. Recolonization pathways of common cormorants and genetic breaks between Atlantic and Pacific kittiwake colonies were reconstructed with genome-wide SNP data emphasizing the importance of glacial history and oceanic barriers [58].

SNPs also plays a critical role to play in aid of conservation management. For instance, low genetic diversity in threatened birds such as the Australasian bittern or Madagascar pochard was determined using SNPs, and this gave rise to properly planned strategies such as habitat re-creation or controlled breeding [36, 69]. Even in wildlife forensics, SNP panels came handy to trace seized bird species from feathers or degraded tissue for tighter action against illicit trade [70]. Despite their utility, SNP markers face challenges as their biallelic nature provides less information per marker compared to multiallelic microsatellites. SNP discovery requires high initialinvestments. SNP discovery may miss rare or divergent alleles not represented in the reference genome, often referred to as reference bias.

## 19.4 Environmental DNA (eDNA) Application in Avian Monitoring

Monitoring bird biodiversity is essential for conservation, but traditional methods (e.g., visual surveys, mist-netting, acoustic recordings) are time-consuming, biased and often miss elusive species. eDNA analysis provides a non-invasive, efficient alternative by detecting genetic material shed into the environment, initially popularized in aquatic systems and now adapted for aquatic birds [137].

**Principles of eDNA Monitoring** DNA consists of DNA from the skin, feathers, faeces and secretions released into water, soil, or air. It is extracted and analyzed using qPCR, metabarcoding, or NGS to identify species. In aquatic settings, avian eDNA persistence depends on temperature, UV, microbial activity and pH [9]. Detection efficiency is influenced by species size, behaviour and density, with waterfowl being ideal candidates [124]. Universal primers targeting mitochondrial genes (12S, 16S, COI) enable multi-species detection in biodiversity studies [137]. High-throughput platforms such as Illumina MiSeq and Oxford Nanopore facilitate massive-scale avian eDNA surveillance as shown in Fig. 19.5.

**Fig. 19.5** Workflow of SNP discovery and genotyping applied to aquatic bird species identification. (Source: Author)

### 19.4.1 Applications of Avian Monitoring with eDNA

Environmental DNA (eDNA) has proven to be a sensitive and non-invasive method for monitoring avian biodiversity, particularly in aquatic and wetland ecosystems. It can identify rare or elusive species that would otherwise remain undetected in normal surveys. For example, the Japanese yellow bunting, a critically endangered species, was detected by eDNA signatures in forest water pools [137]. Stopover habitats of migrants and bird communities may be monitored using estuary and wetland eDNA sampling, guiding conservation planning along migratory flyways.

eDNA also facilitates disease monitoring through the detection of avian hosts of zoonotic diseases such as avian influenza viruses in fecally contaminated water [47]. eDNA monitoring on a recurring frequency also assists in habitat assessment with results of bird assemblage structure, use and effectiveness of wetland restoration and management interventions.

**Case Studies: Airborne eDNA for Urban Bird Monitoring**
New breakthroughs have demonstrated avian eDNA may be recovered from air filters within urban park settings to track species like pigeons (Columba livia), sparrows (Passer domesticus) and even raptors [77]. It is a non-invasive method that is potential for tracking birds within the urban environment.

**Limitations**
The primary threat involves fast eDNA degradation preventing detection. Contamination or stochastic detection in low-abundance samples can lead to false positives. Taxonomic resolution is constrained by the absence of complete avian barcodes. Avian eDNA protocols are not established yet, resulting in non-consistency of results.

## 19.5 Genomic tools for Conservation Management

Aqua bird species are increasingly exposed to environmental stressors such as habitat loss, pollution and climate change, population declines and loss of genetic diversity. Standard conservation practices hardly detect cryptic species and inchoate population declines, for which genomic tools are used to measure genetic variation, connectivity and adaptation markers. Such evidence forms the basis of the determination of Evolutionarily Significant Units (ESUs) and Management Units (MUs) towards focused conservation [3, 141].

**Importance of Genomics in Avian Conservation**
Genomic tools provide increased insight into species biology to guide conservation through maintenance of genetic diversity, control of inbreeding, and determination of population connectivity. Genomics is also used to assess adaptive reaction to environmental stresses and taxonomic uncertainty in migratory waterfowl such as was found in Pacific seabird research [60].

### *19.5.1 Habitat Suitability and Genetic Connectivity*

Waterbirds, sentinel species of ecosystems, demand concerted use of habitat suitability models (HSMs) and genetic connectivity analysis. HSMs forecast species occurrence from environmental information, whereas genetic connectivity can evaluate how far the traits of a landscape control gene movement [118, 120]. Ideal conservation practice combines both to facilitate productive populations in landscapes.

**Modelling Approaches**

(a) **MaxEnt:** Presence-only habitat model of suitability [104]. Applied in the black-necked stilt case [111], it defines the important wetland areas to be conserved.
(b) **Species Distribution Models (SDMs):** These models, e.g., presence-absence data, supplement habitat prediction with algorithms like Generalized Linear Models (GLMs), Boosted Regression Trees (BRT) and Random Forests [32]. GLMs, e.g., were used to identify dabbling ducks' habitat preference [46], whereas Random Forests used to predict black-faced spoonbills' habitat quality [149].

**Genetic Connectivity in Aquatic Birds**

Genetic connectivity is most crucial to long-term species survival, especially in the context of habitat fragmentation and loss. Even with extensive movement, aquatic birds are generally constrained by several genetic barriers to divergence. Recent genomic tools, such as mitochondrial DNA and genome-wide SNPs, detect fine-scale genetic structuring [85]. Conservation measures need to maintain genetic linkedness with habitat corridors and maintain migratory corridors.

**Integrating Habitat Suitability and Genetic Connectivity**

One of the strong means of conserving birds is the synergy of genomic data and habitat suitability models (HSMs). The complementariness enables the determination of connectivity hotspots necessary to preserve genetic exchange between the broken-up populations, as evidenced in a study on the black stork where connectivity-planning enhanced movement corridors [74]. Through the projection of genetic patterns onto ecological matrices, it is simpler to rank significant habitats for conservation, say, in the case of the Australasian bittern [40]. Integration makes adaptive management possible where conservation strategy varies as a function of contemporary ecological and genetic feedback exemplified by whooping crane reintroductions [26].

### *19.5.2 Conservation Breeding Programs and Genetic Health*

Conservation breeding is to be conducted to conserve aquatic avian under climatic change, habitat loss, hunting and pollution threats. Conservation schemes have evolved from population keeping to genetic management programs to sustain evolutionary potential. There is need for genetic fitness with high genetic diversity and low inbreeding to achieve disease resistance, reproductive accomplishment and adaptation [38].

**Principles of Conservation Breeding** Conservation breeding programs, e.g., the Madagascar pochard (*Aythya innotata*), include ex-situ management for protection against extinction and for reintroduction. Genetic studies, including SNP markers, inform breeding strategy by identifying genetic lineages and breeding different individuals separately to develop a healthy, genetically diverse population [69].

**Genetic Objectives in Conservation**
Genetic conservation is designed to conserve allelic diversity by protecting rare alleles, avoid genetic drift that leads to random allele frequency changes, reduce inbreeding by not breeding close relatives and provide even genetic representation across founders. All these are obtained through technologies such as molecular genotyping, pedigree reconstruction and genome-wide analysis. Contemporary breeding schemes combine these with dynamic conservation approaches to enable evidence-based decision-making in the preservation of genetic health and adaptability potential in at-risk avian populations. Some of the conservation methods employed are:

**Pedigree and Studbook Management**
Breeding management uses computer programs such as SPARKS and PMx to handle detailed pedigree information and breed for the best pairings. For instance, molecular pedigree reconstruction in the African penguin (*Spheniscus demersus*) is used to reduce inbreeding [25].

**Re-introduction and Post-release Monitoring**
Genetic monitoring at release guarantees the effective reintroduction of captive-reared animals to wild populations. Genetic evaluation of reintroduced white-faced ibis (*Plegadischihi*), for example, ascertained minimal genetic differentiation, witnessing the effectiveness of the breeding program [119] as shown in Table 19.3.

### *19.5.3 Monitoring Genetic Diversity in Wild Aquatic Bird Populations*

The genetic diversity of wild aquatic bird populations must be monitored to measure their capacity for evolution, their resistance to environment change and general conservation value. Genetic variation constitutes the basis for population viability and robustness, disease resistance, breeding success and adaptive capacity to novel environmental conditions [38]. Aquatic birds inhabiting wetland ecosystems, which occur in productive and most often threatened wetland habitats, are subjected to increasing impacts of climate change, human disturbance and habitat degradation. It

**Table 19.3** Conservation breeding in aquatic birds

| Species | Region | Conservation threats | Molecular tools used | Conservation actions | References |
|---|---|---|---|---|---|
| Whooping Crane (*Grus americana*) | North America | Severe population bottleneck in the 1940s, reduced genetic diversity | Microsatellites, mtDNA | Evaluation of relatedness, assisted reproductive technologies (ART), cryopreservation. | [55] |
| Hawaiian Duck (*Anas wyvilliana*) | Hawaii | Hybridization with invasive mallards (*Anas platyrhynchos*) | mtDNA, nuclear DNA markers | Identification of pure species, selective captive breeding | [35] |
| Madagascar Teal (*Anas bernieri*) | Madagascar | Wetland degradation, small population size | Molecular analysis (e.g., genetic monitoring) | Founder representation, population viability assessments, reintroduction programs | [37] |
| Conservation Breeding in India: Aquatic Avian Species | | | | | |
| Sarus Crane (*Antigone antigone*) | Uttar Pradesh & Gujarat, India | Population decline due to anthropogenic pressures | Microsatellites, mtDNA | Genetic bottleneck assessment, improved founder representation, inter-center genetic exchange | [90] |
| White-bellied Heron (*Ardea insignis*) | Arunachal Pradesh, India & Bhutan | Critically low global population (<250 individuals) | Non-invasive sampling (feathers, droppings), baseline genetic studies | Habitat monitoring, genetic rescue planning | Wildlife Institute of India,WII, ongoing |
| Greater Adjutant Stork (*Leptoptilos dubius*) | Assam & Bihar, India | Range contraction, population decline | Genetic screening | Establishing unrelated breeding pairs, community-based conservation, chick rescue | Community-led projects; conservation NGOs; ongoing |

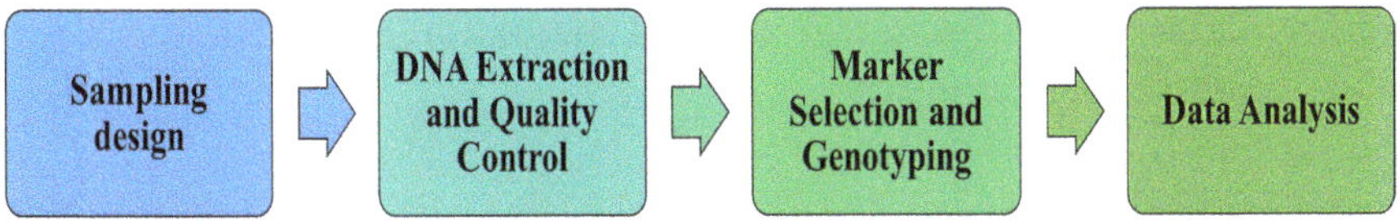


**Fig. 19.6** Workflow for Genetic Diversity Monitoring (Source: Author)

is crucial that their genetic variation is tracked closely to enable evidence-based conservation planning and management.

**Importance of Genetic Diversity**

Genetic diversity is the degree of variation of a species' genetic traits and plays a significant role in reproductive fitness, survivability, and long-term adaptability [4]. It enhances a population's ability to survive environmental variation and, apart from that, reduce threats from inbreeding and evolutionary stability. Genetic monitoring in aquatic birds can reveal population structure, dynamics of gene flow and impacts of fragmentationso providing well-informed conservation action.

**Methodology for Genetic Diversity Monitoring**

1. **Sampling Design:** Sampling site and time should be carefully selected. Sampling must be able to capture every aspect of the environment and geography of the population.
2. **Integrity Control of DNA**: DNA is isolated according to routine procedures and integrity confirmed by spectrophotometry and gel electrophoresis.
3. **Marker Selection and Genotyping:** The markers are chosen according to the G-genomic characteristics of the species. After PCR amplification and sequencing or genotyping.
4. **Data Analysis:** Genetic diversity is quantified using heterozygosity, allelic richness, F-statistics and Bayesian clustering algorithms (e.g., STRUCTURE analysis) as shown in Fig. 19.6

**Applications in Population Management**

Genetic monitoring discriminates between various population subunits, usually referred to as ESUs or MUs, which demand special conservation treatment. Great crested grebe (Podiceps cristatus), for example, had geographically structured populations that influence conservation at the regional level [72]. This study is relevant to dispersal effects, migratory connectivity, and habitat fragmentation. Different conservation zones were established in the case of the Andean flamingo (Phoenicoparrusandinus) according to the genetic results that were generated [79]. The integration of genetic monitoring and ecology allows early detection of declines, evaluation of conservation action and evidence-based management decisions [125].

**Case Studies: eDNA for Migratory Ducks in North America**

eDNA assays have been used to tag the presence and estimate genetic diversity of migratory ducks along flyways for non-invasive management and monitoring [42].

## 19.6 Advances in Next-Generation Sequencing (NGS) and Bioinformatics

Next-Generation Sequencing (NGS) revolutionized avian genomics because it enables high-throughput, fast DNA and RNA sequencing. Methods like Whole Genome Sequencing (WGS), RAD-seq and RNA-seq are now widely employed to study genetic diversity, reveal adaptive characteristics and evaluate population structure in waterbirds [30, 87]. RAD-seq makes it possible to achieve low-cost SNP detection genome-wide, for example, whereas RNA-seq translates gene expression in reaction to environmental stress.

NGS platforms produce enormous amounts of data, which are analyzed with sophisticated bioinformatics pipelines such as sequence alignment (e.g., using BWA or Bowtie), variant calling (e.g., using GATK) and functional annotation (e.g., Blast2GO). These allow for the identification of low-frequency alleles, phylogenetic inference and the inference of past demography.

Conservationally, this method is useful in the identification of evolutionarily significant units (ESUs), tracking genetic health and gene mapping for traits such as disease resistance and salt tolerance. For instance, genomic scans can detect loci that are under selection in plants that live in saline wetlands or estuary ecosystems. Interfusion with GIS-based ecological information also enhances habitat-specific conservation strategies [1, 71]. With declining sequence costs, the integration of NGS with machine learning and ecological modeling can potentially make the conservation efforts for aquatic birds more responsive and predictive.

### *19.6.1 Whole-Genome Sequencing and Comparative Genomics*

Whole-Genome Sequencing (WGS) offers high-resolution analysis of genetic composition, which identifies genes that participate in osmoregulation, thermoregulation and water-proofing feathers of aquatic birds. Variant identification and structural genome analysis by Illumina (short-read) and PacBio/ONT (long-read) platforms are offered [148]. Comparative Genomics contrasts aquatic bird genomes against reference species to identify conserved and lineage-specific genes with molecular adaptation to aquatic environments [34]. These approaches allow genome assembly, annotation and the detection of evolutionary signatures such as gene loss, duplication and transposable element dynamics [31, 52, 56].

### 19.6.2 High-Resolution Genomic Data for Adaptive and Functional Insights

Whole-genome sequencing (WGS) provides an avenue for producing high-quality reference genomes in aquatic bird species to explore aquatic adaptation mechanisms at the genetic level. By employing next-generation sequencing platforms like Illumina, PacBio and Oxford Nanopore, scientists can produce well-contiguous assemblies with deep depth, which reveal genomic variants such as single nucleotide polymorphisms (SNPs), structural variants and copy number variations (CNVs) [31, 80]. Such resolution data sets are critical in the identification of genetic causes of aquatic birds' phenotypic traits, adaptive ecology, and evolutionary processes.

**Genomic Insights into Aquatic Adapations**
WGS data make possible the identification of genomic regions responsible for the functional adaptations in aquatic birds. These include, for instance, lipid metabolism genes, osmoregulation genes, feathers waterproofing and diving physiology genes which can be identified and allocated to the bird's varied survival strategies in the environment. Functional annotation tools like MAKER, BUSCO andInterProScan are used to predict genes and orthologs and demographic models like PSMC (Pairwise Sequentially Markovian Coalescent) and MSMC (Multiple Sequentially Markovian Coalescent) to infer population histories and evolutionary dynamics [34, 148]. All these genomic findings are significant in addressing aquatic birds' adaptation to environmental change.

**Case Study: Great Crested Grebe (*Podiceps cristatus*) [148]**
The Great Crested Grebe is a good example of the way WGS has clarified functional genetic adaptations. Through the genome sequencing of this bird, scientists found genetic variants related to major aquatic adaptations:

- **Lipid Metabolism**: Genes involved in lipid metabolism like APOE and CYP2E1 were subject to positive selection, indicating effective energy storage during extended dives [86].
- **Feather Structure**: Genes Genes that contribute to feather structure and water repellency, including KRT8 and KRT18, were shown to be critical in staying dry and afloat when diving [13].

**Transport of Oxygen** Mutations in the HBB (haemoglobin subunit beta) and HBA (haemoglobin subunit alpha) genes were identified that enable transport and storage of oxygen in diving birds that live in low-oxygen habitats, as shown in Table 19.4 [17, 23].

**Table 19.4** Key genetic adaptations in the great crested grebe

| Gene | Function | Adaptation |
|---|---|---|
| *APOE* | Lipid metabolism | Efficient fat storage for energy during dives |
| *KRT8, KRT18* | Feather structure | Waterproofing to maintain buoyancy and dryness |
| *HBB, HBA* | Oxygen transport | Enhanced oxygen storage for deep and prolonged dives |

These genomics results of aquatic adaptation provide improved understanding of species adaptability to environmental stress like climate change and habitat destruction. Additionally, use of sophisticated demographic models like PSMC allows reconstructing such past population history of the species holding valuable data on potential population bottlenecks and adaptability of aquatic birds.

## *19.6.3 Transcriptomics and Functional for Conservation*

In recent decades, the use of transcriptomics and functional genomics has also offered new opportunities to conservation biology. They allow scientists to bypass traditional genetic markers and explore gene expression and functional characters when exposed to different environmental stresses [121]. By studying gene expression patterns and regulatory strategies, scientists are able to reveal more about organism-environment interaction with stressors such as habitat destruction, global warming and illness. This level of consciousness on a molecular basis is required in order to differentiate adaptive traits, make population health estimates and inform targeted conservation approaches. Lastly, transcriptomics and functional genomics are an important tool for closing genotype to phenotype in a way that has direct utility for species persistence and ecosystem resilience [133].

### 19.6.3.1 RNA-Seq Applications in Understanding Stress Response Mechanisms

RNA-Seq technology facilitates high-throughput measurement of transcript abundance under environmental conditions and in tissues. Transcriptomic analysis has revealed cellular and physiological mechanisms of stress in aquatic birds under salinity changes, temperature fluctuations and heavy metals pollution [134].

For instance, transcriptomic signatures of petroleum hydrocarbon exposure in snow goose (Anser caerulescens) are characterized by upregulation of genes related to oxidative stress, apoptosis, and DNA repair [134]. HISAT2, STAR and DESeq2 computational pipelines and software facilitate differential gene expression analysis, co-expression network reconstruction and GO and KEGG functional enrichment analyses. Such discoveries are crucial in the determination of molecular biomarkers of environmental stress that can be used in early-warning systems of ecosystem monitoring [75].

### 19.6.3.2 Immune System Genomics for Disease Susceptibility Assessments

Immunogenomics is also at the center of explaining the genetics of resistance and susceptibility to diseases, especially in wildlife populations like aquatic birds

[95]. Such species usually carry zoonotic diseases like avian influenza with catastrophic impacts on human and wildlife population health. The application of genomic tools, especially immune transcriptomics, enables scientists to study the hereditary aspect of immune reactions in aquatic birds and carries important implications for susceptibility to disease and resistance [144].

**Key Immune Genes and Pathways in Aquatic Birds**

Water birds have sophisticated immunity systems that consist of innate and adaptive elements that enable them to be highly efficient in fighting their pathogenic environments. The core immune factors such as Toll-like receptors (TLRs), cytokines and Major Histocompatibility Complex (MHC) genes are the focal point of their immune defense mechanisms.

1. **Toll-like Receptors (TLRs):** They belong to the innate immune system and have been shown to be involved in the detection of pathogens. TLRs recognize pathogen-associated molecular patterns (PAMPs) and activate downstream immune responses. Genetic differences within TLR genes, specifically TLR4 and TLR7, have been associated with viral infection susceptibility in aquatic birds, including avian influenza [95].
2. **Cytokines**: These cytokines are involved in immune regulation and inflammation. Pro-inflammatory cytokines such as IL-6 and TNF-$\alpha$ may be modulated by the host during infection so that the host can elicit an effective immune response [144].
3. **Major Histocompatibility Complex (MHC) Genes**: MHC genes are at the core of the adaptive immune system, presenting pathogen-derived peptides to T cells. MHC gene polymorphism is generally associated with increased resistance to disease through more versatile immune responses [121].

**Genomic Insights into Disease Susceptibility**

Studies in aquatic bird populations, especially free-ranging waterbirds, have described the function of polymorphisms in immune gene polymorphisms in regulating vulnerability to diseases. As an illustration, these studies have determined that polymorphisms in TLR4 correlate with variant avian influenza virus (AIV) susceptibility, with individual alleles playing a significant role in heightened resistance to infection.

One such research in migratory waterbirds, for example, the mallard duck (*Anas platyrhynchos*), demonstrated how polymorphisms in TLR4 and TLR7 have been associated with immune response variation against the influenza A virus in a way that some of the variants were correlated with diminished viral load and increased immune clearance [132]. This finding is critical to establishing how specific genetic variations influence resistance to disease in wild populations and illuminates how immune system diversity can influence persistence of pathogens in the environment as shown in Table 19.5.

**Table 19.5** Role of TLR gene polymorphisms in AIV resistance in mallard ducks

| Gene | Polymorphism | Impact on Disease Resistance | Pathogen |
|---|---|---|---|
| TLR4 | Single Nucleotide Polymorphism (SNP) | Certain alleles confer increased resistance to AIV | Avian Influenza Virus (AIV) |
| TLR7 | Polymorphism | Increased viral clearance in specific variants | Avian Influenza Virus (AIV) |

**Table 19.6** Disease management strategies based on immune genomics

| Strategy | Application |
|---|---|
| Genomic Surveillance | Monitoring pathogen spread and identifying resistant populations |
| Targeted Vaccination | Vaccinating high-risk populations based on immune genomic insights |
| Conservation Prioritization | Prioritizing conservation efforts for populations with low immune gene diversity |

Such genomic information are highly beneficial to wildlife managers and conservation biologists in prioritizing populations for disease surveillance, vaccination drives, and conservation efforts. Knowledge of the genetic susceptibility to disease resistance in the context of avian influenza allows high-risk groups to be identified and insights into intervention measures that target disease transmission.

**Practical Implications for Disease Management**

The incorporation of immune system genomics in wildlife disease management allows for more proactive and specific interventions to contain diseases. For example:

- **Vaccination Programs**: Genomic data regarding immune gene polymorphisms allow for the identification of specific populations which are to be targeted with vaccination to achieve maximal resource allocation and improve the efficacy of vaccination [45].
- **Monitoring Disease Resistance**: Genomic surveillance is able to track disease-resistant allele dispersal across populations and identify early warning signs of future disease risk and imminent epidemics [2].
- **Conservation Prioritization** Identifying populations that have lowered genetic diversity in immune-related genes can help conservationists prioritize these populations for genetic management or translocation to more genetically stable populations, as illustrated through experiments with amphibian populations [122] as discussed in Table 19.6.

Incorporation of immune system genomics into disease susceptibility testing can prove to be a valuable tool with which to control and minimize infectious disease impact on water bird populations. By ascertaining meaningful immune gene polymorphisms, scientists can ascertain populations more vulnerable to disease and alter measures to safeguard those species. Also, genomic monitoring and cross-species alignments expose useful insights into how pathogens drive immune system evolution, supporting pre-emptive conservation strategies to prevent disease outbreaks and preserve biodiversity.

### *19.6.4 Epigenetics in Response to Environmental and Anthropogenic Stressors*

Epigenetics, that is, the investigation of heritable variations in gene expression that are not explained by changes in the underlying DNA sequence, is involved in aquatic bird response to environmental and human-induced stressors. The stressorspollution, habitat lossand climatic changehave the potential to initiate adaptive epigenetic modifications providing populations with a fast but reversible way of responding to environmental stress. DNA methylation, histone modification and non-coding RNA expression are the best described epigenetic processes of which DNA methylation is best described in environmental research [135, 138].

#### 19.6.4.1 DNA Methylation, Cytosine Base Methylation, Plays a Part in Gene Expression by Down-Regulating Gene Function

The alteration makes it a suitable biomarker for establishing the impact of environmental pollutants and climate stressors on water birds. New techniques such as Whole-Genome Bisulfite Sequencing (WGBS) and Reduced Representation BisulfiteSequencing (RRBS) are available for studying the complex methylation patterns in aquatic bird species' genomes and generating informative insights on how aquatic birds respond to environmental changes [16, 88].

**Case Study: Black-Headed Gull (*C. ridibundus*) and Pollutants**
Experiments on black-headed gull (*C. ridibundus*) populations inhabiting areas with high levels of pollutants have exhibited higher DNA methylation in the promoter regions of detoxification genes such as CYP1A and GST. Such genes are crucial for detoxifying pollutants and toxins, and their overexpression by methylation may be a sign of an adaptive response to pollution. These results identify the promise of DNA methylation as a potential future biomarker for early pollution stress with a non-invasive approach to monitoring the health of aquatic bird populations [117, 14].

#### 19.6.4.2 Climate Stress and Methylation Shifts

Similarly, the alleviation of stressors including temperature changes together with nutritional deprivation, induced by climate change, has been shown to alter the methylation of the stress regulation genes. For example, elevated methylation levels of the NR3C1 and HSP70 genes have been brought about by temperature changes.

The former is involved in the response to the stress hormone cortisol and is referred to as a heat shock protein gene. These studies indicate that methylation of the DNA performs the function of aiding the organism in enduring environmental shifts and could allow for the possibility of being biologically passed on to make adaptations for fluctuant conditions in the DNA of future organism generations. [5, 129].

#### 19.6.4.3 Epigenomic Tools for Conservation and Monitoring

For the incorporation of epigenomic data into conservation efforts, researchers commonly employ one or more bioinformatics algorithms for calling differentially methylated regions (DMRs) and evaluating their significance. A variety of analytical frameworks, such as MethylKit and MethGo, are often used to analyse methylation data at scale, and the small differences in gene expression that relate to environmental stressors. Now, by adding epigenomic information to conservation predictions, scientists can improve their predictions of how aquatic bird populations will fare based on ongoing environmental challenges. This can supplement tactics such as:

1. **Monitoring Population Health:** Non-invasive epigenetic biomarkers for assessing environmental stress and disease risk in free-ranging wild populations.
2. **Risk Assessments:** Identification of DMRs in important functional genes which may facilitate better risk assessments especially in high environmental degradation areas.
3. **Long-term Conservation Planning:** Understanding inheritance of stress-induced epigenetic alterations will be useful for predicting future population trends and for designing interventions to protect endangered species.

Epigenetics of DNA methylation provides deep understanding of the reaction of aquatic birds to increasing stressors of pollution, global climate change, and habitat destruction. Conservationists will be able to make more accurate estimates of population health and resilience by knowing how environmental stressors cause heritable epigenetic changes in genes. With advancing molecular tools, epigenetic information will increasingly contribute to data-driven, long-term conservation management of aquatic birds and the protection of biodiversity in a time of unprecedented environmental change.

### *19.6.5 Integrating Genomic Approaches into Aquatic Bird Conservation*

Wetland birds are under increasing pressure from numerous factors, including habitat loss, pollution and climate change. Although these traditional conservation methods are helpful, they may not include the depths of genetic changes that will impact how aquatic birds interact with these problems, respond to stress, adapt to change or recover their populations after decline. Increasingly, genomics is becoming an important not-so-secret weapon. Genomic studies of various aquatic bird species are allowing researchers to find genetic patterns that explain the specific genetic patterns behind the birds' responses. These patterns could help us better understand their population structure, movement among breeding areas and potentially long-term, successful survival. For instance, genomics can recognize previously unimaginable genetic bottlenecks or inbreeding that is not possible to quantify in terms of bird numbers alone.

Genomics can help wiser conservation provided it's employed judiciously. Directed conservation can be used for certain populations of birds that carry specific genes to survive in extraordinary environments like high salinity or changing water levels. Genomics enables us to learn about the effect of pressures on diversity from learning about a single species to learning about complete bird societies within wetlands. Overall, applying genomics to conservation for aquatic birds enables conservationists to bridge the gap from observations on the surface to deep genetic consciousness. The shift in strategy is required immediately to stay ahead of global conservation goals and make ecosystems stronger against continuously changing circumstances.

## 19.7 Future Perspectives in Aquatic Avian Genomics

With increasing pressure on waterbird populations, conservation through the use of genomics and real-time observation to safeguard and conserve waterbirds is progressing. Technologies such as environmental DNA (eDNA) sampling and portable sequencers allow researchers to now detect species presence, population dynamics and pathogen emergence in the field by applying sequencer-based analysis. The new technologies are especially critical in terms of tracking cryptic or migratory species in large and inaccessible areas. Besides, multi-omics approaches combining genomics, transcriptomics and metabolomics are creating a more integrative understanding of how species get acclimated to environmental stressors.

Beyond technical considerations, ethics and law are now paramount in genomic conservation. Projects need to use 'rules of engagement' such as the Nagoya Protocol and show equitable benefit-sharing, and comply with any data sovereignty laws which may be in effect. This is particularly true of rare species, and areas where active community-based custodial management is occurring. When TEK is incorporated into and used effectively as part of genomic conservation, conservation outcomes are maximized. TEK uses years of experience with species behaviour, migration and habitat use to improve genomic interpretations and provide new insights into sampling localities. Co-developing research with communities is shown to improve the scientific quality of research, build lasting connections, help realize better science, and further develop grounded and sustainable custodial conservation models.

In the following years, we will direct attention toward integrating high-resolution molecular data with field ecology and predictive models in an ethical and inclusive research context. This combination is important for developing conservation approaches that are scientifically sound, socially responsible and suitable for the local context.

## 19.8 Conclusion

Aquatic birds are crucial to the environment because of their various ecological roles. They are the vital indicators of ecosystem health as their population, genetic diversity and behaviour reflects the kind of environment they are living in. Their survival constantly being challenged by habitat loss, pollution, climate change, and human disturbances. Genomic tools have transformed our ability to study, monitor and conserve these species. From DNA barcoding for accurate species identification to microsatellites and SNPs for understanding population structure, each tool offers unique insights into genetic diversity, adaptation, and connectivity. Environmental DNA (eDNA) has emerged as a non-invasive method for detecting species presence, even in remote or sensitive habitats, while next-generation sequencing and bioinformatics are enabling the discovery of fine-scale genomic signatures of adaptation to aquatic life. Genomic tools help scientists in keeping aquatic bird populations genetically healthier and better prepared for future environmental changes. By combining genetic data with field knowledge, we can design more precise and effective conservation plans. In future, improved genomic methods and strong collaboration between scientists, conservationists, and policymakers will be key to ensuring these birds continue to thrive in our ecosystems.

## References

1. Albert TJ, Molla MN, Muzny DM, Nazareth L, Wheeler D, Song X et al (2007) Direct selection of human genomic loci by microarray hybridization. Nat Methods 4(11):903–905
2. Albery GF, Becker DJ, Brierley L, Brook CE, Christofferson RC, Cohen LE, Dallas TA, Eskew EA, Fagre A, Farrell MJ, Glennon E, Guth S, Joseph MB, Mollentze N, Neely BA, Poisot T, Rasmussen AL, Ryan SJ, Seifert S, Sjodin AR et al (2021) The science of the host-virus network. Nat Microbiol 6(12):1483–1492
3. Allendorf FW, Funk WC, Aitken SN, Byrne M, Luikart G (2022) Conservation and the genomics of populations, 2nd edn. Oxford University Press
4. Allendorf FW, Luikart G, Aitken SN (2013) Conservation and the genetics of populations. Wiley
5. Anastasiadi D, Díaz N, Piferrer F (2017) Small ocean temperature increases elicit stage-dependent changes in DNA methylation and gene expression in a fish, the European sea bass. Sci Rep 7:12401
6. Antil S, Abraham JS, Sripoorna S, Maurya S, Dagar J, Makhija S et al (2023) DNA barcoding, an effective tool for species identification: a review. Mol Biol Rep 50(1):761–775

7. Avise JC (2004) Molecular markers, natural history and evolution, 2nd edn. Sinauer Associates
8. Baker AJ, Dittmann DL (2018) Phylogenetic analysis of avian populations using β-fibrinogen intron 7. Mol Phylogenet Evol 120:1–13
9. Barnes MA, Turner CR (2016) The ecology of environmental DNA and implications for conservation genetics. Conserv Genet 17(1):1–17
10. Barrowclough GF, Cracraft J, Klicka J, Zink RM (2016) How many kinds of birds are there and why does it matter? PLoS One 11(11):e0166307
11. Baumgart SL, Sereno PC, Westneat MW (2021) Wing shape in waterbirds: morphometric patterns associated with behavior, habitat, migration and phylogenetic convergence. Integr Org Biol 3(1):obab011
12. Beja-Pereira A, Oliveira R, Alves PC, Schwartz MK, Luikart G (2009) Advancing ecological understandings through technological transformations in noninvasive genetics. Mol Ecol Resour 9(5):1279–1301. https://doi.org/10.1111/j.1755-0998.2009.02699.x
13. Bininda-Emonds ORP, Cardillo M, Jones KE, MacPhee RDE, Beck RMD, Grenyer R, Price SA, Vos RA, Gittleman JL, Purvis A (2007) The delayed rise of present-day mammals. Nature 446(7135):507–512
14. Blom MP, Peona V, Prost S, Christidis L, Benz BW, Jønsson KA, ... Irestedt M (2024) Hybridization in birds-of-paradise: Widespread ancestral gene flow despite strong sexual selection in a lek-mating system. Iscience 27(7)
15. Bora FD, Popescu M (2024) The evolutionary dynamics of troglobionts: isolation, divergence and speciation. Extreme Life Biospeol Astrobiol 16(1):22–25
16. Bossdorf O, Richards CL, Pigliucci M (2008) Epigenetics for ecologists. Ecol Lett 11(2): 106–115
17. Bouckaert RR (2010) DensiTree: making sense of sets of phylogenetic trees. Bioinformatics 26(11):1372–1373
18. Bowser AK, Diamond AW, Addison JA (2013) From puffins to plankton: a DNA-based analysis of a seabird food chain in the Northern Gulf of Maine. PLoS One 8(12):e83152
19. Brenna-Hernandez AE, Niemann JM (2019) Genetic diversity of aquatic birds: applications of microsatellite markers. Aquat Conserv Mar Freshwat Ecosyst 29(3):389–400
20. Bribiesca-Contreras F, Parslew B, Sellers WI (2021) Functional morphology of the forelimb musculature reflects flight and foraging styles in aquatic birds. J Ornithol 162:779–793
21. Brumfield RT, Beerli P, Nickerson DA, Edwards SV (2003) The utility of single nucleotide polymorphisms in inferences of population history. Trends Ecol Evol 18(5):249–256
22. Campagna L, Toews DP (2022) The genomics of adaptation in birds. Curr Biol 32(20): R1173–R1186
23. Cannarozzi G, Schneider A, Gonnet G (2007) A phylogenomic study of human, dog and mouse. PLoS Comput Biol 3(1):e2
24. Chen Q, Lin H, Zheng C, Mudrik EA, Kashentseva TA, Cheng Y et al (2025) Understanding the past to preserve the future: genomic insights into the conservation management of a critically endangered waterbird. Mol Ecol 34(2):e17606
25. Clucas B, Marzluff JM (2012) Attitudes and actions toward birds in urban areas: human cultural differences influence bird behavior. The Auk 129(1):8–16
26. Converse SJ, Strobel BN, Barzen JA (2019) Reproductive failure in the eastern migratory population: the interaction of research and management
27. Dawson DA, Ball AD, Spurgin LG, Martín-Gálvez D, Stewart IRK, Horsburgh GJ et al (2013) High-utility conserved avian microsatellite markers enable parentage and population studies across a wide range of species. BMC Genomics 14:176
28. Dixit T (2024) A synthesis of coevolution across levels of biological organization. Evolution 78(2):211–220
29. Dodman T, Beaune D, Bichet C, Cadiou B, Carrié A, Diame A et al (2023) Seabird conservation handbook for West Africa (for Mauritania, Senegal, The Gambia, Guinea-

Bissau, Guinea, Sierra Leone & Cabo Verde). BirdLife International Partnership Secretariat in Africa, Dakar, Senegal
30. Ekblom R, Galindo J (2011) Applications of next generation sequencing in molecular ecology of non-model organisms. Heredity 107(1):1–15
31. Ekblom R, Wolf JBW (2014) A field guide to whole-genome sequencing, assembly and annotation. Evol Appl 7(9):1026–1042
32. Elith J, Leathwick JR, Hastie T (2008) A working guide to boosted regression trees. J Anim Ecol 77(4):802–813
33. Ellegren H (2004) Microsatellites: simple sequences with complex evolution. Nat Rev Genet 5(6):435–445
34. Ellegren H (2014) Genome sequencing and population genomics in non-model organisms. Trends Ecol Evol 29(1):51–63
35. Engilis Jr A (2002) The development of an archive for Central Valley bird records and specimens. CVBC Bulletin 5:70-71
36. Ewart KM, Pavlova A, Joseph L, Jelinek A, Donnellan SC, Murphy SA, Sunnucks P (2021) Genome-wide SNPs reveal conservation units in the endangered Australasian Bittern (*Botaurus poiciloptilus*). Emu Austral Ornithol 121(2):125–135
37. Fiolleau T, Roca R, Cloché S, Bouniol D, Raberanto P (2020) Homogenization of geostationary infrared imager channels for cold cloud studies using Megha-Tropiques/ScaRaB. IEEE Trans Geosci Remote Sens 58(9):6609–6622
38. Frankham R, Ballou JD, Briscoe DA (2018) Conservation genetics
39. Gauthier G, Giroux JF, Lochefort L (2006) The impact of goose grazing on arctic and temperate wetlands. Acta Zool Sin 52:108–111
40. Gilbert NA, Ferguson PF (2019) Disturbance, but not the right kind: Exurban development does not create habitat for shrubland birds. Wilson J Ornithol 131(2):243–259
41. Gill FB (2007) Ornithology, 3rd edn. W.H. Freeman
42. Goldberg CS, Turner CR, Deiner K, Klymus KE, Thomsen PF, Murphy MA, Taberlet P (2016) Critical considerations for the application of environmental DNA methods to detect aquatic species. Methods Ecol Evol 7(11):1299–1307
43. Goldstein DB, Schlötterer C (1999) Microsatellites: evolution and applications. Oxford University Press
44. Green AJ, Elmberg J (2014) Ecosystem services provided by waterbirds. Biol Rev 89(1): 105–122
45. Grueber CE, Wallis GP, Jamieson IG (2013) Genetic drift outweighs natural selection at toll-like receptor (TLR) immunity loci in a re-introduced population of a threatened species. Mol Ecol 22(17):4470–4482
46. Hagy HM, Straub JN, Schummer ML, Kaminski RM (2014) Annual variation in food densities and factors affecting wetland use by waterfowl in the Mississippi Alluvial Valley. Wildfowl 436–450
47. Haramoto E, Malla B, Thakali O, Kitajima M (2020) Use of human and animal viral markers to detect fecal contamination in environmental waters in Japan. Sci Total Environ 735:139523
48. Hebert PDN, Cywinska A, Ball SL, deWaard JR (2003) Biological identifications through DNA barcodes. Proc R Soc Lond B Biol Sci 270(1512):313–321
49. Hebert PD, Stoeckle MY, Zemlak TS, Francis CM (2004) Identification of birds through DNA barcodes. PLoS Biol 2(10):e312
50. Hernandez Camacho FB (2024) Signatures of selection, gene flow and genetic drift onthe genomes of a recent avian radiation
51. Hoffman JI, Amos W (2005) Microsatellite genotyping errors: Detection approaches, common sources and consequences for paternal exclusion studies. Mol Ecol 14(2):599–612
52. Jarvis ED, Mirarab S, Aberer AJ, Li B, Houde P, Li C, Ho SY, Faircloth BC, Nabholz B, Howard JT, Suh A, Weber CC, da Fonseca RR, Li J, Zhang F, Li H, Zhou L, Narula N, Liu L, Ganapathy G et al (2014) Whole-genome analyses resolve early branches in the tree of life of modern birds. Science 346(6215):1320–1331

53. Jenni L, Winkler R (2020) The biology of moult in birds. Bloomsbury Publishing
54. Johnson RN (2011) Conservation genetics and wildlife forensics of birds (Chap. 15). In: Huffman
55. Jones SG, Gordon DH, Phillips GM, Richardson BR (2005) Avian community response to a golf-course landscape unit gradient. Wildl Soc Bull 33(2):422–434
56. Kapusta A, Suh A, Feschotte C (2017) Dynamics of genome size evolution in birds and mammals. Proc Natl Acad Sci U S A 114(8):E1460–E1469
57. Kear J (ed) (2005) Ducks, Geese and Swans: Species accounts (Cairina to Mergus), vol 2. Oxford University Press
58. Keller D, Holderegger R, van Strien MJ, Bolliger J (2015) Spatial scale affects landscape genetic analysis of a wetland bird (*Acrocephalus arundinaceus*). Mol Ecol 24(3):602–615
59. Kerr KCR, Lijtmaer DA, Barreira AS, Hebert PDN, Tubaro PL (2009) Probing evolutionary patterns in neotropical birds through DNA barcodes. PLoS One 4(2):e4379
60. Kerr KCR, Stoeckle MY, Dove CJ, Weigt LA, Francis CM, Hebert PDN (2007) Comprehensive DNA barcode coverage of North American birds. Mol Ecol Notes 7(4):535–543
61. Kersten O, Burri R, Fumagalli M, Griesser M, Sæther B-E (2021) Complex population structure of the Atlantic puffin revealed by whole genome analyses. Nat Commun 12:1–11
62. Khan AA, Khan S, Al-Shehri AM (2012) Patterns of genetic diversity and population structure of the threatened Houbara and Macqueen's bustards as revealed by microsatellite markers. Genet Mol Res 11(3):2444–2451
63. Kiran, Singh D, Kour A, Yadav R, Dahiya P, Kumar R, Sharma K (2023) The avian gut microbial community: driving factors and role in wild birds. In: Latest trend in zoology and entomology science, vol 14. Akinik Publications., ISBN: 978-93-5570-676-8, pp 15–31
64. Kushlan JA (1976) Feeding behavior of North American herons. Auk
65. Lavretsky P, McCracken KG, Peters JL (2016) Phylogenetics and genomics reveal polyphyly in the mallard complex (*Aves: Anas*). Auk 133(2):197–210
66. Lavretsky P, McInerney NR, Mohl JE, Peters JL (2024) Human-induced range expansions result in a recent hybrid zone between sister species of ducks. Genes 15(6):651
67. Lees AC, Haskell L, Allinson T, Bezeng SB, Burfield IJ, Renjifo LM et al (2022) State of the world's birds. Annu Rev Environ Resour 47(1):231–260
68. Ležalová-Piálková R (2011) Molecular evidence for extra-pair paternity and intraspecific brood parasitism in the Black-headed Gull. J Ornithol 152(2):291–295
69. Li H, Lee J, Ratsimbazafy H, Székely T (2020) Genetic diversity and conservation of the critically endangered Madagascar pochard (*Aythya innotata*). Conserv Genet 21(3):519–531
70. Lim KC, Lim PE, Chong VC, & Loh KH (2015) Molecular and morphological analyses reveal phylogenetic relationships of stingrays focusing on the family Dasyatidae (Myliobatiformes). PLoS One 10(4):e0120518
71. Liu L, Yu L, Edwards SV (2009) Coalescent methods for estimating phylogenetic trees. Mol Phylogenet Evol 53(1):320–328
72. Liu Y, Wang N, Li J (2018) Population structure and gene flow in great crested grebes (*Podiceps cristatus*) based on microsatellite analysis. Avian Res 9:6
73. Lončar V, Kralj J, Stronen AV, Grgurević M, Pavlinec Ž, Jurinović L et al (2024) High genetic diversity yet weak population genetic structure in European common terns. Sci Rep *14*:Article 29173
74. López-López P, García-Ripollés C, Urios V (2007) Population size, breeding performance and territory quality of Bonelli's eagle Hieraaetus fasciatus in eastern Spain. Bird Study 54(3): 335–342
75. Love AC (2017) The sub-lethal impacts of waste water effluent and its components on Gammarus pulex (Doctoral dissertation). University of Portsmouth
76. Loyau A, Schmeller DS (2009) Polymorphic microsatellites identified by cross-species amplifications in the European Coot *Fulica atra*. J Ornithol 150(4):703–707

77. Lynggaard C, Bertelsen MF, Jensen CV, Johnson MS, Frøslev TG, Olsen MT, Bohmann K (2022) Airborne environmental DNA for terrestrial vertebrate community monitoring. Curr Biol 32(3):701–707.e5
78. MacDonald EC (2025) Sitting Ducks: The physiological and behavioural responses of incubating common eiders (*Somateria mollissima*) to heat stress in the Canadian Arctic (Master's thesis). University of Windsor (Canada)
79. Marconi V, Caziani SM, Derlindati EJ (2011) Population genetic structure of Andean flamingos from Argentina inferred from mitochondrial DNA sequences. Auk 128(4):699–710
80. Mardis ER (2008) The impact of next-generation sequencing technology on genetics. Trends Genet 24(3):133–141
81. Mariyappan M, Rajendran M, Velu S, Johnson AD, Dinesh GK, Solaimuthu K et al (2023) Ecological role and ecosystem services of birds: a review. Int J Env Clim Change 13(6):76–87
82. Maron JL, Estes JA, Croll DA, Danner EM, Elmendorf SC (2006) An introduced predator alters Aleutian Island plant communities by thwarting herbivory. Ecol Lett 9(8):937–944
83. Martin CA, Sheppard EC, Ali HA, Illera JC, Suh A, Spurgin LG, Richardson DS (2024) Genomic landscapes of divergence among Island bird populations: Evidence of parallel adaptation but at different loci? Mol Ecol 33(12):e17365
84. Mazar Barnett J (2001) Comentariosbibliográficos: Aves del mundo (del Hoyo et al.: Handbook of the birds of the world. Volume 4. Sandgrouse to cuckoos). El Hornero 16(01)
85. McClure CJ, Rolek BW, Hayes TI, Hayes CD, Thorstrom R, Curti M, Anderson DL (2017) Successful enhancement of Ridgway's Hawk populations through recruitment of translocated birds. The Condor: Ornithological Applications 119(4):855–864
86. McCormack JE, Faircloth BC, Crawford NG, Gowaty PA, Brumfield RT, Glenn TC (2012) Ultraconserved elements are novel phylogenomic markers that resolve placental mammal phylogeny when combined with species-tree analysis. Mol Phylogenet Evol 63(3):625–637
87. McCormack JE, Hird SM, Zellmer AJ, Carstens BC, Brumfield RT (2013) Applications of next-generation sequencing to phylogeography and phylogenetics. Mol Phylogenet Evol 66(2):526–538
88. McNew SM, Taff CC, Vitousek MN (2024) Manipulation of a social signal affects DNA methylation of a stress-related gene in a free-living bird. J Exp Biol 227(15):jeb246819
89. Miller CV, Bossu CM, Saracco JF, Toews DP, Rushing CS, Roberto-Charron A et al (2024) Genomics-informed conservation units reveal spatial variation in climate vulnerability in a migratory bird. Mol Ecol 33(1):e17199
90. Mirande CM, Harris JT (2019) Crane conservation strategy. International Crane Foundation, Baraboo, Wisconsin, USA
91. Morin PA, Luikart G, Wayne RK, the SNP workshop group (2004) SNPs in ecology, evolution and conservation. Trends Ecol Evol 19(4):208–216
92. Muzio FM, Rubega MA (2024) What do we really know about the water-repellency of feathers? J Avian Biol 2024(11–12):e03259
93. Navalón G, Bright JA, Marugán-Lobón J, Rayfield EJ (2019) The evolutionary relationship among beak shape, mechanical advantage and feeding ecology in modern birds. Evolution 73(3):422–435
94. Nebel S et al (2005) Shorebird foraging behavior in estuarine mudflats. Mar Ecol Prog Ser
95. Nie L, Cai SY, Shao JZ, Chen J (2018) Toll-like receptors, associated biological roles andsignaling networks in non-mammals. Front Immunol 9:1523
96. Nisha SJ, Uma G, Sathishkumar R, Prakash VSG, Isaac R, Citarasu T (2025) Optimization and characterization of bioactive secondary metabolites from Streptomyces sp CMSTAAHL-4 isolated from mangrove sediment. BMC Microbiol 25(1):57
97. Nudds RL, Bowlby JA (2010) Functional morphology of waterfowl beaks. J Avian Biol
98. Nwagu KE, Ibeh AG, Enwere EN, Udechukwu CD, Azumini P (2024) The role of ecogenomics in environmental and forensic discoveries. Bio-Research 22(1):2223–2233

99. Ogidi OI, Akpan UM (2022) Aquatic biodiversity loss: impacts of pollution and anthropogenic activities and strategies for conservation. In: Biodiversity in Africa: potentials, threats and conservation. Springer Nature Singapore, Singapore, pp 421–448
100. Pande A, Rawat N, Sivakumar K, Sathyakumar S, Mathur VB, Mondol S (2018) Cross-species screening of microsatellite markers for individual identification of snow petrel *Pagodroma nivea* and Wilson's storm petrel *Oceanites oceanicus* in Antarctica. PeerJ 6:e5243
101. Paxton KL, Van Riper C III, Theimer TC, Paxton EH (2013) Molecular genetic approaches to conservation of migratory birds in the face of global climate change. In: Lepczyk CA, Warren PS (eds) Urban bird ecology and conservation. University of California Press, pp 131–149
102. Pearce JM, Bollinger KS (2003) Morphological and genetic evidence for seasonal movement of northern pintails in Alaska. Waterbirds 26(4):384–393
103. Pereira SL, Baker AJ (2006) A mitogenomic timescale for birds detects variable phylogenetic rates of molecular evolution and refutes the standard molecular clock. Mol Biol Evol 23(9): 1731–1740
104. Phillips SJ, Anderson RP, Schapire RE (2006) Maximum entropy modeling of species geographic distributions. Ecol Model 190(3-4):231–259
105. Primmer CR, Møller AP, Ellegren H (1996) A wide-range survey of cross-species microsatellite amplification in birds. Mol Ecol 5(3):365–378
106. Raković M, Neto JM, Lopes RJ, Koblik EA, Fadeev IV, Lohman YV, Aghayan SA, Boano G, Pavia M, Perlman Y, Kiat Y, Ben Dov A, Collinson JM, Voelker G, Drovetski SV (2019) Geographic patterns of mtDNA and Z-linked sequence variation in the Common Chiffchaff and the 'chiffchaff complex'. PLoS One 14(1):e0210268
107. Ralls K, Ballou JD (2013) Captive breeding and reintroduction. In: Levin SA (ed) Encyclopedia of biodiversity, vol 1, 2nd edn. Academic, Waltham, pp 662–667
108. Ratnasingham S, Hebert PDN (2007) BOLD: the barcode of life data system (www.barcodinglife.org). Mol Ecol Notes 7(3):355–364
109. Razgour O, Forester B, Taggart J, Bekaert M, Juste J, Ibáñez C, Puechmaille S, Novella-Fernandez R, Alberdi A, Manel S (2019) Considering adaptive genetic variation in climate change vulnerability assessment reduces species range loss projections. Proc Natl Acad Sci USA 116:10418–10423
110. Reed JM, Elphick CS, Oring LW (1998) Life-history and viability analysis of the endangered Hawaiian stilt. Biol Conserv 84(1):35–45
111. Reynolds C (2019) Waterbirds: Birds of a feather flock together. Ostrich 90(4):iii-iv
112. Ribak G, Weihs D, Arad Z (2005) Submerged swimming of the great cormorant Phalacrocorax carbo sinensis is a variant of the burst-and-glide gait. J Exp Biol 208(20):3835–3849
113. Richards EJ, McGirr JA, Wang JR, St. John ME, Poelstra JW, Solano MJ et al (2021) A vertebrate adaptive radiation is assembled from an ancient and disjunct spatiotemporal landscape. Proc Natl Acad Sci U S A 118(20):e2011811118
114. Rijke AM (2023) The structure and functions of the contour feathers of water birds revisited. In: Birds-conservation, research and ecology. IntechOpen
115. Rincon-Sandoval M, De-Kayne R, Shank SD, Pirro S, Ko'ou A, Abueg L et al (2024) Ecological diversification of sea catfishes is accompanied by genome-wide signatures of positive selection. Nat Commun 15(1):10040
116. Romanov MN, Tuttle EM, Houck ML, Modi WS, Chemnick LG, Korody ML et al (2009) The value of avian genomics to the conservation of wildlife. BMC Genomics 10:1–19. BioMed Central
117. Rui QY, Li X, Zhang HB, Guo XM, Zheng N, Zhao L, Guo LQ, Li PH, Yue JJ (2020) Research on the relationship between environmental chemical pollutant exposure and epigenetics. Zhonghua Lao Dong Wei Sheng Zhi Ye Bing Za Zhi 38(3):237–240
118. Runge MC, Sauer JR, Avery ML, Blackwell BF, Koneff MD (2009) Assessing allowable take of migratory birds. J Wildl Manage 73(4):556–565
119. Salzburger W (2012) Introgressive hybridization between color morphs in a population of cichlid fishes twelve years after human-induced secondary admixis

120. Santini L, Benítez-López A, Maiorano L, Čengić M, Huijbregts MA (2021) Assessing the reliability of species distribution projections in climate change research. Divers Distrib 27(6): 1035–1050
121. Sarvari P, Sarvari P, Ramírez-Díaz I, Mahjoubi F, Rubio K (2022) Advances of epigenetic biomarkers and epigenome editing for early diagnosis in breast cancer. Int J Mol Sci 23(17): 9521
122. Savage AE, Zamudio KR (2016) Adaptive tolerance to a pathogenic fungus drives major histocompatibility complex evolution in natural amphibian populations. Proc R Soc B Biol Sci 283(1827):20153115
123. Schreiber RW, Burger J (2002) Biology of Marine Birds. CRC Press
124. Schütz R, Deiner K, Fronhofer EA, Altermatt F (2020) Taxonomic and functional distance from a focal species predict community patterns in stream eDNA metabarcoding. Mol Ecol 29(24):5061–5076
125. Schwartz MK, Luikart G, Waples RS (2007) Genetic monitoring as a promising tool for conservation and management. Trends Ecol Evol 22(1):25–33
126. Seeb JE, Carvalho G, Hauser L, Naish K, Roberts S, Seeb LW (2011) Single-nucleotide polymorphism (SNP) discovery and applications of SNP genotyping in nonmodel organisms. Mol Ecol Resour 11(s1):1–8
127. Shafer ABA, Wolf JBW, Alves PC, Bergström L, Bruford MW, Brännström I et al (2015) Genomics and the challenging translation into conservation practice. Trends Ecol Evol 30(2): 78–87
128. Smith SD, Johnson KP, Allen JM (2019) Myoglobin intron 2 as a phylogenetic marker in avian species: insights into evolutionary divergence. J Avian Biol 50(5):e01983
129. Sonna LA, Fujita J, Gaffin SL, Lilly CM (2002) Invited review: effects of heat and cold stress on mammalian gene expression. J Appl Physiol 92(4):1725–1742
130. Sonsthagen SA, Wilson RE, McCracken KG, Talbot SL (2019) Population genomics of the Common Eider (*Somateria mollissima*): evidence for genetic structure and local adaptation across a widespread sea duck. Ecol Evol 9(4):2106–2121
131. Srinivasan S, Chhatre SS, Guardado JO, Park KC, Parker AR, Rubner MF et al (2014) Quantification of feather structure, wettability and resistance to liquid penetration. J R Soc Interface 11(96):20140287
132. Staley MM (2016) Mechanisms and evolution of disease resistance in birds (Doctoral dissertation, Auburn University). Auburn University
133. Steiner CC, Putnam AS, Hoeck PE, Ryder OA (2013) Conservation genomics of threatened animal species. Annu Rev Anim Biosci 1:261–281
134. Suarez-Bregua P, Álvarez-González M, Parsons KM, Rotllant J, Pierce GJ, Saavedra C (2022) Environmental DNA (eDNA) for monitoring marine mammals: challenges and opportunities. Front Mar Sci 9:Article 987774
135. Sutherland JE, Costa M (2003) Epigenetics and the environment. Ann N Y Acad Sci 983:151–160
136. Taberlet P, Coissac E, Hajibabaei M, Rieseberg LH (2012) Environmental DNA. Mol Ecol 21(8):1789–1793
137. Ushio M, Fukuda H, Inoue T, Makoto K, Kishida O, Sato K et al (2018) Environmental DNA enables detection of terrestrial mammals from forest pond water. Mol Ecol Resour 17(6):e63–e75
138. Vaissière T, Sawan C, Herceg Z (2008) Epigenetic interplay between histone modifications and DNA methylation in gene silencing. Mutat Res 659(1–2):40–48
139. Wang Y, Liu Z, Zhang H, Xu Y (2024) Genetic adaptation in avian species to rapid environmental changes. Int J Mol Evol Biodivers 14(1):45–62
140. Watson H, Videvall E, andersson MN, Isaksson C (2017) Transcriptome analysis of a wild bird reveals physiological responses to the urban environment. Sci Rep 7(1):44180
141. White PJ, Searle JB (2007) Genetic diversity: sources, threats and conservation. In: Levin SA (ed) Encyclopedia of biodiversity, vol 3, 2nd edn. Academic Press, pp 179–189

142. Winker K (2021) An overview of speciation and species limits in birds. Auk 138(2):ukab006
143. Wissen JGM, Taylor MA, Hall BK (2007) Aquatic adaptations in the limbs of amniotes. In: Fins into limbs: evolution, development and transformation, pp 310–322
144. Wlaźlak S, Pietrzak E, Biesek J, Dunislawska A (2023) Modulation of the immune system of chickens: a key factor in maintaining poultry productiona review. Poult Sci 102(8):102785
145. Yadav R, Gupta RK, Kumar R, Kaur T (2022a) Assessment of toxicity of lead and nickel on the biochemical and immunological parameters of earthworm, eudrilus eugeniae. Agric Sci Dig 44(5):922–929
146. Yadav R, Gupta RK, Kumar R, Kaur T, Moun R (2022b) Biochemical response of earthworm, *Eisenia fetida* to heavy metals toxicity. J Appl Nat Sci 14(3):990–998
147. Yodha K, Singh D, Kour A, Yadav R, Kumar R (2024) Species diversity and abundance of birds in ber (*Zizyphus mauritiana*) crop. Asian J Conserv Biol 13(1). https://doi.org/10.53562/ajcb.83683
148. Zhang G, Li C, Li Q, Li B, Larkin DM, Lee C, Storz JF, Antunes A, Greenwold MJ, Meredith RW, Ödeen A, Cui J, Zhou Q, Xu L, Pan H, Wang Z, Jin L, Zhang P, Hu H, Yang W et al (2014) Comparative genomics reveals insights into avian genome evolution and adaptation. Science 346(6215):1311–1320
149. Zhang H, Ma Z, Zhang L, Tang C (2019) Modeling habitat suitability of black-faced spoonbills (*Platalea minor*) in East Asia using ensemble and random forest approaches. Ecol Indic 102:632–641

# Chapter 20
# Assessing Wetland Bird Diversity and Conservation in Dighal Wetlands (IBA Site), Haryana, India: Ecological Insights and Implications

**Parul, Parmesh Kumar, and Mayur Mukut Murlidhar Sharma**

**Abstract** An important bird and biodiversity area (IBA), the Dighal Wetlands of Haryana, India, sustains a rich variety of wetland avifauna especially during the winter migration. This chapter discusses a wide-range of two-year study that dealt with the avian diversity and season variations as well as feeding habits in the wetlands. The research recorded a total number of 90 bird species in wetlands that existed in 62 genera, 23 families, and 12 orders. A total of over 61 percent species identified was represented with the Charadriiformes, Anseriformes and Pelecaniformes orders. Anatidae family turned out to be the most diversified including 19 species highlighting the open water habitat conditions in the wetlands. Most of the wetland bird community was a blend of migratory birds, 62.2% of all the recorded species in the wetland were winter migrants. It implies that the Dighal Wetlands play an important role for stopover and spending winter by the birds migrating through the Central Asian Flyway. Bar-headed Goose (*Anser indicus*), Northern Shoveler (*Spatula clypeata*), and Northern Pintail (*Anas acuta*) were the common species, while Mallard (*Anas platyrhynchos*) and Peregrine Falcon (*Falco peregrinus*) were seen occasionally. Of the recorded species, 23 percent were rare not only some of them required a specific habitat and were thus of conservation concern. Wetland bird community consisted mainly of carnivorous birds, then came the omnivorous, insectivorous, and herbivorous. Foraging methods used were also documented such as grazing, surface feeding, upending, diving and tactile methods such as scything and foot trembling. These feeding patterns had a close correlation with the characteristics of the habitat, characteristics such as depth, vegetation cover,

Parul (✉)
Department of Zoology, Kurukshetra University, Kurukshetra, Haryana, India

P. Kumar
Department of Zoology, Institute of Integrated & Honors Studies, Kurukshetra University, Kurukshetra, Haryana, India

M. M. M. Sharma
J1 Research Collaborator, Department of Surgery, University of Illinois at Chicago, Chicago, IL, USA

 
R. Kumar et al. (eds.), *Ecology, Conservation of Aquatic Birds*, Wetlands: Ecology, Conservation and Management 15, https://doi.org/10.1007/978-3-032-17125-2_20

and prey abundance. Even though the Dighal Wetlands have ecological importance, the anthropogenic activities which are posing a threat are overfishing, pollution, growth and spread of invasive weeds, habitat fragmentation and new infrastructures. This paper highlights an urgent need to establish overall management strategies to control water level, reduce pollution, promote sustainable resource exploitation, and engage local communities in conservation efforts. This is necessary to safeguard ecological integrity of this important wetland. Thus, Dighal Wetlands represent an vital habitat of migrating and resident wetland birds, and offer valuable insights concerning biodiversity protection and ecosystem services in a changing environment.

**Keywords** Conservation · Dighal wetland · Foraging guild · Threats · Wetland birds

## 20.1 Introduction

Wetlands are active ecotones of land and aquatic habitats, and are world-wide regarded as some of the most productive and most ecologically significant environments [30]. With an average coverage of close to six percent of the world surface, these ecosystems provide a comprehensive range of critical services, such as flood regulation, groundwater recharge, carbon sequestration, and nutrient cycling; erosion control, water purification, climate change mitigation [4, 8, 21]. In addition to being hydrologically and biogeochemically important, wetlands are uniquely important as habitats of a great variety of flora and fauna, and avifauna in particular use wetlands in breeding, foraging, roosting, social interactions, and as a stopover resting sites during annual migration [5].

Wetland birds, species that rely wholly or partially on wetland habitats throughout their life cycle, occupy a wide range of ecological niches and trophic levels, making them integral to the structure, function, and resilience of wetland ecosystems [9, 16, 26]. These wetland avian communities encompass diverse taxonomic groups including waterfowl, waders, kingfishers, raptors, and certain passerines. Many functions as flagship or umbrella species, often serving as symbols for wetland conservation and facilitating the protection of co-occurring taxa [15, 29]. Their high mobility and responsiveness to environmental changes make them effective bioindicators for assessing ecological integrity and habitat quality [3, 20].

India, owing to its diverse physiographic and climatic regimes, supports a rich diversity of avian fauna, with more than 1350 bird species recorded nationwide [14, 22]. Among these, over 310 species are associated with wetland habitats [15]. However, wetlands across the country, as elsewhere, are increasingly threatened by various anthropogenic pressures including land-use change, agricultural intensification, eutrophication, proliferation of invasive species, hydrological modifications, and infrastructural development. Collectively, these pressures degrade ecological function and jeopardize the survival and reproductive success of bird populations [1, 6, 13, 18, 23]. Therefore, regular and systematic monitoring of wetland bird assemblages has become a conservation imperative, offering critical insight for the design of evidence-based management strategies [17, 19, 24].

Despite their ecological significance, wetlands in Haryana, particularly those within ecologically sensitive district like Jhajjar, remain underrepresented in long-term ornithological assessments. In this context, the Dighal Wetlands located approximately 60 km west of Delhi, comprise a complex of freshwater marshes, seasonal jheels, shallow ponds, and interspersed agricultural lands. Situated along the Indo-Gangetic migratory corridor and within the Central Asian Flyway (CAF), Dighal Wetlands, provide a vital wintering ground and migratory stopover for a diverse array of wetland bird species. Since it has been shown as an Important Bird and Biodiversity Area (IBA) by BirdLife international underscores its importance regionally and globally in relation to conservation. However, Dighal Wetlands are more likely to be threatened by rapid land-use changes, changing water regimes, spreading invasive flora, excessive use of natural resources, and the growth of human-made infrastructure. They endanger both the ecological functioning of the wetland mosaic and the sustainability of bird assemblages in such wetlands. Against this backdrop, there is a need to have an in-depth site based ecological research to evaluate the pattern of biodiversity, ascertain the conservation priority as well as design policy interventions.

This chapter opens up the findings of a two-year investigative field study (April 2021–March 2023), conducted to determine the diversity, seasonal pattern, foraging guilds and the conservation status of wetland bird species within Dighal Wetlands. At the same time, the results provide key insights regarding the use of this habitat by resident as well as migratory species and the role of the Dighal Wetlands as part of larger regional and global migratory networks. The study proposes active conservation measures on a long-term basis to conserve such a biologically rich and, yet an endangered wetland ecosystem.

## 20.2 Case Study: Dighal Wetlands, Haryana (Source Author)

The present case study provides the findings of a two-year systematic survey (April 2021–March 2023), aimed to chronicle the diversity, abundance, seasonal variation, and foraging habits of wetland birds in the Dighal Wetlands, located in the Jhajjar district of Haryana, North India. The wetlands represent an important ecology site in Central Asian Flyway, as the areas are critical stopover and wintering locations of a large number of resident and migratory bird species.

The Dighal Wetlands have been located geographically at 28°22′N and 76°18′E and include a geographical area of about 131.5 hectares with several interconnected water bodies in the village Dighal of Beri Tehsil. The wetlands occur in the Aravalli plains in the form of a mosaic of perennial ponds and marshes, seasonally flooded fields, and basins fed with agricultural runoff with some basins holding water year-round because of local water management practices and influxes during monsoons.

There agricultural landscape is cultivated with rice paddies, wheat, maize, sugarcane and mustard. Other tree species like the *Eucalyptus*, *Syzygium*, *Acacia* and *Ziziphus* offer canopy cover and nesting sites. Examples of wetland vegetations are emergent hydrophytes such as *Typha* and *Phragmites*, floating plants such as *Eichhornia*. This complex mosaic of aquatic, semi aquatic and cultivated habitats promotes rich birds species diversity and maintains the ecological richness of the wetland complex.

Semi-arid subtropical climate partitions the research place into three unique seasons cool dry season (October–February), hot dry season (March–June) and the monsoon season or wet period (July–September). The average annual rainfall occurs about 444 mm per annum, most of it during the monsoon season. The climate is within the average range of 6 °C in winter and 45 °C in summer.

## 20.3 Systematic Adopted Methodology

Systematic bird surveys were conducted across of eight selected sites within complex of the Dighal Wetlands over a span of two years (April 2021 to March 2023), to evaluate the diversity, abundance, seasonal trends, and foraging behaviours of wetland birds. Point count and direct observation techniques were utilized based on a bi-weekly survey routine. At each wetland location, four to six observation points were established along the edges, situated at least 250 meters apart to avoid overlapping detection areas. Each site underwent surveying 48 times throughout the study period. A five-minute acclimation period was implemented before data collection to reduce the impact of observer on bird behaviour, followed by a ten-minute observation session at every point. Birds were directly observed and counted with Nikon 10x50 field binoculars during the most active hours (06:00–10:00 hrs or 16:00–18:00 hrs), only under suitable weather conditions (i.e., absence of rain, fog, and heavy winds). If in-field identification became difficult, photographs were captured using a Nikon D500 DSLR camera for later confirmation.

Species identification was carried out using conventional field guides [2, 10]. The taxonomic details including order, family, and both common and scientific names, adhered to Praveen and Jayapal [22]. Conservation status of recorded bird species was evaluated according to WPA [25], CITES [7], and IUCN Red List [11].

## 20.4 Some Important Concepts

### *20.4.1 Species Composition and Taxonomic Structure*

During the study period, a total of 90 wetland bird species inhabiting wetlands were identified, representing 62 genera, 23 families, and 12 orders. The order Charadriiformes was the most diverse, consisting of 24 species and making up 27% of the entire wetland bird community. This was followed by Anseriformes (19 species, 21%) and Pelecaniformes (12 species, 13%). Collectively, these three orders accounted for 61.11% of the wetland avian assemblage. In contrast, Falconiformes was the least represented order, with only one species documented. Other orders with low representation included Phoenicopteriformes, Podicipediformes, Suliformes, and Accipitriformes, each contributing very little to the total diversity. A comparative assessment of the diversity of wetland birds at Dighal relative to Haryana and India is detailed in Table 20.1.

### *20.4.2 Species Abundance and Frequency of Occurrence*

Wetland bird species were quantitatively classified into four abundance categories: common, fairly common, uncommon, and rare, based on the frequency of field observations. The analysis revealed distinct abundance trends within the bird community, with 23 species (26%) deemed common, 18 species (20%) categorized as fairly common, 28 species (31%) as uncommon, and 21 species (23%) classified as rare.

The Bar-headed Goose (*Anser indicus*) emerged as the most frequently observed species, with a total of 4510 sightings, indicating its prevalence and dominance throughout the study area. Trailing behind were the Northern Shoveler (*Spatula clypeata*) and Northern Pintail (*Anas acuta*), recorded at 3630 and 1987 sightings, respectively. Conversely, several species exhibited infrequent appearances, with only two to three sightings throughout the study period. These included the Demoiselle Crane (*Grus virgo*), Common Snipe (*Gallinago gallinago*), Brahminy Kite (*Haliastur indus*), Common Kingfisher (*Alcedo atthis*), Water Pipit (*Anthus spinoletta*), Bronze-winged Jacana (*Metopidius indicus*), Indian Thick-knee (*Burhinus indicus*), Baillon's Crake (*Zapornia pusilla*), and Western Marsh Harrier (*Circus aeruginosus*). Notably, the Mallard (*Anas platyrhynchos*) and the Peregrine Falcon (*Falco peregrinus*) were the rarest, each having just one sighting during the two-year study, implying a possibility of transient occurrence or infrequency in this area.

**Table 20.1** Diversity of the wetland birds (comparative analysis with essential morphological characteristics by family) in the Dighal wetlands, Haryana, India

| Sr. no. | Order | Wetland avian family | Key morphological features | Diversity of wetland bird species | | |
|---|---|---|---|---|---|---|
| | | | | Dighal Wetlands | Haryana [12] | India [15] |
| 1. | Anseriformes | Anatidae (Ducks and Geese) | Webbed feet, broad bills, dense waterproof feathers; strong flyers and swimmers. | 19 | 24 | 43 |
| 2. | Charadriiformes | Scolopacidae (Curlew, Snipes and Sandpipers) | Long legs with slender bills for probing; agile fliers with cryptic plumage | 13 | 21 | 37 |
| | | Laridae (Gulls and Terns) | Seabirds with webbed feet, hooked bills, and long wings; typically, grey-white with black markings | 2 | 9 | 30 |
| | | Charadriidae (Plovers and Lapwings) | Small to medium-sized, compact bodies with short bills and strong legs; often inhabit open areas | 4 | 13 | 19 |
| | | Burhinidae (Thick-knees) | Large-eyed, nocturnal, terrestrial waders with cryptic coloration and robust bills | 1 | 2 | 2 |
| | | Recurvirostridae (Avocets and Stilt) | Slender, long-legged birds; avocets with upcurved bills and stilts with straight, thin bills | 2 | 2 | 3 |
| | | Jacanidae (Jacanas) | Tropical waders with elongated toes and claws for walking on floating vegetation; often brightly colored | 2 | 2 | 2 |
| 3. | Pelecaniformes | Ardeidae (Bitterns, Herons and Egrets) | Long-legged, long-necked waders with straight, spear-like bills; neck folded in flight; adept at still hunting | 8 | 14 | 20 |
| | | Threskiornithidae (Ibises and Spoonbills) | Decurved (ibis) or spatula-shaped (spoonbill) bills; social feeders in wetlands | 4 | 4 | 4 |
| 4. | Gruiformes | Gruidae (Cranes) | Tall, elegant birds with long legs and necks; powerful flyers known for elaborate dances | 2 | 3 | 6 |
| | | Rallidae (Rails, Coots and Swamphens) | Medium-sized marsh birds with laterally compressed bodies and lobed feet | 5 | 10 | 17 |
| 5. | Podicipediformes | Podicipedidae (Grebes) | Small to medium-sized diving birds with lobed feet, pointed bills, and striking breeding plumage. | 1 | 4 | 5 |
| 6. | Ciconiiformes | Ciconiidae (Storks) | | 3 | 6 | 9 |

| | | | | | | |
|---|---|---|---|---|---|---|
| | | | Large wading birds with long legs and stout bills; broad wings; silent communicators relying on bill-clattering. | | | |
| 7. | Accipitriformes | Accipitridae (Harriers and Hawks) | Diurnal raptors with hooked beaks, powerful talons, and keen eyesight; broad wings for soaring | 2 | 6 | 11 |
| | | Pandionidae (Osprey) | Specialized fish-eater with reversible outer toes and spiny footpads | 1 | 1 | 1 |
| 8. | Suliformes | Phalacrocoracidae (Cormorants) | Aquatic birds with long necks, hooked bills, and webbed feet; often seen drying wings | 3 | 3 | 3 |
| | | Anhingidae (Darters) | Snake-like necks, spear-shaped bills; partially wettable plumage | 1 | 1 | 1 |
| 9. | Phoenicopteriformes | Phoenicopteridae (Flamingos) | Tall waders with down-curved bills adapted for filter feeding; pinkish plumage due to carotenoid-rich diet. | 1 | 2 | 2 |
| 10. | Falconiformes | Falconidae (Falcons) | Fast-flying raptors with pointed wings, hooked beaks with a tomial tooth; excellent vision. | 1 | 1 | 1 |
| 11. | Coraciformes | Coracidae (Bee-eaters and Rollers) | Brightly colored, slender birds with long beaks; adept aerial insect hunters. | 1 | 3 | 3 |
| | | Alcedinidae (Kingfishers) | Stout birds with large heads and dagger-like bills; skilled fish catchers | 3 | 3 | 12 |
| 12. | Passeriformes | Motacillidae (Pipits and Wagtails) | Ground-dwelling insectivores with slender bodies and long tails | 6 | 6 | 9 |
| | | Hirundinidae (Swallows and Martins) | Streamlined aerial foragers with short bills and forked tails | 5 | 5 | 7 |

### 20.4.3 Family-Level Diversity

In term of avian families, Anatidae emerged as the most diverse group, represented by 19 species and exhibiting the highest relative diversity index (RDi) value of 21.11. This was followed by Scolopacidae (13 species, RDi = 14.44) and Ardeidae (8 species, RDi = 8.88). The Motacillidae family included 6 species (RDi = 6.66), indicating a moderate level of diversity. In contrast, seven families, namely Phoenicopteridae, Podicipedidae, Anhingidae, Burhinidae, Pandionidae, Falconidae, and Meropidae, each have only one species, resulting in an RDi value of 1.11, which emphasizes their limited occurrence in the wetland bird assemblages.

### 20.4.4 Migratory Status and Seasonal Occurrence

Out of the 90 wetland bird species documented in the Dighal wetlands, 30 species (33.3%) were classified as residents, 4 species (4.5%) as summer migrants, and 56 species (62.2%) as winter migrants. This indicates that migrant species constitute a substantial proportion (66.6%) of the wetland avifauna, underscoring the critical role of the Dighal wetlands as a vital seasonal habitat and foraging ground, particularly during the winter season. Migratory species followed distinct arrival and departure patterns, with winter migrants arriving between August and September, peaking from November to March, and departing by March. Summer migrants were observed from April to September, with staggered arrival and departure times.

### 20.4.5 Habitat Use and Foraging Strategies

In the Dighal Wetlands, the choice of habitat was significantly shaped by the accessibility of essential resources such as food, nesting sites, and protective cover. Various bird groups highlighted their ecological adaptation in terms of specific preferable behaviour. As an example, members of the Anatidae family (ducks and geese) preferred deep water bodies with the submerged plants, and they avoided tall emergent vegetation much, indeed. Their foraging activity highlighted their dependence on the plant supplies that are present below and above water surfaces making them specialized herbivores when it comes to their food preferences and preferences towards certain habitats. Waders that tended to shallow waters and marshy habitats, on the other hand, like the Asian Openbill, Painted Stork and Purple Heron, were in turn less inclined to mixed-species groups in these habitats. The feeding behaviours reflected in the cormorants tend to have a broader ecological niche as they adapted to utilised shallow and even deeper waters.

Some species became adapted closely to floating plants, like water hyacinth and lotus patches. These habitats were widely used by birds like the Purple Swamphen and Jacanas as feeding and shelter. At the edges of ponds, the kingfishers were frequently observed skulking along on bankside buildings upon the look-out, to steal fish and water-insects, signifying a dependence on open waters and perching-places in their vicinage. A number of those species of birds shifted their feeding to nearby farmland, especially the waders, shorebirds and wagtails which exploited the seasonal influx of invertebrates into newly ploughed and flooded fields. Tree species such as *Acacia arabica*, *Azadirachta indica*, *Zizyphus jujuba*, *Eucalyptus* spp., *Cassia tora*, *Ficus bengalensis*, *Ficus religiosa*, *Dalbergia sissoo*, and *Prosopis juliflora* located in and around wetlands provided essential roosting and nesting habitats for a variety of wetland birds, including herons, egrets, storks, ibises, cormorants, kites, and kingfishers. This underscores the significance of the vegetation around as nourishing avian diversity in the wetland ecosystem.

The study identified four primary foraging guilds based on dietary patterns: carnivores, omnivores, insectivores, and herbivores. The carnivore guild was the most diverse, consisting of 39 species, followed by omnivores (27 species) and insectivores (18 species). Herbivores, mainly represented by ducks and geese, were the least diverse group, although they held ecological importance.

During the study, a total of 21 distinct foraging techniques were observed, each demonstrating adaptation to certain ecological niches and resource types (Table 20.2). Dabbling was the most common technique, utilized by 18 species, followed by pecking (17 species), picking (16 species), and probing (14 species). Other methods like upending (11 species), diving, and aerial catching (8 species each) were moderately shown. Less frequently noted techniques included foot trembling (6 species) and grazing (5 species). Additionally, various species exhibited specialized foraging behaviors, including filter-feeding in flamingos, stabbing in Oriental Darters, and lateral sweeping of the bill in Eurasian Spoonbills. These specialized foraging methods reflect a high degree of resource partitioning, mitigating interspecific competition and facilitating the coexistence of a diverse bird community within the wetland habitat.

### 20.4.6 Conservation Status and Global Population Trends

A comprehensive evaluation of the conservation status of recorded wetland bird species, based on the IUCN Red List [11], revealed that three species namely, Common Pochard (*Aythya ferina*), Sarus Crane (*Grus antigone*), and River Tern (*Sterna aurantia*) are classified as Vulnerable (VU). Additionally, four species are designated as Near Threatened (NT), including the Ferruginous Pochard (*Aythya nyroca*), Woolly-necked Stork (*Ciconia episcopus*), Eurasian Curlew (*Numenius arquata*), and Black-tailed Godwit (*Limosa limosa*). The majority of the recorded species, totaling 83 bird species (92%), are classified as Least Concern (LC).

In terms of international conservation frameworks, two wetland bird species- the Painted Stork (*Mycteria leucocephala*) and Peregrine Falcon (*Falco peregrinus*) are listed in Appendix I of CITES [7]. Furthermore, six species, including the

**Table 20.2** Certain foraging techniques used by wetland birds in the study area

| Foraging technique | Description | Number of species | Representative species |
|---|---|---|---|
| Dabbling | Feeding at or just below the water surface by tipping forward | 18 | Northern Shoveler (*Spatula clypeata*), Indian Spot-bill (*Anas poecilorhyncha*) |
| Pecking | Picking food items from surfaces or shallow water | 17 | Black-winged Stilt (*Himantopus himantopus*), Red-wattled Lapwing (*Vanellus indicus*) |
| Picking | Grabbing visible prey or plant matter directly with the bill | 16 | Common Moorhen (*Gallinula chloropus*), Purple Swamphen (*Porphyriopoliocephalus*) |
| Probing | Inserting bill into mud or shallow water to locate prey | 14 | Common Snipe (*Gallinagogallinago*), Eurasian Curlew (*Numenius arquata*) |
| Upending | Tipping the body vertically to feed underwater in shallow zones | 11 | Gadwall (*Mareca strepera*), Eurasian Wigeon (*Marecapenelope*) |
| Diving | Submerging the body completely to chase prey underwater | 8 | Little Grebe (*Tachybaptus ruficollis*), Indian Cormorant (*Phalacrocorax fuscicollis*) |
| Catching (Aerial/ Dive) | Capturing flying or swimming prey, either in air or water | 8 | Pied Kingfisher (*Ceryle rudis*), Osprey (*Pandion haliaetus*) |
| Foot Trembling | Stirring the substrate with feet to flush out prey | 6 | Common Redshank (*Tringatotanus*), Black-tailed Godwit (*Limosalimosa*) |
| Grazing | Feeding on grasses or aquatic plants by walking or swimming slowly | 5 | Bar-headed Goose (*Anser indicus*), Greylag Goose (*Anseranser*) |
| Filtering | Using specialized bill structures to filter plankton or small organisms | Few | Greater Flamingo (*Phoenicopterus roseus*) |
| Stabbing | Spearing prey with a rapid thrust of the bill | Few | Oriental Darter (*Anhinga melanogaster*) |
| Sweeping (Lateral) | Swinging bill side-to-side to detect and catch prey | Few | Eurasian Spoonbill (*Platalealeucorodia*) |

Knob-billed Duck (*Sarkidiornis melanotos*), Sarus Crane (*G. antigone*), Demoiselle Crane (*Anthropoides virgo*), Osprey (*Pandion haliaetus*), Western Marsh Harrier (*Circus aeruginosus*), and Brahminy Kite (*Haliastur indus*), appear in Appendix II of CITES. All recorded avian species are protected under Schedule II of [25], with the exception of ten species: Common Pochard, Sarus Crane, Demoiselle Crane, Eurasian Spoonbill, River Tern, Common Greenshank, Brahminy Kite, Osprey, Western Marsh Harrier, and Peregrine Falcon which are listed in Schedule I.

An analysis of the global population trends of recorded wetland bird species, according to IUCN [11], showed that within the study area, 24 species are experiencing increases in global populations, 29 species are witnessing declines, 12 species are stable, and 25 species have unknown population trends. A significant observation was that nine species including the Bar-headed Goose, Common Pochard, Northern Shoveler (*Spatula clypeata*), Indian Spot-billed Duck (*Anas poecilorhyncha*), Northern Pintail (*Anas acuta*), Little Grebe (*Tachybaptus ruficollis*), Black-tailed Godwit, Ruff (*Calidris pugnax*), and Common Sandpiper (*Actitis hypoleucos*), which are global population decreases, were commonly sighted in the study area during the study period.

### *20.4.7 Anthropogenic and Environmental Threats*

Field observations have revealed several human-induced pressures adversely affecting avian diversity and ecological health of wetlands:

- **Intensive Fishing and Bycatch**: Overfishing and non-selective fishing gear disrupt food dynamics and led to unintended bird catches.
- **Entanglement Hazards**: Fishing nets and aerial wires create risks of collision and entrapment, particularly for larger and low-flying avian species.
- **Pollution and Waste Accumulation**: The release of plastic waste, polythene, and household refuse diminishes habitat conditions and posing risks of ingestion and entanglement for wetland birds [27, 28].
- **Eutrophication and Organic Pollution**: Contributions from animal waste and public defecation increase nutrient levels, trigger algal blooms, and cause oxygen depletion.
- **Construction and Habitat Fragmentation**: Infrastructure initiatives result in loss of habitat and increase levels of disturbance.
- **Invasive Aquatic Vegetation**: The spread of water hyacinth (*E. crassipes*) diminishes open-water areas, hindering foraging for diving and fish-eating birds.

## 20.5 Future Prospects and Implications

The Dighal Wetlands should be conserved using an integrated approach involving habitat conservation, sustainable use of resources and community participation. The most important tactics are:

- **Habitat Protection:** Strengthen environmental laws and create buffer zones to protect essential habitats.
- **Sustainable Fisheries Management**: Adopt non-invasive, community-led fishing methods to decrease bycatch and lessen disruptions.
- **Pollution Control**: Enhance waste management practices and reduce agricultural runoff to boost water quality.
- **Community Participation**: Encourage local stewardship through educational campaigns and conservation initiatives.

- **Monitoring and Research**: Implement long-term bird monitoring systems to guide adaptive management strategies.

## 20.6 Conclusion

The Dighal Wetlands are an essential habitat to resident as well as migratory wetland birds especially those that find it a stopover point in migration. The chapter has highlighted the diverse bird population in the wetlands and identified the major trends ecologically and, also, explored the main threats facing the species. The findings demand the need of effective conservation and management measures that balance out habitat against human pressures in terms of development. There is a remote possibility to develop the long-term sustainability of Dighal Wetlands as a breeding and feeding ground for both migratory and resident wetland birds.

## References

1. Ali ASM, Kumar SR, Arun PR (2013) Waterbird assemblage in rural ponds of Samakhiali region, Kutch District, Gujarat, India. Bird Popul 12:12–18
2. Ali S, Ripley SD (1987) Handbook of the birds of India and Pakistan together with those of Bangladesh, Nepal, Bhutan and Sri Lanka. Oxford University Press, Delhi, India
3. Amat JA, Green AJ (2010) Waterbirds as bioindicators of environmental conditions. In: Hurford C, Schneider M, Cowx I (eds) Conservation monitoring in freshwater habitats. Springer, Dordrecht. https://doi.org/10.1007/978-1-4020-9278-7_5
4. Bassi N, Kumar MD, Sharma A, Pardhasaradhi P (2014) Status of wetlands in India: a review of extent, ecosystem benefits, threats and management strategies. J Hydrol Reg Stud 2:1–19. https://doi.org/10.1016/j.ejrh.2014.07.001
5. Buckton S (2007) Managing wetlands for sustainable livelihoods at Koshi Tappu. Danphe 16(1):12–13
6. Butchart SHM, Collar NJ, Stattersfield AJ, Bennun LA (2010) Conservation of the world's birds: the view from 2010. In: Handbook of the Birds of the World, vol 15, pp 13–68
7. CITES (2012) Convention on international trade in endangered species of Wild Fauna & Flora. CITES, Geneva, Switzerland. Downloaded on 20 October 2022. https://www.cites.org
8. Clarkson BR, Sorell BK, Reeves PN, Champion PD, Partridge TR, Clarkson BD (2003) Handbook for monitoring wetland condition. Coordinated monitoring of New Zealand wetlands. A Ministry for the Environment Sustainable Management Fund Project
9. Green AJ, Elmberg J (2014) Ecosystem services provided by waterbirds. Biol Rev 89(1): 105–122. https://doi.org/10.1111/brv.12045
10. Grimmett R, Inskipp C, Inskipp T (2011) Birds of the Indian subcontinent. Oxford University Press, London
11. IUCN (2024) The IUCN red list of threatened species. Version 2024. iucnredlist.org
12. Kalsi RS, Sharma SC, Choudhary JR (2019) Birds of Haryana. Unique Publications, Panipat, India
13. Kafle G, Marcus C, Chaidhary JR, Pariyar H, Adhikari H, Bohara S, Chaudhari U, Ram A, Regmi B (2008) Status of threats to waterbirds of Rupa Lake, Pokhara, Nepal. J Wetland Ecol 1(1/2):9–12

14. Kiran, Singh D, Kour A, Priya, Delu V, Kumar R (2022) Community composition and status of avian diversity at campus and agricultural landscapes of Chaudhary Charan Singh Haryana Agricultural University, Hisar (Haryana). J Appl Nat Sci 14(4):1130–1140. https://doi.org/10.31018/jans.v14i4.3784
15. Kumar A, Sati JP, Tak PC, Alfred JRB (2005) Handbook on Indian wetland birds and their conservation. Zoological Survey of India, Kolkata
16. Kumar P, Gupta SK (2013) Status of wetland birds of Chhilchhila wildlife sanctuary, Haryana, India. J Threat Taxa 5(5):3969–3976. https://doi.org/10.11609/JoTT.o3158.3969-76
17. Kumar P, Sharma A (2019) Wetland birds assemblages in man-made sacred ponds of Kurukshetra, India. Proc Zool Soc 72(1):61–73
18. Kumar P, Rai D, Gupta SK (2016) Wetland bird assemblage in rural ponds of Kurukshetra, India. Waterbirds 39(1):86–98. https://doi.org/10.1675/063.039.0111
19. Lee PF, Ding TS, Hsu FH, Geng S (2004) Breeding bird species richness in Taiwan: distribution on gradients of elevation, primary productivity and urbanization. J Biogeogr 31(2):307–314
20. Morrison ML (1986) Bird populations as indicators of environmental change. In: Johnston RF (ed) Current ornithology. Springer, Boston, pp 429–451
21. Prasad SN, Ramachandra TV, Ahalya N, Sengupta T, Kumar A, Tiwari AK, Vijayan VS, Vijayan L (2002) Conservation of wetlands of India – a review. Trop Ecol 43(1):173–186
22. Praveen J, Jayapal R (2023) Taxonomic updates to the checklist of birds of India and the South Asian region. Indian Birds 18(5):131–134. http://www.indianbirds.in/india
23. Renigald JL, Mahendran C, Sureshkumar S, Pramod P (2007) Birds of Singanallur lake, Coimbatore, Tamil Nadu. Zoo's Print Journal 22(12):2944–2948
24. Sundar KSG, Kittur S (2013) Can wetlands maintained for human use also help conserve biodiversity? Landscape scale patterns of bird use of wetlands in an agricultural landscape in North India. Biol Conserv 168:49–56. https://doi.org/10.1016/j.biocon.2013.09.016
25. The Wildlife (Protection) Amendment Act (2022). http://www.indiaenvironmentportal.org.in/files/file/wildlife%20protection%20amendment%20act%202022.pdf
26. Yashmita-Ulman, Singh M (2022) Avifaunal diversity in unprotected wetlands of Ayodhya District, Uttar Pradesh, India. J Threatened Taxa 14(8):21561–21578. https://orcid.org/0000-0001-7428-3187
27. Yadav R, Gupta RK, Kumar R, Kaur T (2022a) Assessment of toxicity of Lead and nickel on the biochemical and immunological parameters of earthworm, Eudrilus eugeniae. Agric Sci Dig 44(5):922–929
28. Yadav R, Gupta RK, Kumar R, Kaur T, Moun R (2022b) Biochemical response of earthworm, *Eisenia fetida* to heavy metals toxicity. J App Nat Sci 14(3):990–998
29. Yodha K, Singh D, Kour A, Yadav R, Kumar R (2024) Species diversity and abundance of birds in ber (*Zizyphus mauritiana*) crop. Asian J Conserv Biol 13(1). https://doi.org/10.53562/ajcb.83683
30. Zedler JB, Kercher S (2005) Wetland resources: status, trends, ecosystem services, and restorability. Annu Rev Environ Resour 30(1):39–74. https://doi.org/10.1146/annurev.energy.30.050504.144248

# Chapter 21
# Avian Biodiversity in South Asia: The Role of Persistent Organic Pollutants, Regulatory Measures, and Conservation Challenges

**Sharda Kalra, Kapish Samota, Anil Ghanghas, Devinder Singh, and Asha Poonia**

**Abstract** South Asia, home to an exceptional range of avian biodiversity (1.5% and 12% of total flora and faunal diversity with 4494 species of birds), is presently facing increasing threats due to anthropogenic factors, with Persistent Organic Pollutants (POPs) playing a critical role in biodiversity loss. This chapter explores the primary anthropogenic pressures, on the avian diversity of the region, focusing on the impact of POPs on avian species across South Asia. The chapter delve into POPs, including dioxins, PCBs, and PBDEs, which pose significant threats to avian species by disrupting endocrine, neural, and reproductive systems, with species-specific susceptibilities, varying bioaccumulation patterns, and impacts on long-term breeding and survival, especially in predatory and aquatic birds. The chapter also critically assesses the efficacy of regulatory frameworks like the Stockholm Convention and regional policies in curbing POPs. Additionally, the chapter highlights conservation challenges, including habitat fragmentation, inadequate monitoring of pollutant impacts, socio-economic barriers, industrial growth, agricultural dependence on pesticides, and regulatory non-compliance. Overall, the chapter aims to provide an integrated understanding of the intersection between POP pollution, regulatory measures, and conservation efforts to protect avian biodiversity in South Asia. It further outlines strategic recommendations for strengthening policy implementation and fostering sustainable conservation practices.

**Keywords** Aves · Biodiversity · Persistant organic pollutants (POPs) · Toxicity · Wildlife conservation

---

S. Kalra · K. Samota · A. Poonia (✉)
Faculty of Life Science, Department of Zoology, Chaudhary Bansi Lal University, Bhiwani, Haryana, India
e-mail: asha.poonia@cblu.ac.in

A. Ghanghas
Department of Law, Chaudhary Devi Lal University, Sirsa, Haryana, India

D. Singh
Department of Laws, Panjab University, Chandigarh, India



R. Kumar et al. (eds.), *Ecology, Conservation of Aquatic Birds*, Wetlands: Ecology, Conservation and Management 15, https://doi.org/10.1007/978-3-032-17125-2_21

## 21.1 Introduction

South Asia is the home of approximately 15.5 of floral and 12% of faunal of global biodiversity. The animal diversity comprises of 933 species of mammals, 923 reptiles, 4494 birds, 332 species of amphibians and 342 species of riverine fishes. There are 6652 higher plants, 66 conifers and cycads, 764 ferns, and 39,875 flowering plant species that make up the floral diversity [1]. South Asia is most biodiverse region in the world due to their varied ecosystems, which range from semi-arid regions and coastal marshes to the Himalayan temperate-alpine highlands and evergreen woods.

India is the largest of the eight countries in terms of both population and land area. These nations share a rich transboundary environment that includes the high-altitude rangelands of Nepal, Bhutan, and India; the Sundarbans of Bangladesh and India; the Terai woodlands of India, Nepal, and Bhutan; and the numerous river basins that emerge from the Hindu Kush Himalayas. It is necessary to preserve and manage this ecosystem through policies. Recently, it is involved in the Kunming-Montreal Global Biodiversity Framework [2].

### *21.1.1 Biodiversity Hotspots in South Asia*

South Asia contain three biodiversity hotspots includes Western ghat, Indo Burma region and the Himalayas (Fig. 21.1):

**Western Ghat and Sri Lanka** The Western Ghats and Sri Lanka cover most of the deciduous and rain forests, which are located along the western edge of the Indian peninsula. UNESCO has identified at least 325 globally vulnerable species of fish, amphibians, reptiles, birds, plants, and wildlife in the area [3, 4].

**Indo-Burma Region** The Indo-Burma Region is located in the northeastern part of India and spans 2,373,000 sq. km. The Saola, the Grey-shanked Douc, the Leaf Deer, the Annamite Striped Rabbit, the Large-antlered Muntjac, and the Annamite Muntjac are the six large animal species that have been discovered in this region in the last 12 years. Due to over-harvesting and extensive habitat destruction, the majority of the endemic freshwater turtle species are in risk of getting extinct found in this hotspot [5].

**The Himalayan** Northeast India, Bhutan, and the central and eastern regions of Nepal comprises all part of the Himalayas. The Wild Asian Water Buffalo, One-horned Rhino, and other indigenous species are among the 163 endangered species known to exist in this region [6].

**However**, the Andaman and Nicobar Island **is** one of the 34 globally identifies biodiversity hotspots located in Indian Ocean. The Nicobar pigeon, Andaman Scops-owl, Nicobar Bulbul, Nicobar Parakeet, arcondam Hornbill and Andaman Drongo are found in these regions [7].

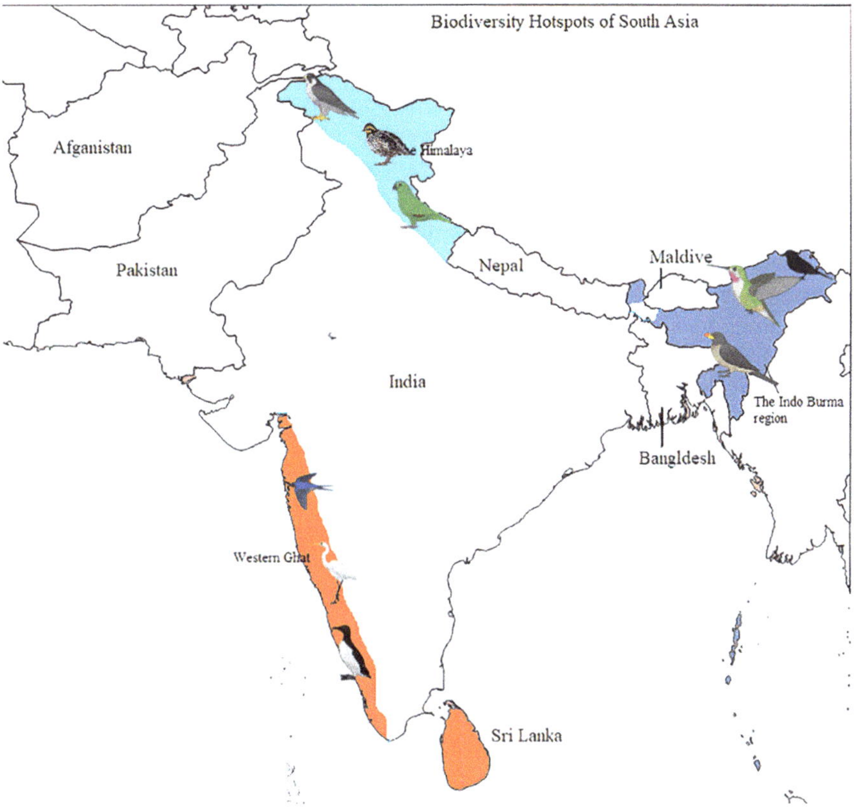


**Fig. 21.1** Showing the biodiversity hotspots region in south Asia

## *21.1.2 Key Threats to Biodiversity*

South Asia's biodiversity plays significant role in ecological stability, cultural heritage, and economic sustenance, although it faces large number of threats from anthropogenic pressures, including habitat loss, pollution, climate change and unsustainable exploitation [8–10].

Habitat Loss and Fragmentation due rapid urbanization, deforestation, agricultural expansion in western ghat and eastern Himalayas reducing ecosystem integrity and decline species biodiversity [9].

Climate change is also a considerable risk in south asia. Temperature fluctuation, precipitation pattern and natural calamity lead to disturbance in life cycle of many in seasonal breeders animals and aquatic diversity [10–14].

Overexploitation and illegal wildlife trade constitute significant threats to biodiversity, particularly in the South Asian countries, where the unsustainable extraction of wild animals and their derivatives is a primary driver of species decline. The poaching of endangered species has profound implications for ecosystem structure and function. Species such as the Indian pangolin (*Manis crassicaudata*), Bengal tiger (*Panthera tigristigris*), and Asian elephant (*Elephas maximus*) are especially susceptible to poaching due to the high commercial value of their body parts. The depletion of these keystone and flagship species disrupts ecological balance and exacerbates the risk of population extinction [9].

Pollution from industrial and agricultural drainage affects the terrestrial as well as aquatic diversity. Many pesticides and other chemicals Leaked into aquatic system lead to eutrophication (accumulation of nutrient specially nitrogen and phosphorus in aquatic system lead to overgrowth of algae on the water surface called as algal bloom) and biomagnifications (bioaccumulation of chemicals and transport from one trophic level to higher trophic level through food chain) through trophic chain adversely affect the whole ecosystem [11]. Biomagnifications have higher impacts on top predator species. Persistent organic molecules (POPs) are the chemicals eliminated through anthropogenic activities and characterized by resistant to degradation in environment, long range persistence, lipophilic and have adverse toxicological impact on wildlife especially on avifauna. Due to their ability to resistance to degradation, accumulated in tissues and biomagnified in higher trophic level causing potential harm to top predatory bird species [15].

## 21.2 Persistent Organic Pollutants (POPs and Their Mechanisms of Toxicity

Persistent organic molecules (POPs) are the chemicals eliminated through anthropogenic activities and characterized by resistant to degradation in environment, long range persistence, lipophilic and have adverse toxicological impact on wildlife especially on avifauna. Due to their ability to resistance to degradation, accumulated in tissues and biomagnified in higher trophic level causing potential harm to top predatory bird species [12, 13].

### *21.2.1 Types and Classes of Persistant Organic Pollutant*

In 1995, the United Nations Environment Programme expanded its research on Persistent Organic Pollutants (POPs), initially concentrating on a group known as the "Dirty Dozen." This group included twelve highly toxic and persistent chemicals: given below. Although many of these pesticides are no longer utilized in agriculture, they are continue to be used in some developing countries [12]. On the basis of origin and their characteristics, the UNEP categorized these POPs into various types:

1. Pesticides: DDT (Dichloro diphenyl trichloroethane), toxaphene, chlordane, aldrin, mirex, dieldrin, heptachlor, endrin [12–14]

2. Industrial pollutants: Polychlorinated biphenyls (PCBs), Hexachlorobenzene [12–14]
3. Co-Products and contaminants: Dioxin and Furans [12–14]

Over time, the list of POPs has grown to include additional hazardous substances, such as tributyltin (TBT) which is an organometallic compound, polycyclic aromatic hydrocarbons which causes cancer and some brominated flame retardants [12–14]. Among these polybrominated diphenyl ethers (PBDEs), organochlorine pesticides (OCPs), dibenzofurans (PCDD/Fs) and polychlorinated dibenzo-p-dioxins and polychlorinated biphenyls (PCBs), which have long lasting hazardous impact on biodiversity [12–14]. The POPs comprised 21 organochlorine pesticides, 18 polychlorinated biphenyls including six non-dioxin-like PCBs and 12 dioxin-like PCBs, 17 polychlorinated dibenzo-p-dioxins and dibenzo-p-furans (PCDD/Fs), perfluorooctane sulfonate (PFOS) and five classes of polybrominated diphenyl ether (PBDE) [12–14].

### 21.2.2 Mechanism of Toxicity in Birds

The Persistent organic pollutants (POPs) are toxic chemicals that are impervious to breakdown, have the ability to bioaccumulation (accumulation of toxic chemicals through food chain), and can be transported across international borders via air, water, and migratory species [13]. These pollutants often settle far from their original sources and build up in terrestrial and aquatic ecosystems. Since many bird species migrate, their movement patterns are an important consideration when assessing exposure to these pollutants and monitoring environmental quality. The breakdown and metabolic byproducts of POPs, which include a variety of anthropogenic organic and elemental elements, have been detected in the tissues of exposed birds. Wild birds often inhabit agricultural and forested areas where pesticides are widely used, exposing them to chemicals through inhalation, dermal contact, or ingestion. Pesticides can be ingested indirectly via preening contaminated feathers or consuming treated crops, weeds, and insects. Birds may also consume pesticide-coated seeds or mistake granules for food or grit, leading to higher exposure. Considering these exposure routes and feeding behaviors, pesticide toxicity should be evaluated based on direct effects and inherent toxicity. With aerial usage of pesticides, oral and dermal exposure gradually took precedence over vital exposure mechanisms such as inhalation and preening [16, 17]. The Magnitude of OC concentrations has been seen increased in the order of sediments, green mussels, earthworms, frogs, lizards, fishes, bird eggs, bats, birds tissues. This study revealed that organochlorine compound is more concentrated in birds than the bats [17]. Many studies have been takes place in birds in respect of persistent organic pollutants as given in Table 21.1.

Variations in pesticide toxicity between birds and mammals are determined by several physiological and anatomical factors [16, 18]. Birds, with their elevated body temperature (~42 °C compared to ~37 °C in mammals), require a faster intake of

**Table 21.1** Various bird species on which effect of persistent organic pollutants has been studied

| Sr. no. | Scientific name | Location | Pollutants | References |
|---|---|---|---|---|
| 1 | Migratory birds | Asia | (OCPs) | [13, 62] |
| 2. | Egret | Pakistan | (OCPs) | [66] |
| 3. | Corvus splendens | India | (OCPs) | [35] |
| 4. | Common red shank, gull-billed tern, pintail snipe, and spotted owlet | India | (OCPs) | [17] |
| 5. | Migratory birds | South India | PCBs and OCPs | [17] |
| 6. | Crow | India | (OCPs) | [63] |
| 7. | Shikra (*Accipiter badius*), white ibis (*Pseudibispapillosa*) | | | [67] |
| 8. | Domestic pigeon (*Columba livia var. domestica*), Black-tailed gull (*Larus crassirostris*), Grey heron (*Ardea cinerea*), Pacific loon (*Gaviapacifica*), Little egrets (*Egrettagarzetta*), Great egrets (*A. alba*) | South Korea | DDT, PCBs, PBDEs, CHLs, HCHs, HCB, HBCDs, PeCB | [59] |
| 9. | White-tailed eagles (*Haliaeetus albicilla*) | Belgium | PCBs, PBDEs and OCPs | [58] |
| 10. | Colonial nesting birds | India | HCH, Endosulfan, DDT, Heptachlor epoxide, Dieldrin | [61] |
| 11 | Spotted owlet (*Athene brama*) and Black kite (*Milvus migrans*) | Pakistan | PBDEs and OCPs | [36] |
| 12 | House sparrows (*Passer domesticus*) | Norway | PCBs, PBDEs and OCPs | [21] |
| 13 | Bermuda petrel (*Pterodroma cahow*) | Canada | PCBs, PBDEs and OCPs and PAH | [23] |
| 14 | Fairy terns (*Gygis alba*), Sooty terns (*Onychoprionfuscatus*) and common noddies (*Anousstolidus*) | Indian Ocean | CHL, DDT, PBDEs, HCBs, PCBEs, CHBs, and HCHs | [26] |
| 15 | Yellow-rumped Warblers (*Setophaga coronata*) | Canada | Methyl-mercury | [28] |
| 16 | Arctic breeding shorebirds | Alaska | Mercury | [30] |
| 17 | Charadrius plovers | China | Mercury | [31] |
| 18 | White-tailed eagles (*Haliaeetus albicilla*) Goshawks (*Accipiter gentilis*) and Golden eagles (*Aquila chrysaetos*) | Norwey | PCBs, DDE and PBDEs | [33, 58] |
| 19 | Sarus crane (*Grus antigone)* | India | Eldrin | [65] |
| 20 | Egyptian Vulture (*Neophron percnopterus*), White-backed Vulture (*Gyps bengalensis*) and Griffon Vulture (*Gyps fulvus)* | Ahmadabad, India | OCPs and PCBs | [34, 64] |
| 21 | Sarus Crane (*Grus antigone)*, Black Ibis (*Pseudibispapillosa)*, White | India | HCH, DDE, DDT | [34] |

(continued)

**Table 21.1** (continued)

| Sr. no. | Scientific name | Location | Pollutants | References |
|---|---|---|---|---|
| | Ibis (*Threskiornismelanocephalus*) and Painted Stork (*Mycteria leucocephala*) | | | |
| 22 | Sarus Cranes(*Grus antigone*), Blue Rock Pigeons and Collared Doves | India | Dielrine | [65] |
| 23 | Flamingo and Red-wattled lapwing | India | HCH, DDE, DDT | [64] |
| 24 | Griffon vulture (*Gyps fulvus*) and Egyptian vulture (*Neophron percnopterus*) | Spain | OCPs and PCBs | [59] |
| 25 | Sparrow hawk (*Accipiter nisus*) and Barn owl (*Tyto alba*) | Belgium | PCBs, PBDEs and OCPs | [60] |

food to sustain their metabolism, potentially leading to greater pesticide consumption from contaminated food sources. Being oviparous, birds eliminate lipophilic pesticides through their eggs, creating an exposure pathway that affects their offspring during early development [16, 18]. Additionally, birds have a relatively smaller liver-to-body size ratio than mammals, and their urinary excretion system involves voiding urine into the cloaca, where reabsorption of metabolites can occur via the cloacal and intestinal walls. Moreover, the presence of the coccygeal mesenteric vein, which connects the hepatic portal and renal veins, enables the direct transport of pesticides and their metabolites from the gastrointestinal tract to both the liver and kidneys, enhancing the role of renal metabolism in birds [16, 18]. Comparative studies have often demonstrated greater toxicity of cholinesterase (ChE)-inhibiting pesticides, such as organophosphates and carbamates, in birds compared to mammals, whereas pyrethroids generally exhibit higher toxicity in mammals. These interspecies differences are primarily associated with variations in the activity of metabolic enzymes, including esterases and oxidases, as well as differences in the sensitivity of toxicological target sites specific to each pesticide [16–18].

POPs harm birds by disrupting their immune, neurological, and endocrine systems, which impacts growth, development, and reproduction. Variability in the aryl hydrocarbon receptor are associated with species-specific sensitivity, and the various POP classes may have helpful or antagonistic effects [19]. Those pollutant which causes disruption of endocrine system in birds was initially recognized through their association with eggshell thinning in bald eagles, which led to a significant decline in chick viability and population numbers [20]. However, the effects of these chemicals exposure in avian species are highly variable, influenced by factors such as species-specific sensitivity, timing, duration, and intensity of exposure, as well as reproductive strategies [21]. Recocial species are particularly susceptible during embryonic development, while altricial species exhibit a more prolonged sensitivity throughout their life stages. Endocrine disruptive chemicals primarily affect the reproductive, thyroid, and stress endocrine systems, though alterations in stress response and immune function are often more complex to interpret [21].

Fluorinated hydrocarbons compounds such as polytetrafluoroethylene including Teflon, perfluorooctane sulfonic acid (PFOS) and perfluorooctanoic acid (PFOA) exert nephrotoxic effects by diffusing passively into renal tubules [21, 22]. Once inside, they disrupt mitochondrial function, leading to compromised energy production required for active excretion [21, 22]. This mitochondrial damage also increases oxidative damage due to the leakage of free radicals and results in cellular apoptosis, progressively impairing kidney function [21, 22].

Persistent organic pollutants (POPs) exhibit lipophilic properties which enable them to cross plasma membranes and accumulation inside adipose tissues [21]. The animal contact with 2,3,7,8-tetrachlorodibenzo-p-dioxin (TCDD) causes significant hepatic toxicity, characterized by lipid accumulation, hepatocyte hypertrophy, infiltration of inflammatory cells, hyperplasia, and elevated levels of hepatic fatty acids, triglycerides, and serum alanine aminotransferase (ALT) [21, 22]. Additionally, exposure to polychlorinated biphenyl (PCB) causes the non-alcoholic fatty liver disease (NAFLD) associated with the reduction in hepatic antioxidant responses [21, 22]. Variety of birds have negatively impactd due to POPs as given in Table 21.1.

## 21.3 Avian POP Contamination in South Asia: Key Case Studies

South Asia, recognized as a biodiversity hotspot hosting a vast array of diverse flora and fauna, faces escalating threats due to rapid industrialization, agricultural intensification, and urban expansion [14]. The environment is substantially exposed to persistent organic pollutants (POPs) as a result of these human activities. These pollutants, characterized by their resistance to degradation, bioaccumulate and biomagnify within food chains, posing a severe threat to biodiversity, with avian species being particularly vulnerable to their toxic effects [13–15]. Some key case studies highlighting the impacts of POP contamination on avian populations in South Asia:

### *21.3.1 The Impact of DDT on Birds of Prey in India*

India is the second largest producer and larger consumer of pesticides all over the south Asia [14, 33]. Although various type of pesticide like HCH, Dieldrine etc. but DDT comprise two third of total consume. Dichlorodiphenyltrichloroethane (DDT), initially lauded for its insecticidal properties and awarded a Nobel Prize, is now recognized for its significant environmental and health impacts [33, 34]. Initially deployed to combat malaria and typhus vectors during World War II, its post-war application as a broad-spectrum pesticide led to widespread environmental contamination [33, 34]. Chronic DDT exposure has been causally linked to neurotoxicity, endocrine disruption, and reproductive dysfunction in both humans and wildlife. Carson's *Silent Spring* (1962) significantly contributed to raising public awareness, resulting in the cessation of DDT use in most developed nations by the 1970s and 1980s. The Stockholm Convention on Persistent Organic Pollutants, 2004 classified DDT among the most hazardous substances, leading to its global restriction [14, 33, 34]. However, the World Health Organization (WHO) continues to advocate for its restricted use in malaria vector control programs in tropical regions, despite associated risks to human and wildlife populations, including top predators [8, 11, 14]. Raptor species, characterized as top predatory avifauna, occupy critical trophic top positions within ecological systems, functioning as bioindicator organisms [33, 34]. Their positioning in complex food web hierarchies enables them to serve as sensitive ecological monitors, providing empirical insights into ecosystem dynamics and environmental health status [8, 11, 34].

Dichlorodiphenyltrichloroethane (DDT), a persistent organochlorine compound, exemplifies biomagnification phenomena through trophic levels. Its molecular properties facilitate bioaccumulation within lipophilic tissues of higher-order consumers. Consequently, raptor populations experienced substantial demographic decline during periods of extensive agricultural chemical deployment, simultaneously emerging as pivotal biosentinel species for environmental contamination assessment. The biochemical characteristics of these avian predators—specifically their lipid-rich physiological composition and long-lived nature—render them exceptional biomonitoring organisms for persistent environmental pollutants, enabling sophisticated ecological risk assessment and environmental surveillance strategies. The order of accumulation of various pesticides hexachlorocyclohexane (HCH) > DDT > heptachlor epoxide > dieldrin revealed that HCH and DDT are prominent among others. Recent study showed there is high concentration of DDTs and Organo chlorine were observed in birds recorded from south India [33–35].

Population Declines of Threatened Raptor Species: Several threatened species, such as the Steppe Eagle (*Aquila nipalensis*), White-rumped Vulture (*Gyps bengalensis*) and Shikra (*Accipiter badius*), have experienced dramatic population declines [18, 33, 35]. These raptors occupy the apex of the trophic level, making them particularly vulnerable to the environmental contaminants [33–35]. The precipitous decline in Gyps vulture populations in South Asia especially in India and Pakistan has been attributed to the veterinary drug diclofenac, highlighting the risks of releasing new chemicals into the environment without thorough testing on non-target species [13, 35]. Additionally, there have been numerous reports of Sarus cranes succumbing to pesticide-treated seeds. The disappearance of the Pallas's Fish Eagle from Keoladeo National Park and other wetlands may be linked to the extensive elimination of pesticides in the surrounding catchment areas [35, 36, 38]. High concentration of DDT have been observed in the fatty deposits of Vulture, Kite and Crow populations in Lucknow, further underscoring the pervasive nature of this contaminant in the environment [36–38].

Eggshell Thinning and Reproductive Impacts: One of the most critical impacts of DDT was its ability to cause thinning of eggshells in birds, primarily due to the accumulation of DDT and its metabolites in the birds' bodies, which interfered with calcium metabolism [36, 38]. Thinner eggshells resulted in higher rates of egg breakage and reduced reproductive success. Additionally, the breakdown products of dicofol, another pesticide, may include DDE, which has been identified as a major cause of reproductive failure in several bird species [18, 38]. As the Great Indian Bustard feeds on insects, often in agricultural fields, it is likely accumulating high levels of pesticides, contributing to the high incidents of egg infertility observed in many areas [17, 38].

Behavioral Changes and Ecosystem Impacts: DDT exposure has also been linked to changes in bird behavior and physiological effects, which could indirectly affect reproduction and survival rates [18, 39]. The decline of predatory bird has had rippling effects on ecosystems, altering food chains and affecting all biodiversity [19, 39]. A recent study showed that naturally occurring fishes which are primary food for the south Indian people have been recorded with maximum residual limit of HCH(hexachlorocyclohexane) perspective provided by Food and agricultural organization and world health organization for fish products [17, 39]. The diminishing effect of organochlorine pesticide were also observed in wild fishes [17, 39].

### 21.3.2 POPs Exposure in Wetland Birds in Bangladesh

POPs have been found extensively in the environment of Bangladesh, and wetland birds are among the most vulnerable species to their effects. The 12 POPs assessed are predominantly derived from industrial processes or employed in agriculture and vector control for public health purposes [12, 13]. Among these, nine compounds are pesticides traditionally applied to agricultural crops or utilized for vector management in public health initiatives [17, 18, 35]. By the late 1970s, Bangladesh had implemented bans or imposed stringent restrictions on the use of these nine pesticides and polychlorinated biphenyls (PCBs). However, recent evidence suggests that certain POPs, including Heptachlor, Dieldrin, DDT, and Chlordane, continue to be used in specific regions of the country [13, 18, 40]. These substances are reportedly deemed essential for maintaining agricultural productivity and addressing public health challenges.

Wetlands serve as transitional zones between land and water, which can be either permanent or temporary, still or flowing, and may encompass freshwater, brackish water, or even marine environments. These ecosystems are vital for sustainable livelihoods, playing crucial roles in nutrient and carbon cycling, hydrological processes, soil formation, supporting natural-resource-based livelihoods, and enhancing crop resilience against pests, diseases, droughts, and floods [18, 40]. In Bangladesh, the Sundarbans stands out as the most prominent saltwater wetland, featuring an extensive mangrove forest [17, 41]. Conversely, Tanguar Haor and Hakaluki Haor are distinguished as exceptional freshwater wetland ecosystems [35, 42]. These ecosystem harbors many birds diversity including threatened species. Studies have shown that wetland birds in Bangladesh, such as Ganges River Dolphin (*Platanista gangeticus*), Pallas's Fishing Eagle (*Haliaeetus leucoryphus*) and the Spoon-billed Sandpiper (*Eurynorhynchus pygmeus*), have high levels of POPs in their tissues [40–42]. These birds have higher concentration of POPs due to top predators and feeding on contaminated fish and other aquatic organisms [40–42]. The prominants POPs were and per-fluoro-octane-sulfonate (PFOS) in finfish (0.1–3.86 ng/g ww) and perfluorooctanoic acid (PFOA) in shellfish (0.07–2.39 ng/g ww) [41] have been recorded in Bangladesh [11, 41]. The coastal region in Bangladesh are highest contaminated with PFOA in water and PFOS in sediment [42, 43].

POP exposure in wetland birds in Bangladesh has also been linked to changes in their behavior, including altered feeding patterns, social behavior, and migration routes [18, 42]. For example, a study on the Pallas's Fishing Eagle in Bangladesh found that exposure to POPs was associated with changes in their diet, with exposed birds preferentially feeding on fish with higher levels of POPs [40–42]. The government of Bangladesh has implemented several measures including the adoption of environment friendly agricultural practices and the establishment of protected areas to lower the POPs usage [18, 25, 41]. Further research is necessary to understand the fullycomprehend of POP exposure in wetland birds in Bangladesh and to develop effective strategies for mitigating its effects [41, 42].

### 21.3.3 Seabirds and Coastal POP Pollution in Sri Lanka

There is a lack of systematic studies measuring POPs concentrations in seabirds or their prey in Sri Lanka [12, 13]. Shorebirds are widely considered as extraordinary endurance athletes because of their amazing long-distance migrations, which are an essential part of their intricate life cycles [9, 18]. However, they are especially vulnerable to environmental contaminants because of their dependency on invertebrate prey in coastal wetland environments [35, 48]. Large pollutant influxes from nearby land and aquatic systems frequently affect these environments [35]. Evaluating the health of shorebirds during migration poses significant challenges due to the limited opportunities to capture and examine individuals [23, 24]. Furthermore, understanding the origins of pollutants encountered by migratory shorebirds and their impacts at physiological, biological, and behavioral levels remains an important but underexplored area of research [23, 25]. Recent advancements, particularly in seabird and songbird studies, have successfully integrated fieldwork with controlled laboratory experiments, offering promising avenues for future investigations [24, 25]. The seabird sooty terns (*Onychoprion fuscatus*) eggs reported high concentration of Persistant organic pollutants includes CHL with 0.21 ng/g wet mass [26], PCB with 1.5 ng/g wm and PBDE with 1.1 ng/g wm [26] than others common noddies (*Anous stolidus*) and fairy terns (*Gygis alba*) reported from Atoll region of Indian Ocean [26].

An innovative approach by capturing wild passerines, administering environmentally relevant doses of specific contaminants, and equipping the birds with tracking devices to monitor the behavioral effects on both exposed and control groups [18, 35]. Applying similar methodologies to migratory shorebirds could provide valuable insights into the health implications of toxicant exposure during their extensive migrations [27, 28].

Despite the ecological significance of shorebirds, there is a notable paucity of data quantifying their risk from environmental pollution [27–29]. Shorebirds, which primarily consume higher invertebrates like crustaceans and mollusks, generally positioned at lower trophic level in contrast to apex predators such as raptors or piscivorous birds [29]. Consequently, they are presumed to accumulate lower levels of contaminants. However, current thresholds used to evaluate pollutant impacts are often derived from studies on piscivores and may not be appropriate for shorebirds [29, 31]. The lower contaminant thresholds for shorebirds to account for their unique ecological and physiological traits and have species-specific role [30, 31]. This knowledge gap is critical to determining risk thresholds and developing effective conservation and monitoring strategies for shorebird populations, which continue to decline globally [17, 31]. Tributyltin (TBT) is an organotin compound, which is a toxic persistant pollutant can remains long time duration in the environment [30, 31]. It acts as reproductive hormones which adversely affects the invertebrates such as masculinity characters appear in females and impaired reproductive system [30–32].

## 21.4 Regulatory Frameworks Addressing POPs in South Asia

South Asia, with its rapid industrialization and agricultural dependence, faces challenges in managing POPs. Regulatory efforts in the region are driven by both international obligations and national initiatives, though implementation varies across countries (Fig. 21.2).

### *21.4.1 International Regulatory Frameworks in South Asia*

#### 21.4.1.1 BRS Convention (Basel, Rottardam and Stockholm Convention) [12]

BRS Convention are multilateral environmental agreements (MEAs) of the Basel, Rotterdam, and Stockholm conventions with a shared objective to protect the environment from the toxic impacts of pollutants [12]. Biennial conferences are convened to review the conventions' regulations, evaluate scientific research, and address emerging environmental health concerns. The 12th Conference (COP-12)

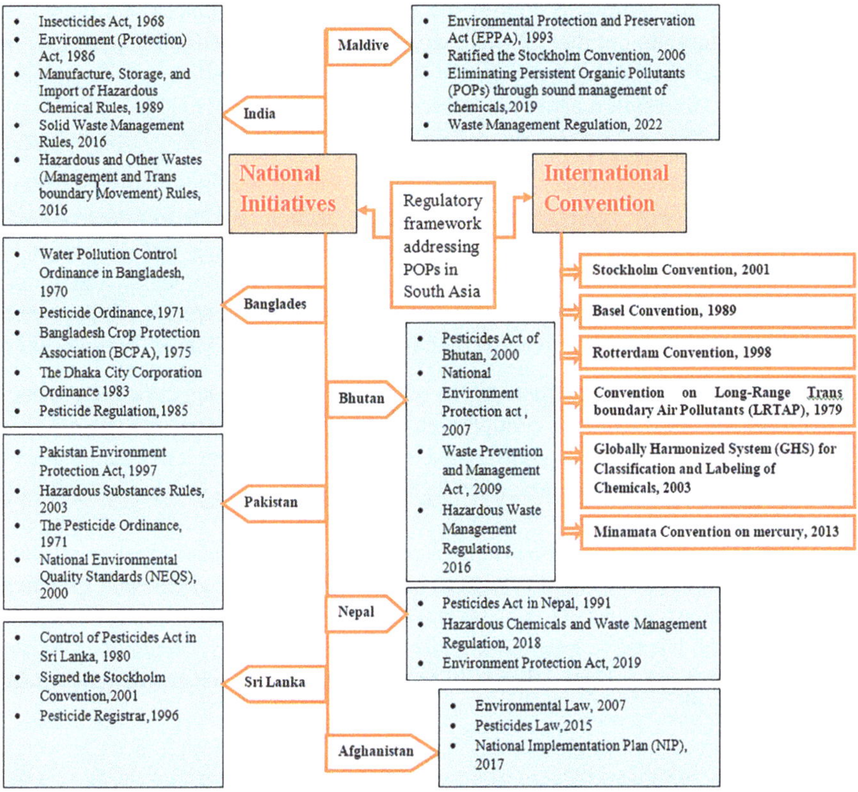


**Fig. 21.2** Showing the regulatory framework addressing Persistent organic pollutants (POPs) in South Asia

has approved the theme *"Make Visible the Invisible: Sound Management of Chemicals and Wastes"* and is scheduled to take place from April 28 to May 19, 2025, in Geneva, Switzerland. Below is a scientific articulation of these conventions and the synergistic approaches to their implementation [10, 12].

**Basel Convention [12]**

The Basel Convention is most comprehensive legal framework to regulate the international movement of harmful chemicals and their proper disposal management [12]. It aims to mitigate risks to human and environmental health associated with improper waste handling by regulating harmful waste throughout its entire lifecycle, encompassing production, transportation, utilization, and disposal [12].

**Rotterdam Convention [12]**

The Rotterdam Convention on the basis of Prior Informed Consent (PIC) serves as an international mechanism to prevent the hazardous chemicals [10, 12]. Through acquiring exporting countries' approval, this convention ensures that no hazardous chemicals mentioned in the convention list are imported into their country [10, 12]. It thereby empowers nations to control the entry of these substances and enhances global health and environmental protection. A Conference of Plenipotentiaries in Rotterdam, Netherlands, approved the Rotterdam Convention on September 10, 1998. The Convention came into force on 24 February 2004 [12].

**Stockholm Convention [12]**
The Stockholm Convention was established to prevent the production, use, and trade of persistant organic pollutants that endanger human health and ecosystems [12]. Through its provisions, the treaty mandates the phasing out of persistent organic pollutants, addressing their release, storage, and environmental contamination [10, 12]. The agreement went into effect on May 17, 2004, after being approved on May 22, 2001, at the Conference of Plenipotentiaries in Stockholm, Sweden [12].

The Conferences of the Parties (COP) of all three conventions initiated a **synergies process** by recognizing overlapping objectives and operational challenges. This synergy process improves the coordinated implementation at regional, national and global level [12]. While maintaining the legal autonomy of each treaty, this approach enhances coherence in policy guidance, streamlines administrative functions, and optimizes resource use [12]. The synergies process has led to transformative reforms within the secretariats of the conventions, fostering more integrated and effective implementation strategies [12]. Stakeholders, including participant countries, regional centers, intergovernmental organizations, and non-governmental organizations (NGO), have a critical role in advancing coherence in chemicals and waste management [12].

#### 21.4.1.2 Convention on Long-Range Transboundary Air Pollutants (LRTAP) [44]

The Convention on Long-Range Transboundary Air Pollution seeks to promote international efforts to mitigate atmospheric contamination that traverses national boundaries [44]. It emphasizes the progressive reduction and prevention of pollutants that impact air quality over extended distances. Similarly, the Protocol addressing Persistent Organic Pollutants, effective from 23 October 2003, focuses on managing substances known for their long-lasting and harmful environmental effects [44]. The Protocol aims to minimize, control, or entirely eliminate the release, emissions, and dispersion of these hazardous compounds to safeguard ecological and human health [44].

#### 21.4.1.3 Globally Harmonized System (GHS) for Classification and Labelling of Chemicals [45]

The Globally Harmonized System (GHS) provides a unified framework for categorizing chemical substances based on their potential risks to health, physical properties, and environmental impacts [45]. This system establishes standardized guidelines to ensure consistency in the assessment and communication of chemical hazards worldwide [45].

#### 21.4.1.4 International Convention on the Control of Harmful Anti-fouling Systems on Ships [46]

The convention aims to prevent the use of toxic organotin compounds in anti-fouling paints to protect marine ecosystems [46]. Anti-fouling systems or bottom paints, used to prevent biofouling on ships, can release organotins into marine environments, causing severe ecological harm due to their toxicity, persistence, and bioaccumulative properties [46]. The Convention, entering into force on 17 September 2008, establishes a global regulatory framework to eliminate these substances and promote sustainable maritime practices. It emphasizes international cooperation to ensure compliance and enforcement among member states, fostering accountability in the shipping industry [46]. Furthermore, the Convention encourages the development of sustainable anti-fouling technologies to balance between protecting marine biodiversity and maintaining the operational efficiency of vessels [46]. It marking a significant milestone in global efforts to reduce environmental impacts from shipping activities.

#### 21.4.1.5 The Vienna Convention and the Montreal Protocol [47]

These conventions have been playing important role in preventing the release of persistent organic pollutants that contribute to ozone layer degradation. Through the systematic reduction and eventual elimination of ozone-depleting substances, many of which are also classified as POPs due to their longevity and toxicity, these agreements have reduced environmental contamination [47]. By encouraging the transition to safer alternatives, they have indirectly minimized the accumulation of harmful compounds in ecosystems. Their global implementation showcases the effectiveness of international cooperation in addressing pollutants that threaten both atmospheric health and broader ecological systems [47].

#### 21.4.1.6 Minamata Convention on Murcury [48]

It is an international organization to protects the human and the environment from the hazardous impact of metal mercury which have become a danger due to their emission in the environment [48]. This Convention was established in October, 2013 at the Conference of Plenipotentiaries, held in Kumamoto, Japan, after its fifth meeting in Geneva, Switzerland [48].

## *21.4.2 Regional and Indian Laws to Regulate POPs*

### 21.4.2.1 Monitoring and Management of Persistent Organic Pollutants (POPs in Asia (1996) [49]

UNU-IAS (United Nations University Institute for the Advanced Study of Sustainability), Shimadzu Corporation has initiated program to strengthening research on POPs in Asian region for proper management environmental contaminants based on scientific methods which is internationally regulated by Stockholm convention [49].

### 21.4.2.2 The Bangladesh Environmental Conservation Acts (1995) [50]

The main aim of this act was to conserve the environment from the pollution and improve quality standard. A national's wetland policy has been implemented to conserve the wetland s for the biodiversity and principle to use the wetland resources along with sustainable development [50].

Water pollution control ordinance (1970) has been implemented to improve the quality of water and prevent the hazardous chemicals drainage into the water system [50].

Environmental pollution control Ordinance (1977) has also implemented for the improve the overall quality and prevent the pollutions in air, water and soil [50].

### 21.4.2.3 Control of Pesticide Act No.33, 1980 and National Environmental Act, Sri Lanka [51]

These acts prevent and regulate the use of persistent organic pollutants in Sri Lanka. Before these, there is continuous use of the chemicals for the agriculture, forest and industries. After implementation, there is restriction to use of some chemicals and some chemicals are permanently banned [51].

### 21.4.2.4 Regulation of Persitant Organic Pollutants (POPs) Rule (2018) [12, 51]

The regulation of Persistent Organic Pollutants (POPs) Rule has been refers to ministry of Environment, Forest and Climate Change of India for the prohibition of use, trade, manufacture, import and exports of seven most hazardous chemicals includes [12, 18, 35]: Chlordecone, Pentacholorobenzene, Hexabromobiphenyl, Hexabromodiphenyl ether and Heptabromodiphenyl ether (Commercial octa-BDE), Hexabromocyclododecane, Tetrabromodiphenyle ether and pentabromodiphenyl ether (Commercial penta-BDE) and Hexachlorobutadine [12, 18, 35, 50].

However India also followed the Stockholm convention to implement and prohibit the trade of the hazardous chemicals. On 13th January, 2006, India has become the part of Stockholm convention which operative in 2004 [12, 51].

Prevention, control, and abatement of environmental pollution: As per the Environment Protection Acts (1986) Chapter III deals with the prohibition of emission of any environmental pollutants in excess. The United Nations conference on the human environment take these step to make environment free from pollution in Stockholm held in June, 1972 [12, 51].

Apart from these, Water pollution act, 1974 to prevent the water pollution, the Air pollution act, 1981 to regulate the emission of harmful substances in air and the Noise pollutionact, 1999 to manage the sound to standard level [51].

#### 21.4.2.5 Hazardous and Other Waste Rule (Management and Transboundary Movement), 2016

The hazardous and other waste management rule, 2016 and Hazardous waste management rule, 2008 ensure the safe handling, use, trade and proper disposal of the waste substance emission from industries, urban development and destruction implemented by govt. of India under the ministry of environment, forest and Climate Change. The acts amended in 2000, 2002 and finally drafted in 2008 as Hazardous Waste (Management, Handling and Tran boundary Movement) Rules, 2008 [50, 52].

Besides these, Pesticide regulations, 1985 in Bangladesh, Pesticides Ordinance, 1971 in Pakistan, Pesticide law, 2015 in Afghanistan, Environmental protection and preservation act, 1993 in Maldives, Pesticides act, 2000 in Bhutan and Pesticides act, 1991 in Nepal were major rules to regulate the pesticides production and usage in south Asian countries as shown in Fig. 21.2.

## 21.5 Challenges of POPs Management

Lack of Awareness is the main challenge that is associated with the use of hazardous chemicals leads to various harmful impacts on human health, environment and biodiversity. Promoting education through campaign and other program may help in aware the people about the effects associated with the use of pollutants. Research on POPs is limited, so there are insufficient mechanisms for the disposals of pollutants which ultimately enter into the food chains and negatively impact the biodiversity. Lack of proper equipments facility for proper disposals of the hazardous chemicals and in handling and lack of trained peoples are also the key factors for the dispersing the pollutants. Lack of unified regulation: there is lack of stringent regulation for the management of the hazardous pollutants. This inconsistency hampers the regional legislation. However slow ratification of international treaties such as Malaysia and Indonesia are not ratified with the Stockholm convention for

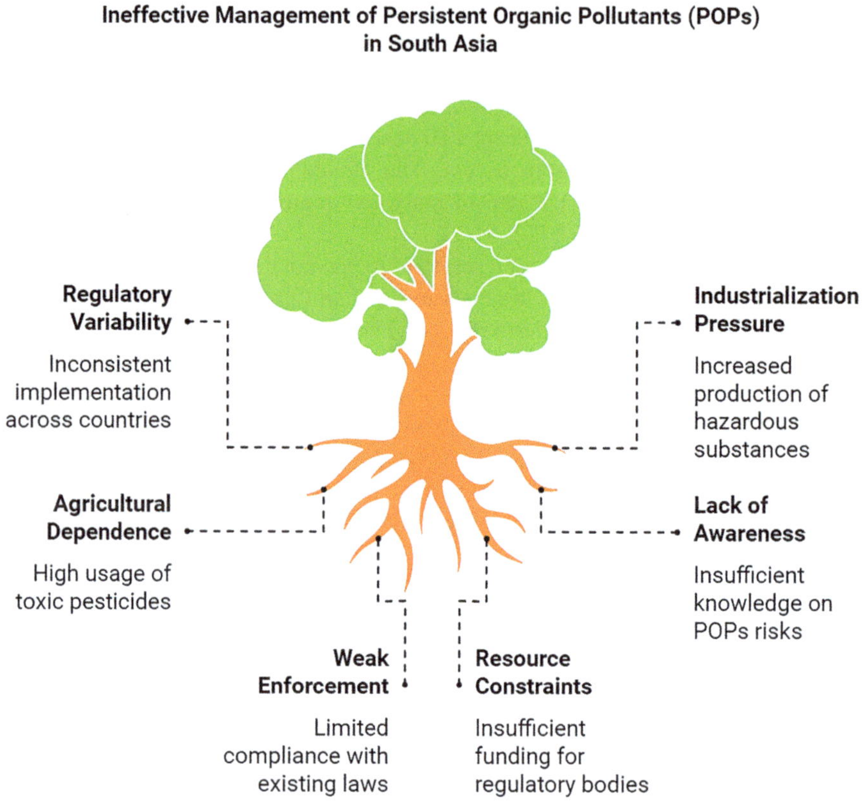


**Fig. 21.3** Showing major challenges associated with the management of Persistant Organic Pollutants in South Asia

the regulations and management of POPs. Traditional practice of agriculture often needs pesticides for the proper harvesting of crop yield. These pesticides are entered into the pest insect bodies and eaten by birds. These chemicals are bioaccumulated in food chain through birds [17, 53]. Rapid industrialization and urbanization has led to elimination of wastes includes solid waste, liquid and gases which accumulated in the environment and affect the biodiversity especially top predators(as shown in Table 21.1).These are some challenges associated with POPs management depicted in Fig. 21.3.

## 21.6 Role of Local Community in Regulation and Management of Persistent Organic Pollutant

Through unofficial pressure on surrounding industries to reduce their emissions, local people play a critical role in managing and reducing pollution. This has a particularly significant effect when official regulations are inadequate or nonexistent. Their action frequently reveals a desire to strike an equilibrium between financial benefits that support the community's welfare and wealth and environmental quality [54, 55]. As evidenced in Jacksonville, where citizen protests impacted EPA actions and grant allocations, they activate local initiatives, contest regulatory choices, and promote environmental justice [55]. They can help in spreading awareness of the detrimental effects of POPs on the environment such as warning neighbors about risk of open burning of waste or educating farmers about potential hazards of insecticides based on POP. New strategies and initiative can also be made in collaboration of local community for better implementation of rules and regulation to prevent the persistant organic pollutants [55, 56]. The disposal of hexachlorobenzene (HCB) in Sydney showed the important role of community engagement in promoting environmental justice and shaping policy decisions [9, 18, 55].

## 21.7 Mitigation Strategies for Avian POP Exposure

A comprehensive approach focusing on legislation, monitoring, and habitat management is necessary for reducing the POPs effects on birds. To mitigate their effects on avian species, the Stockholm Convention has already selected 12 important POPs for global prohibition. There are following strategies that can help to reduce the pollutants exposure to birds [18, 35, 56].

1. Through strengthening legislation nationally and internationally, we can improve the current scenario of prevent birds from the hazardous chemicals. Through proper manufacturing, use and disposal of chemicals by new research methods [56].
2. Enhancing environmental monitoring can also help in reducing the pollutants through regular monitoring of pollutants concentration in air, water and soil. Assessing the pollutants concentrationin bird's tissue like egg and feathers can help in monitoring and early warning of ecological risks [57].
3. Encouraging sustainable practices like using of eco-friendly pesticides in agriculture, recycling of wastes and proper disposal of pollutant before releasing in the environment [35].
4. Promoting habitat conservation such as wetlands, forest, bird's sanctuaries and national parks can help in improving the overall health of biodiversity [41, 42].

5. Education and public awareness can also help in reducing the production, use, import and export of hazardous chemicals. By regular awareness campaign in collaboration of local community can help in reducing the spread of pollutants. More research still needed for appropriate method to deal with the pollutant harmful for the better health of environment [35, 56].

## 21.8 Future Perspective and Policy Recommendations

To enhance current policies addressing persistent organic pollutants (POPs), incentive-based sustainable practices should be introduced to encourage active participation from industries, the agricultural sector, and local communities. Stricter regulations with severe penalties are essential to ensure the proper elimination and disposal of hazardous pollutants. Additionally, personnel involved in handling and disposing of POPs must receive comprehensive training on their severe environmental and health impacts. Personnel involved in handling and disposing of POPs must receive proper training on strategies and their severe environmental and health impacts.

Restoring critical habitats such as wetlands, forests, and grasslands, which support avian biodiversity, is vital. Measures must be implemented to restrict the release of harmful substances into water systems that form the dietary base for terrestrial animals and birds. Protecting primary avian food sources from contamination can mitigate risks of bioaccumulation and biomagnification, safeguarding bird populations from the adverse effects of POPs [17, 68].

Future strategies should emphasize long-term monitoring of birds as bioindicators, use of molecular biomarkers and remote sensing to detect contamination. Migratory species can serve as sentinels of transboundary pollution, highlighting the need for global cooperation. Promoting green evolution and safer alternatives to agrochemicals will further reduce risks, linking POP control directly to avian conservation and broader biodiversity protection. The Global Monitoring Plan uses temporal contaminant trends to evaluate the effectiveness of the Stockholm Convention on Persistent Organic Pollutants. However, these trends are also influenced by climate variability and trophic interactions, making it essential to distinguish regulatory impacts from ecological factors to better guide future policies [68–70]. The standard of hazardous chemical should be reevaluated to reflects the changes in sensitivity to climate change whether they meet the criteria of Stockholm convention. Further research is required to clarify how accelerated melting and contaminant re-mobilization influence accumulation and exposure in marine and freshwater systems. Integrated multimedia assessments, considering contaminant redistribution (particulate-bound vs. dissolved) and food web accumulation, are essential to quantify these impacts systematically [71].

## 21.9 Conclusion

Persistent organic pollution (POPs) are harmful chemicals which are bioaccumulated as well as biomagnified in food chain. The main sources of these pollutants are pesticides for agriculture, industrial and some domestic elimination includes solid waste, liquid and gases. Primarily, top predators includes raptors are highly affected birds species such as seabirds. South asia is most biodiverse region all over the worlds and harbors rich diversity of fauna and flora. However, it is also not avoided from the harmful effects of persistent organic pollutants. Greater flamingo (*Phoenicopterus roseus*), White-rumped vulture (*Gyps bengalensis*), Black kite (*Milvus migrans*) etc. with many other species are vulnerable due to elevated level of DDT and PCBs in agriculture and industrial sectors. Migratory species have prominent role in the dispersal of these toxic pollutants. The hazardous effects on birds includes reproductive impairments (egg shell thinning), neurotoxicity, endocrine toxicity (hormonal imbalance). The current legislations are not sufficient to regulate the emission, production and trade of these harmful and toxic chemicals. There are limited studies on the harmful impacts of these pollutants. More efficient research are still needed for the proper disposal and mitigation of these pollutants.

## References

1. South Asian Cooperative Environment Programme (SACEP) (2016). http://www.sacep.org/pdf/new-publication/South-Asia-Biodiversity-Status-Trend-and-Challenges.pdf
2. World Bank (2023) Supporting global biodiversity conservation with broader species coverage. https://blogs.worldbank.org/en/opendata/supporting-global-biodiversity-conservation-broader-species-coverage
3. Conservation international (2022) Biodiversity hotspots. https://www.conservation.org/priorities/biodiversity-hotspots
4. IUCN (2011) The status and distribution of freshwater biodiversity in the Western Ghats, India. https://iucn.org/resources/publication/status-and-distribution-freshwater-biodiversity-western-ghats-india
5. IUCN (2015) On the verge of extinction: a look at endangered species in the Indo-Burma Hotspot. https://iucn.org/content/verge-extinction-a-look-endangered-species-indo-burma-hotspot
6. World Wildlife Fund (2024) Eastern Himalayas. https://www.worldwildlife.org/places/eastern-himalayas
7. Kerala Biodiversity (2023) Island biodiversity in India. https://keralabiodiversity.org/wp-content/uploads/2023/02/Island_biodiversity_hotspots.pdf
8. Tan YL, Chen JE, Yiew TH, Habibullah MS (2022) Habitat change and biodiversity loss in South and Southeast Asian countries. Environ Sci Pollut Res 29(42):63260–63276
9. Fisher AC (2015) Key aspects of species extinction: habitat loss and overexploitation. In: Environmental resources and applied welfare economics. Routledge, pp 59–69
10. UNEP (2024) UNEP at the climate COP29. https://www.unep.org/unep-climate-cop29
11. Islam R, Kumar S, Karmoker J, Kamruzzaman M, Rahman MA, Biswas N, Tran TKA, Rahman MM (2018) Bioaccumulation and adverse effects of persistent organic pollutants (POPs) on

ecosystems and human exposure: a review study on Bangladesh perspectives. Environ Technol Innovation 12:115–131
12. Stockholm Convention (2023) Stockholm Convention on Persistent Organic Pollutants (POPs). https://www.pops.int/TheConvention/ThePOPs/AllPOPs/tabid/2509/Default.aspx
13. Abbasi NA, Malik RN, Frantz A, Jaspers VLB (2016) A review on current knowledge and future prospects of organohalogen contaminants (OHCs) in Asian birds. Sci Total Environ 542: 411–426
14. UNEP (2024) Persistant Organic Pollutant(POPs) and Pesticides. https://www.unep.org/cep/persistent-organic-pollutants-pops-and-pesticides
15. Haraguchi K, Koizumi A, Inoue K, Harada KH, Hitomi T, Minata M, Tanabe M, Kato Y, Nishimura E, Yamamoto Y, Watanabe T (2009) Levels and regional trends of persistent organochlorines and polybrominated diphenyl ethers in Asian breast milk demonstrate POPs signatures unique to individual countries. Environ Int 35(7):1072–1079
16. Katagi T, Fujisawa T (2021) Acute toxicity and metabolism of pesticides in birds. J Pestic Sci 46(4):305–321
17. Senthilkumar K, Kannan K, Subramanian A, Tanabe S (2001) Accumulation of organochlorine pesticides and polychlorinated biphenyls in sediments, aquatic organisms, birds, bird eggs and bat collected from South India. Environ Sci Pollut Res 8:35–47
18. Walker CH (1983) Pesticides and birds—mechanisms of selective toxicity. Agric Ecosyst Environ 9(2):211–226
19. Hao Y, Zheng S, Wang P, Sun H, Matsiko J, Li W, Li Y, Zhang Q, Jiang G (2021) Ecotoxicology of persistent organic pollutants in birds. Environ Sci: Processes Impacts 23(3): 400–416
20. U.S. Fish and Wildlife Service (2020). https://www.fws.gov/midwest/eagle/Nhistory/biologue.html. May 2020
21. Dimmen MV (2014) Levels and effects of persistent organic pollutants (POPs) on circulating thyroid hormones in house sparrows (Passer domesticus) from Leka, Norway (Master's thesis). Institutt for Biologi
22. Kruzel TA (2020) Kidney disease. In: Textbook of natural medicine. Churchill Livingstone, pp 1507–1517. https://www.sciencedirect.com/science/article/pii/B9780323430449001904; https://doi.org/10.1016/B978-0-323-43044-9.00190-4
23. Campioni L, Oró-Nolla B, Granadeiro JP, Silva MC, Madeiros J, Gjerdrum C, Lacorte S (2024) Exposure of an endangered seabird species to persistent organic pollutants: assessing levels in blood and link with reproductive parameters. Sci Total Environ 930:172814
24. Richard FJ, Gigauri M, Bellini G, Rojas O, Runde A (2021) Warning on nine pollutants and their effects on avian communities. Glob Ecol Conserv 32:e01898
25. Ma Y, Choi CY, Thomas A, Gibson L (2022) Review of contaminant levels and effects in shorebirds: knowledge gaps and conservation priorities. Ecotoxicol Environ Saf 242:113868
26. van der Schyff V, Yive NSCK, Polder A, Cole NC, Tatayah V, Kylin H, Bouwman H (2021) Persistent organic pollutants in sea bird eggs from the Indian Ocean's Mascarene Basin. Sci Total Environ 771:145348
27. Eng ML, Stutchbury BJ, Morrissey CA (2019) A neonicotinoid insecticide reduces fueling and delays migration in songbirds. Science 365(6458):1177–1180
28. Seewagen CL, Ma Y, Morbey YE, Guglielmo CG (2019) Stopover departure behavior and flight orientation of spring-migrant yellow-rumped warblers (Setophaga coronata) experimentally exposed to methylmercury. J Ornithol 160:617–624
29. Colwell MA (2010) Shorebird ecology, conservation, and management. University of California Press
30. Perkins M, Ferguson L, Lanctot RB, Stenhouse IJ, Kendall S, Brown S, Gates HR, Hall JO, Regan K, Evers DC (2016) Mercury exposure and risk in breeding and staging Alaskan shorebirds. Condor 118(3):571–582

31. Su T, Lin X, Huang Q, Jiang D, Zhang C, Zhang X, Dingle C, Goodale E, Que P, Wang R, Liu Y (2020) Mercury exposure in sedentary and migratory Charadrius plovers distributed widely across China. Environ Sci Pollut Res 27:4236–4245
32. Bandara KRV, Chinthaka SDM, Yasawardene SG, Manage PM (2021) Modified, optimized method of determination of Tributyltin (TBT) contamination in coastal water, sediment and biota in Sri Lanka. Mar Pollut Bull 166:112202
33. Eulaers I, Covaci A, Herzke D, Eens M, Sonne C, Moum T, Schnug L, Hanssen SA, Johnsen TV, Bustnes JO, Jaspers VL (2011) A first evaluation of the usefulness of feathers of nestling predatory birds for non-destructive biomonitoring of persistent organic pollutants. Environ Int 37(3):622–630
34. Dhananjayan V, Jayakumar S, Muralidharan S (2011) Assessment of exposure and effect of organochlorine pesticides on birds in India. Res Rev AJ Toxicol 1(3):17–23
35. Malik A, Dharaiya N, Espín S (2018) Is current information on organochlorine exposure sufficient to conserve birds in India? Ecotoxicology 27:1137–1149
36. Abbasi NA, Arukwe A, Jaspers VL, Eulaers I, Mennilo E, Ibor OR, Frantz A, Covaci A, Malik RN (2017) Oxidative stress responses in relationship to persistent organic pollutant levels in feathers and blood of two predatory bird species from Pakistan. Sci Total Environ 580:26–33
37. Ratcliffe DA (1967) Decrease in eggshell weight in certain birds of prey. Nature 215(5097): 208–210
38. Muralidharan S, Jaya Kumar S, Dhananjayan V (2016) Status of pesticide contamination in birds in India. AR Rahmani, Islam, MZ and Kasambe
39. Muralidharan S, Dhananjayan V, Jayanthi P (2009) Organochlorine pesticides in commercial marine fishes of Coimbatore, India and their suitability for human consumption. Environ Res 109(1):15–21
40. Lamsal P, Pant KP, Kumar L, Atreya K (2015) Sustainable livelihoods through conservation of wetland resources: a case of economic benefits from Ghodaghodi Lake, western Nepal. Ecol Soc 20(1)
41. Makowski C, Finkl CW (eds) (2018) Threats to mangrove forests: hazards, vulnerability, and management, vol 25. Springer
42. Alam ABMS, Badhon MK, Sarker MW (2015) Biodiversity of TanguarHaor: A Ramsar site of Bangladesh volume III: fish. IUCN, International Union for Conservation of Nature, Bangladesh Country Office, Dhaka, Bangladesh. pp xii+ 216
43. Habibullah-Al-Mamun M, Ahmed MK, Raknuzzaman M, Islam MS, Negishi J, Nakamichi S, Sekine M, Tokumura M, Masunaga S (2016) Occurrence and distribution of perfluoroalkyl acids (PFAAs) in surface water and sediment of a tropical coastal area (bay of Bengal coast, Bangladesh). Sci Total Environ 571:1089–1104
44. US Departtment and State (1979) Convention on long-range transboundary air pollution. https://www.google.com/search?q=Convention+on+Long-Range+Transboundary+Air+Pollutants+(LRTAP)&oq=Convention+on+Long-Range+Transboundary+Air+Pollutants+(LRTAP)&gs_lcrp=EgZjaHJvbWUyBggAEEUYOTIJCAEQABgNGIAEMggIAhAAGBYYHjIICAMQABgWGB7SAQkzMTYxajBqMTWoAgiwAgE&sourceid=chrome&ie=UTF-8
45. United Nation Institute for Training and Research (UNITAR) (2024). https://unitar.org/sustainable-development-goals/planet/our-portfolio/globally-harmonized-system-classification-and-labelling-chemicals#:~:text=The%20GHS%20is%20an%20internationally,labels%20and%20safety%20data%20sheets
46. International Maritime Organization(IMO) (2008) International Convention on the Control of Harmful Anti-fouling Systems on Ships. https://www.imo.org/en/About/Conventions/Pages/International-Convention-on-the-Control-of-Harmful-Anti-fouling-Systems-on-Ships-(AFS). Retrieved on 10.12.2024
47. UNEP (1985). The Vienna Convention for the protection of the Ozone Layer. https://ozone.unep.org/treaties/vienna-convention. Retrieved on 10.12.2024

48. United Nation (2013) Minimata Convention on Mercury. https://minamataconvention.org/en. Retrieved on 10.12.2024
49. United Nation (2016) Monitoring and Management of Persistent Organic Pollutants (POPs) in Asia, Department of Economic and Social Affairs. https://sdgs.un.org/partnerships/monitoring-and-management-persistent-organic-pollutants-pops-asia
50. The Central Pollution Control Board (2018) Retrieved on 10.12.2024. https://cpcb.nic.in/openpdffile.php?id=UmVwb3J0RmlsZXMvNjU4XzE1MjU2Nzk4NTJfbWVkaWFwaG90bzMwMDE5LnBkZg==
51. Control of Pesticides Act, No. 33 of 1980. Parliament of the democratic socialist republic of Sri Lanka. https://www.srilankalaw.lk/YearWisePdf/1980/CONTROL%20OF%20PESTICIDES%20ACT,%20NO.%2033%20OF%201980.pdf. Retrieved on 10.12.2024
52. The Central Pollution Control Board. Hazardous and other Wastes (Management & Transboundary Movement) Rules, 2016. https://cpcb.nic.in/rules/. Retrived on 11.12.2024
53. Bradbury SP, Coats JR (1989) Comparative toxicology of the pyrethroid insecticides. In: Reviews of environmental contamination and toxicology, pp 133–177
54. Pargal S, Hettige H, Singh M, Wheeler D (1997) Formal and informal regulation of industrial pollution: comparative evidence from Indonesia and the United States. World Bank Econ Rev 11(3):433–450
55. Čapek SM (1992) Environmental justice, regulation, and the local community. Int J Health Serv 22(4):729–746
56. Devendrapandi G, Liu X, Balu R, Ayyamperumal R, Arasu MV, Lavanya M, Reddy VRM, Kim WK, Karthika PC (2024) Innovative remediation strategies for persistent organic pollutants in soil and water: a comprehensive review. Environ Res:118404
57. Medina G (2024) Persistent organic pollutants. https://doi.org/10.1016/b978-0-12-824315-2.01149-0
58. Eulaers I, Covaci A, Hofman J, Nygård T, Halley DJ, Pinxten R, Eens M, Jaspers VL (2011) A comparison of non-destructive sampling strategies to assess the exposure of white-tailed eagle nestlings (Haliaeetus albicilla) to persistent organic pollutants. Sci Total Environ 410:258–265
59. Hong SH, Shim WJ, Han GM, Ha SY, Jang M, Rani M, Hong S, Yeo GY (2014) Levels and profiles of persistent organic pollutants in resident and migratory birds from an urbanized coastal region of South Korea. Sci Total Environ 470:1463–1470
60. Jaspers VLB, Covaci A, Voorspoels S, Dauwe T, Eens M, Schepens P (2006) Brominated flame retardants and organochlorine pollutants in aquatic and terrestrial predatory birds of Belgium: levels, patterns, tissue distribution and condition factors. Environ Pollut 139(2):340–352
61. Jayakumar S, Muralidharan S, Dhananjayan V (2020) Organochlorine pesticide residues among colonial nesting birds in Tamil Nadu, India: a maiden assessment from their breeding grounds. Arch Environ Contam Toxicol 78(4):555–567
62. Kunisue T, Watanabe M, Subramanian A, Sethuraman A, Titenko AM, Qui V, Prudente M, Tanabe S (2003) Accumulation features of persistent organochlorines in resident and migratory birds from Asia. Environ Pollut 125(2):157–172
63. Muñoz-Arnanz J, Cortés-Avizanda A, Donázar-Aramendía I, Arrondo E, Ceballos O, Colomer-Vidal P, Jiménez B, Donázar JA (2024) Levels of persistent organic pollutants (POPs) and the role of anthropic subsidies in the diet of avian scavengers tracked by stable isotopes. Environmental Pollution 343, p.123188
64. Misra V (1989) Monitoring and surveillance of pesticidal pollution of Mahala water reservoir with special reference to its avifauna (Doctoral dissertation, Ph.D thesis). Rajasthan University, Jaipur
65. Muralidharan S (1993) Aldrin poisoning of Sarus cranes (Grus antigone) and a few granivorous birds in Keoladeo National Park, Bharatpur, India. Ecotoxicology 2:196–202
66. Sanpera C, Ruiz X, Jover L, Llorente G, Jabeen R, Muhammad A, Boncompagni E, Fasola M (2003) Persistent organic pollutants in little egret eggs from selected wetlands in Pakistan. Arch Environ Contam Toxicol 44:0360–0368

67. Dhananjayan V (2013) Accumulation pattern of persistent organochlorine pesticides in liver tissues of various species of birds from India. Environ Sci Pollut Res 20:3149–3156
68. Yodha K, Singh D, Kour A, Yadav R, Kumar R (2024) Species diversity and abundance of birds in ber (*Zizyphus mauritiana*) crop. Asian J Conserv Biol 13(1). https://doi.org/10.53562/ajcb.83683
69. Kiran, Singh D, Kour A, Priya, Delu V, Kumar R (2022) Community composition and status of avian diversity at campus and agricultural landscapes of Chaudhary Charan Singh Haryana Agricultural University, Hisar (Haryana). J Appl Nat Sci 14(4):1130–1140. https://doi.org/10.31018/jans.v14i4.3784
70. Borgå K, McKinney MA, Routti H, Fernie KJ, Giebichenstein J, Hallanger I, Muir DC (2022) The influence of global climate change on accumulation and toxicity of persistent organic pollutants and chemicals of emerging concern in Arctic food webs. Environ Sci Process Impacts 24(10):1544–1576
71. Hung H, Halsall C, Ball H, Bidleman T, Dachs J, De Silva A, Hermanson M, Kallenborn R, Muir D, Sühring R, Wang X (2022) Climate change influence on the levels and trends of persistent organic pollutants (POPs) and chemicals of emerging Arctic concern (CEACs) in the Arctic physical environment–a review. Environ Sci Process Impacts 24(10):1577–1615

# Chapter 22
# A Comprehensive Review of Hybridization in Waterbirds of the Indian Subcontinent

**Sarika Kumari, Lakshay Jangra, and Pankaj Sharma**

**Abstract** Research has just started on hybridization, even though it's important to the nature of waterbirds across the ecologically rich Indian subcontinent. This comprehensive chapter aims to collect available information about hybridization events in waterbirds in this region and look at what they mean for species evolution, ecology and protection. Many members of the Anatidae, Ardeidae and Laridae families have been observed undergoing both natural and anthropogenic hybridization. Experts see overlapping breeding areas, changes in habitats and climate-related range changes as the main causes of interspecific breeding. Several hybridization events create infertile offspring or destroy one of the species' genetic assets, yet others help populations become more adaptive and result in the emergence of possible species. Hybrid pairs that have been seen in Indian wetlands are covered in the review, as well as examples of uncertainty in appearance, shared behaviors and the role of molecular tools. There is a special focus on how hybridization influences conservation, mainly with highly endangered species, where gene mixing could endanger the whole population. In addition, hybrid zones are considered important for observing evolutionary changes. Though waterbird hybridization is very important for the environment in India, little scientific work has been done, and what is available is mostly anecdotal. Its findings show that reliable hybridization assessment requires careful monitoring, genetic analysis and the use of all taxonomic approaches. Practical conservation strategies and the balance of wetlands in the region depend on our ability to see these patterns.

S. Kumari
Department of Biotechnology, College- Uka Tarsadia University, Maliba Campus, Bardoli, Gujarat, India

L. Jangra
Department of Zoology, Chaudhary Charan Singh Haryana Agricultural University, Hisar, India

P. Sharma (✉)
Department of Ophthalmology and Visual Sciences, University of Illinois Chicago, Chicago, IL, USA
e-mail: sharmap@uic.edu



R. Kumar et al. (eds.), *Ecology, Conservation of Aquatic Birds*, Wetlands: Ecology, Conservation and Management 15, https://doi.org/10.1007/978-3-032-17125-2_22

**Keywords** Biodiversity · Conservation · Hybridization · Indian subcontinent · Introgression · Waterbirds · Wetlands

## 22.1 Introduction

Hybridization is the interbreeding of genetically different populations or species. It is an evolutionary process that occurs along a continuum, ranging from infrequent mating between closely related individuals to more frequent and intricate gene flow events [22]. Waterbirds (such as ducks, geese, herons, and gulls) are the most likely bird taxa to undergo interspecific hybridization; research shows that more than 60% of waterfowl species do so, with mallards serving as important "gene flow hubs" across several species. In the fields of biodiversity and evolutionary biology, hybridization plays a paradoxical role. While it may provide essential raw materials for adaptation and potentially initiate speciation through introgression and adaptive hybrid zones, hybridization can also endanger species integrity through genetic swamping or outbreeding depression. This dynamic balance makes hybridization a fundamental and intriguing phenomenon in evolutionary studies [62].

The Indian subcontinent is a biogeographically unique area in terms of bird diversity. Home to more than 1350 to 1450 bird species, including approximately 150 endemic species, the subcontinent comprises 13–15% of all bird species worldwide [32]. The convergence of three major biogeographical realms (Indomalayan, Palearctic, and Afrotropical) and the subcontinent's diverse habitats, ranging from dense tropical forests and vast wetlands to rain-scarce deserts and glacier-capped Himalayas, make it a melting pot of avian genomes and evolutionary interaction zones. India is also home to a large number of migratory waterbirds as part of the Central Asian Flyway, increasing the likelihood of interspecific interactions and possible hybridization [34].

This chapter attempts to summarize the current understanding of waterbird hybridization on the Indian subcontinent. It also includes confirmed and suspected cases. These cases are across families such as Anatidae (ducks and geese), Ardeidae (herons and egrets), and Laridae (gulls and terns). The chapter analyzes the contributions of anthropogenic and natural drivers, such as overlapping breeding ranges, habitat modification, and climate-driven range shifts. Finally, it evaluates the methodological approaches used to identify hybrids, ranging from integrative molecular techniques to morphological scrutiny. Additionally, the chapter explores the conservation implications and evolutionary consequences of hybridization, particularly in relation to rare or declining species, where introgression could offer adaptive advantages or pose genetic risks.

## 22.2 Taxonomic Groups Affected by Hybridization in India

In the Indian sub-continent, hybridization of waterbirds has been mainly observed among three of the most prominent avian families, viz. Anatidae (ducks, geese, and swans), Ardeidae (herons, egrets and bitterns) and Laridae (gulls and terns). The diversity of these families makes up a large percentage of the waterbirds in the region and in a bid to co-exist, these often co-occur in mixed species assemblages throughout wetlands interacting interspecifically and possibly completing hybridization.

### *22.2.1 Anatidae (Ducks, Geese, Swans)*

Hybridization is best documented and most wide spread in the family Anatidae both in a worldwide context and in India [28]. Those who belong to this family are described to exhibit complex courtship dynamics, they are very aggressive interspecifically (sexually), and have a tendency to overlap breeding sites [27, 28, 30]. Suspected and confirmed hybrids have been reported in the Indian environment especially of Anas species, including the Northern Pintail (*Anas acuta*), Common Teal (*Anas crecca*) and wild and feral/domestic mallards (*Anas platyrhynchos*), especially the agricultural wetlands and floodplains in the north [38]. The hybridization of wild ducks with domesticated forms has caused reservations about the issue of maladaptive trait introgression and genotype of the wild type [21, 48].

### *22.2.2 Ardeidae (Herons, Egrets, Bitterns)*

In the family Ardeidae, hybridization is relatively infrequent and recorded, especially where the nesting colonies are the same and the plumage is similar. As an example, hybrid specimens that are purported to be of Intermediate Egret (*Ardea intermedia*) and Great Egret (*Ardea alba*) are incidentally reported on Indian wetlands, but are rather anecdotal and not confirmed genetically [47]. With such slight morphological variations between a number of Ardeid species particularly in non-breeding plumage, there is the possibility of hybrid individuals to pass un-identified again indicating the need of more systematic field and genetic studies.

### *22.2.3 Laridae (Gulls, Terns)*

Laridae family- which includes gulls and terns, is well known at the global level on including hybridization, particularly in large and mixed colonies [42]. They have reported hybridization between closely related large white-headed gulls (e.g. Caspian Gull (*Larus cachinnans*), Steppe Gull (*Larus barabensis*), and Heuglin's Gull (*Larus heuglini*)) in the northwestern and coastal areas in the Indian subcontinent during the wintering season [41, 42, 47].The hybrids among these species can be rather difficult to recognize or identify because the taxonomy is complicated and the plumage stages overlap, so they can appear intermediate and difficult to characterize. To accomplish this, it is necessary to conduct a molecular analysis.

Despite these reported or suspected instances, there have been few systematic studies on hybridization among these families in India. The majority of reports are founded on field observation, photos, or morphological exams, and very few studies use molecular diagnostics. There are additional factors which complicate the way in which we can determine even the magnitude of hybridization within these groups, namely, the frequent misidentification, lack of long-term monitoring. A further necessity is in the elaborated research, taking into consideration behavioral, ecological, and genetic information to prove the hybrid status and learn the ecological circumstances, as to the realization of such events.

## 22.3 Drivers and Mechanisms of Hybridization

The process of hybridization between waterbird species is instigated by a complicated interaction between ecological, behavioral, anthropogenic, and genetic factors (Fig. 22.1). All these drivers tend to work synergistically and as a result, this enhances a high likelihood of interspecific mating both in the natural and the adjusted landscapes. Some of these factors have been observed to enable

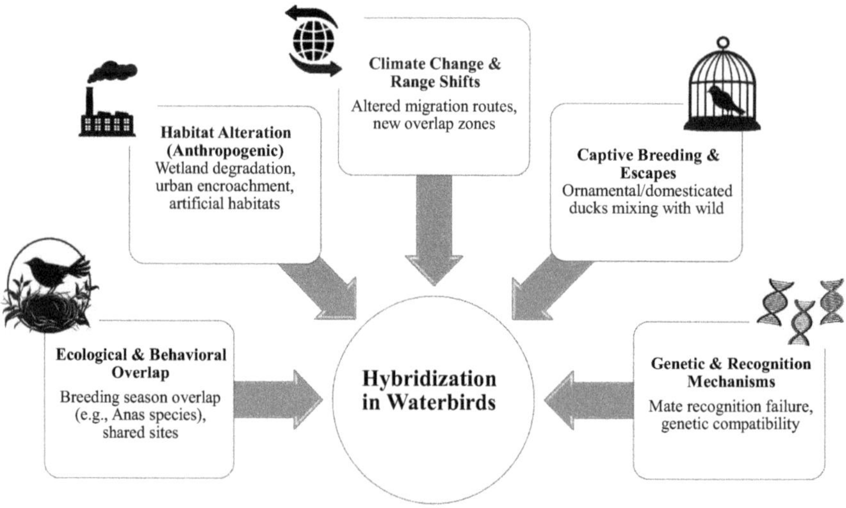


**Fig. 22.1** Drivers and mechanisms of hybridization in waterbirds

hybridization events in the Indian sub-continent especially in wetland areas that are ecologically sensitive.

The significance of ecological and behavioral aspects is crucial in the incidence of the hybridization. However, the overlap of breeding seasons between species of closely related waterbirds is one of the main mechanisms, since the resulting simultaneous breeding activity introduces more chances of interspecific contact. As a case in point, several Anas species (dabbling ducks) overlap in their breeding seasons in the winter and early springtime in wetlands of northern India and this is possibly enhancing hybrid pairings where they co-exist [17, 46, 51].

The other driving force is habitat alteration, especially as resultant of wetland degradation, pollution and urban encroachment [2, 54]. Due to the destruction or modification of the natural habitats, species are generally restricted to new, smaller, and fragmented or artificial habitats where they share, and closely related ones with those [64]. Such new or dense ecological environments may be disruptive to ordinary mate choice activities and escalate hybridization risks [4, 20, 53]. As an example, the birds that were displaced by degradation of floodplains can converge into man-made water bodies such as irrigation tanks, or sewage lagoons, and this results into a distorted interaction between species.

Hybridization is also taking place because of climate change and range movement [13]. The geographical habitats of numerous species are also evolving and lastly, allopatric species are overlapping due to the migration or reproduction seasons as the climate changes [13, 45, 58]. Expansion and shifting of wetland species in the north or upwards in altitude direction due to climate variability is already witnessed in India and this could establish new hybrid zones particularly in the wetland in the foothills of the Himalayas and northeast riverine ecosystems.

Captive breeding and escapes are major factors among those produced by humans and may be acting in certain duck populations [48]. The keeping of ornamental or domesticated ducks in community parks, zoos, and domestic collections is practiced widely and these ducks are at times released, which can hybridize with wild populations. This has been especially apparent in domesticated *Anas platyrhynchos* (commonly referred to as mallards) which can interbreed with wild congeners including spot-billed ducks (*Anas poecilorhyncha*) or northern pintails (*Anas acuta*). This sometimes results in genetic introgression [35, 48]. Due to this spread of artificial wetlands, such as reservoirs, sewage ponds, and aquaculture tank: new habitats have been formed, which in turn has resulted in having an array of waterbird species being attracted to new and old habitats [29, 60]. Although these systems can provide a short-term sense of safety, they are artificial contact zones and would have never existed in the natural ecological state. This enhances the possibility of cross-species meetings and further, hybridization [42].

Lastly, a pairing between mates is inherent in genetic compatibility and failure of mate recognition systems which can result in hybridization in the presence of both a reduction in ecological and behavioral barriers. Relatively similar species are similar enough genetically to raise offspring that are viable and even fertile. Visual and acoustic cues are used in the recognition of species especially in the birds in most instances. Should these cues become vague, hets mating could ensue as a result of charismatic plumage resemblance, change in behavior prompted by habitat, or low population densities [25, 52, 61].

## 22.4 Identification and Detection of Hybrids

The troubles of detecting hybrid individuals in wild bird populations are of both technical and interpretive nature especially in waterbirds that species can dramatically overlap in physical appearance. A major challenge with field identification is that hybrids often have an intermediate or mosaic map of characteristics of plumage, which tend to lie between the parent species in coloration, feather patterning, size or bill shape. This creates the visual form of identification as subjective and vulnerable to misclassification, particularly in wide-ranged and disparate genera like Anas (ducks) and Larus (gulls) [38, 42]. Further, plumage may change seasonally and the non-breeding and juvenile plumages serve to confuse the identification of the diagnostic characters.

Behavioral and vocalization may occasionally be useful in the detection of hybrids, e.g. in species with unusual courtship behavior, vocal calls or flocking. There can also be aberrant mating behavior or uncharacteristic format of vocalizing seen in the hybrids as presented in some species in Anatidae and Laridae in which the hybrids are male and they show poor mating success or aberrant mating preferences [22, 46]. Nonetheless, use of behavioral characteristics are restricted because of the fluctuations of the environment, knowledge of the observer, and the possibility of hybrids to assume the habits of either of its parents, or exhibit entirely new behaviors.

The increasing sensitivity of molecular tools has become critical in the identification of hybrids after the restrictions that morphological and behavioral evaluation brought. Molecular methods that can be used to identify the occurrence of introgression, direction of hybridization, and distinguish F1 hybrid and backcross generations include mitochondrial DNA (mtDNA) sequencing, microsatellite analysis, and lately genome-wide single nucleotide polymorphism (SNP) genotyping [4, 5, 59]. As an example, the mallard and spot-billed duck ancestors in Asia have been examined using nuclear and mitochondrial based markers to prove complicated levels of hybridization that could not have been verified by observational data alone [35] (Fig. 22.2).

Besides these gains, there is the lack of consolidated basis of hybrid identification in most ecological researches. This is due to the fact that having only one source of evidence (e.g. visual, behavioral, or even genetic) causes the incomplete or incomplete evaluation of the qualified person. Therefore, more and more focus is placed on the implementation of integrative taxonomy that unifies morphological, genetic, ecological and behavioural information into a strong basis to detect and classify hybrids [15, 43]. These integrative methods are even more important in areas where avian diversity is high such as the Indian subcontinent, and hybrid zones are relatively poorly documented.

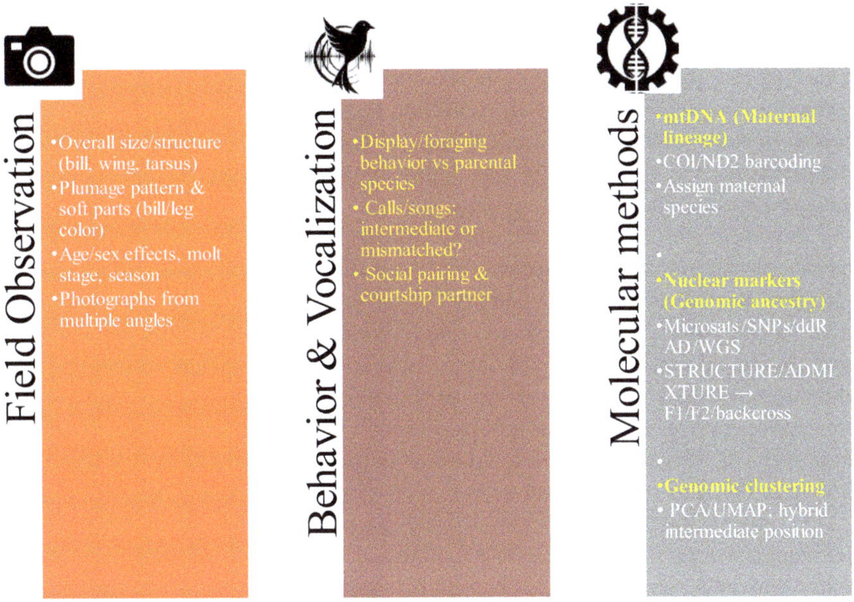


**Fig. 22.2** Identification and detection of waterbird hybrids

This study however aims to conclude that the proper identification of hybrids matters not only concerning comprehension of evolutionary processes but also in conservation planning. The presentation of misidentification can make the actual situation of the threatened species to become unclear or there can be a low representation of genetic introgression. A well-coordinated holistic approach to monitoring is warranted to achieve the elucidation of species edges and population management in the context of hybridization that may be a natural phenomenon or occur as a result of human interference due to the disturbance.

## 22.5 Evolutionary and Ecological Outcomes of Hybridization

Depending upon the frequency, directionality and effect of interspecific gene flow, hybridization in waterbirds can result in a number of evolutionary and ecological consequences. These effects include the positive process of adaptive introgression and even speciation and the negative effects of genetic swamping and out breeding depression. Such results are fundamental in deciphering the influence of hybridization on the formation of the biodiversity and ecological dynamics.

Genetic swamping of the gene pool in a rare or endemic species in repeated backcrosses with a more numerous closely related species is one of the most controversial effects of hybridization [48]. In this process, there are chances of eventuated genetic or demographic extinction of the less common species particularly in small or isolated populations. India There is a particular concern in India regarding endemic or declining waterbird taxa vulnerable to hybridization with very numerous congeners, including hybridisation between wild spot-billed ducks (*Anas poecilorhyncha*) and domesticated mallards (*Anas platyrhynchos*) [35]. Under those circumstances the introgression can undermine the evolutionary uniqueness and localized adaptations of the indigenous species.

Conversely, introgression, the stable transfer of genetic material between the species of one group into the gene pool of another can also be a source of genetic variation that drives adaptive evolution. There is a possibility that population can attain useful alleles through introgressive hybridization, which helps in adapting a new or altered environment [24, 56]. As one example, genes underlying new plumage coloration, immune response or migratory behavior have been introgressed between species in other avian groups, possibly increasing ecological-stress tolerance survival. Nevertheless, these adaptive impacts have not been well examined among Indian waterbirds, which depicts a very important research gap.

The other significant impact of hybridization is hybrid zone referred to as a geographical area in which the distributions of two species overlap and hybrids with common occurrence are thus present [8]. Such areas are de facto a laboratory to watch evolution in motion since ecological relatedness is a natural meal ticket in the process of reproductive isolation, repression of hybridization, or even specification of hybrids [1, 8, 23]. Such zones can be changeable and temporary in migratory waterbirds, influenced by movements in breeding or wintering areas under the influence of climate change and habitat change. Hybrid zones are quite published in formations where most of the temperate regions productions are involved (e.g. gulls and ducks), however, these zones have not been catalogued in India as a result of inadequate long-term scrutinizing in region.

Ecological and evolutionary dynamics are also affected by the fitness consequences of hybrids. Hybrid vigor (heterosis) In special circumstances, hybrid vigor may lead to superior fitness so that hybrids are more fit than the parent species in survivorship, male or female reproductive capacity, or adaptability [11, 49]. It may result in demographic or ecological dominance of hybrids in one or some environments. Though, it may also be the case that vice versa, instead of outbreeding depression, it does the opposite, hybrids pay a price in decreased fitness, because the complex of genes adapted together are broken and in the ecological sense are not matched [18]. Hybrid fitness landscape is consequently a determining factor between the innovation of evolution and the decline of demography associated with hybridization.

The ecological viewpoint can create biodiversity and ecological niches through hybridization. Hybrids can also very well be filling in new and/or intermediary ecological niches and this aspect affects the actions of competition and resource use, as well as species interaction within the wetland community [53]. Within bird assemblages this can lead to changes in population structure or parental replacement. Long-term hybridization can led to the fuzziness of species boundaries, taxonomic difficulties and the impairment of taxonomic priorities [59].

## 22.6 Implications for Conservation Management

Hybridization presents an evolutionary potential and severe conservation dilemmas particularly with rare or endangered species of water birds. Species management, legal protection and policy formulation are simplified by hybridization especially in the Indian subcontinent that has a high biodiversity and a high social-ecological pressures on conservation. Specially of worry is the hybridization that happens with the threatened wild populations and the more usual or dedicated members, which would lead to genetic watering down factions which are already susceptible.

The case of the critically endangered Baer Pochard (*Aythya baeri*) is one of the striking examples: its population worldwide decreased significantly because of habitat loss and hybridization with another widespread duck, the Ferruginous Duck (*Aythya nyroca*) [36, 55, 63]. Sharing of morphological features as well as shared ranges in the non-breeding seasons puts the possibility of hybrid pairings at risk, and this can not be felt unless genetic screening is done. This process of genetic introgression can potentially not only decrease the genetic uniqueness of *A. baeri* but also inhibits a proper estimation of the population, subsequently complicating the conservation planning [37].

Likewise, Indian Skimmer (*Rynchops albicollis*), endangered because of habitat fragmentation, and restricted breeding success may experience the risk of potential hybridization in exotic wetlands in the presence of gulls or terns [6, 16]. There is also a documented instance of hybridization with Indian Skimmer though it is not greatly prevalent but such a threat is high in modified or destroyed habitat where the interaction of species is distorted. To illustrate, these cases point out that hybridization must not only be addressed as an immediate conservation risk especially on reproductively viable subsets of endangered species.

Genetic diluting, a phenomenon of genetic backcrossing which produces diluted pure genetic lineages is one of the major dangers of hybridization. Low frequency hybridization events may have long term evolutionary impacts on narrow distribution or endemic species [48]. And the challenge is compounded by the absence of regular molecular observation in most Indian wetlands whose individuals present underreporting and misidentification of hybrid people.

The complications that hybrids bring about are usually handled improperly by legal and conservation frameworks. The major international and national laws which are aimed at conservation of distinct species are the Wildlife Protection Act (1972) of India and IUCN Red List appraisals [12]. The status of hybridization and introgression is still uncertain and therefore there are gaps or ambiguities in policy of managing hybrid populations [4]. An example is that, the hybrids might not be considered as granted protection under species-specific acts, although they are hybrids of endangered taxonomical units. This provokes really challenging questions of their control and in particular breeding programs and habitat reconstructions.

The importance of reducing intensity of wetlands must be given priority in order to reduce the impact of the above challenges. The preservation of natural wetlands involves the preservation of the ecological processes upon which it is based, so they are less prone to inducing unnatural species interactions than artificial or man-made habitats [14, 19, 31]. The wetland reclamation, reservoir constructions, and pollution are all of anthropogenic origin, which places the waterbird species in homes shared and suboptimal, raising the risk of interspecific breeding [31, 44]. Therefore, the preservation and the recovery of natural wetlands environments can be a secondary step in minimizing risks of hybrids.

## 22.7 Regional Case Studies

Another commonly cited instance of hybridization in **India** is the genus Anas of ducks with largescale hybridization in the North and central wetlands. Field ornithologists in Keoladeo National Park (Bharatpur, Rajasthan), a Ramsar site, and a UNESCO World Heritage wetland, have noted the suspected hybrids between Spot-billed Duck (*Anas poecilorhyncha*) and Mallard (*Anas platyrhynchos*), in particular during winter when migratory mallards adding in on Central Asia used winters overlap with the local residents [40, 65]. They are intermediate septentrional hybrids with intermediate plumage (greenish) and a greenish head patch and changeable bill patterns and can be confused with pure species in the field. Even though the genetic confirmation is still pending in most reports, there is evidence of such hybrid individuals frequently photographed and posted in citizen science websites such as eBird and the journal of Indian birds.

Similar interspecific competition phenomena have been observed in Chilika Lake, Odisha, the largest coastal lagoon in India and also one of the primary wintering areas of migratory ducks, between Northern Pintails (*Anas acuta*), the Gadwalls (*Anas strepera*), and Common Teals (*Anas crecca*). Numbers and density of birds are high and the roosting and foraging habit of these birds overlaps during non-breeding season which also causes an increase in chances of hybridization particularly in the changed environmental conditions [7, 50]. Morphological intermediates between Gadwalls and Mallards have been reported in some populations in Chilika but fully worked up molecular analysis to establish hybridization is yet to be done.

The Rann of Kutch in Gujarat, a seasonal saline wetland which supports large flocks of flamingos, cranes and ducks, has also been cited as another probable location where hybridization may occur. Hybridization in the gulls (*Larus* spp.) has been reported world-wide in such coastal locations [41], although Caspian Gull (*Larus Cachinnans*) and Heuglin Gull (*Larus heuglini*) hybrids have been suspected on winter surveys in Kutch [47]. Such hybrids are difficult to identify since they have variable stages of plumages, morphometrics overlaps, and there is limited availability of comparative molecular samples of Indian populations.

The existence of hybrid geese in Nepal is unconfirmed, but there have been unconfirmed reports of hybrid geese especially between the Greylag and Bar-headed Geese in Nepal, especially in the Koshi Tappu Wildlife Reserve where large concentrations of waterfowl occur. The two species are reported to co-occur in migration and wintering, and the courtship among the two species has been reported once in a while [10]. In the event that the hybrid offspring is viable it might hamper conservation efforts in the bar-headed goose a culturally and ecologically important species in South Asia.

**In Bangladesh**, Tanguar Haor wetland complex in the northeast and Hakaluki Haor system are also suggested as locations of possible occurrence of hybrid ducks, particularly in the post-monsoon season when migratory and resident species mix. Nonetheless, there are still no formal studies that report hybridization in the area, a factor that portrays general lack in ornithological studies and wetland surveillance in Northeast South Asia [3, 9, 39].

**In Sri Lanka**, hybridization is less often seen but it could happen in such wetlands as Bundala National Park and Anawilundawa Sanctuary, where egrets, herons and ducks are abundant in large populations. There are anecdotal descriptions of a bird between the two egret generations, possibly a hybrid of the Intermediate (*Ardea intermedia*) and Great Egrets (*Ardea alba*), but no published morphological or genetic examination has been made, nor were field reports published [33].

## 22.8 Future Prospectus and Knowledge Gaps

The study of hybridization in waterbirds in the Indian subcontinent is disjointed, anecdotal and under documented in the literature, even as the ecological and evolutionary importance of hybridization has risen to prominence around the world. One critical shortcoming is the fact that the majority of reported hybrid cases are composed of either field-observed or photographed evidence with little or no corroborating morphological measurement or molecular verification [38, 42]. Such anecdotal records, as useful though incomplete first indicators, may yield to misidentification and are particularly vulnerable to this condition in species that are highly variable within their species or those in which plumage characteristics are shared. This is even complicated by the fact that no protocols have been put in place when checking these regularly and there are also few ornithologists who have been trained on how to identify hybrids correctly.

The lack of molecular analyses of hybridization in Indian water birds is one of the most serious research gaps. Although DNA-based markers including mitochondrial markers, microsatellites, and genome-wide SNP have become very useful in hybrid detection across the globe [35, 59], these markers are hardly used in the South Asian setting. Only a few Indian facilities hold genetic reference databases of indigenous waterbird species and publicly available DNA barcode libraries of bird taxa remain in early stages of development [34]. Such undeveloped molecular infrastructure can restrict the chance to verify suspected hybrids and to apprehend the degree and course of inter-specific gene flow.

To fill these gaps, it is necessary to observe ecological monitoring programs that will last long and will specialize in the discovery of hybrid zones, individual phenotype tracking over time and correlating the occurrence of the hybrid with environmental factors such as habitat change, climate change, and human impact [13]. These should be incorporated within the bird census schemes already in execution such as the Asian Waterbird Census and carried out together with the protective area managers and conservation NGOs.

It is also necessary to come up with an integrated hybrid detection protocol. This would include standardized photographic records, morphology scoring, behavioural observation and sampling of genetic study. The possible existence of a centralized reporting site where ornithologists, birders, and forest officials could report possible hybrids to be studied further can be linked with the open databases such as eBird, India Biodiversity Portal, etc. Moreover, community involvement especially people residing along important wetlands may be useful in determining the hybrids and their behaviour and movement patterns hence facilitating participatory conservation [57].

Ornithologists, taxonomists, geneticists, and conservation managers will play an important role in developing an integrated framework of studying hybridization. Technology transfer, development of genetic labs and sharing data in south Asia might be encouraged by international cooperations and capacity-building programs. Moreover, hybridization ought to be factored in species recovery plans and IUCN evaluations, particularly in the case of endangered species whereby the practice of hybridization can jeopardize the genetic integrity of the target species [4, 26].

## 22.9 Conclusion

Waterbird interbreeding is a phenomenon of ecological importance but little studied in the Indian sub-continent. As this review has pointed out, hybridization has twofold role—on the one hand, it works in the evolutionary context as it introduces a genetic variety, thus enables adaptation, or in some extreme instances, leads to speciation. Alternatively, it poses serious conservation problems particularly in such cases as rare, endemic or declining species which are under the threat of genetic mixing and demographic imbalance. The diverse, dynamic wetlands of South Asia, located between the foothill slopes of the Himalayas and the outer lagoons, are in a very special place, as species overlaps, and both natural vegetation and human-made hybridization are possible.

But there is still a lack of insight on hybridization in the area, due to overdependence on anecdotal accounts, scant molecular evidence and the lack of organized surveillance systems. The majority of reported hybrids have never been verified and the existence and effect of hybrid individuals is highly underrepresented in the conservation frameworks. Such ignorance is a setback in addressing the issue of biodiversity management, especially now that environmental modifications are changing species patterns and relationships. In order to overcome these hurdles, integrative, and area specific studies which integrate field ecology, genetics and conservation planning are required with a sense of urgency. The development of standard hybrid detection procedures, enhancement of the identification capacity of the species and engaging the local communities in monitoring of the wetlands is a significant step ahead. Furthermore, hybridization needs to be taken as not only a complicated or mistaken problem of taxonomy but as an inevitable part of the evolutionary landscape bounded to be learned and handled in a context. With collaboration and multidisciplinary research, supported by conservation objectives in mind, we will have a more comprehensive view on the role of hybridization in waterbirds and we will be in a better position to preserve biodiversity, the avifaunal and wetland heritage of the region on a long-term perspective.

## References

1. Abbott R, Albach D, Ansell S, Arntzen JW, Baird SJ, Bierne N, Boughman J, Brelsford A, Buerkle CA, Buggs R (2013) Hybridization and speciation. J Evol Biol 26(2):229–246
2. Ahmad M, Ahmad WS, Ahmad SN, Jamal S, Saqib M (2024) Tracing the roots of wetland degradation in India: A systematic review of anthropogenic drivers, ecological consequences and conservation strategies. GeoJournal 89(1). https://doi.org/10.1007/s10708-024-10997-9
3. Ahsan MF, Khan MMH, Chowdhury SH, Jaman MF, Khan MAR, Sultana S, Islam MA, Islam F, Shovon TA, Roy S (n.d.) Red List of Bangladesh. Retrieved July 20, 2025, from https://www.researchgate.net/profile/K-M-Rahman-2/post/Can-anyone-provide-examples-with-references-of-countries-that-use-the-IUCN-method-for-species-conservation-assessments/attachment/5a62f085b53d2f0bba4c8262/AS%3A584778805968897%401516433540935/download/ShowFile.pdf
4. Allendorf FW, Leary RF, Spruell P, Wenburg JK (2001) The problems with hybrids: setting conservation guidelines. Trends Ecol Evol 16(11):613–622
5. Anderson EC, Thompson E (2002) A model-based method for identifying species hybrids using multilocus genetic data. Genetics 160(3):1217–1229
6. Ankit K, Ahmad M, Ranjan V, Kumar S, Hussain SA, Gopi GV (2024) Observations on Indian Skimmer Rynchops albicollis Swainson, 1838 (Aves: Charadriiformes: Laridae) breeding colonies in Middle Ganges stretch, India. J Threat Taxa 16(2):24737–24745
7. Avifauna of Chilika, Odisha: Assessment of Spatial and Temporal Changes (2020) In: Balachandran S, Pattnaik AK, Gangaiamaran P, Katti T (eds) Wetlands: Ecology, Conservation and Management. Springer International Publishing, pp 335–363. https://doi.org/10.1007/978-3-030-33424-6_13
8. Barton NH, Hewitt GM (1985) Analysis of hybrid zones. Annu Rev Ecol Syst 16:113–148
9. Bird JP, Lees AC, Chowdhury SU, Martin R, Haque EU (2010) A survey of the critically endangered spoon-billed sandpiper Eurynorhynchus pygmeus in Bangladesh and key future research and conservation recommendations. Forktail 26(26):1–8
10. Birds of Nepal: their status and conservation especially with regards to watershed perspectives. (2020). In H. S. Baral & C. Inskipp, Hindu Kush-Himalaya watersheds downhill: landscape ecology and conservation perspectives (pp. 435–458). Springer International Publishing. https://doi.org/10.1007/978-3-030-36275-1_22
11. Burke JM, Arnold ML (2001) Genetics and the fitness of hybrids. Annu Rev Genet 35(1): 31–52. https://doi.org/10.1146/annurev.genet.35.102401.085719
12. Choudhary S (2024) The Wildlife (Protection) Act, 1972: a comprehensive study on wildlife conservation in India. Available at SSRN 4848859. https://papers.ssrn.com/sol3/papers.cfm?abstract_id=4848859
13. Chunco AJ (2014) Hybridization in a warmer world. Ecol Evol 4(10):2019–2031. https://doi.org/10.1002/ece3.1052
14. Davidson NC (2014) How much wetland has the world lost? Long-term and recent trends in global wetland area. Mar Freshw Res 65(10):934–941

15. Dayrat B (2005) Towards integrative taxonomy. Biol J Linn Soc 85(3):407–417
16. Debata S, Kar T, Palei HS, Swain KK (2019) Breeding ecology and causes of nest failure in the Indian Skimmer *Rynchops albicollis*. Bird Study 66(2):243–250. https://doi.org/10.1080/00063657.2019.1655526
17. Devineau O, Guillemain M, Johnson AR, Lebreton J (2010) A comparison of green-winged teal Anas crecca survival and harvest between Europe and North America. Wildl Biol 16(1):12–24. https://doi.org/10.2981/08-071
18. Edmands S (2007) Between a rock and a hard place: evaluating the relative risks of inbreeding and outbreeding for conservation and management. Mol Ecol 16(3):463–475. https://doi.org/10.1111/j.1365-294x.2006.03148.x
19. Finlayson C, D'Cruz R (2005) Inland water systems. In: Ecosystems and human well-being: current state and trends. Island Press, pp 551–583. https://researchoutput.csu.edu.au/en/publications/inland-water-systems
20. Fitzpatrick BM, Johnson JR, Kump DK, Smith JJ, Voss SR, Shaffer HB (2010) Rapid spread of invasive genes into a threatened native species. Proc Natl Acad Sci 107(8):3606–3610. https://doi.org/10.1073/pnas.0911802107
21. Frei D, Reichlin P, Seehausen O, Feulner PGD (2023) Introgression from extinct species facilitates adaptation to its vacated niche. Mol Ecol 32(4):841–853. https://doi.org/10.1111/mec.16791
22. Grant PR, Grant BR (1992) Hybridization of Bird species. Science 256(5054):193–197. https://doi.org/10.1126/science.256.5054.193
23. Harrison RG (1990) Hybrid zones: windows on evolutionary process. https://www.cabidigitallibrary.org/doi/full/10.5555/19910191041
24. Hedrick PW (2013) Adaptive introgression in animals: examples and comparison to new mutation and standing variation as sources of adaptive variation. Mol Ecol 22(18):4606–4618. https://doi.org/10.1111/mec.12415
25. Irwin DE, Bensch S, Price TD (2001) Speciation in a ring. Nature 409(6818):333–337
26. Jackiw RN, Mandil G, Hager HA (2015) A framework to guide the conservation of species hybrids based on ethical and ecological considerations. Conserv Biol 29(4):1040–1051. https://doi.org/10.1111/cobi.12526
27. Johnsgard PA (1960a) Comparative behaviour of the Anatidae and its evolutionary implications. Wild 11(11):31–45
28. Johnsgard PA (1960b) Hybridization in the Anatidae and its taxonomic implications. Condor 62(1):25–33
29. Johnsgard PA (1961) The taxonomy of the Anatidae—a behavioural analysis. Ibis 103:71–85. https://digitalcommons.unl.edu/johnsgard/29/
30. Johnsgard PA (1963) Behavioral isolating mechanisms in the family Anatidae. https://digitalcommons.unl.edu/johnsgard/23/
31. Kingsford RT (2000) Ecological impacts of dams, water diversions and river management on floodplain wetlands in Australia. Austral Ecol 25(2):109–127. https://doi.org/10.1046/j.1442-9993.2000.01036.x
32. Kiran, Singh D, Kour A, Priya, Delu V, Kumar R (2022) Community composition and status of avian diversity at campus and agricultural landscapes of Chaudhary Charan Singh Haryana Agricultural University, Hisar (Haryana). J Appl Nat Sci 14(4):1130–1140. https://doi.org/10.31018/jans.v14i4.3784
33. Kotagama SW, Ratnavira G (2010) An illustrated guide to the birds of Sri Lanka. Field Ornithology Group of Sri Lanka
34. Kumar N, Gupta U, Jhala YV, Qureshi Q, Gosler AG, Sergio F (2020) GPS-telemetry unveils the regular high-elevation crossing of the Himalayas by a migratory raptor: Implications for definition of a "Central Asian Flyway". Sci Rep 10(1):15988
35. Lavretsky P, McInerney NR, Mohl JE, Brown JI, James HF, McCracken KG, Fleischer RC (2020) Assessing changes in genomic divergence following a century of human-mediated

secondary contact among wild and captive-bred ducks. Mol Ecol 29(3):578–595. https://doi.org/10.1111/mec.15343
36. Liang W, Lei J, Ren B, Cao R, Yang Z, Wu N, Jia Y (2022) The impacts of a large water transfer project on a waterbird community in the receiving dam: a case study of miyun reservoir, China. Remote Sens 14(2):417
37. Liu D, Zhou Y, Fei Y, Xie C, Hou S (2021) Mitochondrial genome of the critically endangered Baer's Pochard, Aythya baeri, and its phylogenetic relationship with other Anatidae species. Sci Rep 11(1):24302
38. McCarthy EM (2006) Handbook of avian hybrids of the world. Oxford University Press. https://books.google.com/books?hl=en&lr=&id=MwInO7z_Y3oC&oi=fnd&pg=PR13&ots=cFdrBevF1Y&sig=5nf4SnmJp5FcIrgYS632WCJh3FQ
39. Nahar S, Rahman MH, Islam MS, Alam MA, Lupa ST, Amin A (2025) Fish biodiversity at Haor wetland of Bangladesh: A comprehensive review. Int J Aquac Fish Sci 11(1):001–009
40. Narwade SS, Jathar GA, Rahmani AR (2005) Bibliography of the Birds of South. https://www.academia.edu/download/90017955/BUCEROS_15.pdf
41. Olsen KM (2010) Gulls of Europe, Asia and North America. Bloomsbury Publishing. https://books.google.com/books?hl=en&lr=&id=5PTUBAAAQBAJ&oi=fnd&pg=PA2&dq=Olsen,+K.+M.,+%26+Larsson,+H.+(2003).+Title:+Gulls+of+Europe,+Asia+and+North+America.&ots=eF8im-vbLC&sig=QtiC6j0Wc7C7fdNMa5HJ2aOye74
42. Ottenburghs J, Nisbet I (2025) Hybridization in terns: a review. Mar Ornithol 53(1):83–89
43. Padial JM, Miralles A, De La Riva I, Vences M (2010) The integrative future of taxonomy. Front Zool 7(1):16. https://doi.org/10.1186/1742-9994-7-16
44. Paracuellos M, Tellería JL (2004) Factors affecting the distribution of a waterbird community: the role of habitat configuration and bird abundance. Waterbirds 27(4):446–453
45. Parmesan C, Yohe G (2003) A globally coherent fingerprint of climate change impacts across natural systems. Nature 421(6918):37–42
46. Randler C (2002) Avian hybridization, mixed pairing and female choice. Anim Behav 63(1):103–119
47. Rasmussen PC, Anderton JC (2005) Birds of south Asia: The Ripley guide, vol 2. https://wgbis.ces.iisc.ac.in/biodiversity/sahyadri_enews/newsletter/issue57/bibliography/7-birds-of-south-asia.pdf
48. Rhymer JM, Simberloff D (1996) Extinction by hybridization and introgression. Annu Rev Ecol Syst 27(1):83–109. https://doi.org/10.1146/annurev.ecolsys.27.1.83
49. Rieseberg LH, Archer MA, Wayne RK (1999) Transgressive segregation, adaptation and speciation. Heredity 83(4):363–372
50. Samal KT (2017) Aquatic bird diversity with reference to selected wetlands of Koraput, Odisha, India (PhD Thesis, Central University of Odisha). https://idr.cuo.ac.in/bitstream/handle/123456789/100/09_chapter-2.pdf?sequence=9&isAllowed=y
51. Sangster G, Collinson M, Helbig AJ, Knox AG, Parkin DT, Prater T (2001) The taxonomic status of green-winged teal Anas carolinensis. Br Birds 94:218–226
52. Secondi J, Bretagnolle V, Compagnon C, Faivre B (2003) Species-specific song convergence in a moving hybrid zone between two passerines. Biol J Linn Soc 80(3):507–517
53. Seehausen O, Takimoto G, Roy D, Jokela J (2008) Speciation reversal and biodiversity dynamics with hybridization in changing environments. Mol Ecol 17(1):30–44. https://doi.org/10.1111/j.1365-294x.2007.03529.x
54. Sievers M, Hale R, Parris KM, Swearer SE (2018) Impacts of human-induced environmental change in wetlands on aquatic animals. Biol Rev 93(1):529–554. https://doi.org/10.1111/brv.12358
55. Song Y, Mo X, He M (n.d.) Future Climate and Land Use Change Jointly Shape the Suitable Habitats of Endangered Waterbirds. Retrieved July 20, 2025, from https://papers.ssrn.com/sol3/papers.cfm?abstract_id=5336372
56. Suarez-Gonzalez A, Lexer C, Cronk QCB (2018) Adaptive introgression: a plant perspective. Biol Lett 14(3):20170688. https://doi.org/10.1098/rsbl.2017.0688

57. Sutherland WJ, Newton I (2004) Bird ecology and conservation: A handbook of techniques. Oxford University Press. https://books.google.com/books?hl=en&lr=&id=icwTDAAAQBAJ&oi=fnd&pg=PR7&dq=Sutherland,+W.+J.,+Newton,+I.,+%26+Green,+R.+E.+(2004).+Title:+Bird+Ecology+and+Conservation:+A+Handbook+of+Techniques.&ots=XEu3ZiG9UE&sig=iUjm9qI-TnBY-MzNVvu2bfe_oeo
58. Taylor SA, Larson EL (2019) Insights from genomes into the evolutionary importance and prevalence of hybridization in nature. Nat Ecol Evol 3(2):170–177
59. Toews DPL, Brelsford A (2012) The biogeography of mitochondrial and nuclear discordance in animals. Mol Ecol 21(16):3907–3930. https://doi.org/10.1111/j.1365-294x.2012.05664.x
60. Tong A, Ouyang H, Zhou Y, Li Z (2025) Multidimensional Bird habitat network resilience assessment and ecological strategic space identification in international Wetland City. Land 14(6):1166
61. Uy JAC, Borgia G (2000) Sexual selection drives rapid divergence in bowerbird display traits. Evolution 54(1):273–278. https://doi.org/10.1111/j.0014-3820.2000.tb00027.x
62. Vedder D, Lens L, Martin CA, Pellikka P, Adhikari H, Heiskanen J, Engler JO, Sarmento Cabral J (2022) Hybridization may aid evolutionary rescue of an endangered East African passerine. Evol Appl 15(7):1177–1188. https://doi.org/10.1111/eva.13440
63. Yadav SK, Lamichhane BR, Subedi N, Kadariya R, Paudel U, Shahi R, Pokharel N (2024) Diversity, abundance, and trend of water birds outside the protected area of Western Lowlands, Nepal. The Ring 46(1):3–18. https://doi.org/10.2478/ring-2024-0001
64. Yodha K, Singh D, Kour A, Yadav R, Kumar R (2024) Species diversity and abundance of birds in ber (*Zizyphus mauritiana*) crop. Asian J Conserv Biol 13(1). https://doi.org/10.53562/ajcb.83683
65. Zhang Y, Bao Q, Cao Z, Bian Y, Zhang Y, Cao Z, Chen G, Xu Q (2023) Chinese domestic ducks evolved from mallard duck (Anas Platyrhynchos) and spot-billed duck (A. zonorhyncha). Animals 13(7):1156

# Chapter 23
# Navigating Coexistence: The Dynamics of Migratory Birds, Aquatic Bird Communities, and Human Influences in a Changing Wetland Landscape

**Aazad Prakash Ojha, Oyungerel Orosoo, and Ponnusamy Thillai Arasu**

**Abstract** Migratory birds play a vital role in maintaining ecological balance, health, and biodiversity. However, their existence and survival are increasingly threatened by habitat loss and human interference in wetlands worldwide. This review explores the complex dynamics of migratory birds, aquatic bird communities, and human influence in three globally significant wetlands: the Pantanal wetland in South America, Okavango Delta wetland in Africa, and Sundarbans wetland in Bangladesh and India. By comparing the ecological characteristics, avian diversity, and conservation challenges across these wetlands, we highlight the importance of understanding the coexistence of migratory and resident bird species in the face of anthropogenic pressure and changing landscapes. Human activities such as deforestation, agricultural expansion, and tourism development have reshaped these wetlands, affecting hydrological patterns, habitat heterogeneity, and resource availability for birds. Migratory and resident species have adapted to these changes through niche partitioning strategies, including spatial, temporal, and behavioral adaptations facilitated by the dynamic nature of these wetlands. However, the delicate balance of coexistence is threatened by ongoing habitat degradation and climate change. To navigate the coexistence and ensure the long-term conservation of these wetlands and their avian diversity, we propose adaptive management approaches that integrate ecological goals with sustainable human activities. These strategies include fire management, water flow regulation, mangrove restoration, and

A. P. Ojha
Department of Zoology, Udit Narayan, P.G. College, Padrauna, Kushinagar, Uttar Pradesh, India
e-mail: aazad.ojha@hotmaill.com

O. Orosoo (✉)
National Pingtung University of Science and Technology, Pingtung, Taiwan, People's Republic of China

P. T. Arasu
Department of Chemistry, College of Natural and Computational Sciences, Wollega University, Nekemte, Ethiopia

 
R. Kumar et al. (eds.), *Ecology, Conservation of Aquatic Birds*, Wetlands: Ecology, Conservation and Management 15, https://doi.org/10.1007/978-3-032-17125-2_23

community-based conservation initiatives. By engaging local communities, promoting cross-border collaboration, and prioritizing research on avian population dynamics and habitat responses, we can work towards preserving these invaluable wetland ecosystems as sanctuaries for migratory and resident birds. Future research should focus on long-term bird population monitoring, habitat responses, and economic analyses of conservation benefits to strengthen strategies.

**Keywords** Anthropogenic pressures · Aquatic bird communities · Coexistence · Conservation strategies · Habitat degradation · Migratory birds · Wetland landscape

## 23.1 Introduction

The once predictable and harmonious arrival of migratory birds (MB) in coastal wetlands is now increasingly disrupted due to habitat loss and human interference. The MB play a vital role in maintenance of ecological balance, health and biodiversity because they participate in processes such as pollination, seed dispersal, and pest control [12, 61, 68]. Their existence and survival depend on the wetland ecosystems worldwide. These wetlands are home to different bird species, including MB; however, these ecosystems are currently threatened. During long-distance travel, MB interacts with wetland habitat as well as local avian populations. This interaction shows coexistence with other species. Therefore, by understanding their impact of human activities on wetlands and avian diversity, coexistence and conservation challenges become important.

Beside importance of wetland ecological role for MB, they are facing various threats from human activities and climate change. It has led to a decline of biodiversity from global wetlands, e.g., over 50% biodiversity has got decline from the wetlands worldwide [46]. This suggests the need of conservation planning to address human impact in these habitats. In addition, a critical gap exists to understand the dynamic interactions between MB, aquatic birds, and anthropogenic influence on changing landscapes. Existing studies have examined loss of habitat, climate change effects, and avian ecology. However, there is a need to adapt a comprehensive approach for protecting these ecosystems, their biodiversity and coexistence among various challenges.

We are going to address these gaps by conducting the review study of complex dynamics of migratory and local bird coexistence within three renowned large wetlands of the world i.e. the Pantanal wetland (PW) in South America, the Okavango Delta wetland (ODW) in Africa, and the Sundarbans wetland (SW) in Bangladesh and India. To address these aims, this chapter is structured into seven sections. Following this introduction, we will first explore the ecological significance of three important wetlands that serve as critical habitats for MB, and provide an overview of MB and their ecological role. Next, we provide a global perspective on how human activities affect the coexistence of wetlands and avian species. Next, we analyzed the coexistence of migratory and resident birds. We then navigate coexistence through conservation strategies and challenges. Finally, we address sustainable avian coexistence through adaptive management, focusing on conservation and management strategies that promote harmonious coexistence. By presenting illustrative examples from globally significant wetlands and synthesizing current research, this chapter provides a thorough review and insights for informed conservation and management actions in these critical ecosystems.

## 23.2 Overview of Wetlands as Critical Habitats

Wetlands are diverse ecosystems that serve as critical habitats for MB and several other species. To begin with, these ecosystems remain temporarily or permanently flooded and play a basic role in maintaining life on earth. For example, wetland systems provide a wide range of ecosystem services, including homes and resources for the MB [40]. More specifically, in the North American Prairie Pothole area, wetlands are important breeding sites that support the continent's waterfowl populations [43], Likewise, the Great Basin of the North American Pacific Flyway is essential for aquatic avian species [21]. Despite these ecological benefits, recent studies have highlighted that increased environmental and human activity pressure threatens wetland sustainability. This calls attention to the urgent need to conserve biospheric ecosystems for the continued support of biodiversity.

### *23.2.1 Migratory Birds: Overview and Ecological Roles*

MB are well adapted to long-distance flying journeys from breeding to wintering sites and vice versa. Such distant migration is possible due to the presence of anatomical, physiological and behavioral characters such flight muscles, pneumatic bones, metabolism, inbuilt navigation systems etc. These characters enable them to exploit seasonal resources optimally. Their navigation system allows them to orient themselves to the Earth's magnetic field, celestial signals, and geographical indicators. These adaptations help in their large-scale movement and worldwide distribution, with the highest MB diversity observed in the Northern Hemisphere [61].

The availability of resources and distribution level play an important role in determining the selection of single or multiple wintering sites for MB. Therefore, some individuals exhibit long post-migratory movements during the non-breeding season [62]. According to Somveille et al. [61], one in five bird species has isolated breeding and overwintering distribution. Thus, the selection of wintering sites and their long-distance movement strategies determine their ecological importance in different regions and enable them to occupy diverse ecological niches and habitats. Therefore, understanding their wintering grounds is crucial for effective conservation planning. An overview of wetland ecosystems and MB's ecological role has shown in Table 23.1.

**Table 23.1** Overview of wetland ecosystems and migratory birds' ecological roles

| Aspect | Details |
|---|---|
| Wetland ecosystems | Dynamic ecosystems that provide critical habitats for migratory birds and other organisms [40]. |
| Ecosystem services of Wetlands | Shelter, food resources, and other ecological benefits essential for sustaining migratory bird populations [40]. |
| Environmental pressures | Anthropogenic and environmental pressures threatening the stability of wetland ecosystems. |
| Migratory birds (MB) | Adapt to long-distance migration between breeding and wintering grounds using physiological and behavioral adaptations [61]. |
| Navigation and adaptations | Use Earth's magnetic field, celestial cues, and geographical landmarks for navigation [61]. |
| Movement strategies | May occupy single or multiple wintering sites depending on resource availability and disturbances [62]. |
| Global distribution | Highest diversity observed in the Northern Hemisphere [61]. |
| Ecosystem contributions | Pollination, seed dispersal, nutrient recycling, and habitat modification [65]. |
| Nutrient cycling role | Transporting seeds and nutrients across regions enhances plant diversity and ecosystem health [33]. |
| Environmental indicators | MB serve as indicators of environmental health, reflecting changes in habitat, climate, and pollution [60]. |
| Conservation significance | MB sites are important for conservation and ecological planning [61]. |

The role of the MB in the global ecosystem and biodiversity is vital, as it supports its processes and functions. They establish ecological connections with habitats and provide benefits to humans through their services in many locations [60]. For example, they help in pollination, seed dispersal, nutrient recycling, and modification of habitat [65]. Their role in nutrient cycling by distributing seeds and nutrients supports plant diversity and ecosystem health [33]. Additionally, they act as environmental health indicators because of their sensitivity to changes in habitat quality, climate change, and pollution levels. Thus, because of these above discussed ecological functions MB considered as ecologically significant group of birds.

### *23.2.2 Global Wetland Diversity: A World of Avian Adaptations and Pressures*

To understand dynamics of bird communities, MB and human impact on navigating coexistence, we need to compare different global wetland ecosystems. For this purpose, we selected and explored three important wetlands of the world e.g. Pantanal Wetland (PW) in South America, Okavango Delta Wetland (ODW) in South Africa, and Sundarbans Wetland (SW) in Bangladesh and India (Fig. 23.1). We performed review of existing studies to collect information on these wetlands. These wetlands were selected because of their distinct characteristics such as having

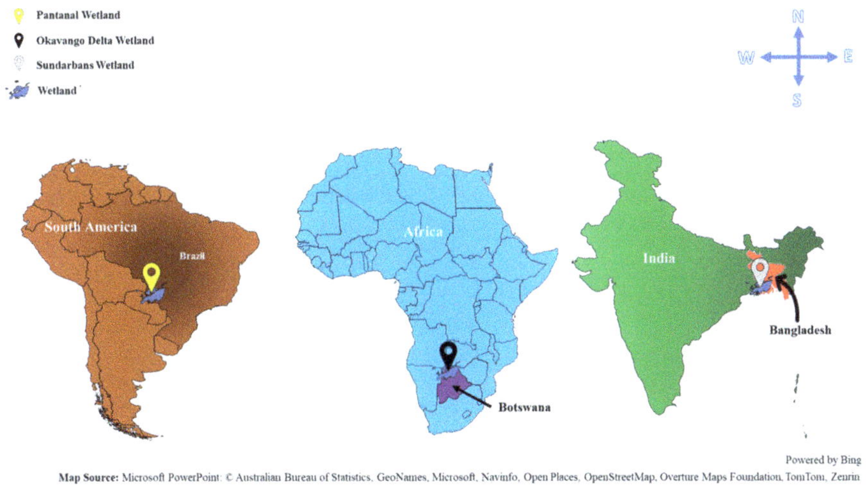


**Fig. 23.1** Map showing three wetlands under study: Pantanal (Yellow), Okavango (Black), and Sundarbans (Grey) (Map source: Microsoft PowerPoint; Base Map data: © Microsoft powered by Bing)

large areas, geography and habitat quality, and experiencing various anthropogenic threats. For example, The PW has balanced agricultural expansion with avian diversity, The ODW is located in an arid landscape facing emerging threats from tourism and drought conditions, and The SW mangrove forest's coastal wetland experiences anthropogenic pressure and climate change impact, which are suitable for understanding the complex dynamics of avian communities and human influence scenarios. In the following subsections, we describe these three wetlands and their comparative analyses.

#### 23.2.2.1 The Pantanal, South America

The PW, the world's largest freshwater tropical wetland, covers approximately 140,000 km$^2$ of the Upper Paraguay River Basin in Brazil (South America). The diverse mosaic of habitats, such as lowland floodplains, gallery forests, grasslands, and seasonally flooded areas, distinguishes them from other wetlands. Its dense vegetation supports a rich terrestrial and aquatic biota. [1, 23, 30]. Its extensive area provides a vital habitat for these MB, further highlighting the importance of conserving this invaluable ecosystem [34]. Thus, this unique wetland has many MB as winter guests, and some key species are shown in Table 23.2.

This wetland faces conservation challenges, such as deforestation, agricultural extension, and hydropower plant construction. These activities threaten wetland ecosystems and cause loss of habitat and biodiversity. Beside this, economic pressures from outside drive development projects to benefit external groups rather than

**Table 23.2** Comparison of key characteristics and human influences across globally significant wetlands

| Wetland name | Ecosystem type | Key Migratory Birds | Major Human Influences |
|---|---|---|---|
| Pantanal, South America | Tropical Floodplain mosaic of forests, grasslands, and wetlands | Southern Pochard (*Netta peposaca*)<br>Chilean Flamingo (*Phoenicopterus chilensis*)<br>Neotropic Cormorant (*Phalacrocorax olivaceus*)<br>White-faced Ibis (*Plegadis chihi*)<br>Wood Stork (*Mycteria americana*)<br>Jabiru Stork (*Jabiru mycteria*)<br>Various species of ducks<br>Source: (Antas [2]; Junk et al. [32]; Pinho et al. [50]) | Agriculture<br>Expansion<br>Ranching<br>Infrastructure |
| Okavango Delta, Africa | Inland Delta<br>Alluvial fan with permanent marshlands and seasonally flooded plains | African Darter (*Anhinga rufa*)<br>Green-backed Heron (*Butorides striata*)<br>Slaty Egret (*Egretta vinaceigula*)<br>Squacco Heron (*Ardeola ralloides*)<br>African Spoonbill (*Platalea alba*)<br>Saddle-billed Stork (*Ephippiorhynchus senegalensis*)<br>Greater Flamingo (*Phoenicopterus roseus*)<br>Fish Eagles, Cranes, Kingfishers, Ducks<br>Source: (Francis et al. [13]; Ramberg et al. [54]; Bartlam-Brooks et al. [3]; Hudson et al. [28]) | Emerging<br>Tourism<br>Water<br>Abstraction<br>(Potential) |
| Sundarbans, Asia | Mangrove Forest<br>Coastal mangrove ecosystem with tidal rivers and mudflats | Black-tailed Godwit (*Limosa limosa*)<br>Lesser Sand Plover (*Charadrius mongolus*)<br>Brahminy Kite (*Haliastur indus*)<br>Masked Finfoot (*Heliopais personata*)<br>Storks, ducks, Kingfishers, Waders etc. | Deforestation<br>Aquaculture<br>Overfishing<br>Pollution |

local communities [23, 31, 59]. In summary, conservation efforts are required to balance ecological preservation and sustainable development to protect this unique and biodiverse habitat.

#### 23.2.2.2 The Okavango Delta, Botswana, Africa

The ODW in northern Botswana is the largest inland freshwater wetland in South Africa, and is recognized as a biodiversity hotspot. This site also recognized as a World Heritage Site by UNESCO, is important habitat for MB [26]. This is an alluvial fan which gets its water from the Okavango River. This river originates in the highlands of Angola, and it is fed by the Cubango and Cuito rivers before flowing in Namibia into Botswana [47]. This hydrological system makes permanent swamps and seasonally flooded plains which sustain a wide range of wildlife [4, 13]. This seasonal flooding is important in maintaining ecological balance by supporting terrestrial and aquatic habitats in this wetland. This system provides essential resources for plant and animal life, mainly for MB that depend on this wetland for their survival [60]. Given its ecological importance, ODW remains a vital biodiversity resource. Therefore, it is necessary to understand hydrological and ecological factors for conservation efforts, particularly in the face of climate change and habitat loss.

Previous studies documented 444 bird species, with majorly being waterbirds. Among them, sixteen species were breeding, and four species were threatened (one endangered and three vulnerable species) [18]. The key MB species are listed in Table 23.2.

The ODW faces many conservation challenges such as upstream water extraction, climate change, and land-use changes, such as urbanization and tourism. Growing tourism in this area has affected the environment, caused habitat loss, and waste management issues [42, 49]. Moreover, water resources are becoming limited due to population growth and agricultural development that leads to competition for limited water resources [18] and also threaten flooding regimes, which are important for delta biodiversity. Moreover, conservation efforts are becoming complicated due to the absence of long-term monitoring and comprehensive management strategies [10, 13]. In summary, ODW is under threat due to environmental and anthropogenic pressures.

#### 23.2.2.3 Sundarbans, Bangladesh and India (Asia)

SW is the world's largest coastal mangrove wetland situated in Bangladesh and India. It spreads over approximately 10,000 km$^2$ across Bangladesh and India. This wetland comprises tidal rivers, mudflats, and diverse biodiversity, and crucial habitat for species such as Royal Bengal tigers, various herons, kingfishers, and mudskippers [16, 35, 52]. This ecosystem is characterized by a subtropical monsoonal climate with significant annual rainfall and frequent cyclonic storms. Additionally, sediment deposition from the major rivers like the Ganges, Brahmaputra, and Meghna Rivers makes diverse salinity gradients that enrich the biodiversity of the region [15, 19]. Therefore, SW is not only ecologically significant, but also a dynamic system affected by climatic and hydrological factors.

This wetland system is also a critical habitat for avian biodiversity and approximately 598 bird species have been documented, of which 68% are migratory [39]. During migration, the seasonal flood and nutrient-rich intertidal zones provide refuge and feeding grounds when food availability in their native habitats declines for MB [52]. These birds rely on SW for both feeding and breeding purposes,

**Table 23.3** Comparative Overview of Key Features and Conservation Needs Across Global Wetlands

| Feature | Pantanal, South America | Okavango Delta, Africa | Sundarbans, Asia |
|---|---|---|---|
| Area covered | 140,000 km$^2$ (Harris et al. [23]; Alho and Silva [1]) | 15,000 km$^2$ (Francis et al. [13]; Heyden et al. [26]) | 10,000 km$^2$ (Gopal and Chauhan [19]; Ghosh et al. [15]) |
| Key hydrological feature | Seasonal flooding from Upper Paraguay River | Okavango River floods originating in Angola Highlands [47] | Ganges-Brahmaputra-Meghna River system creating diverse salinity gradients [52] |
| Biodiversity hotspot | Yes | Yes | Yes |
| No. of bird species | ~600 species (Antas [2]; Junk et al. [32]; Pinho et al. [50]) | ~444 species (Bartlam-Brooks et al. [3]; Gondo and Kolawole [18]) | ~598 species (Mallick [39]; Rahman [52]) |
| Migratory Bird proportion | High (~70%) (Antas [2]; Kantek et al. [34]) | High [28] | High (68%) [39] |
| Conservation status | Threatened by agriculture and deforestation (Junk and Cunha [31]; Schulz et al. [59]) | Facing threats from climate change, water abstraction, and tourism (Mbaiwa [42]; Darkoh and Mbaiwa [10]) | Pressured by aquaculture, overfishing, and climate change (Sarkar [58]; Roy et al. [56]) |
| Urgent conservation needs | Sustainable development, habitat preservation | Integrated water management, community-based conservation | Resource management, ecological damage mitigation |

making this wetland as a key conservation site for avian species [19, 39]. SW harbors several key avian species (see Table 23.2) such as the eastern curlew, lesser adjutant stork, and various ducks and shorebirds [35, 52]. Given its ecological importance, conservation efforts in SW are required for maintaining the balance of this biodiverse habitat.

Besides the above-mentioned ecological importance, SW face conservation challenges due to human activities and environmental changes. For example, agricultural land expansion and aquaculture encroachment on mangrove areas have led to habitat loss [56]. Additionally, chemical runoff from agricultural practices deteriorates water quality and affects aquatic species [58]. Moreover, local biodiversity was depleted by unsustainable fishing and harvesting and also disrupted ecological balance [58]. These pressures not only threaten biodiversity, but also affect the local communities' livelihoods which depend on fisheries and ecotourism [24, 57]. Given these escalating threats, we need urgent conservation strategies to promote sustainable resource management and mitigate ecological damage [6, 24].

The Tables 23.2 and 23.3 summarize the comparative analysis of three wetland ecosystems under study, about their wetland types, key avian species, and major human influences. See Sect. 23.2.2 for detailed descriptions of each wetland. Besides this, a comparative overview of key features and conservation needs is provided in Table 23.3.

#### 23.2.2.4 Global Wetland Diversity Summary

The PW, ODW, and SW wetlands, which are distinct in their geographical locations and ecological characteristics, share similarities in their importance for MB and the conservation challenges they face. All these three wetlands are critical habitats for diverse avian species including migratory birds. However, they experience increased pressure from human activities and environmental changes.

The PW faces threats from deforestation and agricultural expansion, while ODW with upstream water extraction and growing tourism. Despite these challenges, each wetland maintains its own unique ecological importance. In conclusion, these wetlands show global importance of conservation for MB populations and balance between human activities and ecological preservation.

## 23.3 Human Influences Reshaping Wetland Landscapes and Avian Coexistence

Wetlands are facing many problems caused by human activities and making it more difficult for different plants and animals to survive. These issues are concerning because they not only damage wetland habitats, but also disrupt coexistence between migratory and local birds with humans. For instance, rapid urbanization, infrastructure development, land use pattern changes and encroachment by human activities affected the community composition of birds. To solve these problems, we need strategies that protect nature as well as allowing economic growth [41, 45]. In the following sections, we will look more closely at how human activities affect wetlands.

### *23.3.1 Human Influence on Wetlands*

Human activities significantly reshape wetland landscapes and affect avian coexistence in the PW, ODW, and SW wetlands. These changes occurred due to a combination of ecological, hydrological, and anthropogenic factors. The human activities impact is particularly evident in changes in water flow patterns and vegetation composition [37]. Consequently, many bird species faced habitat loss and increased competition for resources. Therefore, conservation efforts in these regions must find the complex interaction between human needs and ecological preservation for long-term survival of diverse avian communities.

As discussed earlier, PW is experiencing various ecological threats like deforestation, agricultural practices, and hydropower construction, affecting local livelihoods and ecological integrity [59]. Additionally, sediment loads from intense land use and climatic factors affect rivers, habitats, and wildlife [1, 38]. In 2020, wildfires in an area of approximately 19, 890 $km^2$ affected millions of faunal species [63]. Research by Alho and Silva [1] showed habitat fragmentation restricted dispersal of birds and reduced species richness and community composition. In light of these challenges, we need urgent conservation approaches to mitigate habitat destruction and land use practices. Addressing these issues is essential for preserving both the biodiversity and livelihoods of communities that depend on their resources.

Likewise, flooding and fire regimes have affected the ODW's bird species, which reflects broader ecosystem responses to climate change and human threat [5]. While human influence in the ODW is less direct than in the PW, they remain significant. Beside this, upstream water utilization and proposed hydropower projects in Angola pose a major threat by reducing flood volumes and disrupting the delta's flood-pulse dynamics [44]. In addition, tourism growth and grazing pressure also introduce ecological pressures in this area [9]. These human influences show a complex balance between conservation efforts and economic activities in the region. Thus, sustainable water management and land-use policies are becoming essential for preserving the ODW's ecosystem and the diverse bird species.

Thus, as we discussed above about PW and ODW, similar influences have been observed in SW [31]. In SW, population pressure, aquaculture, and deforestation for shrimp farming and settlement cause a 1–2% annual decline in mangrove cover in some areas [17]. These changes reduced nesting sites for arboreal species such as brown-winged kingfishers [53]. Due to growing population demand and colonial policies, historical tidal forests are converting into agricultural land that has further disrupted the ecosystem [55]. The mangrove ecosystem is degrading due to reduced freshwater inflows and increased salinity also affecting avian habitats [25]. Besides these human impacts, climate change also exerted an impact with sea level rise and frequent cyclones. Projections show a 1-meter sea-level rise could submerge up to 75% of the mangrove forest, displacing tidal-dependent bird species [19]. Additionally, pollution from agricultural runoff introduces contaminants that threaten fish-eating birds such as the White-bellied Sea eagle [8]. Although some avian species, such as herons, have adapted to certain conditions, this trend indicates declining biodiversity [29]. Urgent conservation measures are required to restore mangrove cover, regulate aquaculture, and ensure sustainable water management.

### 23.3.2 Human Impact on Avian Co-existence

Human activities have caused hydrological changes, habitat fragmentation, and climate-induced stressors in the PW, ODW, and SW wetlands, impacting avian coexistence. Study has shown that wetland bird diversity depends on habitat heterogeneity where human activities degrade by simplifying landscapes [36, 66]. Avian species in these wetlands depend on natural disturbance e.g. floods in the PW and ODW, and tides in the SW—to sustain ecological balance. However, human disruptions caused the decline of specialist species while favoring opportunistic species. For example, in the PW, cattle grazing can mimic natural grazing by capybaras, thereby supporting wading birds [14]. In SW, aquaculture ponds create artificial habitats which benefit heron species while disadvantage forest-dependent other species [48]. These shifts show how human influence reshapes avian communities, affecting migratory birds'wintering grounds. Addressing these imbalances requires sustainable conservation strategies to restore natural flood cycles, protect mangrove habitats, and regulate land use to preserve avian diversity.

## 23.4 Coexistence

Coexistence, in the context of human-wildlife interactions, is defined as a sustainable and dynamic state in which humans and wildlife adapt to share landscapes. Resources with human interactions are effectively governed to ensure that wildlife populations persist in socially legitimate ways to maintain tolerable risk levels [22, 51]. There is a complex coexistence observed between humans and birds in this area, for example, avian species often seen coexist with humans by sharing land, thereby affecting agricultural practices. Research has shown that birds can increase the production of woody crops, although their effects on herbaceous crops may differ. Non-lethal strategies, including the use of deterrent devices and modifications in sowing practices, have shown efficacy in reducing crop losses, which is essential for promoting harmonious coexistence [27]. In areas such as the PW, birds have more tolerance to human disturbance, mainly in areas near small farms. This indicates behavioral adaptation plays important role in coexistence [64].

### *23.4.1 Strategies Used by Migratory Birds for Coexistence*

MB use niche partitioning approach to avoid competition for resources and shows remarkable coexistence in wetland ecosystems. In the PW, seasonal flooding is common that causes temporal and spatial niche partitioning, allowing these birds to share resources. Greenberg et al. [20] observed that Wilson's phalarope (*Phalaropus tricolor*) and golden plover (*Pluvialis dominica*) exhibited segregation in their foraging sites during stopovers, with plovers favoring shallow mudflats and phalaropes utilizing deeper pools. This spatial segregation reduces competition for food sources and enables resource utilization efficiently during migration. Thus, the resource partitioning strategy shows adaptations in different environmental conditions, as well as maximizes the use of resources in the PW.

While in ODW, behavioral adaptation and selection of habitat are the main strategies used by MB. For example, Bino et al. [7] observed how little stint

(*Calidris minuta*) (Palaearctic waders) and common sandpiper (*Actitis hypoleucos*) partition their feeding zones during peak floods. Sandpipers forage along plant-covered margins, although little stints explore open shallows to minimize competition. This spatial segregation facilitates the optimal provisioning for MB during stopovers, thus support avian diversity in this area. These adaptive foraging behaviors show how MB optimizes habitat use during this migration to ODW.

In SW, MB shows habitat partitioning or niche partitioning on a small scale, which is influenced by tidal cycles. For example, according to Datta et al. [11], species such as whimbrel (*Numenius phaeopus*) and black tailed godwit (*Limosa limosa*) separate their foraging times and depths to co-exist with each other. While godwits forage on exposed mudflats during low tide, whimbrels prefer the higher intertidal zones, which minimizes their overlap in the winter. This strategic habitat partitioning helps preserve the fragile ecological balance in SW and benefits both resident bird (RB) and MB species. This partitioning illustrates the impact of tidal cycle on resource allocation and how they enable birds to coexist. These observed patterns of niche partitioning in wetlands show how MB reduces interspecific-competition, promotes efficient resource utilization, and ecosystem stability. Thus, conservation of wetlands is crucial for maintaining its functions which sustain avian diversity during the migration period.

### 23.4.2 Coexistence with Resident Species

The coexistence of migratory and RB species in the PW relies on complementary interactions, although competition may occur under resource constraints. Migratory waders benefit from the foraging trails of resident southern lapwings (*Vanellus chilensis*), which disturb insects in grazed areas and enhance prey availability [14, 67]. However, competition increases when resources are limited, such as after post-wildfire habitat loss, where reduced wetland extent forces migratory and resident species, such as the Jabiru stork (*Jabiru mycteria*), into overlapping niches, increasing the pressure on shared food sources [63]. PW heterogeneity supports this eco-balance that enables varied strategies among migratory flocks and facilitates coexistence between them and resident species. These complementary and competitive dynamics show the importance of maintaining habitat heterogeneity to sustain avian biodiversity.

In the ODW, this coexistence involves balancing competition and facilitation. For example, migratory White-faced whistling duck (*Dendrocygna viduata*) competes with resident Wattled cranes (*Grus carunculatus*) for aquatic vegetation during flood recessions [36]. However, cranes benefit from ducks' grazing which exposes hidden prey and fosters a complementary relationship that reduces direct competition. Fish eagles (resident predator) sometimes prey on migratory flocks, but their predation rate is low which shows stable coexistence. ODW's pulsed flooding sustains this balance, creating sequential niches that migratory and RB exploit.

In SW, migratory and resident species often coexist through niche complementarity, although habitat loss threatens this balance. Migratory herons, such as the Grey heron (*Ardea cinerea*), share artificial aquaculture ponds with resident Little cormorants (*Microcarbo niger*), where both species exploit abundant fish but minimize conflict by using different hunting techniques—herons wade, while cormorants dive [48]. However, mangrove loss shrinks forested habitats, pushing residents like Mangrove pitta (*Pitta megarhyncha*) into competition with migratory passerines for cover, disrupting coexistence. The tidal dynamics and mangrove structure of SW facilitate interactions between migratory and resident species; however, ongoing habitat degradation poses significant challenges in maintaining this ecological balance.

### 23.4.3 Mechanisms of Coexistence

MB in the PW, ODW, and SW wetlands coexist via niche partitioning. Niche partitioning involves spatial, temporal, or behavioral adaptations that are facilitated by changing water levels. Flooding in the PW and ODW and tidal movements in the SW create diverse habitats. These varied environments allow different bird species to use specific resources at different times which in turn reduces competition [36]. Such niche partitioning minimizes the overlap between migratory species and also promotes biodiversity by enabling efficient resource use during temporary visits. These niche partitioning patterns show the crucial role of water dynamics in the coexistence of MB across wetland ecosystems.

The Coexistence in wetlands ranges from competition to cooperation which depends on resource availability. Species compete with one another when resources are limited. However, they can show positive biotic interaction if resources are abundant in general. But this may not always be true, for example, we observed five species including both migratory and resident species were sharing water resources with positive interaction with limited food resources at arid landscape in Barli village, Jodhpur (Rajasthan), India in 2019 (see Fig. 23.2). Additionally, resident species benefit MB by stabilizing wetland ecosystems [14]. Thus, such coexistence among resident and MB can maintain wetland ecosystems by providing a stable environment. This interaction between competition and cooperation shows the complex relationships that allow migratory and resident species to coexist in sensitive ecosystems. An overview and summary of the mechanism of coexistence have been provided in Table 23.4 and Table 23.5 respectively.

### 23.4.4 Summary

The PW, ODW, and SW wetlands show complex patterns of coexistence between migratory and RB species, driven by niche partitioning and dynamic ecological

**Fig. 23.2** Five species sharing resources when resources become limited. (Source: Author)

**Table 23.4** Brief description of different mechanism for coexistence

| Mechanism | Description | References |
|---|---|---|
| Niche partitioning | Spatial, temporal, or behavioral adaptations facilitated by changing water levels (flooding and tides) to reduce competition. | [36] |
| Competition | Occurs when resources are scarce, leading to overlap in resource use between migratory and resident species. | |
| Cooperation/ facilitation | Foraging activities of one group (migratory or resident) can benefit the other by making prey more accessible or stabilizing ecosystems. | [14] |

**Table 23.5** Summary of Coexistence Mechanisms in Pantanal, Okavango Delta and Sundarbans wetland ecosystem

| Wetland | Coexistence among migratory birds | Coexistence with residents |
|---|---|---|
| Pantanal | Spatial niche partitioning, e.g., mudflats vs. pools | Facilitation via resident foraging trails |
| Okavango delta | Behavioral adaptations, e.g., feeding zone segregation | Competition and facilitation, e.g., prey exposure |
| Sundarbans | Tidal-driven temporal segregation | Niche complementarity, e.g., different hunting methods |

interactions. MB share stopover and wintering sites through spatial, temporal, and behavioral partitioning. They reduce competition and optimize resource use. Meanwhile, interactions with resident species oscillate between competition and facilitation are influenced by wetland hydrology and habitat diversity. These interactions maintain a delicate balance that sustains biodiversity across diverse ecosystems. To ensure the continued coexistence of migratory and resident species in wetlands, proactive conservation efforts are necessary to preserve the habitat diversity and ecological integrity. By protecting these ecosystems, we can ensure that they remain essential sanctuaries for bird biodiversity, thus benefiting both migratory and resident species.

## 23.5 Navigating Coexistence: Conservation Strategies and Management Approaches

Effective conservation strategies and management approaches are essential to sustaining avian coexistence in the PW, ODW, and SW wetlands while accommodating human livelihoods. These three wetlands support complex interactions between migratory and RB species. However, as we discussed earlier, growing pressure from agriculture, water management, and climate change threatens these ecosystems and disrupts their coexistence. Recent studies have shown the importance of developing adaptive strategies that balance ecological preservation with sustainable human activities. Different strategies for navigating coexistence have been summarized in flow chart (See Fig. 23.3). The following sections explore successful conservation initiatives, ongoing challenges, and future directions for navigating the coexistence of these ecosystems.

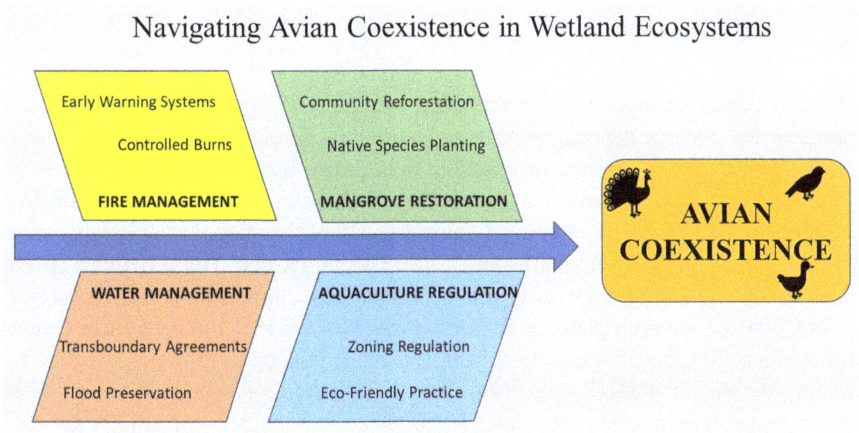


**Fig. 23.3** Flowchart showing conservation strategies implemented for conserving avian species and wetlands under study

### 23.5.1 Pantanal: Fire Management and Sustainable Grazing

To conserve bird diversity in PW, it is necessary to address the challenges posed by wildfires and excessive grazing through integrated fire management and sustainable grazing practices. Recent large-scale fires discussed earlier affected nearly 20% of the wetland area. This disrupted bird habitats and highlighted the need for effective conservation strategies. Tomas et al. [63] recommended an integrated approach to fire management that combines controlled burns with early detection systems to reduce the risk of uncontrolled fires that endanger MB stopover sites and resident nesting areas. This approach restores the flood-pulse dynamics that maintain habitat heterogeneity, supporting diverse bird communities.

Sustainable grazing is another critical strategy to maintain avian coexistence in the PW region. Garcia et al. [14] demonstrated that regulated cattle grazing mimics natural disturbances caused by capybaras, preserving open habitats for migratory waders such as the Southern lapwing (*Vanellus chilensis*) while supporting resident species. In addition, overgrazing by domestic and wild grazer species harms the environment and interrupts coexistence. To address this issue, many researchers have recommended the use of rotational grazing systems and other management strategies. These targeted approaches can balance livestock needs and conservation plans. Community-based conservation initiatives that engage local farmers in fire and grazing management increase coexistence by aligning human livelihoods with ecological aims. Nonetheless, two primary challenges persist: securing long-term funding, and effectively enforcing regulations. Addressing these issues is essential for successful conservation efforts in this region. This might require exploring new funding sources and improving local governance structures.

### 23.5.2 Okavango Delta: Water Management and Ecotourism Regulation

Effective conservation of avian coexistence within the ODW requires planned water management and regulated ecotourism. The delta's flood-dependent ecosystem is threatened by upstream water abstraction and climate-induced drought. Bino et al. [7] advocate for transboundary agreements with Angola to regulate the water flow from the Okavango River, thereby ensuring seasonal flooding that maintains stop-over sites for MB and foraging grounds for resident species. These policies aim to preserve the hydrological integrity that is essential for supporting avian diversity.

In ODW, ecotourism plays an important role in economy, but it requires careful management to reduce its ecological impact. Kingsford et al. [36] suggest that, by limiting the visitor numbers and controlling infrastructure development can reduce habitat fragmentation, migratory path disturbance and resident breeding sites. The income from ecotourism can support habitat restoration and monitoring efforts, as seen in community-driven programs that conserve water channels and habitat for species such as African fish eagle (*Haliaeetus vocifer*). It is crucial to balance the economic significance of ecotourism with its ecological effects, and adaptive management frameworks are recommended to modify strategies based on bird population dynamics and flood data.

### *23.5.3 Sundarbans: Mangrove Restoration and Aquaculture Limits*

Restoring mangroves and regulating aquaculture are vital for maintaining avian coexistence in SW. Mangrove loss due to sea level rise and aquaculture expansion

**Table 23.6** Overview of major threats, conservational approaches, community engagement and outcomes within Pantanal, Okavango Delta and Sundarbans wetland ecosystem

| Wetland | Major threats | Conservation strategies | Community engagement | Conservation outcomes |
|---|---|---|---|---|
| Pantanal | Wildfires, overgrazing | Integrated fire management, sustainable grazing | Ranchers in fire/ grazing mgmt. | Restored habitats, increased bird diversity |
| Okavango | Water abstraction, droughts | Transboundary water agreements, ecotourism control | Local guides monitoring | Hydrological stability, reduced habitat loss |
| Sundarbans | Mangrove loss, aquaculture | Mangrove restoration, aquaculture zoning | Villagers leading reforestation | Enhanced wetland integrity, biodiversity protection |

**Table 23.7** Overview of primary threats and action implemented with evidence/initiatives in Pantanal, Okavango Delta and Sundarbans wetland and ongoing challenges

| Wetland | Primary threats | Conservation strategies | Key evidence/ initiatives | Ongoing challenges |
|---|---|---|---|---|
| Pantanal | Wildfires, overgrazing | Integrated fire management (controlled burns, early warnings); sustainable/rotational grazing | 2020 wildfire impact; studies by Tomas et al. [63] and Garcia et al. [14] | Funding, enforcement, habitat degradation |
| Okavango | Water abstraction, climate-induced droughts, unregulated ecotourism | Transboundary water management; capping tourist numbers; infrastructure restrictions | Water flow agreements with Angola [7]; ecotourism controls [36] | Balancing economic benefits with ecological needs |
| Sundarbans | Mangrove loss, aquaculture expansion | Large-scale mangrove restoration; zoning for sustainable aquaculture | Reforestation projects [11]; aquaculture zoning [48] | Cross-border policy harmonization, enforcement issues |

threatens the delicate balance between migratory and resident species. Datta et al. [11] suggested large-scale mangrove restoration using native species such as *Avicennia* and *Rhizophora* to rebuild habitats for forest-dependent residents and migratory stopovers. Community-based reforestation projects have been successful in securing tidal areas, improving niche availability, and ensuring avian diversity. A summary of major threats, conservation challenges and management approaches has been represented in Tables 23.6 and 23.7.

Management of aquaculture expansion is critical for sustaining wetland integrity in SW. Mukherjee and Ray [48] recommended zoning regulations to limit pond expansion and preserve natural wetlands for migratory waders with sustainable fish farming. Incentives for eco-friendly practices support both human livelihoods and avian diversity such as polyculture systems that reduce habitat conversion as demonstrated by Grey herons (*Ardea cinerea*) and Little cormorants (*Microcarbo niger*) coexisting in managed ponds. However, implementation remains a challenge, and cross-border collaboration between India and Bangladesh is needed to harmonize policies and protect this shared ecosystem.

Tables 23.6 and 23.7 clearly summarize the unique conservation challenges and management approaches across the three regions, making it easier for readers to grasp their differences and commonalities.

## 23.6 Sustaining Avian Coexistence Through Adaptive Management

Adaptive management strategies that integrate ecological goals with human livelihoods are required to sustain avian coexistence. Conservation efforts in the PW, OSW, and SW (see Tables 23.6 and 23.7) have shown that maintaining habitat heterogeneity and regulating anthropogenic pressure can protect avian biodiversity. For example, combining fire management, water management, and habitat restoration with sustainable practices can help these ecosystems to serve as sanctuaries for migratory and resident species. Kingsford et al. [36] emphasize adaptive management—using real-time data on bird populations and habitat conditions—to refine these approaches, ensuring flexibility amid climate variability. For long-term success, this requires community involvement, implementation mechanisms, and cross-border collaboration to preserve globally significant wetlands.

Thus, community engagement is crucial. In the PW, ranchers participate in fire control, in the ODW, local guides monitor tourism impacts, and in the SW, villagers lead reforestation. These efforts align conservation with human needs, fostering the coexistence between humans and wildlife. Challenges include funding shortages, policy enforcement, and climate change, all of which amplify habitat loss [63]. Integrated landscape approaches, which combine protected areas with sustainable use zones, offer a promising framework, as evidenced by pilot projects in each wetland.

## 23.7 Future Directions

Future research should focus on the long-term monitoring of bird populations and habitat responses along with economic analyses of conservation benefits to strengthen these strategies. By balancing human activities with ecological needs, these wetlands can remain sanctuaries for migratory and RB, sustaining their rich coexistence in the future. Conservation alternatives must include such strategies to protect critical wetland ecosystems like the PW, ODW, and SW. These strategies must prioritize ecosystem health and involve local community engagement. Potential conservation measures considered were: (1) Use of controlled fire management and grazing practices, (2) Establishing water usage agreements, and (3) Starting mangrove reforestation programs: Adaptive management strategies should be adopted to make adjustments to the strategy, if necessary. Future research should experiment with: (1) Long-term monitoring of avian populations; (2) Habitat change assessment; (3) Assessing the impact of wetland conservation on local economies: With a moderation of development encroaching upon these wetlands, such ecological systems can continue to exist as critical microhabitats that host various ecosystems through time. Strategy, if necessary. Future research should experiment with: (1) Long-term monitoring of avian populations; (2) Habitat change assessment (3) Assessing the impact of wetland conservation on local economies: With a moderation of development encroaching upon these wetlands, such ecological systems can continue to exist as critical microhabitats that host various ecosystems through time.

## 23.8 Conclusion

Navigating coexistence in the PW, ODW, and SW requires conservation strategies and management approaches that preserve wetland dynamics and engage the local communities. Fire and grazing regulations, water flow agreements, and mangrove restoration resolve specific threats, while adaptive management ensures resilience. Many species employ different approaches for maintaining coexistence, such as niche partitioning, sharing habitat, and synchronized activities to survive under highly pressurized environments or ongoing threats. By mitigating these environmental challenges and adopting the right conservation strategies, we can maintain harmonious coexistence among aquatic bird species and humans.

## References

1. Alho C, Silva J (2012) Effects of severe floods and droughts on wildlife of the Pantanal wetland (Brazil)—a review. Animals 2:591–610. https://doi.org/10.3390/ani2040591

2. Antas P (1994) Migration and other movements among the lower Paraná River valley wetlands, Argentina, and the South Brazil/Pantanal wetlands. Bird Conserv Int 4:181–190. https://doi.org/10.1017/S0959270900002768
3. Bartlam-Brooks HLA, Bonyongo MC, Harris S (2013) How landscape scale changes affect ecological processes in conservation areas: external factors influence land use by zebra (Equus burchelli) in the Okavango Delta. Ecol Evol 3(9):2795–2805. https://doi.org/10.1002/ece3.676
4. Bauer P, Gumbricht T, Kinzelbach W (2006) A regional coupled surface water/groundwater model of the Okavango Delta, Botswana. Water Resour Res 42. https://doi.org/10.1029/2005WR004234
5. Becker F, Boyes R, Wittmer H, Inman V, Marsland S (2023) Landscape-wide pulse events predict trait-based responses among wetland birds in perennial channels of a dryland wetland. Ecol Appl:e2931. https://doi.org/10.1002/eap.2931
6. Biller D, Sánchez-Triana E (2013) Enlisting markets in the conservation and sustainable use of biodiversity in South Asia's Sundarbans. Int J Soc Ecol Sustain Dev 4(3):71–86. https://doi.org/10.4018/jsesd.2013070106
7. Bino G, Wolski P, Murray-Hudson M, Mosepele K, Kingsford RT (2020) Flood variability and waterbird responses in the Okavango Delta, Botswana. Ecohydrol 13(7):e2240. https://doi.org/10.1002/eco.2240
8. Chowdhury GW, Koldewey HJ, Duncan E, Napper IE, Niloy MNH (2017) Anthropogenic threats to fish-eating birds in the Sundarbans. Oryx 51(4):689–696. https://doi.org/10.1017/S003060531600045X
9. Cumming GS, Gaidet N, Ndlovu M (2013) Large herbivores and ecosystem engineering in the Okavango Delta. J Arid Environ 89:87–94. https://doi.org/10.1016/j.jaridenv.2012.09.010
10. Darkoh M, Mbaiwa J (2014) Okavango delta—a Kalahari oasis under environmental threats. J Biodivers Endanger Species 2:1–6. https://doi.org/10.4172/2332-2543.1000138
11. Datta D, Chattopadhyay RN, Guha P (2022) Impact of climate change and anthropogenic pressure on the Sundarbans mangroves and its biodiversity. Reg Environ Chang 22(2):67. https://doi.org/10.1007/s10113-022-01918-5
12. Figueres C (2018) United nations framework convention on climate change (UNFCCC). Yearb Int Coop Environ Dev 1998–99. https://doi.org/10.4324/9781315066547-11
13. Francis R, Bino G, Inman V, Brandis K, Kingsford R (2021) The Okavango Delta's waterbirds—trends and threatening processes. Glob Ecol Conserv 30:e01763. https://doi.org/10.1016/J.GECCO.2021.E01763
14. Garcia LC, Pott A, Damasceno-Junior GA (2021) Cattle grazing as a management tool for wetland birds in the Pantanal. Wetlands 41(8):108. https://doi.org/10.1007/s13157-021-01503-5
15. Ghosh A, Schmidt S, Fickert T, Nüsser M (2015) The Indian Sundarban mangrove forests: history, utilization, conservation strategies and local perception. Diversity 7:149–169. https://doi.org/10.3390/D7020149
16. Ghosh P (2015) Conservation and conflicts in the Sundarban Biosphere Reserve, India. Geogr Rev 105(4):429–440. https://doi.org/10.1111/j.1931-0846.2015.12101.x
17. Giri C, Pengra B, Zhu Z, Singh A, Tieszen LL (2007) Monitoring mangrove forest dynamics of the Sundarbans using multi-temporal satellite data. Estuar Coast Shelf Sci 73(1–2):91–100. https://doi.org/10.1016/j.ecss.2006.12.008
18. Gondo R, Kolawole OD (2019) Sustainable water resources management: issues and principles of water governance in the Okavango Delta, Botswana. Int J Rural Manag 15(2):198–217. https://doi.org/10.1177/0973005219865369
19. Gopal B, Chauhan M (2006) Biodiversity and its conservation in the Sundarban mangrove ecosystem. Aquat Sci 68:338–354. https://doi.org/10.1007/s00027-006-0868-8
20. Greenberg R, Marra PP, Wooller MJ (2020) Stable isotope ecology of migratory shorebirds in tropical wetlands: insights from the Pantanal. J Ornithol 161(3):743–755. https://doi.org/10.1007/s10336-020-01767-8

21. Haig S, Murphy S, Matthews J, Arismendi I, Safeeq M (2019) Climate-altered wetlands challenge waterbird use and migratory connectivity in arid landscapes. Sci Rep 9. https://doi.org/10.1038/s41598-019-41135-y
22. Haight JD, Larson KL, Clark JAG, Lewis JS, Hall SJ (2023) Social-ecological drivers of metropolitan residents' comfort living with wildlife. Front Conserv Sci 4:1248238. https://doi.org/10.3389/fcosc.2023.1248238
23. Harris M, Tomas W, Mourão G, Da Silva C, Guimarães E, Sonoda F, Fachim E (2005) Safeguarding the Pantanal wetlands: threats and conservation initiatives. Conserv Biol 19. https://doi.org/10.1111/j.1523-1739.2005.00708.x
24. Hasanuzzaman AFM, Sayeed MAB, Rahaman S, Huq KA (2010) Implications of climate change for fisheries and aquaculture in the Sundarbans region of Bangladesh. Khulna Univ Stud:73–82. https://doi.org/10.53808/kus.si.sesb.2010.73-82-ls
25. Henderson ACG, Das S, Ghosh T, Panizzo VN, Moorhouse H, Roberts LR, Walton RE, Zheng Y, Bass AM, McGowan S (2021) The Indian Sundarbans: biogeochemical dynamics and anthropogenic impacts. Newcastle University. https://doi.org/10.1007/978-3-030-68980-3_15
26. Heyden VDS, Cwecwe Y, Sado T, Von Brandis R, Creer S, Skelton P, Neef G, Mosie I, Miya M, Grevesse T (2023) Environmental DNA biomonitoring in biodiversity hotspots: a case study of fishes of the Okavango Delta. Environ DNA 5(6):1720–1731. https://doi.org/10.1002/edn3.494
27. Huang C, Zhou K, Huang Y, Fan P, Liu Y, Lee T (2023) Insights into the coexistence of birds and humans in cropland through meta-analyses of bird exclosure studies, crop loss mitigation experiments, and social surveys. PLoS Biol 21. https://doi.org/10.1371/journal.pbio.3002166
28. Hudson MM, Mendelson J, Mosepele K, Ramberg L, Vanderpost C, Wolski P (2010) Okavango Delta: floods of life. Raison Publishers. Accessed at https://policycommons.net/artifacts/1376449/okavango-delta/1990711/
29. Islam MZ, Rahmani AR (2008) Potential and existing Ramsar sites in India. Oxford University Press
30. Ivory S, McGlue M, Spera S, Silva A, Bergier I (2019) Vegetation, rainfall, and pulsing hydrology in the Pantanal, the world's largest tropical wetland. Environ Res Lett 14. https://doi.org/10.1088/1748-9326/ab4ffe
31. Junk W, Brown M, Campbell I, Finlayson M, Gopal B, Ramberg L, Warner B (2006) The comparative biodiversity of seven globally important wetlands: a synthesis. Aquat Sci 68:400–414. https://doi.org/10.1007/s00027-006-0856-z
32. Junk WJ, da Silva CJ, da Cunha CN (2011) The Pantanal: ecology, biodiversity and sustainable management of a large neotropical seasonal wetland. Pensoft Publishers. https://pure.mpg.de/pubman/faces/ViewItemFullPage.jsp?itemId=item_1505043
33. Kabir MH, Shawon RAR, Haque MK, Shyama IJ, Islam MA, Saila MSN, Hossain MJ, Rahman MM (2024) Enhancing conservation strategies for wild and migratory bird habitats in South Asia: a comprehensive review. Asian J Environ Ecol. https://doi.org/10.9734/ajee/2024/v23i1522
34. Kantek DLZ, Melo RC, Miyazaki SS, Castilheiro WFF, Filho MS (2020) Aves aquáticas da estação ecológica de Taiamã: variação sazonal da estrutura da comunidade e a importância das áreas protegidas no Pantanal. Biodiversidade Bras – BioBrasil 10(3):24–40. https://doi.org/10.37002/biobrasil.v10i3.1513
35. Khan MS, Abdullah S, Salam MA, Mandal TR, Hossain R (2021) Review assessment of biodiversity loss of Sundarban forest: highlights on causes and impacts. Indones J For Res 8(1):85–97. https://doi.org/10.20886/ijfr.2021.8.1.85-97
36. Kingsford RT, Bino G, Porter JL (2021) Declining waterbird populations in wetlands: the role of habitat loss and hydrological change. Glob Ecol Conserv 30:e01787. https://doi.org/10.1016/j.gecco.2021.e01787
37. Kiran, Singh D, Kour A, Priya, Delu V, Kumar R (2022) Community composition and status of avian diversity at campus and agricultural landscapes of Chaudhary Charan Singh Haryana

Agricultural University, Hisar (Haryana). J Appl Nat Sci 14(4):1130–1140. https://doi.org/10.31018/jans.v14i4.3784
38. Louzada R, Bergier I, Assine M (2020) Landscape changes in avulsive river systems: case study of Taquari River on Brazilian Pantanal wetlands. Sci Total Environ 723:138067. https://doi.org/10.1016/j.scitotenv.2020.138067
39. Mallick JK (2023) A review and field verification of avian diversity and habitat use in the greater Sundarbans of India, Bangladesh and their hinterland. Probe-Animal Sci 5(1):1778. https://doi.org/10.18686/pas.v5i1.1778
40. Maltby E, Acreman MC (2011) Ecosystem services of wetlands: pathfinder for a new paradigm. Accessed from https://www.tandfonline.com/doi/pdf/10.1080/02626667.2011.631014
41. Mao Q, Liao C, Wu Z, Guan W, Yang W, Tang Y, Wu G (2019) Effects of land cover pattern along urban-rural gradient on bird diversity in wetlands. Diversity. https://doi.org/10.3390/D11060086
42. Mbaiwa JE (2011) Hotel companies, poverty and sustainable tourism in the Okavango Delta, Botswana. World J Entrep Manag Sustain Dev 7(1):47–58. https://doi.org/10.1108/20425961201000030
43. McKenna O, Mushet D, Kucia S, McCulloch E (2021) Limited shifts in the distribution of migratory bird breeding habitat density in response to future changes in climate. Ecol Appl 31. https://doi.org/10.1002/eap.2428
44. Mendelsohn J, Vanderpost C, Ramberg L, Murray-Hudson M, Wolski P, Mosepele K (2010) Okavango Delta: floods of life. Raison Publishers. Accessed from https://www.researchgate.net/profile/John-Mendelsohn/publication/232560418_Okavango_Delta_Floods_of_Life/links/65588af0b86a1d521bf25b0e/Okavango-Delta-Floods-of-Life.pdf
45. Mereta ST, Lemmens P, De Meester L, Goethals PLM, Boets P (2021) The relative importance of human disturbance, environmental and spatial factors on the community composition of wetland birds. Water 13(23):3448. https://doi.org/10.3390/w13233448
46. Mishra A, Kumar S, Patra R, Kumar A, Sahu H, Chandra N, Pande C, Alshehri F (2023) Physicochemical parameters of water and its implications on avifauna and habitat quality. Sustainability. https://doi.org/10.3390/su15129494
47. Moses O (2024) Projected changes in rainfall and temperature using CMIP6 models over the Okavango River Basin, Southern Africa. Theor Appl Climatol 155(6):5337–5351. https://doi.org/10.1007/s00704-024-04950-6
48. Mukherjee S, Ray S (2023) Ecological impacts of aquaculture on avian diversity in the Indian Sundarbans. Environ Monit Assess 195(4):489. https://doi.org/10.1007/s10661-023-11025-6
49. Musora O (2018) Employees' perceptions of environmental impacts of tourism activities in the Okavango Delta, Botswana. Int J Hosp Tour Manag 2(1):13. https://doi.org/10.11648/j.ijhtm.20180201.13
50. Pinho JB, Aragona M, Hakamada KYP, Marini MÂ (2017) Migration patterns and seasonal forest use by birds in the Brazilian Pantanal. Bird Conserv Int 27(3):371–387. https://doi.org/10.1017/s0959270916000290
51. Pooley S, Vasava A, Bhatia S (2020) Rethinking the study of human-wildlife coexistence. Conserv Biol 35(3):784–793. https://doi.org/10.1111/cobi.13653
52. Rahman MM (2023) Impact of climate change: whispering from the Sundarbans mangrove of Bangladesh. SSRN Electron J. https://doi.org/10.2139/ssrn.4449671
53. Rahman MM, Giedraitis VR, Lieberman LS, Akhtar MT, Haque MM (2011) Environmental impacts of shrimp farming in the Sundarbans mangrove forest. Environ Monit Assess 183(1–4):291–304. https://doi.org/10.1007/s10661-011-1922-8
54. Ramberg L, Hancock P, Lindholm M, Meyer T, Ringrose S, Slíva J, As J, Post C (2006) Species diversity of the Okavango Delta, Botswana. Aquat Sci 68:310–337. https://doi.org/10.1007/s00027-006-0857-y
55. Richards JF, Flint EP (1990) Long-term transformations in the Sundarbans wetlands forests of Bengal. Agric Hum Values 7(2):17–33. https://doi.org/10.1007/BF01530433

56. Roy S, Ghosh T, Hewavithana DK (2024) Transformation of coastal wetlands in the Sundarban Delta (1999–2020). Environ Monit Assess 196(8). https://doi.org/10.1007/s10661-024-12901-x
57. Rudra A, Chattopadhyay A (2019) Environmental change of coastal Sundarbans: impact on livelihood and standard of living status of indigenous people. Environ Qual Manag 29(2): 77–84. https://doi.org/10.1002/tqem.21667
58. Sarkar SK (2022) The imperiled Sundarban: wetland of international importance. Elsevier eBooks. https://doi.org/10.1016/b978-0-12-821139-7.00092-1
59. Schulz C, Whitney B, Rossetto O, Neves D, Crabb L, De Oliveira E, Lima P, Afzal M, Laing A, De Souza Fernandes L, Da Silva C, Steinke V, Steinke E, Saito C (2019) Physical, ecological and human dimensions of environmental change in Brazil's Pantanal wetland: synthesis and research agenda. Sci Total Environ 687:1011–1027. https://doi.org/10.1016/j.scitotenv.2019.06.023
60. van der Schalie R, van der Vliet M, Albergel C, Dorigo W, Wolski P, De Jeu R (2022) Characterizing natural variability in complex hydrological systems using passive microwave-based climate data records: a case study for the Okavango Delta. Hydrology and Earth System Sciences 26(13):3611–3627. https://doi.org/10.5194/hess-2021-637
61. Semmens DJ, Diffendorfer JE, López-Hoffman L, Shapiro CD (2011) Accounting for the ecosystem services of migratory species: quantifying migration support and spatial subsidies. Ecol Econ 70(12):2236–2242. https://doi.org/10.1016/j.ecolecon.2011.07.002
62. Somveille M, Manica A, Butchart SH, Rodrigues AS (2013) Mapping global diversity patterns for migratory birds. PLoS One 8. https://doi.org/10.1371/journal.pone.0070907
63. Teitelbaum CS, Bachner NC, Hall RJ (2023) Post-migratory nonbreeding movements of birds: a review and case study. Ecol Evol 13(3). https://doi.org/10.1002/ece3.9893
64. Tomas WM, Berlinck CN, Chiaravalloti RM, Faggioni GP, Strüssmann C, Libonati R, Abrahão CR, Alvarenga GC, Bacellar AEF, Batistella FR, Bernardino TC, Botelho AL, Camilo AR, Campos RC, Carniello MA, Cezar SF, Costa FR, Deus FF, Dias RA, Morato RG (2021) The 2020 wildfires in the Pantanal: a disaster for biodiversity and ecosystem services. Environ Conserv 48(4):245–252. https://doi.org/10.1017/S0376892921000278
65. Tryjanowski P, Kosicki J, Hromada M, Mikula P (2019) The emergence of tolerance of human disturbance in Neotropical birds. J Trop Ecol 36:1–5. https://doi.org/10.1017/S0266467419000282
66. Whelan CJ, Şekercioğlu ÇH, Wenny DG (2015) Why birds matter: from economic ornithology to ecosystem services. J Ornithol 156(S1):227–238. https://doi.org/10.1007/s10336-015-1229-y
67. Yodha K, Singh D, Kour A, Yadav R, Kumar R (2024) Species diversity and abundance of birds in ber (*Zizyphus mauritiana*) crop. Asian J Conserv Biol 13(1). https://doi.org/10.53562/ajcb.83683
68. Yodha K, Singh D, Kour A, Yaday R, Dahiya P, Kumar R (2023) Estimation of predatory bird damage to winter guava crop. Indian J Ecol 50(5 (SI)):1787–1793. https://doi.org/10.55362/IJE/2023/4138
69. Zhang W, Wei J, Xu Y (2023) Prioritizing global conservation of migratory birds over their migration network. One Earth. https://doi.org/10.1016/j.oneear.2023.08.017

# Chapter 24
# Ecological Attributes of Nano-Based Antimicrobial Agents for Disease Control in Wild Aquatic Bird Populations

**Narender Mohan, Pooja Verma, Jyoti Ahlawat, Neelam Rani, and Sonika Jakhar**

**Abstract** Wild aquatic birds are vital to aquatic ecosystems, fulfilling roles as pollinators, seed dispersers, and facilitators of nutrient cycling. Yet, their populations face significant threats from infectious diseases such as avian influenza and botulism, which risk ecological disruption and biodiversity loss. Conventional disease management strategies often fall short due to inefficacy or unintended environmental consequences, prompting the exploration of innovative alternatives. This chapter examines the potential of nano-based antimicrobial agents including silver, copper, and zinc nanoparticles as targeted solutions for disease control in these populations. By taking advantage of their distinctive physicochemical traits, such as a high surface-area-to-volume ratio and the ability to specifically target pathogens, these agents fight microbes through various mechanisms including the production of reactive oxygen species, disruption of cell membranes as well as interference with the genetic material. Their advantages over traditional antimicrobials include enhanced efficacy, reduced resistance development and broad-spectrum activity. However, ecological concerns stick with their environmental behaviour including long-distance transport, aggregation, persistence, and bioaccumulation risks within aquatic food webs. These factors offer emphasis to possible impacts on non-target species as well as ecosystem integrity, ultimately, demanding thorough ethical evaluation and risk assessment frameworks. The chapter emphasizes an urgent need for inter-disciplinary research to address gaps in long-term ecological effects, species-specific vulnerabilities, and the expansion of globally synchronized regulatory standards. By balancing innovation with environmental

N. Mohan (✉) · P. Verma · S. Jakhar
Division of Quality and Basic Sciences, ICAR-IIWBR Karnal, Karnal, Haryana, India

J. Ahlawat
Department of Plant Biotechnology, G.D. Goenka University, Sohna Rural, Haryana, India

N. Rani
Department of Biotechnology, Guru Jambheshwar University of Science and Technology, Hisar, Haryana, India



R. Kumar et al. (eds.), *Ecology, Conservation of Aquatic Birds*, Wetlands: Ecology, Conservation and Management 15, https://doi.org/10.1007/978-3-032-17125-2_24

stewardship, nano-based antimicrobials could come forward as a sustainable tool for safe-guarding avian health and aquatic ecosystem resilience.

**Keywords** Aquatic system · Antimicrobial agents · Ecology · Nanotechnology · Sustainable agriculture

## 24.1 Introduction

Wild aquatic birds, such as waterfowl, shorebirds, and seabirds, are essential components of aquatic ecosystems worldwide. Their existence and activities are crucial for nutrient cycling, supporting ecosystem functions, and preserving biodiversity. Wild aquatic birds are crucial to the health of aquatic ecosystems. By foraging in wetlands, coastal areas, and open waters, they help control insect populations, disperse seeds, and even contribute to the movement of nutrients across ecosystems [108]. For example, shorebirds help maintain the balance in mudflats and coastal habitats by feeding on small invertebrates and acting as indicators of environmental health [21]. Seabirds, with their long migratory patterns, also play a role in connecting different ecosystems, which can help with the distribution of nutrients between ocean and land [14]. In essence, these birds are not just pretty to look at but are integral players in the environmental processes that keep ecosystems functioning smoothly.

Nevertheless, these bird populations face growing threats from the spread of infectious diseases, including avian influenza, botulism, and a range of bacterial and fungal infections [15, 86]. Outbreaks of such diseases can lead to significant population declines, disrupt migration patterns, and damage ecosystem stability. Moreover, due to the indefinite migration of the bird species and the interconnectivity of aquatic environments, diseases through infected birds can spread rapidly across various regions and habitats affecting diverse species [46]. In such cases, even traditional methods of controlling diseases in wild bird populations such as culling infected individuals, habitat modification, and quarantine could fail in their effectiveness and come with ethical as well as ecological concerns. Also, these approaches may not fully address the underlying causes of disease spread. Thus, new and innovative solutions are needed to protect both the birds as well as the environments they inhabit. One hopeful possibility is the employment of nano-based antimicrobial agents in disease control in order to offer potentially more targeted, sustainable, and efficient alternative. For illustration, several studies have indicated that specific antimicrobial-activated metal nanostructures have been effective in fighting deadly, contagious diseases [33, 61]. The utilization of potential nanoparticles could be an effective means of reducing the broadcast of harmful diseases in wild aquatic birds either through direct exposure or ecological contamination [35].

In spite of that, nano-based antimicrobial agents in sensitive habitats like wetlands and aquatic environments have raised numerous serious ecological considerations such as their behaviour with the living forms, persistence in the ecosystems and impacts on the environment. In view to the above mentioned concerns, understanding their behaviour, persistence as well as ecological impacts is crucial to guarantee that the use of these agents will not damage the non-targeted species. Moreover, the concerns about their accumulation in food chains, food webs, in aquatic organisms, and potential toxicity to other wildlife are under evaluation to avoid any interference in the ecological balance [35]. For illustration, fishes occupy the highest position in the aquatic food chain and can accumulate large amounts of certain heavy metals from water, food, or sediments. These harmful metals are then taken up by superior aquatic organisms or wild seabirds [36]. Reports by Hosseini et al. [37] and Asztemborska et al. [9] have also suggested that lower trophic-level organisms show the probability to become a source of nanoparticles for higher organisms. Therefore, it is noteworthy to examine the flow of nanomaterials through aquatic food chains as well as their potential to biomagnify because it is vital to evaluate the environmental risks and also, to determine their harmful ecological impacts for a more sustainable future. Moreover, a comprehensive risk assessment is also required to determine the nanomaterials that pose significant threats to ecosystems. In that order, sustainable management practices (green nanotechnology, biodegradable nanoparticles, comprehensive risk assessments and strong regulatory frameworks) could guarantee that nanotechnology contributes positively to environmental, societal, and economic sustainability without compromising the health of ecosystems as well as humankind [43].

This chapter aims to investigate the sustainable characteristics of nano-based antimicrobial agents, particularly focusing on the potential role in controlling deadly diseases in wild aquatic bird populations. Also, the incorporation of advanced technologies with wildlife conservation could help in minimizing environmental threats through sustainable management practices.

### *24.1.1 Wild Aquatic Bird Populations and Their Importance*

Wild aquatic bird populations include ducks, swans, geese and various shorebirds. They contribute in maintaining the balance of aquatic ecosystems. They act as seed dispersers, pollinators and predators of invertebrates and small fishes [16, 17, 28]. In addition, their migratory patterns contribute to the transmission of nutrients across large geographical areas. Wild aquatic birds also help in regulating other species by playing a key role in the food chain. Furthermore, they serve as indicators of ecological health and help in signalling environmental changes such as shifts in water quality or the impacts of climatic changes [21]. Hence, it is necessary to assess their ecological role while assessing any interference that could have an effect on their populations.

### *24.1.2 Diseases in Wild Aquatic Bird Populations and Their Prevention*

Wild aquatic bird populations undergo numerous health issues due to numerous deadly diseases such as avian influenza (bird flu), avian botulism, and avian cholera can spread rapidly through both migratory and resident bird populations, often with devastating consequences. It is well-accepted that these avian infections (e.g. botulism, salmonellosis, cholera, influenza, and west nile encephalitis) can adversely impact populations of other wild bird species [38]. These hazardous diseases not only threaten the birds' survival but can also lead to declines in biodiversity, disrupt the local ecosystems, and even pose zoonotic risks to humans as well as domesticated animals [99]. The spread of disease worsens by environmental factors like shifts in water temperature, habitat destruction, and pollution. All of these factors further weaken birds' immune systems. Traditional methods of disease management, such as culling, or quarantine measures, have been met with ethical concerns in which infected or suspected animals are killed and their carcasses are destroyed to prevent the spread of diseases [31]. But these methods are not always effective at controlling outbreaks. Therefore, there is a growing need for innovative, environmentally sustainable, and non-invasive approaches to disease control in these vulnerable populations.

### *24.1.3 Introduction of Nano-Based Antimicrobial Agents as a Potential Solution*

Nano-based antimicrobial agents employ a promising strategy for tackling the increasing threat of contagious diseases in wild bird populations. These agents are prepared using nanoparticles demonstrating superior antimicrobial effectiveness when compared to their larger and bulk forms. Nanoparticles are exceptionally small (ranging in size from 1–100 nm) and can be made of various materials such as metals, polymers, ceramics, or lipids. The nano-scale size, larger surface area and ability to interact with microbial cell membranes make them highly capable in suppressing the growth of bacteria, viruses, and fungi [1]. Nanoparticles can be altered to target specific pathogens enabling a more focused and efficient manner of disease management. For instance, polymer-based nanoparticles linked with mannose-specific lectins have shown stronger binding to carbohydrate receptors on the cell surface of *Helicobacter pylori*, a pathogenic bacterium [30]. Therefore, nano-based antimicrobial agents should be added to water supplies, food, or environmental surfaces for wild aquatic birds in order to prevent diseases in a more sustainable way.

## 24.2 Nano-Based Antimicrobial Agents

Nano-based antimicrobial agents are nanoparticles made from metals such as silver (Ag), copper (Cu), and zinc (Zn). They have distinctive characteristics that make them extremely effective in deactivating several harmful pathogens. The size, surface area, and ability of the nanoparticles to co-operate with microbial cells on a molecular level deliver higher antimicrobial abilities [1, 59, 63, 71]. They are engineered at nano-scale because of their unique biological, chemical and physical properties that exhibit inhibitory effects against various pathogens [13, 61]. Moreover, nano-based antimicrobial agents engage with pathogens and inhibit their growth by disrupting their cell membranes through secretion of metal ions and also, by releasing reactive oxygen species (ROS) [87, 104].

### *24.2.1 Different Types of Nano-Based Antimicrobial Agents*

Some different types of nano-based antimicrobial agents are as follows:

#### 24.2.1.1 Silver Nanoparticles (AgNPs)

Silver nanoparticles are one of the most extensively studied and employed nanomaterials. Silver has unique antimicrobial qualities and its ability to suppress the growth of microbes is further enhanced by its high reactivity in nanoparticle form. AgNPs have anti-fungal, anti-inflammatory, anti-viral and anti-platelet capabilities [12, 79, 82]. Moreover, the green synthesis of AgNPs deals environmental-friendly and sustainable method that not only gratifies the needs of the investigators but also minimizes the potential environmental hazards [74].

#### 24.2.1.2 Copper Oxide Nanoparticles (CuO NPs)

CuO nanoparticles possess broad-spectrum antimicrobial activity. They are skilled of efficiently eradicating environmental pollutants due to their simple preparation process, cost-effectiveness and non-toxic nature [11]. They are predominantly effective against bacteria and fungi. They are also used in casings for therapeutic devices, textiles, and other exteriors. Moreover, green synthesis is regarded as a vital constituent in generating CuO NPs as it eradicates the carbon-based solvents from artificial systems and also, eliminates the harmful elements generated by chemical reactions [6].

#### 24.2.1.3 Zinc Oxide Nanoparticles (ZnO NPs)

Zinc oxide (ZnO) is presently classified as a Generally Recognized as Safe (GRAS) element by the Food and Drug Administration and is employed as a food additive [26]. It has also been reported that ZnO NPs exhibit antimicrobial effectiveness against a wide-ranging spectrum of microorganisms including bacteria, fungi, and viruses. As a result, they are often unified into polymeric matrices to improve the antimicrobial performance and functional properties of packaging ingredients [25]. In addition, ZnO NPs are widely used in personal care items, medical equipment and food packing.

#### 24.2.1.4 Titanium Dioxide Nanoparticles ($TiO_2$ NPs)

$TiO_2$NPs can act as photo-catalysts under UV light releasing reactive oxygen species (ROS). ROS damage microbial cells because of their high reactivity and can interact with biological molecules causing damage such as membrane lipid peroxidation, DNA fragmentation, and oxidation of amino acids and active sites in proteins [54, 73]. The ROS-induced photo-toxicity of $TiO_2$ makes it a highly effective nanoparticle with strong antibacterial and antiviral capabilities, commonly applied in purification of drinking water and therapeutic sterilization [68, 78, 100]. Additionally, $TiO_2$ NPs show great potential for breaking down chemical pollutants in both surface and groundwater [54].Furthermore, they are often used in wound dressings and disinfecting surfaces.

#### 24.2.1.5 Gold Nanoparticles (AuNPs)

Gold nanoparticles, though more commonly used in diagnostic applications, have also been found to perform antifungal activities depending on size, shape, and concentrations [83, 84, 109]. It has been reported by Ahmad et al. [3] that AuNPs displayed excellent size-dependent antifungal activity and bio-cidal action against *Candida* isolates. Moreover, green synthesis methods hold significance as they represent an environmentally friendly strategy that utilizes natural biological resources and steers clear of harmful substances to produce numerous distinct nanoparticles [90]. For instance, eco-friendly production of cobalt ferrite nanoparticles from extracts of grape peel, pulp, as well as honey-assisted synthesis of cobalt-zinc ferrite nanoparticles have been reported [95, 96].

### *24.2.2 Mechanisms Utilized by Nano-Based Antimicrobial Agents*

Nano-based antimicrobial agents act through several mechanisms, depending on their size, shape, and surface properties. The mechanism of action utilised by nano-based antimicrobial agents have been shown in Fig. 24.1. These include:

#### 24.2.2.1 Release of Reactive Oxygen Species (ROS)

Nanoparticles such as AgNPs and CuO NPs can induce the generation of ROS when they come into contact with microorganisms. ROS can damage essential biomolecules in the pathogen (e.g., proteins, lipids, as well as DNA) resulting in cell death. But, an excessive and unregulated production of reactive oxygen species (ROS) can trigger oxidative stress, which disrupts the cell's ability to sustain normal redox-regulated physiological functions [88].

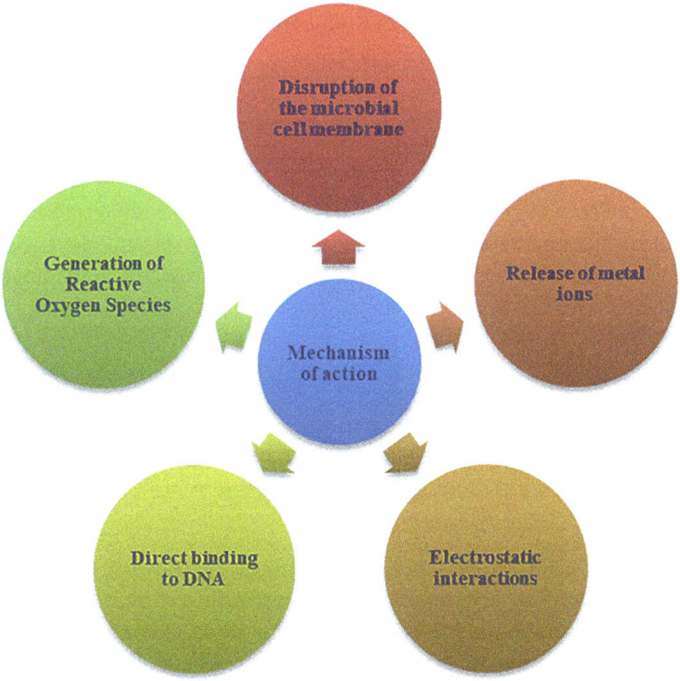


**Fig. 24.1** Mechanism of action of nano-based antimicrobial agents

#### 24.2.2.2 Disruption of the Cell Membrane

Nano-materials can physically interact with microbial cell membranes causing its disruption. This can lead to leakage of all the cellular contents, loss of membrane integrity, and ultimately, cell death [64, 67].

#### 24.2.2.3 Release of Metal Ions

Silver and copper nanoparticles can release metal ions ($Ag^+$ and $Cu^{2+}$), which are toxic to microbial cells. These toxic ions can bind to microbial proteins, DNA, and other vital cellular components, disrupting cellular function and leading to antimicrobial activity [45, 66].

#### 24.2.2.4 Binding to DNA

Some nanoparticles, like silver nanoparticles, can interact directly with microbial DNA. This can prevent DNA replication and transcription, inhibiting the growth and reproduction of microorganisms. It has been reported that exchanges of gold ions with cellular DNA leads to bacterial biofilm inhibition and causes its subsequent damage [69]. Comparable findings have been reported for RNA, where binding interferes with ribosomal activity resulting in bacterial cell death and a notable decrease in biofilm formation [20].

#### 24.2.2.5 Electrostatic Interactions

Electrostatic interactions between nanoparticles and the negatively charged groups on bacterial surfaces cause membrane disruption and cytoplasmic leakage. The cell surfaces of bacteria have a more negative charge as compared to mammalian cells which enables enhanced electrostatic interactions with positively charged substances [56].

### *24.2.3 Advantages of Using Nano-Materials Over Conventional Antimicrobial Agents*

Nanoparticles (NPs) have small size and larger surface area which empowers greater interaction with microbial cells making them more effective as antimicrobial agents even at lower concentrations [23]. Moreover, the ratio of surface area to volume increases due to reduction in the size of the nanoparticles [23, 107]. These substances, then, enter into the microorganisms through their cell walls or membranes by active transport or facilitated diffusion. Additionally, their high surface area-to-volume ratio improves their interactions with membranes, making it easier for them to destroy microbial structures [29, 70]. Furthermore, it enables them to eliminate the internal components of cells that are protected from larger molecules [75]. Moreover, the broad-spectrum activity of nano-based antimicrobial agents allow them to target strains of bacteria that are resistant to multiple drugs, silver nanoparticles are the most commonly used metallic nanomaterials [18, 44]. Mechanism of action of nano-based antimicrobial agents are enlisted in Fig. 24.1.

The mechanism of action implemented provides resistance to microbes which are significantly lower than that of conventional antibiotics [49, 61]. Furthermore, nanoparticles have been shown to be less harmful to human cells when used in low concentrations because their interactions with biological systems are diminished at lower exposure levels [58]. As a result, it is safer for therapeutic purposes like surface disinfection or wound healing. Similarly, the development of metallic nanoparticles through green synthesis has been brought to light by advances in nanotechnology and nano-biotechnology. This approach utilizes various biological substrates like plants, fungi, bacteria and alga. Also, they are recognized for being eco-friendly, straight-forward, cost-effective, and constantly reproducible [10, 39, 62].

## 24.3 Ecological Considerations in Wild Aquatic Bird Populations

### *24.3.1 Aquatic Ecosystems and the Role of Wild Aquatic Birds*

Aquatic ecosystems are environments dominated by water, covering freshwater lakes, rivers, marshes, wetlands, and coastal marine regions. These ecosystems feature distinctive plant and animal life that is adapted to living in or near water. Wild birds are vital for maintaining ecological balance in such environments. Wild aquatic birds, such as ducks, geese, swans, and seabirds, are involved in various ecological processes:

#### 24.3.1.1 Nutrient Cycling

Wild aquatic birds contribute to nutrient dynamics through their droppings, which provide essential nutrients like nitrogen and phosphorus to aquatic plants and microorganisms [2]. Additionally, in coastal areas seabirds process bulk quantities of food in a small space and help in transferring nutrients from the aquatic zone to the terrestrial zone [24, 55].

#### 24.3.1.2 Seed Dispersal

These aquatic birds help in the spread of aquatic plants by eating seeds and transporting them across different areas. Moreover, several herbivorous waterbirds that eat macrophytes could potentially act as significant carriers for seed dispersal due to their abundance near aquatic ecosystems [19].

#### 24.3.1.3 Predation and Control

Birds play a vital role in the food web, serving both as predators of smaller organisms like insects, fish, and invertebrates and as prey for higher organisms. Studies have revealed that bird predation can significantly reduce populations of certain species. For example, it has been found to lower the density of over-wintering spiders in northern spruce forests [8]. Additionally, numerous studies conducted in both terrestrial and aquatic environments have documented a decline in population of invertebrates due to extreme bird predation [105].

#### 24.3.1.4 Pollination

Specific bird species (hummingbirds, waterbirds, nector feeding birds etc.) also contribute to the pollination of plants in wetland ecosystems.

### *24.3.2 Spreading of Diseases in Wild Aquatic Bird Populations and Their Ecological Impact*

Wild aquatic bird populations are susceptible to a variety of diseases, both viral and bacterial, which can spread rapidly through their colonies. These diseases can be transmitted through water, direct contact, and migration routes. Numerous aquatic bird species are migratory, covering vast distances between their breeding and wintering sites, which contributes to the widespread transmission of diseases across broad regions. For instance, wild birds may contract enteric pathogens like *Salmonella* and *Campylobacter* spp. by consuming untreated sewage or waste. These pathogens can then be transmitted to humans either directly or indirectly through contamination of commercial poultry farms [32, 72]. Moreover, water sources and wetlands are often shared by large numbers of migratory birds from different regions and species leading to increase in the likelihood of disease transmission in these concentrated areas. In addition to all of that, pollution, habitat destruction, and climate change can weaken bird immune systems which make them more susceptible to diseases and less able to recover from outbreaks [48]. Thus, the ecological impact of disease in these populations can be significant as some diseases, such as avian influenza or avian botulism, can lead to high mortality rates in bird populations, ultimately, disrupting breeding success and overall population dynamics. The declines in bird populations may lead to unevenness in the food web, ultimately, affecting plant growth, insect populations and numerous animal species that depend on birds for food or other ecological services [34]. Furthermore, diseases that excessively affect certain bird species may change community dynamics which can cause reductions in some species while others may flourish, potentially leading to shifts in ecosystem functions.

### *24.3.3 Potential Risks and Benefits of Nano-Based Antimicrobial Agents*

The utilization of nano-based antimicrobial agents comes with both probable benefits and risks (Fig. 24.2).

#### 24.3.3.1 Benefits

**Controlling Disease Outbreaks** Antimicrobial treatments minimize the spread of infectious agents in cases of severe disease outbreaks and also, help in protecting susceptible bird populations from rapid decline [5].

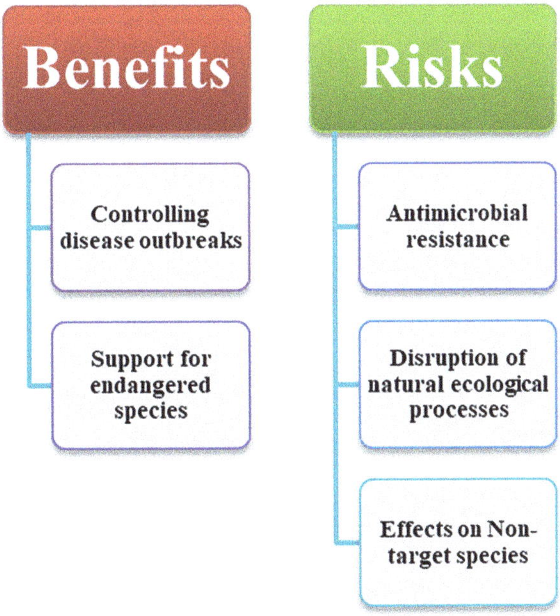


**Fig. 24.2** Benefits and risks of nano-based antimicrobial agents

**Support for Endangered Species** Some species that are already under threat due to environmental pressures could be benefitted from nano-based antimicrobial intervention to help boost their survival during disease outbreaks. Furthermore, applying nanotechnology in wildlife management holds potential for improving disease prevention, diagnosis, and treatment, as well as aiding in animal monitoring, promoting better health and nutrition, and boosting reproductive performance [42].

#### 24.3.3.2 Risks

##### Antimicrobial Resistance

The most noteworthy concern is the expansion of antimicrobial resistance (AMR). Overuse or misuse of antimicrobial agents in wildlife could lead to the evolution of resistant pathogens, which would not only affect bird populations but could also spread to other wildlife and even human populations [4].

##### Disruption of Natural Ecological Processes

Wild ecosystems rely on a miscellaneous array of microorganisms that contribute to essential functions such as nutrient cycling, decomposition as well as synergetic relationships with plants and animals [89].Therefore, the introduction of antimicrobial agents alters the composition of microbial communities leading to reduction in biodiversity and disrupting these essential ecological processes. Beneficial microbes in the environment or living on the birds could be affected, leading to unforeseen ecological consequences.

##### Non-target Effects

Even though the targeted pathogens have received the majority of research attention, the unintended consequences on non-target species are increasing concerns. Nanoparticles disturb aquatic ecosystems in unpredictable ways which has an impact on fish invertebrates and microorganisms. Concerns have also been raised about the use of nanoparticles in pest control as they cause potential damage to beneficial insects [50]. Moreover, these agents might cause unintentional harm to other species including birds, plants, or invertebrates in the ecosystem [81].

## 24.4 Influence of Nano-Based Antimicrobial Agents on Environment and Its Sustainability

Nanotechnology has several applications in environmental clean-up, medicinal approaches, and sustainable agriculture. Its effects on aquatic ecosystems, however, continue to be an important research topic. Although nanoparticles are extremely valuable, they might cause a risk to the environment and its inhabitants. Thus, to prevent any ecological imbalance, nanoparticles must be appropriately evaluated.

### *24.4.1 Persistence of Nanoparticles in Aquatic Ecosystems*

The potential of nanoparticles is influenced by their surface charge in aquatic environments as well as the water's physical and chemical characteristics, including temperature, salinity, and pH. To determine how durable nanoparticles are in water, scientists examine their mobility and propensity to interact with other materials.

#### 24.4.1.1 Mobility and Transport

Because of their nanosize, nanoparticles are very mobile in aquatic environments. It enables them to travel great distances via sediments, water columns, or other aquatic life [76, 106].

#### 24.4.1.2 Aggregation

In water, nanoparticles can occasionally group together to form clusters that impair their mobility. It impacts sediment-dwelling organisms in water resources by speeding up their sedimentation rate and enabling them to congregate in benthic areas [94].

#### 24.4.1.3 Persistence

Depending on their makeup and characteristics, certain nanoparticles can endure for longer periods of time [106]. For example, compared to organic-based nanoparticles, which break down more quickly, inorganic nanoparticles might be more persistent. Furthermore, nanoparticles may continue to affect ecosystems for years or even decades. To guarantee a beneficial effect on various ecosystems, it is necessary to assess the nanoparticles' capacity to endure in aquatic settings [97].

### *24.4.2 Accumulation of Nanoparticles in the Food Chain and Their Impacts on Ecosystems*

Many organisms, including planktons, small invertebrates, fish, and top predators, consume nanoparticles. Thus, their ability to bioaccumulate within aquatic food chains raises serious concerns:

#### 24.4.2.1 Bioaccumulation

Smaller organisms (phytoplanktons, zooplanktons) and benthic organisms may absorb nanoparticles through their digestive tract, gills, or skin in an aquatic environment. As a result, nanoparticles may accumulate in the food chain and get transmitted to higher predators such as fish-eating birds or mammals. I n results in causing deadly effects on their health [22, 85].

#### 24.4.2.2 Trophic Transfer

The transmission of nanoparticles through the food chain may result in the absorption of toxic substances in higher trophic levels leading to toxicity in top predators and possibly, humans who consume contaminated sea-food [85, 94].

#### 24.4.2.3 Adverse Effects on Non-targeted Species

Nanoparticles could also have direct toxic effects on non-target species in aquatic ecosystems. For example, some nanoparticles have been shown to affect the growth, behaviour, reproduction, and survival of aquatic invertebrates, such as crustaceans, or fish species [101]. The nanoparticles may interfere with cellular processes, cause oxidative stress, or damage cell membranes.

### *24.4.3 Biodegradability and Eco-Toxicity of Nano-Based Agents*

#### 24.4.3.1 Biodegradability

Some nanoparticles are inherently biodegradable, breaking down into less harmful substances over time due to microbial or chemical processes. For instance, certain organic or biodegradable nanoparticles (e.g., lipid-based nanoparticles or those made from biopolymers) may degrade more easily in natural environments [60]. However, many nanoparticles, especially those made from metals (such as silver or gold), may persist in the environment for extended periods, raising concerns about their long-term effects on ecosystems. The biodegradability of nanoparticles is influenced by factors such as the material's chemical composition, surface functionalization, and the surrounding environmental conditions [51]. Metal nanoparticles may oxidize over time liberating toxic ions into the environment leading to increase in their toxicity levels [77].

#### 24.4.3.2 Ecological Toxicity

Nanoparticles present significant ecological risks. It enables them to engage in specific interactions with cells and biological molecules that are not possible for larger particles. Furthermore, the functional groups located on the surface of nanoparticles affect their interaction with the biological systems and also, add to their toxicity. Certain metal-based nanoparticles can release metal ions into water resulting in disruption of the normal functioning of organisms [103]. Additionally, smaller nanoparticles because of their larger surface area exhibit higher toxicity [91].

## 24.5 Risk Assessment and Ethical Concerns

The unintended consequences of using nanotechnology in nature without fully comprehending its long-term effects on biodiversity are one of the major ethical concerns. Furthermore, the health of aquatic organisms could be directly impacted by a variety of nanotechnological applications [42].

### *24.5.1 Strategies for Risk Assessment*

To know the possible risks of nanotechnology and guarantee its application in wildlife management, risk assessment is highly crucial. These consist of:

#### 24.5.1.1 Toxicology Testing

Nanomaterials may have a variety of effects on ecosystems, food chains, or individual species, therefore, their toxicity needs to be carefully examined. To protect against any negative effects brought on by their bioaccumulation or biomagnification, their interactions with living systems should to be carefully considered [102].

#### 24.5.1.2 Environmental Persistence

Nanoparticles may contribute to the environmental stability of wildlife birds. One crucial phase in managing long-term risks is the production to disposal of nanoparticles, [106]. Therefore, it is essential to evaluate the interactions of nanoparticles with soil, water, and air.

#### 24.5.1.3 Dosage and Exposure

A controlled exposure and accurate dosage is very crucial for nano-based antimicrobials to avoid over-exposure in certain vulnerable species [98]. The risk assessment strategies can also evaluate variable environmental conditions and guarantee that nano-based solutions do not adversely interrupt non-targeted species.

## 24.6 Future Guidelines and Research Gaps in Nanotechnology for Control of Diseases

Nanotechnology can control the spread of numerous diseases in wildlife bird populations by recommending precise, targeted and resourceful explanations. Some of the **future guidelines** (Fig. 24.3) for the betterment of these antimicrobial agents include:

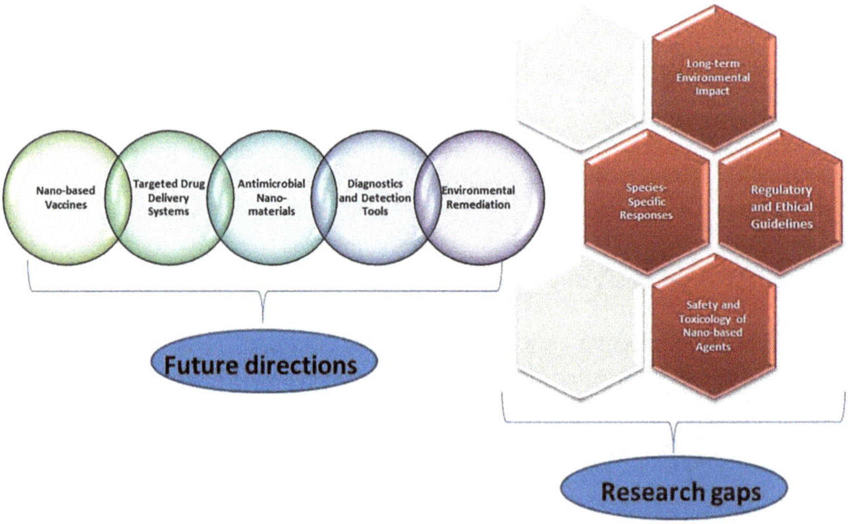


**Fig. 24.3** Future directions and research gaps in usage of nano-based antimicrobial agents

### 24.6.1 Nano-Based Vaccines

Nanoparticles can be used as potential delivery systems for vaccines providing controlled release into systems and boosting immune responses. Designing specific nano-vaccines that are targeted at avian diseases could help in controlling the transmission of zoonotic infections [53]. Recent research explores the use of various nanomaterials such as liposomes, dendrimers and nano-structured viruses to effectively carry antigens and trigger protective immunity [7].

### 24.6.2 Targeted Drug Delivery Systems

Nanotechnology allows transport of therapeutic agents directly at the infection site. This targeted approach reduces the volume of drug required, their side-effects leading to improving the efficiency of treatments [92]. In wild seabird's disease management, this could be critical for treating many diseases.

### 24.6.3 Antimicrobial Nanoparticles

Metal-based nanoparticles have strong antimicrobial properties [27]. These can inhibit bacterial or fungal infections in bird populations either by directly applying those to affected areas or by incorporating them into water bodies or vegetation [47].

### 24.6.4 Diagnostics and Detection Tools

Nanotechnology is focused on the improvement of highly sensitive diagnostic tools for early detection of deadly diseases in wild seabird populations. Biosensor-based nanoparticles can rapidly distinguish pathogens in blood, saliva or faecal samples [80]. In order to control and prevent epidemics, early detection of deadly diseases is necessary.

### 24.6.5 Environmental Remediation

Nanoparticles are also under investigation for their potential to rid an environment from foci where in a particular population was disseminating the pathogen [52]. For example, nanoparticles might be tasked with sanitising waterways of nuisance bacteria or a disease-causing virus to protect wild population of seabirds from environmental diseases [65].

Despite of so many attractive features of nanotechnology certain **research gaps** persist to be fulfilled for its safe implication in disease prevention (Fig. 24.3).

### *24.6.6 Long-Term Environmental Impact*

The nano-based treatments are being extensively studied, but still there is a lot to know about the long-term environmental effects of nanoparticles. The bioaccumulation of nanoparticles in different ecosystems, such as soil, water, and food chains require advanced research [41]. To avoid unanticipated ecological consequences, it is essential to gain a deeper understanding about nanoparticles and their interaction with different environments.

### *24.6.7 Species-Specific Responses*

Wild waterbird communities differ widely in their physiological reactions to novel improved therapies. More studies are needed to narrow down the metabolic pathways of various species and their talk with nano-based antimicrobials. For instance, a specific treatment may be effective in one species and not another. And, so specific approaches are required for different groups of wildlife [93].

### *24.6.8 Safety and Toxicology of Nano-Based Agents*

Nanoparticles have shown antimicrobial properties and their potential toxicity to wildlife, both in the short and long term, remains poorly understood [57]. Further research on the toxicity of nano-based agents to various animal species is important to ensure that these technologies do not unintentionally harm the wildlife bird populations they are intended to protect.

### *24.6.9 Regulatory and Ethical Guidelines*

As nanotechnology is advancing, there is a clear need for improvement of regulatory frameworks and ethical guidelines for its potential use [40]. Further research is obligatory to create protocols for testing, deploying, and monitoring nano-based interventions in wild aquatic bird populations ensuring that the technology is both safe and effective.

## 24.7 Future Perspectives

There are still some important research gaps regarding the long-term environmental impact, species-specific effects and toxicity of nanoparticles despite significant progress. Therefore, further research to completely understand the effects of nano-based antimicrobial agents on aquatic bird populations as well as the environment is necessary before deploying any nano-based formulation. Moreover, systematic risk assessments should be conducted to investigate the long-term effects of these agents on wildlife as well as ecosystems. It is necessary to integrate nano-based antimicrobial agents into advanced wildlife management initiatives that include techniques such as survey of deadly diseases, habitat restoration, monitoring of different populations, and immunization campaigns. Additionally, in order to guarantee that nanotechnology is used "ethically" with a focus on sustainability and minimal environmental impact, regulatory organizations should evaluate all the objectives and ethical issues. Furthermore, collaboration among wildlife managers, researchers, conservationists, local communities, and policymakers will be necessary to ensure that nanotechnology is in line with sustainability.

## 24.8 Conclusion

Nanotechnology offers numerous innovative solutions for managing wildlife diseases, including nano-based vaccines, targeted drug delivery systems, antimicrobial nanomaterials, and highly sensitive diagnostic tools. These technologies can enable more efficient, precise, and less aggressive interventions in health management of wild bird populations. Ethical concerns surrounding the impact of nanotechnology on different ecosystems are important. The likelihood of long-term ecological impacts and unintentional harm to non-targeted species emphasizes the necessity of systematic risk assessments and preventative actions.

## References

1. Aderibigbe BA (2017) Metal-based nanoparticles for the treatment of infectious diseases. Molecules 22(8):1370
2. Adhurya S, Das S, Ray S (2020) Guanotrophication by waterbirds in freshwater lakes: a review on ecosystem perspective. In: Mathematical analysis and applications in modeling: ICMAAM 2018, Kolkata, India, January 9–12, pp 253–269
3. Ahmad T, Wani IA, Lone IH, Ganguly A, Manzoor N, Ahmad A et al (2013) Antifungal activity of gold nanoparticles prepared by solvothermal method. Mater Res Bull 48(1):12–20
4. Ahmed SK, Hussein S, Qurbani K, Ibrahim RH, Fareeq A, Mahmood KA, Mohamed MG (2024) Antimicrobial resistance: impacts, challenges, and future prospects. J Med Surg Public Health 2:100081

5. Al-Awsi GRL, Alameri AA, Al-Dhalimy AMB, Gabr GA, Kianfar E (2023) Application of nano-antibiotics in the diagnosis and treatment of infectious diseases. Braz J Biol 84:e264946
6. Alhalili Z (2022) Green synthesis of copper oxide nanoparticles CuO NPs from eucalyptus Globoulus leaf extract: adsorption and design of experiments. Arab J Chem 15(5):103739
7. Aljabali AA, Obeid MA (2020) Inorganic-organic nanomaterials for therapeutics and molecular imaging applications. Nanosci Nanotechnol Asia 10(6):748–765
8. Askenmo C, Von Brömssen A, Ekman J, Jansson C (1977) Impact of some wintering birds on spider abundance in spruce. Oikos 28:90–94
9. Asztemborska M, Jakubiak M, Stęborowski R, Chajduk E, Bystrzejewska-Piotrowska G (2018) Titanium dioxide nanoparticle circulation in an aquatic ecosystem. Water Air Soil Pollut 229:1–9
10. Bandeira M, Giovanela M, Roesch-Ely M, Devine DM, da Silva Crespo J (2020) Green synthesis of zinc oxide nanoparticles: a review of the synthesis methodology and mechanism of formation. Sustain Chem Pharm 15:100223
11. Bashiri M, Rezanezhad M, Tavakkoli-Moghaddam R, Hasanzadeh H (2018) Mathematical modeling for a p-mobile hub location problem in a dynamic environment by a genetic algorithm. Appl Math Model 54:151–169
12. Beyene HD, Werkneh AA, Bezabh HK, Ambaye TG (2017) Synthesis paradigm and applications of silver nanoparticles (AgNPs), a review. Sustain Mater Technol 13:18–23
13. Bhattacharya R, Mukherjee P (2008) Biological properties of “naked” metal nanoparticles. Adv Drug Deliv Rev 60(11):1289–1306
14. Booty J (2023) The effects of shorebirds on the erodibility, properties and ecosystem functioning of intertidal muddy sediments (Doctoral dissertation). University of East Anglia
15. Boroomand Z, Mayahi M, Hosseini H, Valadbeigi S (2019) Detection and isolation of H9N2 subtype of avian influenza virus in house sparrows (Passer domesticus) of Ahvaz, Iran. Arch Razi Institute 74(4):439–444
16. Buij, R., Melman, T. C., Loonen, M. J., & Fox, A. D. (2017). Balancing ecosystem function
17. Chang ER, Zozaya EL, Kuijper DPJ, Bakker JP (2005) Seed dispersal by small herbivores and tidal water: are they important filters in the assembly of salt-marsh communities? Funct Ecol 19:665–673
18. Chen J, Yang L, Chen J, Liu W, Zhang D, Xu P et al (2019) Composite of silver nanoparticles and photosensitizer leads to mutual enhancement of antimicrobial efficacy and promotes wound healing. Chem Eng J 374:1373–1381
19. Clausen P, Nolet BA, Fox AD, Klaassen M (2002) Long-distance endozoochorous dispersal of submerged macrophyte seeds by migratory waterbirds in northern Europe—a critical review of possibilities and limitations. Actaoecologica 23(3):191–203
20. Cui Y, Zhao Y, Tian Y, Zhang W, Lü X, Jiang X (2012) The molecular mechanism of action of bactericidal gold nanoparticles on Escherichia coli. Biomaterials 33(7):2327–2333
21. Danchenko SA (2010) Shorebirds assemblages as indicators of ecological status and sediment quality in the Ria Formosa lagoon (Master’s thesis). Universidade do Algarve (Portugal)
22. Dang F, Huang Y, Wang Y, Zhou D, Xing B (2021) Transfer and toxicity of silver nanoparticles in the food chain. Environ Sci Nano 8(6):1519–1535
23. Dinesh R, Anandaraj M, Srinivasan V, Hamza S (2012) Engineered nanoparticles in the soil and their potential implications to microbial activity. Geoderma 173:19–27
24. Ellis JC (2005) Marine birds on land: a review of plant biomass, species richness, and community composition in seabird colonies. Plant Ecol 181:227–241
25. Espitia PJP, Otoni CG, Soares NFF (2016) Zinc oxide nanoparticles for food packaging applications. In: Antimicrobial food packaging. Academic, pp 425–431
26. Espitia PJP, Soares NDFF, Coimbra JSDR, de Andrade NJ, Cruz RS, Medeiros EAA (2012) Zinc oxide nanoparticles: synthesis, antimicrobial activity and food packaging applications. Food Bioprocess Technol 5:1447–1464
27. Fan X, Yahia LH, Sacher E (2021) Antimicrobial properties of the Ag, Cu nanoparticle system. Biology 10(2):137

28. Ferraro DO, Oesterheld M (2002) Effect of defoliation on grass growth. A quantitative review. Oikos 98(1):125–133
29. Franci G, Falanga A, Galdiero S, Palomba L, Rai M, Morelli G, Galdiero M (2015) Silver nanoparticles as potential antibacterial agents. Molecules 20(5):8856–8874
30. Gao W, Thamphiwatana S, Angsantikul P, Zhang L (2014) Nanoparticle approaches against bacterial infections. Wiley Interdiscip Rev Nanomed Nanobiotechnol 6(6):532–547
31. Garner MG, Dubé C, Stevenson MA, Sanson RL, Estrada C, Griffin J (2007) Evaluating alternative approaches to managing animal disease outbreaks–the role of modelling in policy formulation. Veterinaria Italiana 43(2):285–298
32. Georgopoulou I, Tsiouris V (2008) The potential role of migratory birds in the transmission of zoonoses. Vet Ital 44(4):671–677
33. Goodman CM, McCusker CD, Yilmaz T, Rotello VM (2004) Toxicity of gold nanoparticles functionalized with cationic and anionic side chains. Bioconjug Chem 15(4):897–900
34. Gutierrez-Arellano C, Mulligan M (2018) A review of regulation ecosystem services and disservices from faunal populations and potential impacts of agriculturalisation on their provision, globally. Nat Conserv 30:1–39
35. Hauser-Davis RA, Monteiro F, Rocha RCCD, Lemos L, Cardoso MD, Siciliano S (2020) Titanium as a contaminant of emerging concern in the aquatic environment and the current knowledge gap regarding seabird contamination
36. Henry F, Amara R, Courcot L, Lacouture D, Bertho ML (2004) Heavy metals in four fish species from the French coast of the Eastern English Channel and Southern Bight of the North Sea. Environ Int 30(5):675–683
37. Hosseini M, Rahmanpour SH, Mashinchian MA (2015) Heavy metal ions on titanium dioxide nano-particle: biomagnification in an experimental aquatic food chain
38. Hubálek Z (2021) Pathogenic microorganisms associated with gulls and terns (Laridae). J Vertebr Biol 70(3):21009–1
39. Huston M, DeBella M, DiBella M, Gupta A (2021) Green synthesis of nanomaterials. Nanomaterials 11(8):2130
40. Isibor PO (2024) Regulations and policy considerations for nanoparticle safety. In: Environmental nanotoxicology: combatting the minute contaminants. Springer Nature Switzerland, Cham, pp 295–316
41. Isibor PO, Oyegbade SA, Oni JG, Ahmed WW, Abiodun EO, Oyewole OA (2024) Nanoparticles in food chains: bioaccumulation and trophic transfer. In: Environmental nanotoxicology: combatting the minute contaminants. Springer Nature Switzerland, Cham, pp 203–233
42. Jawre S, Jadav KK, Tiwari DK, Arora N, Kumar R, Mamatha D et al (2024) Nanotechnology in wildlife management. In: Nanotechnology theranostics in livestock diseases and management. Springer Nature Singapore, Singapore, pp 411–423
43. Kamali M, Persson KM, Costa ME, Capela I (2019) Sustainability criteria for assessing nanotechnology applicability in industrial wastewater treatment: current status and future outlook. Environ Int 125:261–276
44. Kang XQ, Qiao Y, Lu XY, Jiang SP, Li WS, Wang XJ et al (2019) Tocopherol polyethylene glycol succinate-modified hollow silver nanoparticles for combating bacteria-resistance. Biomater Sci 7(6):2520–2532
45. Karlsson HL, Cronholm P, Hedberg Y, Tornberg M, De Battice L, Svedhem S, Wallinder IO (2013) Cell membrane damage and protein interaction induced by copper containing nanoparticles—importance of the metal release process. Toxicology 313(1):59–69
46. Kiran, Singh D, Kour A, Yadav R, Dahiya P, Kumar R, Sharma K (2023) The avian gut microbial community: driving factors and role in wild birds. In: Latest trend in zoology and entomology science, vol 14. Akinik Publications., ISBN: 978-93-5570-676-8, pp 15–13
47. Konieczny J, Rdzawski Z (2012) Antibacterial properties of copper and its alloys. Arch Mater Sci Eng 56(2):53–60

48. Koprivnikar J, Forbes MR, Baker RL (2007) Contaminant effects on host-parasite interactions: atrazine, frogs, and trematodes. Environ Toxicol Chem Int J 26(10):2166–2170
49. Kumar M, Curtis A, Hoskins C (2018) Application of nanoparticle technologies in the combat against anti-microbial resistance. Pharmaceutics 10(1):11
50. Kumari K, Singh P, Bauddh K, Sweta, Mallick S, Chandra R (2019) Implications of metal nanoparticles on aquatic fauna: a review. Nanosci Nanotechnol Asia 9(1):30–43
51. Lead JR, Batley GE, Alvarez PJ, Croteau MN, Handy RD, McLaughlin MJ et al (2018) Nanomaterials in the environment: behavior, fate, bioavailability, and effects—an updated review. Environ Toxicol Chem 37(8):2029–2063
52. Li Q, Chen X, Zhuang J, Chen X (2016) Decontaminating soil organic pollutants with manufactured nanoparticles. Environ Sci Pollut Res 23:11533–11548
53. Li Y, Xiao Y, Chen Y, Huang K (2021) Nano-based approaches in the development of antiviral agents and vaccines. Life Sci 265:118761
54. Ma H, Brennan A, Diamond SA (2012) Photocatalytic reactive oxygen species production and phototoxicity of titanium dioxide nanoparticles are dependent on the solar ultraviolet radiation spectrum. Environ Toxicol Chem 31(9):2099–2107
55. Mariyappan M, Rajendran M, Velu S, Johnson AD, Dinesh GK, Solaimuthu K et al (2023) Ecological role and ecosystem services of birds: a review. Int J Environ Clim Change 13(6): 76–87
56. Matsuzaki K (2009) Control of cell selectivity of antimicrobial peptides. Biochim Biophys Acta 1788(8):1687–1692
57. Méndez-Vilas A (2011) Nanoparticles and their potential application as antimicrobials. In: Science against microbial pathogens: communicating current research and technological advances. FORMATEX
58. Mironava T, Hadjiargyrou M, Simon M, Jurukovski V, Rafailovich MH (2010) Gold nanoparticles cellular toxicity and recovery: effect of size, concentration and exposure time. Nanotoxicology 4(1):120–137
59. Mody VV, Siwale R, Singh A, Mody HR (2010) Introduction to metallic nanoparticles. J Pharm Bioall Sci 2(4):282–289
60. Mondéjar-López M, García-Simarro MP, Navarro-Simarro P, Gómez-Gómez L, Ahrazem O, Niza E (2024) A review on the encapsulation of "eco-friendly" compounds in natural polymer-based nanoparticles as next generation nano-agrochemicals for sustainable agriculture and crop management. Int J Biol Macromol 280:136030
61. Mubeen B, Ansar AN, Rasool R, Ullah I, Imam SS, Alshehri S et al (2021) Nanotechnology as a novel approach in combating microbes providing an alternative to antibiotics. Antibiotics 10(12):1473
62. Nadaroglu H, Güngör AA, Ince S (2017) Synthesis of nanoparticles by green synthesis method. Int J Innov Res Rev 1(1):6–9
63. Nisha, Anshu, Rani S, Kumar S, Kumar R, Kumar A (2025) Recent technologies and strategies for bioremediation of toxic and hazardous waste. In: Dubey KK, Shah MP, Yadav A (eds) Challenges and sustainable solutions in bioremediation. CRC Press, pp 1–16. https://doi.org/10.1201/9781003407317-1
64. Ogunsona EO, Muthuraj R, Ojogbo E, Valerio O, Mekonnen TH (2020) Engineered nanomaterials for antimicrobial applications: a review. Appl Mater Today 18:100473
65. Okeke ES, Chukwudozie KI, Nyaruaba R, Ita RE, Oladipo A, Ejeromedoghene O et al (2022) Antibiotic resistance in aquaculture and aquatic organisms: a review of current nanotechnology applications for sustainable management. Environ Sci Pollut Res Int 29(46):69241–69274
66. Palza H (2015) Antimicrobial polymers with metal nanoparticles. Int J Mol Sci 16(1): 2099–2116
67. Parandhaman T, Das A, Ramalingam B, Samanta D, Sastry TP, Mandal AB, Das SK (2015) Antimicrobial behavior of biosynthesized silica–silver nanocomposite for water disinfection: a mechanistic perspective. J Hazard Mater 290:117–126

68. Prasad GK, Agarwal GS, Singh B, Rai GP, Vijayaraghavan R (2009) Photocatalytic inactivation of Bacillus anthracis by titaniananomaterials. J Hazard Mater 165(1–3):506–510
69. Radzig M, Koksharova O, Khmel I, Ivanov V, Yorov K, Kiwi J et al (2019) Femtosecond spectroscopy of Au hot-electron injection into TiO2: evidence for Au/TiO2 plasmonphotocatalysis by bactericidal Au ions and related phenomena. Nanomaterials 9(2): 217
70. Ramalingam B, Parandhaman T, Das SK (2016) Antibacterial effects of biosynthesized silver nanoparticles on surface ultrastructure and nanomechanical properties of gram-negative bacteria viz. Escherichia coli and Pseudomonas aeruginosa. ACS Appl Mater Interfaces 8(7): 4963–4976
71. Rani S, Sharma S, Singh D, Sheoran R, Kumar R (2025) Improved production of anti-cancer compounds from medicinal plants through molecular and biotechnological mutagenesis strategies. In: Chen JT (ed) Biotechnology, multiple omics, and precision breeding in medicinal plants. CRC Press, pp 287–297
72. Reed KD, Meece JK, Henkel JS, Shukla SK (2003) Birds, migration and emerging zoonoses: West Nile virus, Lyme disease, influenza A and enteropathogens. Clin Med Res 1(1):5–12
73. Riley PA (1994) Free radicals in biology: oxidative stress and the effects of ionizing radiation. Int J Radiat Biol 65(1):27–33
74. Rogach AL (2000) NanocrystallineCdTe and CdTe (S) particles: wet chemical preparation, size-dependent optical properties and p
75. Rudramurthy GR, Swamy MK, Sinniah UR, Ghasemzadeh A (2016) Nanoparticles: alternatives against drug-resistant pathogenic microbes. Molecules 21(7):836
76. Saleh N, Kim HJ, Phenrat T, Matyjaszewski K, Tilton RD, Lowry GV (2008) Ionic strength and composition affect the mobility of surface-modified Fe0 nanoparticles in water-saturated sand columns. Environ Sci Technol 42(9):3349–3355
77. Sengul AB, Asmatulu E (2020) Toxicity of metal and metal oxide nanoparticles: a review. Environ Chem Lett 18(5):1659–1683
78. Seo JW, Chung H, Kim MY, Lee J, Choi IH, Cheon J (2007) Development of water-soluble single-crystalline TiO2 nanoparticles for photocatalytic cancer-cell treatment. Small 3(5): 850–853
79. Shankar SS, Rai A, Ahmad A, Sastry M (2005) Controlling the optical properties of lemongrass extract synthesized gold nanotriangles and potential application in infrared-absorbing optical coatings. Chem Mater 17(3):566–572
80. Shanker R, Singh G, Jyoti A, Dwivedi PD, Singh SP (2020) Nanotechnology and detection of microbial pathogens. In: Animal biotechnology. Academic, pp 593–611
81. Shanmugam K, Sakthivel E, Rajendran J, Jeyaraman P (2024) Nanoparticles in aquatic ecosystems: origins, destiny, and ecological consequences. Environ Sci Arch 3(1):111–124
82. Sharma MMM, Kapoor D, Loyal A, Kumar R, Sharma P, Husen A (2025) Plant response to silver nanoparticles in terms of growth, development, production, and protection: an overview. In: Husen A (ed) Plant response to silver nanoparticles. Springer, Singapore, pp 1–22. https://doi.org/10.1007/978-981-97-7352-7_1
83. Sharma MMM, Kapoor D, Loyal A, Kumar R, Sharma P, Husen A (2024a) Effect of gold nanoparticles on seed germination, plant growth, and plant protection. In: Husen A (ed) Plant response to gold nanoparticles. Springer, pp 155–175. https://doi.org/10.1007/978-981-97-7174-5_8
84. Sharma MMM, Kapoor D, Loyal A, Kumar R, Sharma P, Husen A (2024b) Mitigation of drought stress in plants using gold nanoparticles. In: Husen A (ed) Plant response to gold nanoparticles. Springer, pp 193–209. https://doi.org/10.1007/978-981-97-7174-5_10
85. Shi Q, Wang CL, Zhang H, Chen C, Zhang X, Chang XL (2020) Trophic transfer and biomagnification of fullerenol nanoparticles in an aquatic food chain. Environ Sci Nano 7(4):1240–1251
86. Skarin H, TevellÅberg A, Woudstra C, Hansen T, Löfström C, Koene M et al (2013) The workshop on animal botulism in Europe. Biosecur Bioterror 11(S1):S183–S190

87. Slavin YN, Asnis J, Hńfeli UO, Bach H (2017) Metal nanoparticles: understanding the mechanisms behind antibacterial activity. J Nanobiotechnol 15:1–20
88. Snezhkina AV, Kudryavtseva AV, Kardymon OL, Savvateeva MV, Melnikova NV, Krasnov GS, Dmitriev AA (2019) ROS generation and antioxidant defense systems in normal and malignant cells. Oxid Med Cell Longev 2019(1):6175804
89. Sokol NW, Slessarev E, Marschmann GL, Nicolas A, Blazewicz SJ, Brodie EL et al (2022) Life and death in the soil microbiome: how ecological processes influence biogeochemistry. Nat Rev Microbiol 20(7):415–430
90. Soltys L, Olkhovyy O, Tatarchuk T, Naushad M (2021) Green synthesis of metal and metal oxide nanoparticles: principles of green chemistry and raw materials. Magnetochemistry 7(11):145
91. Sukhanova A, Bozrova S, Sokolov P, Berestovoy M, Karaulov A, Nabiev I (2018) Dependence of nanoparticle toxicity on their physical and chemical properties. Nanoscale Res Lett 13:1–21
92. Suri SS, Fenniri H, Singh B (2007) Nanotechnology-based drug delivery systems. J Occup Med Toxicol 2:1–6
93. Swaisgood RR, Shepherdson DJ (2005) Scientific approaches to enrichment and stereotypies in zoo animals: what's been done and where should we go next? Zoo Biol 24(6):499–518
94. Tangaa SR, Selck H, Winther-Nielsen M, Khan FR (2016) Trophic transfer of metal-based nanoparticles in aquatic environments: a review and recommendations for future research focus. Environ Sci Nano 3(5):966–981
95. Tatarchuk T, Danyliuk N, Shyichuk A, Kotsyubynsky V, Lapchuk I, Mandzyuk V (2021b) Green synthesis of cobalt ferrite using grape extract: the impact of cation distribution and inversion degree on the catalytic activity in the decomposition of hydrogen peroxide. Emerg Mater 5:89–103
96. Tatarchuk T, Shyichuk A, Sojka Z, Gryboś J, Naushad M, Kotsyubynsky V et al (2021a) Green synthesis, structure, cations distribution and bonding characteristics of superparamagnetic cobalt-zinc ferrites nanoparticles for Pb (II) adsorption and magnetic hyperthermia applications. J Mol Liq 328:115375
97. Thakur A, Kumar A (2023, August) Ecotoxicity analysis and risk assessment of nanomaterials for the environmental remediation. Macromol Symp 410(1):2100438
98. Thanigaivel S, Vickram AS, Anbarasu K, Gulothungan G, Nanmaran R, Vignesh D et al (2021) Ecotoxicological assessment and dermal layer interactions of nanoparticle and its routes of penetrations. Saudi J Biol Sci 28(9):5168–5174
99. Tsiodras S, Kelesidis T, Kelesidis I, Bauchinger U, Falagas ME (2008) Human infections associated with wild birds. J Infect 56(2):83–98
100. Tsuang YH, Sun JS, Huang YC, Lu CH, Chang WHS, Wang CC (2008) Studies of photokilling of bacteria using titanium dioxide nanoparticles. Artif Organs 32(2):167–174
101. Turan NB, Erkan HS, Engin GO, Bilgili MS (2019) Nanoparticles in the aquatic environment: usage, properties, transformation and toxicity—a review. Process Saf Environ Prot 130:238–249
102. Uddin MN, Desai F, Asmatulu E (2020) Engineered nanomaterials in the environment: bioaccumulation, biomagnification and biotransformation. Environ Chem Lett 18(4): 1073–1083
103. Vasyukova IA, Zakharova OV, Chaika VV, Pikula KS, Golokhvast KS, Gusev AA (2021) Toxic effect of metal-based nanomaterials on representatives of marine ecosystems: a review. Nanobiotechnol Rep 16:138–154
104. Wang L, Hu C, Shao L (2017) The antimicrobial activity of nanoparticles: present situation and prospects for the future. Int J Nanomedicine 12:1227–1249
105. Whelan CJ, Wenny DG, Marquis RJ (2008) Ecosystem services providedby birds. Ann N Y Acad Sci 1134(1):25–60
106. Wigginton NS, Haus KL, HochellaJr MF (2007) Aquatic environmental nanoparticles. J Environ Monit 9(12):1306–1316

107. Wong MS, Sun DS, Chang HH (2010) Bactericidal performance ofvisible-light responsive titaniaphotocatalyst with silver nanostructures. PLoS One 5(4):e10394
108. Yodha K, Singh D, Kour A, Yadav R, Kumar R (2024) Species diversity and abundance of birds in ber (*Zizyphus mauritiana*) crop. Asian J Conserv Biol 13(1). https://doi.org/10.53562/ajcb.83683
109. Zhang Y, ShareenaDasari TP, Deng H, Yu H (2015) Antimicrobial activityof gold nanoparticles and ionic gold. J Environ Sci Health Part C 33(3):286–327

# Index




R. Kumar et al. (eds.), *Ecology, Conservation of Aquatic Birds*, Wetlands: Ecology, Conservation and Management 15, https://doi.org/10.1007/978-3-032-17125-2

## M



## N



## O



## P



## R



## S



## T



## U

**W**

GPSR Compliance
The European Union's (EU) General Product Safety Regulation (GPSR) is a set of rules that requires consumer products to be safe and our obligations to ensure this.

If you have any concerns about our products, you can contact us on

ProductSafety@springernature.com

In case Publisher is established outside the EU, the EU authorized representative is:

Springer Nature Customer Service Center GmbH
Europaplatz 3
69115 Heidelberg, Germany

www.ingramcontent.com/pod-product-compliance
Ingram Content Group UK Ltd.
Pitfield, Milton Keynes, MK11 3LW, UK
UKHW020742021026
476198UK00003B/15
* 9 7 8 3 0 3 2 1 7 1 2 4 5 *